LEAF OPTICAL PROPERTIES

Plant leaves collectively represent the largest above-ground surface area of plant material in virtually all environments. Their optical properties determine where and how energy and gas exchange occurs, which in turn drives the energy budget of the planet, and defines its ecology and habitability. This book reviews the state-of-the-art research on leaf optics. Topics covered include leaf traits, the anatomy and structure of leaves, leaf color, biophysics and spectroscopy, radiometry, radiative transfer models, and remote and proximal sensing. A physical approach is emphasized throughout, providing the necessary foundations in physics, chemistry, and biology to make the context accessible to readers from various subject backgrounds. It is a valuable resource for advanced students, researchers, and government agency practitioners in remote sensing, plant physiology, ecology, resource management, and conservation.

STÉPHANE JACQUEMOUD is Professor of Remote Sensing at the Université de Paris. He has held positions in the Department of Physics (1995–2004) and the Department of Earth, Environmental and Planetary Sciences (2004 to today). He currently works at the Institut de Physique du Globe de Paris. His research focuses on remote sensing of natural surfaces in the visible/infrared domain, and its applications in geophysics, the environment, and exobiology.

SUSAN USTIN is Distinguished Professor of Environmental Resource Science at the University of California, Davis. Her research focuses on the detection of plant health and plant identification to better understand the functioning and composition of ecosystems using imaging spectroscopy. She received an honorary doctorate from the University of Zurich in 2012, and became a fellow of the American Geophysical Union in 2017.

LEAF OPTICAL PROPERTIES

STÉPHANE JACQUEMOUD

Université de Paris

SUSAN USTIN

University of California, Davis

CAMBRIDGE
UNIVERSITY PRESS

CAMBRIDGE
UNIVERSITY PRESS

University Printing House, Cambridge CB2 8BS, United Kingdom

One Liberty Plaza, 20th Floor, New York, NY 10006, USA

477 Williamstown Road, Port Melbourne, VIC 3207, Australia

314–321, 3rd Floor, Plot 3, Splendor Forum, Jasola District Centre,
New Delhi – 110025, India

79 Anson Road, #06–04/06, Singapore 079906

Cambridge University Press is part of the University of Cambridge.

It furthers the University's mission by disseminating knowledge in the pursuit of
education, learning, and research at the highest international levels of excellence.

www.cambridge.org
Information on this title: www.cambridge.org/9781108481267
DOI: 10.1017/9781108686457

First published 2019

Printed in the United Kingdom by TJ International Ltd. Padstow Cornwall

A catalogue record for this publication is available from the British Library.

Library of Congress Cataloging-in-Publication Data
Names: Jacquemoud, Stéphane, 1965– author. | Ustin, Susan L., 1943– author.
Title: Leaf optical properties / Stéphane Jacquemoud, Université de Paris, Susan Ustin,
University of California, Davis.
Description: Cambridge, United Kingdom ; New York, NY : Cambridge University Press, 2019. | Includes
bibliographical references and index.
Identifiers: LCCN 2019008517 | ISBN 9781108481267 (hardback : alk. paper)
Subjects: LCSH: Foliar diagnosis. | Leaf – Optical properties.
Classification: LCC QK865 .J34 2019 | DDC 572/.2–dc23
LC record available at https://lccn.loc.gov/2019008517

ISBN 978-1-108-48126-7 Hardback

Additional resources for this publication at www.cambridge.org/leafopticalproperties.

Contents

Color plate section can be found between pages 210 and 211

Preface

Plant leaves are the main surfaces of phyto-elements in most plant communities. They are all around us, but contrary to flowers, only a few books have focused on them: in their book *Leaves*, Prance and Sandved (1985) paint a fascinating picture of their formation, characteristics, and uses. Their book presents more than 300 breathtaking photos of leaves taken from their travels around the world. At the age of 87, Vitale (1997) released her first book, *Leaves: In Myth, Magic & Medicine*, based on research she conducted over almost 30 years. She provides authentic portraits of 110 leaf specimens all grown in North America, mixing delightful short stories of where they originate, how they inspired poets and myth-makers, or how they were used as medicines. Among the many subjects that these authors cover is leaf color, but only in passing. The interaction of light with plant leaves, which results in leaf color, is nevertheless of interest to the botanist, forester, geographer, biophysicist, biochemist, ecologist, hydrologist, agronomist, and others. Because leaves collectively represent the largest surface area of plant material in virtually all environments, they drive the energy budget of the planet and define its ecology and its habitability. It is worthwhile remembering that the photosynthetic function of leaves is essential for life on Earth (Vogelmann and Gorton, 2014).

Lee's *Nature's Palette: The Science of Plant Color* (Lee, 2007) is one of the first books entirely devoted to plant color, taking the reader through its social history, ecology, evolution, and biochemistry. It includes flowers, leaves, fruits and seeds, stems, and roots. His approach is, however, mainly that of a chemist or a biologist. The book *Photoprotection in Plants – Optical Screening-based Mechanisms* (Solovchenko, 2010) deals with the optical screening of excessive and potentially harmful solar radiation by plant leaves. Photoprotection is important for juvenile and senescing plants as well as when under environmental stresses. Visual plant defenses (camouflage, mimicry and aposematism via coloration, morphology, and even movement) against herbivores is the central topic of *Defensive (anti-herbivory) Coloration in Land Plants* (Lev-Yadun, 2016). That book develops the author's current understanding on defensive plant coloration and related issues.

Leaf optical properties have been the subject of hundreds of studies, most since the middle of the last century. Despite their obvious importance in many scientific domains including plant physiology, ecology, remote sensing, environmental physics, or image synthesis, a reference book that covers the entire subject of how light interacts with plant leaves has not previously been published. This book reviews the state-of-the-science of leaf optics. Because of the depth of this subject, we have restrained from adding sections on canopy and landscape-scale processes, because of the impractical length of such a book and because there are other books that address these subjects.

This book was written for the interested scientists who work across the wide range of disciplines cited above. Many of these readers do not share a common understanding of these properties, therefore we have included brief descriptions of the fundamental discoveries in leaf optics in the solar domain, thermal infrared, and microwaves and how these relate to other disciplines. The first four chapters provide descriptions of the basic component parts of plants and their relevant biochemical composition, then how light is absorbed or scattered by leaf constituents in different wavelength regions, and finally, how light interactions are measured. These are the building blocks that determine what materials light can interact with. All of these are presented in the context of how light interacts with biological and physical processes. Chapters 5 to 7 describe the optical properties of leaves and the sources of variability; these are the physical rules that control the interactions. Chapters 8 to 10 review leaf reflectance models, their uses and limitations, and provide a detailed review of the most widely used leaf optics model, PROSPECT, and a three-dimensional ray tracing model, RAYTRAN. Chapters 11 and 12 include information extraction methods and their application to a wide range of applications. A short conclusion is followed by references and then several appendices that provide more detail on several of the physical and optical processes, mathematical methods, and available datasets.

Symbols

A	absorptance
A	one-sided area of a fresh leaf (m^2)
c	velocity of light in a vacuum (2.998×10^8 m s^{-1})
C	capacitance of a medium
d_p	penetration depth of light (m)
dw	leaf dry mass (kg)
E	irradiance (W m^{-2})
$F_{\parallel}$	parallel component of the Fresnel equations
$F_{\perp}$	perpendicular component of the Fresnel equations
fw	leaf fresh mass (kg)
h	Planck's constant (6.626×10^{-34} J s)
I	radiant intensity (W sr^{-1})
I_c	upward collimated radiant flux
I_d	upward diffuse radiant flux
I_p	intensity of the polarized component of light
I_u	intensity of the unpolarized component of light
J_c	downward collimated radiant flux
J_d	downward diffuse radiant flux
k	specific absorption coefficient (m^{-1})
k_B	Boltzmann constant (1.38×10^{-23} J K^{-1})
ℓ	leaf thickness (m)
L	radiance (W m^{-2} sr^{-1})
L_e	spectral radiance of a blackbody (W $m^{-2}\mu m^{-1}$ sr^{-1})
M	radiant emittance (or radiant exitance) (W m^{-2})
$\widetilde{n}$	complex refractive index, $\widetilde{n} = n_r + i\ n_i$
n_r	real part of the complex refractive index (or refractive index noted n)
n_i	imaginary part of the complex refractive index
N_A	Avogadro number (6.02214×10^{23} mol^{-1})
p	degree of polarization of light
q	energy of a photon (J or eV)
Q	magnitude of the polarization ellipse
Q	radiant energy (J)
R	reflectance

R	resistance of a medium
r_{ij}	average reflectivity from medium *i* to medium *j*
s	scattering coefficient of a medium (m^{-1})
$\vec{S}$	Stokes vector
t_{ij}	average transmissivity from medium *i* to medium *j*
T	transmittance
T	absolute temperature (K)
U	orientation of the polarization ellipse
v	velocity of light in a medium ($m\ s^{-1}$)
V	circularity of the polarization ellipse
Γ	reflection coefficient in the microwaves
$\widetilde{\epsilon}$	relative dielectric permittivity, $\widetilde{\epsilon}_r = \epsilon_r + i\epsilon_i$
ϵ_0	vacuum permittivity (8.854×10^{-12} F m^{-1})
ϵ_{eff}	effective dielectric constant of a medium
ϵ_i	imaginary part of the relative dielectric permittivity (or loss factor noted ϵ'')
ϵ_r	real part of the relative dielectric permittivity (or dielectric constant noted ϵ')
θ_c	critical angle (° or rad)
θ_B	Brewster's angle (° or rad)
λ	wavelength (m)
ν	frequency (Hz)
ξ	polarization azimuth
ρ_d	leaf density (dry mass per unit volume) (kg m^{-3})
σ	wavenumber (m^{-1})
σ_e	extinction coefficient of a medium (m^{-1})
σ_k	absorption coefficient of a medium (m^{-1})
σ_s	scattering coefficient of a medium (m^{-1})
τ	fraction of light transmitted through a medium
Φ	radiant flux (or radiant power) (J s^{-1} or W)
χ	shape of the polarization ellipse
ψ	leaf water potential (Pa)
ω	angular frequency (rad s^{-1})
Ω	solid angle (sr)

Acknowledgments

We owe acknowledgments to many people for their help and advice on this book. Actually, given the ten or more years we have worked on and thought about it, we owe thanks and gratitude to more people than we can remember. Of course, we owe our families, especially our respective spouses, Christelle Gée and James Doyle, for their patience in accepting the loss of many long weekends and vacations while we worked, and despite this, their continued cooperation and support throughout these years. We thank and appreciate our friends and colleagues for their generous encouragement. Many of them have read various chapters and drafts, sometimes several versions, and their comments and suggestions have helped us tremendously. Last but not least, we thank our MSc and PhD students and postdoctoral fellows, who have helped elucidate many aspects of leaf optics, and who are cited throughout the book. Their contributions to this field have contributed greatly to our understanding. While we have received extensive aid and cooperation in reaching completion of this book, any errors, misinterpretations, or other mistakes are our own.

1
A Brief History of Leaf Color

The purpose of this introductory chapter is to provide a general survey to readers from various backgrounds about how we have thought about leaf properties related to their interactions with light. For example, questions such as "is it colored because of how light contacts the surface or because some colors of light are absorbed by particular materials?" These questions aroused curiosity about how the nature of interactions with light influences leaf properties, such as observations of leaf color differences on the upper and lower foliar surfaces or why leaves change color in the fall. Investigations from Aristotle up to the 19th century focused on the causes of leaf color and its variation and how these relate to how leaves function. Finally, we introduce some of the earliest studies on the physical mechanisms for the color patterns observed. As the chemical properties of leaves became known, researchers began to show close linkages between the three-dimensional structures that result from their anatomical and morphological patterns, the patterns of light absorption and scattering across the electromagnetic spectrum, and the associated physiological processes related to carbon, water, and nutrient budgets. The modern study of leaf optics really began with the 20th century. The evolving understanding of the botanical characteristics of leaves provides a context to understanding their optical processes.

For a long time, plants have been considered primarily from the point of view of utility and medicinal use. Historically, Assyrians, Egyptians, Chinese, and Indians recorded information about the external and internal characteristics of plants, but their descriptions were often fragmented or enigmatic. Greek philosophers contributed to the early development of botany as a science. It appears likely that the study of leaf optics began with a desire to understand the color of foliage. Little is known about Aristotle's work on the nature of plants (384–322 BC). In the *De Coloribus*, he identified four colors corresponding to the four elements: earth (black), air (white), fire (yellow), and water (blue). Aristotle described the changes in the color of plants during their development, maturity, and decay, although the passages at issue are not easy to interpret:

> . . . stagnant waters, and hence also sap in plants, are yellow green (χλωρον) at first, but next, when darkened by the rays of the sun, they become grass green (ποωδεζ) . . . When grass green water, including the sap in plants, is mixed with the rays of the sun, it is to some extent darkened . . . Furthermore, as the black in plants gradually weakens, grass green reverts to yellow green again.

Around 300 BC, Aristotle's student, Theophrastus (371–286 BC), produced the first work describing plant leaves in his encyclopaedia *De Historia Plantarum*, which we know from the Latin translation in the second half of the 15th century by Teodoro Gaza. For instance, he noticed differences in color between the upper and lower surfaces of some species (Hort, 1916):

Now all leaves differ as to their upper and under surfaces; and in most trees the upper surfaces are greener and smoother, as they have the fibres and veins in the under surfaces, even as the human hand has its "lines" but even the upper surface of the leaf of the olive is sometimes whiter and less smooth.

(book 1, chap. 10)

In the French translation, Suzanne Amigues remarks that the symmetry of the two comparisons is apparent (Amigues, 2003). Generally, "the upper surfaces are greener and smoother" than the lower surfaces. The upper surface of the olive leaf is a little rough and gray-green, the lower surface silky and silver-green: it is greener and less rough above than below. Compared to the upper surface of other species which are green and smooth, the olive leaf is "whiter and less smooth". Three centuries later, Pliny the Elder textually repeats the same words in his *Naturalis Historia* (Pliny the Elder, 1855):

These trees [elm, lime, olive, white poplar and willow] also present in their leaves the same difference that is to be observed in those of all the rest: the underside, which looks toward the ground, is of a green grassy colour, and has a smooth surface; while the veins, the callous skin, and the articulations, lie upon the upper face, the veins making incision in the parts beneath, like those to be seen upon the human hand. The leaf of the olive is whiter above, and not so smooth.

(book 16, chap. 36)

The late Renaissance period coincides with renewed interest in botanical medicines in the early botanic gardens of Leiden (1577), Montpellier (1593), and Heidelberg (1597) and their botanical descriptions in herbals, for example, the *De Historia Stirpium* of Leonhart Fuchs (1542) or the *Kitab-i hasha'ish* ("The book of herbs") published in 1595 (Figure 1.1). The latter is actually a Persian translation of the *De materia medica* of the Hellenistic scholar Dioscorides written in the 1st century AD.

In chapter 8 of the Notebooks of Leonardo da Vinci (Richter, 1970), *Botany for Painters and Elements of Landscape Painting*, probably written between 1513 and 1515, we find the earliest reference on the interaction of light with plant leaves. Although this chapter is less famous than his chapter 7, *On the Proportions and in the Movements of the Human Figure*, it is considered of similar biological significance. Da Vinci attempts to provide scientific explanations for why things look as they do and sets up rules to guide artists in representing trees. In particular, he explains how the colors of leaves should look in sunlight and in shade (Figure 1.2). The proportions of light and shade depend on the position of the leaf in relation to the Sun and the viewer. Seen from below, a concave leaf surface will be partly in shade and partly transparent, while the upper exposed surface is in light.

Da Vinci also noted that light reflected from the darkest leaves approximated the color of the atmosphere (Figure 1.3) because light on the illuminated portion combines the blueness of the atmosphere with the dark hue of the leaf to reflect a blue color. Yellow-green leaves do not reflect blue but combine the reflected blue of the atmosphere with the yellow of the leaf to appear yellow-green.

In the 17th and 18th centuries, scientists began to study the origin of green color in plants, which is widespread in the plant kingdom. In *Experiments and Considerations Touching Colours*, the famous British scientist Robert Boyle raised the question of leaf color and especially its changes throughout the seasons (Boyle, 1664):

First I have been willing to leave unmentioned the most part of those phenomena of colours, that Nature presents us of her own accord such as the different colours that [ ] appear upon the fading of flowers and leaves, [ ] etc. together with a thousand other obvious instances of the changes of colours.

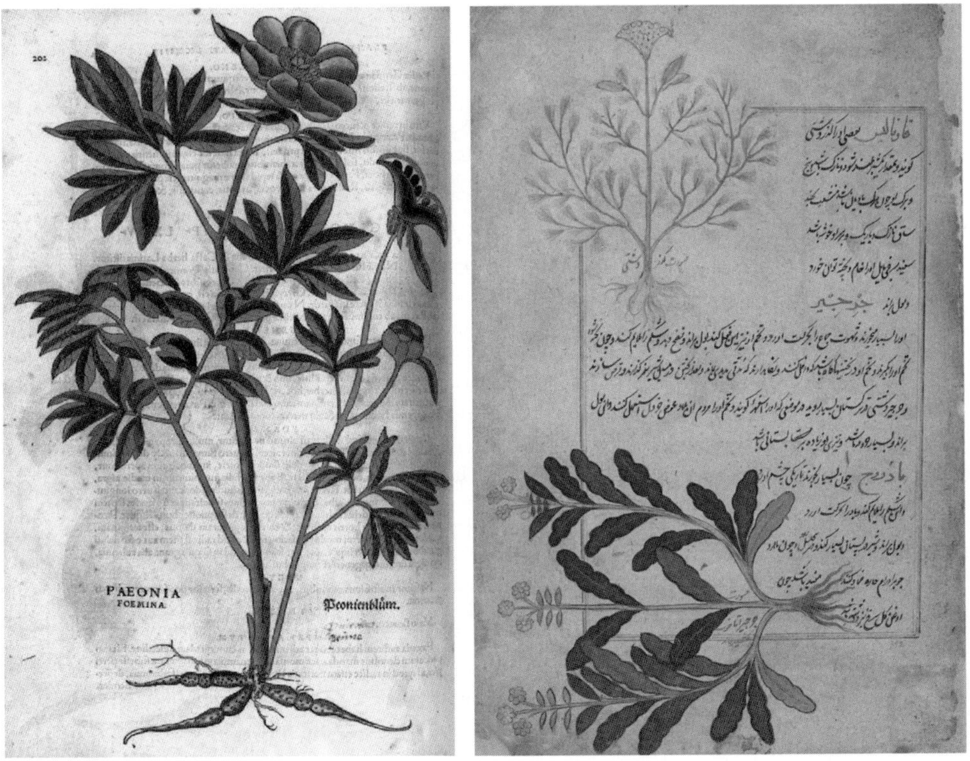

Figure 1.1 Illustrations of leaf structures: (left) from the *De historia Stirpium*; and (right) from the *Kitab-i hasha'ish* depicting bifacial leaf color. (A black and white version of this figure will appear in some formats. For the color version, please refer to the plate section.)

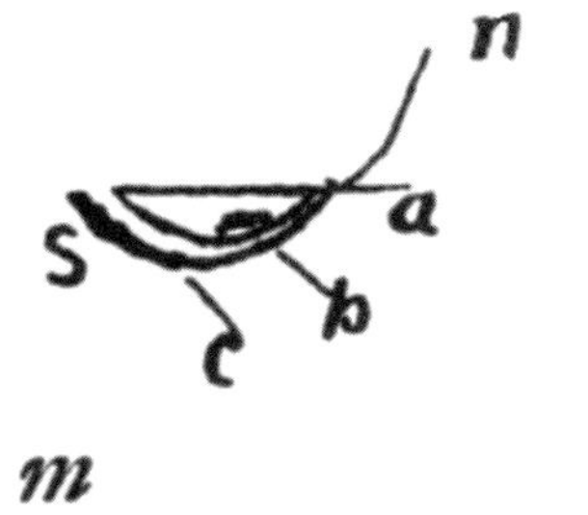

Figure 1.2 Proportion of light and shade in a leaf as seen by Da Vinci: "Sometimes a leaf has three accidents [of light] that is: shade, lustre [reflected light] and transparency [transmitted light]. Thus, if the light were at *n* as regards the leaf *s*, and the eye at *m*, it would see *a* in full light, *b* in shadow and *c* transparent" (Richter, 1970).

Later, he ponders why the two sides of some leaves display a notable disparity of colors that is revealed "when a breath of wind passes though them". However, Boyle does not consider light, but only thinks of colored matter. His theory of colors is quite different from that of Descartes, Newton, or Hooke, whose approaches were more physical: in his famous book *Optiks*, Newton (1704) explained the color of objects in terms of light interacting with them: "These colours arise from hence, that some natural bodies reflect some sorts of rays, other sorts more copiously than the rest".

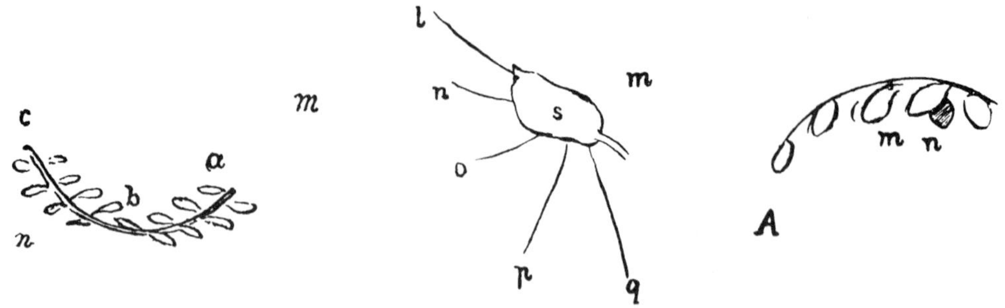

Figure 1.3 The gradations of shade and color in leaves as seen by Da Vinci: "If the light comes from *m* and the eye is at *n* the eye will see the colour of the leaves *a b* all affected by the colour of *m* – that is of the atmosphere; and *b c* will be seen from the underside as transparent, with a beautiful green colour verging on yellow. If *m* is the luminous body lighting up the leaf *s* all the eyes that see the underside of this leaf will see it of a beautiful light green, being transparent. In very many cases the positions of the leaves will be without shadow [or in full light], and their underside will be transparent and the right side lustrous [reflecting light]" (Richter, 1970).

Newton used prisms to prove that white light was actually made up of waves of different colors; then he showed that objects appear to be certain colors because they absorb and reflect different amounts and wavelengths of light. Newton surprisingly barely mentioned plant leaves or flowers in his book. For instance, Delaval (1774) only found one observation concerning the degradations of the green in the plants that fade. He was convinced that Newton's results obtained on transparent natural bodies, like glass, water, and air, were applicable to opaque bodies, and that a lack of experiments prevented him from discovering the origin of color in the animal and plant worlds.

In parallel, the chemists of the 17th and 18th centuries prepared "infusions of plants into several sorts of liquors" to extract their coloring principles or they burned leaves in the open air, assuming that the ash that remained after combustion contained these substances. Some, like the German physicians Johann Joachim Becher, in *Physica Subterranea* (Becher, 1669), and Georg Ernst Stahl, in *Opusculum Chymico-Physico-Medicum* (Stahl, 1715), attribute the green color of leaves to iron. While he was appointed counselor of mines for Saxony, in the heart of Germany, Johann Friedrich Henckel published *Flora Saturnisans* where he studied the chemical similarities between plants and minerals (Henckel, 1760). However, he could not make up his mind whether leaf color was due to iron or copper. He noted that M. L'Abbé de Vallemont had heard that tree leaves in the vicinity of gold mines have a golden color that is produced by ground exhalations. And to prove it, he relates this passage from Boyle: "Folia arborum saepiùs aureo colore obducta inveniri ab auri fodinarum exhalationibus metallicis". However, Lemery (1706) wonders about the presence of iron in plants: does it naturally occur or is it formed when plants are burnt to ashes? He is inclined to favor the first hypothesis:

Le fer étant répandu en abondance dans toutes sortes de terres, & pouvant être aisément dissous par les premières liqueurs salines qui l'arrosent [ ]; ces liqueurs montant ensuite par la chaleur du Soleil dans les tuyaux des plantes pour les nourrir & les faire croître: ces liqueurs, dis-je, portent naturellement avec elles le fer donc elles se sont chargées.[1]

[1] Iron is abundantly present in all kinds of soils, and is easily dissolved by the first saline liquors that water it []; these liquors then go up by the heat of the Sun in the tubes of the plants to feed them and to make them grow: these liquors, I say, naturally carry with them iron they are laden with.

By vitrifying this ash, M. le Comte de Mouroux produced a stable dark green glass (Mouroux, 1770). He confirmed that the elements which cause the green color are independent of the body in which they are embedded. However, other authors showed that the color of the glass might be rather due to the temperature applied during the vitrification process. Mouroux also mentions a new theory expressed by Benjamin Franklin:

> I have been rather inclined to think that the fluid fire as well as the fluid air, is attracted by plants in their growths and becomes consolidated with the other materials of which they are formed and makes a great part of their substance.
>
> *(Franklin, 1751)*

and concludes that, with the help of this *fire*, plants receive a vital principle and the development of color. Note that although these theories were supported by several similar experiments, the results sometimes contradict each other (e.g., Delafolie, 1774; Achard, 1778; Morozzo, 1782).

The natural variation of colors observed in flowers and leaves was also attributed to the mixing of sulfurs with different salts contained in the sieve cells. In a discourse read at the Royal Society of London on 3 May 1677, the English physician Nehemiah Grew seems to confirm these early observations (Grew, 1682):

> I suppose therefore, that not only green, but all the colours of plants, are a kind of precipitate, resulting from the concurrence of the saline parts of the air, with the saline and sulphurious parts of the plant.

He explains that he could extract the green color of leaves with olive oil, but not with water. At the instigation of the Academy (understood to be the French *Académie Royale des Sciences* created in 1666), a complete chemical analysis of 1400 plant species was undertaken to know their nature, properties, and uses. They were found to contain common substances, in particular sulfurs, which at that time were suspected by chemists to be the cause of color. Thus Geoffroy (1707) notes: "le vert qui est la couleur la plus ordinaire des feuilles, est vraisemblablement l'effet d'une huile raréfiée dans les feuilles, & mêlée avec les sels volatiles & fixes de la sève".[2] He also explains why leaves turn red in autumn by the beginning of the cold season: the pores and channels of the sieve cells stop flowing, so the sap is retained in the leaves where it turns sour; the acid that is produced destroys the alkali that is behind the green color, and the remaining sulfurs cause the red color. Senebier (1783), who was struck by the co-occurrence between leaf greenness and carbon fixation under the action of solar radiation, hypothesized that carbon was the cause of the green color. In 1809, he was the first to show that green matter was located in the parenchyma cells. Sulfur, iron, copper, gold, or carbon: at the end of the 18th century, physicists (or alchemists) were actually still in disagreement about the origin of color in leaves and other plant organs. But is chemistry the right way to unveil the secrets of plants?

The dominant influence of light on leaf color was recognized for the first time by the 17th-century English naturalist and botanist John Ray who noted that only light affects the green color of plants and that leaves turn white in darkness (Ray, 1686); if the plants that have been grown in the dark under an opaque vase are returned to the light, the leaves soon lose their white hue, and eventually assume their natural color; the rapidity with which they become green, and the intensity of their color, will be in proportion to the amount of light to which they are exposed. Ray made sure that this phenomenon was due neither to the deprivation of the air nor to the influence of heat. The evolution of current ideas on

[2] Green, which is the most ordinary color of the leaves, is probably the effect of a rarefied oil in the leaves, which is mixed with the volatile and fixed salts of the sap.

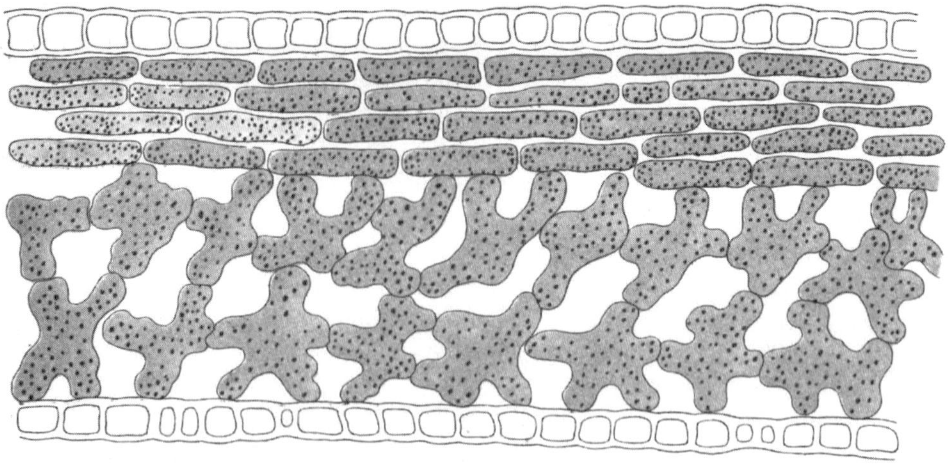

Figure 1.4 Cross-section of a European dogs tooth (*Erythronium dens-canis*) leaf. The epidermal cells are colorless; the upper mesophyll is made of elongated cylindrical cells that contain "granular" and "gelatinous" chlorophyll; the cytoplasm is either red or colorless, as it contains erythrophyll, the red coloring matter of leaves, or not; the white areas are due to the interposition of a thin air layer between the epidermis and the upper mesophyll (Morren, 1858a). (A black and white version of this figure will appear in some formats. For the color version, please refer to the plate section.)

leaf color is inseparable from photosynthesis and the discovery of leaf pigments (for a comprehensive review, see Govindjee and Krogmann, 2004, and Hill, 2012). Joseph Priestley, an English chemist, initiated the experimental study of photosynthesis by discovering oxygen (termed dephlogisticated air) in 1774 and published his findings the same year (Priestley, 1774). He showed that oxygen could be produced by plants and support respiration in animals. He was followed by Jan Ingen-Housz who showed that light was essential to photosynthesis (Ingen-Housz, 1779), Jean Senebier who discovered that CO_2 was required for photosynthetic growth (Senebier, 1783), and Nicholas-Théodore de Saussure who highlighted the role of water (de Saussure, 1804). Finally, the German physicist Julius Robert von Mayer defined photosynthesis in his second publication using the principle of conservation of energy (Mayer, 1845). He showed that light energy from the Sun was stored as chemical energy in products formed during photosynthesis. The theory of native metals (iron, copper, gold, etc.) or non-metallic elements occurring in the native state (carbon, sulfur, etc.) to explain leaf color lost momentum for the benefit of green substances, first called "gelatinous green matter", green starch", "viridine", "resin", or "chromule". These substances can be found in the form of granules or in the amorphous state (Figure 1.4).

The word "chlorophyll" came from French in the early 19th century. It was made up from the Greek words *chloros* "light green" and *phyllon* "leaf" by two French pharmacists, Pelletier and Caventou (1817). It is ironic to note that they did not make a big deal of this name: "nous n'avons aucun droit pour nommer une substance connue depuis longtemps, et à l'histoire de laquelle nous n'avons ajouté que quelques faits; cependant nous proposerons, sans y mettre aucune importance le nom de chlorophylle".[3] Originally, all pigments were referred to as

[3] We have no right to name a substance known for a long time, and to the history of which we have added only a few facts; however, we will propose, without putting any importance on it, the name chlorophyll.

chlorophyll. The term anthocyanin (*anthos* being Greek for flower, and *kyanos* for blue) has been used since Marquart (1835) to represent the coloring matter that is responsible for the various colors found in flowers, fruits, and autumn foliage. Note that it has been long called erythrophyll (*eruthros* being Greek for red). Berzelius (1837a, 1837b) named the yellow pigments obtained from the autumn leaves xanthophylls (*xanthos* being Greek for yellow). According to these authors, anthocyanin and xanthophyll were considered as a counterpart to chlorophyll and assumed to be the result of a metamorphosis of chlorophyll under the effect of solar radiation (Phipson, 1858). It seems that chemists had trouble extracting and purifying leaf pigments (Filhol, 1865). For instance, without pure chlorophyll extract they could not make a judgment on the nature of chlorophyll: is it a pure substance or a mixture of several? Verdeil (1851) announced that chlorophyll was related to the coloring matter of blood, and as such, it contained iron. Fremy (1860, 1865) has ascertained that it was composed of two coloring principles, one a yellow called phylloxanthin, the other a blue called phyllocyanin. Later, Sorby (1872) identified two chlorophyll pigments, but the identification of different leaf pigments was only resolved in the 20th century. Scientists ran into another snag with the evolution of leaf color in the fall that occurs during leaf senescence (Macaire-Princep, 1828; Berzelius, 1837c; Morren, 1858a,1858b, 1858c), but also periodically in evergreen leaves (Mohl, 1837, 1838). Morren (1858a), who reviewed all the past and current theories explaining plant color, lamented the fact that this subject had only interested scientists in passing, but had rarely given rise to specific studies. In the middle of the 19th century, the main pigments that produce leaf colors – green, yellow, red, brown, etc. – are named, although their molecular formulas were not identified.

The second half of the 19th century is devoted to the more difficult study of their structure and evolution and, in connection with this, their relationship with the physical environment of growing plants. Once achieved, scientists rapidly started to study their intrinsic optical properties, in particular the absorption and fluorescence emission of chlorophyll. Brewster (1834) focused the Sun's light by a lens and he studied its dispersion by an English laurel (*Prunus laurocerasus*) leaf solution obtained by *absolute alcohol* (pure ethanol) extraction. The light transmitted through the solution was analysed by a prism. He wrote:

> we shall observe a spectrum of the most beautiful kind. In place of seeing the green space with a portion of blue on one side and yellow on the other, as the Newtonian theory would lead us to expect, we perceive a spectrum divided into several coloured bands of unequal breadths, and having their colours greatly changed by absorption. . .

as illustrated in Figure 1.5.

In the same article, Brewster described for the first time a very remarkable phenomenon, which he designated as *internal dispersion* and which is a major discovery in plant physiology: chlorophyll fluorescence emission. He experimentally showed an emission of red light in an alcohol extract of laurel (*Prunus laurocerasus*) leaves: "In making a strong beam of the Sun's light pass through the green fluid, I was surprised to observe that its colour was a brilliant red, complementary to the green". Even if leaf optical properties naturally included blue and red fluorescence, fluorescence is a world apart which has already given rise to many books and hundreds of dedicated articles, due to the close link between fluorescence and photosynthesis. For this reason, we will only mention it occasionally in this book. Twenty years after Brewster, Stokes (1852a) made a new experiment of light dispersion

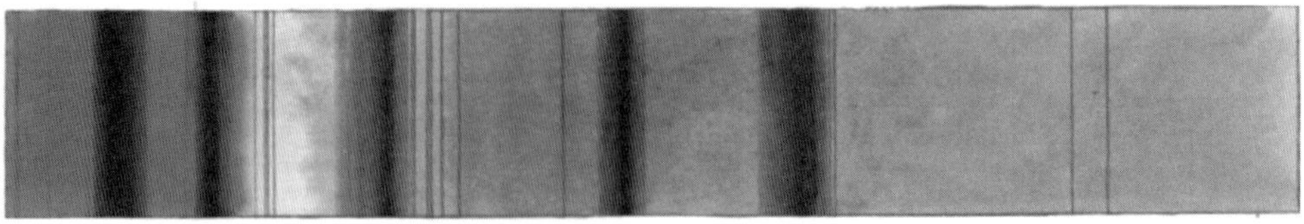

Figure 1.5 Decomposition of light by a solution of an English laurel green leaf showing several main absorption bands (from Brewster (1834), *Transactions of the Royal Society of Edinburgh*. Copyright © 1834 with permission from Cambridge University Press). (A black and white version of this figure will appear in some formats. For the color version, please refer to the plate section.)

with a solution of common nettle (*Urtica dioica*) leaf. He used a candle as an illumination source and found five absorption bands situated in the blue (No. 5), in the green-yellow (No. 4), in the yellow (No. 3), and in the red (Nos. 1 and 2). He also probably mentions the first experiment dedicated to the measurement of a leaf transmittance spectrum:

It should be noted that although the absorption produced by leaf-green is best studied in a solution, its leading characters may be observed very well by merely placing a green leaf behind a slit, as near as possible to the flame of a candle, and then viewing the slit through a prism.

Brewster (1855) translated Newton's statements for a leaf as follows:

> The leaf of a plant, for example, appeared green in the white light of day, because it had the property of reflecting green light in greater abundance than any other. When the leaf was placed in homogeneous red light, it no longer appeared green, because there were no green rays in the red [ ] The green leaf, for example, stops or absorbs the red, blue and violet rays of the white light which falls upon it, and reflects and transmits only those which compose its peculiar green.

He and other authors undertook an extensive investigation of absorption spectroscopy and demonstrated linkages between vegetal colors and the absorption spectrum of plant leaf extracts (e.g., Ångström, 1853, 1854; Salm-Horstmar, 1854, 1855a, 1855b, 1856; Harting, 1855; Landrin, 1864; Hagenbach, 1870; Sorby, 1871a; Chautard, 1872; Schönn, 1872; Palmer, 1877a, 1877b; Timiriazeff, 1903; Ursprung, 1918). In particular, they studied the evolution over time of leaf pigment content and therefore of leaf color (e.g., Sorby, 1871b, 1884; Martin and Thomas, 1887; Gauthier, 1906).

Leaf extrinsic optical properties were first measured at the dawn of the 20th century, to answer the question of storage of solar energy in green plants. Scientists entered the era of quantitative data through the parallel development of measuring instruments. It is interesting to note that the earliest papers on leaf optical properties addressed the question of radiative energy exchange between plant leaves and their surrounding environment. The French physicist Edmond Becquerel, who discovered evidence of radioactivity, was the first to attack the question of storage of solar energy in green plants (Becquerel, 1868). Following this early work, Timiriazeff (1903), in a long article titled *The cosmical function of the green plant*, showed that the absorptance of direct sunlight by leaves, called the *economic coefficient*, was approximately 25%, an obviously underestimated value. As cited by Ansari and Loomis (1959), the earliest information regarding leaf temperatures dates from the mid-1870s when Askenasy (1875) held a mercury thermometer against the surface of thin leaves of *Sempervivum* and observed that leaves in sunlight were 4 to 5°C warmer than the surrounding air. More accurate determinations have been made by means of type-T (copper-constantan) or type-K (chromel-alumel) thermocouples and potentiometers. The thermocouple

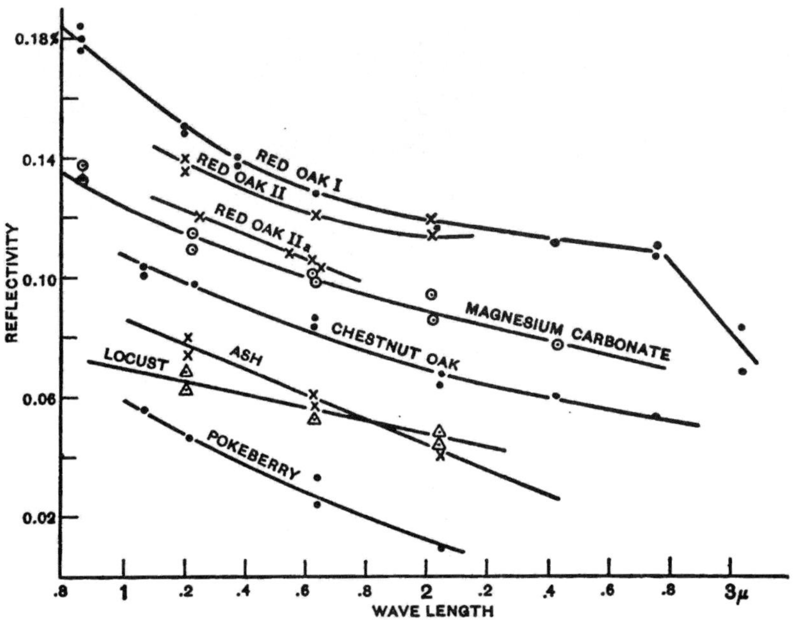

Figure 1.6 Reflecting power of green leaves (from Coblentz (1913), *Bulletin of the Bureau of Standards*. Copyright © 1913 with permission from the National Institute of Standards and Technology).

may be threaded through the mesophyll, inserted into the midrib, or glued to the underside of the leaf under experiment. Nevertheless, a good contact between leaf and junction is required to accurately measure leaf temperature. This article opened the way to new fields of applications in leaf radiative budget (e.g., Maquenne, 1880; Brown, 1905; Brown and Escombe, 1905; A.M.S., 1909; Clum, 1926).

In his studies of the *reflecting power* of matte surfaces, Coblentz (1913) published the first measurements of the reflection of visible radiation from leaf surfaces, undertaken in May 1908. At that time he used a bolometer, a mirror spectrometer, and a fluorite prism. The green leaves of nine plant species were illuminated at an angle of incidence of 45° with a Nernst glower, an obsolete device for providing a continuous source of infrared radiation from 2 to 14 micrometers. The curves of *reflecting power* show a regular decrease in reflection of energy of between 0.9 μm and 3.0 μm, which is true overall (Figure 1.6). However, these curves are difficult to interpret due to the poor spectral resolution, and the low near-infrared values do not seem to be realistic. Coblentz also provided the transmittance of plant leaves for the first time: about 20% of the energy, a definitely overestimated value, was found to pass through common lilac (*Syringa vulgaris*) and black locust (*Robinia pseudoacacia*) leaves at 600 nm.

The description of leaf optical properties in terms of physical processes, that is absorption of light by photosynthetic pigments and diffusion at the air–cell wall interfaces (Figure 1.7), is often attributed to Willstätter and Stoll (1918), who shrewdly understood that photosynthesis could not be reduced to a biochemical reaction.

Using a prism spectrophotometer and magnesium carbonate ($MgCO_3$) as a photometric standard surface, Shull (1928, 1929) measured the reflectance spectrum of the upper and lower surfaces of

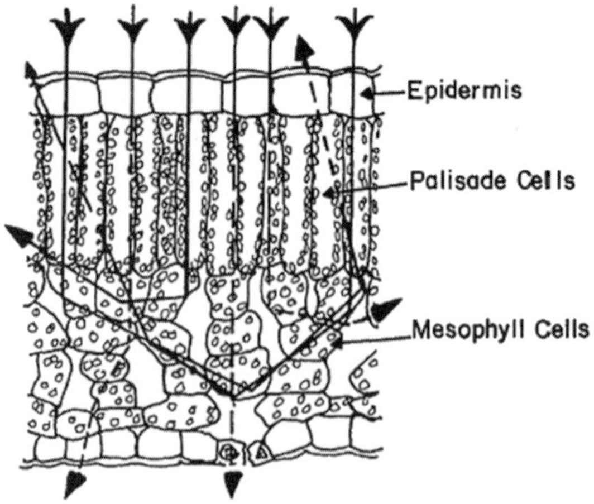

Figure 1.7 Schematic drawing depicting the Willstätter and Stoll (1918, p. 123) theory on the pathway of light through a dorsiventral leaf of *Acer negundo* (adapted from Sinclair et al. (1973), *Agronomy Journal*. Copyright © 1973 with permission from the American Society of Agronomy).

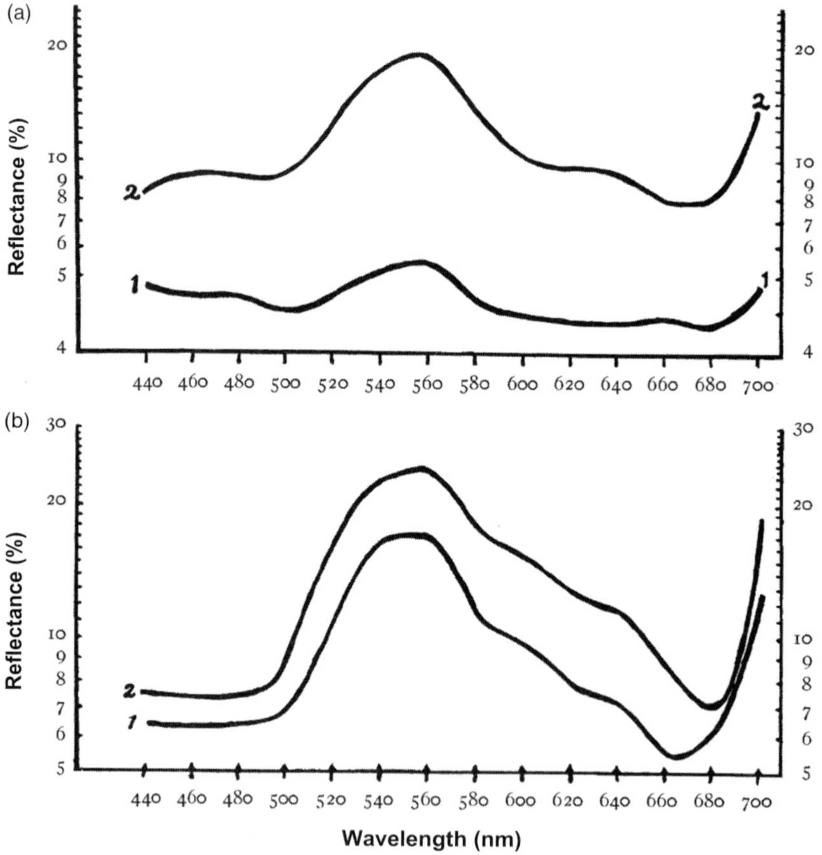

Figure 1.8 Reflection curves for leaves of: (a) rhododendron (1) and jonquil (2); (b) violet (1) and iris (2) (from Shull (1929), *Botanical Gazette*. Copyright © 1929 with permission from the University of Chicago Press).

a number of leaves from 430 nm to 700 nm (Figure 1.8). His results clearly indicate a maximum reflectance at approximately 550 nm (green) and a minimum reflectance between 660 nm and 680 nm (red) corresponding approximately to the maximum absorption band of chlorophyll. Unfortunately the spectra stop before the near-infrared plateau. The work of Shull initiated a series of articles on the monochromatic reflection of plant leaves. As reported by Billings and Morris (1951), it took almost 20 years and the improvement of monochromatic spectrophotometers before one could measure reflectance spectra continuously from the visible to the near infrared. The results of Rabideau et al. (1946) showed a sharp rise in reflectance starting at about 675 nm and continuing to a plateau of about 50% extending from about 750 nm to 850 nm, the limit of their observations. The advent, around 1940, of sensitive photographic film in the near infrared had already highlighted a much higher reflectance of plant leaves in this spectral range.

The applications of such studies are numerous and cover many scientific disciplines, from plant physiology (photosynthesis and photomorphogenesis) to remote sensing in the optical domain (environmental studies, precision farming, and ecology). Most papers have focused on the leaf spectral properties (hemispherical reflectance and transmittance) in connection with their biochemical content (chlorophyll, water, dry matter, etc.) and their anatomical structure. For instance, plant stress resulting from an insect attack or a nitrogen deficiency induces degradation of the leaf chlorophyll content, which has repercussions on the leaf optical properties: the reflectance and transmittance increase over the whole visible spectrum. This relation between cause and effect allows the estimation of leaf biochemistry – the chlorophyll content in this particular case – by establishing empirical relationships between the variable of interest and the leaf reflectance or transmittance, or better still, by directly using a physical model. It is now well established that leaf reflectance and transmittance are closely related to the biochemical content and anatomical structure, which depend on the plant species and, of course, on many environmental factors. Quantitative relations between these optical properties and these biophysical characteristics were empirically established in the second half of the 20th century: for example, stress can involve degradation of the chlorophyll pigments which, in turn, will cause an increase in reflectance and transmittance in the visible. In parallel, radiative transfer models were used to simulate these physical processes and estimate leaf biochemical composition.

2

Leaf Biophysics

This chapter aims to describe the basic anatomy of common groups of higher plants. including the tissues of leaves, their main cell types, and the biochemical constituents that characterize their functional properties. It should provide enough detail on the main construction of plant leaves and how major groups of plants are distinguished based on anatomy, morphology, cell type distribution, and biochemistry. The three-dimensional structure and arrangement of the organelles, cells, and tissues in the leaf are critical to understanding the photon transport in leaf tissue and how these traits relate to the physiological processes of photosynthesis, respiration, and transpiration.

The conquest of land by plants started about 450 million years ago. Today, they are the dominant life form on Earth. Green plants are those that obtain most of their energy from sunlight via the photosynthetic process, and include familiar organisms such as trees, herbs, shrubs, grasses, ferns, mosses, and green algae (Figure 2.1 and Figure 2.2). Here, we focus on higher plants (vascular plants) that have lignified tissues for conducting water, minerals, and photosynthetic products through the plant; these represent the most widespread group of land plants. Since the optics of ferns have given rise to very few remote sensing studies, spermatophytes (seed plants) are central to this book. These include angiosperms (flowering plants possessing seeds enclosed in carpels, ~ 260 000 species) and gymnosperms (plants possessing naked seeds, ~ 700 species). The coniferous gymnosperms are woody trees and shrubs (pines, firs, spruces, sequoias, junipers, cypresses, larches, cedars, etc.) and most, with the exception of the *Podocarpus* family, have needle-shaped leaves. Other groups of gymnosperms include cycads, ginkgo, and the *Gnetophyta* (*Gnetum* and *Welwitschia*) with broader leaf blades. Among the angiosperms, the major distinction is between monocotyledons (or monocots, ~ 35 000 species) and dicotyledons (or core eudicots, ~ 165 000 species in 7–10 clades). These differences are particularly important in terms of foliar anatomy and, as seen later, in terms of optical properties. For instance, monocots represent nine clades with diverse growth forms, including grasses (wheat, maize, barley, bamboo, etc.), orchids, aquatic herbs, and trees (*Pandanus, Dracaena*). Eudicots, with eight clades, also have highly diverse growth forms including herbs, parasitic plants, shrubs, vines, lianas, and large trees (oak, beech, southern birch, hazel, etc.).

Although the absolute number of plant species is large (~ 298 000 species, of which ~ 216 000 have been described and catalogued, see Mora et al., 2011), the number of dominant species in ecosystems, that is, those that structurally dominate the canopy layer, is much smaller, with the number of gymnosperm species being the smallest, followed by dicotyledonous evergreen species, deciduous woody species, and herbaceous species representing the largest dominance in terms of biodiversity. This reduces the number

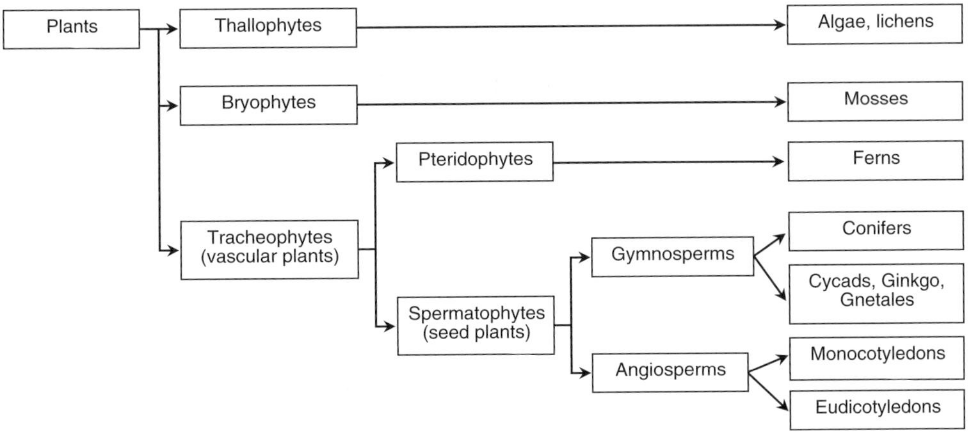

Figure 2.1 Simplified classification of plants. The principal cultivated species belong to the group *Spermatophytes* (Lecointre et al., 2007; www.tolweb.org/tree/).

Figure 2.2 A few examples of plant types. (A black and white version of this figure will appear in some formats. For the color version, please refer to the plate section.)

of species to study, especially if some convergence of optical properties exists among plant species based on an assumption that there is an integration of leaf form and function that has occurred over evolutionary time. There is renewed interest in classifying species into groups relating to function rather than to taxonomy. Plant functional traits are the features (morphological, physiological, and phenological) that represent ecological strategies and determine how plants respond to environmental factors. Their measurement in many species from different habitats and microclimatic conditions requires standardization that has been described in the literature (Pérez-Harguindeguy et al., 2013).

2.1 Leaf Anatomy

Variation in internal leaf structure has been a source of scientific fascination since the late 1600s. The anatomical organization of plant leaves is generally conservative within a species and remains largely unchanged even under widely differing environmental conditions, but leaf anatomy can display large differences between species and even within species for leaves grown under various environmental conditions, including intensity and quality of light. All leaves share the general tissue organization of epidermis, mesophyll, and vascular structures, and are generally flattened or compressed in their dorsiventral orientation consistent with their primary function of harvesting light for photosynthesis. However, the characteristics of specific tissues vary widely with species characteristics, environmental conditions, especially due to the light environment, and availability of nutrients and water. For example, there are three primary pathways for photosynthesis, each with different anatomical characteristics. The most widespread and common form is termed C_3, indicating that the first product of photosynthesis is a three-carbon sugar. The other two pathways are C_4 (forming four-carbon sugars) and CAM (Crassulacean Acid Metabolism, a specialized type of C_4 photosynthesis adapted to arid conditions in which the stomata are open at night to take up CO_2 but the conversion to C_4 sugars occurs during the day). C_4 and CAM pathways are often found in harsh environments like arid, saline, or wetland ecosystems. These pathways and corresponding anatomical characteristics are described in greater detail afterwards.

Conifers have the widest geographic distribution within the gymnosperm group, although restricted to the northern hemisphere. They have foliage that is characterized as needle-shaped, scale-like, or strap-shaped. Leaves are generally persistent for several growing seasons. A cross-section through a fresh pine needle illustrates the tissue organization typically described in botanical literature (Figure 2.3). The needle or strap-shaped leaves are typically schlerophyllous, with small cells, thick cell walls, and little air space between mesophyll cells, and with stomata located below the epidermal surface, which altogether contribute to limiting the exchange of gases with the surrounding environment. The thick cross-section produces a low surface-to-volume ratio which also aids in reducing water loss in dry climates. Angiosperms have more differentiated and structurally diverse leaf tissues than gymnosperms. Monocots and eudicots (Figure 2.3) have characteristically different leaf anatomical structures, although some prominent features common to both groups emerge.

2.1.1 Epidermis

The epidermis, also called the dermal tissue, is the specialized outermost cell layer of the leaf, which is typically one cell thick. It regulates exchange of water and carbon dioxide between the interior of the leaf and the air and provides mechanical strength to the leaf. The epidermis is made of several kinds of specialized differentiated cells, including guard cells that form the stomatal complex and regulate and allow gas exchange with the interior, subsidiary cells, and epidermal hairs (trichomes).

The cuticle is a non-cellular protective layer that is produced by the epidermal cells and that covers the entire leaf surface (Figure 2.4). It may be covered with wax that can have many forms from smooth to complex three-dimensional multicellular structures. It is among the most resistant parts of the plant to decomposition so that it has a high potential for preservation, as evidenced by fragments extracted from soil cores. Therefore it reflects the cell pattern of the epidermis and can be used for

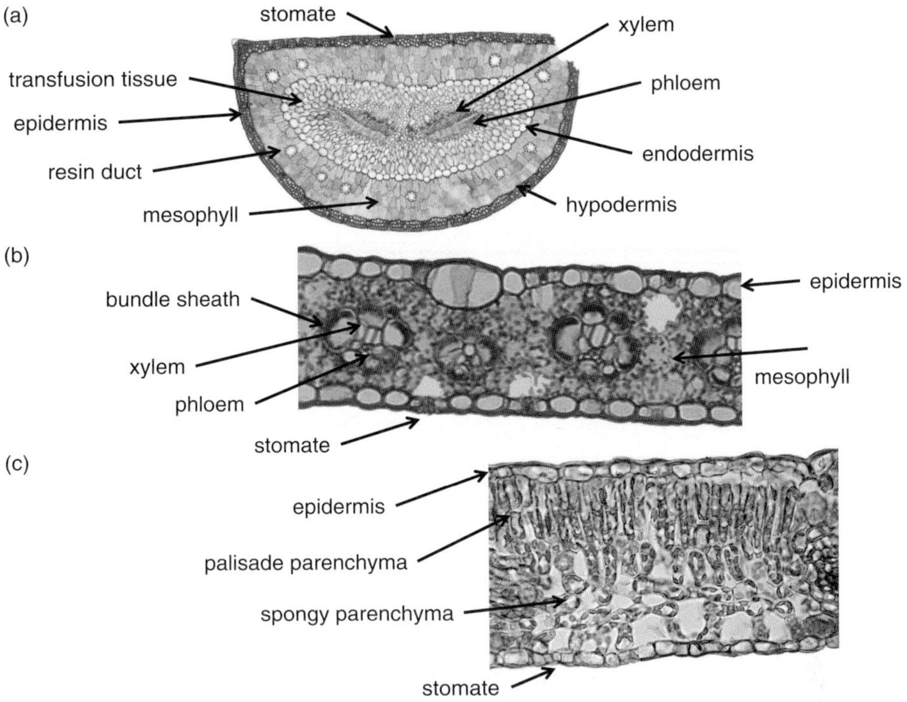

Figure 2.3 Micrograph cross-sections of the anatomy of (a) a conifer needle (© Philippe Labrot/http://nirgal .net/); (b) a monocot (maize) leaf (© Larry Klotz, Department of Biological Sciences, SUNY Cortland); and (c) a dicot (beech) leaf.

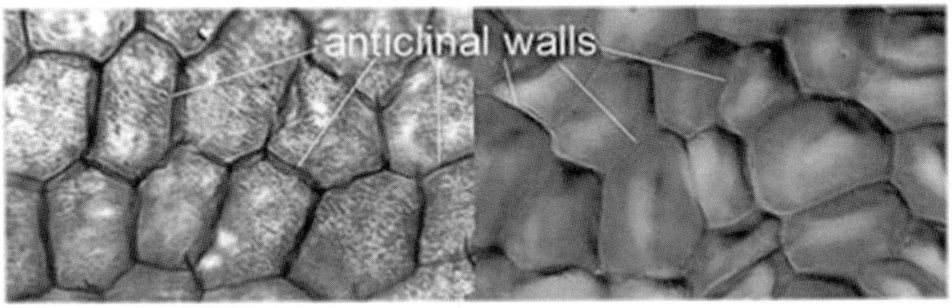

Figure 2.4 Pteridosperm cuticles from the Upper Carboniferous (left) and Upper Permian (right), clearly show relief from the anticlinal walls (www.uni-muenster.de/GeoPalaeontologie/palaeobotanik/).

identifying plant fossils and reconstructing phylogenies and paleoclimate/ecology (e.g., Wolfe and Upchurch, 1986, 1987; Upchurch, 1995).

The cuticle serves many functions. It provides a permeability barrier for water and other molecules, which limits evaporation and gas exchange. It restricts contamination from external water; therefore it is a significant barrier to herbicide or pesticide absorption. The micro- and nanostructures of the cuticle confer defensive properties that prevent penetration by dust, atmospheric pollutants, and microorganisms including virus particles, bacterial cells, and the spores or growing filaments of fungi (Muller and Riederer, 2005). In addition to these leaf functions, many plants, such as those of the sacred lotus (*Nelumbo nucifera*), exhibit ultra-hydrophobic epicuticular wax crystals and self-cleaning properties that were described by Neinhuis and Barthlott (1997).

2.1.1.1 Gymnosperms

The surface of a needle is covered by a thick cuticle, made of cutin, a three-dimensional polymer of hydroxy fatty acids, embedded in waxy substances. The waxes and other cuticular materials differ in chemical composition between species, causing differences in the luster (glossiness vs. glaucousness or bloom) of leaves that affects the light-scattering properties. The outer leaf layer is composed of a single layer of epidermal cells with high phenolic content and thick lignified cell walls. The epidermal cells are interrupted by stomata that are arranged in rows along the needle's length close to the location of the terminal xylem cells. Stomata are located below the leaf surface, which consists of guard cells around a cavity that regulates the exchange of water and carbon dioxide into and out of the leaf. In the next layer, a sclerified hypodermis may be present, which is formed from the ground parenchyma and located between the epidermis and the mesophyll in some conifer species. The cell walls in this layer may also contain a waxy material, suberin, which limits the transverse exchange of water out of the underlying mesophyll to the stomata.

2.1.1.2 Angiosperms

As with gymnosperms, the epidermis, usually made of a single cell layer, covers the whole surface of the leaf blade. The epidermis protects the leaf from dehydration, insect and animal predation, UV (ultraviolet) radiation, and mechanical stresses. It consists of thick-walled epidermal cells without chloroplasts, which are coated on the outer side with a waxy cuticle of thickness and structure that vary depending on the species (Figure 2.5). The epidermis forms the boundary separating the plant's inner ground tissue from the external world; the waxy cuticle prevents water loss and gas exchange except where stomata occur.

The epidermal surfaces of plant organs, such as stems and leaves, are often ornamented with unicellular to complex multi-cellular outgrowth structures called trichomes, the most common type of which is hairs, but which also include other structures like scales (Figure 2.6). All the major groups of higher plants have the capacity to produce such structures. Since they originate from epidermal tissues, they proceed to develop through growth, differentiation or cell division. Trichomes grow out of the plane of the epidermal cells and can be short or long and stringy, star-shaped, dendritic, hooked, and many other shapes. They can be gland-tipped with sticky, toxic, or irritating substances that provide protection from herbivory. Flavonoids and other phenolic compounds are common in the epidermal cells and in leaf hairs and are implicated in UV protection. In some species, such as desert plants, both waxes and hairs can be sufficiently dense to make leaves appear white instead of green, which provides a more favorable energy balance for these species (see Section 12.1).

Hairs can be sparse or dense on the leaf surface. Between a glabrous leaf, which lacks hairs, and a tomentose leaf, which is densely covered with short woolly hairs, nature displays a wide variety of pubescent leaves. Leaves and hairs occur in many forms. They can be hirsute (covered with coarse hairs), hispid (covered with stiff or rough hairs), downy (covered with fine soft hairs), or strigose (having stiff, straight, closely appressed hairs), etc. as shown in Figure 2.7.

In the epidermis, as with conifers, the stomata regulate the exchange of gases and water vapor between the outside air and the interior of the leaf. Monocots often have stomata on both laminar

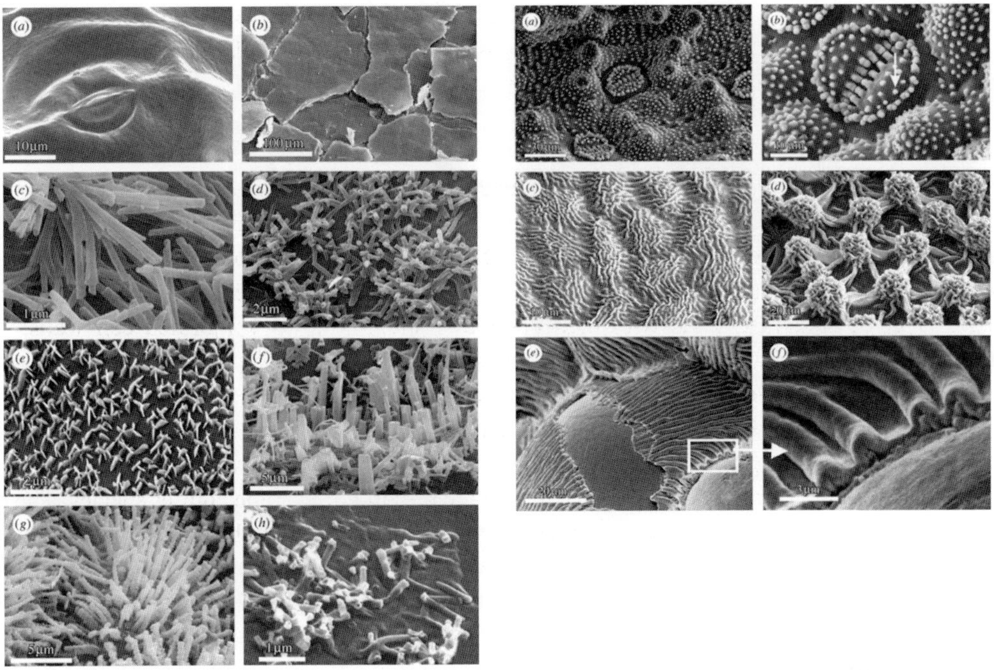

Figure 2.5 Scanning electron microscopy (SEM) micrographs of cell surface. Left panel: structuring by epicuticular waxes: (a) a thin wax film found on many plant surfaces as in *Hydrocotyle bonariensis*; (b) a crust of wax plates on a leaf of *Crassula ovata*; (c) β-diketone wax tubules on *Eucalyptus gunnii* leaves; (d) nonacosan ol tubules on *Thalictrum flavum* glaucous leaves; (e) wax platelets on *Robinia pseudoacacia* leaves arranged in rosettes; (f) simple rodlets on a leaf of *Brassica oleracea*; (g) transversely ridged rodlets on a leaf of *Sassafras albidum*; and (h) mechanically isolated waxes from a leaf of *Thalictrum flavum* on a glass surface. Right panel: Structuring by subcuticular insertions and cuticle folding: (a, b) The cell surface roughness of the common horsetail, *Equisetum arvense*, is induced by subcuticular Silicon-dioxide crystals (arrow in (b)); (c) *Schismatoglottis neoguineensis*, the structure of the cell surface is caused by cuticle folding; (d) *Alocasia macrorhiza* cells are flat (tabular) on lower leaf surface, and the cuticle in the central field forms node-like exposed structures; and (e, f) *Aztekium ritteri*, with part of the cuticle removed, shows that surface structuring is induced by the cuticle and not by the underlying cell wall (f) shown in detail (from Koch and Barthlott (2009), *Philosophical Transactions A*. Copyright © 2009 with permission from The Royal Society Publishing).

faces (bifacial) while dicots typically have stomata that are restricted to the dorsal (lower or abaxial) face. The stomatal apparatus consists of a pair of guard cells that can open or close and subsidiary cells that connect to guard cells by plasmodesmata. A substomatal cavity lies below the stomata and provides the actual substrate where gas exchange occurs. The stomatal aperture describes the state of opening or closing of the stomatal guard cells. In contrast to other epidermal cells, the guard cells of the stomata uniquely contain chlorophyll in functional chloroplasts, which regulate stomatal opening. The relative openness of the stomata depends on several interacting plant and environmental properties. Opening the guard cells is a turgor-driven process, requiring an increase in cations within the guard cells relative to the surrounding epidermal cells, so that the guard cells have a positive gradient for water absorption. Photosynthesis in the guard cells creates the osmotic gradient that causes water to diffuse into the guard cells to balance the water potential. The guard cells have thickened cell walls

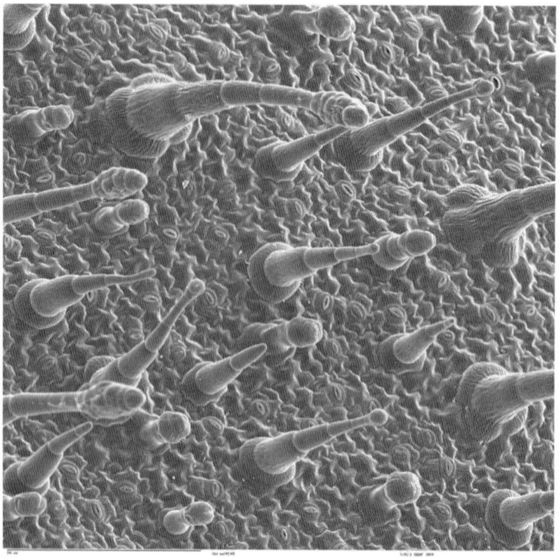

Figure 2.6 SEM image of *Nicotiana alata* upper leaf surface, showing trichomes and a few stomata (© Louisa Howard, Dartmouth electron microscope facility, Wikimedia Commons, http://commons.wikimedia.org/wiki/File:Leaf_epidermis.jpg).

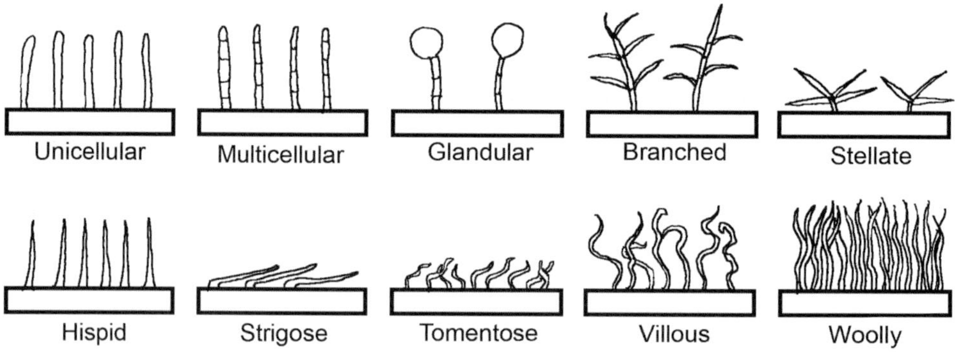

Figure 2.7 Examples of various types of leaf hair arrangements (by courtesy of Nathan Doyle).

on the inner surface that cause the two guard cells to pull apart when they are at full turgor, opening the stomata. This basic functioning is modified by the light intensity and quality, temperature gradients between the leaf and air, absolute humidity, CO_2 concentration, and water potential gradients between the air and soil. Although the leaf stomata respond dynamically to changing environmental conditions, plant hormones, particularly kinetin and abscisic acid, also affect the stomatal aperture, which contribute a longer-term memory of past stress conditions into the stomatal aperture system. One way in which these properties are manifest in optical properties is that as stomata close, the leaves continue to absorb sunlight; therefore, more energy must be dissipated as kinetic heat and less as latent heat. This temperature change creates the potential to measure changes in water status through changes in thermal infrared emissions.

2.1.2 Mesophyll

The mesophyll is the tissue that includes all cells in the leaf that are between the upper and lower epidermis. It is primarily composed of parenchyma cells, the ground tissue of the leaf. Some parenchyma cells contain chloroplasts and are termed chlorenchyma; these produce the green color of the leaf. Other tissues found within the mesophyll include the vascular cells, which are the cells that transport water and nutrients from the soil to the leaf or those that transport carbohydrates (the products of photosynthesis) to the rest of the plant.

2.1.2.1 Gymnosperms

The mesophyll is the layer beneath the epidermis that is composed of undifferentiated thin-walled chloroplast-containing parenchyma cells that form a compact tissue (with few air spaces) (Figure 2.3). This layer is usually one to a few cells thick, except where the stomata are connected to the vascular system. Resin canals are also located in this tissue, which allows resin to be extruded from leaf or stem surfaces, sometimes coating them and changing surface reflectance, if abundant. The innermost layer of the leaf mesophyll tissue is the single-layered endodermis that surrounds the central vascular bundle. The suberized cell walls of the endodermis force water to pass through the living cells, providing another site for the control of water transport.

2.1.2.2 Angiosperms

The mesophyll tissue is composed of a homogeneous layer of chloroplast-containing parenchyma (also termed chlorenchyma) in monocots and a differentiated one in eudicots, which have the primary responsibility for photosynthesis (Figure 2.3). Most eudicot leaves display two types of ground tissues: the palisade parenchyma and the spongy mesophyll. The palisade parenchyma that is located just below the upper (adaxial) epidermis is most composed of one layer but this layer can comprise several rows of tightly packed elongated palisade cells, with the long axis perpendicular to the leaf surface in vertically aligned cells. Small air cavities may exist between adjacent palisade cells, which would increase the area of cell walls exposed to the air. These cells contain many chloroplasts and are the primary site of photosynthesis in the leaf. Typically, leaves grown in the shade, compared to leaves grown in full sunlight, have fewer layers of palisade parenchyma and less cell wall thickening and are thus thinner. The spongy mesophyll is located between the palisade parenchyma and the lower (abaxial) epidermis and is composed of loosely packed, irregularly shaped thin-walled cells that are separated by large intercellular air spaces. Although contributing to photosynthesis these cells generally have significantly fewer chloroplasts than the palisade layer. This tissue promotes the exchange of gases through the large air spaces that connect to the stomata on the abaxial surface. Finally, in some species, the palisade parenchyma is present on both faces of the leaf blade. This bilateral symmetry is often found in species where a combination of wind and canopy architecture results in similar illumination on both leaf surfaces.

2.1.3 Vascular System

The vascular system is composed of two types of tissues located in the mesophyll, the xylem, and the phloem, which respectively transport water to the leaves or carbohydrates to the rest of the plant. The cells are long and narrow to facilitate these transport functions or are specialized with thick cell walls to provide physical support for the leaf, termed fibers and tracheids. Between the xylem and phloem cells is the vascular cambium, a meristematic tissue that differentiates and produces new cells as needed for plant growth.

2.1.3.1 Gymnosperms

The leaves of conifers display a central vascular bundle which transports fluids between the leaf and the roots, with the xylem transporting water and minerals from the roots to the leaves along water potential gradients. Xylem fibers and tracheids are nonliving lignified cells that provide rigidity and structural support because of the cellulose, hemicelluloses, lignin, and other materials deposited in the secondary cell walls. Water transport along potential energy gradients is possible through the tracheids because of this structural support. The phloem transports photosynthetic assimilate from the mesophyll to other plant organs through thin-walled sieve tubes and companion parenchyma cells. These cells have only primary cell walls, which are composed of cellulose, hemicelluloses (e.g., xyloglucan), pectin, glycoproteins, and enzymes. The tissue located between the vascular tissue and the endodermis is termed the transfusion tissue.

2.1.3.2 Angiosperms

The vascular system of the leaf is composed of a network of primary, secondary, and tertiary veins throughout the mesophyll of the lamina and their distribution pattern defines the architecture of the venation. The vascular conducting system is continuous through the plant. The vascular bundles are composed of the midrib and the vein system. The xylem is usually located on the adaxial (upper) side of the leaf with the phloem on the abaxial (lower) side and this conducting tissue is surrounded by strengthening tissue including sclerenchyma fibers and sclerids with thickened secondary walls. Xylem cells are composed of fibers and vessels, with secondary wall thickening and the phloem is composed of sieve cells and companion cells.

In the leaf veins of numerous angiosperms, the vasculature is surrounded by a ring of compactly arranged bundle-sheath cells. They conduct compounds both into the leaf, particularly during leaf development, and out of the leaf, during export of photosynthates and during senescence. Compared to C_3 plants the bundle-sheath cells of C_4 plants have expanded physiological functions. The leaves of all C_4 plants are composed of two types of photosynthetic cells: an inner layer made of bundle-sheath cells, and an outer layer made of mesophyll cells. This arrangement, known as Kranz anatomy, ensures a quick transfer of the malic acid to the bundle-sheath cells, which have chloroplasts that lack grana but are rich in starch, consistent with their function in creating and storing the carbohydrate products from the Calvin cycle. The bundle-sheath cell walls in C_4 leaves are suberized to reduce conductance of CO_2 into the mesophyll, enhancing the potential for the dark reactions of photosynthesis. Many grasses exhibit C_4 photosynthesis, including several agricultural crops like maize, sorghum, and sugar cane. C_4 species are found in environmentally stressed but diverse habitats including saline marshes and deserts, alpine tundra, and grasslands and are common in hot and disturbed habitats (Sage, 2005). These anatomical differences, combined with biochemical differences, such as the low nitrogen concentrations typical of C_4 leaves, create differences in the optical properties of C_3 and C_4 species. The evolution of C_4 species arose independently four times in *Cyperaceae* (26 genera and 1350 C_4 species), about 710 times in grasses (4600 C4 species), and 31 times in dicots, most commonly in the *Chenopodaceae* and *Amaranthaceae* with 800 species and 1200 additional species from 15 families.

2.2 Leaf Shape and Venation

It has long been noted that leaf size and shape are closely related to the physical environment and requirements of mechanical structure. Raunkiaer (1934) defined leaf size classes and a leaf dimension index to show relationships to climatic variables like solar energy and soil moisture. More recently, Roth-Neblestick et al. (2001) reviewed the literature on the relationship between leaf form and function. Taylor (1975) used a leaf dimension index to relate to environmental variables that let him examine the optimization of leaf form for energy balance and photosynthetic production. Ideally, understanding how relationships between leaf form and function are expressed in their optical properties would allow a new basis for nondestructive monitoring of plant responses to environmental conditions.

In angiosperms, leaf shape and major venation patterns are tightly linked in development (Dengler and Kang, 2001). Hickey (1973) identified six major types of venation (Figure 2.8). The veins generally arise pinnately (single mid-vein from which smaller veins diverge on either side), palmately (several main veins, each having many small branches with a common point of origin), or in parallel (parallel veins of nearly equal size that branch from near the base; smaller transverse commissural veins form the network). In monocots, vein transport occurs along the long axis of the leaf in a parallel array of primary veins. Parallel venation is also characteristic of gymnosperms. Dichotomous or dendritic (open ended) venation are found in ginkgo (*Ginkgo biloba*) and some ferns and conifers (Roth-Nebelsick et al., 2001). In dicots, leaf venation is hierarchical, with smaller secondary veins branching from larger primary veins, and with higher-order veins forming a reticulate network such that most cells are within a few cells of a vein. This type of venation reduces the transport distance between cells.

External environmental factors like edaphic, geologic (e.g., Deroin and Deroin, 1996), altitudinal (e.g., Ehleringer, 1988; Richardson and Berlyn, 2002c; Richardson et al., 2003a), water availability, or other environmental gradients generally induce some phenotypic plasticity in leaf development. Typically, stress-inducing environmental conditions reduce leaf area and specific leaf area (e.g., Gratani et al., 2006; Sardans et al., 2006) and correspond to changes in the biochemical composition of the leaf, particularly pigments and water content.

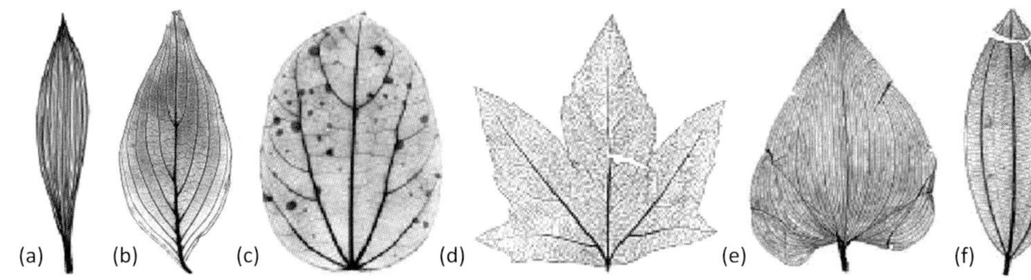

Figure 2.8 Six categories of leaf venation: (a) parallelodromous (e.g., *Maianthemum canadense*); (b) eucamptodromous (e.g., *Cornus officinalis*); (c) basal actinodromous (e.g., *Arcangelisia lemniscata*), (d) palinactinodromous (e.g., *Platanus racemosa*); (e) campylodromous (*Maianthemum dilatatum*; and (f) basal acrodomous (*Miconia*) (adapted from Leaf Architecture Working Group (1999), Smithsonian Institution, Washington).

Table 2.1 *Chemical composition of a plant leaf*

		Cuticle	Vacuole	Cell wall	Chloroplast	Cytoplasm
Water			●			●
Polysaccharides	Cellulose, hemicellulose, pectin			●		
	Starch				●	
	Sucrose, raffinose, fructan					●
Photosynthetic pigments	Chlorophylls, carotenoids				●	
Polyphenols	Lignin			●		
	Anthocyanins and other flavonoids		●			
Proteins			●	●	●	●
Waxes and oils		●				

2.3 Leaf Biochemical Composition

A stoichiometry of the component chemistry in living plants is available in Sterner and Elser (2002). Plant growth is closely coordinated with available resources in the critical zone (Field et al., 1992) and metabolic processes that require investment in biochemical synthesis to be proportional to meet basic metabolic activities. Reich et al. (1997) expanded this analysis, showing that diverse species from tropical and temperate forests to alpine tundra and desert had similar relationships between leaf structure and function and plant growth in all biomes. In a study of 2548 species, Wright et al. (2004) found strong linear relationships between maximum assimilation, nitrogen, phosphorous, specific leaf mass, and leaf longevity. Thus there is a strong case for stoichiometry among species that maintains plant chemistry within limited ranges to achieve a balance between the energy costs for producing different classes of compounds such as pigments, nitrogen, water, and carbohydrates and their integrated functions (Table 2.1). The apparent convergence of leaf characteristics found by Wright et al. (2004) and expanded to 100 000 species in Kattge et al. (2011, 2015) supports the idea that leaf optical properties reflect the integration of these forms (Ustin and Gamon, 2010). Cellulose, hemicellulose, lignin, sugars, amino acids, plant hormones, and other components that comprise a small fraction of the composition of the cell walls are discussed under the generic term of dry matter (Fourty et al., 1996a).

2.3.1 Water

The quantitative determination of water content in leaf tissues is of great concern in plant physiology, ecology, forestry, agriculture, etc. Water represents from 50% to 90% of the mass of a fresh leaf. It is mainly located in the cell vacuole and cytoplasm. Part of it is also contained in the chloroplasts (Carter and McCain, 1993). The same authors obtained water profiles and showed an asymmetric distribution: for instance in fresh sunlit leaves, maximum water density (water mass per leaf volume)

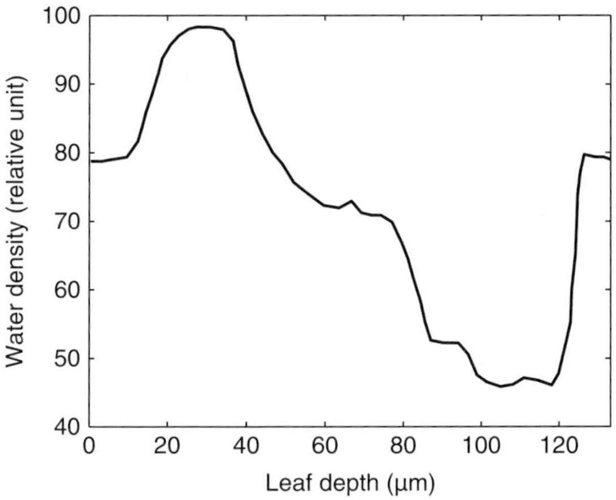

Figure 2.9 Relative density (water mass per leaf volume) of total water at a given depth from the adaxial surface of a maple (*Acer platanoides*) leaf. Leaf relative water content is approximately 100% (from Carter and McCain (1993), *Remote Sensing of Environment*. Copyright © 1993 with permission from Elsevier).

occurred in the upper two-thirds of the palisade parenchyma, while high densities were observed in the epidermal layers near the upper and lower surfaces (Figure 2.9).

The availability of water controls many plant functions. Water stress may occur seasonally as soil moisture reserves are depleted, and diurnally when transpiration exceeds the rate at which water is supplied from the roots to the leaves. It can also be induced in plants in order to improve the quality of harvestable parts of some crops such as fruits and vegetables. In such a situation, transport of water from the roots to the leaves is reduced or stopped because of a decrease in stomatal conductance (due to partial or full stomatal closure), which also limits the uptake of CO_2 for photosynthesis and thus decreases primary productivity. This results in, at best, reversible inhibition of growth, but prolonged exposure to water stress due to drought can lead to irreversible cell damage, loss of chlorophyll that eventually leads to leaf abscission (Bowman, 1989; Carter, 1991; Hadjiloucas et al., 1999), and plant mortality if the water deficit continues beyond the point when plants can resume growth, termed the permanent wilting point. At the leaf level wilting caused by loss of water content changes both the optical properties related to water absorption (see Section 6.2.2) and refraction from internal leaf light scattering (see Section 6.1.1). At the canopy level such optical changes in leaf reflectance and absorption are enhanced by wilting and changes in the leaf angle distribution for short duration water deficits, followed by abscission and loss of foliage for drought occurring over longer time intervals. If new growth occurs during partial stress periods, the plant will produce smaller leaves due to lack of sufficient turgor pressure to fully expand the leaf area. There are several ways of measuring leaf water content and physiological variables that characterize leaf water status. Leaves are preferably collected during early morning to ensure initially high water content and they are placed immediately in plastic bags to minimize water loss and returned to the laboratory for measurements.

2.3.1.1 Water Density

Water density (WD) (kg m^{-3}) is the weight of the water per its unit volume. It depends on temperature but also on dissolved salt content. The maximum value for pure liquid water occurs at about 4°C, at which it is equal to 999.972 kg m^{-3}. When the temperature changes, either greater or less than 4°C, the density decreases from this value. Since water is the "standard" by which other liquids are measured, the specific gravity of water at about 4°C is therefore equal to 1. Note that the density of ocean water at the sea surface is about 1027 kg m^{-3}.

2.3.1.2 Leaf Water Potential

Leaf water potential (ψ) (Pa) is considered to be the most physiologically meaningful indicator to determine plant water stress commonly used by plant physiologists. It is the sum of all driving energy gradients that move water. These result from the turgor potential, osmotic potential, and matrix potential (tensile force between water and solid materials in the cell). The water potential is the difference between the chemical potential for leaf water (with solutes) and pure water divided by the partial molar volume of water. It is the driving force for water movement from the soil to the stomata on the leaf surface and into the surrounding air mass (Tyree and Hammel, 1972; Tyree and Jarvis, 1982). To measure it, one places a leaf attached to a petiole inside a pressure chamber. Then pressurized gas is slowly added to the chamber; as the pressure increases at some point water will be pushed to the cut end of the petiole. The pressure that is required to force water to the cut surface is equal and opposite to the water potential of the leaf. Predawn measured water potential in crops typically ranges between 0 and −7.0 MPa. At the plant level, ψ also includes the gravitational potential related to the upward force required to equal the downward force of gravity. While unimportant in most herbaceous species, it becomes important for water transport to foliage and the evapotranspiration rates in tall trees.

2.3.1.3 Fuel Moisture Content

The measure of fuel moisture content (FMC) (g g^{-1}) is expressed as a percentage of the oven dry weight of the leaf. It is critical to predicting wildfire risk in natural ecosystems, such as forests and open landscapes (wildfire risk assessment also includes topography, wind speed, air temperature, fuel type, fuel biomass, live to dead fuel, etc.) (Danson and Bowyer, 2004; Dauriac, 2004). To determine FMC from field measurements, mature leaves of the most abundant species or an indicator species are collected, preferably during early morning to ensure high water content. The samples are weighed while fresh, and then oven-dried and weighed again to calculate FMC as

$$FMC = \frac{fw - dw}{dw} \tag{2.1}$$

where fw is the fresh leaf mass (g) and dw the oven-dry leaf mass (g).

2.3.1.4 Leaf Water Content

Leaf water content (LWC) (g g^{-1}), or gravimetric water content (GWC), is expressed as a percentage of leaf fresh weight, which ranges between 0 and 100%:

$$LWC = \frac{fw - dw}{fw} \tag{2.2}$$

2.3.1.5 Equivalent Water Thickness

Equivalent water thickness (EWT) (g cm^{-2}) is the mass of water per unit leaf area:

$$EWT = \frac{fw - dw}{A} \tag{2.3}$$

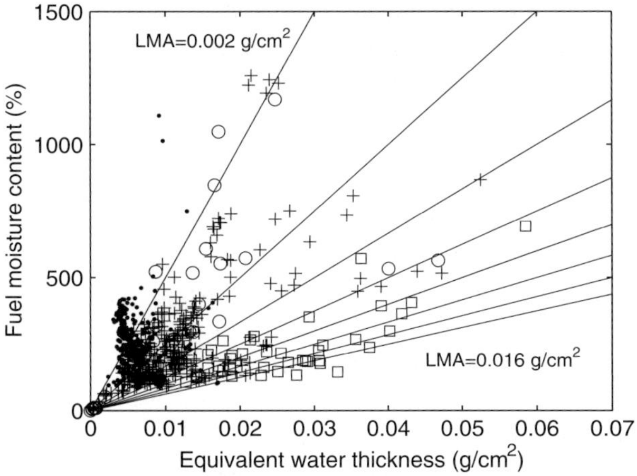

Figure 2.10 Relationship between equivalent water thickness (EWT) and fuel moisture content (FMC) for four datasets: plus (330 samples; Hosgood et al., 1995), circles (32 samples; Jacquemoud and Baret, 1990), squares (41 samples; Asner and Martin, 2009), dots (322 samples; Féret et al., 2008). The leaf mass per area (LMA) isolines plotted from 0.002 to 0.016 g cm^{-2} at 0.002 g cm^{-2} intervals (adapted from Danson and Bowyer (2004)).

where fw is the fresh leaf mass (g), dw the oven-dry leaf mass (g), and A the one-sided leaf area (m^2). EWT is normally expressed in g cm^{-2} but, since the density of pure water is 1 g cm^{-3}, it is also frequently given in cm. The second term of Eq. 2.3 is called leaf mass per area (LMA) (see Section 2.4). EWT and FMC are two different ways to define leaf water content but they are not directly related unless LMA is known (Ceccato et al., 2001). Indeed, by dividing Eq. 2.1 into Eq. 2.3, Danson and Bowyer (2004) noticed that FMC was related both to the amount of water and dry matter in a leaf:

$$FMC = \frac{EWT}{LMA} \tag{2.4}$$

These relationships are illustrated graphically by plotting LMA isolines against EWT and FMC (Figure 2.10). The data points illustrate FMC, LMA, and EWT for leaves of different species. That figure also shows that FMC and EWT are perfectly correlated when LMA is constant. Note that at the pixel scale, EWT is the hypothetical thickness of water in a pixel when it is distributed over the area of the pixel.

2.3.1.6 Relative Water Content

Relative water content (RWT) (g g^{-1}) is the ratio of the actual leaf water content to the maximum water content at full turgor. To obtain this particular observation state, petioles are cut underwater and the leaves allowed to absorb water until full turgidity is reached (determined by reweighing the leaves over time until no change in weight is found, at which point the sample is accepted to be at full turgor):

$$RWC = \frac{fw - dw}{tw - dw} \tag{2.5}$$

where *tw* is the leaf mass in the fully turgid state (g). It has been demonstrated to be inversely related to leaf water potential ($1/\psi$), which controls plant response to water stress. However, it should be noted that different species may have the same *RWC* at different water contents in their leaves because of variance introduced by different turgid and dry matter weights of species in nature (Ceccato et al., 2001). Additionally, the same *RWC* in different species (or for different environmental conditions) does not signify a specific stress level in the leaf. For a given leaf, there is generally a one-to-one relationship between ψ and *RWC*, which can be graphed as a pressure-volume curve (Tyree and Hammel, 1972). From this curve it is then possible to graphically estimate the water potential at zero turgor (the point when leaves wilt), and the osmotic and matrix potentials. These water potential components define the drought status of the leaf. Because of the correlation between water potential and *fw*, an optical property sensitive to small changes in *EWT* can indicate the hydration state of vegetation.

2.3.2 Polysaccharides

The six-carbon sugar, glucose, is the primary product of photosynthesis. However, the plant cell wall is primarily composed of structurally complex molecules which are compactly arranged and extremely resistant to degradation. A review of cell wall chemistry is found in Carpita and McCann (2000). The wall polysaccharides are often classified into cellulose, hemicelluloses, and pectin (Vincent, 1999). They are represented in almost all cell walls in varying proportions. They are made up of particular kinds of sugars, which are linked to each other by glycosidic bonds in characteristic patterns to form polymer chains. The chemical structure of the three above-mentioned polysaccharides is presented in Figure 2.11. Cellulose occurs as a crystalline polymer of β-D-glucose units with the formula $(C_6H_{10}O_5)_n$. It represents approximately 50% of the total biomass and 100 billion tons are produced per annum. Thus it is the most abundant organic molecule on Earth (Alberts, 2004). Hemicelluloses are a family of polysaccharides that are more complex than sugars and less complex than cellulose. They are the second most abundant polysaccharide in nature after cellulose. They include xylans (the most abundant form of hemicellulose), xyloglucans, arabinoxylans, mannans, glucomannans, galactomannans, galactans, etc. Besides glucose, they are heteropolymers containing many different neutral sugar monomers of xylose, arabinose, mannose, galactose, etc. Pectin is the third group of structural polysaccharides characteristically found in primary cell walls of plant leaves. It is composed of a chain of galacturonic acid and other molecules. It plays an important role in plant growth by allowing cell wall extension (Harolt et al., 2010). In the cell wall, the cellulose fibers are intertwined with hemicellulose molecules in a matrix of pectin.

Starch is a primary storage product of photosynthesis in leaves. In most plants, a large fraction of the carbon assimilated during the day is stored transiently as starch in the chloroplasts for use during the subsequent night. Then it provides a continued supply of sugars to sustain metabolism in the leaf and for export to sink organs. If leaves are placed in darkness for 24 hours, the starch disappears. Starch is a carbohydrate consisting of a large number of glucose units joined by glycosidic bonds (Figure 2.12). This architecture allows the formation of a dense semi-crystalline insoluble granule (Zeeman et al., 2007).

Note that other storage forms are synthesized in the leaves of some species: sucrose (disaccharide composed of glucose and fructose), raffinose (trisaccharide composed of galactose, fructose, and glucose), and fructan (polymer of fructose). The cytosol of photosynthetic cells contains several enzymes required for the important regulatory interconversion of starch to these molecules.

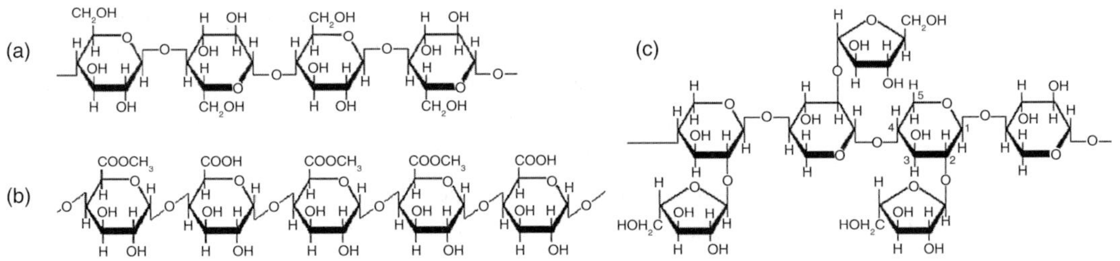

Figure 2.11 Chemical structure of: (a) cellulose; (b) pectin; and (c) arabinoxylan.

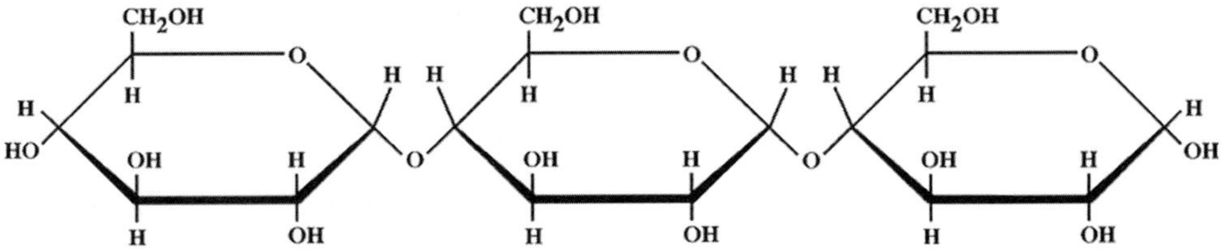

Figure 2.12 Chemical structure of starch.

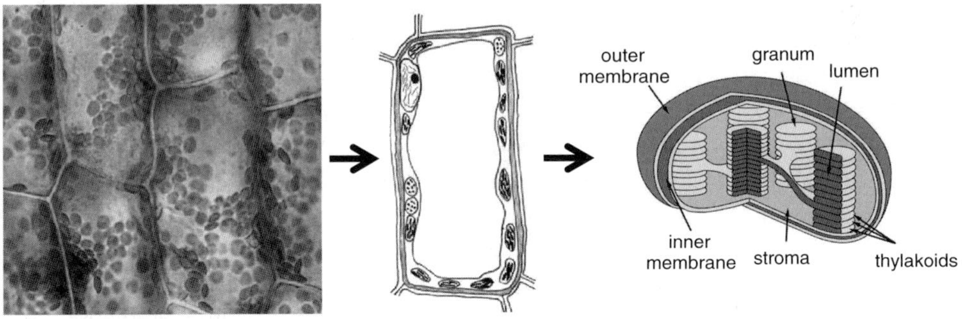

Figure 2.13 Distribution and structure of the chloroplast.

2.3.3 Photosynthetic Pigments

All photosynthetic pigments are located in the chloroplasts, which are membrane-bound organelles distributed in the cytoplasm of leaf mesophyll cells (Figure 2.13). Absorption of light in the visible spectrum by plant photosynthetic pigments produces a unique spectral reflectance signature that is responsible for plant color.

Light is captured in the process of photosynthesis (Govindjee and Krogmann, 2004) that provides the energy for plants to synthesize carbohydrate (organic matter) from CO_2, through a series of electron transfers that occur on the stacks of the thylakoid membranes in chloroplasts. The global photosynthesis reaction is

$$CO_2 + 2H_2O \xrightarrow{h\nu} CH_2O + O_2 + H_2O \tag{2.6}$$

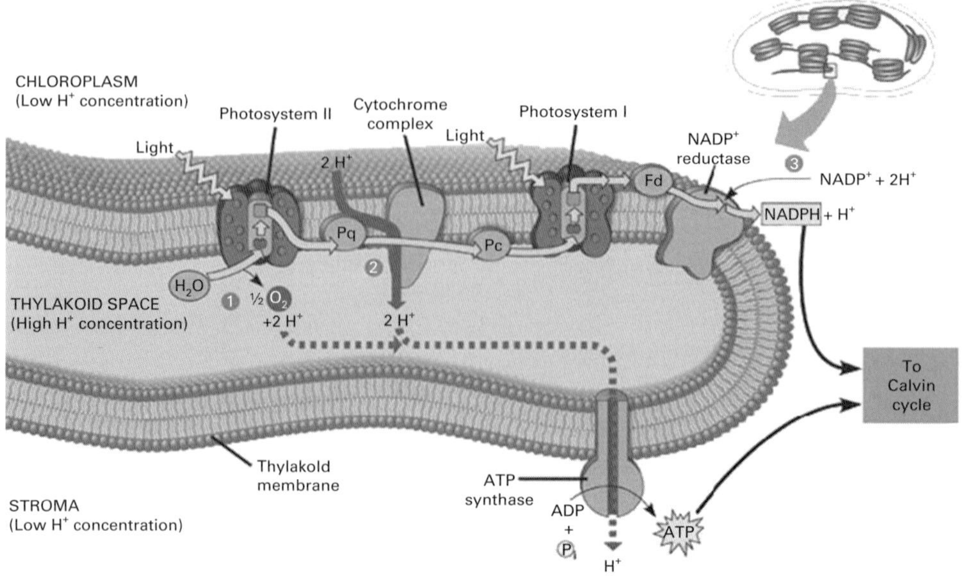

Figure 2.14 Light-dependent reactions of photosynthesis at the thylakoid membrane.

Figure 2.14 schematically illustrates the sequence of carriers that captures light energy and transfers it to the dark reactions of the Calvin cycle. In the intact chloroplast membrane (the thylakoid), the pigment-protein complexes are organized into two photosystems: the photosystem I (PSI) and photosystem II (PSII), which are also termed P700 and P680, respectively, for the wavelengths that define their absorption maxima *in vivo*. The photosystem antenna harvest light and transfer energy through a series of energy conversion reactions termed the Z-Scheme. This simplified diagram shows the arrangement of the electron carriers on the thylakoid, while the term Z-scheme refers to the energy state as the electron is passed through the carriers. The antenna system increases the effective cross-section of light absorption and the efficiency of energy capture and transfer to the reaction center. The size and composition of the antenna is flexible and leaves developing under low light intensity generally have more antenna pigment molecules per reaction center than those developed under high light intensity.

Photosynthesis occurs in two phases. The first, called the light phase, absorbs light energy which is captured by exciting an electron to a higher orbital state (shown as the lightning strikes on the figure). One photon is captured by each reaction center and the combined energy, minus transfer loss, is the chemical energy that is required to reduce ADP to ATP and $NADP^+$ to NADPH. The second, called the dark phase, uses this energy to transform atmospheric CO_2 into organic carbon during the Calvin cycle. The dark phase of photosynthesis, termed in full, the Calvin-Benson-Bassham Cycle, is a series of reactions that utilize the energy from the light phase (ATP and NADPH) to convert CO_2 and water into organic carbohydrate, specifically a 3-carbon sugar, glyceraldehyde-3-phosphate. The primary enzyme involved in catalyzing this conversion is ribulose-1,5-bisphosphate carboxylase oxygenase (RuBisCO). RuBisCO is the most abundant enzyme on Earth and comprises 50 to 60% of the soluble protein in plant leaves (Evans, 1989). It is the only enzyme in nature able to fix CO_2 into carbohydrate molecules, thus its activity is essential to life. The thylakoid nitrogen concentration is

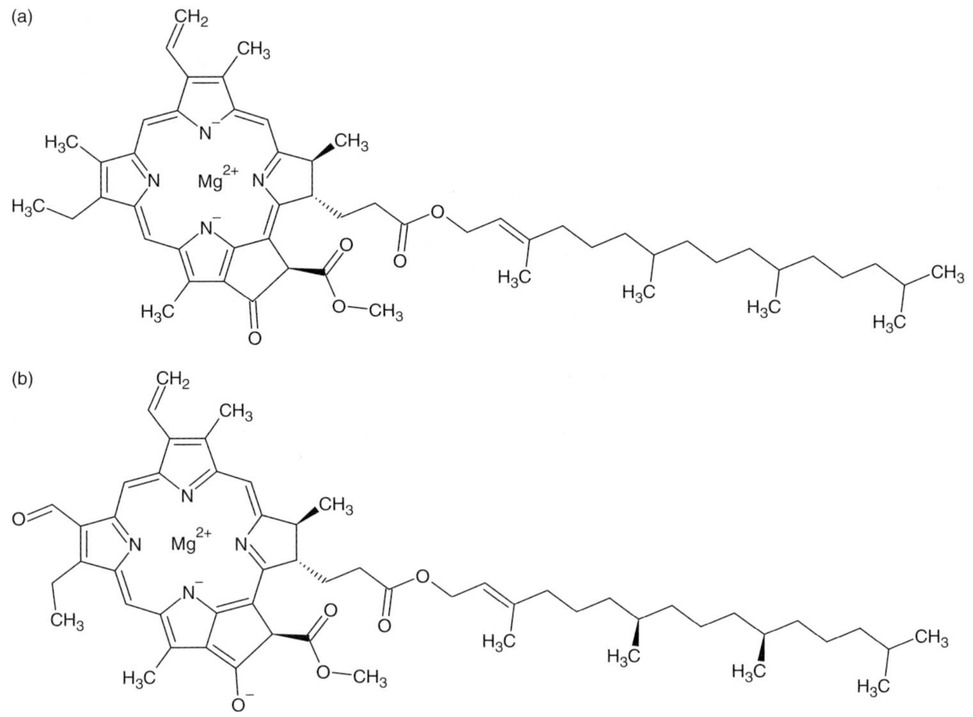

Figure 2.15 Chemical structure of (a) chlorophyll *a* and (b) chlorophyll *b*.

generally roughly proportional to the chlorophyll content (50 mol thylakoid N mol^{-1} Chl) and at the species level there are strong linear relationships between nitrogen and both RuBisCO and chlorophyll (e.g., Field and Mooney, 1986; Evans, 1989).

As seen earlier, chlorophyll is the most important photosynthetic pigment, in quantity and in quality, which explains the green color of vegetation. On average, one chloroplast contains 6.7×10^8 molecules of chlorophyll, which corresponds to a concentration of about 0.24 M in the thylakoids, the membranes on which the light reactions of photosynthesis occur (the dark phase reactions occur in the stroma of the chloroplast). Chlorophyll is a chlorine molecule with a central porphyrin ring containing a magnesium atom. In the leaves of higher plants, it exists in two forms, chlorophyll *a* as the major pigment and chlorophyll *b* as an accessory pigment. They only differ by a substituent attached to the pyrrole ring on the porphyrin ring opposite the phytol tail (Figure 2.15). On chlorophyll *a* the substituent is a simple methyl ($–CH_3$) group; chlorophyll *b* has an aldehyde (–C(=O)H) functional group at the same location. Their chemical formulas and molar masses are given in Table 2.2.

Besides chlorophyll *a* and chlorophyll *b*, the photosynthetic antenna (the organized association of pigments that capture photons and transfers energy to the chlorophyll *a* reaction center) contain other membrane-bound accessory pigments that include β-carotene, lutein, and the xanthophyll cycle pigments (Lichtenthaler, 1987). These molecules broaden the wavelength range over which plants can capture light and serve other functions in photosynthesis. Although the contribution of these molecules to leaf mass is negligible (approximately one-thousandth), they play an essential role at the beginning of the food chain that supports all life on Earth. Carotenoids are probably the most widely

Table 2.2 *Chemical formula and molar mass of chlorophylls and carotenoids.*

Name	Chemical formula	PubChem CID	Molar mass (g mol^{-1})
Chlorophyll *a*	$C_{55}H_{72}O_5N_4$ Mg	6433192	893.49
Chlorophyll *b*	$C_{55}H_{70}O_6N_4$ Mg	6450186	907.47
α-carotene	$C_{40}H_{56}$	4369188	536.87
Lutein	$C_{40}H_{56}O_2$	5281243	568.88
β-carotene	$C_{40}H_{56}$	5280489	536.87
Zeaxanthin	$C_{40}H_{56}O_2$	5280899	568.88
Antheraxanthin	$C_{40}H_{56}O_3$	5281223	584.87
Violaxanthin	$C_{40}H_{56}O_4$	448438	600.85
Neoxanthin	$C_{40}H_{56}O_4$	5281247	600.87

distributed class of pigments found in nature, occurring in microorganisms, plants, and animals (Young, 1991). In higher plants, they are biosynthesized and accumulate in the chloroplasts. Like the chlorophylls, they are found in both photosystems in the thylakoid membrane where they are non-covalently linked to membrane-associated proteins.

Carotenoids are defined by their chemical structure, which is formally derived from a 40-carbon polyene chain. This chain is considered to be the backbone of the molecule, which is terminated by cyclic end-groups (rings) and may be complemented with oxygen-containing functional groups. They are divided into oxygen-free carotenes (α-carotene and β-carotene) and different forms of oxygen-containing xanthophylls (lutein, zeaxanthin, violaxanthin, neoxanthin, antheraxanthin), such as one or several hydroxy or epoxy groups. Introduction of hydroxyl functional groups (–OH) into α-carotene gives rise to lutein, while β-carotene is the precursor of zeaxanthin, antheraxanthin, and violaxanthin, all recognized as contributors to photoregulation in chloroplasts. As for neoxanthin, which is intermediate in the biosynthesis of the plant hormone abscisic acid, it is produced from violaxanthin (Figure 2.16). The chemical formulas and molar masses of carotenoids are given in Table 2.2. The percentage composition (by % weight) varies between species and plant conditions within the range 25–40% for β-carotene, 40–57% for lutein, 9–20% for violaxanthin, and 5–13% for neoxanthin (Lichtenthaler, 1987; Demmig-Adams et al., 1996).

Carotenoids increase the spectrum over which light can be absorbed for photosynthesis and perform other functions like attraction for pollinators or seed dispersal, not all of which are fully understood. The role of β-carotene in photosynthetic systems is two-fold: first, it serves as an accessory light-harvesting pigment, its main function, trapping light energy and transferring it to chlorophyll molecules by singlet–singlet energy transfer. Second, it is involved in photoregulation of light by dissipating excess absorbed energy and avoiding oxidative damage to the photosynthetic system under excess UV light or freezing temperatures (Young, 1991; Dey and Harborne, 1997). Lutein and neoxanthin apparently only act as accessory pigments in light absorption. The interconversion of violaxanthin and zeaxanthin through an epoxidation reaction, where antheraxanthin is an intermediate, provides a photo-protective mechanism called the xanthophyll cycle that aids dissipation of excess energy (Demmig-Adams, 1990; Graßmann, 2005). It does this by increasing the conjugation of double bonds between violaxanthin and zeaxanthin. On illumination, the level of violaxanthin in the leaf decreases while the levels of antheraxanthin and zeaxanthin increase.

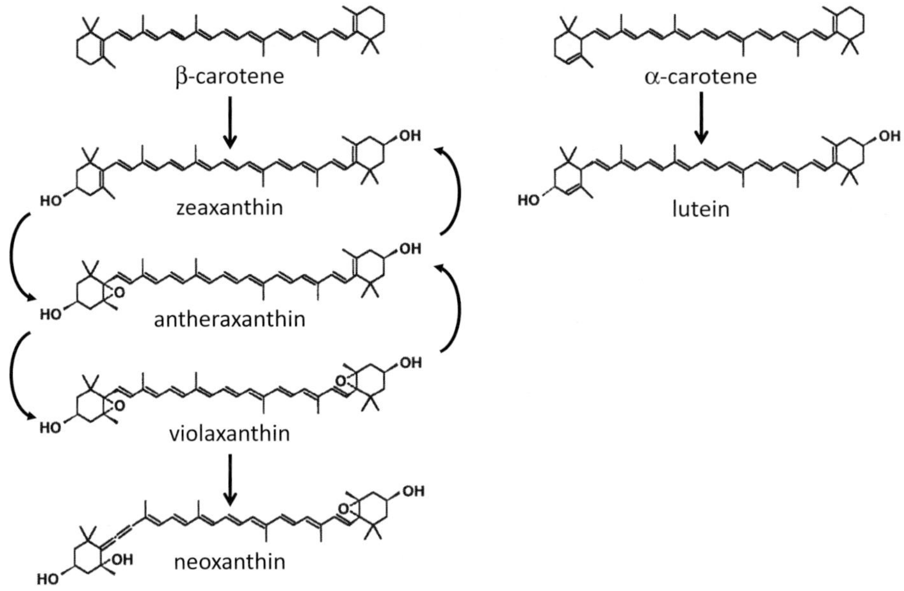

Figure 2.16 Carotenoid biosynthetic pathway in higher plants showing the xanthophyll cycle (adapted from Niyogi et al. (1997), *Proceedings of the National Academy of Sciences*. Copyright © 1997 with permission from National Academy of Sciences).

A return to low light reconverts zeaxanthin back to violaxanthin, at the expense of antheraxanthin and zeaxanthin. In dark-adapted leaves, violaxanthin levels are usually high compared to antheraxanthin and zeaxanthin levels. The consequence of increasing the conjugated state of the double bond change is that more high energy light is reflected and less absorbed in the low light state. Control of the epoxidation state provides a mechanism for rapid response to changing light levels and the energetic state of the leaf.

In summary, photosynthetic pigments provide a mechanism for active regulation at all stages of energy capture and storage (Figure 2.17). As a component of the photosystem antenna, chlorophyll *b* almost totally transfers its absorbed energy to the chlorophyll *a* in the reaction center. Carotenoids absorb in the blue-green and transfer most of this energy on a femtosecond (1 fs = 10^{-15} s) time scale to an energy state of chlorophyll *a* and *b*, with the remaining energy dissipated as thermal energy (Dutton et al., 1943). One of the important regulatory mechanisms is "nonphotochemical quenching," a mechanism to dissipate excess energy before it is transferred to the reaction center (Demmig et al., 1987; Gilmore et al., 1995).

2.3.4 Phenolic Compounds

Phenolic compounds, or polyphenols, represent a large and diverse group of related chemicals (~ 8000 known structures), the composition of which is highly variable, both qualitatively and quantitatively. They account for 40% of organic carbon circulating in the biosphere so they are the most abundant secondary metabolites in plants (Cheynier, 2005). Polyphenols contained in the leaf

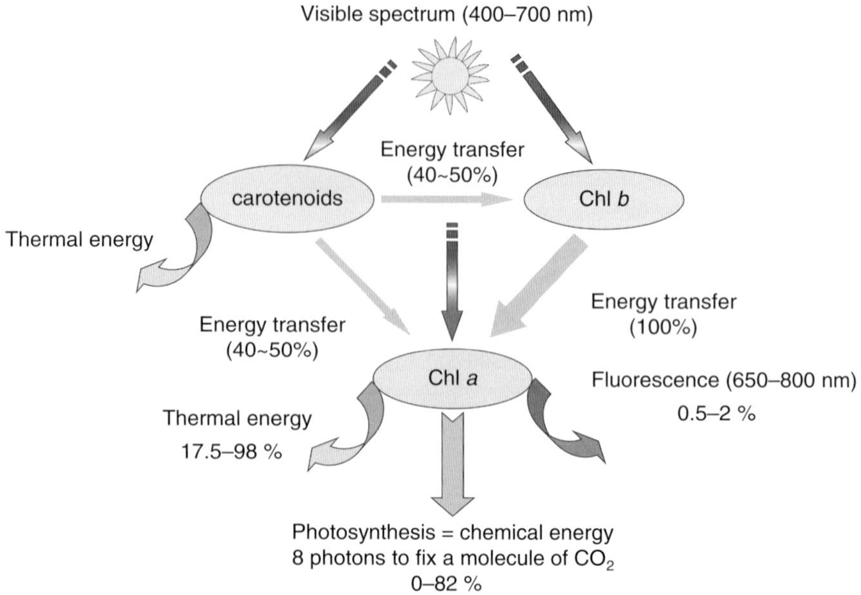

Figure 2.17 Photosynthetic pigments: excitation energy transfer.

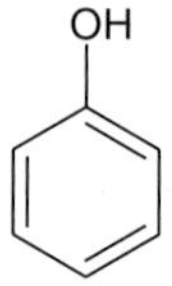

Figure 2.18 Chemical structure of phenol showing the phenyl group ($-C_6H_5$) bonded to a hydroxyl group (–OH).

epidermis perform several important functions: they are thought to serve as antioxidants or to protect chlorophyll in the mesophyll layers from high light intensity by absorbing UV (strongly) and blue light (weakly). Some polyphenols form protective compounds that are not directly active in metabolism, such as tannins and flavonoids, but provide a defense from herbivores and pathogens, while others have less obvious roles. Some are ubiquitous, whereas others are restricted to specific families or species. Despite a huge number of studies devoted to these molecules in general, understanding the full extent of different roles of polyphenols is therefore still incomplete. They show a great diversity of structures, ranging from rather simple molecules containing one or more phenol units (Figure 2.18) to high-molecular-weight polymer structures (Figure 2.21). Polyphenols are generally classified into soluble compounds (flavonoids, of which the cytoplasmic pigment anthocyanin is an example, phenolic acids, phenylpropanoids, quinones, etc.) and non-soluble compounds (lignins, condensed tannins, cell-wall bound hydroxycinammic acids, etc.). Lovelock et al. (1992) established a good relationship between soluble phenolic compounds and leaf mass per area (LMA). When the plant cells are damaged, the phenolic compounds, which are highly unstable, rapidly transform into various products by enzymatic reaction.

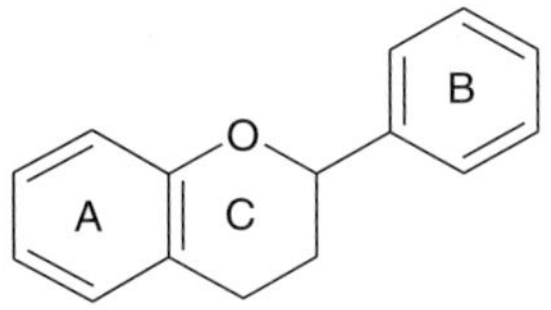

Figure 2.19 General flavonoid structure.

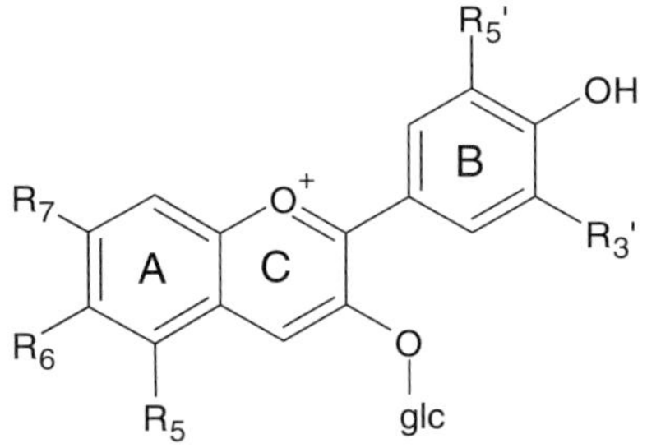

Figure 2.20 Chemical formula of the anthocyanin + glucose complex.

2.3.4.1 Flavonoids

Flavonoids are the largest group of water soluble polyphenols (~ 4500 identified molecules) (Figure 2.19). They contain 15 carbon atoms and are characterized by two aromatic cycles (A- and B- rings) linked by a 3-carbon bridge (C-ring that has an oxygen atom). Flavonoids can undergo modifications of their aromatic cycles, including hydroxylations, methylations, glycosylations, acylations, or prenylations (Pourcel et al., 2007). This particular structure contributes to their great diversity. Six major sub-groups are found in higher plants: flavones, flavonols, flavanones, chalcones, anthocyanins, and isoflavonoids (Winkel-Shirley, 2001). Condensed tannins, also called proanthocyanidins because they release anthocyanidins, form a heterogeneous group of flavonols that are widespread in the plant kingdom and are generally characterized by their capacity to precipitate proteins (Heil et al., 2002).

Flavonols are the most ubiquitous flavonoids in leaves, and the main representatives are quercetin and kaempferol. They accumulate in the vacuole of epidermal cells and have both a screening function in UV-B protection and an antioxidant function in photoprotection.

Anthocyanins are plant pigments closely associated with the colors in the autumn foliage of deciduous plant species (see Chapter 1). They are synthesized via the flavonoid pathway in the cytoplasm of the palisade and/or spongy mesophyll cells of leaves and are sequestered into the vacuole (Gould et al., 2009). They are often observed during early leaf development and under environmental stresses, but they are not directly involved in photosynthesis. They are thought to protect the photosystems from excess light (Merzlyak and Chivkunova, 2000; Gitelson et al., 2001), along with other functions, for example, protection from herbivory and pathogen infection (Close and Beadle, 2003). Under stress (e.g., high temperatures) or during early leaf development (Hoch et al.,

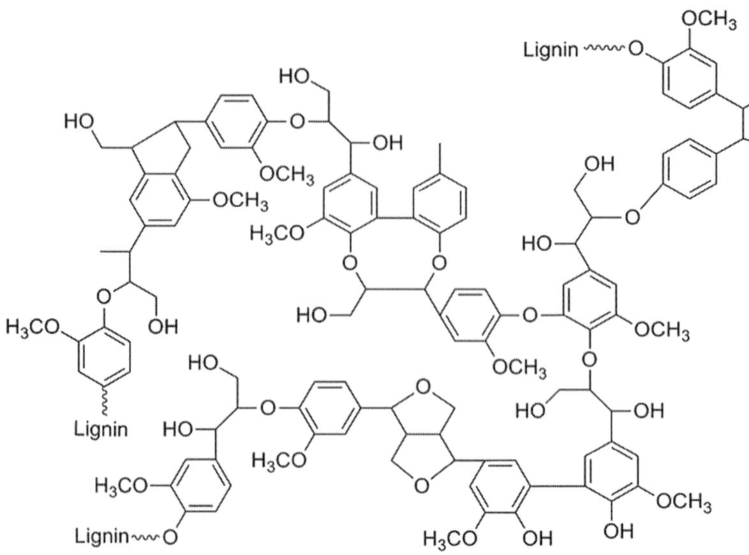

Figure 2.21 Chemical structure of lignin.

2001) they become more visible, consequently measurement of leaf anthocyanins may be another indicator of physiological state (Curran et al., 1991). Figure 2.20 presents their chemical structure. The glucose molecule that is covalently attached to the C ring increases their solubility in water. Anthocyanin pigmentation is largely due to the positive charge on the C ring: when that charge is neutralized, anthocyanins lose their pigmentation and become colorless.

Isoflavonoids are secondary metabolites that can be divided into isoflavones and pterocarpans (Appendix B). The leaves of red clover (*Trifolium pratense*), as well as those of many other species of the *Fabaceae* family, are known to contain high concentrations of these compounds. The isoflavones found in such leaves include biochanin A, formononetin, daidzein, genistein, pratensein, prunetin, pseudobaptigenin, calycosin, methylorobol, afrormosin, texasin, irilin B, irilone, etc. (Saviranta et al., 2010). The pterocarpans include bitucarpin A and B, erybraedin A and B, dolichin A and B, glyceollin, etc.

2.3.4.2 Lignin

Lignin, the core constituent of wood, is a large heterogeneous phenylpropanoid macromolecule formed with different methoxylated monomers that vary somewhat across taxonomic groups (Figure 2.21). It is very abundant on Earth. It contributes to secondary wall thickening in plant leaves and is important in the decomposition of leaves and in the quality of soil organic matter. Lignin makes cell walls resistant to compression, conferring rigidity to the cellular walls. These molecules resist bacterial degradation and the non-decomposed residues of cell wall constituents are the main components of soil organic matter. The rate of breakdown of cellulose and lignin residues also plays a role in the availability of nutrients to support new plant growth in the soil complex, thus knowledge of their abundance provides important evidence on the quality of the environment for plant growth.

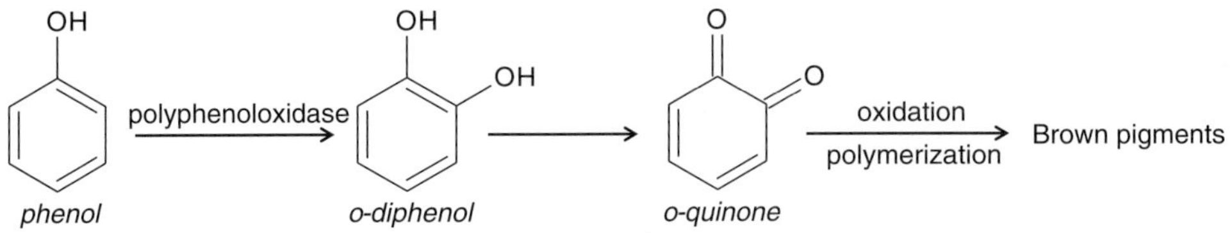

Figure 2.22 Diagram showing the oxidation and polymerization of polyphenoloxidase to brown pigments (from Cardenas et al. (1970), *Remote Sensing of Environment*. Copyright © 1970 with permission from Elsevier).

2.3.4.3 Condensed Tannins

Condensed tannins are found in nearly all plant families and comprise the next largest fraction of dry matter in leaves after cellulose. They play an important function in regulating the decomposition and nitrogen cycling of leaves. They also show antibiotic properties. For instance, young leaves of cotton (*Gossypium hirsutum*) may be more resistant to verticillium wilt than older leaves due to their high concentration of condensed tannins (see Section 7.2.1).

2.3.4.4 Brown Pigments

Brown pigments are the colored residues of degraded pigments during the process of leaf senescence. As seen earlier, phenolic compounds are unstable and quickly oxidize when plant tissues are damaged. Anthocyanins are colored but do not contribute significantly in browning reactions. Leaf browning is a very complex process which results from an increase in enzymatic activity. As leaves senesce and decay, they degrade into colored substances described as brown pigments (Cardenas et al., 1970; Pourcel et al., 2007). This chemical reaction sequence has been studied in lettuce (*Lactuca sativa* L.) leaves (Degl'innocenti et al., 2005). The current biochemical model for plant tissue browning considers the metabolism of phenylpropanoids: phenylalanine ammonia-lyase (PAL), an enzyme present in higher plants, catalyzes the first reaction in the phenylpropanoid pathway leading to the biosynthesis of phenolic compounds; in response to stress such as tissue wounding, pathogenic attack, light, or low temperatures, these compounds are oxidized by the action of polyphenol oxidase (PPO), a copper-containing enzyme located in chloroplasts and also widely distributed in plants (Boeckx et al., 2015). After cellular destruction, PPO comes into contact with phenolic compounds. The hydroxylation of phenols produces o-diphenols, and the oxidation of o-diphenols with molecular oxygen from air produces o-quinones, which in turn spontaneously polymerize to form brown pigments (Figure 2.22). PAL and PPO are ubiquitous enzymes of plants. Their activity is low in non-stressed leaves – during the normal life cycle of plants, a slow browning reaction known as functional browning occurs – which increases in the presence of both biotic and abiotic stresses. The extent of browning in leaves can be assessed visually. Note that degradation processes continue beyond the death of the leaf as part of the larger process of decomposition into humus and the organic components of the soil. These breakdown products have not been well studied, although understanding the nutrient changes could contribute to better understanding of nutrient cycling and soil fertility.

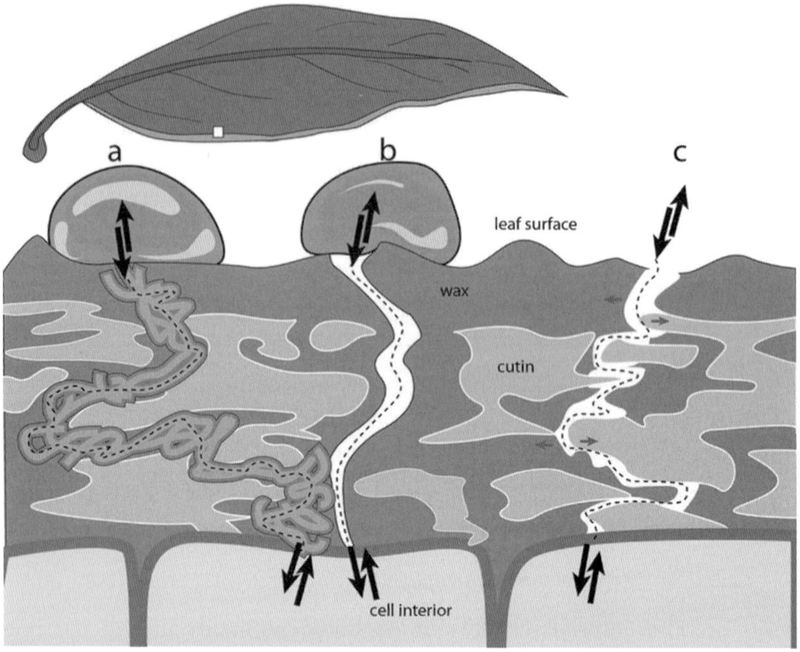

Figure 2.23 Leaf cuticle structure showing how chemical entities are selectively allowed to pass: (a) water, hydrated calcium, and other ions; (b) carbohydrates and other hydrophillic compounds; and (c) hydrophobic compounds (www.asknature.org/).

2.3.5 Waxes and Oils

The cuticle, or cuticular membrane, is the most superficial layer of the leaf. It is a protective barrier between these organs and their environment and a prerequisite of terrestrial plant life. It helped vascular plants to establish themselves on dry land around 400 million years ago. The cuticle is actually not a simple layer but is composed of epicuticular waxes and cutin structured in a complex matrix of long-chain hydrophobic compounds (Figure 2.23) that occupy approximately the outer 0.1–10 μm of the plant surface. It ends at the cell wall underlying the epidermis. In leaves, the weight of an isolated cuticle ranges from 400 to 800 μg cm^{-2}. Details on the chemical, structural, and ecological properties and diversity of plant cuticles are found in several books (e.g., Kolattukudy, 1976; Juniper and Jeffree, 1983; Hamilton, 1995; Kerstiens, 1996; Riederer and Müller, 2006). From a physiological point of view, the main function ascribed to the cuticle is to minimize water loss. However, from a more general point of view, this role in the regulation of plant water is accompanied by other core functions; the cuticle also protects the leaf against physical (UV irradiation), chemical (pollution), and biological (herbivores and pathogens) aggressions. For many cuticular functions it is not known which component, wax or cutin, is important in which physical properties, or if they act in unison. However, the epicuticular wax is thought to confer resistance to insect herbivory and to constitute a mechanical barrier against insect pests, while cutin is thought to be crucial in the early events of plant infection by many pathogenic fungi.

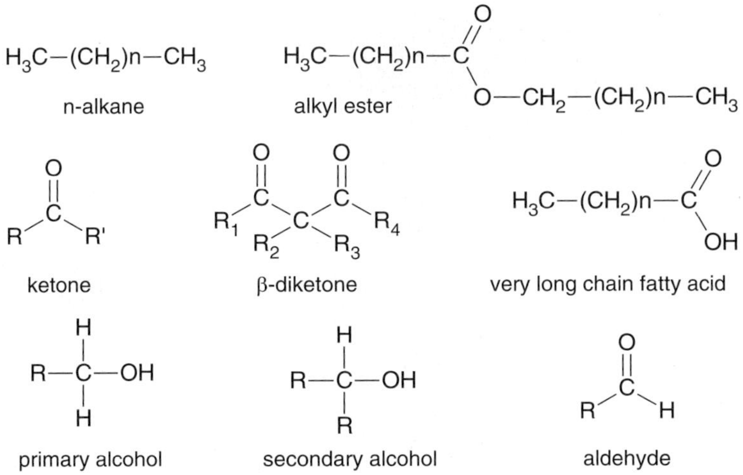

Figure 2.24 Structure of the main components that occur in leaf epicuticular waxes: n-alkane, alkyl ester, ketone, β-diketone, very long chain fatty acid, primary alcohol, secondary alcohol, and aldehyde.

2.3.5.1 Epicuticular Waxes

Epicuticular waxes occur on the surface of all plant leaves, from small quantities on leaves of some temperate annuals (~ 1–5 μg cm^{-2}) to very large quantities on leaves of the Brazilian palm tree (*Copernicia cerifera*), the famous source of carnauba wax (several mg cm^{-2}). They do not form a chemically homogeneous group but they are mixtures of aliphatic hydrocarbons with a variety of substituted groups. Common examples are n-alkanes (paraffins) that range in chain lengths from C_{17} to C_{35} and usually occur as odd-numbered homologues. The hydrocarbon chains may have substituted groups in mid-chain (alkyl esters, ketones, β-diketones, and secondary alcohols) or terminal (very long chain fatty acids, primary alcohols, and aldehydes) positions (Figure 2.24).

R_1 and R_2 have commonly 10 to 20 carbon atoms or more. Several isomers, that is, molecules that have the same chemical formula, may coexist. The unique characteristics of plant cuticles derive both from the numerous organic compounds involved, and the diverse structural arrangements of the components (Holloway, 1982). They come in a variety of forms such as granules, flakes, filaments, crystalline nanotubes, or rodlets (Figure 2.5). Note that intracuticular waxes are located inside the cutin matrix.

2.3.5.2 Cutin

Cutin is the main chemical component of the plant cuticle (40–80% by weight). It is essentially a layered structure which consists largely of omega hydroxy acids and their derivatives, which are interlinked via ester bonds to form a polyester polymer of indeterminate size (Heredia, 2003). The 16-hydroxy hexadecanoic acid (16 atoms of carbon) and 18-hydroxy octadecanoic acid (18 atoms of carbon) are common monomers of cutin (Appendix B). Additional hydroxy or epoxy substituents may be present in the mid-chain position, for example, 9,16- or 10,16-dihydroxy hexadecanoic acid, 18-hydroxy-9,10-epoxy octadecanoic acid, or 9,10,18-trihydroxy octadecanoic acid.

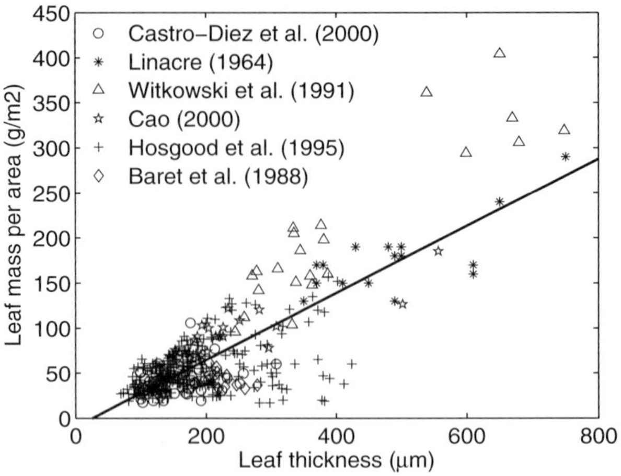

Figure 2.25 Regression between leaf mass per area *LMA* and leaf thickness ℓ (431 samples) (after Linacre, 1964 (16 samples); Witkowski et al., 1991 (22 samples); Cao, 2000 (22 samples); Hosgood et al., 1994 (319 samples); Castro-Díez et al., 2000 (52 samples)).

2.3.5.3 Oils

Epidermal exuberances, for example, hair cells and spines, which enhance light scattering and reduce the heat load on the leaf energy budget, can contain essential oils and other fluids. Monoterpenes and sesquiterpenes are the main constituents of oils that have numerous ecological functions in the plant kingdom: their antioxidant properties are well known, but they also act as allelopathic agents, repellants, or attractants in plant–plant or plant–pathogen–herbivore interactions (Graβmann, 2005). For instance, the Holm oak (*Quercus ilex*) is considered to be one of the strongest monoterpene (α-pinene, β-pinene, and sabinene) emitting species in the Mediterranean basin, the emission of which is controlled by light intensity and, to a lesser extent, by ozone concentration. Aromatic and cyclic compounds such as flavonoids and terpenoids may also be present in plant cuticles and leaf hair glands in smaller amounts (e.g., Tulloch, 1976; Baker, 1982; Bianchi, 1995). Some triterpenes, such as ursolic acid and oleanolic acid, have been identified in the epicuticular wax of Madagascar periwinkle (*Catharanthus roseus*) and Tasmanian bluegum (*Eucalyptus globulus*) leaves, respectively.

2.4 Dry Matter

Air can be considered as a mixture of dry air and water vapor; similarly, one can consider that a plant leaf is made of dry matter and water. The leaf mass per area (LMA) (g cm^{-2}), also called specific leaf mass (SLM) or specific leaf weight (SLW) in older literature, is defined as the ratio of the oven-dry leaf mass dw (g) over the one-sided area of a fresh leaf A (cm^2):

$$LMA = \frac{dw}{A} \tag{2.7}$$

In laminar leaves, *LMA* can also be considered the product of two physical properties of leaves, namely the leaf density ρ_d defined as the dry mass per unit volume (g cm^{-3}) and the leaf thickness ℓ (cm) (e.g., Witkowski and Lamont, 1991; Niinemets, 1999; Poorter et al., 2009):

$$LMA = \frac{dw}{A} = \frac{\ell \times dw}{\ell \times A} = \ell \times \rho_d \tag{2.8}$$

Thus the leaf mass per area correlates positively with leaf thickness (Figure 2.25). They are both lower in deciduous than in evergreen species.

The specific leaf area (*SLA*), a variable frequently used in growth models, is simply the reciprocal of *LMA* and therefore is expressed in cm^2 g^{-1}:

$$SLA = \frac{A}{dw} \tag{2.9}$$

SLA is related to gas exchange, potential photosynthetic rate, decomposition rate, relative growth rate, leaf nitrogen concentration, and leaf lifespan across species. Leaf dry matter content (*LDMC*) (mg g^{-1}) is the oven-dry mass of a leaf *dw* (mg) divided by its fresh mass *fw* (g):

$$LDMC = \frac{dw}{fw} \tag{2.10}$$

Note that *LDMC* is simply related to leaf water content (*LWC*) (Eq. 2.2) by

$$LWC = \frac{fw - dw}{fw} = 1 - \frac{dw}{fw} = 1 - LDMC \tag{2.11}$$

It is also possible to relate *LDMC* to *LMA* and equivalent water thickness (*EWT*) (Eq. 2.3):

$$LDMC = \frac{dw}{A} \times \frac{A}{fw} = LMA \times \frac{A}{fw} \tag{2.12}$$

but

$$\frac{fw}{A} = \frac{fw - dw + dw}{A} = EWT + LMA \tag{2.13}$$

therefore

$$LDMC = \frac{LMA}{EWT + LMA} \tag{2.14}$$

In laminar leaves, *LDMC* is also related to *SLA* by a formal relationship involving leaf thickness ℓ and the average leaf density ρ_f defined as the fresh mass per unit volume (g cm^{-3}) (Vile et al., 2005; Pérez-Harguindeguy, 2013):

$$LDMC = \frac{dw}{fw} = \frac{A}{SLA} \times \frac{1}{fw} = \frac{1}{\rho_f \times SLA \times \ell} \tag{2.15}$$

Note that ρ_f is not the density of leaf tissues (tissue fresh mass per tissue volume). Indeed it includes the mass and volume of leaf water, as well as the volume of intercellular air spaces. Assuming that $\rho_f = 1$ g cm^{-3}, Eq. 2.15 simplifies to $LDMC = 1/(SLA \times \ell)$. *LDMC* therefore tends to be inversely related to *SLA* and ℓ. It has been shown to correlate negatively with relative growth rate and positively with leaf lifespan. Garnier et al. (2001) showed that leaves with high *SLA* and low *LDMC* were subject to measurement error unless hydrated prior to measurement. They provided a standard measurement protocol. Specific leaf area (*SLA*) and leaf dry matter content (*LDMC*) are two fundamental leaf functional traits that express meaningful differences in ecological behavior among plant species. They vary greatly between leaves on the same plant, different species, and may be controlled by different environmental variables such as nutrient, moisture, and light regimes.

2.5 Natural Range and Relationships of Leaf Constituents

Leaf construction involves a stoichiometric balance among chlorophyll, nitrogen, water, and other biochemicals, all convolved with leaf structure that is somehow linked with the specific leaf area (or the leaf mass per area). These parameters are therefore not independent, biophysically or ecologically (Asner, 2008). Hundreds of publications provide information about leaf biochemical content: however, most of them deal with one nutrient or stress agent at a time, one plant species, or a family of plants, or plants collected in a given ecosystem; the units in which these contents are expressed are often different, which makes the comparison between datasets difficult; and finally, the display of leaf biochemical constituents varies with the environmental conditions that plants develop under. For instance, in an experiment on British plant species, Rosevear et al. (2001) showed that the proportion of carotenoids increased in plants grown in high irradiance environments compared to chlorophyll, while including xanthophyll cycle pigments increased relative to carotenoids. In deep shade, medium and high light plants were more responsive than those adapted to low light environments. Baltzer and Thomas (2005) also grew plants under different light intensities and nutrient levels. They found chlorophyll and carotenoid concentrations and other anatomical changes increased more under high nitrogen and low light compared to higher light levels. The review by Wright et al. (2004) shows that leaf adaptations have nearly continuous variation but that leaf traits (leaf longevity, leaf mass per area, and nutrient contents) associated with assimilation have high covariance. More recently, the review by Kattge et al. (2011, 2015) on plant traits from a global database of almost 5.6 million entries from 100 000 plant species found the same trait-based patterns. They found up to 75% of the total leaf trait variation occurred within traditional plant functional type groups (PFTs) defined by growth form and leaf longevity, rather than between them. Variability among C:N:P ratios can be as high at individual sites as globally (e.g., Wright et al., 2004; Elser et al., 2010). Osnas et al. (2013) showed strong relationships, among traits after normalization for mass, retain the general structures found with *LMA* and leaf longevity, but are quantitatively different due to the normalization, representing physiological differences. Relationships among traits are not random but co-occur along axes either promoting productivity (high N, high P, high assimilation and respiration rates, low *LMA*, short-lived leaves) or promoting stress tolerance (low N, low P, low assimilation and respiration rates, high *LMA*, and long-lived leaves). Leaf nutrient concentration decreases with plant size, although P decreases faster than N.

Table 2.3 *Basic statistics of four variables studied on ~ 1400 leaves (from Féret et al., 2011).* C_{ab} *is the chlorophyll* a+b *content,* C_{xc} *the carotenoid+xanthophyll content,* EWT *the equivalent water thickness, and* LMA *the leaf mass par area.*

	C_{ab} (μg cm^{-2})	C_{xc} (μg cm^{-2})	*EWT* (cm)	*LMA* (g cm^{-2})
Number of samples	1417	1106	725	1034
Minimum	0.30	0.04	0.0043	0.0008
Maximum	106.72	25.30	0.0713	0.0331
Mean	32.81	8.58	0.0129	0.0077
Standard deviation	18.87	3.95	0.0073	0.0035

There is considerable stoichiometry in plants due to the integration among tissues and organs because of allocation shifts with size (Elser et al., 2010) as a consequence of allocation to structure rather than metabolic activities. Neither Kattge et al. (2011) nor Wright et al. (2004) studies include plant pigments in their databases. A study by Asner and Martin (2009) of 37 rainforest tree species in Kauai, Hawaii, shows marked species differences in pigment composition and spectral differences. Therefore it is not clear at the global scale of species distribution how much variance is genetic and how much is attributable to environmental patterns. In order to better know the natural distribution of foliar biochemical variables and their correlations, Féret et al. (2011) analyzed 17 independent datasets including a wide range of leaf spectral, chemical, and structural properties of 1417 leaves corresponding to about 120 different species from various growing conditions and developmental stages. Because these data were collected for different applications (development of a spectral index, model calibration, or inventory), these datasets differed in their content and format: spectral domain, type of measurements (reflectance and/or transmittance), measured biochemical variables, etc. Table 2.3 provides a list of the range, mean, and standard deviation of the four key plant biochemicals that contribute to the optical properties of leaves.

The distribution of C_{ab}, C_{xc}, and *LMA* can be fitted with a Gaussian distribution, while *EWT* is better represented by a lognormal distribution (Figure 2.26). Although this database, which mainly contains deciduous trees, cereal crops, and fruit trees, and a very few humid tropical species, does not represent all types of vegetation, it is one of the largest datasets compiled to date.

The correlations between chlorophyll *a* and chlorophyll *b* and between total chlorophyll and carotenoids confirm results from the literature (Figure 2.27). The average chlorophyll *a:b* ratio ranges between 2 and 4 in all datasets. Growth conditions and environmental factors modify this ratio (Lichtenthaler, 1987). Under high light conditions, it ranges from 3.2 to 4.0 in sunlit leaves, and under low light conditions it ranges from 2.5 to 2.9 in shaded leaves. Most of the samples in these datasets are shaded leaves. Changes in chlorophyll *a:b* ratios have also been reported for plants exposed to environmental stressors. For example, Rock et al. (1988) noted a disproportional loss of chlorophyll *b* in foliage exposed to air pollution. The chlorophyll:carotenoid ratio is approximately equal to 4 and its variability is greater than the chlorophyll *a:b* ratio: it ranges between 2 and 6, except for some extreme cases.

The LOPEX'93 database developed in the early 1990s by the European Commission to connect plant biophysical properties to plant reflectance and transmission properties (Hosgood et al., 1994),

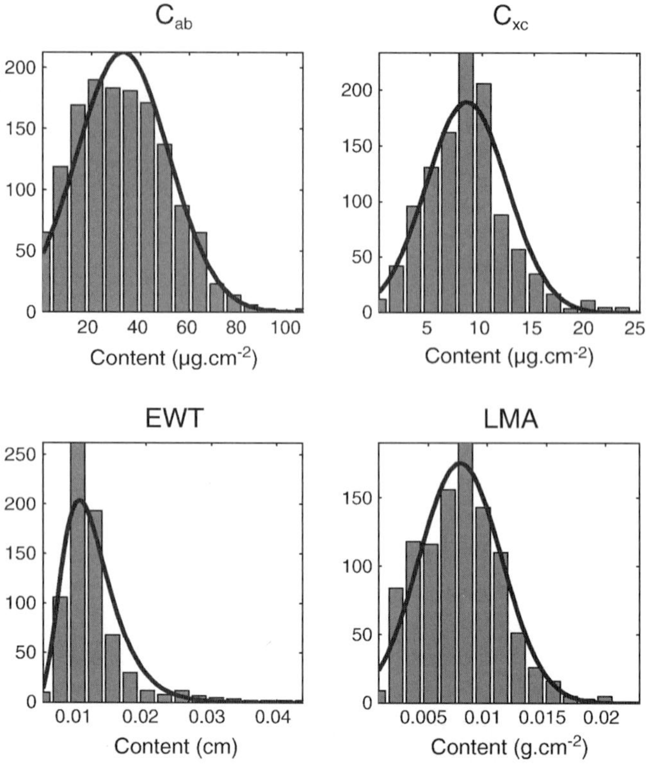

Figure 2.26 Distribution of chlorophyll *a+b* content C_{ab}, carotenoids plus xanthophylls content C_{xc}, equivalent water thickness *EWT*, and leaf mass per area *LMA* (from Féret et al. (2011), *Remote Sensing of Environment*. Copyright © 2011 with permission from Elsevier).

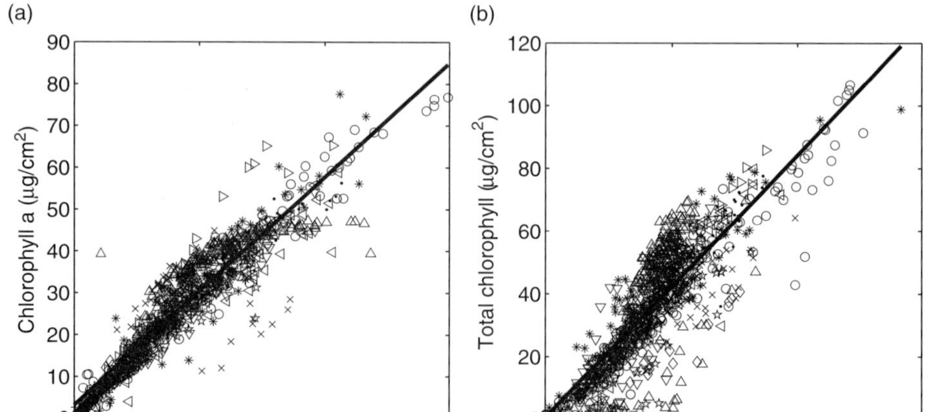

Figure 2.27 (a) Chlorophyll *a:b* ratio: 1430 samples, slope = 2.73 and (b) total chlorophyll:carotenoids ratio: 1332 samples, slope = 4.19.

Table 2.4 *Leaf thickness and biochemical content in the LOPEX'93 database.*

Variable	Unit	Range	Mean	Standard deviation
Leaf thickness	µm	86.4–780.0	194.5	114.9
FMC	%	44.9–92.4	66.4	11.0
EWT	mg cm^{-2}	4.6–40.5	11.5	6.7
LMA	mg cm^{-2}	1.9–13.5	4.5	10.7
Cellulose	mg cm^{-2}	0.31–5.45	1.08	0.72
Hemicellulose	mg cm^{-2}	0.02–3.32	0.8	0.64
Lignin	mg cm^{-2}	0.03–3.05	0.6	0.57
Protein	mg cm^{-2}	0.48–1.72	0.96	0.29
Starch	mg cm^{-2}	0–0.98	0.11	0.15
C	mg cm^{-2}	0.79–6.65	2.53	1.20
N	mg cm^{-2}	0.09–0.33	0.16	0.05

was one of the first to provide an overview of leaf biochemistry (Table 2.4). Finally, some new databases want to exhaustively inventory the phenology and the foliar biochemical properties of an ecosystem (see Appendix J).

2.6 Developmental Evolution of Leaf Constituents

How do pigments adapt to climate and environmental change? The display of pigments in a leaf varies depending on the species, the phenological cycle of healthy plant leaves, and environmental conditions. When they are initially expanding, the chloroplasts are not mature and photosynthesis does not occur or occurs at a reduced efficiency. They start functioning right away since chlorophyll is formed when the leaf is exposed to light. The pigment complexes are light green at this time as the chlorophylls are being converted to functioning molecules from protochlorophyllide. Many immature leaves with non-functioning photosynthetic systems start out as red in color due to the presence of anthocyanins and possibly other flavonoids which provide a protective function. Light absorption by these non-photosynthetic pigments seems to be protective of the developing chloroplasts. As the leaves mature they become deeper green, a characteristic of photosynthetically functioning foliage. At this time the colors of the carotenoid and anthocyanin (in very low quantity) molecules while still present, are masked. But at senescence, chlorophyll is lost before other pigments, creating the intense yellow and red colors of foliage in the fall. For species that undergo cold weather senescence the leaf color becomes dominated by the carotenoid and anthocyanin pigments and leaves turn yellow and red. These pigments protect the photosystems from photoinhibitory light during the mobilization and resorption of essential nutrients by the leaf (Hoch et al., 2001). While these and associated anatomical changes are dramatic and have large impacts on leaf optical properties, the duration of these color changes that occur at the beginning and end of the growth cycle is brief, around one week or so in length. We know many details about the biochemical and physiological bases of autumn colors, that is, how the phenomenon occurs, but we do not know why it occurs, so we do not know what, if any, is the adaptive advantage of colored leaves. The leaves of many deciduous trees change their color in autumn before falling: it is

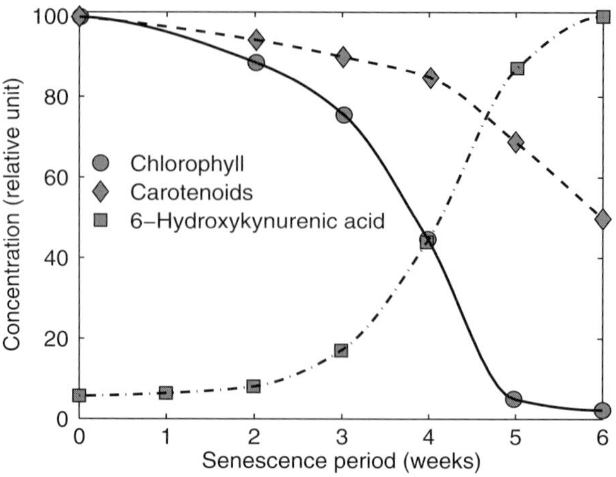

Figure 2.28 Evolution of chlorophyll, carotenoids, and 6-hydroxykynurenic acid during leaf senescence in ginkgo (*Ginkgo biloba*). Relative amounts per unit of fresh leaf weight (redrawn from Matile (2000)).

a striking phenomenon, well known to everybody. Yet nobody has explained its possible adaptive meaning.

2.6.1 Normal Conditions

Foliar pigments are conditioned by light and temperature. This is particularly true in temperate climates where seasons control leaf ontogenesis, from early to senescent stage, but it can be extended to any climate even when annual climatic variations are attenuated. When plants bud, the main foliar pigments are carotenoids and anthocyanins. Leaves contain undifferentiated organelles, the protoplasts, which are transformed into chloroplasts under the action of incident light. Chlorophyll and proteins that form the pigment–protein complex are then synthesized. When the photosynthetic apparatus forms, the leaf changes from a sink to a source state, then photosynthesis produces more carbohydrates necessary for metabolism than are consumed. In most herbaceous and deciduous plants, this change occurs when leaves are about 30 to 60% of their mature size. This maturation can occur later in some tropical species for which the photosynthetic apparatus does not develop before the end of the leaf growth (Woodall et al., 1998). This adaptive mechanism limits the nutritional value of leaves as translocation of nutrients is delayed, thus protecting them from predators. This increases the likelihood that they will reach an advanced developmental stage, allowing them to optimize productivity (Kursar and Coley, 1992).

When the leaf reaches maturity and starts to provide the carbohydrates necessary to plant growth, pigments control its optical properties in the visible domain because of high chlorophyll and carotenoid contents. Several studies detail pigment dynamics during leaf development (e.g., Makrides and Goldthwaite, 1981; Matile, 2000; Hörtensteiner and Feller, 2002). In autumn foliage, these chloroplast pigments degrade after the chlorophylls breakdown and thus they are revealed in the foliage spectrum due to the loss of chlorophyll (Merzlyak and Gitelson, 1995). As seen in Figure 2.28, the chlorophyll content decreases faster than the carotenoid content during the senescence because the latter pigments are more stable (Biswal,

Figure 2.29 Variation of Japanese cherry (*Prunus serrulata*) leaf color during leaf development. (A black and white version of this figure will appear in some formats. For the color version, please refer to the plate section.)

1995). When chlorophyll degrades, it is converted to a colorless form without the magnesium central core. In *Ginkgo biloba*, chlorophyll breakdown is also accompanied by the synthesis of a fluorescent compound, 6-hydroxykynurenic acid, a secondary metabolite of tryptophan which appears to have the effect of an optical brightener. Note that during senescence, anthocyanins are the only pigments synthesized from the colorless flavonols located in the vacuole (Wilkinson et al., 2002).

Variations in the relative concentrations of these pigments explain the basic color of plant leaves, as illustrated by Figure 2.29. It shows Japanese cherry tree (*Prunus serrulata* "Hisakura") leaves collected in the Jardin des Plantes (Paris, France). They are green in spring and summer and yellow and brown in autumn.

2.6.2 Effect of Light

Leaf pigments adapt to the intensity of incident light. In moderate light, a variation of light intensity causes a dissociated adaptation at the chloroplast and leaf scales. Murchie and Horton (1997) studied the response of 22 different species and Bailey et al. (2001) were more particularly interested in the case of *Arabidopsis thaliana*. At the chloroplast scale, an increase in light intensity causes a change in the total chlorophyll content, which represents a decrease in the volumetric concentration but a stable or increase in the content per unit leaf area. These two effects are explained by the increase in foliar volume and the number of chloroplasts. The increase in the light intensity also induces a decrease in the concentration of photosystems in the thylakoid membrane and a decrease in the PSII size. In spite of lower chlorophyll content, one observes an increase in photosynthetic activity because chlorophyll efficiency improves. In the same way, the increase in the chlorophyll *a:b* ratio under high light intensity seems to be in contradiction to the increase in the PSII:PSI ratio (Bailey et al., 2001). It is explained by a reduced LHCP2 (primary source of chlorophyll *b*) size within the PSII. At the leaf scale, the increase in light intensity produces a thickening of the lamina and an increase in the number of chloroplasts, because of an increase in the number of cell layers in the palisade parenchyma, which is rich in chloroplasts. However, Murchie and Horton (1997) showed that these mechanisms were not universal; the phenotypic plasticity of the leaf that is preferentially invested in the production of carbon-rich molecules, like phenolic compounds, rather than nitrogen-rich molecules, like proteins and chlorophyll, when the duration of light exposure increases (Meyer et al., 2006). Leaves also adapt to the carotenoid pool. Demmig-Adams et al. (1996) compared the carotenoid composition of sunlit

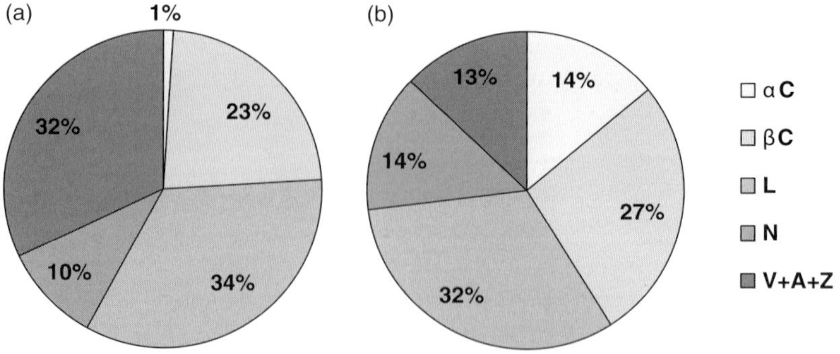

Figure 2.30 Carotenoid composition in shaded (a) and sunlit (b) leaves of lawyer vine (*Smilax australis*) growing in their natural habitat at a coastal site in Australia (αC: α-carotene, βC: β-carotene, L: lutein, N: neoxanthin, V+A+Z: violaxanthin + antheraxanthin + zeaxanthin) (from Demmig-Adams et al. (1996), *The FASEB Journal*. Copyright © 1996 with permission from The Federation of American Societies for Experimental Biology).

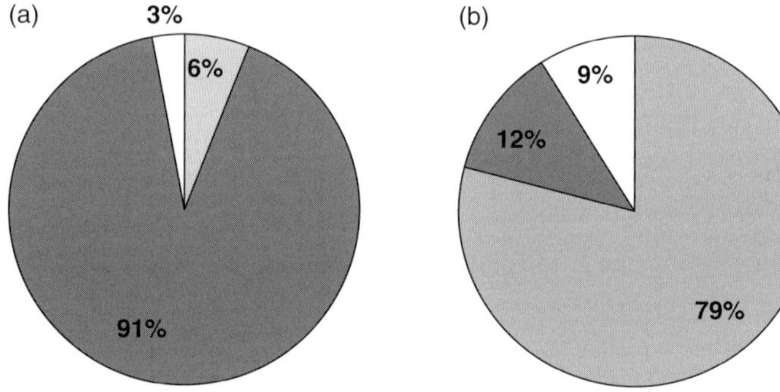

Figure 2.31 Allocation of absorbed light to photosynthesis (P, dark gray) vs. heat dissipation (D, light gray) at midday in shaded (a) and sunlit (b) leaves of lawyer vine (*Smilax australis*) growing in their natural habitat at a coastal site in Australia) (from Demmig-Adams et al. (1996), *The FASEB Journal*. Copyright © 1996 with permission from The Federation of American Societies for Experimental Biology).

leaves and shaded leaves of *Smilax australis*. They showed that the chlorophyll:carotenoid ratio decreases in the case of overexposure to light and they illustrate an important variation in the relative proportions of various carotenoids, with a strong increase in zeaxanthin, violaxanthin, and antheraxanthin, which contributes to the xanthophyll cycle (Figure 2.30).

This result should be related to the resource fraction allocated to photosynthesis that strongly decreases in sunlight for heat dissipation. Thus most tropical understory species usually have a low photosynthetic capacity such that they are not able to use more than 10% of incident solar energy to carry out photosynthesis (Figure 2.31).

A light that is too intense causes photoinhibition, which results in a significant reduction of photosynthetic activity. This photoinhibition performs at several scales; at the cell scale the chloroplasts that are under optimal lighting conditions preferentially move towards the periclinal cell wall to maximize their exposure to incident light. When illumination becomes too strong, they move to rearrange the chloroplasts along the anticlinal cell walls to be oriented sideways to the incident light, which reduces possible damage of the photosynthetic apparatus. These movements are controlled by a system that is sensitive to UV and blue radiation, and less often to red radiation (Trojan and Gabrys, 1996; Wada et al., 2003). At the level of the photosynthetic apparatus, one first observes a dynamic photoinhibition, during which the LHC2 migrates from the PSII towards the PSI and transfers its excitation energy to decrease the formation of active forms of oxygen in the environment of the PSII. This photoinhibition is reversible. If this situation persists, an irreversible chronic photoinhibition leads to the destruction of the photosynthetic pigments by photo-oxidation (Powles, 1984; Hendry et al., 1987; Alves et al., 2002). The start of the xanthophyll cycle is then the principal protection mechanism that dissipates the energy absorbed by chlorophylls at the level of the PSII antenna in the form of heat. Behera and Choudhury (2002) showed that the destruction of photosynthetic pigments induced by overexposure to light was less important for carotenoids than for chlorophylls. This effect is thus comparable to the sequence of pigment degradation in senescence.

3

Spectroscopy of Leaf Molecules

This chapter provides the basis for the absorption and internal scattering properties of leaves derived from theoretical spectroscopy of various chemical components. The absorption of electromagnetic radiation by leaf constituents occurs in different regions of the spectrum. Molecular electronic transitions take place in the ultraviolet (UV) and visible spectrum. This occurs when electrons in a molecule are excited from one energy level to a higher energy level. Transitions between two levels can occur upon the absorption of a photon. The photon energy associated with the infrared range is not high enough to excite electrons to higher orbits, but it may induce vibrational excitation of covalently bonded atoms and groups. Vibrational transitions are thus responsible for absorption in the near infrared, shortwave infrared, and thermal infrared. They can be associated with the stretching or bending of specific bonds. The absorption features that occur in the far-infrared and microwave result from quantum mechanical molecular rotational transitions. Molecular rotations require little energy to excite them. Sometimes these absorptions bands are highly localized. Most of the time however, when the energy level of two modes are close, they can mix. Measuring the absorption coefficients in leaves is particularly challenging because extraction of the chemicals, particularly where reactions are membrane bound, results in changing the chemical bond structure and state of the material, which modifies its refractive index and the relative permittivity, thus changing its absorption properties.

With plant leaves we face two kinds of difficulties: first, as pointed out earlier, the absorption properties of the biochemical constituents which, except for water, are poorly known. This problem is most acute for chemicals that are embedded in membrane bound complexes, such as the photosynthetic pigments. Pigment absorption features are only available *in vitro*, which means that the larger protein complexes in which pigments were imbedded have been lost when measurements are made, thus impacting the bond structure and absorption wavelengths. The chemical bonds are further changed by the polarity of the solvent used in the extraction. In most cases pigment absorption features are only known over the visible spectrum although some absorb in the UV and near infrared. Second, leaves are not homogeneous media. Therefore, a more comprehensive knowledge of absorption and refraction, but also of leaf biochemistry in relation to leaf anatomy, is required before we can apply a purely theoretical approach to compute the refractive index in a complex medium like a plant leaf.

3.1 Theory

3.1.1 The Complex Refractive Index

The optical properties of any kind of medium (extrinsic optical properties) depend on the complex refractive index of its constituents (intrinsic optical properties):

$$\tilde{n} = n_r - in_i \tag{3.1}$$

The real part n_r and the imaginary part n_i are both functions of the wavelength; they are generically called optical constants. The real part is the refractive index n defined as the ratio of velocity of light c (m s^{-1}) in a vacuum to its velocity ω (m s^{-1}) in the medium: $n = c/\omega$ (see Appendix E). In physics, the imaginary part is called the extinction coefficient and is related to light absorption. Experimentalists generally prefer the absorption coefficient k (m^{-1}) that can be expressed in terms of the extinction coefficient as

$$k = \frac{4\pi n_i}{\lambda} \tag{3.2}$$

It should be noted that agreement on the definition of the term "absorption coefficient" (sometimes also referred to as the Lambert absorption coefficient) and on the use of the symbol k is far from universal, and quantities differing from k and/or n_i by various factors have been frequently used to describe absorption. Knowledge of n and k is useful for the theoretical calculation of reflectivity (or transmissivity) values at the interface between two media and for determination of light transmission within a medium, respectively. Each of the foliar constituents (e.g., pigments, water, cellulose) should be individually isolated and measured. Although this is not problematic for water, which has been accurately determined, there is considerable disagreement in the literature as to the correct values of $\tilde{n}$ for the other constituents. For different reasons that will be detailed later, such optical constants are difficult to measure for the photosynthetic pigments that are imbedded in the two photosystems (PSI and PSII) within the chloroplasts, and for the cell wall constituents.

3.1.2 The Relative Dielectric Permittivity

In the terahertz microwave and radio-wave region, the dielectric properties of a material, which basically reflect the electric charge movement inside the material in response to an external electric field, are defined by the relative dielectric permittivity, simply called relative permittivity hereafter. Since there is no standard dedicated symbol representing the relative permittivity, we will use the symbol $\tilde{\epsilon}$ in this book to avoid confusion with the emissivity given by the symbol ε. Note that the word "relative" in front of permittivity implies that the numbers are reported relative to the vacuum permittivity ϵ_0 (8.854×10^{-12} F m^{-1}) and then are dimensionless. Like the refractive index $\tilde{n}$, the relative permittivity of the material is wavelength (or frequency) dependent and has both real and imaginary mathematical parts:

$$\tilde{\epsilon} = \epsilon_r - i\epsilon_i \tag{3.3}$$

The real part ϵ_r (dielectric constant noted ϵ') is related to the stored energy within the medium and the imaginary part ϵ_i (loss factor noted ϵ'') is related to the dissipation of energy within the medium. The refractive index and the relative permittivity are also related:

$$\tilde{\epsilon} = (n_r - in_i)^2 \tag{3.4}$$

Table 3.1 *Relative permittivity and refractive index of pure liquid water.*

Frequency (wavelength)	Temperature	Relative permittivity		Refractive index	
		Real part	Imaginary part	Real part	Imaginary part
1 GHz (30 cm)	0°C	86.90	9.11	9.33	0.49
	20°C	79.96	4.43	8.95	0.25
	40°C	73.09	2.55	8.55	0.15
50 GHz (0.6 cm)	0°C	8.21	14.52	3.53	2.06
	20°C	13.32	22.71	4.45	2.55
	40°C	20.33	28.29	5.25	2.69
300 GHz (0.1 cm)	0°C	4.84	3.49	2.32	0.75
	20°C	5.11	4.97	2.47	1.00
	40°C	5.43	6.24	2.62	1.19

Expanding Eq. 3.4 leads to $\epsilon_r = n_r^2 - n_i^2$ and $\epsilon_i = 2n_r n_i$. Conversion between n and ϵ is then done by

$$n_r^2 = \frac{1}{2}\left(\sqrt{\epsilon_r^2 + \epsilon_i^2} + \epsilon_r\right) \tag{3.5}$$

$$n_i^2 = \frac{1}{2}\left(\sqrt{\epsilon_r^2 + \epsilon_i^2} - \epsilon_r\right) \tag{3.6}$$

Table 3.1 gives the relative permittivity $\widetilde{\epsilon}$ and the complex refractive index $\widetilde{n}$ calculated using Eq. 3.5 and Eq. 3.6 for pure liquid water, at different temperatures. At 20°C (room temperature) and 1 GHz, the literature generally gives approximately 80 for the dielectric constant (real part) of water when the loss factor (imaginary part) is smaller by a factor of 20.

3.1.3 The Kramers–Kronig Relations

The dispersion and loss of electromagnetic radiation in a homogeneous medium are not independent of each other. In other words, the real and imaginary parts of the complex refractive index $\widetilde{n}$ and the complex relative permittivity $\widetilde{\epsilon}$, which are descriptive of light–matter interaction phenomena, are not wholly independent but are connected by a special form of Hilbert transforms: the Kramers–Kronig relations (e.g., Lucarini et al., 2005; Waters et al., 2005). These relations introduced in the 1920s state that the real part, at a given wavelength, can be expressed as an integral of the imaginary part, over all wavelengths, and vice versa. Applied to the frequency-dependent refractive index $\widetilde{n}(\omega)$, they take the form of

$$\Re\left(\widetilde{n}(\omega)\right) = \frac{2}{\pi} P \int_0^{\infty} \frac{\omega' \Im\left(\widetilde{n}(\omega')\right)}{\omega'^2 - \omega^2} d\omega' \tag{3.7}$$

$$\Im\left(\widetilde{n}(\omega)\right) = -\frac{2\omega}{\pi} P \int_0^{\infty} \frac{\Re\left(\widetilde{n}(\omega')\right)}{\omega'^2 - \omega^2} d\omega' \tag{3.8}$$

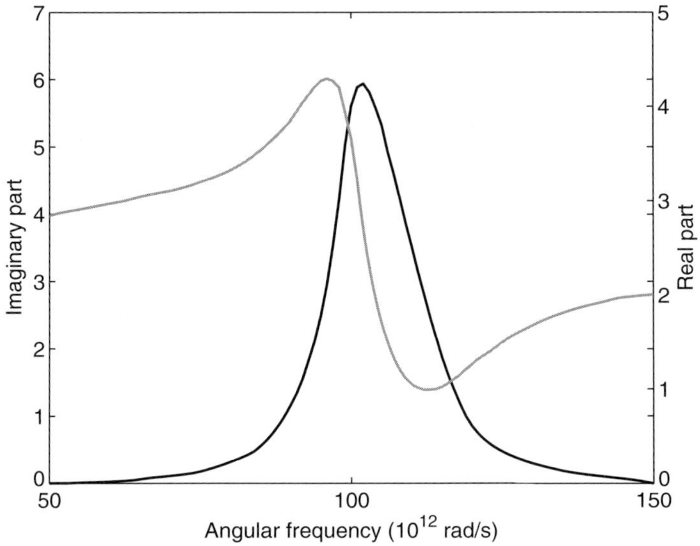

Figure 3.1 Variation of the complex refractive index as a function of the angular frequency. The imaginary part (a) *y*-axis is colored in black, and the real part (b) *y*-axis is colored in gray (redrawn from Elias and Lafait (2006)).

Integrals are taken over all positive frequencies ω' and P stands for the Cauchy principal value of the integral. Implementation of such relations in the case of finite-bandwidth data requires extrapolation of the spectral data beyond the experimental data range and linearity properties of the optical constants. Truncation of the integrals may affect the accuracy of the results. Figure 3.1 illustrates the Kramers–Kronig transform applied to the absorption coefficient k for a simple single electronic dipole oscillator of resonance frequency $\omega_0 = 100 \times 10^{12}$ rad/s. The refractive index n slowly increases before or after the band, while it rapidly decreases close to the band. Results would be similar with the loss factor ϵ'' and the dielectric constant ϵ'.

These relations are useful to determine the real part or the imaginary part of the complex refractive index when it is difficult to experimentally measure either optical variable over a wide range of frequencies. Note that it is generally much more difficult to measure the real part than the imaginary part. As an example, Hale and Querry (1973) compiled values of absorption coefficient $k(\lambda)$ for pure liquid water from the scientific literature and successfully computed its refractive index $n(\lambda)$ in the 200 nm–200 μm spectral region by applying a subtractive Kramers–Kronig analysis. Zolotarev and Demin (1977) extended that work from 0.01 nm (X-rays) to 1 m (radio frequencies) (Figure 3.2).

Just as the complex refractive index is exactly known for pure liquid water, there is a large amount of experimental work devoted to its relative permittivity (Figure 3.3).

In the same way, the refractive index of the thylakoid membrane (Figure 3.4) can be determined by measuring the gradient of unidirectional light in chloroplast suspensions (e.g., Paillotin et al., 1993, 1998; Dobek et al., 1994; Gapinski et al., 1994; Naqvi et al., 2004). Unfortunately the result of this (*in vitro*) experiment cannot be transferred as it is to the (*in vivo*) leaf where light is diffused. Moreover the macroscopic distribution of light within the leaf blade depends on chloroplast concentration. The Kramers–Kronig relations are thus unsuitable to determine the refractive index of these foliar constituents.

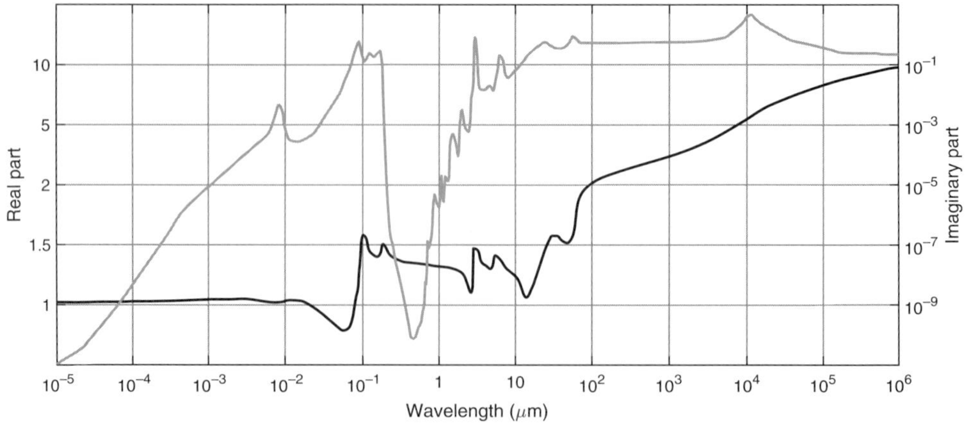

Figure 3.2 Complex refractive index of water between 0.1 Å and 1 m at 25°C (redrawn from Zolotarev and Demin (1977)).

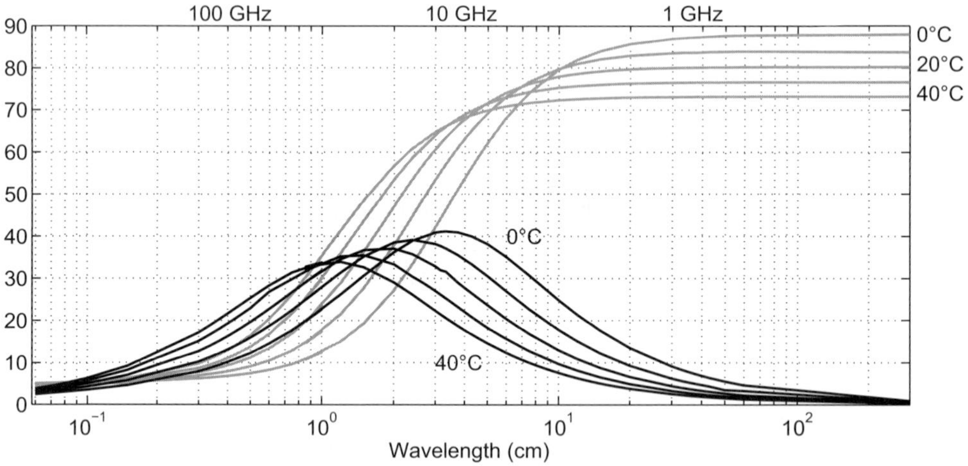

Figure 3.3 Simulated data for the relative permittivity of pure liquid water at 0°C, 10°C, 20°C, 30°C, and 40°C. The real part ϵ' is represented by black lines and the imaginary part ϵ'' is represented by gray lines. The data are initially calculated for a given frequency and temperature using the model derived by Meissner and Wentz (2004).

Recently, Chen and Weng (2012) calculated the wavelength specific complex refractive index for water and dry matter thanks to a Kramers–Kronig constrained variational analysis of optical spectra, which they tested with good results on the LOPEX data and the PROSPECT model (see Chapter 9).

3.1.4 The Effective Medium Theory

The effective medium theory describes the macroscopic properties of a composite or mixed material medium based on the properties and the relative fractions of its components and could contribute to

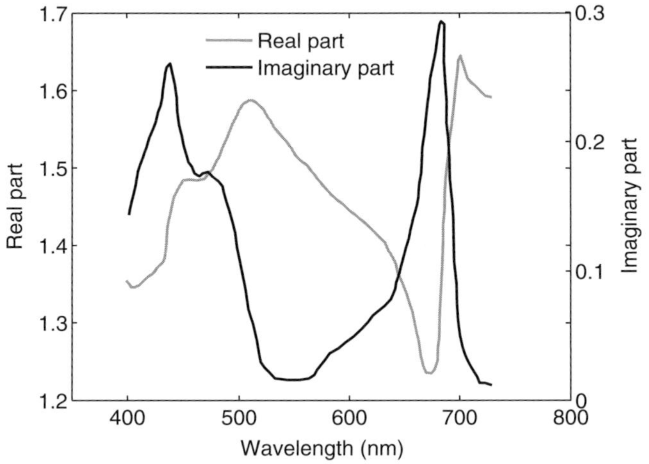

Figure 3.4 The real and imaginary parts of the refractive index, $n = n_r$ (gray line) and n_i (black line), as a function of wavelength, describing light scattering (reflection, diffraction) and absorption, respectively, for the thylakoid membrane. The real part is calculated according to Eq. 3.7 by integration of the imaginary part over the 200–750 nm range. The constant part of the refractive index $n = 1.42$ is obtained from the fit of the experimental data (redrawn from Dobek et al. (1994)).

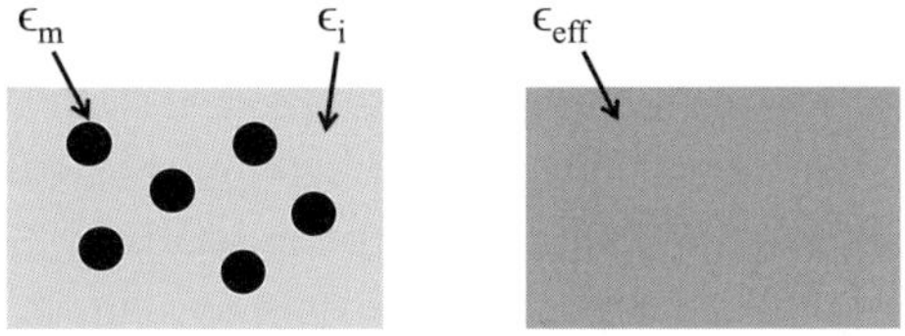

Figure 3.5 Effective medium theory of a two-component medium.

a better determination of the effective permittivity (or effective refractive index) in an inhomogeneous medium. For instance, the Maxwell Garnett theory (e.g., Maxwell Garnett, 1904; Markel, 2016), which yields an expression for the effective dielectric constant of a medium in which small spheres are embedded, could be investigated. Imagine a matrix (dielectric constant ϵ_m) and spherical inclusions (dielectric constant ϵ_i) occupying a volume fraction δ_i ranging from 0 to 1 (Figure 3.5).

At low concentration, the effective dielectric constant of such a complex medium can be written in terms of the permittivities and volume fractions of the individual constituents:

$$\epsilon_{eff}(\omega) = \epsilon_m(\omega)\frac{2\delta_i\left(\epsilon_i(\omega) - \epsilon_m(\omega)\right) + \epsilon_i(\omega) + 2\epsilon_m(\omega)}{2\epsilon_m(\omega) + \epsilon_i(\omega) + \delta_i\left(\epsilon_m(\omega) - \epsilon_i(\omega)\right)} \tag{3.9}$$

Other formulas have been proposed for mixtures of ellipsoidal or arbitrarily shaped particles (Scheller et al., 2010). For instance, the Landau–Lifshitz–Looyenga model considers a given volumetric fraction of particles with unknown shape; by successively adding an infinitesimal amount of particles, the effective permittivity slightly increases, which can be described by a Taylor approximation:

$$\sqrt[3]{\epsilon_{eff}(\omega)} = \delta_i \sqrt[3]{\epsilon_i(\omega)} + (1 - \delta_i) \sqrt[3]{\epsilon_m(\omega)} \tag{3.10}$$

This model has been used to determine leaf water content with terahertz radiation: the leaf is described as an irregular multi-compound mixture, which mainly consists of air, water, and solid plant material. Equation 3.10 computes the effective dielectric constant of a medium consisting of two components so that an extension has to be made in order to match these enhanced requirements. Then an electromagnetic model is used to simulate the backscattering properties of leaves in this wavelength region (see Section 8.4). It is not certain that the effective medium theory would provide satisfactory results in the visible domain because it is normally applied to particles much smaller than the chloroplasts; however, because it has been successfully used in other situations that do not fit the model requirements, it still presents possibilities to improve the calculation of the refractive index of leaves.

3.2 Pigment-Specific Absorption Coefficients

The absorption of light by photosynthetic pigments dominates the optical properties of green leaves in the visible spectrum (400–700 nm).

3.2.1 Chlorophylls

The spectroscopy of chlorophyll can be described in the broadest context of porphyrin optical properties. These molecules, which are characterized by a tetrapyrrolic porphyrin ring, all display an absorption band in the near UV and a system of absorption bands in the visible. The nature of their electronic transitions depends on the symmetry of the ring system. In the case of chlorophyll, the symmetry is low. The magnesium ion that is bound to the tetrapyrrolic ring in a rigid square planar arrangement is capable of a variety of oxidation states allowing for the acceptance or donation of electrons. The absorption peak of chlorophyll *a* in methanol is located at 658 nm and corresponds to a transition from the fundamental electronic level S_0 toward the first excited level S_1 (Figure 3.6). It is also called the Q_y transition because it is mainly polarized along the *y*-axis of the molecule (Platt, 1956). This band has a vibrational satellite at 612 nm. The peak at 572 nm corresponds to a second electronic transition S_2, which is polarized along the *x*-axis and is then called Q_x. The Soret band, the maximum of which is at 426 nm, is a composite band that contains at least three electronic transitions $S_{3,4}$.

In vitro absorption spectra for chlorophylls and several accessory photosynthetic pigments (e.g., carotenoids like the xanthophylls) have been measured in several solvents. For instance, Seely and Jensen (1965) recorded and tabulated the intrinsic optical properties of chlorophyll *a* in 40 solvents. The two major absorption peaks in the blue and in the red of chlorophylls *a* and *b* shift to longer wavelengths with increasing polarity and/or water content of the solvent. Table 3.2 provides some physical properties of the main solvents used for plant pigments. The Polarity index that measures the ability of the solvent to interact with various test solutes increases with solvent polarity (Snyder, 1978 Szalay et al., 1974). Diethyl ether, also known as ethyl ether, is generally considered a "non-polar" solvent compared to ethanol, acetone, or methanol.

Chlorophylls in diethyl ether consequently produce the shortest wavelengths for the blue and red peaks, compared to chlorophylls in ethanol (Figure 3.7). With increasing polarity of the solvent,

Table 3.2 *Specifications of the main solvents used to extract photosynthetic pigments.*

Solvent	Polarity index	Refractive index	Density at 20°C (g/ml)
Diethyl ether	2.8	1.352	0.713
Ethanol	4.3	1.361	0.789
Acetone	5.1	1.359	0.790
Methanol	5.1	1.328	0.791
Water	10.2	1.333	1.000

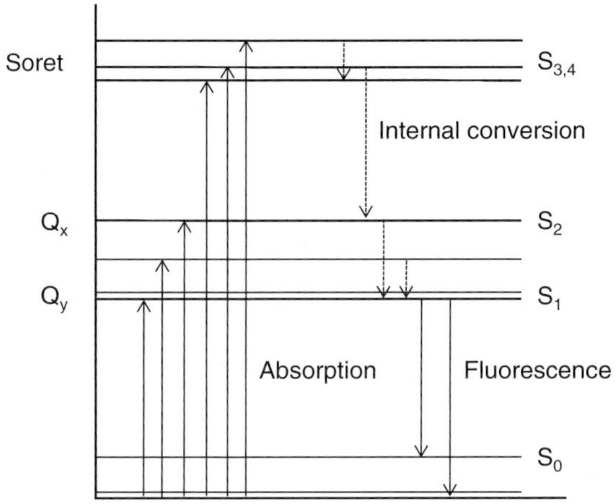

Figure 3.6 Energy diagram of chlorophyll *a*.

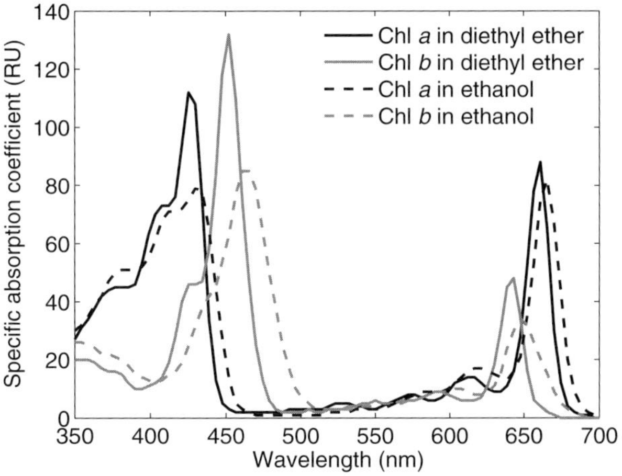

Figure 3.7 Specific absorption coefficients for chlorophyll *a* and *b* in diethyl ether and ethanol (redrawn from Lichtenthaler (1987)).

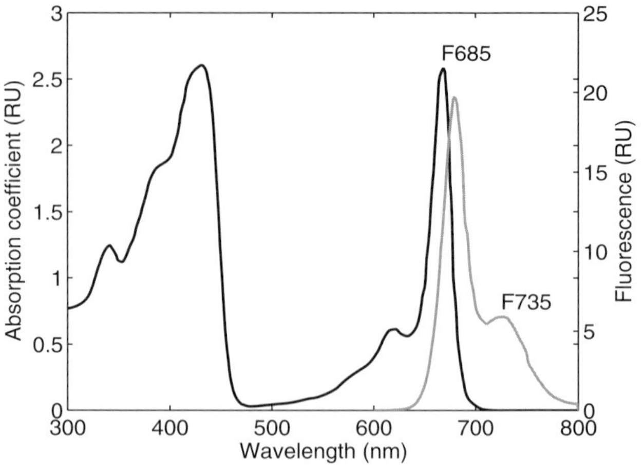

Figure 3.8 Absorption (black line) and emission (gray line) spectra of chlorophyll *a* in solution in methanol.

one can note that the shift is less pronounced in the red than in the blue, that it is larger for chlorophyll *b* than for chlorophyll *a*, and that the absorption peaks become smaller and broader. The absorption spectra of these molecules in a solution of methanol produces absorption peaks close to those measured *in vivo*. Compared to chlorophyll *a*, the first absorption peak of chlorophyll *b* is slightly shifted toward longer wavelengths while the second peak is shifted toward shorter wavelengths. Because the pigment spectrum shifts depending upon the solvent used to extract them, and because the membrane-bound protein complex is removed during extraction, the true *in vivo* configuration of plant pigment absorption coefficients remains uncertain. Moreover, the overlapping wavelengths of these absorption coefficients make their individual identification in leaf reflectance spectra difficult and challenging to predict pigment concentrations from inverse modeling.

Chlorophyll *a* is the only molecule in plant leaves that emits light in the red wavelength region. Fluorescence spectra of other colored materials are generally measured at other wavelengths in diluted and non-turbid solutions. The situation is very different in a leaf where chlorophylls are highly concentrated. Fluorescence corresponds to radiative de-excitation of the lowest vibrational level S_1 (or Q_y) toward the fundamental level S_0, which explains the mirrored emission spectrum compared to the absorption spectrum (Figure 3.8). The emission spectrum has two main peaks at 670 nm and at 725 nm. One can notice that these two spectra overlap in the red but not in the infrared region. The distance between the excitation and emission wavelengths is called the Stokes Shift. If the chlorophyll concentration is high, photons emitted in the red are likely to be reabsorbed.

Several scenarios are presented in Figure 3.9, based on the initial and final states of excitation. At room temperature, the fundamental level S_0 is much more populated than its excited vibrational levels. A transition from the fundamental level S_0 toward the vibrational levels of S_1 (e.g., A_1 or A_2) is the most likely. Absorption occurring from the excited vibrational levels of S_0 (e.g., A_3) produces the same fluorescence emission spectrum as A_1 and A_2 because all the transitions start from the same initial level. This particular case shows that the energy of the emitted light can be higher than the energy of the absorbed light, in other words, the wavelength of emitted photons can be smaller than

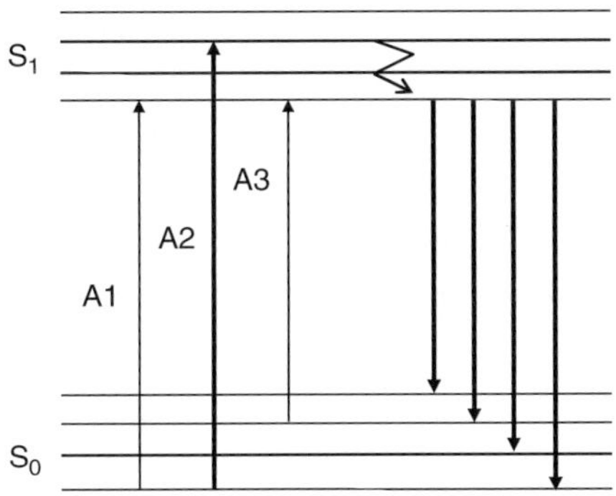

Figure 3.9 Fluorescence emission. The transition A_2 is followed by a fast de-excitation toward the lowest vibrational level of S_1 and by fluorescence emission. Absorptions A_1, A_2, and A_3 induce the same fluorescence emission spectrum.

the wavelength of absorbed photons. However, the emission spectrum will be less intense because the population of chlorophyll *a* molecules in this situation is lower.

3.2.2 Carotenoids

Carotenoids are a large family of water-repelling red, orange, or yellow accessory pigments, which primarily absorb in the blue-green wavelengths and extend the wavelength range for energy absorption by photosynthesis. The carotenoids of higher plants include β-carotene, lutein, zeaxanthin, violaxanthin, and neoxanthin as major components of the photochemically active thylakoids of chloroplasts (Lichtenthaler, 1987). The molecular structure of carotenoids displays a central pattern consisting of alternating single (C–C) and double (C=C) carbon–carbon bonds. These bonds form a system called a conjugated π-electron system: they interact with each other, allowing electrons in the molecule to move freely across these areas of the molecules. The schematic energy diagram of zeaxanthin in Figure 3.10 shows the transitions which are allowed (straight line) and those which are not allowed (dashed line) by one-photon excitation. The transition from the ground state S_0 (also noted $1A_g^-$) to the first singlet excited state S_1 (also noted $2A_g^-$) is optically forbidden. According to selection rules, the first low-lying excited state that is allowed in a one-photon transition from the ground state is the second singlet state S_2 (also noted $1B_u^+$). The strong blue-green light absorption peak by carotenoids originates from this transition. It appears likely that the S_1 level of carotenoids would be involved in the xanthophyll cycle of higher plants (Polívka et al., 1999). As the S_2 to S_1 transition is allowed, the molecules that are excited to the S_2 state can relax to the S_1 state and then produce fluorescence (S_1 to S_0 transition). However, in carotenoids, it is very weak and short, yet long enough to be probed by femtosecond techniques.

Figure 3.11 exhibits typical absorption spectra of carotenoids in methanol. They are characterized by three well separated absorption peaks (zeaxanthin, violaxanthin, and neoxanthin) or two peaks with one shoulder (α-carotene and β-carotene). Oxygen-free carotenoids absorb light at longer wavelengths than oxygen-bearing xanthophylls: 449.2 nm (2.76 eV) and 475.4 nm (2.61 eV) for β-carotene versus 435.6 nm (2.85 eV) and 464.0 nm (2.67 eV) for neoxanthin. As for chlorophylls, the wavelength position of the absorption peaks of carotenoids also depends on the type of solvent and its water content.

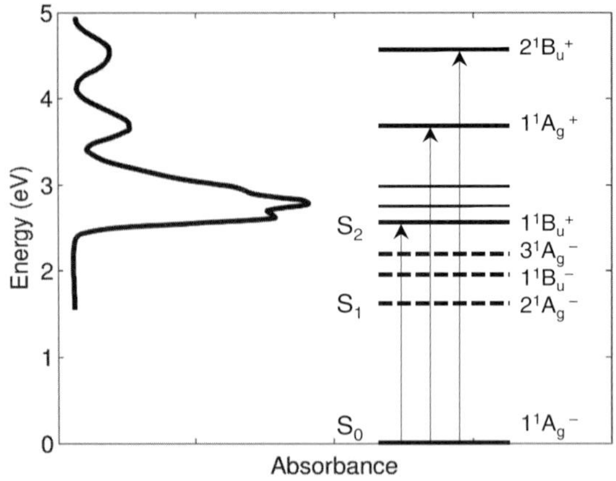

Figure 3.10 Energy diagram of a zeaxanthin molecule (redrawn from Milanowska and Gruszecki (2005)).

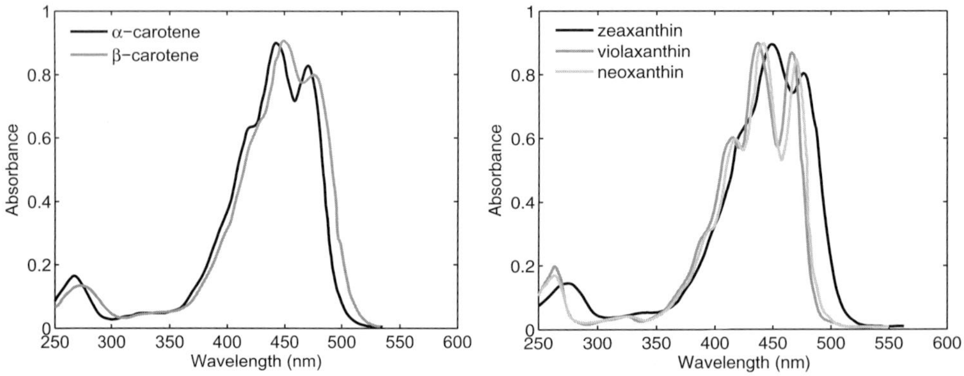

Figure 3.11 Specific absorption coefficients spectra of leaf carotenoids (source: http://lipidbank.jp/).

3.2.3 Anthocyanins and Other Phenolic Compounds

A wide variety of phenolic compounds are found in plant tissues. As described previously, anthocyanins are flavonoid pigments not associated with photosynthesis. They are usually dissolved in the cell sap of plant vacuoles and their color varies from red to blue. Due to the difficulties of preparing crystalline anthocyanin free from impurities, the absorption coefficients found in the literature singularly lack uniformity. They generally display two main absorption bands, one in the UV region (260 to 280 nm) and another one in the visible region (450 to 550 nm). The single absorption maximum of anthocyanins around 520 nm can be detected by changes in spectral reflectance of leaves in the green region. Because these pigments usually cause little change in the red-edge, it can be used as a reference for comparison to photosynthetic pigments (e.g., Curran et al., 1991; Neill and Gould, 1999; Gitelson et al., 2001, 2006). The presence of cinnamic acid acylation is revealed by the presence of a third absorption band in the 310 to 360 nm range (Figure 3.12).

The spectral features of anthocyanins, and hence their colors, are very dependent on pH, temperature, solvent, and the presence of other molecules that may interact with them. The sensitivity to pH is

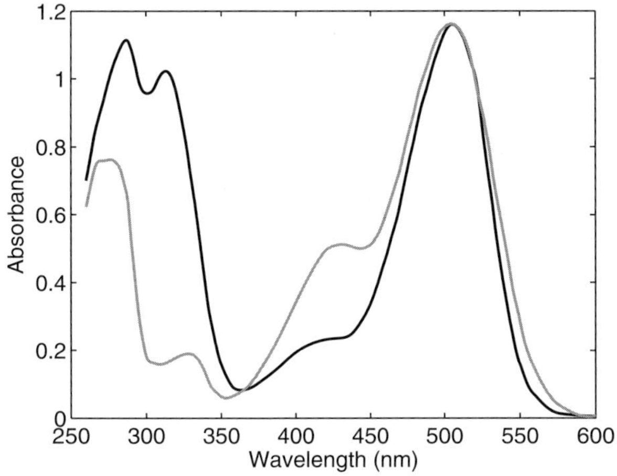

Figure 3.12 Spectral absorbance of two pelargonidin derivatives in pH 1 buffer: 3,5-glycosylated anthocyanins with cinnamic acid acylation (black curve) and 3-glycosylated anthocyanins with no cinnamic acid (gray curve) (redrawn from Giusti and Jin (2008)).

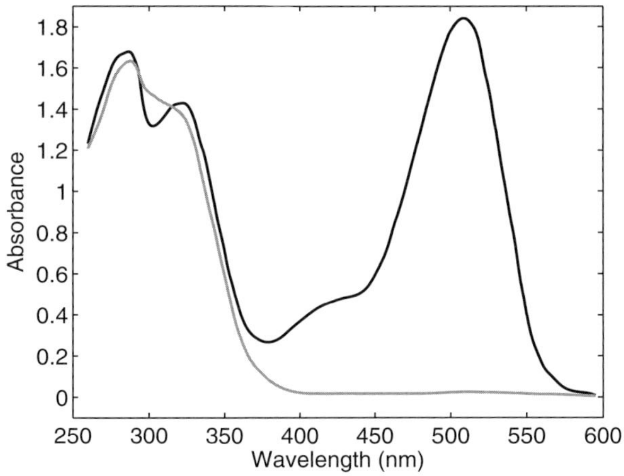

Figure 3.13 Spectral characteristics of purified radish anthocyanins (acylated pelargonidin-3-sophoroside-5-glucoside derivatives) in pH 1.0 (black line) and pH 4.5 (gray line) buffers (redrawn from Giusti and Wrolstad (2001)).

due to a reversible structural change that occurs in the C ring of the molecule (see Figure 11.20). If leaves contain acidic sap (pH ~ 1.0), the C ring of the molecule is positively charged. Anthocyanins strongly absorb light between 450 nm and 550 nm (blue and green light), with a maximum at about 520 nm. Thus they will appear as violet or bluish-red to the eye. If the sap is less acidic (pH ~ 4.5), the positive charge is lost due to hydration of the C ring. Anthocyanins absorb less light in the visible range; they display a duller purple color or are colorless (Figure 3.13). If many natural anthocyanins become colorless in neutral and alkaline solution, a few lose very little color. This property has been associated with the presence of cinnamic acid acylation. Finally, in variegated leaves, the green cells may contain anthocyanin, which is rendered colorless because of local variation in the vacuole pH.

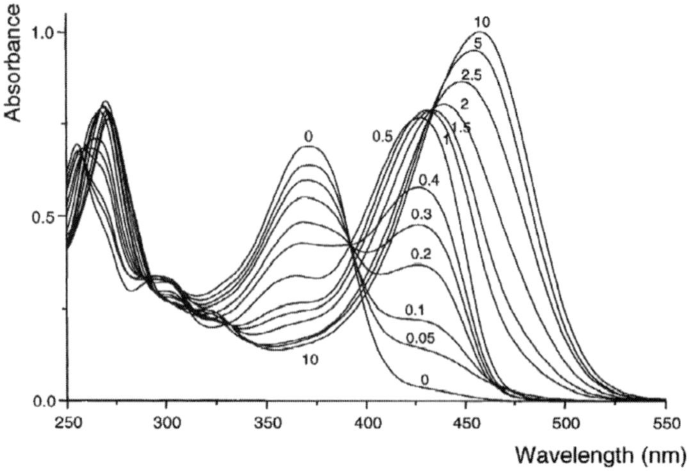

Figure 3.14 Absorption spectra of quercetin in methanol during titration with Al(3). Concentration range: 0–10 μM (from Cornard and Merlin (2002), *Journal of Inorganic Biochemistry*. Copyright © 2002 with permission from Elsevier).

However, since the various colors of tissues containing anthocyanins are due not only to the presence of several anthocyanins but also to changing amounts and mixtures of them, to alterations in the pH of the cell sap, to the ash content of the latter, to co-pigmentation (e.g., addition of tannin), and to the colloidal condition of certain other components of the sap, the *in vivo* determination of anthocyanins is difficult (Blank, 1947). Kokaly and Skidmore (2015) identified phenolic compounds in plants by measuring leaf reflectance at 1.66 μm.

Flavonoids that are ubiquitous fulfill many plant functions. They accumulate primarily in epidermal cells where they protect photosynthetic activity from high sunlight, particularly UV-B radiation (Agati et al., 2013). Quercetin is a common flavonoid, the absorption spectrum of which is shown in Figure 3.14.

Jacquemoud et al. (2000) found a strong absorption peak in the blue spectrum of albino leaves at wavelengths shorter than 450 nm that was attributed to polyphenols, particularly flavonols (Cerovic et al., 2002). These compounds are present in green leaves but are hidden by chlorophylls and carotenoids. Brown pigments occur during leaf senescence. Soon after death, leaves are dry and light brown in color but after winter the color has become much darker brown, even though they are still recognized as leaf detritus. Figure 3.15 displays the "average" absorption spectrum of brown leaves that should be multiplied by a factor to account for different leaf colors.

3.3 Water-Specific Absorption Coefficients

Because of the importance of this molecule in physics, chemistry, and biology, there is a large body of literature on the measurement of the optical properties of water throughout the entire electromagnetic spectrum from gamma rays to radio waves. A water molecule, the structure of which is shown in Figure 3.16, consists of two hydrogen atoms and one oxygen atom. Water exists in three different states: gas (water vapor), liquid (water), and solid (ice). For gaseous water, the H–O–H bond angle is approximately 104.45° and the O–H bond length is approximately 0.95718 Å. The latter value is slightly higher in liquid water due to hydrogen bonding.

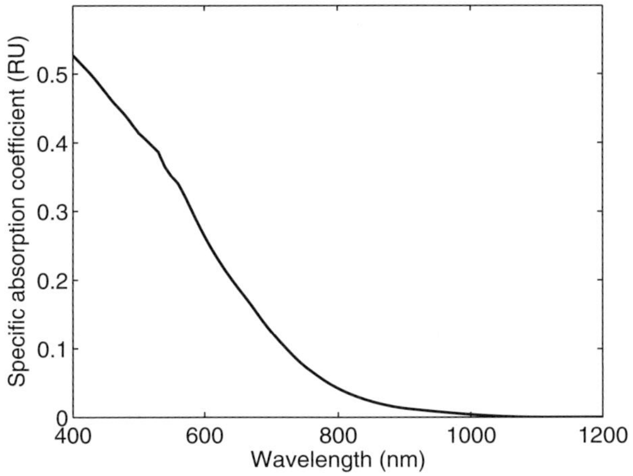

Figure 3.15 Absorption spectrum of brown pigments (F. Baret, personal communication).

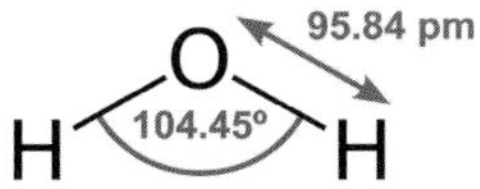

Figure 3.16 Chemical structure of water.

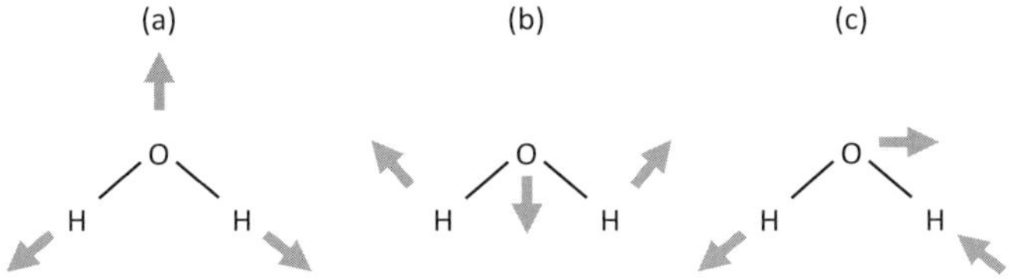

Figure 3.17 Three degrees of vibrational freedom of the molecule of water: (a) Symmetric stretching mode at $\nu_1 = 2.73$ μm; (b) Bending mode at $\nu_2 = 6.27$ μm; and (c) Asymmetric stretching mode at $\nu_3 = 2.66$ μm.

Water spectroscopy is complex: it depends on the phase (gas, liquid, or solid) of the molecule, which changes with temperature. The band structure of condensed phases is much more complex than that of free molecules. The main absorption features, which occur in the infrared, result from vibrational transitions involving various overtones and combinations of three fundamental vibrational transitions (Figure 3.17): ν_1 (H–O–H symmetric stretch), ν_2 (H–O–H bending of the covalent bonds), and ν_3 (H–O–H asymmetric stretch). Water is almost transparent to visible light, a property that is put to good use by photosynthesis: small peaks and shoulders that correspond to overtone and combined overtone bands can be found at 514 nm, 606 nm, 660 nm, 739 nm, and 836 nm. Note that the measurement of water absorption at these wavelengths is still difficult and controversial (Litjens et al., 1999; Fry, 2000; Quickenden et al., 2000). These discrepancies are smaller in the shortwave-infrared where one observes two major water absorption peaks centered near 1470 nm and 1900 nm, and two minor absorption peaks centered near 970 nm and 1200 nm. The absorption feature centered near 970 nm is attributed to a $2\nu_1 + \nu_3$ combination, the one near 1200 nm due to a $\nu_1 + \nu_2 + \nu_3$ combination, the one

Table 3.3 *Vibrational-rotational transitions of gaseous and liquid water at 20°C. Wavelength (μm) = 10^4/wavenumber (cm^{-1}). Note that a and b are integers.*

Liquid		Gas			
Wavelength (μm)	Wavenumber (cm^{-1})	Wavelength (μm)	Wavenumber (cm^{-1})	Assignment	$a+b$
0.401	24940			$av_1 + bv_3$	$a+b=8$
0.449	22270			$av_1 + bv_3$	$a+b=7$
0.514	19460			$av_1 + bv_3$	$a+b=6$
0.606	16500			$av_1 + bv_3$	$a+b=5$
0.660	15150			$av_1 + v_2 + bv_3$	$a+b=4$
0.739	13530			$av_1 + bv_3$	$a+b=4$
0.836	11960			$av_1 + v_2 + bv_3$	$a+b=3$
0.970	10310	1.004	9960	$av_1 + bv_3$	$a+b=3$
1.200	8330	1.272	7861	$av_1 + v_2 + bv_3$	$a+b=2$
1.470	6800	1.536	6510	$av_1 + bv_3$	$a+b=2$
1.900	5260	1.990	5025	$av_1 + v_2 + bv_3$	$a+b=1$
2.870	3490	2.662	3756	v_3, asymmetric stretch	
3.050	3277	2.738	3652	v_1, symmetric stretch	
4.650	2150			$v_2 + L_2$	
6.080	1645	6.270	1595	v_2, bend	
15.0	686.3			L_2, librations	
25.0	395.5			L_1, librations	
55.0	183.4			intermolecular stretch	
200.0	50			intermolecular bend	

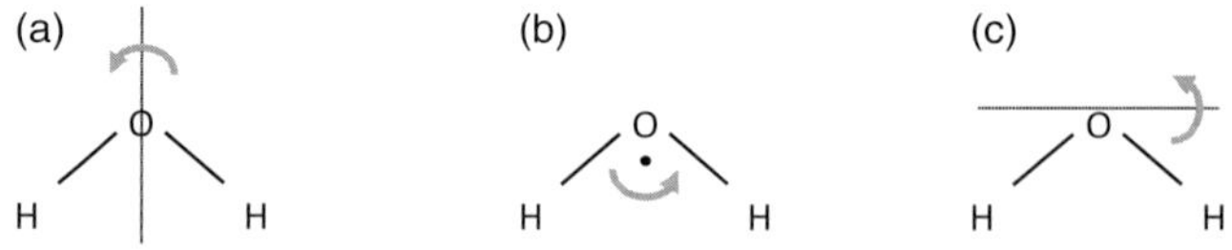

Figure 3.18 Three degrees of rotational freedom of the water molecule: (a) Rotational axis C_2; (b) Rotational axis σ_v; and (c) Rotational axis σ'_v.

near 1470 nm to a $v_1 + v_3$ combination, and the one near 1900 nm to a $v_2 + v_3$ combination (Table 3.3).

The water molecule is asymmetric and has small independent moments of inertia. The three rotational modes represent rotation about the two-fold symmetry axis called C_2 and about two perpendicular mirror planes in and out of the H–O–H plane, called σ_v and σ'_v, respectively, which contain the principal axis (Figure 3.18). In liquid water and ice, rotations tend to be restricted by hydrogen bonds: the small amplitude rotations, or librations, form two absorption bands: a minor one L_1 with a peak at approximately 25 μm and a major one L_2 with a peak at approximately 15 μm (Table 3.3). The infrared spectrum is the result of vibrational overtones and combinations with librations. The rotation-vibration spectrum of water vapor, which spans between approximately 10 cm^{-1} and 40000 cm^{-1}, contains several millions of absorption lines. In liquid water and ice, energy levels are so close to each other that they almost form a continuum.

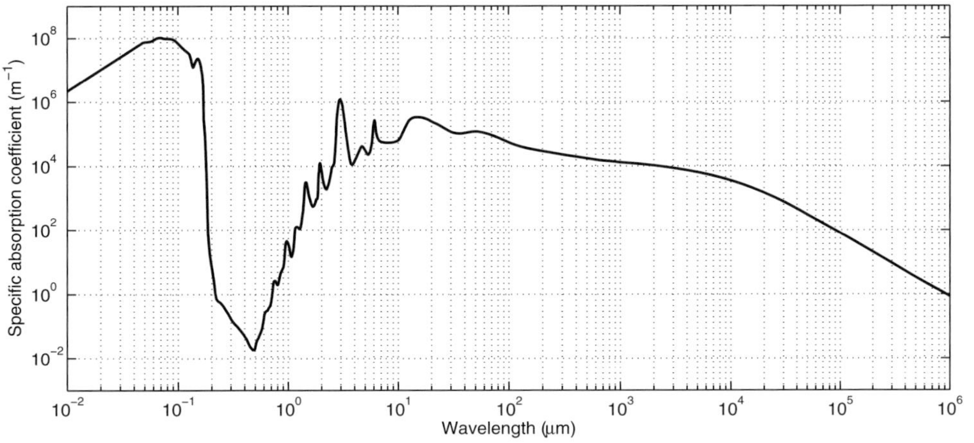

Figure 3.19 Specific absorption coefficient of pure liquid water (redrawn after Segelstein (1981)).

Finally, in the microwave region, the absorption of electromagnetic radiation by liquid water (equivalent to the dielectric loss ϵ'') is due to the shifting of the water dipole in the oscillating electric field of the wave.

Irvine and Pollack (1968) and more recently Wozniak and Dera (2007) reviewed the infrared optical properties of water. The specific absorption coefficient $k(\lambda)$ of water has been measured by many authors. Its determination is in practice a challenge due to the difficulty of obtaining pure distilled water alone and to experimental problems measuring the transmission of liquid through a cell of variable thickness, depending on the small or large value of k. Values of the complex refractive index can be downloaded on the website of Prahl (2001). Figure 3.19 shows the linear coefficient of absorption of pure liquid water for a wide spectral range (from UV light to microwave and radiowave radiation). The absorption spectra of leaf water and pure liquid water are very similar in the near and shortwave infrared (Jacquemoud, unpublished data).

3.4 Cell Wall Constituent-Specific Absorption Coefficients

The biochemical compounds found in leaves are solid, almost insoluble materials, not suitable for conventional spectrophotometric analysis. They can be analyzed *in situ*, but the information obtained is limited because their absorption spectra overlap, or *in vitro* as powdered materials. The search for an isolation method which does not change their structure too much is an unanswered question. Most extraction and purification processes destroy their original form. Depending on the source and the isolation procedure, their spectrum can vary significantly. Then providing specific absorption coefficients for pure constituents is attempting the impossible. Several non-destructive analytical techniques have been developed to determine the absorption spectra: Raman spectroscopy or photoacoustic spectroscopy (PAS) in the UV and visible region, near-infrared spectroscopy (NIRS) in the near-infrared and shortwave-infrared region, and Fourier transform infrared spectroscopy (FTIR) in the thermal infrared region. Since the instrumentation is different in these regions, the techniques have traditionally been considered separately. Note that the National Institute of Standards and Technology, USA, or the National Institute of Advanced Industrial Science and Technology, Japan (http://sdbs.riodb.aist.go.jp/), provide some IR spectroscopy databases, but that there are no published data for many molecules.

The organic compounds (e.g., cellulose, hemicellulose, lignin, structural proteins) that comprise the dry matter of plant cell walls form complex assemblages. For instance, if we consider lignin only, there are several chemical and structural types of this polymer that are characteristic of hardwoods, softwoods, and grasses. Their absorption features are the result of the stretching and bending vibrations of the strong molecular bonds between light atoms, for example, hydrogen atoms, and the atoms of carbon, nitrogen, and oxygen. The fundamental absorption band is located at the energy level (or wavelength) that allows the molecule to rise to higher vibrational states. These compounds actually strongly absorb radiation in the UV ($\lambda \leq 0.4$ μm) and in the middle-infrared ($\lambda \geq 2.5$ μm) region, so that the absorption bands observed in the near infrared and shortwave infrared are relatively weak and broad compared to those observed in the middle-infrared spectrum. They originate as harmonics ($2f$, $3f$, $4f$, etc. where f is the fundamental frequency) and overtones (any frequency higher than the fundamental frequency) of the fundamental stretching frequencies of the C–H, O–H, and N–H bonds (Murray and Williams, 1987; Peterson et al., 1988; Wessman et al., 1988a; Schmidt, 2010). Several chemical compounds absorb at similar wavebands, so that a wavelength is rarely uniquely related to specific molecule (Table 3.4). Curran (1989) cites an example of the strong O–H bond that is a component of the absorption spectra of water, cellulose, sugar, starch, and lignin.

It is difficult to remove cellulose, hemicellulose, and lignin from leaves without degrading their polymeric network superstructure and consequently changing their spectroscopic properties. Singh et al. (2012) measured the UV-visible absorbance spectrum of sal (*Shorea robusta*) leaf lignin in alkaline solution. They attribute the high absorption in the range of 200–220 nm to the $\pi \rightarrow \pi^*$ transition of the aromatic ring. Gould (1982) obtained UV and visible absorption spectra of cell walls both on powdered material and *in situ* using photoacoustic spectroscopy. The absorption spectrum of cellulose does not display any pronounced features and is characterized by a practically monotonic decrease as a function of the wavelength (Figure 3.20). Lignin shows a single major, relatively unstructured absorption band at about 280 nm, which is due to the phenolic structure of the free and etherified hydroxyl group (Figure 3.20). One observes an absorption tail extending to the visible between 400 nm and 600 nm. Occasionally, additional minima or maxima occur at other wavelengths affecting lignin color, as the result of oxidation of the polymer.

The infrared absorption spectra of five biochemical compounds found in leaves are plotted in Figure 3.21 as $ln(1/R)$, where R is the reflectance acquired from pure powdered material measured in laboratory with a spectrophotometer (see Section 4.5.4). Oak wood lignin has peaks at 1420 and 1940 nm ascribed to O–H bonds, and 1690 and 2100 nm ascribed to aromatic C–H bonds and phenolic hydroxyls, respectively (Ziechmann, 1964). As noted by Curran (1989), absorption features in leaves are broadened by multiple scattering and often interfere with one another. He cites an example where the first overtones of the N–H and O–H stretch overlap for most of their width.

Most fundamental molecular vibrations revealed by FTIR spectroscopy fall in the spectral region ranging from 2.5 to 25 μm (4000 to 400 cm^{-1}). Maréchal and Chanzy (2000) examined hydrogen bonding in cellulose chains; about one-third formed strong bonds with adjacent chains that formed various conformational structures. The attenuated total reflection (ATR) spectrum of cellulose

Table 3.4 *Absorption features in near infrared, shortwave infrared, and thermal infrared that have been related to particular foliar chemical concentrations (after Ziechmann, 1964; Curran, 1989; Elvidge, 1990; McLellan et al., 1991a, 1991b; Ben-Dor et al., 1997; Ribeiro da Luz, 2006; Ribeiro da Luz and Crowley, 2007; Coward, 2010; Agarwal and Atalla, 2010).*

Wavelength (μm)	Wavenumber (cm^{-1})	Bond vibration	Chemical(s)
0.91	10989	C–H stretch, third overtone	protein
0.93	10753	C–H stretch, third overtone	oil
0.97	10309	O–H bend, first overtone	starch
0.99	10101	O–H stretch, second overtone	starch
1.02	9804	N–H stretch	protein
1.04	9615	C–H stretch, C–H deformation	oil
1.12	8929	C–H stretch, second overtone	lignin
1.20	8333	O–H bend, first overtone	cellulose, starch, lignin
1.42	7042	C–H stretch, C–H deformation	lignin
1.45	6897	O–H stretch, first overtone, C–H stretch, C–H deformation	starch, sugar, lignin
1.49	6711	O–H stretch, first overtone	cellulose, sugar
1.51	6623	N–H stretch, first overtone	protein
1.53	6536	O–H stretch, first overtone	starch
1.54	6494	O–H stretch, first overtone	starch, cellulose
1.58	6329	O–H stretch, first overtone	starch, sugar
1.69	5917	C–H stretch, first overtone	lignin, starch, protein, nitrogen
1.78	5618	C–H stretch, first overtone, O–H stretch, H–O–H deformation	cellulose, sugar, starch
1.82	5495	O–H stretch, C–O stretch, second overtone	cellulose
1.90	5263	O–H stretch, C–O stretch	starch
1.94	5154	O–H stretch, O–H deformation	lignin, protein, starch, cellulose
1.96	5102	O–H stretch, O–H bend	sugar, starch
1.98	5051	N–H asymmetry	protein
2.00	5000	O–H deformation, C–H deformation	starch
2.06	4854	N=H bend, second overtone, N=H bend, N=H stretch	protein
2.08	4808	O–H stretch, O–H deformation	sugar, starch
2.10	4762	O–H bend, C–O stretch, C–O–C stretch, third overtone	starch, lignin, cellulose
2.13	4695	N–H stretch	protein
2.18	4587	N–H bend, second overtone, C–H stretch, C–O stretch, C=O stretch, C–N stretch	protein
2.24	4673	C–H stretch	protein
2.25	4651	O–H stretch, O–H deformation	starch

Table 3.4 (*cont.*)

Wavelength (μm)	Wavenumber (cm^{-1})	Bond vibration	Chemical(s)
2.27	4608	C–H stretch, O–H stretch, CH_2 bend, CH_2 stretch	cellulose, sugar, starch
2.28	4587	C–H stretch, CH_2 deformation	starch, cellulose
2.30	4348	N–H stretch, C=O stretch, C–H bend, second overtone	protein
2.31	4329	C–H bend, second overtone	oil
2.32	4310	C–H stretch, CH_2 deformation	starch
2.34	4274	C–H stretch, O–H deformation, C–H deformation, O–H stretch	cellulose
2.35	4255	CH_2 bend, second overtone, C–H deformation, second overtone	cellulose, protein
2.38	4202	COO–CH3 stretch, second overtone	pectin, protein
2.77	3603	O–H stretch	nonacosane
3.04	3288	O–H stretch	water, lignin
3.43	2914	CH_2 stretch	lignin, nonacosane
3.48	2871	CH_3 stretch	nonacosane
3.51	2846	CH_2 stretch	lignin, nonacosane
4.71	2125		water
5.76	1735	C=O stretch	lignin, esters
5.92	1688	C=O stretch	lignin, oleanolic acid
6.12	1633	O=H bend	water
6.79	1472	CH_2 bend	lignin, nonacosane
6.84	1462	CH_2 bend	lignin, nonacosane
8.58	1166	C–O stretch	lignin, esters
9.48	1055	C–O stretch	cellulose, primary alcohols
9.62	1039	COH bend	hemicellulose
9.69	1032	C–O stretch	cellulose, secondary alcohols
9.72	1029		oleanolic acid
10.03	997		oleanolic acid
13.72	729		nonacosane
13.91	719		nonacosane

(Figure 3.22) displays two strong absorption bands at 9.47 μm and 9.70 μm, which they assigned to C–O stretching vibrations of the primary and secondary alcohol functional groups, respectively. Weaker bands appear at 8.62 μm, 9.03 μm, and 11.15 μm. The ATR spectrum of hemicellulose xylan displays a strong absorption band at 9.62 μm that is probably related to COH bending (Figure 3.22).

Derkacheva and Sukov (2005) modeled a set of stable wave number bands (740–1620 cm^{-1}) for isolated softwood lignins that varied in terms of bandwidths and peak intensities, which they then used to separate different lignin molecules. Although many articles have been devoted to the

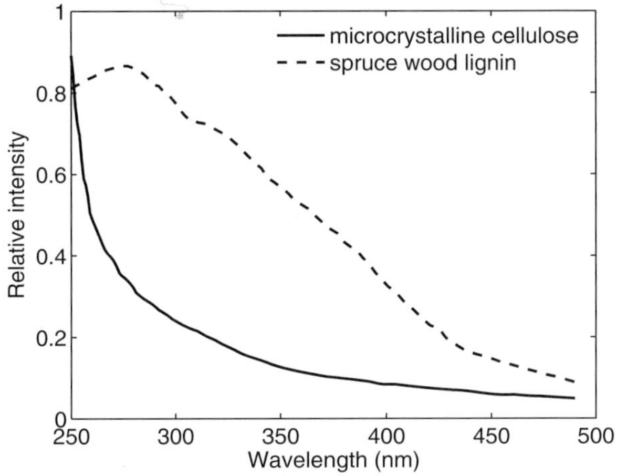

Figure 3.20 Photoacoustic absorption spectrum of microcrystalline cellulose and milled spruce wood lignin powders. All spectra have been divided by the spectrum of a carbon lampblack sample to correct for the intensity spectrum of the illumination (redrawn from Gould (1982)).

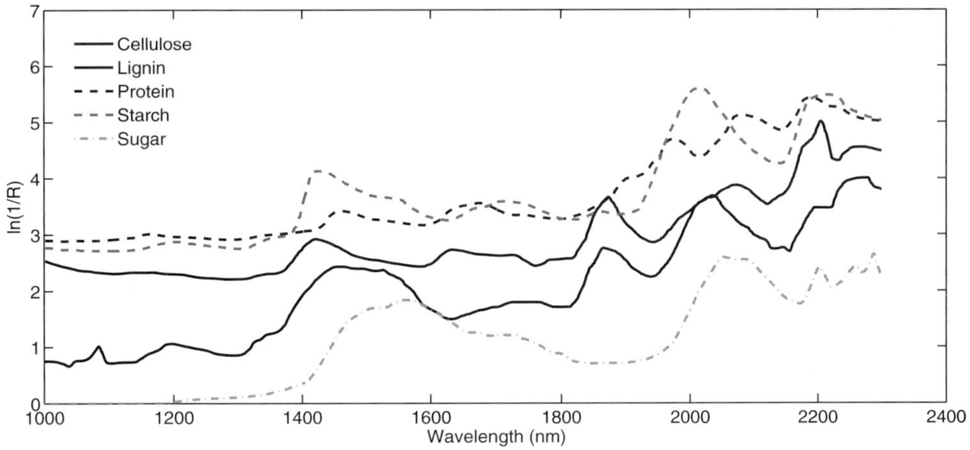

Figure 3.21 Absorption spectra of five biochemical compounds found in leaves (redrawn from Wessman (1990) and Burns and Ciurczak (2007)).

absorption features of cellulose, hemicellulose, and lignin, full band interpretation is still under discussion (Ribeiro da Luz and Crowley, 2007) (Figure 3.23).

Miller (2001) concurs with Curran (1989), concluding that because of the large number of chemical functional groups that are active in the near infrared (e.g., CH, CN, NH, OH, C=O), many of which produce overtone and combination bands that overlap or can be found in different molecular configurations, it becomes difficult or impossible to identify bands that separate them into their chemical constituents. He further concludes that because the NIR spectrum is not random but results from fundamental molecular vibration mechanisms, multivariate calibration methods can provide predictions of these complex relationships.

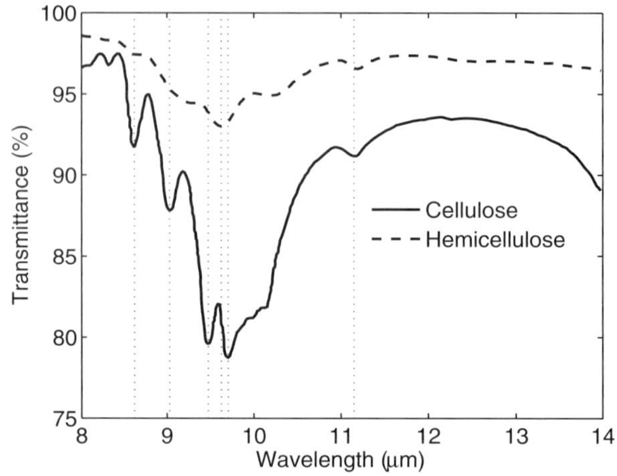

Figure 3.22 ATR spectrum of cellulose and hemicellulose. The vertical lines correspond to the main absorption bands: 8.62 μm, 9.03 μm, 9.47 μm, 9.70 μm, and 11.15 μm for cellulose; 9.62 μm for hemicellulose (redrawn from Ribeiro da Luz and Crowley (2007)).

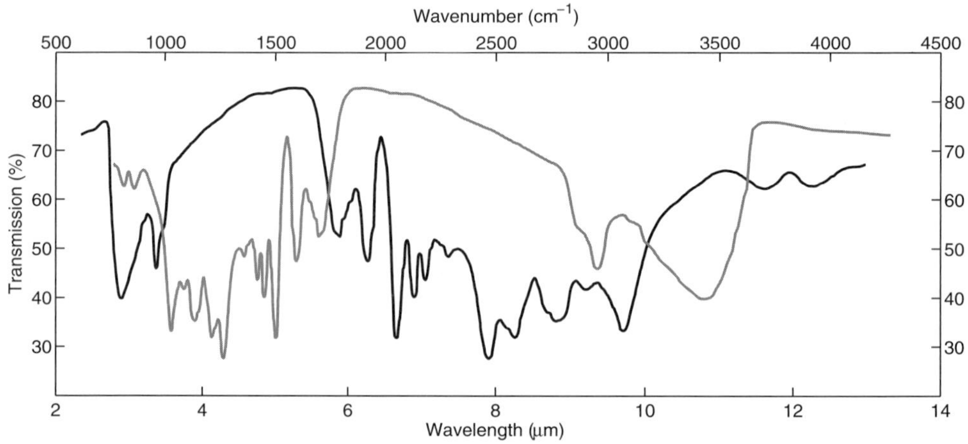

Figure 3.23 Infrared transmission spectra of spruce wood lignin expressed as a function of the wavelength (μm, black line) and the wavenumber (cm^{-1}, gray line) (redrawn from Ziechmann (1964)).

3.5 Other Minor Constituent-Specific Absorption Coefficients

In addition to the classes of organic compounds identified in the cell wall, the cytoplasm includes soluble proteins, enzymes, sugars, and starch granules. Because these compounds represent categories of molecules, the absorptions vary within groups and between groups giving rise to complex overlapping absorptions, making individual spectral identification difficult or impossible (Figure 3.21). Of these the only element that has received significant attention has been nitrogen. Nitrogenous molecules are found in a wide range of compounds and all parts of the cell, thus its identification based on spectroscopy has had limited success. To date, most approaches are empirical, using various statistical approaches.

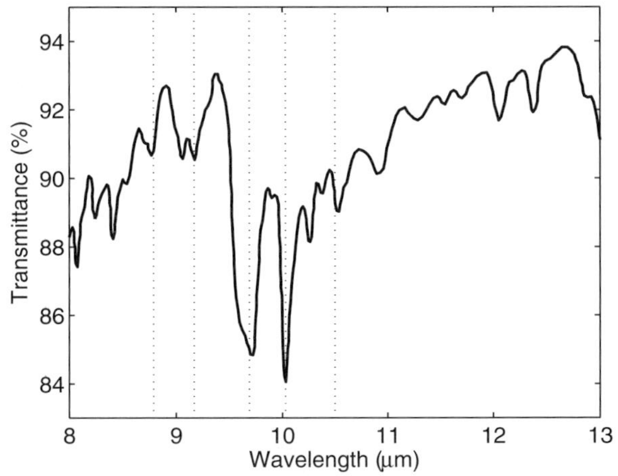

Figure 3.24 ATR spectrum of oleanolic acid. The lines correspond to the main absorption bands at 8.79 μm, 9.17 μm, 9.69 μm, 10.03 μm, and 10.52 μm (redrawn from Ribeiro da Luz and Crowley (2007)).

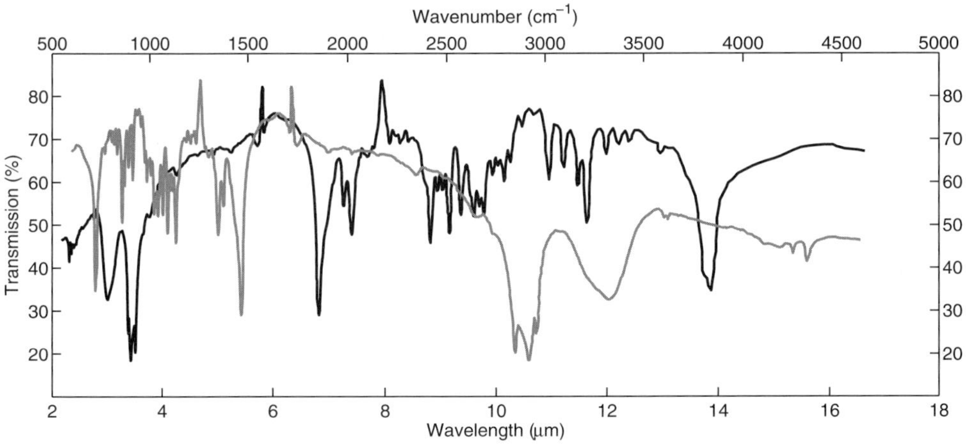

Figure 3.25 FTIR spectra of synthesized, racemic nonacosan-10-ol as a function of wavelength (μm, black line) and the wavenumber (cm^{-1}, gray line) (redrawn from Coward (2010)).

The spectroscopy of epicuticular waxes, which cover leaf surfaces and are expected to display distinctive spectral features in the thermal infrared, has been particularly studied in the past decade. The phase behavior of leaf cuticular waxes from English Ivy (*Hedera helix*) and English Walnut (*Juglans regia*) was studied using FTIR equipped with a horizontal attenuated total reflectance (HATR) appliance (Merk et al., 1998). They observed changes in the symmetric stretching mode of CH_2 groups (2856–2848 cm^{-1}) indicative of orthorhombic arrangements of alkyl chains. Dubis et al. (1999, 2001) also used FTIR with an HATR on several crop and tree species to examine epicuticular waxes and were able to identify long-chain aliphatic compounds, mainly hydrocarbons and carbonyl groups at 1717–1738 cm^{-1} and aromatic esters in the 709–766 cm^{-1} range. Fundamental vibration modes of various molecular functional groups produce characteristic spectral absorption features that could serve to "fingerprint" many compounds (Ribeiro da Luz, 2006; Ribeiro da Luz and Crowley, 2007). Such functional groups and related spectral features include hydroxyl (OH) in alcohols and acids, carbonyl (C=O) in esters, ketones, aldehydes and acids, and methyl (CH_3) and methylene (CH_2) in alkanes. Figure 3.24 displays the transmission spectrum of oleanolic acid and Figure 3.25 that of nonacosan-10-ol, one of the main components of crystalline epicuticular wax nanotubes.

3.6 Refractive Index of Leaf Constituents

Discontinuities in the refractive index between air and the main leaf constituents, that is, water, dry matter, and chlorophyll, cause reflection and refraction phenomena that contribute to diffuse light within plant leaves. That information is consequently required to properly calculate their optical properties. The refractive index is wavelength dependent but, with a few exceptions, spectral variation is unknown because of the challenges to obtain it. Therefore in most studies, the refractive index is provided at a given wavelength. Several physically based direct or indirect methods have been used to infer this property according to the nature (solid, liquid, gas) of the material, including ellipsometry, refractometry, microscopy, and spectrophotometry. There is no ambiguity for the refractive index of air that can be taken as $n = 1.0$ with satisfactory accuracy, unlike uses in atmospheric sciences. Nonetheless, the refractive index of pure liquid water (Figure 3.26) as well as many other properties are a continuing puzzle for scientists, and there are several formulations of the index of refraction equation as a function of temperature, pressure, and the wavelength of light (e.g., Hale and Querry, 1973; Djurišić and Stanić, 1998; Huibers, 1997; Robinson et al., 2000).

Since leaf sap mainly consists of water and inorganic ions, its refractive index should be similar to water. Portable hand refractometers have been used in the field to determine a correlation between the refractive index and the osmotic potential of leaf sap (Slavík, 1959; Shimshi and Livne, 1967). It is found to range between 1.34 and 1.36 for different species, with some values very close to those of pure liquid water in the visible spectrum. Our knowledge of the refractive

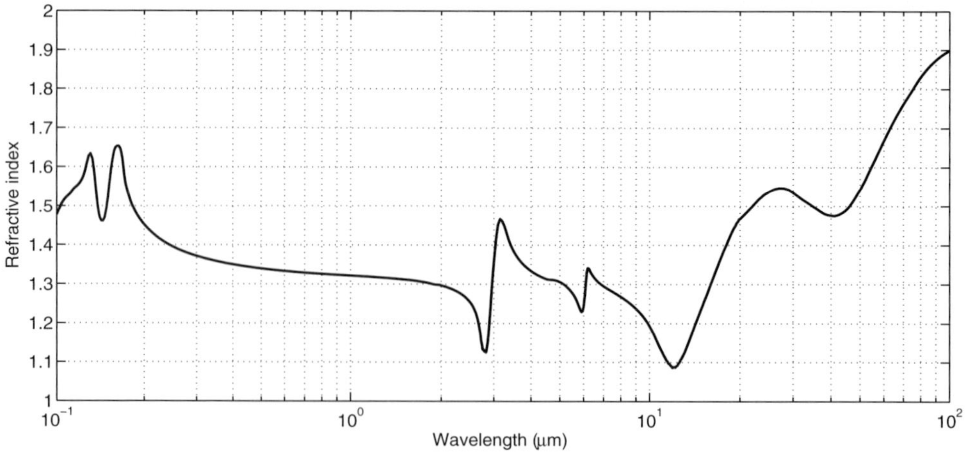

Figure 3.26 Index of refraction of pure liquid water (redrawn from Segelstein (1981)).

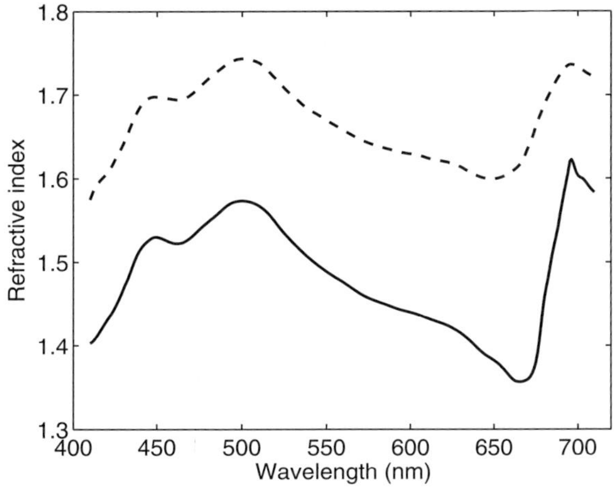

Figure 3.27 Normal (dashed line) and tangential (solid line) components of the refractive indices of photosynthetic membranes (redrawn from Paillotin et al. (1998)).

index of plant tissues is actually very limited. Although the chemical properties of the plant cuticle have been studied in detail (Martin and Juniper, 1970; Cutler et al., 1982; Kerstiens, 1996), its optical properties have received very few actual measurements. As far as chlorophyll is concerned, Kumar and Silva (1973) take the value $n = 1.52$ published by Charney and Brackett (1961). This value is in the range of the refractive indices of photosynthetic membranes provided by Paillotin et al. (1998), as illustrated in Figure 3.27.

A series of experimental studies were performed at the beginning of the 20th century. Senn (1908) (cited by Castle, 1933) measured the refractive indices of entire palisade cells and obtained $n = 1.47$ for green bean (*Phaseolus vulgaris*), $n = 1.49$ for faba bean (*Vicia faba*), and $n = 1.48$ for dandelion (*Taraxacum officinale*). At the Osborn Botanical Laboratory, Brown (1920) determined the refractive index of cell walls, cuticle, vessels, etc. and showed that the different cell membranes of a tissue had characteristic values. n may be accurately determined using microscopic techniques by immersing leaf pieces in a series of oils of different refractive index. In oblique illumination, the outlines of the tissue appear dark on the side toward which the diaphragm is shifted if its refractive index is higher than that of the liquid; and they appear dark on the side away if it is lower. The refractive index of different wax mixtures, including carnauba wax, a substance obtained from the leaf surface of a species of palm tree native to north-eastern Brazil, carnaubeira palm (*Copernicia prunifera*), was investigated by Waller and Seibert (1955) who found $n = 1.46$. By measuring the reflectance of leaf pieces whose intercellular spaces had been infiltrated with oils of different refractive index, Woolley (1975) showed that the refractive index of the antidermal walls of the abaxial epidermis of soybean soybean (*Glycine max*) leaves changed with degree of hydration: at 800 nm, it is higher in oven-dried leaves (between 1.525 and 1.545) than in fully hydrated leaves (between 1.405 and 1.415). Note that leaves are collected before mid-morning to ensure opened stomata and consequent ease of infiltration. For dry leaves, these values are comparable to $n = 1.557$ given by Manabe et al. (2001) for cellulose, one of the main components of cell walls. Recently, Sultanova et al. (2009) have fitted experimental data for cellulose with the Sellmeier dispersion formula, an empirical relationship between refractive index and wavelength (http://refractiveindex.info/). Moreover, the average

Table 3.5 *Plant refractive indexes found in the literature. When the wavelength is not mentioned, indexes generally correspond to values averaged over the whole visible range.*

Medium	Index of refraction	Wavelength	Reference
Air	1.00		
Water	1.33	700 nm	Segelstein (1981)
Cell sap	1.36	589 nm	Charney and Brackett (1961)
Hair	1.47–1.49	800 nm	Woolley (1975)
Cutin	1.5		Vanderbilt et al. (1991)
Wax (bayberry)	1.436		Vanderbilt et al. (1991)
Wax (jojoba)	1.465		Vanderbilt et al. (1991)
Wax (carnauba)	1.467–1.472		Waller and Seibert (1955), Allen et al. (1969), Vanderbilt et al. (1991)
Wax (n-alkane)	1.436–1.451		
Wax (secondary alcohol)	1.458		
Cuticle	1.45		Seyfried and Fukshansky (1983)
Crystalline cellulose	1.557		Manabe et al. (2001)
Cell wall	1.52	700–1300 nm	Kumar and Silva (1973)
Cell wall	1.333–1.472	850 nm	Gausman et al. (1974a)
Hydrated cell wall	1.42	800 nm	Woolley (1975)
Non-hydrated cell wall	1.51	800 nm	Woolley (1975)
Cellulose	1.47		Belov et al. (2011)
Cellulose	1.480–1.461	437–1052 nm	Sultanova et al. (2009)
Chloroplast	1.415–1.420	589 nm	Charney and Brackett (1961)
Whole leaf	1.38–1.42	400–1100 nm	Rvachev and Guminetskii (1966)
Whole leaf	1.47–1.48		Woolley (1971)
Whole leaf	1.43		Vanderbilt and Grant (1985)

refractive index of plant cell walls may vary from one species to another: $n = 1.333$–1.436 for cucumber, $n = 1.414$–1.472 for blackeye pea, $n = 1.333$–1.454 for tomato, and $n = 1.333$–1.472 for string bean leaves (Gausman et al., 1974a). By measuring the reflection of polarized light at the Brewster angle, Rvachev and Guminetskii (1966) determined the refractive index of several leaves as a whole and arrived at values of approximately 1.38 to 1.42 in the visible near-infrared (VIS-NIR). In their ray tracing model, Kumar and Silva (1973) assume that $n = 1.52$ for cell walls, a value that notably differs from $n = 1.37$ published by Charney and Brackett (1961), while Seyfried and Fukshansty (1983) use $n = 1.425$ to simulate light gradients in plant tissues. Kumar and Silva (1973) and Yamada and Fujimura (1991) assume that $n = 1.36$ in the cytosol and the vacuole, a value close to the refractive index of water. Table 3.5 provides examples of refractive indexes in the published literature. Most of these are average values over the visible domain, although the indices should vary spectrally. Current data demonstrates considerable overlap in the values for leaf constituents.

The first whole spectrum from a maize (*Zea mays*) leaf is published by Allen et al. (1969), who used a radiative transfer model. In the earliest versions of PROSPECT (Jacquemoud and Baret, 1990), albino (400–800 nm) and dry (800–2500 nm) maize leaves, regarded as a compact leaf ($N = 1$),

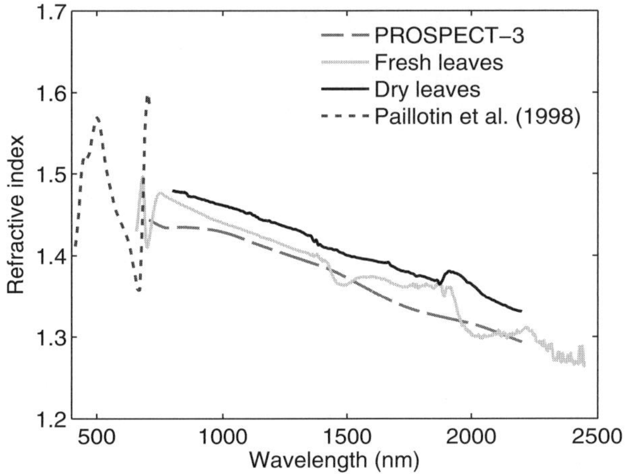

Figure 3.28 Refractive index of plant leaves computed by different ways (redrawn from Féret et al. (2008)).

provided a similar profile. The refractive index ranges between 1.43 and 1.5 in the VIS and then regularly decreases to 1.27 in the MWIR and it is presumed to be the same for all leaves. As shown by Féret et al. (2008), such an index is adapted from non-absorbent leaves so that there is good reason to think that a fresh green leaf is different (Figure 3.28). Recent work takes into account the differences in chloroplast and cytosol refractive indexes to calculate leaf optical properties (Krekov et al., 2009).

4

Measurement of Leaf Optical Properties

This chapter provides a background on measurements of optical properties. First, we review the terminology used to describe electromagnetic radiation, starting from definition of terms used in describing electromagnetic radiation, blackbody radiation, solar spectrum, and radiometric units (radiance, irradiance, etc.). The fate of photons intersecting a surface like a leaf is explained in terms of properties that either depend on or are independent of the illumination and viewing angles; terminology for directional dependence observations, for example, for bidirectional measurements, attenuated total reflection, and polarization are described. Measurement of polarized and unpolarized bidirectional reflectance distribution function (BRDF), bidirectional reflectance factor (BRF), bidirectional transmittance distribution function (BTDF), and bidirectional transmittance function (BTF) are explained. Experimental measurements of leaf optical properties are often difficult to make consistent; here progress in characterizing optical properties is reviewed as are descriptions of how the measurements are conducted. The term SI stands for *Système International d'unités* (International System of Units).

4.1 Terminology

4.1.1 Electromagnetic Radiation

Electromagnetic radiation can be described in the form of an electromagnetic wave that has an electric field **E** and a magnetic induction **B**. These two-field components oscillate in phase perpendicular to each other in the direction of energy propagation **C** (Figure 4.1).

Contrary to other types of waves that need gas, liquid, or solid material to move through, electromagnetic waves can travel through empty space and their speed, often abbreviated as c, is constant in a vacuum (2.998×10^8 m s^{-1}). In media other than a vacuum, electromagnetic waves travel at different speeds, a phenomenon known as dispersion. Electromagnetic radiation is characterized by five quantities that are time dependent and likely to provide part of the information about the target:

- The direction of propagation **C**
- The intensity I
- The wavelength λ
- The polarization
- The phase

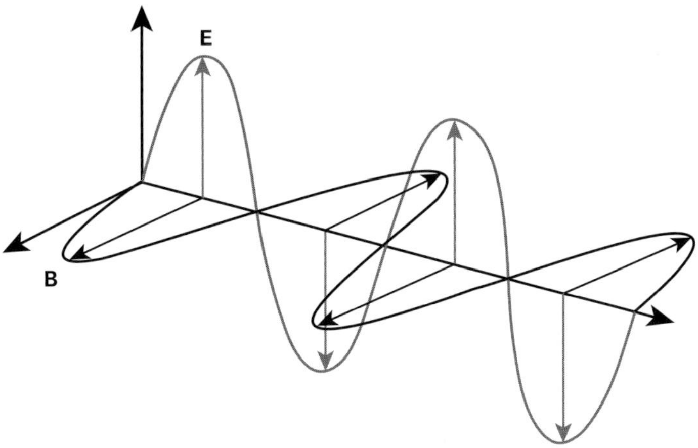

Figure 4.1 Electromagnetic wave: coupled oscillation of the electric field and the magnetic field (© CCRS/CCT).

Although the SI unit of wavelength is the meter (m), the wavelength dimension may be expressed in various units that are convenient for the purpose. The nanometer unit is 10^{-9} m and the micron is 10^{-6} m, which are convenient for wavelengths in the visible (VIS) and shortwave infrared (SWIR) parts of the electromagnetic spectrum. The wavenumber of a monochromatic ray of electromagnetic radiation, conveniently used in the midwave and longwave infrared, is the number of cycles per wavelength (or the number of wavelengths per unit distance):

$$\sigma = 1/\lambda \tag{4.1}$$

It has dimensions of reciprocal length, so its SI unit is m^{-1} (or cm^{-1}). Frequency is defined as a number of cycles per unit time. It is related to the wavelength by the speed of light in vacuum:

$$v = c/\lambda \tag{4.2}$$

The SI unit for v is the hertz (Hz ≡ s^{-1}). The energy of a photon is inversely proportional to wavelength:

$$q = hv = hc/\lambda \tag{4.3}$$

where h is Planck's constant (6.626×10^{-34} J s). This relationship means that light consisting of high/low energy photons has a short/long wavelength (e.g., blue/red light). It is often used to compute the change in energy resulting from photon absorption and to predict the frequency of the light emitted for a given energy transition (see Chapter 3). Since energy is inversely proportional to wavelength, energy is directly proportional to wavenumber. Thus a 3000 cm^{-1} photon has three times the energy of a 1000 cm^{-1} photon. Energy should be expressed in joule (J) but a commonly used unit for photons is the electron-volt (eV) equal to approximately 1.602×10^{-19} J. Consequently, we have:

$$hc = \frac{6.626 \times 10^{-34}}{1.602 \times 10^{-19}} 2.998 \times 10^{8} = 1.2398 \times 10^{-6} \text{ eV m} = 1.2398 \text{ eV } \mu\text{m} \tag{4.4}$$

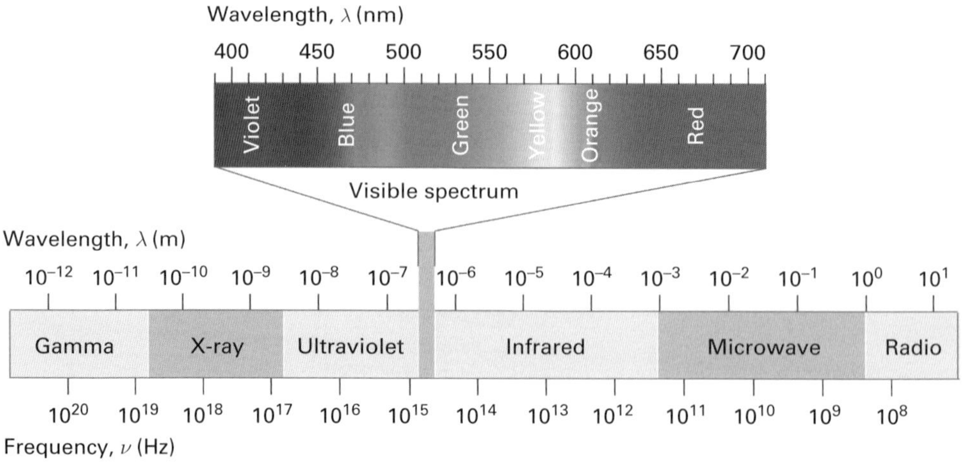

Figure 4.2 Electromagnetic spectrum (by courtesy of Averill and Eldredge, 2011). (A black and white version of this figure will appear in some formats. For the color version, please refer to the plate section.)

The range of all possible electromagnetic radiation, called the electromagnetic spectrum, is divided into several wavelength domains. Figure 4.2 displays on a logarithmic wavelength scale, the regions of the electromagnetic radiation spectrum from the short wave, high frequency gamma rays to the very long wave, low frequency radio waves. The human eye has an extremely limited range of sensitivity between 400 nm and 700 nm beyond which are the unseen regions of the spectrum (Gates, 1967). The ultraviolet is beyond the short wavelength limit of the visible and the infrared is beyond the long wavelength limit of vision. There are no precisely defined boundaries in the spectrum; rather the naming conventions depend on the domain of application (astronomy, remote sensing, meteorology, communications, etc.) or on the response range of various detector types.

A new international standard has been published by the International Standards Organization: the document ISO 21348 *Space Environment (natural and artificial) – Process for determining solar irradiances*, that describes the process for representing solar irradiances (www.iso.org/). It is applicable to measurement sets, reference spectra, empirical models, theoretical models, and solar irradiance proxies or indices that provide solar irradiance products representing parts or all of the solar electromagnetic spectrum. From highest energy to lowest energy, the forms of light typically considered range from gamma rays and X-rays; ultraviolet, visible, and infrared radiation; microwaves to radio frequencies (Table 4.1).

Because the Earth's atmosphere is opaque to gamma rays, X-rays, part of the ultraviolet, and radiation between 14 μm and 1 mm, most Earth observation satellites operate in three wavelength domains: visible, infrared, and microwave. Leaf optical properties have been studied mainly in the visible and the infrared, which is often divided into the following commonly used sub-division scheme: near infrared (NIR) from 700 nm to 1.0 μm, shortwave infrared (SWIR) from 1 μm to 3 μm, midwave infrared (MWIR) from 3 to 8 μm, longwave infrared (LWIR) from 8 μm to 12 μm, very longwave infrared (VLWIR) from 12 μm to about 25 μm, and farwave infrared (FWIR) from 25 μm to 1 mm that includes terahertz radiation. Finally, the thermal infrared (TIR) region, in which sensors provide a measurement based on thermal emission of objects near Earth temperatures, refers to electromagnetic waves ranging between 3 μm and 20 μm. These sub-categories commonly used in remote sensing differ from those listed in Table 4.1 and can quickly become confusing due to the domain specific terms!

Table 4.1 *ISO 21348 definitions of solar irradiance spectral categories (by courtesy of Space Environment Technologies (http://www.spacewx.com/)).*

Spectral category	Spectral sub-category	Notes	Wavelength range (SI prefixes)	Frequency or wave number or energy range (SI prefixes)
Gamma-rays			10 fm ≤ λ < 1 pm	
X-rays		Hard X-rays	1 pm ≤ λ < 0.10 nm	
	SXR	Soft X-rays	0.10 nm ≤ λ < 10 nm	
Ultraviolet	VUV	Vacuum Ultraviolet	10 nm ≤ λ < 200 nm	
	EUV	Extreme Ultraviolet	10 nm ≤ λ < 121 nm	123.9842 eV ≥ E > 10.2466 eV
	H Lyman-α	Hydrogen Lyman-α	121 nm ≤ λ < 122 nm	
	FUV	Far Ultraviolet	122 nm ≤ λ < 200 nm	10.1626 eV ≥ E > 6.1992 eV
	UVC	Ultraviolet C	100 nm ≤ λ < 280 nm	12.3984 eV ≥ E > 4.4280 eV
	MUV	Middle Ultraviolet	200 nm ≤ λ < 300 nm	6.1992 eV ≥ E > 4.1328 eV
	UVB	Ultraviolet B	280 nm ≤ λ < 315 nm	4.4280 eV ≥ E > 3.9360 eV
	NUV	Near Ultraviolet	300 nm ≤ λ < 400 nm	4.1328 eV ≥ E > 3.0996 eV
	UVA	Ultraviolet A	315 nm ≤ λ < 400 nm	3.9360 eV ≥ E > 3.0996 eV
Visible		Purple	360 nm ≤ λ < 450 nm	3.4440 eV ≥ E > 2.7552 eV
		Blue	450 nm ≤ λ < 500 nm	2.7552 eV ≥ E > 2.4797 eV
		Green	500 nm ≤ λ < 570 nm	2.4797 eV ≥ E > 2.1752 eV
		Yellow	570 nm ≤ λ < 591 nm	
		Orange	591 nm ≤ λ < 610 nm	
		Red	610 nm ≤ λ < 760 nm	
Infrared	IR-A	NIR	760 nm ≤ λ < 1.4 μm	13,157.89 cm^{-1} ≥ σ > 7,142.86 cm^{-1}
	IR-B	Middle Infrared	1.4 μm ≤ λ < 3 μm	7,142.86 cm^{-1} ≥ σ > 3,333.33 cm^{-1}
	IR-C	Far Infrared	3 μm ≤ λ < 1 mm	3,333.33 cm^{-1} ≥ σ > 10.00 cm^{-1}
Microwaves	W		3 mm ≤ λ < 5.35 mm	100.0 GHz ≥ ν > 56.0 GHz
	V		5.35 mm ≤ λ < 6.52 mm	56.0 GHz ≥ ν > 46.0 GHz
	Q		6.52 mm ≤ λ < 8.33 mm	46.0 GHz ≥ ν > 36.0 GHz
	K		8.33 mm ≤ λ < 27.5 mm	36.0 GHz ≥ ν > 10.9 GHz
	X		2.75 cm ≤ λ < 5.77 cm	10.9 GHz ≥ ν > 5.2 GHz
	C		4.84 cm ≤ λ < 7.69 cm	6.2 GHz ≥ ν > 3.9 GHz
	S		5.77 cm ≤ λ < 19.3 cm	5.2 GHz ≥ ν > 1.55 GHz
	L		19.3 cm ≤ λ < 76.9 cm	1.55 GHz ≥ ν > 0.39 GHz
	P		76.9 cm ≤ λ < 1.33 m	0.39 GHz ≥ ν > 0.225 GHz
Radio	EHF	Extremely High Frequency	1 mm ≤ λ < 10 mm	300.0 GHz ≥ ν > 30.0 GHz
	SHF	Super-High Frequency	10 mm ≤ λ < 10 cm	30.0 GHz ≥ ν > 3.0 GHz
	UHF	Ultra-High Frequency	10 cm ≤ λ < 1 m	3000 MHz ≥ ν > 300 MHz
	VHF	Very High Frequency	1 m ≤ λ < 10 m	300 MHz ≥ ν > 30 MHz
	HF	High Frequency	10 m ≤ λ < 100 m	30 MHz ≥ ν > 3 MHz

4.1.2 Polarization

Polarization is a fundamental characteristic of electromagnetic radiation, alongside intensity, frequency, and coherence. As seen above, both the electric field and magnetic field are oscillating in an electromagnetic wave but in different (orthogonal) directions. By convention, the state of polarization is described by reference to the direction of the electric field in a plane defined by the direction of propagation of the light wave and the surface normal, over one period of the oscillation. It may be oriented in a single direction (linear polarization) or rotate as the wave travels (circular or elliptical polarization). When there is no clearly defined state of polarization because the electric field vector varies randomly with time, the light is said to be unpolarized or randomly polarized. This is typically the case for sunlight. Five parameters need to be measured to define the state of polarization:

(1) The intensity of the unpolarized component (I_u)
(2) The intensity of the polarized component (I_p)
(3) The shape of the polarization ellipse (χ) where $\tan\chi = \pm b/a$ and a and b are the semi-major and semi-minor axes of the polarization ellipse
(4) The polarization azimuth (ξ) given by the orientation of the polarization ellipse semi-major axis to a reference direction
(5) The direction of rotation, or handedness, of the electric vector ($\tan\chi$)

The total intensity of the beam is defined as $I = I_u + I_p$ and the degree of polarization of light is expressed as

$$p = \frac{I_p}{I_u + I_p} = \frac{I_p}{I} \tag{4.5}$$

Thus, $I_p = p\,I$ is the intensity of the polarized component of the light and $I_u = (1-p)I$ is the intensity of the unpolarized component. Due to the difficulty in measuring some of these parameters directly, it is advantageous to use an alternative description of polarization, one which is based on observable quantities. Stokes (1852b) conveniently introduced four quantities which are functions only of observables of the electromagnetic wave. They are now known as the Stokes parameters that vary with the wavelength of the electromagnetic radiation and are dependent on the geometry of the source and sensor position. They are generally combined into the Stokes vector:

$$\mathbf{S} = \begin{pmatrix} s_0 \\ s_1 \\ s_2 \\ s_3 \end{pmatrix} = \begin{pmatrix} I \\ Q \\ U \\ V \end{pmatrix} = \begin{pmatrix} I_{\|} + I_{\perp} \\ I_{\|} - I_{\perp} \\ I_{\backslash} - I_{/} \\ I_{+} - I_{-} \end{pmatrix} \tag{4.6}$$

where $I_{\|}$, $I_{\perp}$, $I_{\backslash}$, and $I_{/}$ are the intensities measured at linear polarizing angles of 0°, 90°, 45°, and 135°, respectively, and I_+ and I_- with left and right circular filters (Wu et al., 2013; Savenkov, 2015). Then I is the total intensity of the light beam, Q is the difference between the intensity of the horizontal polarized component ($I_{\|}$) and that of the vertical linearly polarized component ($I_{\perp}$), U is the difference between the intensity of the linearly polarized component oriented at a 45° angle ($I_{\backslash}$) and that of the linearly polarized component oriented at a –45° angle ($I_{/}$), and V is the difference between the intensity of the left circularly polarized component (I_+) and that of the right circularly polarized component (I_-). One can define the degree of (full) polarization as

$$p = \frac{\sqrt{Q^2 + U^2 + V^2}}{I} \tag{4.7}$$

The shape of the polarization ellipse is given by

$$\chi = \frac{1}{2} \arctan \frac{V}{\sqrt{Q^2 + U^2}} \tag{4.8}$$

and the polarization azimuth is given by

$$\xi = \frac{1}{2} \arctan \frac{U}{Q} \tag{4.9}$$

While scattered light may be described by the four components of the Stokes vector, significant progress has been achieved toward understanding only the first component I, the intensity of the scattered light. According to Fresnel's Equations (Appendix E), if unpolarized light is incident on the interface between two dielectric materials with different refractive indices n_1 and n_2, there is a particular angle of incidence, called Brewster's angle (θ_B), at which the reflected beam is completely linearly polarized (only light polarized perpendicular to the beam plane has a reflected component) and the transmitted beam is partly polarized (all the components polarized parallel to the beam plane and some of the components polarized perpendicular to the beam plane are transmitted through the surface) (Figure 4.3). This angle is easily calculated by $\theta_B = \arctan \frac{n_2}{n_1}$. For example, at the boundary between air ($n_1 = 1.00$) and water ($n_2 = 1.33$), $\theta_B = 53°$, while at the boundary between air and leaf surface ($n_2 = 1.45$), the angle is higher at $\theta_B = 55.4°$.

To better understand the scattering behavior of rough surfaces that are known to be depolarizing, the Mueller matrix M may be measured. The Mueller matrix is a four-by-four real matrix that describes the linear transformation of an incident Stokes vector $\mathbf{S}$ into a reflected (or transmitted) Stokes vector $\mathbf{S}'$:

$$\mathbf{S}' = M\ \mathbf{S} \tag{4.10}$$

It is written as

$$M = \begin{pmatrix} m_{11} & m_{12} & m_{13} & m_{14} \\ m_{21} & m_{22} & m_{23} & m_{24} \\ m_{31} & m_{32} & m_{33} & m_{34} \\ m_{41} & m_{42} & m_{43} & m_{44} \end{pmatrix} \tag{4.11}$$

where the scattering coefficients m_{ij} depend on wavelength, incident and scattered angles for the experimental configuration, the optical properties of the reflecting surface, and the orientation of the reflecting surface with respect to the plane of incidence, if the surface is anisotropic (Krishnan and Nordine, 1994). Talmage and Curran (1986) and recently Peltoniemi et al. (2015) published review articles on the use of polarized light in remote sensing.

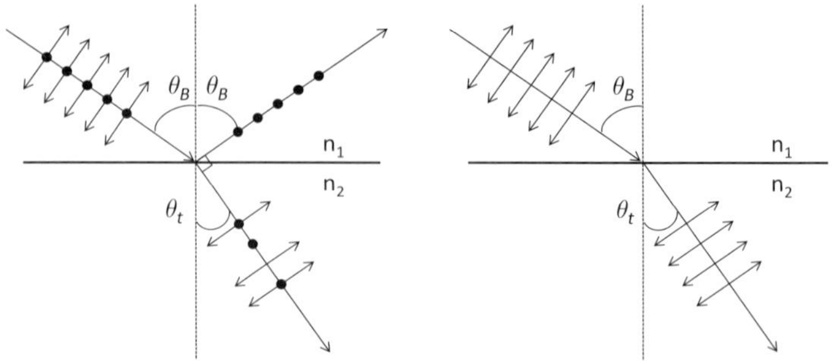

Figure 4.3 Polarization of light at Brewster's angle. Only the parallel (arrows) and perpendicular (discs) electric fields are represented. The incident light is unpolarized, the reflected light is completely polarized, and the transmitted light is partially polarized.

4.1.3 Radiometric Units

This section itemizes the definitions, symbols, and units that are helpful in radiometry and photometry (e.g., Meyer-Arendt, 1968; Sliney, 2007). One can point out that several expressions may be used for similar quantities, which makes things more complex than in other areas of optics.

4.1.3.1 Solid Angle

The intersection of the cone with a sphere of radius 1 defines a surface whose area is equal to the solid angle (Ω) subtended by the cone (Figure 4.4). It is measured in units of steradian (sr) following the SI nomenclature. We often need to know the amount of energy radiated in an element of the solid angle $d\Omega$ that is calculated by

$$d\Omega = \frac{dS}{r^2} = \frac{(rd\theta)(r\,\sin\theta d\phi)}{r^2} = \sin\theta d\theta d\phi \tag{4.12}$$

where dS is the surface of intersection and r the radius of the sphere.

The solid angle of a cone with apex angle 2θ is the area of a spherical cap on a unit sphere:

$$\Omega = \int_{\theta=0}^{\alpha}\int_{\phi=0}^{2\pi} \sin\theta d\theta d\phi = 2\pi[-\cos\theta]_0^{\alpha} = 2\pi(1 - \cos\alpha) \tag{4.13}$$

For small angles, $\cos\alpha = 1 - \frac{\alpha^2}{2!} + \frac{\alpha^4}{4!} + \cdots$. therefore $\Omega = \pi\alpha^2$. The solid angle corresponding to the whole space equals 4π sr.

4.1.3.2 Radiant Energy

The radiant energy (Q) is the energy traveling in the form of electromagnetic waves. When this radiation hits an object, it may be partly or completely absorbed, and transformed into another form. The standard unit is the joule (J).

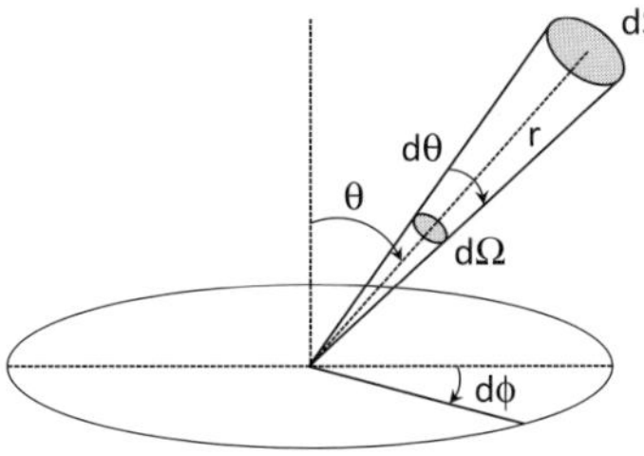

Figure 4.4 The solid angle.

4.1.3.3 Radiant Flux

The radiant flux (Φ) (or radiant power) is the radiant energy (Q) transferred per unit time. The standard unit is joule per second (J s^{-1}) or watt (W). Note that it is measured in the same units as power in general:

$$\Phi = \frac{dQ}{dt} \text{ or } Q = \int \Phi dt \tag{4.14}$$

4.1.3.4 Radiant Intensity

The radiant intensity (I) is the radiant flux (Φ) proceeding from a point source per unit solid angle in the direction being considered. It is a measure of the intensity of electromagnetic radiation. The standard unit is watt per steradian (W sr^{-1}):

$$I = \frac{d\Phi}{d\Omega} \tag{4.15}$$

4.1.3.5 Irradiance

The irradiance (E) is the radiant flux (Φ) incident upon a surface per unit area. The standard unit is watt per square meter (W m^{-2}):

$$E = \frac{d\Phi}{dA} \tag{4.16}$$

Spectral irradiance characterizes the light at a single wavelength or frequency and is expressed in (W m^{-2} nm^{-1}). In some applications like photosynthesis, it is more convenient to express light in terms of a quantum flux. Irradiance (E) is related to quantum flux (N) by the following equation:

$$E = N \times N_A \times \frac{hc}{\lambda} \tag{4.17}$$

where N_A is the Avogadro number (6.02214×10^{23} mol^{-1}), h is the Planck constant (6.626×10^{-34} J s), c is the speed of light in a vacuum (2.998×10^8 m s^{-1}), and λ is the wavelength (m). The standard unit is mol photons per square meter and second (mol (photon) m^{-2} s^{-1}). If the wavelength is given in nm, it must be first converted into meters by multiplying by 10^{-9}. For example, 1 µmol m^{-2} s^{-1} of 400 nm light is equivalent to 0.3 W m^{-2}, while for 700 nm it is 0.17 W m^{-2} (Evans et al., 2004).

4.1.3.6 Fluence Rate

The fluence rate (F) (or scalar irradiance) is the radiant flux (Φ) incident on a sphere of unit cross-section. The standard unit is watt per square meter (W m^{-2}). Although irradiance and fluence rate are expressed in the same units, the distinction between the two physical quantities is important in biology, particularly in UV photobiology (Rupert and Latarjet, 1978; Björn and Vogelmann, 1996). Fluence rate is of particular interest for the characterization of the natural environment of plants because the latter with few exceptions are three-dimensional rather than flat objects:

$$E = \frac{d\Phi}{da} \tag{4.18}$$

4.1.3.7 Radiance

The radiance (L) is the radiant flux (Φ) that leaves a surface per unit solid angle and unit projected area of that emitting surface. Radiance is the physical quantity that corresponds closely to the visual brightness of a surface:

$$L = \frac{d^2\Phi}{dA\cos\theta d\Omega} \tag{4.19}$$

where Φ is the radiant flux (W) emitted, A the area of the source (m^2), Ω the solid angle (sr) subtended by the observation, and θ the angle between the normal to the surface and the specified direction. The term $\cos\theta$ transforms area dA to projected area. If the amount of flux $d^2\Phi$ that leaves an element of area dA in solid angle $d\Omega$ in all directions is proportional to $\cos\theta$, the surface is said to obey Lambert's cosine law and is often referred to as a Lambertian, or perfectly diffuse radiating (or reflecting or transmitting) surface. Because the projected area is also proportional to $\cos\theta$, the radiance of such a perfectly diffuse surface is independent of the direction of view. The standard unit is watt per square meter per steradian (W m^{-2} sr^{-1}). Spectral radiance characterizes the light at a single wavelength or frequency and is expressed in (W m^{-2} sr^{-1} nm^{-1}).

4.1.3.8 Radiant Emittance

The radiant emittance (M) (or radiant exitance) is the radiant flux per unit area that emerges from a surface. The standard unit is watt per square meter (W m^{-2}).

4.1.4 Blackbody Radiation

One distinguishes between natural energy sources (the Sun, the Earth and other planetary surfaces, and the atmosphere) defined as passive radiation and artificial energy sources (LiDAR, Radar) defined as active sources. All bodies emit and absorb electromagnetic radiation. A blackbody is an ideal object that absorbs all the radiation that contacts it. In this case there is no reflected or transmitted flux and therefore the existing flux only consists of emitted flux. In the laboratory, one can simulate a blackbody as a large cavity with a small entrance hole (Planck radiator). The

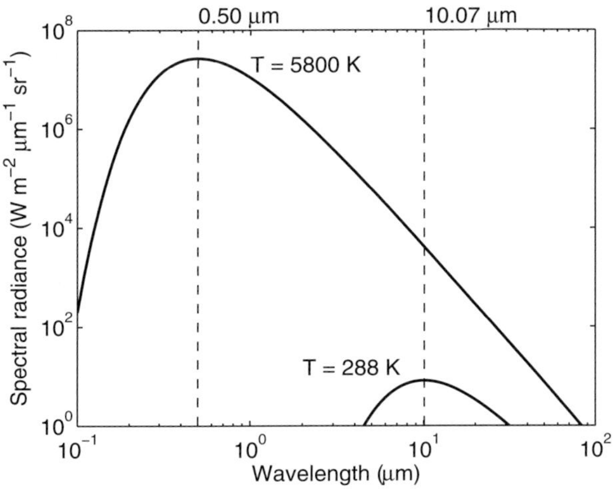

Figure 4.5 Blackbody spectral radiance calculated by Planck's law for the Sun (5800 K) and Earth (288 K) at mean emitting temperatures.

absorption of incident radiation increases its internal energy and induces the emission of unpolarized radiation over a broad spectral distribution depending on its temperature. The maximum amount of energy radiated occurs at a particular wavelength, the position of which shifts with the temperature of the source, as seen in Figure 4.5. If the blackbody is a perfectly diffusing surface, the energy radiated into space, per unit area, would equal the absorbed energy for thermal balance. The formulation of the blackbody spectral radiance in unit of power per unit of surfacc area per unit of wavelength per unit of solid angle (W m^{-2} μm^{-1} sr^{-1}) is given by the famous Planck's law (Schanda, 1986; Gates, 2003):

$$L_e(\lambda, T) = \frac{2hc^2}{\lambda^5\left(e^{hc/k\lambda T} - 1\right)} \tag{4.20}$$

where λ is the wavelength (m), T is the absolute temperature (K), h is the Planck constant (6.626×10^{-34} J s), c is the speed of light in a vacuum (2.998×10^8 m s^{-1}), and k is the Boltzmann constant (1.38×10^{-23} J K^{-1}). Figure 4.5 presents the variation of $L_e(\lambda, T)$ as a function of the wavelength for two particular temperatures: the surface temperature of the Sun ($T = 5800$ K) and the average temperature of the Earth ($T = 288$ K). One can show that spectra of different temperatures never cross each other.

According to Planck's law, the blackbody spectral radiance $L_e(\lambda, T)$ is a continuous function of the wavelength that tends toward 0 when $\lambda \to 0$ or $\lambda \to \infty$ and that presents a maximum value at a wavelength λ_{max} determined by the surface temperature of the blackbody. It is obtained by equating the first derivative of $L_e(\lambda, T)$ with respect to the wavelength at zero (see Appendix C for details):

$$\frac{dL_e(\lambda, T)}{d\lambda} = 0 \Rightarrow \lambda_{max} T = 2.897 \times 10^{-3} \text{mK} \tag{4.21}$$

It is clear that $\lambda_{max} = 499$ nm when $T = 5800$ K (the Sun), that $\lambda_{max} = 10$ μm when $T = 288$ K (the Earth), and that the cosmic background radiation corresponding to a blackbody temperature of about 2.726 K has a maximum of emission in the microwave at 1.06 mm. Equation 4.21 is better known as Wien's displacement law that predicts the shift in λ_{max} with change in temperature of the radiating body. Let us now integrate $L_e(\lambda, T)$ over the whole spectrum and over the hemisphere into which the radiation is emitted. First we integrate over the wavelength from zero to infinity to obtain the total radiance in units of power per unit of surface area per unit of solid angle (W m^{-2} sr^{-1}) (see Appendix C for details):

$$L_e(T) = \int_0^\infty L_e(\lambda, T)\, d\lambda = \frac{2\pi^4 k^4}{15h^3c^2} T^4 \tag{4.22}$$

For the integration over the hemisphere we assume that the emission from a plane radiating surface into the hemisphere is proportional to the projected surface area and that it is isotropic, that is, $L_e(T)$ does not depend on direction:

$$M(T) = \int_\Omega L_e(T) \cos\theta d\Omega = L_e(T) \int_0^{\pi/2} \int_0^{2\pi} \cos\theta \sin\theta\, d\theta\, d\phi = \pi L_e(T) \tag{4.23}$$

This distribution leads to the famous Stefan–Boltzmann Law:

$$M(T) = \frac{2\pi^5 k^4}{15h^3c^2} T^4 = \sigma T^4 \tag{4.24}$$

where σ is the Stefan-Boltzmann constant (5.67×10^{-8} W m^{-2} K^{-4}). The total radiated power of a blackbody surface is shown to be proportional to the fourth power of the temperature.

4.2 What to Measure?

When studying the optical properties of an object, it is necessary to distinguish between those which depend on the illumination and viewing directions, for example, the reflection factor, and those which do not depend on the angles, for example, the index of refraction. The properties of the first are called *apparent* and the second, *inherent*. In physical models, inherent properties are used to simulate the apparent properties, and *vice versa*. For example, the index of refraction and the absorption coefficient of a medium are used in geometric optics approximations to calculate the factors of reflection and transmission of the medium (see Chapter 3). Palmer (1997) notes that there has been a continuing dialog over terminology, particularly between the suffixes *–ance* and *-ivity*: the suggested usage seems to employ the suffix *-ivity* (i.e., transmissivity, absorptivity, and reflectivity) for properties of a pure material (inherent optical properties) and the suffix *-ance* for the characteristics of a specimen or sample (apparent optical properties).

Whenever a plant leaf is exposed to radiant energy Φ_i from a source such as the Sun, this energy has three modes of dissipation: an amount of energy Φ_r is reflected and another Φ_t is transmitted (Figure 4.6). The energy absorbed by a leaf Φ_a is transformed into another form of energy: thermal emission, conduction to internal tissue, evapotranspiration, and photochemical reactions. The apparent optical properties of leaves characterize their capacity to absorb, reflect, or transmit light.

In Physics, the law of conservation of energy states that the total amount of energy in an isolated system remains constant. In consequence, these four terms are connected by

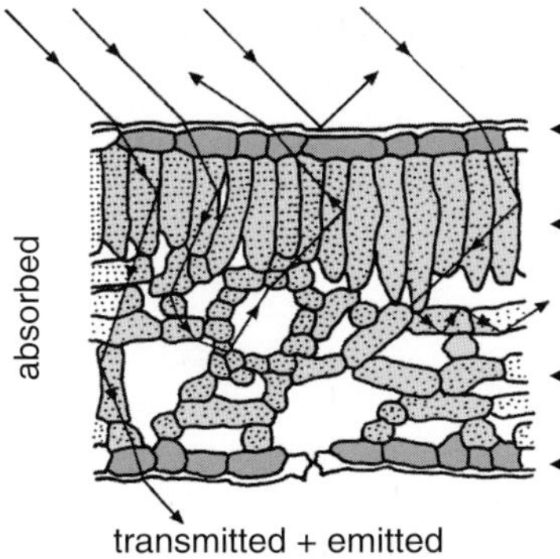

Figure 4.6 Reflected, emitted, absorbed, and transmitted radiation from a leaf.

$$\Phi_i = \Phi_r + \Phi_t + \Phi_a \tag{4.25}$$

The factor of reflection or reflectance (R), the factor of transmission or transmittance (T), and the factor of absorption or absorptance (A) are defined by

$$R = \frac{\Phi_r}{\Phi_i} \tag{4.26}$$

$$T = \frac{\Phi_t}{\Phi_i} \tag{4.27}$$

$$A = \frac{\Phi_a}{\Phi_i} \tag{4.28}$$

and then

$$R + T + A = 1 \tag{4.29}$$

The apparent optical properties R, T, and A depend on the wavelength of the radiation (spectral variation) and the geometries of illumination and observation (directional variation). They quantify what is commonly called the color or the appearance of an object. Thus, healthy leaves are green because their spectral reflectance peaks in the visible wavelengths at around 550 nm. Mirrors are bright objects because their directional reflectance is strong in a specific direction and practically null in the others. Nicodemus et al. (1977) proposed a precise nomenclature of these apparent optical properties. To study the spectral variation of the optical properties of plant leaves, directional-hemispherical reflectance (DHR) and directional-hemispherical transmittance (DHT) are commonly measured. That means that incident light comes from a well-defined direction but that the reflected and transmitted light rays are integrated over the whole hemisphere.

4.2.1 Bidirectional Reflectance/Transmittance Distribution Function

Studies on leaf optical properties use physical concepts such as the BRDF. Although strictly speaking it can never be measured directly, the BRDF is essential to characterize the intrinsic bidirectional

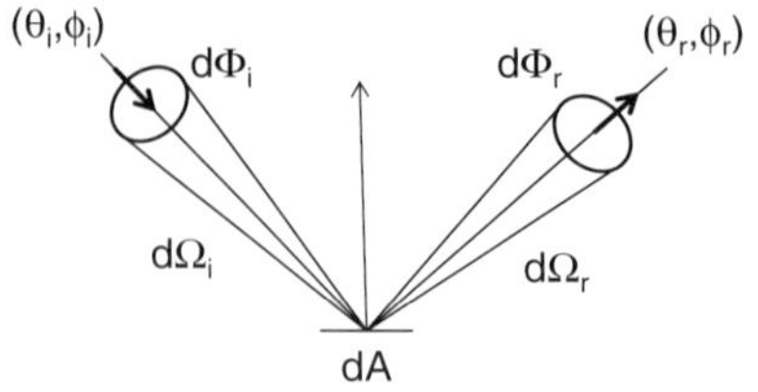

Figure 4.7 Geometry of incident and reflected elementary beams.

optical properties of surfaces. It was first introduced by Nicodemus et al. (1977) and a recent review of this nomenclature in remote sensing is available in Schaepman-Strub et al. (2006). The BRDF is a physical concept implementing infinitesimal flux quantities. As in Figure 4.7, consider an elementary surface dA uniformly irradiated in the direction (θ_i, ϕ_i).

If $d\Phi_i$ (W) is the incident radiant flux confined in the solid angle element $d\Omega_i$, the irradiance dE_i (W m^{-2}) is defined as

$$dE_i = \frac{d\Phi_i}{dA} \tag{4.30}$$

Assume now that this elementary surface is observed in the direction (θ_r, ϕ_r) in the solid angle element $d\Omega_r = \sin\theta_r d\theta_r d\phi_r$. If $d\Phi_r$ (W) is the reflected radiant flux, the radiance dL_r (W m^{-2} sr^{-1}) is given by

$$dL_r = \frac{1}{dA\cos\theta_r}\frac{d\Phi_r}{d\Omega_r} \tag{4.31}$$

The BRDF (sr^{-1}) of the surface is then defined as

$$BRDF(\theta_i, \phi_i, \theta_r, \phi_r, \lambda) = \frac{dL_r(\theta_i, \phi_i, \theta_r, \phi_r, \lambda)}{dE_i(\theta_i, \phi_i, \lambda)} = \frac{d\Phi_r}{d\Phi_i \cos\theta_r d\Omega_r} \tag{4.32}$$

According to this definition, the BRDF corresponds to the incident and reflected radiation that is included in solid angle elements. In practice, they are finite and both illumination and measurement are imbedded in a cone. Then one should define a new quantity called biconical rather than bidirectional, which corresponds to an angular integration of the BRDF on the corresponding fields. However, for the sake of simplicity, the term BRDF will be used hereafter to indicate biconical measurements. There is actually no possible confusion since the considered incident and reflected solid angles are small. In the case of thin and translucent surfaces, one can also define the BTDF, also expressed in (sr^{-1}) by replacing the numerator of Eq. 4.32 with the radiance dL_t emitted toward the lower hemisphere:

$$BTDF(\theta_i, \phi_i, \theta_r, \phi_r, \lambda) = \frac{dL_t(\theta_i, \phi_i, \theta_r, \phi_r, \lambda)}{dE_i(\theta_i, \phi_i, \lambda)} \tag{4.33}$$

The definition of the bidirectional scattering distribution function (BSDF) generalizes this concept for all conditions of diffusion (Asmail, 1991): it is the ratio of the scattered radiance to incident

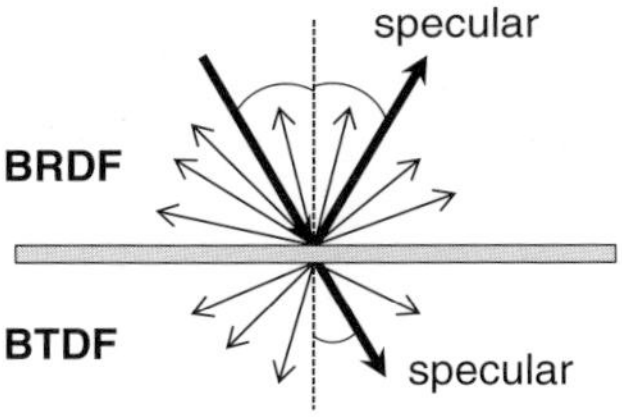

Figure 4.8 Concept of BRDF and BTDF.

irradiance at a given wavelength. Therefore it may be split into reflected and transmitted components, which are treated separately as the BRDF and the BTDF, as seen in Figure 4.8. For many surfaces, the specular reflection and transmission is much larger than the diffuse scattering. Most often the BSDF is used to name the general mathematical function which, as a probability density function, describes the way in which the light is scattered by a surface into a particular direction.

4.2.2 Bidirectional Reflectance Factor

In practice, the BRDF is difficult to measure. Instead, the bidirectional reflectance factor (BRF) defined as the ratio of the radiance reflected by a target surface and the radiance reflected by a Lambertian surface in the same direction is commonly used (Appendix D):

$$BRF(\theta_i, \phi_i, \theta_r, \phi_r, \lambda) = \pi \quad BRDF(\theta_i, \phi_i, \theta_r, \phi_r, \lambda) \tag{4.34}$$

4.2.3 Directional-Hemispherical Reflectance Factor

The BRDF quantifies the energy that is reflected in a particular direction. To assess the amount of light reflected in a whole hemisphere, one calculates the directional-hemispherical reflectance factor (DHRF) which is related to the BRDF by the following equation (Nicodemus et al., 1977):

$$DHRF(\theta_i, \phi_i, \lambda) = \iint BRDF(\theta_i, \phi_i, \theta_r, \phi_r, \lambda) \cos\theta_r d\Omega_r \tag{4.35}$$

where the integral covers the upper hemisphere. The same equation applies for the directional-hemispherical transmittance factor (DHTF) where the BRDF is replaced by the BTDF and the upper hemisphere by the lower one.

4.2.4 Attenuated Total Reflectance

Introduced in the 1960s, attenuated total reflection (ATR) spectroscopy is today the most widely used Fourier transform infrared (FTIR) technique to study materials that are difficult to analyze by classical methods, for example, plant leaves that are too thick for transmission analysis because most of the infrared radiation is totally absorbed (Ribeiro da Luz and Crowley, 2007). The measuring principle is simple: a sample of low refractive index ($n_2 < n_1$) is placed in close contact with a transparent plate of high refractive index, for instance, zinc selenide or diamond crystals ($n_1 = 2.43$ at 10 μm). When the

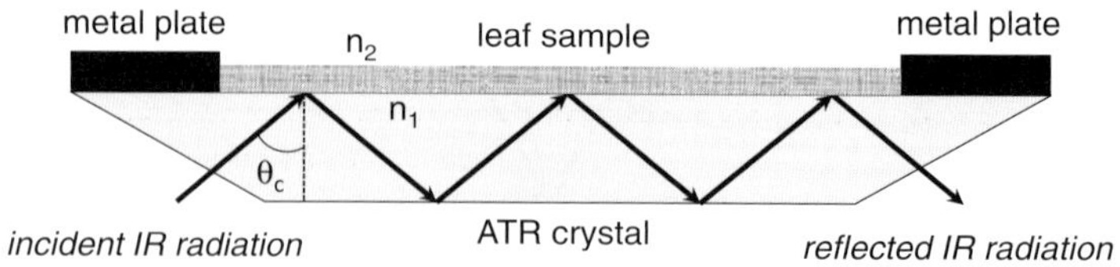

Figure 4.9 Schematic illustration of the principle of the infrared ATR spectroscopic technique: n_1 and n_2 are the refractive indices of the ATR crystal and the leaf sample, respectively. Note that condition $n_2 < n_1$ is required for the total internal reflection to occur at the ATR crystal–sample interface.

angle of incidence exceeds a critical angle θ_c, radiation propagating in such an optically dense medium undergoes total internal reflection at the interface, as illustrated by Figure 4.9.

θ_c is determined by

$$\theta_c = \sin^{-1}\left(\frac{n_2}{n_1}\right) \tag{4.36}$$

The system is then designed so that the incident infrared beam impinges on the plate at an angle greater than θ_c. According to Descartes' law, the incident beam should be totally reflected at the crystal–sample interface and directed toward the detector. In fact, internal reflection creates an evanescent wave which extends at a short distance beyond the surface of the crystal into the sample of lower index. In regions of the infrared spectrum where absorption is high, the intensity of the evanescent wave is attenuated exponentially with distance from the boundary. The distance, which is on the order of one or two micrometers in a plant leaf, makes ATR generally insensitive to sample thickness. The resulting spectrum is similar to a transmission spectrum with some differences in peak intensities because of the variable penetration. The penetration depth of the infrared beam is defined as the distance required for the electric field amplitude to fall to e^{-1} of its value at the surface and is given by the following equation:

$$d_p = \frac{\lambda}{2\pi\left(n_1^2 \sin^2\theta - n_2^2\right)^{1/2}} \tag{4.37}$$

where λ is the wavelength of light, θ is the angle of incidence of the infrared beam relative to the perpendicular from the surface of the crystal, and n_1 and n_2 are the indices of refraction of the plate and the sample, respectively. Equation 4.37 shows that d_p increases at longer wavelengths (lower wavenumbers), that it decreases as the angle of incidence increases, and that it is lowest for the crystal with the highest refractive index. If $n_2 = 1.5$ for plant cuticles and $n_1 = 2.43$ for ZnSe crystals, the critical angle equals $\theta_c = 38.12°$. Assume that we exceed it, for instance, $\theta = 45°$, to observe a purely ATR spectral result. The penetration depth in leaves is about 0.47 μm at 2500 nm (4000 cm^{-1}) to 1.90 μm at 10 μm (1000 cm^{-1}).

4.3 Measurement of Leaf Color

Color is the most natural way to describe leaf optical properties, but certainly not the simplest. The common terms used to describe plant colors are often subjective and make comparisons difficult. So it

Table 4.2 *Main color spaces used in leaf color science. The color coordinates are directly measured using a digital camera, a colorimeter, or they are derived from spectra acquired using a spectrometer.*

Color space	Description	Reference
XYZ	The original CIE 1931 color space. Y is a measure of the brightness of a color	Spomer et al. (1988), Yoshimura et al. (1991a), Comar et al. (2012)
xyY	Normalized chromaticity values, derived from the CIE XYZ color space. Y is the same as in XYZ	Ruth et al. (1991), Bacci et al. (1996, 1998), Buschmann et al. (2012)
L*a*b*	CIE 1976 color space. L* is a nonlinear scaling of L, normalized to a reference white point	Kwolek (1982), Meir et al. (1992), Madeira et al. (2003), Döring et al. (2009), Kendal et al. (2013)
LCH	Cylindrical version of the L*a*b* color space	Deák et al. (2007)
RGB	Color matching by combinations of monochromatic primary colors Red (R), Green (G) and Blue (B)	Andrieu et al. (1992), Nakatani and Kawashima (1994), Gay et al. (2008), Tian et al. (2010), Yuzhu et al. (2011), Lunadei et al. (2012), Yang et al. (2014)
sRGB	Standard created by HP and Microsoft in 1996 and adopted by major manufacturers for use on monitors, printers, and the internet	Shibghatallah et al. (2013)
HSI	Model derived from the RGB color space and representing every color with three components: hue (H), saturation (S), intensity (I)	Dillen et al. (2012)
Munsell	Introduced in 1915 and revised twice since then, this color system uses hue (H), chroma (C), and lightness value (V). Oldest color model in common use today	Zhang et al. (1998)

is important to quantitatively measure them. The International Commission on Illumination (CIE for Commission Internationale de l'Éclairage) has defined a number of standard color spaces, more correctly known as models, which are widely used. A color space is a completely specified scheme for describing the color of light, ordinarily using three numerical values called coordinates. In 1931, the CIE studied human color perception (see Section 12.5.3) and developed a standard, called the CIE XYZ. This standard defines a three-dimensional space where three values, called tristimulus values, define a color. In the decades since that initial specification, the CIE has developed several additional color space specifications that attempt to provide alternative color representations that are better suited to some purposes than XYZ (Table 4.2).

To avoid the utilization of a three-dimensional diagram for the color representation, the relative trichromatic coefficients x and y were calculated as follows:

$$\begin{cases} x = \dfrac{X}{X+Y+Z} \\ y = \dfrac{Y}{X+Y+Z} \\ z = \dfrac{Z}{X+Y+Z} = 1 - x - y \end{cases} \tag{4.38}$$

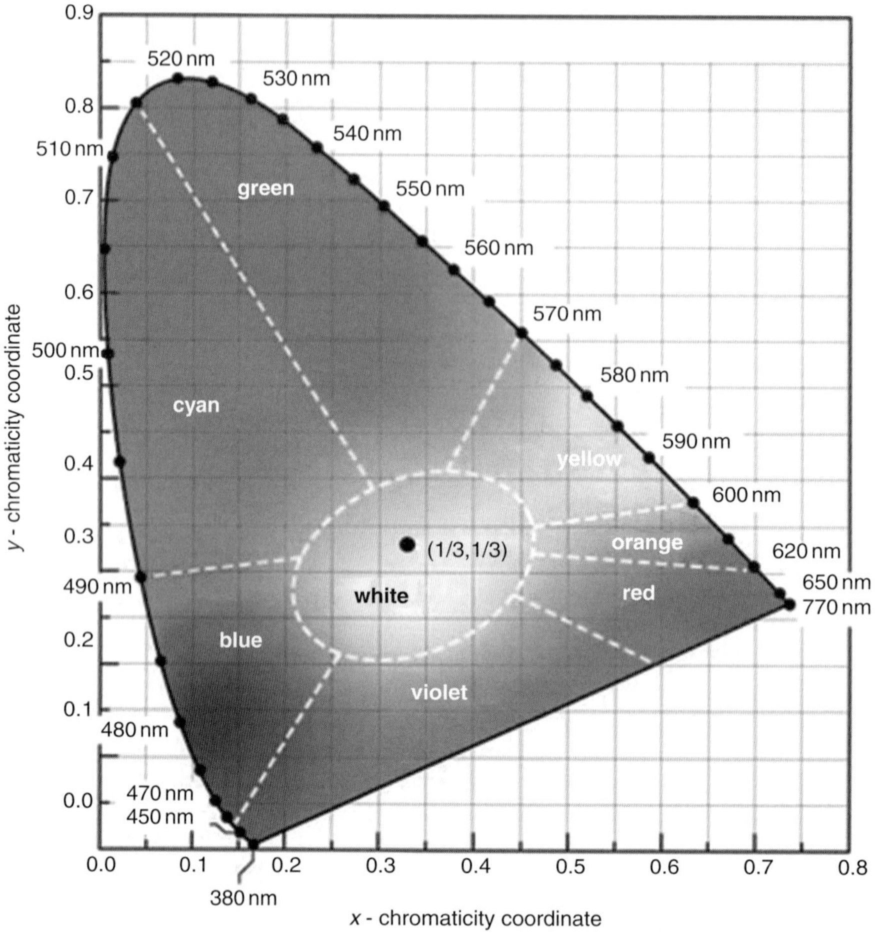

Figure 4.10 CIE 1931 (x, y) chromaticity color diagram relating coordinates and wavelengths (from Schubert (2006) Copyright © with permission from Cambridge University Press). (A black and white version of this figure will appear in some formats. For the color version, please refer to the plate section.)

Since $x + y + z = 1$, they permit the description of every color by means of two components in a chromaticity diagram (Figure 4.10). Monochromatic colors are located on the perimeter and white light is located in the center of the diagram. Color saturation decreases toward the center of the diagram.

Then the line drawn from the white point through the (x, y) point intersects the curved line (red line) which is labeled with wavelength values (Figure 4.11).

The CIE 1976 color space, also called the L*a*b* color space, is a standardized (device-independent) non-linear transformation of the RGB color space. It was created in an effort to get a perceptually uniform color space that could be correlated with the human perception of colors. It has linear measures of lightness (L^*) on a scale of 0 (black) to 100 (white) and two chromaticity coordinates (a^* and b^*) that are suitable for use in mathematical models. As show in Figure 4.12, the a^* dimension represents a spectrum from green (negative) to magenta (positive) and the b^* dimension represents a spectrum from blue (negative) to yellow (positive). The area near the origin (very low absolute values of a^* and b^*) represents grays, that is, colors of little or no chromaticity.

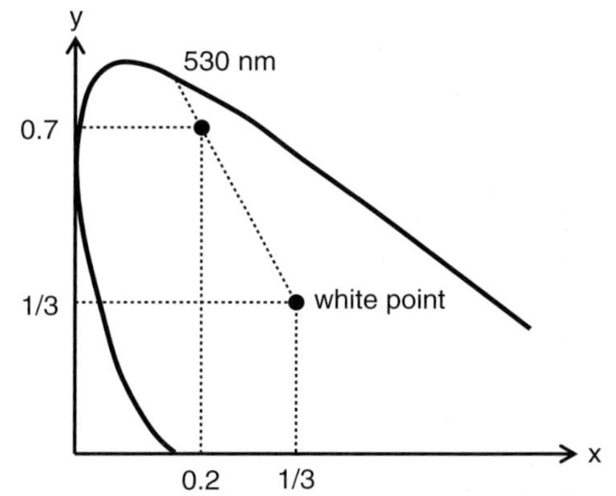

Figure 4.11 Determination of the dominant wavelength (~ 530 nm) corresponding to the value $(x, y) = (0.2, 0.7)$ extending the line centered on the white point about $(x, y) = (1/3, 1/3)$ (redrawn from Shibghatallah et al. (2013)).

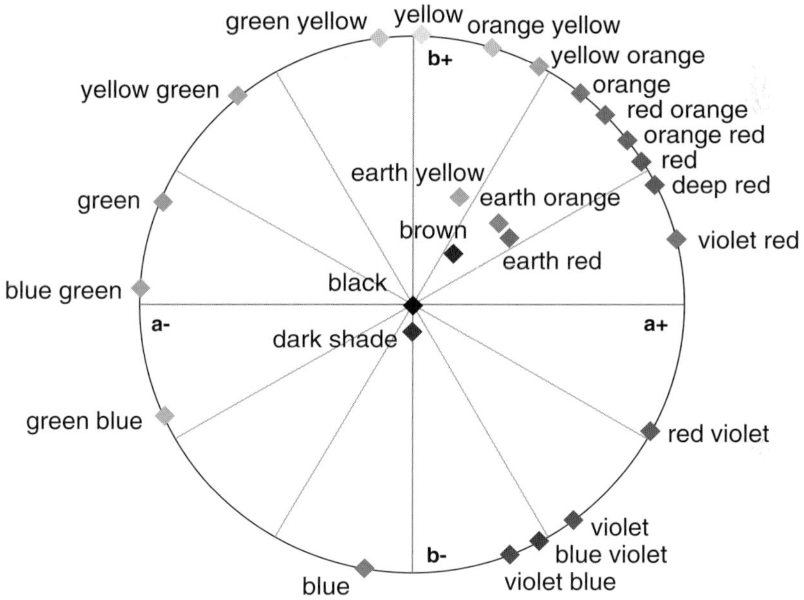

Figure 4.12 The CIELAB a*b* plane. (A black and white version of this figure will appear in some formats. For the color version, please refer to the plate section.)

The L^*, a^*, and b^* are calculated as

$$\begin{cases} L^* = 116\left(\dfrac{Y}{Y_n}\right)^{1/3} - 16 \\ a^* = 500\left(\left(\dfrac{X}{X_n}\right)^{1/3} - \left(\dfrac{Y}{Y_n}\right)^{1/3}\right) \\ b^* = 200\left(\left(\dfrac{Y}{Y_n}\right)^{1/3} - \left(\dfrac{Z}{Z_n}\right)^{1/3}\right) \end{cases} \tag{4.39}$$

a^* and b^* are Cartesian coordinates. LCH (also denoted as CIE L*C*h or L*u*v*) is a cylindrical version of the L*a*b* color space, which has the advantage of providing means for measuring important functions such as: i) the metric hue angle, $h = \arctan(b^*/a^*)$, which is recommended by the CIE as the psychometric correlate of the visually perceived attribute of hue; and ii) the chroma or "saturation," $C^* = \sqrt{a^{*2} + b^{*2}}$. Finally, the degree of color difference $\Delta E = \sqrt{\Delta L^2 + \Delta a^2 + \Delta b^2}$ expresses the difference between two color samples by a single function number.

Color matching studies carried out in the 1920s showed that colored samples could be matched by combinations of monochromatic primary colors red (700 nm), green (546.1 nm), and blue (435.8 nm). Digital images (scanner, digital photograph, etc.) are typically saved as a rectangular grid of pixels in RGB format (linear values ranging from 0 to 1). Adding certain proportions of each primary color allows simulating a huge number of colors. They can be represented in a cubic space, but differences in color in this space are not closely related to human perception of color difference. Moreover, RGB values are typically not standardized; they depend on the instrument used to capture the image so that there are actually several common RGB working spaces (sRGB, Adobe, Apple, PAL/SECAM, Wide Gamut, etc.). As a consequence, digital images are often transformed into different color spaces more suited to a particular application. The process of transforming coordinate values in one color space (RGB) to corresponding values in another color space (XYZ) involves three linear equations:

$$\begin{bmatrix} X \\ Y \\ Z \end{bmatrix} = M \begin{bmatrix} R \\ G \\ B \end{bmatrix} \tag{4.40}$$

where M is a 3-by-3 transform matrix that changes from one RGB working space to another. Digital images may be standardized using a reference object such as the ColorChecker chart in the image. In a second step, (x, y) coordinates are produced using Eq. 4.38. Note that it is possible to convert XYZ values to RGB values using the inverse matrix of M:

$$\begin{bmatrix} R \\ G \\ B \end{bmatrix} = M^{-1} \begin{bmatrix} X \\ Y \\ Z \end{bmatrix} \tag{4.41}$$

The conversion from RGB to HSI or from HSI to RGB begins with normalizing RGB values:

$$\begin{cases} r = \dfrac{R}{R+G+B} \\ g = \dfrac{G}{R+G+B} \\ b = \dfrac{B}{R+G+B} \end{cases} \tag{4.42}$$

Each normalized H, S, and I components are then obtained by

$$\begin{cases} h = \cos^{-1}\left\{\dfrac{0.5\left((r-g)+(r-b)\right)}{\sqrt{(r-g)^2+(r-b)(g-b)}}\right\} & h \in [0,\pi] \text{ for } b<g \\ h = 2\pi - \cos^{-1}\left\{\dfrac{0.5\left((r-g)+(r-b)\right)}{\sqrt{(r-g)^2+(r-b)(g-b)}}\right\} & h \in [\pi,2\pi] \text{ for } b>g \\ s = 1 - 3\min\{R,G,B\} & \\ i = \dfrac{R+G+B}{3\times 255} & \end{cases} \tag{4.43}$$

For convenience, h, s, and i values are converted in the ranges of (0,360), (0,100), and (0,255), respectively by: $H = h \times 180/\pi$, $S = s \times 100$, and $I = i \times 255$.

Leaf color can be described using three main approaches: visual systems, colorimeters, and spectrometers. Visual systems that involve comparison with color charts under controlled illumination will be presented in Section 12.3.1. Colorimeters are widely used in industrial and food applications. Tristimulus colorimetry is based on the three component theory of color vision, which states that the eye possesses receptors for three primary colors (red, green, blue) and that all colors are seen as mixtures of these three primary colors. Therefore a colorimeter is an instrument that uses edge band filters, or some similar system, to provide the red, green, and blue components of a color. For instance, digital cameras may be considered as non-contact colorimeters. Their quality has improved markedly in recent years, as well as their affordability and availability. Therefore they are a suitable tool for lab or field data collection as long as the quality of the images is suitable for plant color measurements (Kendal et al., 2013). The spectrophotometer breaks light up into a spectrum, using a grating or similar system and determines the reflection or transmission spectrum of an object. Spectrophotometric methods then numerically characterize the appearance of the color under a specified "Standard Illuminant" (e.g., D65 daylight). The colorimetric components can be produced in different space colors from the spectral data recorded in the visible band. For instance, the tristimulus values for a color with a spectral power distribution $I(\lambda)$ are given in terms of the standard observer by

$$\begin{cases} X = \int_0^\infty I(\lambda)\bar{x}(\lambda)d\lambda \\ Y = \int_0^\infty I(\lambda)\bar{y}(\lambda)d\lambda \\ Y = \int_0^\infty I(\lambda)\bar{z}(\lambda)d\lambda \end{cases} \tag{4.44}$$

where λ is the wavelength of the monochromatic light (nm) and $\bar{x}(\lambda)$, $\bar{y}(\lambda)$, and $\bar{z}(\lambda)$ are the CIE's color matching functions (CMF) that describe the chromatic response of the observer (Figure 4.13). Their measurements are valid for 2° (1931) and 10° (1964) fields of view, respectively. Note that in 1978, Vos made additional corrections to Judd's (1951) revision of the CIE 1931 2°Color matching functions, and that there are most recent proposals that have yet to be ratified by the CIE.

As previously, the RGB values for a color with a spectral power distribution $I(\lambda)$ are given in terms of the standard observer by

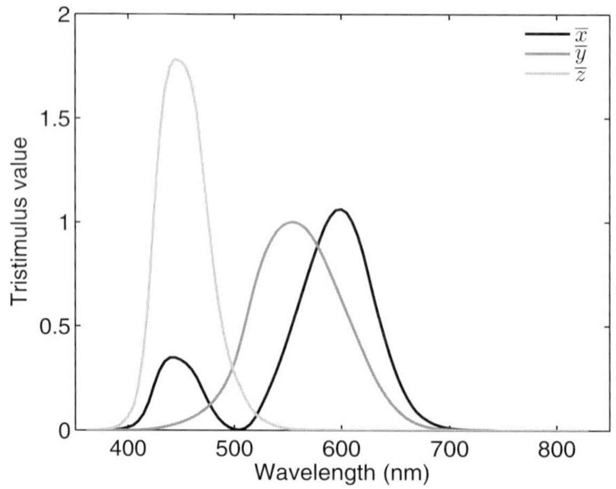

Figure 4.13 The CIE 1931 XYZ color matching functions. Note that the $\overline{y}(\lambda)$ color matching function is identical to the eye sensitivity function $V(\lambda)$.

$$\begin{cases} R = \int_0^\infty I(\lambda)\overline{r}(\lambda)d\lambda \\ G = \int_0^\infty I(\lambda)\overline{g}(\lambda)d\lambda \\ B = \int_0^\infty I(\lambda)\overline{b}(\lambda)d\lambda \end{cases} \tag{4.45}$$

where λ (nm) is the wavelength of the monochromatic light and $\overline{r}(\lambda)$, $\overline{g}(\lambda)$, and $\overline{b}(\lambda)$ are the CIE's color matching functions that describe the chromatic response of the observer (Figure 4.14).

Despite the variety of methods in the practice of color measurement, many instruments are not convenient to use since their results are difficult to interpret due to the fact that color is usually expressed differently from human color sensation.

4.4 Measurement of Leaf BRDF/BTDF

The bidirectional optical properties of an object describe the variation in reflectance, transmittance, and absorptance with illumination and observation directions. Measuring these properties finds application in various fields including industrial control, planetary studies, remote sensing, medical diagnosis, metrology of optical surfaces, synthesis imagery, etc. Some photo-goniometers adapted for field measurements over different types of surfaces, including vegetation canopies, soil, water, and snow, are specially developed for remote sensing applications. Others are designed for laboratory measurements of small plants, urban materials, soils, etc. (Von Schönermark et al., 2004). There are no commercial goniometers adapted to the measurement of leaf BRDF (or worse, leaf BTDF), but also of other translucent targets like fabric, paper, insect integuments, or human skin, so that scientists have generally developed their own instruments. Measurements are generally performed in the visible/near-infrared region, in one to several wavebands, and in polarized or non-polarized conditions.

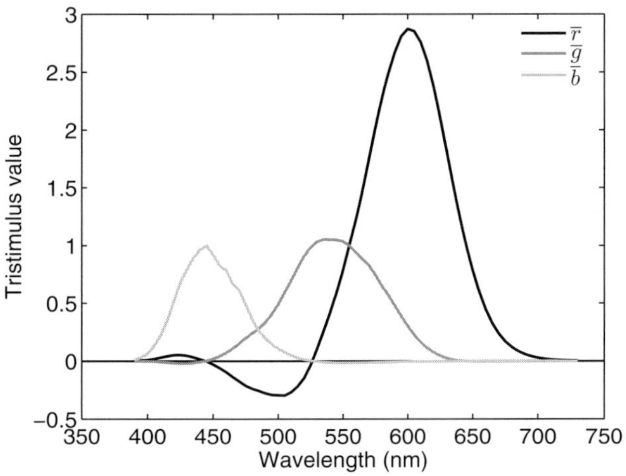

Figure 4.14 Stiles & Burch 2°Color matching functions based on measurements made by ten observers.

To calculate the BRDF of a sample, it is necessary to measure its brightness and the incident illumination. In order to simplify calibrations, there are several methods requiring information about the sample or a reference sample (Briottet et al., 2004). The first simplification relies on the measurement of the BRDF of a sample, the DHRF (defined in Section 4.2.3) of which is known. Knowledge of it allows expressing irradiance as a function of sample reflectance; in consequence the measurement of irradiance is unnecessary. Let ρ be the directional-hemispherical reflection factor:

$$\rho = \int BRDF(\theta_v, \varphi_v)\ \cos\theta_v\ d\Omega_v = \int \frac{L(\theta_v, \varphi_v)}{E}\ \cos\theta_v\ d\Omega_v \tag{4.46}$$

Then one can calculate the irradiance:

$$E = \frac{1}{\rho} \int L(\theta_v, \varphi_v)\ \cos\theta_v\ d\Omega_v \tag{4.47}$$

That can be replaced in the definition of the BRDF to obtain

$$BRDF(\theta_v, \varphi_v) = \rho \frac{L(\theta_v, \varphi_v)}{\int L(\theta_v, \varphi_v)\ \cos\theta_v\ d\Omega_v} \tag{4.48}$$

This method can be used, for instance, to estimate the BRDF of a calibration sample when the directional hemispherical reflection factor is known. The second method consists of measuring the BRDF of a reference panel, a painted $BaSO_4$ panel, or a Spectralon® standard of known reflectance:

$$BRDF_{ref} = \frac{L_{ref}}{E} \tag{4.49}$$

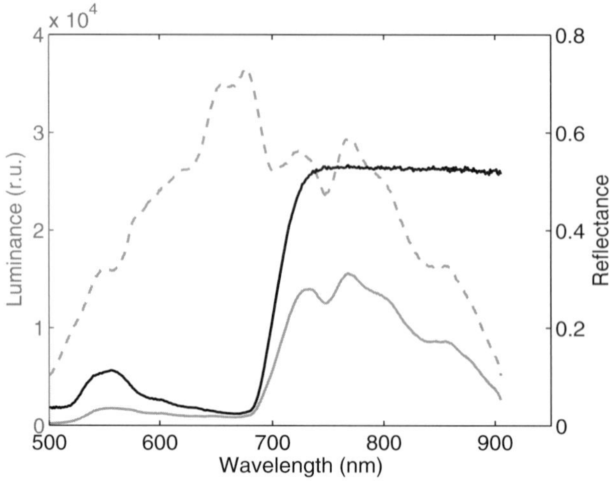

Figure 4.15 Spectral reflectance of a Johnson grass (*Sorghum halepense*) leaf in the visible and near-infrared region measured for illumination zenith angle $\theta_s = 40°$, viewing zenith angle $\theta_v = 20°$, and relative azimuth angle $\varphi = 60°$. Left axis, gray color: luminance signal measured on the reference panel (dashed line) and the leaf (gray line). Right axis, black color: leaf BRF (black line).

$$BRDF_{leaf} = \frac{L_{leaf}}{E} \tag{4.50}$$

If the irradiance is the same in both equations, then the leaf BRDF is the amount of flux received from a sample surface divided by the amount of flux which would be received from an ideal, hypothetical, perfectly diffuse, perfectly white standard can be written as

$$BRDF_{leaf} = \frac{L_{leaf}}{L_{ref}} BRDF_{ref} \tag{4.51}$$

where L_{leaf} and L_{ref} are measured with the same illumination (Figure 4.15).

As said earlier, illumination must be well-collimated and the size of the solid angle of the viewing cone must be small to approach the real BRDF. These two methods avoid absolute calibration. Assume now that the luminance L of the sample represented as a voltage V, corrected from the dark level of the instrument V_{dark}, is proportional to the rough signal provided by the apparatus:

$$L(\theta_v, \varphi_v) = K\Big(V(\theta_v, \varphi_v) - V_{dark}\Big) \tag{4.52}$$

with K as a calibration factor. By substituting Eq. 4.52 into Eq. 4.48, one obtains as the leaf BRDF:

$$BRDF_{leaf}(\theta_v, \varphi_v) = \rho \frac{V_{leaf}(\theta_v, \varphi_v) - V_{dark}}{\int \Big(V_{leaf}(\theta_v, \varphi_v) - V_{dark}\Big) \cos\theta_v \, d\Omega_v} \tag{4.53}$$

with the first method and

$$BRDF_{leaf}(\theta_v, \varphi_v) = \frac{V_{leaf}(\theta_v, \varphi_v) - V_{dark}}{V_{ref}(\theta_v, \varphi_v) - V_{dark}} BRDF_{ref}(\theta_v, \varphi_v), \tag{4.54}$$

with the second method. Note that the multiplication factor K disappeared. Measurements of leaf reflection can be taken at the Brewster angle using a polarization photometer. Equation 4.54 also applies to the bidirectional reflectance factor, which is related to the bidirectional reflectance distribution factor by a factor of π, as seen in Eq. 4.34. If $BRF_{\min}$ and $BRF_{\max}$ are the minimum and maximum amount of radiation transmitted by the polarization analyzer, respectively, then one can determine the following variables (Grant, 1985):

$$\begin{aligned} BRF &= (BRF_{\max} + BRF_{\min})/2 \\ BRF_p &= (BRF_{\max} - BRF_{\min})/2 \end{aligned} \tag{4.55}$$

where BRF is the bidirectional factor of the leaf measured without the polarization analyzer and BRF_p represents it polarized component. Then $BRF_u = BRF - BRF_q$ is the non-polarized component and, as seen in Eq. 4.7, the degree of polarization of light can be assessed by

$$p = 100 \times \frac{BRF_p}{BRF} \tag{4.56}$$

In order to sample the BRDF or BTDF, one or several detectors can be mounted on a rotation stage while the light source is fixed in position. Sometimes, however, the measurement set-up is the opposite: the detectors are fixed and the light source is mounted on a rotation stage. The literature is also equally divided into works that involve non-polarized and polarized light, sometimes both in the same study.

4.4.1 Non-Polarized Measurements

Since the leaf is an inhomogeneous object, it scatters incident light in all directions. The early generation of photo-goniometers allows measurement of scattering in the principal plane. By studying the angular distribution of the reflected and transmitted radiation, Seybold (1933a) concluded that the diffusion is anisotropic. Metzner (1957) published the first measurements of BRF and noticed differences that he attributed to the shape of the leaf surface (Figure 4.16).

Between 1960 and 1966, several papers appeared in Russian journals and some were translated into English (Tageyeva and Brandt, 1960a, 1961a). Shul'gin et al. (1960a) built a photo-goniometer and confirmed that leaf reflectance does not follow Lambert's cosine law. They also observed an increasing diffuse component with leaf thickness. This work continued in the United States and in Australia in the early 1970s when three important articles were published. Breece and Holmes (1971) improved the way to acquire leaf bidirectional reflectance at several wavelengths between 375 nm and 1000 nm. They are the first to use the concept of BRDF conceptualized by Nicodemus (1968) three years before, which has since become widespread. They also defined the BTDF, its equivalent in transmission. They noted that the directional distribution of reflected radiation is approximately isotropic in the NIR where little light is absorbed, whereas it is highly anisotropic in the visible with

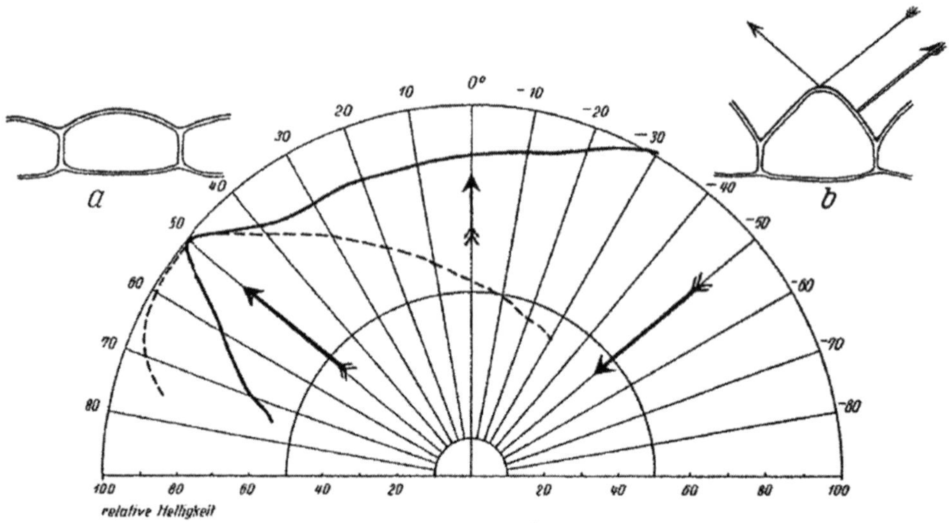

Figure 4.16 Bidirectional reflectance of a *Philodendron surinamense* leaf. Dashed line: adaxial face. Plain line: abaxial face (from Metzner (1957), *Die Kulturpflanze*. Copyright © 1957 with permission from Springer).

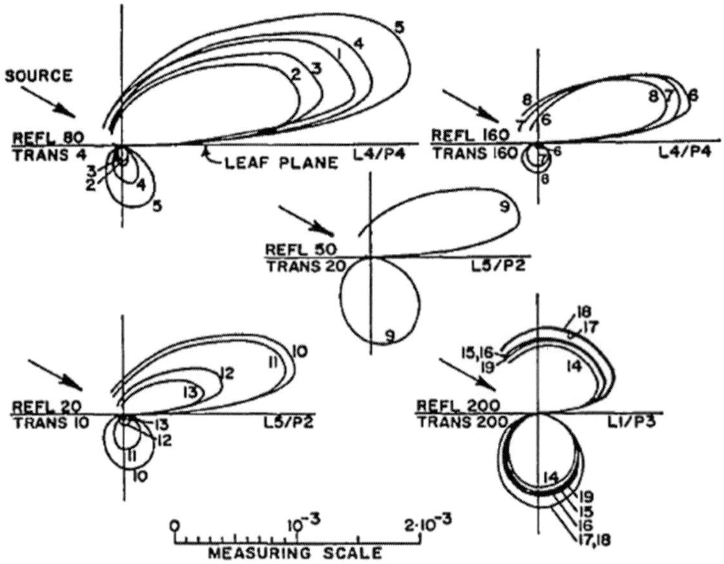

Figure 4.17 Polar plot of soybean leaves with top incidence at $\theta_i = 60°$ in the UV (1), VIS (2–13), and NIR (14–19) (from Breece and Holmes (1971), *Applied Optics*. Copyright © 1971 with permission from The Optical Society of America).

a strong specular peak (Figure 4.17). They also hypothesized that the non-Lambertian component of reflectance which is spread about the specular reflection angle was due to specular reflection by the leaf surface or by the first few layers within the leaf tissue. Finally, while leaf reflectance exhibits an angular dependency on incidence angle, leaf transmittance is nearly Lambertian.

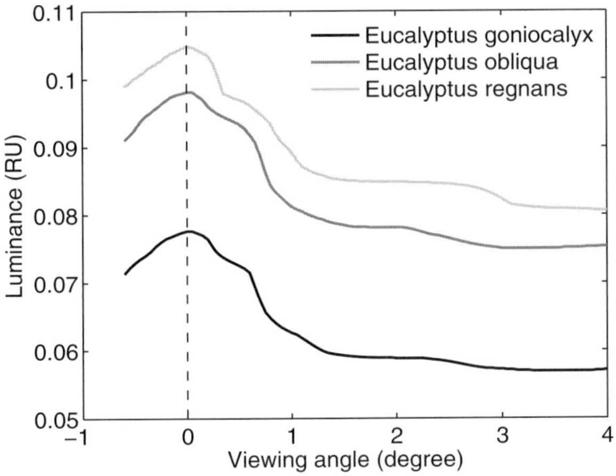

Figure 4.18 Luminance indicatrices of mature isobilateral leaves of three *Eucalyptus* leaves for 45° incidence angle (redrawn from Howard (1969)).

Woolley (1971) provides a thorough study of leaf optical properties and reports that the specular component is up to 10 times more important than the diffuse component in certain directions of observation. He evaluates the contribution of the specular component to reflection in the visible at 50% of the total reflectance and the corresponding solid angle at one steradian. He notes that, for rough leaves, the maximum reflection is not observed exactly in the specular direction but at a higher zenith angle. This phenomenon was already observed by Torrance and Sparrow (1967) on optical surfaces and called "off specular reflection." He also carried out measurements of polarization, identifying the specular contribution in the direction of the maximum reflection.

The hot spot phenomenon is characterized by a local maximum of reflectance in the backward direction, at zero phase angles (angle between the direction of incidence and observation). Obtaining it experimentally is difficult. It was identified on *Eucalyptus* leaves by Howard (1969, 1971a), who noticed an increase of reflectance of approximately 20%. The maximum intensity is observed at normal incidence and the form (peak centered on the null phase) is accentuated when incident angles increase (Figure 4.18). Although Greiner et al. (2007) focus their paper on the measurement and modeling of the specular peak, the hot spot is noticeable in the BRDF.

Studies on leaf bidirectional properties multiplied in the 1980s when the first explanatory models and complete leaf BRDF measurements are produced. Kestner et al. (1988) designed a new photo-goniometer operated at the NASA Goddard Space Flight Center to characterize the diffusion of light by soil and vegetation. They show that reflectance varies with species, leaf orientation, and view angle. Brakke et al. (1989, 1993) used it to illuminate plant leaves at 632.8 nm and from this they proposed an empirical model of directional reflectance that fits the measurements acquired with this instrument. The photo-goniometer built by Pedrini (1991) and his colleagues at the Joint Research Centre is one-of-a-kind: it is designed to measure the directional reflectance and transmittance, but also the directional backward and forward fluorescence. Their measurements of such indicatrices obtained on *Philodendron* (Figure 4.19) and tobacco leaves for different incident angles are unique. They confirm that the fluorescence emission flux is isotropic.

The full light distribution of soybean and maize leaves over a hemisphere was first published by Walter-Shea et al. (1989). Measurements were accomplished using two broad-band sensors in the

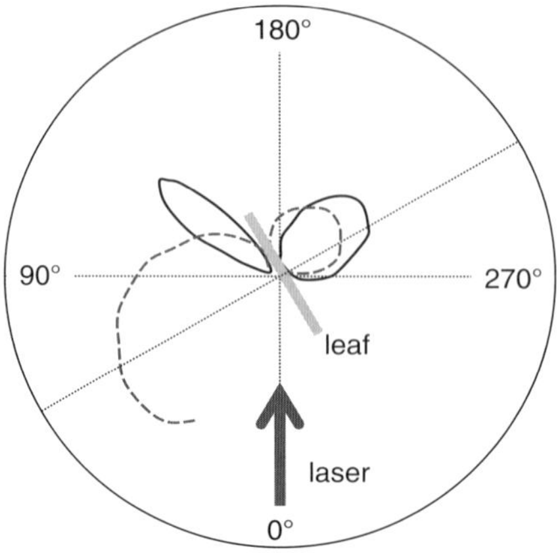

Figure 4.19 Reflection, transmission (black line), backward and forward fluorescence (gray dots) of a *Philodendron* leaf for 60° incidence angle (from Pedrini (1991), personal communication).

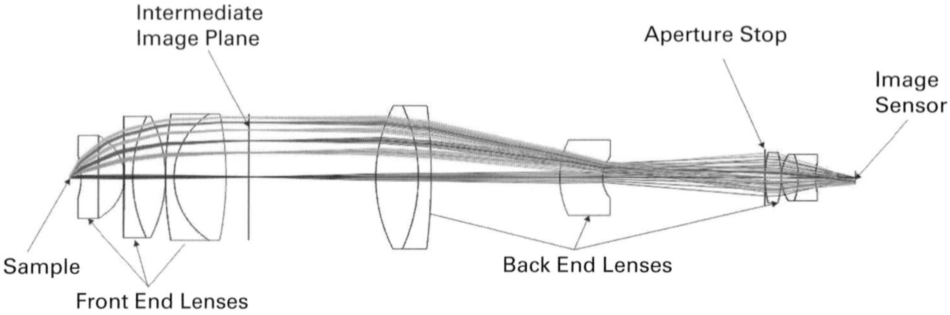

Figure 4.20 Schematic of a conoscope lens (http://eckop.com/optics/opticsadvanced-lenses/how-conoscope-lenses-work/).

VIS and NIR. Few studies have combined both the spectral and directional dimensions, the complementarity of which is crucial to better understand the determinism of leaf surface reflectance. The latest generation of instruments, called spectro-gonio-photometers are able to perform such measurements (e.g., Bousquet et al., 2005; Biliouris et al., 2007; Combes et al., 2007; Lolli et al., 2014). Finally, Comar et al. (2012) used a conoscopic system based on Fourier optics that projects a collimated light beam on the leaf while catching simultaneously all rays reflected by the surface onto a cooled Charge Coupled Device (CCD) matrix, with a very high angular resolution. What is special about a conoscope lens is that it precisely maps incoming angles to positions on the image (Figure 4.20). Table 4.3 presents a comprehensive chronological summary of leaf bidirectional optical measurements.

4.4.2 *Polarized Measurements*

Once again, the first studies were published by Russian scientists (e.g., Shul'gin and Khazanov, 1961; Moldau, 1965; Shul'gin and Moldau, 1965; Pospergelis, 1969), who measured polarization to

Table 4.3 *Articles that present measurements of bidirectional leaf optical properties in non-polarized light. R (reflectance), T (transmittance), PP (principal plane).*

Reference	RT	θ_i [deg]	θ_r [deg]	φ [deg]	λ [nm]	Leaf species
Metzner (1957)	R	50	0→75	PP	VIS	88 species
Shul'gin et al. (1960a)	R	0	10→80	PP	400→600	~ 30 species
Howard (1969)	R	0, 45, 60	$\theta_v = \theta_s$	PP		*Eucalyptus* sp.
Tageyeva and Brandt (1960a)	RT	0→72	0	PP	404, 463, 545, 656	*Ficus elastica, Hibiscus rosa-sinensis, Lactuca sativa*
Breece and Holmes (1971)	RT	0, 15, 30, 45, 60	0→90	PP	19 bands 375→1000	*Glycine max, Zea mays*
Howard (1971a)	R	0, 15, 30, 45, 60	0→90	PP	VIS	*Eucalyptus* sp.
McClendon (1984)	R	60	0, 60	PP	632.8	*Populus alba*
Salisbury et al. (1987)	R	8→50	8→50	PP	400→800	4 species
Brakke et al. (1989)	RT	0, 30, 60	0→70	PP	632.8	*Acer rubrum, Quercus rubra*
Walter-Shea et al. (1989)	RT	20, 45, 70	0→75	10→350	2 bands VIS-NIR	*Glycine max, Zea mays*
Brakke (1992)	RT	0, 30, 60	0→90	PP	590, 910	*Acer rubrum, Liriodendron tulipifera, Quercus rubra*
Brakke et al. (1993)	RT	0, 30, 60	0→90	PP	632.8	*Acer rubrum, Liriodendron tulipifera, Quercus rubra*
Sanz (1994)	RT	variable	0→80	PP	645, 840	*Alocasia macrorhizia, Ipomoea fistulosa*
Okayama (1996)	R	15→90	0	PP	VIS	23 species
Lao et al. (2005)	RT	0, 30, 60	0→90	PP	650, 830	*Euonymus japonicus*
Biliouris et al. (2007,)	R	15, 30, 45, 60, 75	0→90	0→360	350→2500	*Fagus sylvatica*
Combes et al. (2007)	RT	0, 20, 40, 60	0→85	0→360	450→900	*Julans regia, Prunus laurocerasus, Sorghum halepense*
Georgiev et al. (2007)	R	0, 67	0→80	PP	340, 470, 870	*Acacia greggii, Liriodendron tulipifera*

Table 4.3 (*cont.*)

Reference	RT	θ_i [deg]	θ_r [deg]	φ [deg]	λ [nm]	Leaf species
Greiner et al. (2007)	RT	0, 10, 20, 30, 40, 50, 60, 70, 78	0→90	PP	1064	*Acer saccarum, Populus deltoides*
Liu and Lao (2007)	R	0, 30, 60, 75	0→90	PP	650	*Ginkgo biloba*
Okayama and Li (2007)		10→80	0→60	0, 30, 60, 90	VIS	*Camellia japonica, Ulmus* sp.
Georgiev et al. (2009)	R	0, 67	0→80	PP	340, 470, 870	*Liriodendron tulipifera, Robinia pseudoacacia*
Comar et al. (2012)	R	45	0→80	0→360	RGB	*Triticum durum*
Comar et al. (2014)	R	45	0→80	0→360	RGB	*Sorghum halepense, Triticum durum*

separate the specular component from the diffuse component. They observed that the polarized component was wavelength independent and proposed to use the degree of polarization as an indicator of leaf surface physiological status. Two decades later, in her PhD thesis, Grant (1985) develops a portable photometer for the measurement of leaf polarized reflectance at the angle of incidence of 55° to the surface normal (Figure 4.21).

At the particular angle called the Brewster angle, polarization allows separation of surface and subsurface components of reflectance (Vanderbilt and Grant, 1986; Grant et al., 1987a, 1987b, 1993; Kharuk and Yegorov, 1990). These authors test the hypothesis that this method can separate diffuse and specular reflectance. The principal results at the leaf scale and their consequences at the canopy scale are summarized in Vanderbilt et al. (1991). Kharuk and Yegorov (1990) experimentally found that polarization measured over the 1.75 to 2.2 μm wavelengths at the Brewster's angle changed as water stress developed in leaves, due to changing leaf angles. Pollution from industrial dust also decreases polarization, suggesting that plant stress can be monitored from polarization measurements. As these measurements are performed at a single viewing angle or a few angles, Sarto et al. (1989) developed a goniometer where all components are fixed to simultaneously measure the polarized reflectance of a leaf in many directions.

A series of articles published since the 1990s looked at the spectropolarization of plants with ornamental (Shuplyak et al., 1994; Dlugunovich et al., 2001a; Vanderbilt and Daughtry, 2012), agricultural (Shuplyak et al., 1997; Belyaev et al., 2000), or ecological (Krishnan and Nordine, 1994; Atrashevskii et al., 1998, 1999; Shashar et al., 1998; Dlugunovich et al., 2001b; Wu et al., 2013) interest. For instance, Atrashevskii et al. (1998) were interested in maple (*Acer platanoides*) leaves, a species common in the Minsk area, at various stages of development and senescence. More than 1000 distributions were measured between 1993 and 1995 under various ecological

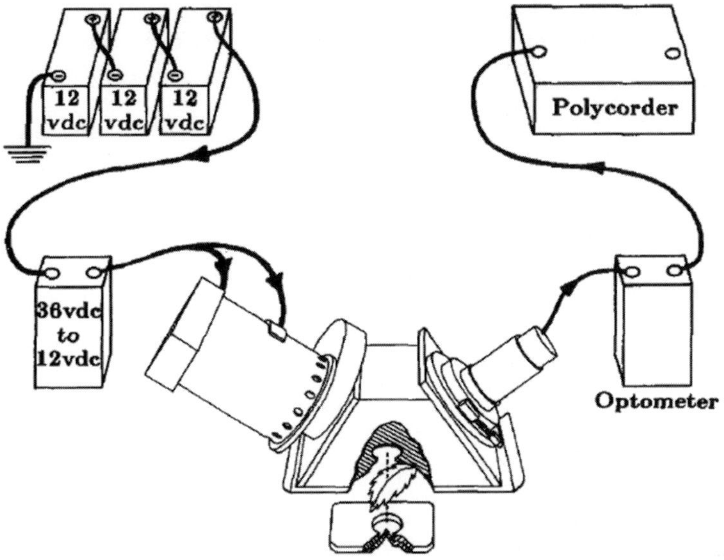

Figure 4.21 Portable photometer used to measure *in situ* polarization of leaves at angles of incidence and reflectance of 55° (from Grant et al. (1993), *Physiologia Plantarum*. Copyright © 1993 with permission from John Wiley & Sons).

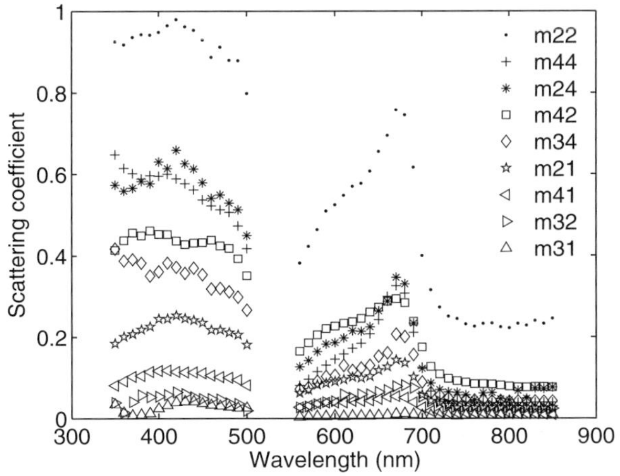

Figure 4.22 Scattering coefficients for thale cress (*Arabidopsis thaliana*) (redrawn from Martin et al. (2010)).

conditions. These authors established that the characteristics of polarization change with the leaf's physiological state. Savenkov et al. (2007) experimentally show that the epidermal cells control the polarization state of light entering the leaf, suggesting that they act as a filter of sunlight to a specific polarization state.

In the past decade, Raven et al. (2002) extended these measurements to the TIR with the intention of discriminating plant types in scenes of natural vegetation. They observe significant differences

Table 4.4 *Articles that present measurements of bidirectional leaf optical properties in polarized light. R (reflectance), T (transmittance), PP (principal plane), PAR (photosynthetically active radiation). Note that a review of specular and diffuse reflections is given in Grant (1987c).*

Reference	RT	θ_i [deg]	θ_r [deg]	φ [deg]	λ [nm]	Leaf species
Shul'gin and Khazanov (1961)	R	20, 45, 60	0→90	PP	PAR	*Camellia japonica, Cineraria hybrida, Gesneria cardinalis,Prunus laurocerasus*
Moldau (1965)	R	0, 30, 60	0→60	PP	495, 725	
Shul'gin and Moldau (1965)	RT	0, 45	0→90	PP	554, 671, 731	*Zea mays*
Egan (1970)	R	30	0→86	PP	350→1000	*Medicago sativa, Solanum tuberosum, Triticum aestivum, Zea mays*
Woolley (1971)	R	15, 45	15→80	PP	VIS	*Glycine max* sp., *Philodendron, Zea mays*
Vanderbilt and Grant (1986)	R	55	55	PP	5 bands	*Quercus prinus, Quercus rubra, Quercus velutina*
Grant et al. (1987a)	R	55	55	PP	5 bands	*Sorghum bicolor*
Grant et al. (1987b)	R	55	55	PP	5 bands	*Coleus blumei*
Woessner and Hapke (1987)	R	0, 60		PP	448, 554, 690, 820	*Trifolium repens*
Sarto et al. (1989)	R	0→82	0→82	0→360	VIS	*Magnolia* sp.
Ma et al. (1990)	RT	30	0→90	PP	632.8	*Prunus laurocerasus, Solanum tuberosum, Zea mays*
Yanovskaya et al. (1991)	R	5→80	0→90	PP	630	*Solanum tuberosum*
Grant et al. (1993)	R	55	55	PP	5 bands	~18 species
Brakke (1994)	RT	0, 55	0→90	PP	590, 725	*Acer rubrum, Quercus alba, Quercus rubra*
Krishnan and Nordine (1994)	R	40		PP	633	*Acer* sp., *Taxus* sp.
Shuplyak et al. (1994)	R	20→80	20→80	PP	400→800	*Kalanchoe daigremontiana*
Shuplyak et al. (1997)	RT	70	0→90	PP	400→850	*Solanum tuberosum*
Atrashevskii et al. (1998, 1999)	R	60, 70	0→90	PP	460, 550, 680, 740	*Acer platanoides*
Belyaev et al. (2000)	RT	70	0→90	PP	380→1080	*Solanum tuberosum*
Dlugunovich et al. (2001a)	R	0→85	5→80	PP	630	5 species
Dlugunovich et al. (2001b)	R	55	35→75	PP	630	*Rhododendron smirnovii*

Table 4.4 (*cont.*)

Reference	RT	θ_i [deg]	θ_r [deg]	φ [deg]	λ [nm]	Leaf species
Raven et al. (2002)	R	60	10→80	PP	300→2500 2 μm→15 μm	*Prunus laurocerasus, Vesbascum thapsus*
Savenkov et al. (2003)	RT	0→24	153→170	PP	630	*Quercus robur*
Goldstein and Cox (2004)	RT	0	18 angles	PP	700→2300	20 species
Savenkov et al. (2004)	RT	0	100→170	PP	630	*Triticum* sp.
Zhumar and Tsaryuk (2005)	R	0→30	40→70	PP	630	*Begonia sp. Hedera helix*
Savenkov et al. (2007)					630	*Chlorophytum* sp.
Voshchula et al. (2007)	R	40→70	40→70	PP	630	*Begonia* sp., *Hedera helix*
Martin et al. (2010)	R	45	135	PP	350→850	*Arabidopsis thaliana, Ficus benjamina, Quercus robur*
Xie et al. (2010)	R		0→60	PP		*Zea mays*
Vanderbilt et al. (2012)	R	56	56	PP	400→2400	*Cannabis sativa*
Wu et al. (2013)	R			PP	437→902	*Sedum spectabile*
Vanderbilt et al. (2014)	R	56	56	PP	400→2400	*Cannabis sativa, Coffea arabica, Ficus* sp., *Philodendron* sp., *Spathiphyllum* sp.

in polarimetric signatures between laurel (*Prunus laurocerasus*) leaves that are glossy and mullein (*Verbascum thapsus*) leaves that are highly pubescent, which was expected due to the high dissimilarity between these two species. Savenkov et al. (2003, 2004) went further in the analysis of polarized data by determining the complete Mueller matrix and anisotropy parameter for oak and wheat leaves. Physical interpretation of these data in terms of light scattering, both at the leaf surface and within the blade, remains a challenge (Savenkov and Muttiah, 2004a). The most recent studies particularly investigate the ellipticity e and the polarization azimuth ξ (Eq. 4.9) to infer information about the leaf cuticle, such as refractive index and surface roughness (Zhumar and Tsaryuk, 2005; Voshchula et al., 2007). Goldstein and Cox (2004) measured monostatic spectropolarimetric reflectance distribution functions of 20 leaf species from 700 to 2300 nm. They determined the Mueller matrix of these samples and showed that leaves are absorbing mirrors. Figure 4.22 displays the spectral variation of the Mueller matrix scattering coefficients of thale cress (*Arabidopsis thaliana*) leaves, which are covered with small hair-like filaments (trichomes) and can be considered as microscopically rough (Martin et al., 2010). The m_{21} values at wavelengths longer than 750 nm are a measure of surface roughness. Table 4.4 includes the main articles published in this field.

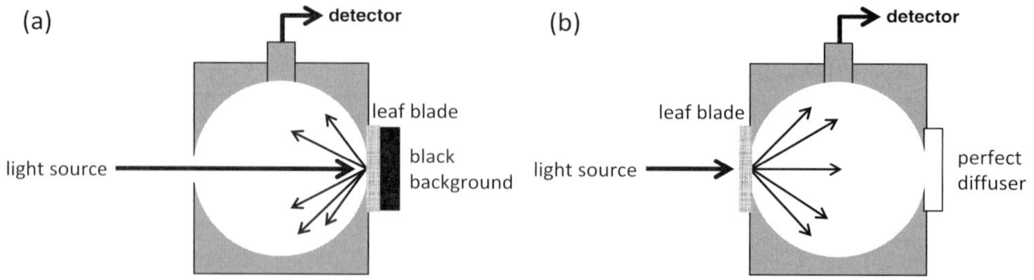

Figure 4.23 Typical measurement of the directional-hemispherical: (a) reflectance; and (b) transmittance.

4.5 Measurement of Leaf DHRF/DHTF

There is a long history of measuring the spectral reflectance and transmittance of plant leaves or organs in the laboratory or in the field using spectrophotometers equipped with an integrating sphere.

4.5.1 Broadleaf Samples: Non-Polarized Measurements

Directional-hemispherical reflectance $R(\lambda)$ and transmittance measurements $T(\lambda)$ are typically carried out using a conventional spectrophotometer equipped with an integrating sphere that averages the signal reflected or transmitted in all directions. Laboratory instruments generally consist of the following components: a) light source; b) monochromator; c) integrating sphere (also called Ulbricht sphere) coated on the inside with a highly reflecting layer over the required wavelength range; d) sample holder; and e) detector. In a single-beam instrument for instance, the radiance of a diffuse reference placed at the exit port of the sphere ($L_{ref,exit}$) is first measured. Then the radiance of a sample placed either at the entrance port ($L_{leaf,entrance}$) or at the exit port ($L_{leaf,exit}$) of the sphere is measured (Figure 4.23). In this case, a plug or a light trap (black cone or velveteen) that absorbs 100% of incident light is positioned just behind the sample. Finally, the specular component may be either naturally included in or rejected from total reflectance by switching between an 8° and a 0° sample holder, respectively. The latter forces part of the incident light to be reflected back to the entrance port.

To a first approximation, total reflectance and transmittance are calculated by

$$R_{leaf}(\lambda) = \frac{L_{leaf,exit}(\lambda)}{L_{ref,exit}(\lambda)} \tag{4.57}$$

$$T_{leaf}(\lambda) = \frac{L_{leaf,entrance}(\lambda)}{L_{ref,exit}(\lambda)} \tag{4.58}$$

and the total absorptance $A_{leaf}(\lambda)$ is easily derived using Eq. 4.29. As stressed by Myers and Allen (1968), the leaf sample must be placed in two separate positions on the integrating sphere to obtain both reflectance and transmittance measurements, which is not trivial. Indeed, apart from the fact that rearranging it is time-consuming, errors may be introduced if the sample is not spatially homogeneous or if the alignment with respect to the incident illumination changes. Carlson and Yarger (1971a) proposed a method for calculating leaf transmittance from leaf reflectance measurements (see Section 4.5.3). Pickering et al. (1992, 1993) got around this difficulty by using two adjacent integrating spheres: the sample is sandwiched between the exit port of one sphere and the entrance port of the other, which permits simultaneous determination of is reflectance and the transmittance. Some corrections that take into account the geometry of the sphere and the optical properties of the reference are generally required to calculate absolute values of reflectance and transmittance. Merzlyak et al. (2002, 2004) claimed that all the spectra published over the last 50 years and indicating an average absorptance of 5–20% in the 780–900 nm region were mistaken due to incomplete collection of the transmitted light by the detection system. As seen before, diffuse transmittance of a leaf is recorded by placing the leaf against the entrance port of an integrating sphere, a few millimeters away from its inner wall. The result is that a fraction of the transmitted light fails to strike the integrating sphere. These authors propose to apply a correction factor that allows one to compensate for that drift and that decreases the absorptance in this region. However, this work is still controversial. Since the reference is not a perfect diffuser, one should multiply Eq. 4.57 by its DHRF (same for the transmittance):

$$R_{leaf}(\lambda) = \frac{L_{leaf,exit}(\lambda)}{L_{ref,exit}(\lambda)} DHRF_{ref}(\lambda) \tag{4.59}$$

Diffuse reference standards exist in several forms. Ideally, one should use a perfect diffuser which is a theoretical perfectly white ($\rho = 1$) and Lambertian (same radiance from any viewing angle, see Appendix D) surface (Walter-Shea and Biehl, 1990; Palmer, 1997). Certain materials approach the ideal over a limited angular and wavelength range. Historically, magnesium oxide (MgO) was used in the visible spectrum. It was replaced first by barium sulfate ($BaSO_4$) and more recently by polytetrafluoroethylene (PTFE), the reflectance properties of which are provided by the manufacturer in the different wavelength regions of use. This fine, white powder, when pressed to a density of 1 g cm^{-3}, is close to ideal over a wide spectral range. Contrary to $BaSO_4$, PTFE is hydrophobic, thus it is washable and insensitive to humidity changes. However, it is not totally Lambertian, showing a fall of BRDF at angles far removed from the specular direction. Figure 4.24 shows the 8° hemispherical reflectance of Spectralon®, a white thermoplastic resin that is highly Lambertian and that reflects more than 99% of incident light from 400 nm to 1500 nm and more than 95% of incident light from 250 nm to 2500 nm. For that reason it is recommended for measurements in the UV-VIS-NIR.

In the TIR, Infragold® provides diffuse reflectance between 92% and 96% in the 2.5–15 μm range. It is very stable is used to calibrate spectrometers in the MIR-FIR (Figure 4.25).

The directional hemispherical reflectance (DHR) of a single leaf should also be corrected from the dark background on which the sample is superposed. There are different models of leaf-background light interactions (Miller et al., 1992; Noble and Crowe, 2001a). For a homogeneous leaf, the reflectance and transmittance of which are similar for both sides of the blade, multiple reflections between the leaf underside and the background produces the following measured apparent (or measured) reflectance (Figure 4.26):

$$\begin{aligned} R_{meas}(\lambda) &= R(\lambda) + T(\lambda)R_b(\lambda)T(\lambda) + T(\lambda)R_b(\lambda)R(\lambda)R_b(\lambda)T(\lambda) + \dots \\ &= R(\lambda) + T^2(\lambda)R_b(\lambda)\Big(1 + R(\lambda)R_b(\lambda) + \dots\Big) \\ &= R(\lambda) + \frac{T^2(\lambda)R_b(\lambda)}{1 - R(\lambda)R_b(\lambda)} \end{aligned} \tag{4.60}$$

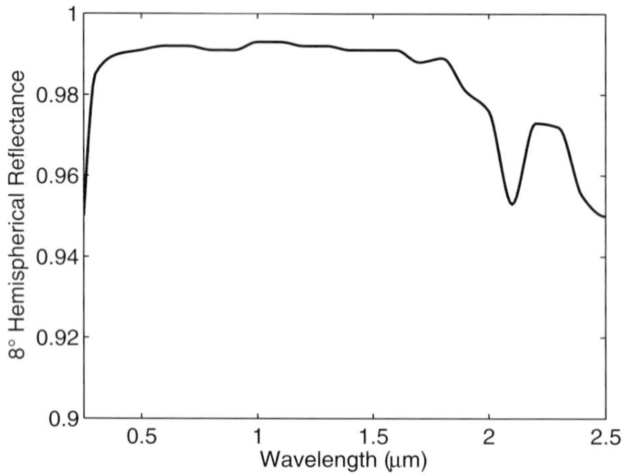

Figure 4.24 Typical 8° hemispherical reflectance of an optical-grade Spectralon® material (SRM-990) (from Labsphere, Inc).

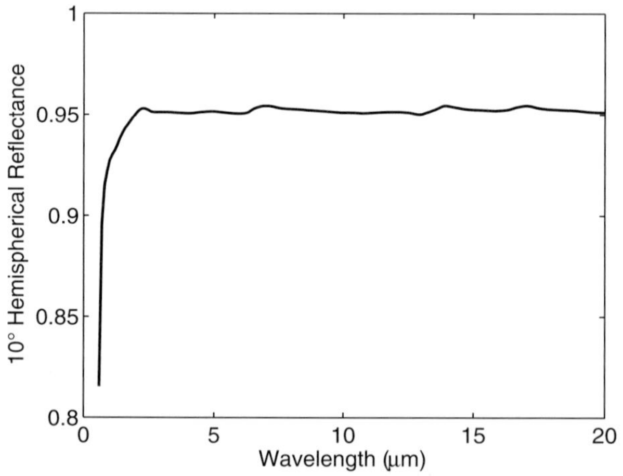

Figure 4.25 Typical 10° hemispherical reflectance of an Infragold® material (from Labsphere, Inc.).

where R_{meas} and T are the measured DHR and DHT, respectively, R_b is the DHR of the dark background, and R is the real leaf DHR. Inferring R from Eq. 4.60 is then straightforward.

A simplified approach is proposed by Lillesaeter (1982), who considers only the first two terms of Eq. 4.60, which becomes

$$R_{meas} = R(\lambda) + T^2(\lambda)R_b(\lambda) \tag{4.61}$$

In addition, the reflectance of optically thick samples $R_\infty(\lambda)$ can be measured by stacking leaves in order to magnify the radiometric signal and to minimize the leaf-to-leaf variability (e.g., Allen and Richardson, 1968; Gausman et al., 1976; Lillesaeter, 1982; Miller et al., 1992; Blackburn, 1999). It is

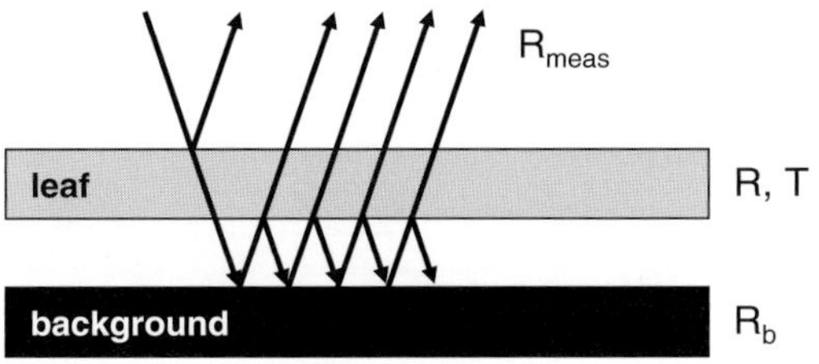

Figure 4.26 Model of leaf-background light interaction.

possible to theoretically link $R_\infty(\lambda)$ to $R(\lambda)$ and $T(\lambda)$. In the Stokes theory, the reflectance of N plates is given by

$$R_N = \frac{b^N - b^{-N}}{ab^N - a^{-1}b^{-N}} \tag{4.62}$$

where a and b depend on the reflectance and transmittance of one leaf (see Chapter 9 for details). By dividing Eq. 4.62 by b^N one obtains

$$R_N = \frac{1 - b^{-2N}}{a - a^{-1}b^{-2N}} \tag{4.63}$$

So that when the number of plates tends to infinity, Eq. 4.63 tends to

$$R_\infty = a^{-1} = \frac{2R}{R^2 - T^2 + 1 + \sqrt{\Delta}} \tag{4.64}$$

where $\Delta = (R + T + 1)(R + T - 1)(R - T + 1)(R - T - 1)$. In this situation, the reflectance of the background does not matter. Allen and Richardson (1968) found no further change in reflectance in the NIR-MWIR when mature cotton leaves had been stacked to a depth of eight or more. Saturation occurs much faster in the VIS.

In certain cases, like a remote field experiment, laboratory measurements are inconvenient. It is, however, possible to attach an integrating sphere interfaced to a portable spectrometer to produce laboratory-like spectrophotometer measurements. For instance, the portable spectroradiometer FieldSpec Pro (Analytical Spectral Devices, Inc., Boulder, CO) attached to the $BaSO_4$ LI-1800-12 external integrating sphere (LI-COR, Inc., Lincoln, NE) via a fiber optic enjoyed considerable popularity over recent years in a variety of field campaigns. The LI-COR 787 halogen lamp can be used to illuminate the target (Figure 4.27). After the LI-1800-12 was discontinued in 2003, integrating spheres manufactured by Labsphere or other companies are now used.

Vanderbilt et al. (1987) designed a lightweight field portable transmissometer that measures the hemispherical transmittance of a leaf sample and could be adapted to various commercial spectrometers. This design could provide a replacement for the LI-1800-12 integrating sphere. Nostell et al. (1999) designed a single beam instrument that rotates the light and can measure reflectance and transmittance from separate integrating spheres that can also be measured when varying the angle of incidence.

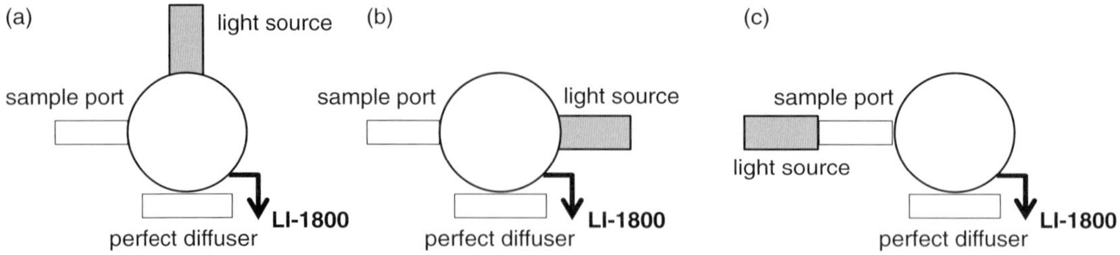

Figure 4.27 The LI-1800 integrating sphere the configured in: (a) the reference mode; (b) the reflectance mode; and (c) the transmittance mode (redrawn from Daughtry et al. (1989)).

4.5.2 Narrow-Leaf Samples

When the diameter of the entrance (exit) port of the integrating sphere is larger than the sample, or when the illuminating beam has a larger cross-sectional area, the measurement of its optical properties constitutes an experimental challenge (Noble and Crowe, 2007; Jernshøj and Hassing, 2009; Yáñez-Rausell et al., 2014a). For example, the diameter of the sample port of the integrating sphere of the LI-COR LI-1800 spectroradiometer is 14.5 mm with an illumination beam diameter of 11.4 mm. Conifer needles, grasses, or small leaves in general (Ramsey and Rangoonwala, 2004), the length and width of which may be smaller than these dimensions, are of concern in the first place. Some spectrophotometers, especially in the TIR, allow samples to be positioned horizontally, and then the infinite reflectance of an optically thick sample is measured (Dawson et al., 1998a, 1998b). In most standard instruments, however, the sample must be maintained in vertical alignment. A technique consists in filling a glass cuvette with leaves and measuring R_∞, but additional corrections are required to account for multiple reflections between the sample and the glass wall and it prohibits measurements of transmittance. An alternative technique consists of arranging the leaves side-by-side into a single layer and fixing them using a transparent tape (Ahern, 1988). If a few gaps may be acceptable for reflectance measurements, this is certainly not the case for transmittance measurements. Moreover, the time required to prepare the sample for measurement can result in some physiological degradation, for example, pigment and water loss that would lead to erroneous interpretation of optical properties. Daughtry et al. (1989) propose a protocol that enables the measurement of conifer needles that are longer than the diameter of the sample port of the integrating sphere. They measure the reflectance and the transmittance of a mat of evenly spaced needles, with gaps between needles approximately one needle width apart; then the sample is removed from the port of the integrating sphere and the needles are coated with an opaque flat black paint; the flux at 680 nm transmitted through the blackened sample allows indirect determination of the gap fraction (the ratio of gap area to total sample area). However, Middleton et al. (1996) and Mesarch et al. (1999) showed that this method led to large underestimates of transmittance values, especially in the VIS, and improved it by evaluating the influence of gaps between sample elements using an image-analysis system. The equations for calculating reflectance R and transmittance T of individual needle are given as

$$R = \frac{1}{1 - GF} \times \frac{I^R_{sample} - I_{empty}}{I_{ref} - I_{empty}} \tag{4.65}$$

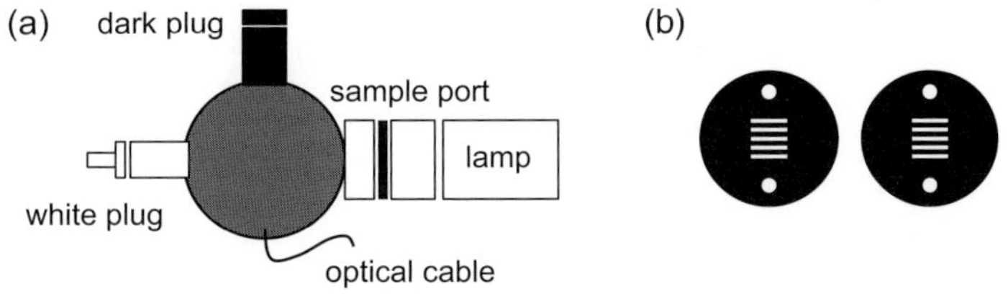

Figure 4.28 Device to measure needle optical properties. (a) LI-COR 1800 integrating sphere coupled with an ASD field spectrometer. Needles are mounted on (b) black anodized carriers and placed in the integrating sphere port to obtain reflectance and transmittance (redrawn from Moorthy et al. (2008)).

$$T = \frac{1}{1 - GF} \times \left(\frac{I^T_{sample}}{I_{ref} - I_{empty}} - GF \times R_{sphere} \right) \tag{4.66}$$

where GF is the gap fraction of the sample, I^R_{sample} and I^T_{sample} are the radiation reflected and transmitted by the sample, I_{ref} is the reference radiation, I_{empty} is the stray light radiation (radiation flux of the empty illuminated sample port in reflectance mode), and R_{sphere} is the reflectance of sphere walls. In his PhD thesis, Harron (2000) designed a black anodized carrier specifically for Jack Pine needles and that eliminates the need for gap fraction estimation (Figure 4.28).

Equation 4.65 and Eq. 4.66 have been adapted to take into account the optical effects of the sample holder and multiple scattering between the measured needles (Malenovsky et al., 2006; Noble and Crowe, 2007; Moorthy et al., 2008; Lukeš et al., 2013; Noda et al., 2013) and to calculate the uncertainties associated with the measurements (Yáñez-Rausell et al., 2014b). For instance, the latter used an optically stable artificial material simulating needle leaves. Some measurements of conifer needles that are shorter than the diameter of the sample port of the integrating sphere (Case 3 in Daughtry et al., 1989) were performed by Williams (1991). The optical properties of needles are paradoxically still obscure, as illustrated by the few studies reported in the recent literature. However, more emphasis on these measurements is needed to resolve these problems because conifers represent a significant fraction of forest ecosystems, providing important ecosystem services including carbon sequestration.

Because performing spectral measurements of needles is difficult for all the reasons set out above, spectral scattering properties of needles, shoots, or branches have been performed (e.g., Rock et al., 1994; Ross et al., 1994; Kaasalainen and Rautiainen, 2007; Thérézien et al., 2007; Masaitis et al., 2013; Yang et al., 2016a; Rautiainen et al., 2018).

4.5.3 Special Case Where only Reflectance Can Be Measured

In the absence of an adapted instrument, a field spectroradiometer can be used to derive both leaf bidirectional reflectance and transmittance by measuring only reflectance alternatively with a black background and a white background. Illumination is provided by one or more lamps carefully positioned around the spectrometer head. For instance, applying Eq. 4.61 over the two contrasting backgrounds of reflectance R_b and R_w, respectively (Lillesaeter, 1982):

$$R^b_{meas} = R(\lambda) + T^2(\lambda)R_b(\lambda) \quad (4.67)$$

$$R^w_{meas} = R(\lambda) + T^2(\lambda)R_w(\lambda) \quad (4.68)$$

A simple transformation gives

$$T^2(\lambda) = \frac{R^w_{meas}(\lambda) - R^b_{meas}(\lambda)}{R_w(\lambda) - R_b(\lambda)}, \quad (4.69)$$

and

$$R(\lambda) = R^b_{meas} - T^2(\lambda)R_b(\lambda) \text{ or } R(\lambda) = R^w_{meas} - T^2(\lambda)R_w(\lambda) \quad (4.70)$$

However, Miller et al. (1992) show that the derived reflectance and transmittance spectra of a single leaf calculated this way are unrealistic because $R(\lambda) + T(\lambda) > 1$ in the NIR region. They find much better results applying the same principle to Eq. 4.69 that includes the multiple reflectances between the leaf underside and the background, and refine these methods for dorsiventral leaves. Major et al. (1993) obtained similar results. This method provides a stopgap solution for field studies, but is strongly dependent on the measurement conditions (Noble and Crowe, 2001a).

4.5.4 *Measurement of Ground and Dried Leaves*

Near-infrared spectroscopy (NIRS) is a rapid and non-invasive analytical technique that is widely used to quantitatively analyze powdered samples. Norris et al. (1976) demonstrated for the first time that forage samples could be analyzed in a few minutes with such technology. Leaves of different species, sometimes mixed with stems or other plant parts, are generally oven-dried for several days at 70°C, ground to a fine powder in a blade grinder, and then oven-dried again before analyzes are conducted. The powder is placed in small aluminum cups for measurement of the DHRF in an upright position (Figure 4.29).

For the sake of simplicity, the DHRF of such a sample is noted R_∞ for infinite reflectance: the same amount of light is reflected whatever the background reflectance $R_b = 0$ or $R_b = 1$. As seen in Appendix F, the absorbance (optical depth) of a semi-transparent sample is given by $\log_{10} \frac{1}{T}$, the base 10 logarithm of $\frac{1}{T}$ according to Beer–Lambert's law. It is proportional to the elemental concentrations in the sample. When the sample is opaque, like in NIR spectroscopy, a pseudo-absorbance expressed as $\log_{10} \frac{1}{R_\infty}$ is often calculated by analogy with transmission spectra because this gives a curve comparable to an absorption curve with peak recordings occurring at wavelengths corresponding to absorption bands in the sample (Norris et al., 1976). However, note that there is no theoretical reason for this logarithmic form and that the pseudo-absorbance only gives an approximation of the absorbance (Griffiths and Dahm, 2008). In practice, although $\log_{10} \frac{1}{R_\infty}$ should not linearly vary as a function of the concentrations c of the absorber, many empirical equations of the form $c \propto \log 1/R_\infty$ have been implemented for various applications.

A much more satisfying treatment has been advanced by Kubelka and Munk (1931). In the case of an infinitely thick medium, the K–M theory leads to a particularly simple solution for the reflectance:

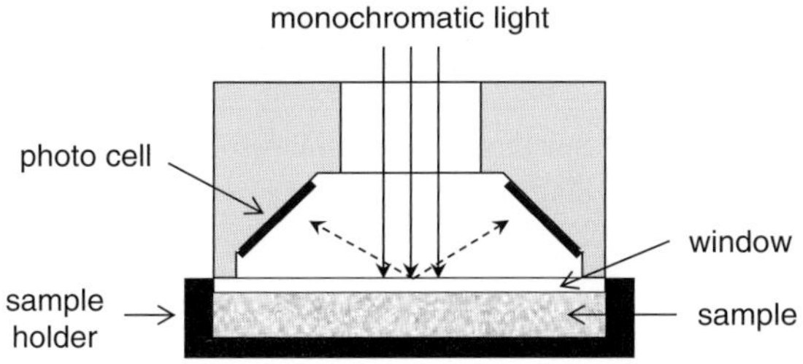

Figure 4.29 Optics for reflectance measurement. Four lead sulfide photo cells surround the sample; the one in the front and the one at the back are not shown (redrawn from Norris et al. (1976)).

$$R_\infty = \frac{1 - \left(K/(K+2S)\right)^{1/2}}{1 + \left(K/(K+2S)\right)^{1/2}} \tag{4.71}$$

with K (m^{-1}) and S (m^{-1}) the absorption and scattering coefficients of the medium, respectively. This relationship may be rewritten in the alternative form:

$$\frac{K}{S} = \frac{(1 - R_\infty)^2}{2R_\infty} = f(R_\infty) \tag{4.72}$$

Since the spectral dependence of S is small compared to K over a limited wavelength range, a plot of $f(R_\infty)$, also known as the K–M function, versus wavelength should display the same features as the absorption spectrum (Birth and Hetch, 1987):

$$c \propto f(R_\infty) \tag{4.73}$$

4.6 Portable Photometers and Other Probes

Laboratory spectroradiometers, equipped with an integrating sphere, can measure leaf directional-hemispherical optical properties in several contiguous wavebands to determine the chlorophyll or water content by using spectral indices (see Chapter 9). However, besides their price, such instruments are often difficult to use in the field. A wide variety of portable spectroradiometers have been used by investigators for outdoor applications (Table 4.5). Geophysical and Environmental Research Corporation (GER) pioneered the science of field spectroscopy 30 years ago, followed by Analytical Spectral Devices (ASD Inc., Boulder, CO) with the widespread FieldSpec®, a high resolution spectroradiometer designed for fast spectral measurements. It actually consists of three separate instruments that cover the 350–975 nm, 976–1770 nm, and 1771–2500 nm wavelength ranges. Ocean Optics, Inc. (Dunedin, FL) marketed the SpectroClip-JAZ-TR, a fully integrated system that measured diffuse reflectance and transmittance of plant leaves from 350 to 1000 nm. The two-channel spectrometer has originally been built for chlorophyll analysis. The UniSpec-SC spectral analysis system, commonly referred to as a leaf reflectometer, has been designed by PP Systems (Amesbury, MA) to

Table 4.5 *Characteristics of commercialized spectroradiometers used to measure leaf optical properties. FWHM = full width at half maximum of an emission line.*

Company	Instrument	Spectral range (nm)	Spectral resolution (FWHM)	Sampling interval
ASD	FieldSpec Pro FR	350–2500	3 nm @ 700 nm, 10 nm @ 1400 and 2100 nm	1.4 nm (350–1050 nm), 2 nm (1000–2500 nm)
	FieldSpec HandHeld	325–1075	3.5 nm @ 700 nm	1.6 nm
Li-COR	LI-1800.22	300–1100		
Ocean Optics	USB2000+	200–1100	~0.1–10.0 nm	
	SpectroClip-JAZ-TR	350–1000	1.3 nm	
PP Systems	UniSpec-SC Spectral Analysis System	350–1100	< 10 nm	3.3 nm
Spectra Vista Corporation	GER 1500	350–1050	3.2 nm	1.5 nm
	HR-768i	350–2500	3.5 nm @ 700 nm, 16 nm @ 1500 nm, 14 nm @ 2100 nm	1.5 nm (350–1000 nm), 7.6 nm (1000–1890 nm), 5 nm)1890–2500 nm)

measure reflectance of individual leaves for ecosystem remote sensing applications. It is a self-contained system operating in the 350–1100 nm wavelength range, which that does not require a separate PC for operation.

As illustrated by Figure 4.30, three scenarios are possible:

1) The fiber optic of the instrument is inserted into the port of an external integrating sphere, for instance the LI-1800-12 or the ASD RTS-3ZC, with an internal halogen light source to measure diffuse reflectance or transmittance. This case has been previously studied in Section 4.5.1.
2) The fiber optic of the instrument is interfaced with special accessories (*Plant Probe* and *Leaf Clip*), with an internal halogen light source. A two-sided, rotating head containing an embedded background panel allows performing both reflectance (black panel face, < 5% reflectance) and transmittance (Spectralon ® white panel face, > 99% reflectance in the VIS-NIR) measurements. Using such an accessory reduces external disturbances such as the illumination angle or unstable light conditions.
3) The fiber optic of the instrument is connected to a gun (*Pistol Grip*) that looks at the leaf vertically, for instance in the laboratory when the source of illumination is a lamp, or in any direction, for instance in the field when the source of illumination is the Sun. The bidirectional reflectance is then measured.

Calibration of these data may be a tough task, which most of the time is not performed correctly. There are very few comparative studies in which reflectance of the same leaves is recorded using different systems or spectrometers, each with a unique spectral resolution and sampling interval. Castro-Esau et al. (2006a) performed an instrument comparison and standardization of sampling procedures in the frame of the Spectral Network (SpecNet) project. According to the instrument,

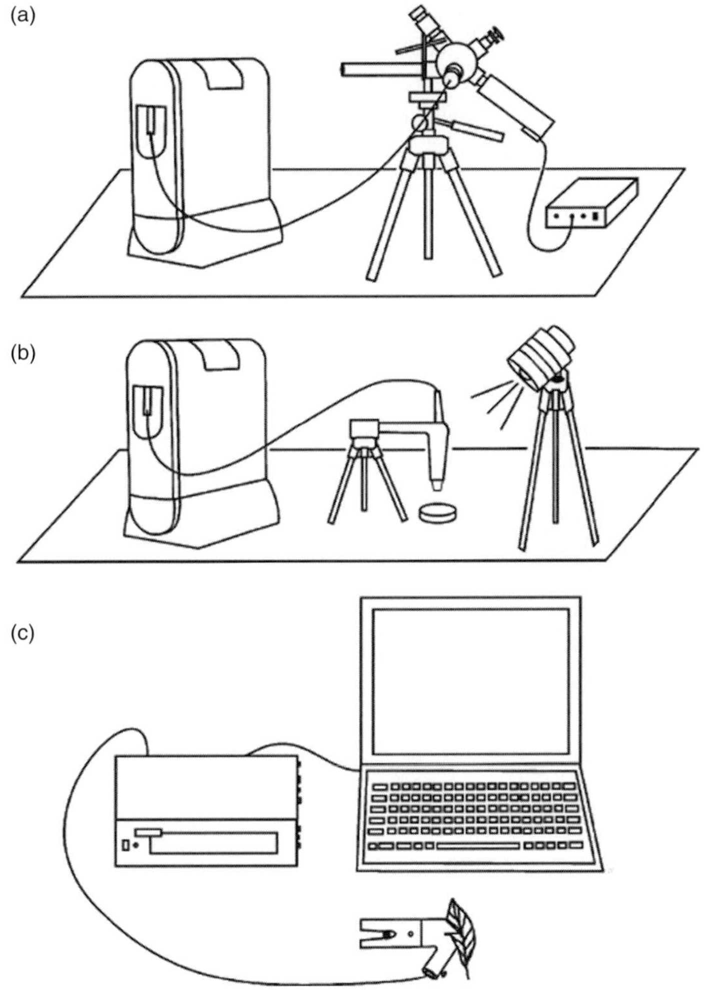

Figure 4.30 Illustration of spectrophotometer set-up: (a) the fiber optic is inserted into the port of an integrating sphere; (b) the fiber optic is connected to a gun; and (c) the fiber-optic is inserted into a leaf clip (from Castro-Esau et al. (2006a), *Remote Sensing of Environment*. Copyright © 2006 with permission from Elsevier).

they observed minor to major shape and amplitude differences between spectra of the same leaves. For some applications, such differences may not be important on a practical level. With the exception of PRI, the spectral indices calculated were similar. For other applications, like the use of physical models, such differences may be critical. Thus, inverting the PROSPECT model on measurement performed using the ASD gun or leaf clip may lead to biased estimations of leaf chemical constituents.

Handheld, battery operated photometers are an easy way to do a rapid, non-destructive determination of the chemical content of intact leaves. The first instruments sensitive to chlorophyll were designed in the 1950s (e.g., Weybrew and Green, 1952; Benedict and Swidler, 1961; Inada, 1963, 1965; Hardwick and Baker, 1973; Wallihan, 1973; Macnicol et al., 1976; Hardacre et al., 1984). Figure 4.31 displays three "historical" chlorophyllmeters. The first two (Yellowmeter and Portable Reflectance Meter) are based on reflectance measurement, the third (Chlorophyllo meter) on transmittance measurements.

To date, the most popular is the SPAD-502 (*Soil Plant Analysis Development*, Konica Minolta, Tokyo, Japan) that has given rise to an abundant literature over the last two decades (Yadava, 1986; Schaper and Chacko, 1991; Markwell et al., 1995; Manetas et al., 1998; Richardson et al., 2002; Yuzhu et al., 2011, among others). The principle on which it works is as follows: the leaf surface is illuminated by two LEDs (light emitting diodes) at 650 nm and 950 nm and the transmitted fraction of light is measured by a Silicon photodiode. The two transmittances are then combined to produce a spectral index, the so-called SPAD unit, which is related to the chlorophyll concentration using a calibration relationship provided by the manufacturer (Figure 4.32).

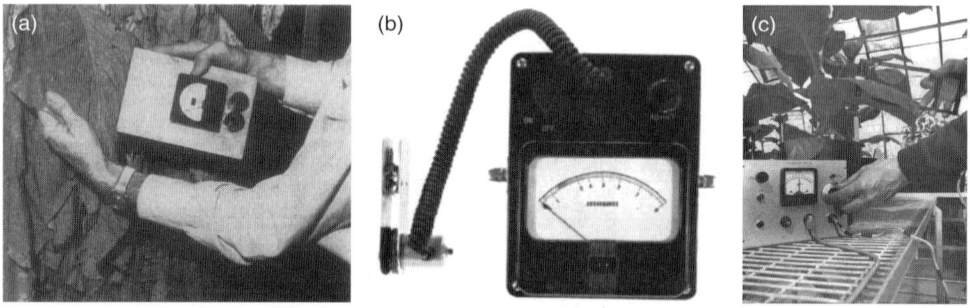

Figure 4.31 Three early photometers designed to measure leaf chlorophyll content: (a) The Yellowmeter (from Weybrew and Green (1952), *Science*. Copyright © 1952 with permission from American Association for the Advancement of Science); (b) The Portable Reflectance Meter (from Wallihan (1973), *Agronomy Journal*. Copyright © 1973 with permission from Oxford University Press); and (c) The Chlorophyllometer (from Macnicol et al. (1976), *Annals of Botany*. Copyright © 1976 with permission from Oxford University Press).

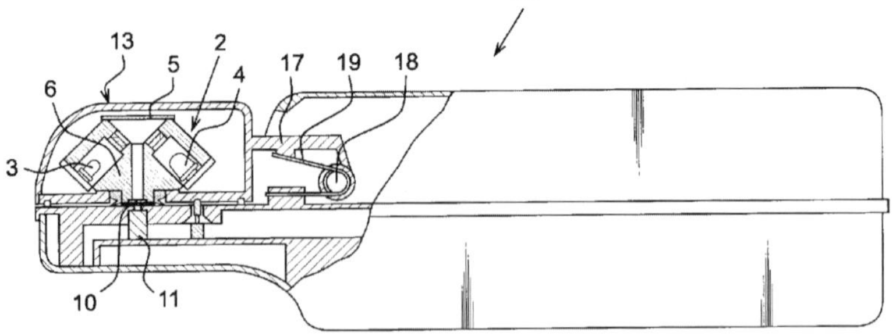

Figure 4.32 Diagrammatic sectional view showing the principle of a chlorophyllmeter apparatus based on transmittance measurements (extracted from the US Patent 5,859,435 "Content measuring apparatus for plant leaf" by Satake et al. (1997)).

Table 4.6 *Characteristics of commercialized photometer used to determine leaf chemical content. R = reflectance, T = transmittance*

Company	Instrument	Area (cm^2)	Wavebands (nm)	Principle
Adetel Group	Hydroleme	n/a	n/a	T
Force-A	Dualex Scientific+	1.88	UV-A, red, NIR	T
FT Green LLC	atLEAF+	n/a	660, 940	T
Hansatech Instruments	CL-01	n/a	620, 940	T
Konica Minolta	SPAD 502	0.06	650, 950	T
Opti-Sciences	CCM-200	0.71	653, 931	T
Photon Systems	N-Pen N 100	variable	565, 760	R
Instruments	PlantPen NDVI 200	variable	660, 740	R
	PlantPen PRI 200	variable	531, 570	R
Spectrum Technologies	CM-100	variable	700, 840	R
	FieldScout CM-1000	variable	700, 840	R
Trimble	GreenSeeker-505	variable	red, NIR	R
Yara International ASA	Hydro N-Tester	n/a	red, NIR	T
Zhejiang Bethlehem Aparatus	TYS-A	n/a	n/a	n/a

Other commercial instruments, like the CCM-200 (Chlorophyll Content Meter, Opti-Sciences, Inc., Hudson, NH), the Dualex Scientific+ (FORCE-A, Orsay, France), the CL-01 (Hansatech Instruments, Norfolk, UK), the Hydro N-Tester (Yara International ASA, Oslo, Norway), or the atLEAF+ (FT Green LLC, Wilmington, DE), are based on the same principle, therefore their readings are highly correlated to those of the SPAD-502 (Table 4.6). Besides chlorophyll, the Dualex Scientific+ measures leaf flavonols and anthocyanins. Instead of the transmittance, the FieldScout CM-1000 (Chlorophyll Meter, Spectrum Technologies, Inc., Aurora, IL) measures the reflectance at 700 nm and 840 nm offering the advantages of increased speed and reduced effort to make each reading: at 30.5 cm, it integrates a 1.35 cm diameter area on a plant leaf. The N-Pen N 100 (Nitrogen), PlantPen PRI 200 (Photochemical Reflectance Index), and PlantPen NDVI 200 (Normalized Difference Vegetation Index) are three other recent reflectance-based devices developed by Photon Systems Instruments (Drasov, Czech Republic) that measure the nitrogen and chlorophyll content of a leaf sample by calculating spectral indices. Similarly, the GreenSeeker-505 (Trimble Navigation, Ltd., Sunnyvale, CA) measures reflectance at two bands in the red and NIR and the configurable Crop Circle ACS-470 (Holland Scientific, Inc., Lincoln, NE) measures three user selected bands offering rapid assessment of crop condition. Except for the CCM-200 that calculates a Chlorophyll Concentration Index (CCI), all instruments deliver values that have no units but that are theoretically proportional (linearly or not) to the amount of chlorophyll in the leaf through a calibration curve. Similar instruments have been developed in laboratories for pigment determination but are not marketed (Han et al., 2009; Sumriddetchkajorn and Intaravanne, 2014).

As for water, techniques based upon the observation of leaf temperature kinetics following changes in leaf energy balance (Buriol et al., 1984a, 1984b; de Parceveaux et al., 1995) or upon the transmission measurement of terahertz radiation (Hadjiloucas et al., 1999) are available to measure leaf water state. Contrary to chlorophyll, there no is commercial portable device devoted to the measurement of water content using optical methods, although there are several ongoing studies. For example, the RAMIS (RAdiomètre portatif de Mesure In Situ) portable field radiometer determines total chlorophyll content, leaf equivalent water thickness, and leaf mass per area by measuring the leaf transmittance at five wavelengths in the VIS-NIR-MIR and inverting a neural network or the PROSPECT leaf optical properties model (Pavan et al., 2004). The latest version of RAMIS is based on reflectance measurements and has been optimized to determine leaf water and dry matter content (Conejo et al., 2010). An optoelectronic device called HYDROLEME for measuring the water content in a plant element gave rise to a recent patent (Galtie and Lescure, 2009). The main application for the HYDEOLEME is monitoring and assessment of fire risk.

Instruments based on transmittance measurements seem to provide better results than instruments based on reflectance measurements because the leaf blade is clamped, which creates a closed chamber around the area to be measured. Therefore they are not affected by sunlight conditions in the field. However, they measure only one spot of one leaf for each measurement, so many measurements must be taken to get a reliable average.

4.7 Measurement of Leaf Absorption Profiles

In most publications, the photons that pass through a plant leaf are counted as total transmittance. Unfortunately, transmittance does not provide any information about light gradients within the mesophyll, although downwelling and backscattered fluxes are both important in photomorphogenesis and photosynthesis. Measurements of light quality and quantity at different depths can be made using a fiber optic microprobe that can be constructed easily from commercially available materials (Vogelmann and Björn, 1984; Vogelmann et al., 1988, 1989, 1991). As illustrated by Figure 4.33, a leaf sample of a few square centimeters is mounted between two aligned Plexiglas slides perforated with a 5 mm diameter opening in the center of each slide. On one side of the entrance there is a monochromatic (laser) or white (halogen or xenon-arc lamp) collimated light source, so that its adaxial surface is illuminated perpendicular to the leaf blade. On the other side is a coaxial fiber-optic microprobe, which is cut by a diamond knife, threaded through the eye of a needle, and firmly glued onto it: the microprobe is then slowly advanced through a vein-free region of the leaf, with a high-resolution stepping motor, to penetrate from the abaxial side toward the irradiated surface.

Light that falls within the acceptance cone of the fiber is easily guided through the central core of the fiber via internal reflection at the core-cladding interface caused by different refractive indices, whereas that that falls outside of the acceptance cone is lost in the cladding (Figure 4.34).

The acceptance cone is bounded by a critical acceptance angle θ_a from a line parallel to the central axis of the fiber:

$$\theta_a = \arcsin \frac{\sqrt{n_1^2 - n_2^2}}{n_0} \tag{4.74}$$

where n_1 is the refractive index of the fiber core, n_2 is that of the surrounding cladding material, and n_0 is that of the surrounding medium, respectively. In most optical fibers, the refractive index of the core is slightly higher than that of the cladding. Note that the acceptance angle varies with the refractive index of the medium in which the fiber is used. Since the microprobe in plant tissues encounters both liquid and gaseous phases, it can be calculated both for water and air. Moreover, such a microprobe captures more light that strikes it normally (0°) than at more oblique directions. The relationship

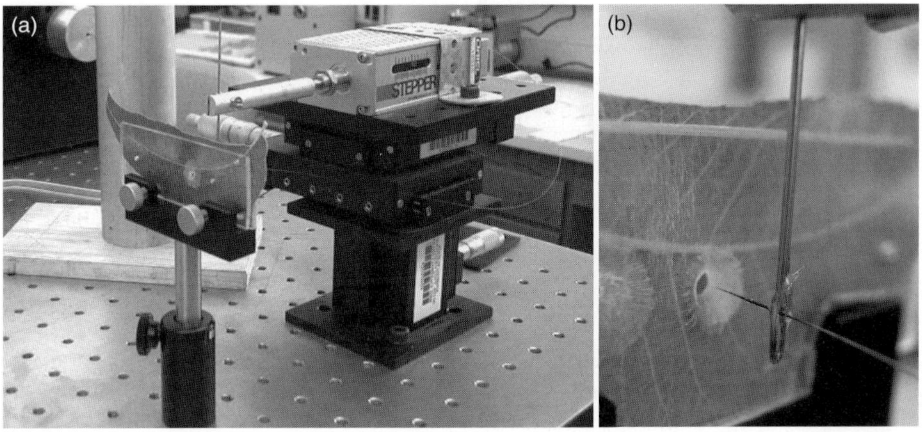

Figure 4.33 (a) Typical fiber-optic micro-sensor used in environmental studies. (b) Mounting the leaf sample (from Qi et al. (2003b). Copyright © 2003 with permission from SPIE).

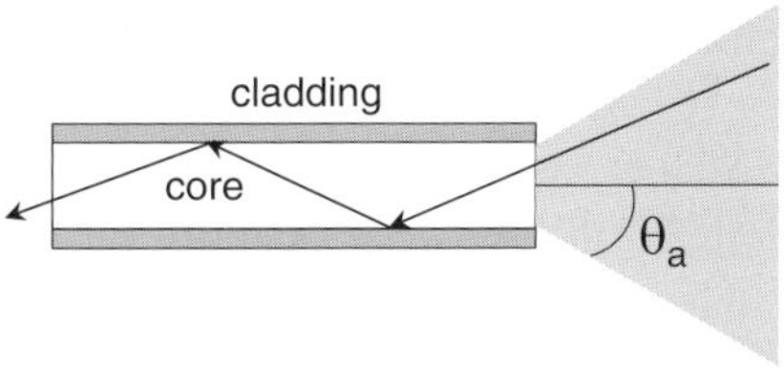

Figure 4.34 Diagram showing how light is guided through a metal-coated optical fiber: (1) acceptance angle, (2–4) radiation introduced at an angle smaller than or equal to the acceptance angle is transmitted, (5–6) radiation introduced at an angle greater than the acceptance angle is not transmitted (redrawn from Pilarski and Rajba (2004)).

between the angle at which light intersects the fiber and the amount that is captured by the fiber is called the acceptance function, which is generally a Gaussian function:

$$S(\theta) = S(0)\exp\left(-\theta^2/2\sigma^2\right) \tag{4.75}$$

Alternatively, to facilitate penetration into tissue, the fiber optic may be inserted through and fixed in a thin-walled hypodermic needle (Figure 4.35).

The light is guided to the opposite end of the fiber to a calibrated spectroradiometer interfaced with a computer. According to the position of the fiber-optic microprobe, one can measure the amount of light that travels through the leaf in a forward or a backward direction. A measurement in a single direction is generally sufficient to measure directional light. As illustrated by Figure 4.36, measurements of light at 0 and 30° give estimates of the amount of light that travels through the leaf in a forward direction, whereas measurement at 150° gives an estimate of the back-scattered light.

Since light within the leaf is not absolutely isotropic, Vogelmann and Björn (1984) recommend performing at least six measurements in different representative directions to measure diffuse light. If we set aside veins, plant leaves exhibit no left or right sides, they only have a top and a bottom surface. Assuming that leaf anatomy is radially symmetrical, the distribution of light has similar

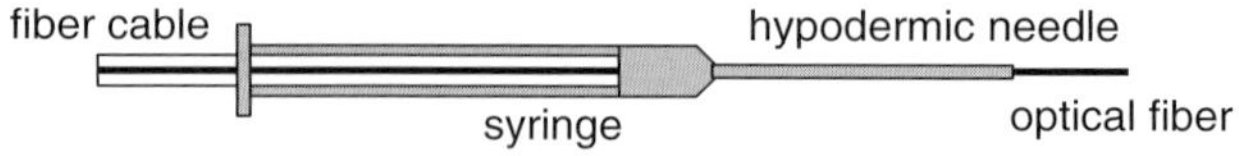

Figure 4.35 Schematic drawing of how single-strand fibers can be fixed in a syringe that can be mounted on a micromanipulator (redrawn from Leadbetter (2005)).

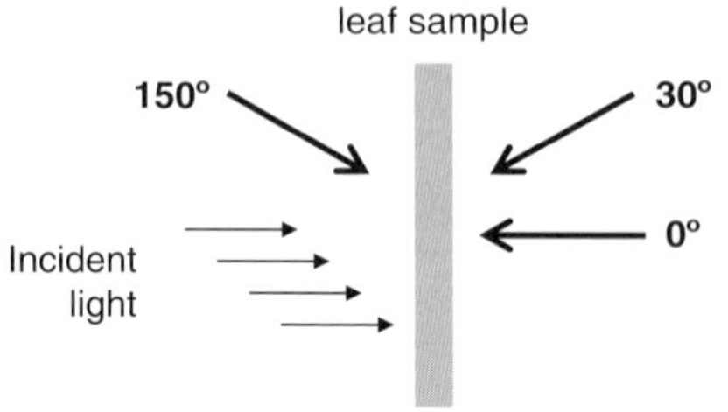

Figure 4.36 Sampling orientations of the fiber-optic microprobe. The adaxial surface of the leaf is positioned so that it faces the light source.

properties. Such measurements are somewhat difficult to carry out experimentally and have given rise to the beginnings of a controversy (Chwirot and Slevin, 1986; Björn, 1986). Note that multiplying the number of angles may destroy leaf tissues and alter light transmission.

Irradiance is the adequate definition for the physical quantity measured by the fiber in one direction. Spherical irradiance (or radiant energy fluence rate), defined as the total radiant flux incident from all directions onto a small sphere divided by the cross-sectional area of that sphere, is the common term used when measuring diffuse light (see Section 4.1.3). SI unit of both terms is W m^{-2}.

4.8 Measurement of Leaf Surface Temperature

Comprehensive literature reviews of leaf temperature can be found in Jackson (1982). In the past 50 years, infrared thermometers have been widely used to measure leaf surface temperature (e.g., Fuchs and Tanner, 1966; Gates, 1962; Gates et al., 1968; Idso et al., 1969; Takiuchi and Hashimoto, 1977). This is a non-contact method that consists in relating the radiation emitted from the leaf M to the surface temperature T_s by the Stefan–Boltzmann law:

$$M = \varepsilon \, \sigma \, T_s^4 \tag{4.76}$$

where ε is the emissivity of the surface and σ is the Stefan–Boltzmann constant (5.67×10^{-8} W m^{-2} K^{-4}). Most thermometers have bandpass filters that allow measurement of energy within the 8–14 μm waveband, which includes the peak of blackbody emission at normal temperatures. Calibration of such instruments can be achieved, for instance by pointing them toward a Leslie cube whose temperatures are measured with a precision thermometer. For the determination of leaf temperatures, infrared thermometers are held within a few millimeters of the surface. They can provide an absolute temperature reading for a blackbody surface with errors in the range of 0.1 to 0.3 K but for precise measurements the emissivity must be either known or determined. The emissivity of a single leaf ranges between 0.95 and 1.00, with most plant leaves being between

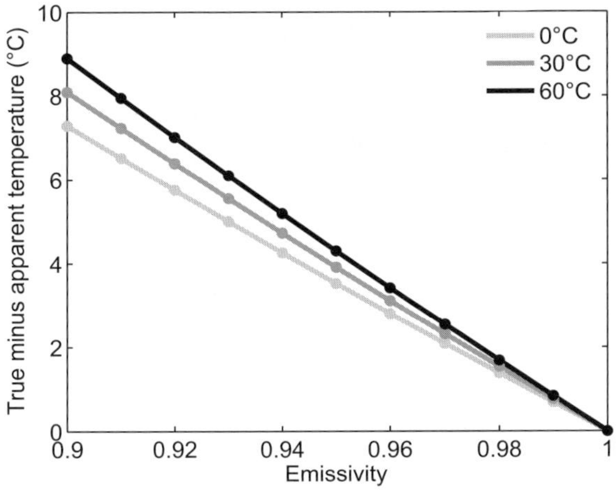

Figure 4.37 The maximum increment to be added to the apparent temperature T obtained by assuming a unit emissivity, to obtain the actual surface temperature T_s as a function of emissivity for two values of the apparent temperature (redrawn from Jackson (1982)).

0.97 and 0.98. Note that it is not representative of the emissivity of a plant canopy because of the multiple internal reflections resulting from the canopy architecture. Assuming that the leaf emissivity is $\varepsilon = 0.97$, Figure 4.37 shows that for an apparent temperature of about 30°C, the maximum error involved by assuming $\varepsilon = 1$ would be about 2°C.

Point measurements are useful only when the temperature distribution over an object is sufficiently known. This is barely the case of a plant leaf. Hashimoto et al. (1984) pioneered the use of a thermographic camera to visualize differences in surface temperature distribution over sunflower (*Helianthus annuus*) leaves. Computer software transforms emitted infrared radiation into thermal images in which temperature levels are indicated by a false-color gradient. Such images are very useful to characterize leaf water stress (Hashimoto et al., 1984; Cohen et al., 2005), UV irradiation (Umezaki and Shimadaa, 1999), transpiration rate and water flow velocity (Garbe et al., 2002), stomatal conductance (Jones, 1999), or plant-virus interactions (Chaerle et al., 1999, 2001; Chaerle and van der Straeten, 2000). Note that leaf surface temperature obtained by an infrared thermometer or a thermographic camera may be calibrated using a thermocouple.

4.9 Measurement of Leaf Electrical Properties

In the 1980s, the development of radar remote sensing, in particular applied to vegetation studies, prompted scientists to develop new techniques for measuring the microwave dielectric constant ϵ of plant leaves. The University of Michigan, Ann Arbor, pioneered such experimental measurements (e.g., Ulaby and Jedlicka, 1984; El-Rayes and Ulaby, 1987; Sarabandi and Ulaby, 1988) in parallel with other teams (e.g., Tan, 1981; Hill et al., 1986). Reviews can be found in Chukhlantsev (2006) and Chung (2007). Many techniques have been investigated to determine the real part (ϵ') and imaginary part (ϵ'') of the complex dielectric constant of plant leaves *in vivo* over the desired frequency band: one can measure the magnitude and phase of the complex reflection coefficient Γ by using an open-ended coaxial probe in contact with the leaf (Figure 4.38) (El-Rayes and Ulaby,

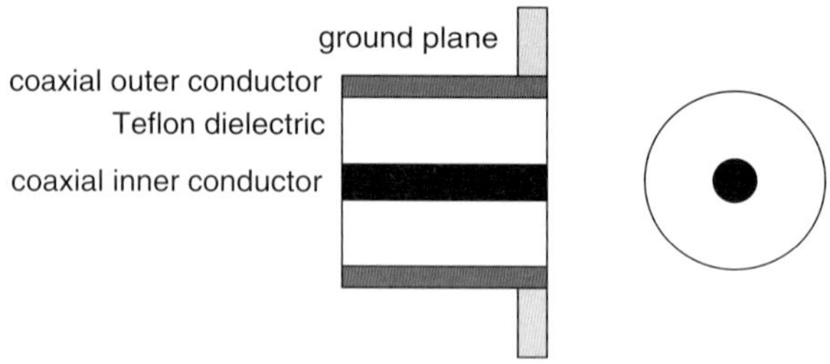

Figure 4.38 Coaxial probe: (a) longitudinal cross-section of the probe tip; and (b) low frequency probe (redrawn from El-Rayes and Ulaby (1987)).

1987). Since the thickness of a single leaf is not sufficient to simulate a slab of infinite electrical thickness, two electrically thick materials with known dielectric constant, for instance metal and Plexiglas, can be alternately placed in the background. The dielectric properties of the leaf can be then deduced from these two measurements. Note that no air gaps should remain between the probe and the leaf sample, but that the contact should not be too strong to avoid crushing the sample, otherwise its electrical properties may be changed.

Another technique consists of measuring the reflection coefficient Γ of a leaf sample placed in a rectangular waveguide system perpendicularly to the direction of propagation. A modulated microwave is amplified and fed into the directional coupler. The waveguide is connected to an analyzer on one end and terminated in a resistive load on the other end (Sarabandi and Ulaby, 1988; Chuah et al., 1995, 1997). The only purpose of the resistive load, which is matched to the characteristic impedance of the waveguide, is to absorb all the energy without causing standing waves. Finally, the sample holder is sandwiched between two thin pieces of styrofoam that keep it in place. Then an explicit expression for the relative complex dielectric constant can be derived from Γ using a simple propagation model. As an example, the second-order solution for Γ is given by (Sarabandi and Ulaby, 1988):

$$\epsilon = \frac{1 - \left(1 - \frac{1}{2}\left(\frac{\lambda}{a}\right)^2 - i\tau k_z\left(\frac{\lambda}{2a}\right)^2 - 2i\frac{k_z}{\tau k^2}\right)\Gamma}{1 + (1 + i\tau k_z)\Gamma} \tag{4.77}$$

with

$$k_z = \sqrt{k^2 - \left(\frac{\pi}{a}\right)^2} = \frac{\pi}{\lambda a}\sqrt{4a^2 - \lambda^2} \tag{4.78}$$

where λ is the free-space wavelength, a is the base of the waveguide cross-section, τ is the thickness of the leaf, and $k = 2\pi/\lambda$. A practical problem in the reflection method is to hold the leaf sample at exactly the input flange of the waveguide. A small offset in the positioning of the slab may induce some errors in the retrieval of ϵ (Chung, 2006). To get around this difficulty, one can measure the magnitude and phase of the field transmission coefficient T instead of Γ (Ulaby and Jedlicka, 1984; Chung, 2007). As before, an explicit expression for ϵ can be derived from T:

$$\epsilon = \frac{\left\{1 + \frac{(\tau k)^2}{8}\left(\frac{\lambda}{a}\right)^2 + i\frac{(\tau k)^2}{2\tau k_z}\left(1 - \frac{1}{2}\left(\frac{\lambda}{a}\right)^2\right)\right\} - \frac{1}{\mathrm{T}}}{\frac{(\tau k)^2}{2\tau k_z} - i\frac{(\tau k)^2}{2\tau k_z}} \tag{4.79}$$

Figure 4.39 Experimental set-up to measure leaf capacitance: the leafy plant moisture sensor (https://agsci.psu.edu/entrepreneur/news/2014/amin-afzal-the-inventor-behind-leafy-ag-springboard-winner).

where all the variables are listed above. This technique is especially suitable for high-loss materials such as moist vegetation.

Finally, the dielectric properties of leaves can be measured using capacitive methods. The leaf is mounted between two conductive parallel-plate electrodes forming a sample capacitor. A potential is applied between the two electrodes and the current that passes through the leaf is measured (Figure 4.39) (Baffioni et al., 1985; Hill et al., 1986, 1987; Broadhurst et al., 1987; Saltas et al., 2007; Afzal and Mousavi, 2008; Afzal et al., 2010).

The sample capacitor can be considered as an equivalent electrical circuit that consists in a capacitance, C, in parallel with a resistance, R. These values, which are wavelength dependent, are the output of the dielectric analyzer and are associated with the real and imaginary part of the complex dielectric permittivity through the relation (Saltas et al., 2007):

$$\epsilon(\omega) = \frac{C(\omega)\, d}{\varepsilon_0\, \pi\, r^2} - i \frac{d}{\omega\, R(\omega)\varepsilon_0 \pi\, r^2} \tag{4.80}$$

where $\omega = 2\pi f$ is the angular frequency, d is the distance between the two electrodes, r is their radius, and ε_0 is the permittivity of the vacuum. Note that in this technique the epidermis is often removed to give a more homogeneous specimen and liquid contact with the electrodes. Like in other wavelength domains, leaf transmittance decreases with leaf thickness in the microwave band (Mätzler and Sume, 1989). Consequently, τ is an influential factor in most of the above-mentioned methods in the determination of ϵ and it should be known *a priori*. Nevertheless, Chuah et al. (1997) reported that there was no noticeable difference between the dielectric constants of the adaxial and abaxial sides of a leaf.

5

Leaf Optical Properties in Different Wavelength Domains

This chapter describes the optical properties of leaves at the epidermis and how, to a large extent, the anatomical and morphological structure of the epidermis moderates, controls, and influences the optical properties of the leaf mesophyll and its functioning. We start with the properties of waxes, hairs, and the three-dimensional surface structures and explain many optical phenomena related to scattering of light away from the leaf such as iridescence and specular reflectance, and how surface roughness interacts with water. We discuss how light is focused into the leaf by the epidermal cells, illustrating how this is critical to leaf functions like exchanges of energy and gases. Next we review the volume scattering of the entire leaf across wavelength domains from X-rays to microwaves, each providing different information about the structure and function of leaves. Finally, we introduce evidence for light gradients within leaves and how these continue to extend our understanding of the integration of form and function within leaves.

5.1 Surface Scattering

5.1.1 Iridescence

Interference-produced iridescence is commonly encountered in some insects, birds, fish, etc. (Vukusic and Sambles, 2003; Berthier, 2006) but quite rare in the plant world where leaf and flower color is mainly due to pigments (Glover and Whitney, 2010). Although it is found in a few plant species (perhaps a few tens), iridescence has puzzled botanists for more than a century (Lee, 1977, 1986, 1997, 2007; McPherson, 2010; Gebeshuber and Lee, 2012; Vignolini et al., 2013; Diah et al., 2014). Plant iridescence was first noticed in 1896 by German botanist Ernst Stahl, who traveled extensively to Ceylon and Java. Studying Willdenow's spikemoss (*Selaginella wildenowii*), a common species of southeastern Asia also called the peacock fern due to it blue hues, he observed "granules" of what could be a reflective "pigment" in the cuticle (Stahl, 1896). The surface of some leaves in the shaded environment of humid tropical forests displays changeable blue, blue-green, purple, or violet hues by direct or diffuse light, depending on the angle of observation (Figure 5.1). When viewed at some angles, this optical effect may vanish and the leaf surface only exhibits the green color of chlorophyll (Fox and Wells, 1971). When dried, the coloration disappears, and when rehydrated, the coloration reappears. In understory plants, the most spectacular iridescence is the electric blue foliage observed in peacock begonia (*Begonia pavonina*), an endemic species native to the mountains of West Malaysia. At a distance, these plants can be mistaken for light blue plastic bags because the color is so unexpected (Blanc, 2002). Iridescent plants are not the exclusive privilege of Asia and other examples can be found in Africa, Central, and South America. Colors other than blue

Figure 5.1 Iridescent leaves of: (a) octopus tree (*Schefflera actinophylla*); (b) peacock fern (*Selaginella willdenowii*); (c) peacock begonia (*Begonia pavonina*); (d) sedge (*Mapania caudata*); (e) *Diplazium cordifolium* from Borneo; and (f) Persian shield (*Strobilanthes dyeriana*) (from *Vertical Garden* by Patrick Blanc, www.verticalgardenpatrickblanc.com/inspiration/leaf-colors). (A black and white version of this figure will appear in some formats. For the color version, please refer to the plate section.)

also exist in nature: for instance, the leaves of *Anoectochilus roxburghii*, an orchid used in Chinese medicine, are characterized by an intricate network of gold veins; *Alocasia cuprea*, which is naturally found only on the island of Borneo, has heavily textured metallic coppery-green iridescent leaves. Finally, some species like *Lindsaea lucida*, a fern, or *Begonia guttata*, an Angiosperm, display a very surprising green iridescence that could be confused with the green chlorophyll pigment color.

Iridescent colors are distinguished from pigment colors in that they have a physical rather than chemical origin: no blue pigment is extractable by solvents. The structural basis for this phenomenon varies across species (Table 5.1). It can originate from the adaxial cuticle, epidermis, or from deeper cellular layers. Graham et al. (1993) showed that light interference in *Danae nodosa* was caused by a helicoidal outer cell wall, comparable to the structure of some insect exoskeletons (Kutschera, 2008). The cell wall in this type of plant has multiple dense layers alternating with arcs of cellulose microfibrils, the orientation of which changes by a fixed angle from one layer to the next (Lee, 2011; Wilts et al., 2014). In the Malaysian tropical rainforest understory herb *Mapania caudata*, Strout et al. (2013) reported that structural leaf coloration may be enhanced by a layering of silica nanoparticles in the adaxial epidermal cell wall. In *Selaginella*, iridescence results from the separation of two reflective lamellae in the outer wall of the adaxial epidermis (Lee and Lowry, 1975; Hébant and Lee, 1984). Graham et al. (1993) showed that adjacent thylakoids in specialized chloroplasts of *Trichomanes elegans* located next to the outer wall of the adaxial epidermis could also produce iridescence. Because of our limited knowledge of leaf iridescence, Gould and Lee (1996) suggest that other physical mechanisms may yet exist.

Iridescence is created when the distance between consecutive ultrastructures is small compared to the wavelength of incident light. A basic understanding of this physical phenomenon is illustrated by interference created from a single thin film of thickness d (Figure 5.2).

It is shown that, whatever the incidence angle, the rays reflected by the film overlap and constructively interfere when their difference in optical path δ is an odd multiple of half the wavelength. In optics, the optical path length through a medium is defined as the product of the

Table 5.1 *List of physical mechanisms that yield iridescence in plants.*

Physical mechanisms	Leaf species	Reference
Parallel lamellae	*Selaginella willdenowii*	Lee and Lowry (1975), Thomas et al. (2010)
	Selaginella willdenowii, Selaginella uncinata	Hébant and Lee (1984)
	Begonia pavonina, Phyllagathis rotundifolia	Gould and Lee (1996)
Helicoidal cell walls	*Danae nodosa*	Graham et al. (1993)
	Diplazium tomentosum, Lindsaea lucida	Gould and Lee (1996)
	Mapania caudata	Strout et al. (2013)
Specialized chloroplasts	*Trichomanes elegans*	Graham et al. (1993)

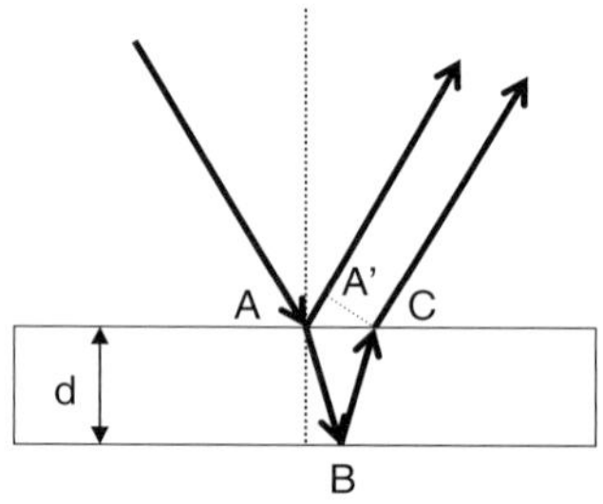

Figure 5.2 Single-layer interference.

geometric length of the path light follows and its index of refraction. Then δ is equal to the product of the geometrical path $AB + BC$ times the refractive index of the film n minus the geometrical path AA':

$$\delta = (AB + BC)n - AA' = 2AB\ n - AA' \tag{5.1}$$

If θ_i and θ_t are the incident and refracted angle, respectively, we have $AB = d/\cos\theta_t$ and $AA' = AC\sin\theta_i$. It is straightforward that $AC = 2d\ \tan\theta_t$. According to the Snell–Descartes law of refraction $\sin\theta_i = n\sin\theta_t$. Equation. 5.1 can then be simplified as

$$\delta = \frac{2d}{\cos\theta_t}\ n - 2d\ \tan\theta_t \sin\theta_i = 2dn\cos\theta_t \tag{5.2}$$

Thus, Eq. 5.2 for constructive interferences becomes

$$2dn\cos\theta_t = (2k+1)\frac{\lambda}{2}\ \text{with}\ k = 1, 2, 3, \ldots \tag{5.3}$$

The thickness d is related to the wavelength of maximum reflectance $\lambda_{\max}$:

$$d = \frac{\lambda_{\max}}{4n\cos\theta_t} \tag{5.4}$$

If the refractive index is smaller than the surrounding medium, a phase shift occurs requiring twice the thickness $D = 2d$ for the same wavelength of constructive interference (Graham et al., 1993). Equation 5.4 becomes

$$D = \frac{\lambda_{\max}}{2n\cos\theta_i} \tag{5.5}$$

So far there has been no clear verification of adaptive advantages for iridescence, like an improvement in net photosynthesis under low light conditions as suggested by Lee and Lowry (1975). For instance, Thomas et al. (2010) showed that the reflectance of iridescent (juvenile, blue) and non-iridescent (mature, green) *Selaginella willdenowii* leaves collected from the same plant was very similar over the 600–700 nm range, which coincides with photosynthetically important wavelengths. A comparison of the two types of leaves indicates that the loss of the iridescence is accompanied by a loss of multi-layer structures. They also mention a possible "scarecrow" effect to avoid predatory insects and herbivores: iridescent leaves may be confusing for some animals, camouflaging their shape and edge. Iridescence is also likely to have functions in pollinator recognition, photoprotection, and photoreception (Glover and Whitney, 2010; Diah et al., 2014). Finally, some fruits and flowers are known to be iridescent, suggesting pollinator or dispersal mechanisms (Vignolini et al., 2013, 2015).

5.1.2 Wax

In opposition to coherent scattering that can produce the phenomenon of iridescence, incoherent Rayleigh scattering can be caused by the presence of an intact crystalline epicuticular wax layer at the surface of glaucous leaves: it typically occurs when particles are much smaller than the wavelength of incident light. In that case, shorter wavelengths (the blue end of the visible spectrum) are preferentially scattered, causing the characteristic bluish gray-green color found in the leaves (or needles) of a number of plant species, including blue spruce (*Picea pungens*), Atlas cedar (*Cedrus atlantica*), chalk dudleya (*Dudleya brittonii*), and blue gum (*Eucalyptus globulus*). When dissolving the wax deposit from the surface, the green leaf color beneath is revealed, but the blue color is not found in the solvent because the scattering structure has been destroyed (Vogelmann, 1993; Vogelman and Gorton, 2014). Depending on the species or variety, the presence of a wax layer at the leaf surface may change the type of light reflection from specular (shiny), found on leaves with smooth surfaces, to diffuse (dull), associated with microscopically rough surfaces (e.g., Barnes and Cardoso-Vilhena, 1996; Pfündel et al., 2006). Jordan et al. (2005) proposed, based on analysis of 123 species, that scleromorphic cuticles and specialized structures below the epidermis in the southern hemisphere family *Proteaceae* protect the mesophyll cells from excess solar UV radiation. The cuticle may also substantially modify the spectral reflectance, transmittance, and absorptance in the visible and, to a lesser extent, in the near infrared. Strongly glaucous leaves, which are characteristic of a number of species, have a waxy glaucescence that has been experimentally removed by lightly wiping their surface with cotton wool and compared to naturally occurring non-glaucous leaves (Cameron, 1970; Clark and Lister, 1975; Eller and Willi, 1977b, 1977c; Reicosky and Hanover, 1978; Mulroy, 1979; Berg, 1985; Close et al., 2007). Glaucous leaves generally reflect more light than non-glaucous leaves at all wavelengths (Figure 5.3). Because ultraviolet-B (280–320 nm) radiation reaching the Earth's surface can cause damage to nucleic acids, proteins, and photosystems of the photosynthetic apparatus, many studies focused on protection by the leaf cuticle in this spectral region (see Section 5.2.2).

Finally, Esteban et al. (2014) linked the changes in color in Spanish juniper (*Juniperus thurifera*) to wax accumulation: they reported that glaucous needles showed higher rates of defoliation, more

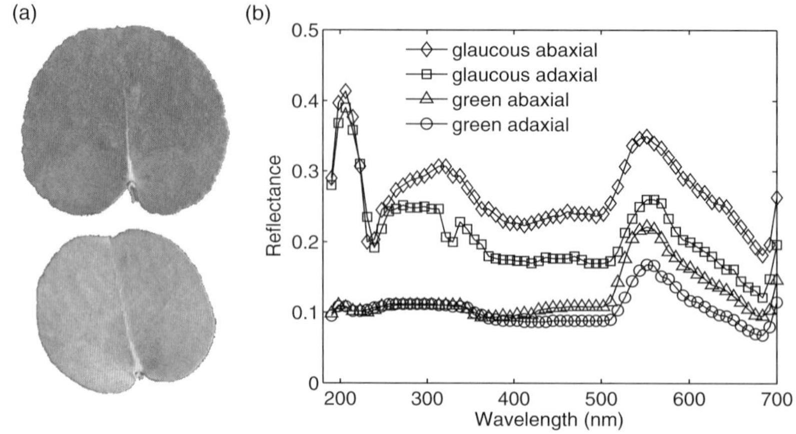

Figure 5.3 (a) Representative green and glaucous juvenile leaves from the urn tree (*Eucalyptus urnigera*). (b) Reflectance from the abaxial and adaxial sides of these leaves (redrawn from Close et al. (2007)).

physiological photoprotective mechanisms, and lower photochemical efficiencies than green phenotypes, suggesting that these may be protective mechanisms in the drought stressed/higher temperature Mediterranean climate summers and cold winters. They used it as a decline indicator.

5.1.3 Pubescence

White poplar (*Populus alba*) leaves, which may display two very different pubescence types on their adaxial and abaxial surfaces, are emblematic of pubescent leaves. In his *Naturalis Historia*, Pliny the Elder noted that "the white poplar has a part-colored leaf, white on the upper side and green beneath," while Figure 5.4 clearly shows the opposite orientation (Bostock and Riley, 1855). In Greco-Roman mythology, this tree species gave the name to the most beautiful of the nymphs, Leuce, a daughter of Oceanus. Pluto, the ruler of the underworld, fell in love with her but she wanted to escape. To keep her near him, he turned her into a white poplar which he placed at the entrance of Hades. The double color represented the duality between the upper and the underworld.

Johnson (1975) assigned three main categories of ecological functions to leaf pubescence, which reflects the selective forces that have been implicated in the evolution of plant species: the physical environment, the biotic environment, and internal physiological controls. In response to biotic factors that are in contact with leaves, including predators, parasites, and pathogens, various forms of surface pubescence may have been selected for as a mechanism of defense or resistance. The physiological significance of pubescence may be considered with respect to its role in material exchanges and metabolic regulation. The first selective force which is mostly related to the leaf energy budget (see Section 12.1) is the only one that is developed hereafter. In windy locations, hairs break up the flow of air across the plant surface and increase the depth of the leaf boundary layer, reducing evapotranspiration and CO_2 flux (Sandquist and Ehleringer, 1997). High altitude plants are also assumed to be particularly pubescent for that reason. However, Körner (2003) reports that this vision, which comes from the presence of representative plants with tomentose leaves in the mountains, like the famous edelweiss, is wrong and that the majority of alpine plants are not distinguished from species grown at lower altitudes. Leaf

Figure 5.4 (a) White poplar (*Populus alba*) (from Lobelius, *Icones*, 1581 (Tidestrom, 1909)). (b) Leaves, right leaf blade shows the lower, abaxial surface and left leaf shows the upper, adaxial surface (www.treebrowser.org/). Note the wide variation in leaf margin and shape characteristics.

pubescence also tends to increase along environmental gradients of decreasing precipitation: therefore it is presented as an adaptative feature of plants growing in hot or arid climates because pubescence greatly reduces the heat load of leaves by increasing the reflectance from their surface, which consequently reduces the amount of absorbed radiation (Shull, 1928, 1929; Billings and Morris, 1951; Gausman and Cardenas, 1969, 1973; Eller, 1977a; Eller and Willi, 1977b, 1977c; Karabourniotis et al., 1992, 1993, 1995; Ntefidou and Manetas, 1996; Manetas, 2003). Both mechanisms contribute to controlling leaf surface temperature (Figure 5.5).

The impact of such an adaptative strategy on the net photosynthetic rate was studied by Ehleringer on two closely related species, brittlebush (*Encelia farinose*), a drought-deciduous shrub with white densely pubescent leaves, and California brittlebush (*Encelia californica*) which has green, glaucous (non-pubescent) leaves (Ehleringer, 1976; Ehleringer et al., 1976a; Ehleringer and Bjorkman, 1978; Ehleringer and Mooney, 1978; Ehleringer and Cook, 1990) (Figure 5.6). Ge et al. (2012) report similar changes from Russian olive (*Eleagnus angustifolius*), white poplar (*Populus al*ba), waxyleaf privet (*Ligustrum quihoui*), and Japanese hop (*Humulus scandens*).

Ehleringer (1984) later examined pubescence in many species of plants from the Sonoran and Mojave Deserts of the USA and reported that dense trichomes increased reflectance across all wavelengths between 400 nm and 3000 nm, decreasing absorption by a factor of three compared to glaucous leaved plants, reducing leaf temperatures and photosynthetic and transpiration rates. He did not find much effect on the boundary layer from dense trichomes in the hot desert species he studied, while Meinzer and Goldstein (1985) found that for frailejones (*Espeletia timotensis*), a rosette plant of the high Andes, the greatest effect was on altering the energy budget by increasing boundary layer thickness, thus increasing leaf resistance to water transport. Smith and Hare (2004) did not find any significant differences in PAR absorptance of sacred datura (*Datura wrigtii*) leaves displaying various trichome types, nor in their temperature, transpiration or stomatal conductance. When the trichomes of *Encelia densifolia* are wet, reflectance is decreased to a level comparable to the glaucous *Encelia* species (Harrington and Clark, 1989). Understanding and modeling the optical properties of hairy

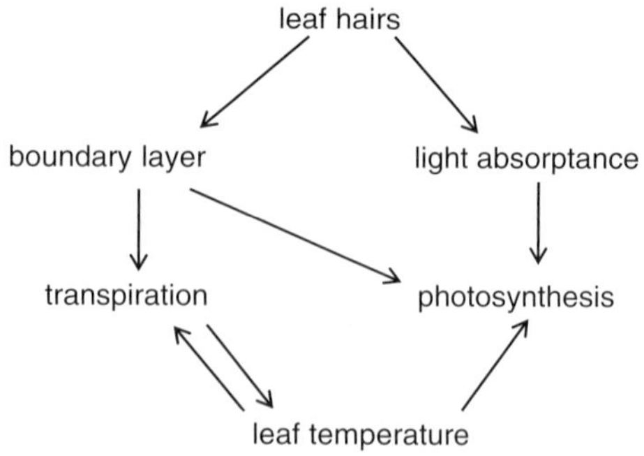

Figure 5.5 Interactions between leaf hairs and leaf physiological processes (redrawn from Ehleringer and Mooney (1978)).

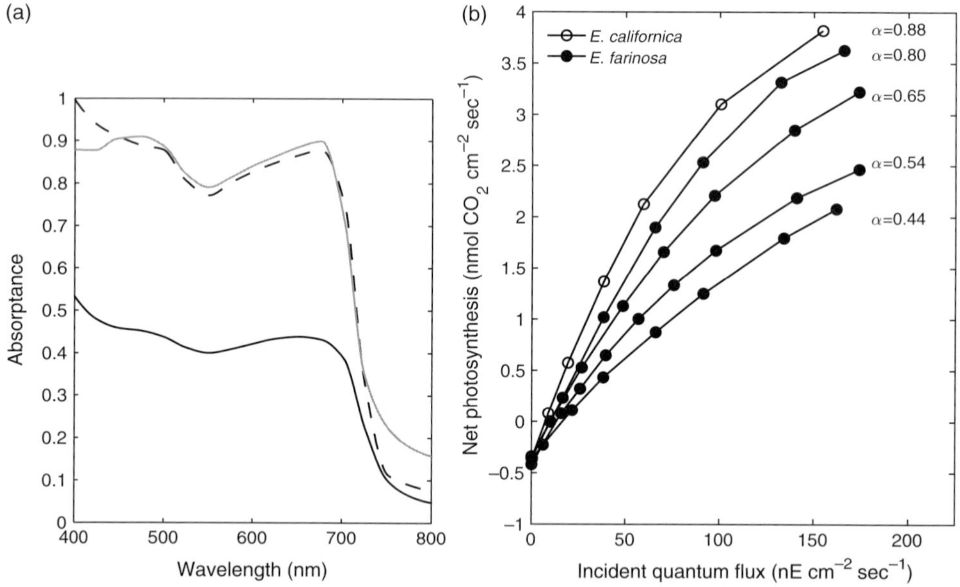

Figure 5.6 (a) Leaf absorptance spectra for brittlebush (*Encelia farinose*) with hairs intact (black solid line) and after the hairs were removed (black dotted line) and the leaf absorptance spectrum for an intact Californian brittlebush (*Encelia californica*) leaf (gray solid line) in the visible and near-infrared spectrum (redrawn from Ehleringer and Björkman (1978)). (b) Net photosynthesis versus incident quantum flux of leaves differing in pubescence and then absorptance, for 325 μbar CO_2 and 30°C (redrawn from Ehleringer and Mooney (1978)).

leaves still remains a question at issue. Vigneron et al. (2005) found a physical explanation for the white color of the edelweiss (*Leontopodium nivale*) bracts, the surface of which is covered by a layer of white hairs of variable thickness and density (Figure 5.7). In the visible, the reflectance of the bracts is high (~ 60%) and nearly constant, which explains their white appearance, but it surprisingly becomes very low in the ultraviolet below 400 nm, dissipating this high energy radiation before it reaches the epidermis. When observed with a light microscope, the filaments that form the hair layer are transparent, so that the color is not determined at this length scale but is instead due to submicrometer photonic structures.

Figure 5.7 Edelweiss (*Leontopodium nivale*) (from Vigneron et al. (2005), *Physical Review E*. Copyright © 2005 with permission from the American Physical Society).

5.1.4 Specular Behavior

5.1.4.1 Discrimination of Specular and Diffuse Reflections

Because of a thick smooth cuticle and high chlorophyll content, cherry laurel (*Prunus laurocerasus*) is a textbook case to observe a strong white specular peak on a green diffuse background that is caused by pigment absorption. Bousquet et al. (2005a) measured the BRDF of a laurel leaf at 98 viewing directions to locate the specular peak, and in 400 wavelengths to see its spectral composition. The data are plotted in polar coordinates in the lower part of Figure 5.8 at three wavelengths (550, 670, and 780 nm) and a given illumination direction ($\theta_s = 60°$). Incident direction is represented by a star and directions of experimental data acquisition are marked by dots. The other values have been interpolated for this representation. Thus one plot shows variation of the BRDF over the whole hemisphere. The first waveband corresponds to the maximum leaf reflection in the visible domain (green), the second to a minimum reflection (red), and the third lies in the near-infrared domain where reflection reaches a high plateau (due to minimum absorption).

The BRDF pattern consists of a thin high peak in the specular direction and a uniform background in all other viewing directions. The peak maximum is about 1.2 sr^{-1} and may vary up to 10% with wavelength. The intensity of the background is smaller than 0.1 sr^{-1} in the visible and about 0.2 sr^{-1} in the near infrared. The peak in the BRDF is thus identified as almost pure specular reflection. In comparison, although diffuse reflection appears very low, it becomes the predominant reflectance once integrated over the whole hemisphere. Indeed, although the specular peak reaches high intensity it remains confined within a small solid angle. The cherry laurel DHRF which has been evaluated by numerical summation is about 0.1 in the red, 0.2 in the green, and 0.5 in the near infrared. If the contribution of specular reflection to the DHRF is constant, it must be less than 0.1 DHRF units. Although small, this could still represent half or more of the reflected light in the visible domain. In a study of leaf diffuse and specular reflectance from 33 species, McClendon (1984) found that most leaves were "shiny" regardless of their visual appearance. In only two cases for leaves with thick hairs did the fraction of diffuse reflectance exceed specular, which was attributed to the convex outer surfaces of the epidermal cells.

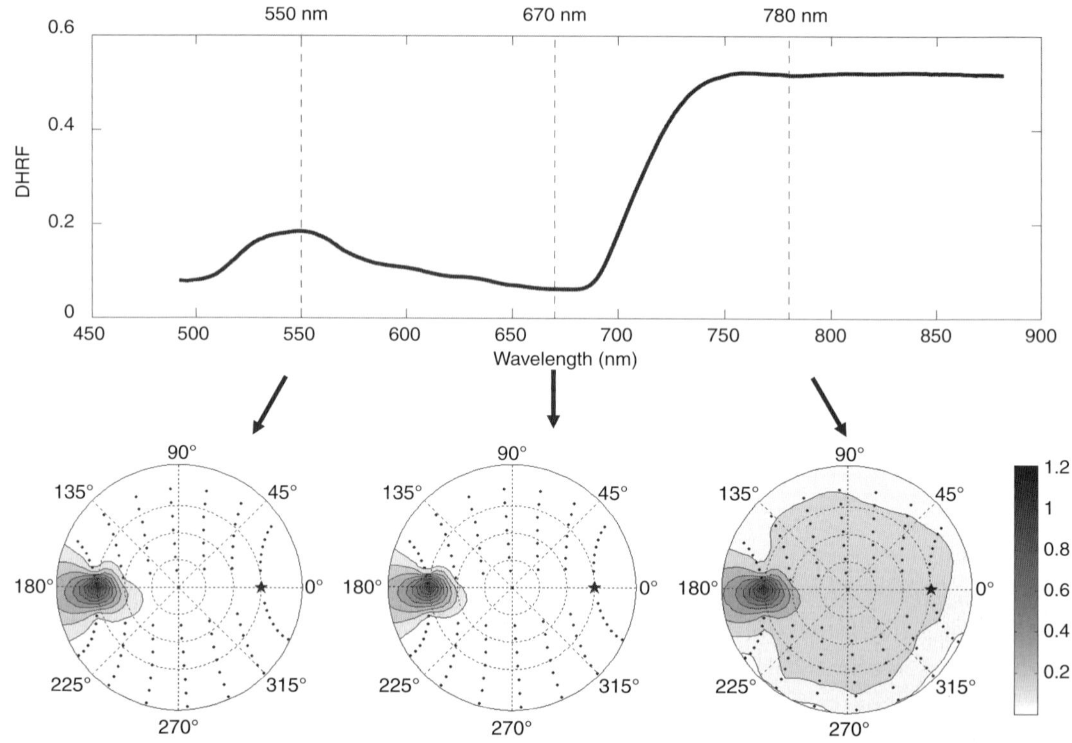

Figure 5.8 Cherry laurel (*Prunus laurocerasus*) DHRF (top) and BRDF (bottom) calculated for $\theta_s = 60°$ at 550 nm (left), 670 (middle), and 780 nm (right). The bar scale is the same for the three plots (from Bousquet et al. (2005), *Remote Sensing of Environment*. Copyright © 2005 with permission from Elsevier).

5.1.4.2 Influence of Surface Roughness

Differences in bidirectional reflectance distribution function (BRDF) patterns are thought to come from specular reflection. For wavelengths of strong absorption, for instance, 680 nm where chlorophyll is at full absorption, specular reflection is highlighted when compared to diffuse reflection, as previously noted by McClendon (1984) and Walter-Shea et al. (1989). The experimental BRDF of the three leaf species, European beech (*Fagus sylvatica*), cherry laurel (*Prunus laurocerasus*), and common hazel (*Corylus avellana*) are presented in Figure 5.9 to illustrate different magnitudes of BRDF and directional patterns. All leaves showed forward scattering, whereas no backscattering was observed.

With its thick cuticle creating a very smooth surface on the adaxial face, cherry laurel displayed a thin high BRDF peak localized in the specular direction (Figure 5.10a). Common hazel was much more Lambertian, with a maximum reflectance measured at very high viewing angles, which may be due to its hairy surface and undulated shape (Figure 5.10b). Finally, European beech showed an intermediate BRDF pattern with a lower reflectance and peak localized over a larger area between 20° and 40°. As noted by Woolley (1971) and Walter-Shea et al. (1989), the maxima of BRDF may appear at a greater angle than expected for a mirror-like reflection. This effect, which is due to leaf roughness (Torrance and Sparrow, 1967), tends to vanish when the latter decreases. As the zenith illumination angle increases, the three samples become more specular and the BRDF peak becomes narrower and higher. For a 60° illumination angle, BRDF maxima are 1.1 sr^{-1} for cherry laurel (*Prunus laurocerasus)*, 0.29

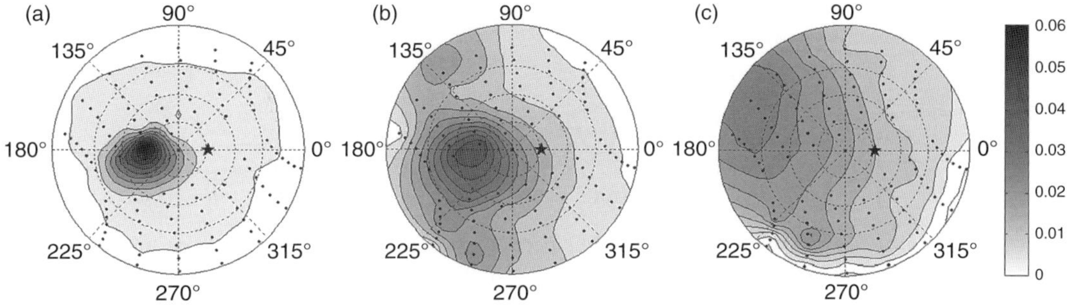

Figure 5.9 (a) BRDF of adaxial leaf faces of cherry laurel (*Prunus laurocerasus*); (b) European beech (*Fagus sylvatica*); and (c) common hazel (*Corylus avellana*), for $\theta_s = 21°$ and $\lambda = 680$ nm. The bar scale is the same for the three plots (from Bousquet et al. (2005), *Remote Sensing of Environment*. Copyright © 2005 with permission from Elsevier).

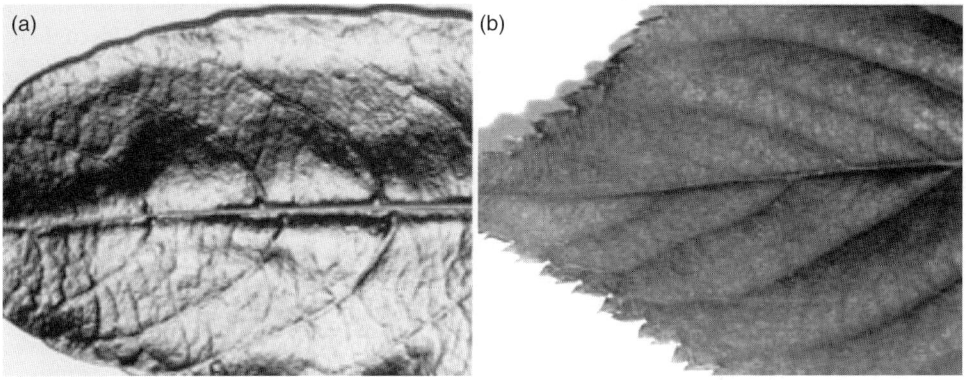

Figure 5.10 Leaf surface differences between cherry laurel (*Prunus laurocerasus*) (a) and common hazel (*Corylus avellana*) (b) leaves produce large differences in specular and diffuse scattering from the leaf surfaces.

sr^{-1} for European beech (*Fagus sylvatica*), and 0.14 sr^{-1} for common hazel (*Corylus avellana*). These differences between the three samples are such that differentiation of these taxa based upon their BRDF shape is possible.

Comar et al. (2012) examined the directionality of wheat leaf reflectance and found strong anisotropy with a strong specular lobe in the forward scattering direction that was most pronounced when the illumination direction was perpendicular to the venation. The specular feature accounted for about 30% in the forward direction and 5% in the backward direction for each of the three visible bands measured. The magnitude of this reflectance calls into question whether leaf chemistry can be retrieved from reflectance spectra without taking the directionality of scattering into account.

5.1.5 Epidermal Cell

As seen in Section 2.1.1, the epidermis is the outermost cell layer of plant leaves. Epidermal cells are structurally and functionally variable. Their shape may affect the way light interacts with a plant leaf. Haberlandt (1914), who studied heliotropism, that is, the ability of plant leaves to track the

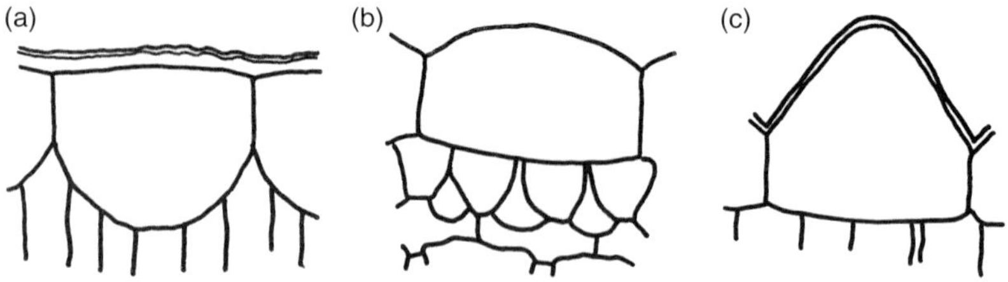

Figure 5.11 Vertical section through the adaxial epidermis of (a) Brazil raintree (*Brunfelsia pauciflora*), (b) wood sorrel (*Oxalis acetosella*), and (c) *Ruellia* sp. (redrawn from Haberlandt (1914)).

movements of the Sun, wondered if they possessed special optical sense-organs. He noticed that most plants had diaheliotropic leaves, that is their adaxial side faces the Sun, and that the lamina could force the petiole to move it into a favorable position to maximize the amount of light available to the leaf (heliotropic equilibrium). The convex cells of the upper epidermis, which are sensitive to the direction of light, seem to act as photoreceptors and also they can focus it on the epidermis itself or somewhere within the mesophyll at the level of the chloroplasts, based on the curvature of their outer wall (Figure 5.11).

Besides the control of leaf orientation, light-focusing properties of epidermal cells may increase the capture of diffuse light, and then the photosynthesis rate of leaves growing in the extreme shade which is prevalent in tropical rain forests. The relevance of such an adaptation was investigated by Bone et al. (1985) with a ray tracing model. Using micrographs of transverse sections, they studied the distribution of refracted light within 15 shaded/sunlit leaves and showed concentration of light close to the chloroplasts both for direct and diffuse environments. This phenomenon is not restricted to tropical understory plants but seems to be widespread among all plants. Martin et al. (1989) experimentally measured the amount of transmitted light within alfafa (*Medicago sativa*) leaves with a fiber-optic probe. Under collimated light they observed concentration of light in the palisade which, locally, can be two or more times that of incident levels (Figure 5.12a). The peaks tend to disappear under diffuse light (Figure 5.12b). This technique is a reliable way to measure focal lengths, but the angular sensitivity of the probe to light may induce errors.

One can compare an epidermal cell to a biconvex (both surfaces are convex) or plano-convex (one of the surfaces is flat) thick lens. In the Gaussian approximation, a collimated beam traveling parallel to the axis of such an ideal lens and passing through it is focused to an infinitesimally small spot on the axis, at a certain distance behind the lens called the focal length f'. It is the algebraic measure of the distance SF' which is positive (Figure 5.13a). In fact, among all parallel incident rays, only those close to the optical axis converge to F'. Light that passes through the lens near its periphery will have a shorter focal length (Figure 5.13b). Such a spherical aberration is characteristic of hemispherical surfaces that concentrate light within a vertical zone. Moreover, because epidermal cells are seldom symmetrical, the focal point is more like an irregularly shaped spot that has a relatively large area (Poulson and Vogelmann, 1990). Another type of aberration is coma which occurs when rays pass through the lens at an angle to the optical axis θ. Rays which pass through the center of the lens of focal length f' are focused at a point with distance $f' \tan\theta$ from the axis (Figure 5.13 c).

Assuming that an epidermal cell is an ideal convex lens, its image focal length f' is calculated using the formula:

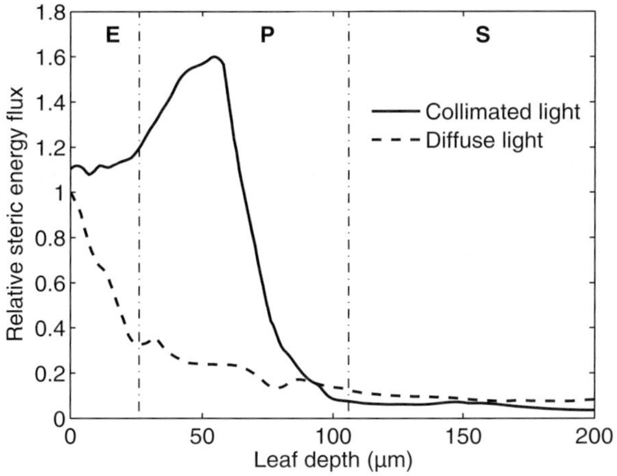

Figure 5.12 (a) Amount of 550 nm light within alfafa (*Medicago sativa*) leaves irradiated with collimated and diffuse light. (b) The boundaries of the upper epidermis (E), palisade (P), and spongy mesophyll (S) are shown at the top of the graph (redrawn from Martin et al. (1989)).

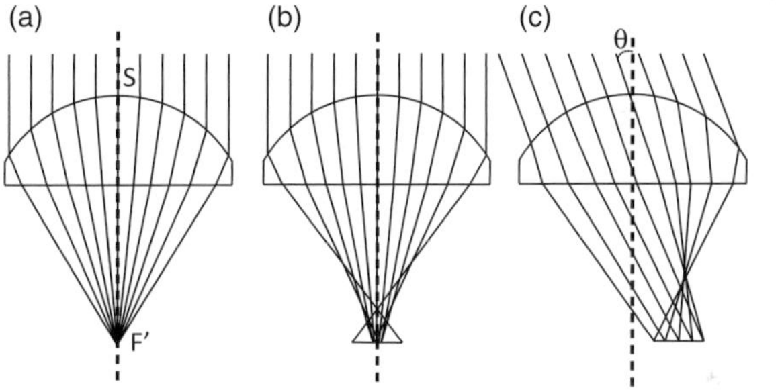

Figure 5.13 Light focusing through: (a) an ideal lens (aspheric); (b) a spherical lens showing spherical aberration; and (c) a spherical lens showing coma aberration.

$$f' = \frac{n_2}{n_2 - n_1} R \tag{5.6}$$

where R is the radius of curvature of the outer epidermal wall, $n_1 = 1.0$ the refractive index of air, and n_2 the refractive index of leaf interior. Given a particular form of epidermal cell, measured from cross-sections of fresh leaves (e.g., Poulson and Vogelmann 1990; Martin et al., 1991; Myers et al., 1994), and the index of refraction of cytoplasm ($n_2 = 1.36$), it is possible to determine the focal length of the epidermal cells using Eq. 5.6, as Vogelmann et al. (1996a) did with about 50 species collected from a wide variety of habitats. They showed that the focal plane was located in any mesophyll tissue, according to leaf anatomy, and that in certain cases it could be situated beyond the leaf blade (Table 5.2). Epidermal cells with a small radius of curvature usually focus light within the upper epidermis or beneath in the palisade. Brodersen and Vogelmann (2007) showed that regardless of whether epidermal cells were lens-shaped or not, both types absorbed more direct light compared to diffuse light.

Focal length of epidermal cells and concentration factors can also be determined directly using high-resolution agarose replicas of the adaxial leaf surfaces. Except for long hairs, the shape of which

Table 5.2 *Focal length of epidermal cells from the adaxial surface of some leaves. E, epidermis; E/P, epidermis-palisade boundary; P, palisade; P/S, palisade-spongy mesophyll boundary; S, spongy mesophyll; •, focal plane lies outside the leaf*

Species	Radius of curvature (μm)	Image focal length (μm)	Location of focal plane	Reference
Trifolium repens	14	53	P	Vogelmann et al. (1996a)
Polygonatum pubescens	17	64	E/P	Vogelmann et al. (1996a)
Oxalis europaea	24	91	P	Poulson and Vogelmann (1990)
Rosa woodsii	26	98	E	Vogelmann et al. (1996a)
Thermopsis montana	26	100	P	Martin et al. (1991)
Oxalis stricta	30	113	P	Poulson and Vogelmann (1990)
Smilacina stellata	37	140	P	Martin et al. (1991)
Lamium amplexicaule	41	155	S	Vogelmann et al. (1996a)
Oxalis montana	47	178	P	Poulson and Vogelmann (1990)
Cerastium vulgatum	57	215	P/S	Vogelmann et al. (1996a)
Arctostaphylos uva-ursi	109	412	•	Vogelmann et al. (1996a)

is poorly preserved by this technique, such replicas are representative of the leaf surface and are very stable. f' is measured from calibrated vertical displacements of the microscope stage between the surface of the cast and the position of maximum focal intensification within the cast. Both methods provide similar results.

Leaf epidermis spectral reflectance, transmittance, and absorptance have also been studied in a variety of plant species, in the ultraviolet (mostly) and the visible (less). Only plants species with an epidermis that can be easily mechanically isolated from fresh turgid leaves, by either direct peeling or by gently scraping off the abaxial side of the leaf, have been investigated. Lin and Ehleringer (1983) compared the optical properties of the adaxial face of flameflower (*Talinum paniculatum*) leaves from 400 nm to 800 nm, before and after the upper epidermis were removed with a dissecting needle (Figure 5.14). They showed that leaf absorptance increased and reflectance decreased by about 3% each over the photosynthetically active wavelengths (400–700 nm), but did not observe any change in the near infrared (700–800 nm). Changes in leaf transmittance were also observed, but to a lesser extent. In cabbage (*Brassica oleracea*), the epidermis of which is covered with a waxy layer, these changes were higher and observable over the whole visible to near-infrared domain.

5.2 Volume Scattering of the Entire Leaf

5.2.1 X-Rays

Different techniques have been used to obtain X-ray images of plant leaves. X-ray imagery is a well-known tool in medicine and art, but is also useful biologically because it generally provides more structural information than conventional techniques like optical microscopy and electron microscopy. When a leaf is placed between an X-ray source, for instance, an X-ray tube with a tungsten anode that is commonly used in medicine, and a photographic sensitive film, the vascular tissue that is

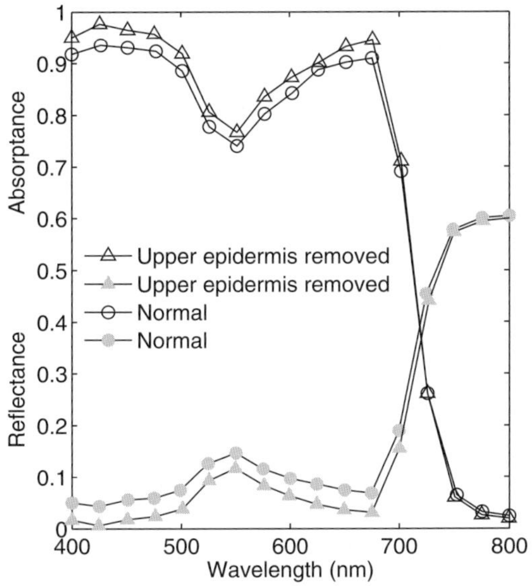

Figure 5.14 Absorptance and reflectance spectra of intact leaves of *Talinum paniculatum* measured with upper leaf surface facing the light source and also with the upper epidermis removed (redrawn from Lin and Ehleringer (1983)).

more attenuated by X-radiation than most of the other tissues produces an image of the veins as a pattern of lighter (less exposed) lines against a darker (more exposed) ground (Wing, 1992). Figure 5.15 is the image of an orchid tree leaf, where the size, shape, and first- and second-order veins of the leaf are clearly observed. If the veins are too thin, or if the cuticle and mesophyll are too thick, the sharpness of the picture may be affected. However, Wing (1992) who studied foliar vein networks of different species – the vascularization of angiosperm leaves is complex and provides useful information in systematic analyses – mentions that almost all investigated species have revealed fourth- or higher-order veins.

Other X-ray image reconstruction techniques have been investigated in botany. Synchrotron radiation, which is generated by particle accelerators and is quasi monochromatic, is the brightest artificial source of hard and soft X-rays. The incident synchrotron beam illuminates a portion of the leaf and the sample is moved relative to the beam to produce an image. Soft X-rays produced by laser plasma sources have been also extensively used. In the spectral region between 2.3 nm and 4.4 nm (0.28–0.53 keV), known as the water-window, the absorption of carbon is 10 times that of water, hence plant leaves can be imaged with good contrast in a natural aqueous environment. Both methods are now widely used in research. The main application in plant leaves is the detection of the presence of trace elements (Cu, Zn, As, Cd, Pb, Hg, Ca, Fe, Ni) from their selective absorption properties. Heavy metals are known to play a critical role in plant metabolism, either as a natural component (like Mg in chlorophyll) or as a pollutant because of their toxicity. Indeed, their accumulation in our environment, which may occur naturally or be due to anthropogenic activities, is a health risk factor because such elements potentially can be transferred to living organisms through fresh water or vegetables (see Section 7.1.3). One of the most promising and natural methods, which have been proposed for detecting and removing contaminants, seems to be the phytoremediation (Kaiser et al., 2005, 2007; Reale et al.,

Figure 5.15 X-ray positive of an orchid tree (*Bauhinia blakeana*) leaf by Albert C. Koetsier who uses X-rays to create black and white photos of plants (www.beyondlight.com/).

2006). Reale et al. (2004) artificially doped common soapwort (*Saponaria officinalis*) leaves with different concentrations of cadmium and imaged them with 1 KeV soft-X-rays. They showed the capacity of this method to quantify the uptake of heavy metals by plants. Reale et al. (2008a) successfully determined the natural sulfur content of maize (*Zea mays*) and garden pea (*Pisum sativum*) leaves and the lead content of samples treated in a solution of lead acetate (PbAc) at three different concentrations. Moreover, they showed higher absorption in the zone where chlorophyll (and hence magnesium) should be present by exposing the white and green regions of a leaf of English ivy (*Hedera helix*) to X-ray radiation, obtained from a plasma source by using an Mg target.

Some plants adapted to soils with heavy metals have mechanisms to accumulate specific metals in their tissues. Ultramafic soils can have high concentrations of nickel, chromium, and cobalt. In California, the Siskiyou Mountains Pennycress (*Thlaspi montanum*) is known to accumulate nickel. Heath et al. (1997) used X-ray imagery and found that Ni was localized in the subsidiary cells of the stomatal complex. All of the cells of the stomatal complex arise from a single leaf meristemoid cell, while Ni seems to be excluded from the guard cells, leading to speculation that the functional significance could relate to water transport and osmotic potential. A recent discovery by Lintern et al. (2013) made the news, when these scientists discovered that eucalyptus trees in the Kalgoorlie region of Western Australia, the roots of which extend tens of meters into the ground, were able to draw up gold particles and to deposit them in their leaves. Using X-ray imagery, they could locate and see the gold in the leaves, which would otherwise have been untraceable.

5.2.2 Ultraviolet (UV)

It was discovered in the mid-1970s that some human-produced chemicals, specifically chlorofluorocarbons, were leading to depletion of the stratospheric ozone layer. In 1984, scientists from the

British Antarctic Survey detected that a large area of the atmosphere above the South Pole was depleted in ozone, termed the ozone hole. This discovery was soon followed by TOMS (Total Ozone Mapping Spectrometer) satellite verification of the ozone hole at both poles during winter to early spring. The loss of ozone has caused an increase in the amount of solar ultraviolet radiation (UV-B, the most dangerous form, from 280 to 315 nm and UV-A from 315 to 400 nm) reaching the Earth, with possible medical and ecological consequences both for animals and plants (National Academy of Sciences, 1982). Exposure to excessive UV-A and UV-B radiation increases the probability of skin cancers, both non-malignant and malignant forms, and cataracts (www.epa.gov/ozone/science/effects/). For instance, Australians, who receive about 15% more UV radiation in the summer months than Europeans, experience the highest rates of skin cancer in the world. Ozone may also adversely affect terrestrial plant life and aquatic ecosystems. Recent years have seen substantial advances in our understanding of the sensitivity of higher plants to UV-B radiation: it may change their structural (e.g., leaf thickness), physiological, and biochemical properties; it may damage the photosynthetic apparatus by altering the chloroplast structure; and different lesions of DNA that controls the synthesis of proteins may also be observed. Sensitivity to UV-B radiation varies considerably within and between plant species, so that the natural capacity of plants to avoid exposure or evolve some mechanisms of defense is still a question at issue. Plant mechanisms of resistance to UV-B radiation fall into three categories: epidermal screening, high antioxidative status, and efficient repair of damage (Nybakken and Bilger, 2007). Changes in the optical properties of outer tissues of the leaf are considered to play a primary role, and much attention has been paid to the cuticle and the epidermis that function like a skin to the plant, providing a barrier between the outside world and the leaf interior (see Section 2.1.1).

Since the early work by Wuhrmann-Meyer and Wuhrmann-Meycr (1941), Lautenschlager-Fleury (1955), Kozlova (1957), or Shul'gin et al. (1960b), leaf optical properties in the ultraviolet have been extensively studied by many authors and are still a fruitful area of research in plant physiology. These authors report that leaves of low altitude plants reflect less than 10% of the UV radiation and transmit hardly anything. As a result they absorb 90% of it (e.g., Caldwell, 1968; Allen et al., 1975; Rodriguez and Gausman, 1977; Lovelock et al., 1992; Yang et al., 1995; Gonzalez et al., 1996; Grant et al., 2003; Qi et al., 2003b; Yoshimura et al., 2010; Bálint et al., 2013). Table 5.3 sums up the published values of leaf reflectance and transmittance both in the UV-B and UV-A domains, but also the optical properties of isolated epidermis and cuticles, the two leaf structures that act as filters of UV radiation, as seen hereafter.

Figure 5.16 shows that reflectance in the ultraviolet is generally not wavelength specific, that is, it displays similar levels as in the visible part of the spectrum. Nevertheless, marked differences among plant genera have been reported.

The literature is full of speculation about how such unfavorable environmental conditions are likely to change leaf optical properties. For instance, UV-B radiation reflection may depend on plant age. Lee and Lowry (1980) studied six tropical rainforest species and showed that the reflectance of young leaves was significantly less than that of mature leaves that contain lower levels of anthocyanins and total phenols. Do mountain plants absorb more light in the ultraviolet? In fact, UV contribution to total solar radiation tends to increase with altitude. As mentioned by Körner (2003), there were several attempts at answering the question of whether high altitude plants reflect more ultraviolet light than low altitude plants. Caldwell (1968) compared the UV-A reflectance of leaves from several species located at about 1800 m elevation (montane plants) and at about 3750 m

Table 5.3 *Statistics on leaf optical properties, expressed in %, in the UV spectral region (UV-B from 280 to 315 nm and UV-A from 315 to 400 nm).*

Tissue	R/T	Domain	N	Min	Max	Mean±Std	Reference
Whole leaf	R	UV-B	123	0.60	24.50	5.93±2.77	Yang et al. (1995), Ruhland and Day (1996), Filella and Penuelas (1999), Bawhey and Grant (2003), Grant et al. (2003), Qi et al. (2003a, 2003b, 2010)
		UV-A	154	1.00	35.10	5.20±4.34	Visser et al. (1997), Qi et al. (2010)
	T	UV-B	48	0.00	0. 31	0.03±0.06	Caldwell (1968), Yang et al. (1995), Filella and Penuelas (1999), Holmes and Keiller (2002), Bawhey and Grant (2003), Grant et al. (2003), Bálint et al. (2013)
		UV-A	6	0.09	0.56	0.28±0.18	Visser et al. (1997)
Isolated epidermis	R	UV-B	2	4.00	4.00	4.00±0.00	Robberecht and Caldwell (1986)
		UV-BA	16	3.60	8.70	5.43±1.23	Gausman et al. (1975b), Rodriguez and Gausman (1977)
	T	UV-B	213	0.02	32.20	5.91±4.56	Lautenschlager-Fleury (1955), Robberecht et al. (1980), Robberecht and Caldwell (1983, 1986), Liakoura et al. (2003), Nybakken et al. (2004a, 2004b)
		UV-A	157	0.40	47.00	7.04±6.82	Caldwell (1968), Robberecht and Caldwell (1983), Nybakken et al. (2004a, 2004b)
		UV-BA	41	0.10	86.10	7.36±15.92	Gausman et al. (1975b), Rodriguez and Gausman (1977), Robberecht and Caldwell (1978)
Isolated cuticle	T	UV-B	81	0.30	59.50	21.73±18.23	Krauss et al. (1997)

elevation (alpine plants). He found no significant difference between these two groups ($R = 0.0300 \pm 0.0143$ for the upper side and $R = 0.0312 \pm 0.0139$ for the lower side of the first group; $R = 0.0247 \pm 0.0054$ for the upper side and $R = 0.0244 \pm 0.0058$ for the lower side of the second group). This was confirmed by Filella and Peñuelas (1999), who studied evergreen oaks (*Quercus ilex*) growing at 200 and 1200 m elevation and alpenrose shrubs (*Rhododendron ferrugineum*) growing at 2200 m elevation: reflectance ranged between 5% in the UV-A region and 12% in the UV-B region, and it was significantly higher in sunlit than in shaded leaves. Finally, Rozema et al. (1997) showed that the absorption of UV-B of leaf extracts increased with increasing elevation.

Where in the leaf blade is UV radiation mostly attenuated? At least 90% of the radiation that does reach the surface of plant leaves is attenuated before reaching the mesophyll tissues. The cuticle, along with other structures like trichomes and glandular hairs, form the first line of defense against high UV radiation. Bawhey and Grant (2003), who measured the reflectance of wild-type and mutant

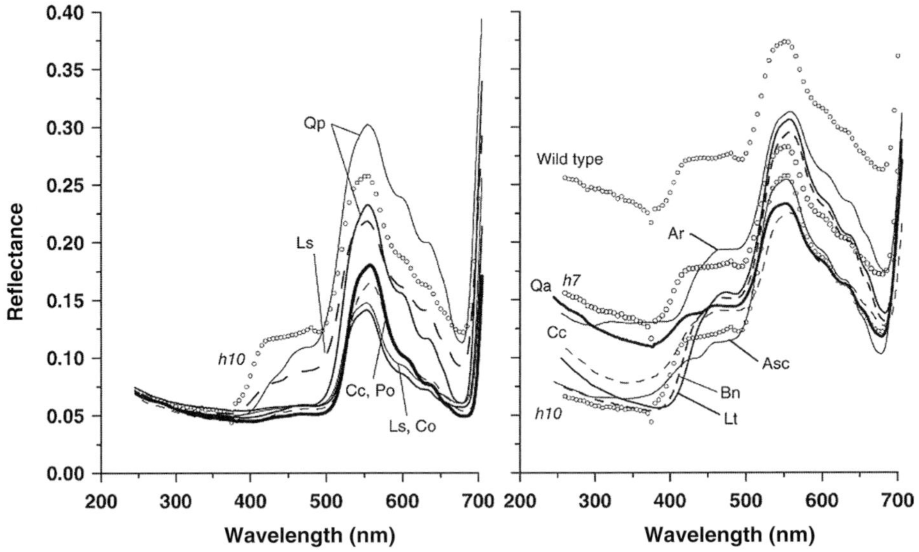

Figure 5.16 Leaf spectral reflectance in the ultraviolet (from Grant et al. (2003), *Agricultural and Forest Meteorology*. Copyright © 2003 with permission from Elsevier).

leaves of sorghum (*Sorghum bicolor*) containing different amounts of wax, showed that they were causally related; greater density of filaments induces greater reflectance. This is confirmed by Grant et al. (2003), who used a larger database containing 20 deciduous tree species: they suggest that the magnitude of the UV reflectance is a function of the shape, diameter, and distribution of epicuticular wax leaf surface structures. This has been also reported for Giant Chalk Dudleya (*Dudleya brittonii*) (Mulroy, 1979) and blue spruce (*Picea pungens*) (Clark and Lister, 1975) or for certain desert species with dense pubescence (Robberecht et al., 1980; Karabourniotis et al., 1992). However, if glaucous leaves are very effective reflectors of both UV and visible radiation, it seems that pubescent leaves are more effective in reflecting visible wavelengths than ultraviolet radiation (Holmes and Keiller, 2002). Turunen et al. (1999) showed that the epidermis of mature primary needles of loblolly pine (*Pinus contorta*) transmitted less than 2.5% of UV radiation but cotyledons had less UV screening, especially for UV-A, while both cotyledons and needles grown without UV exposure had substantial declines in UV-A and UV-B transmittance. To differentiate between UV attenuation by cellular compounds of plant cells as opposed to attenuation by the cuticular membrane and leaf epidermis, transmittance of isolated fresh epidermal samples can be measured. The first systematic analysis was conducted by Lautenschlager-Fleury (1955), who found a consistently low UV-B transmission (i.e., high absorption) for all plants tested. This was later confirmed by others (e.g., Caldwell, 1968; Allen et al., 1975; Gausman et al., 1975b; Caldwell et al., 1983; Robberecht and Caldwell, 1978, 1983, 1986; Flint et al., 1985; Qi et al., 2002). Figure 5.17 depicts the anatomy of the upper side of a dicot leaf; UV-absorbing compounds that occur principally in the cuticle, the vacuole, and the cell walls of the epidermis absorb approximately 90% of UV radiation. Less than 5% of radiation is reflected by the leaf surface and about the same amount reaches the mesophyll tissue. Scattering and absorption within the mesophyll result in less than 1% of the incident UV radiation entering the lower epidermis Also, no detectable UV flux passes through the entire leaf (Caldwell et al., 1983).

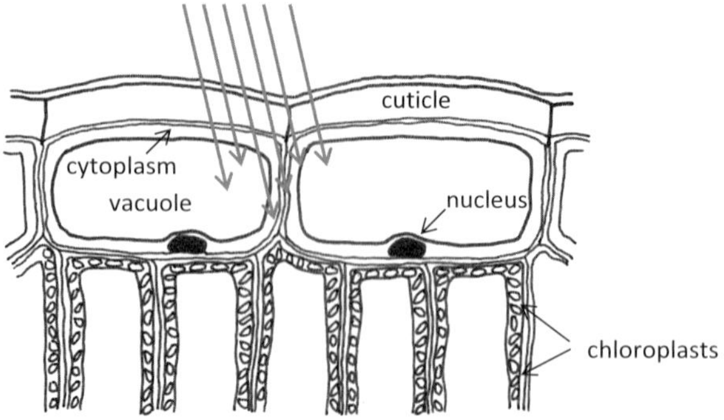

Figure 5.17 Partitioning of UV-B radiation incident on a glabrous leaf surface based on a survey of 100 species (redrawn from Caldwell et al. (1983)).

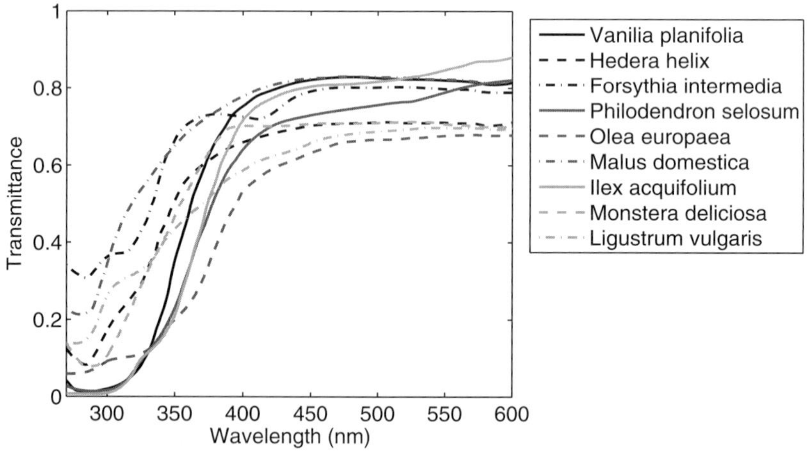

Figure 5.18 Transmittance spectra of isolated cuticular membranes of nine leaves (redrawn from Krauss et al. (1997)).

Krauss et al. (1997) measured the transmittance spectra of isolated leaf cuticular membranes of 27 species in the range from 270 nm to 600 nm. The cuticles that are practically translucent in the visible strongly absorb ultraviolet radiation and attenuation is nearly identical for a wide range of species (Figure 5.18).

Robberecht et al. (1980) measured leaf epidermal transmittance of UV-B radiation (295–320 nm) along a latitudinal gradient of solar ultraviolet radiation. In equatorial and tropical regions characterized by high UV-B radiation, mean epidermal transmittance for the species examined was less than 2%. At higher latitudes, it could exceed 5% (Figure 5.19). Such variation along natural gradients may reflect plant response to UV-B radiation. To determine whether a natural change in UV-B radiation along a latitudinal gradient would lead to a gradient in epidermal transmittance, Nybakken et al. (2004a, 2004b) studied six species collected in northern Norway, southern Norway, and in the French Alps. They conclude a lack of a clear latitudinal gradient in UV-B screening.

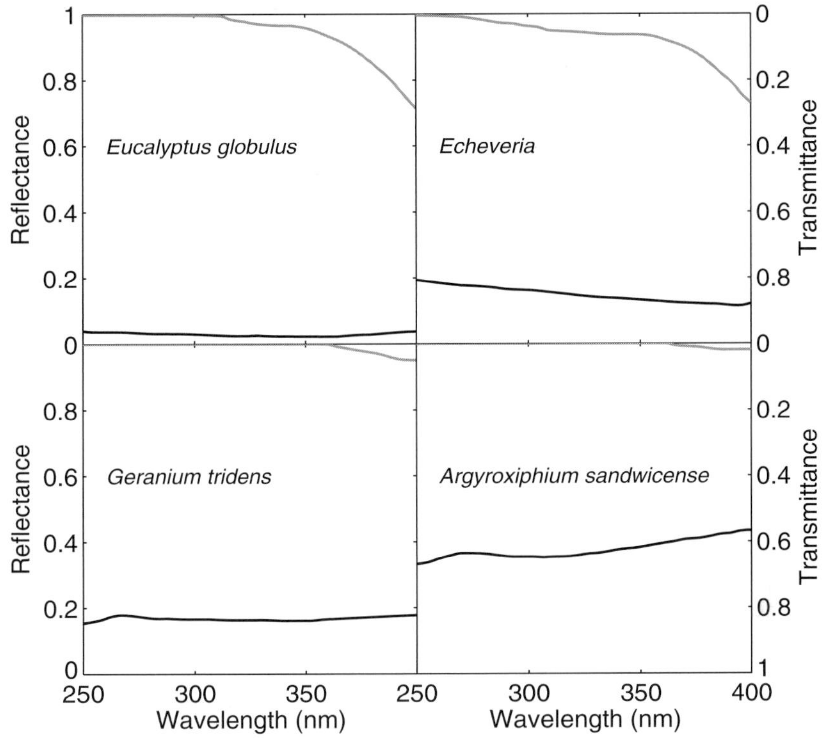

Figure 5.19 Leaf epidermal UV optical properties of four species from exposed habitats in equatorial and tropical latitudes. Three major types of leaf surface characteristics are represented: (1) glabrous (*Eucalyptus globulus*); (2) glaucous (*Echeveria* sp.); and (3) pubescent (*Geranium tridens* and *Argyroxiphium sandwicense*) (redrawn from Robberecht et al. (1980)).

Absorption profiles have also been performed in the ultraviolet region of the spectrum using the techniques described in Section 3.5. The penetration depth of UV radiation into subalpine conifer needles, which are characterized by a thick and lignified epidermis, has been examined by several authors (Bornman and Vogelmann, 1988; Day et al., 1992, 1994; DeLucia et al., 1992). UV-B radiation is systematically attenuated more readily than UV-A radiation, which penetrates less than blue or red photosynthetically active light. Bornman and Vogelmann (1988) and DeLucia et al. (1992) found a higher transmittance in white spruce (*Picea engelmannii*) than Rocky Mountain fir (*Abies lasiocarpa*) at all wavelengths. The depth of penetration in both species also seemed to decrease as foliage matured. Finally the removal of the epicuticular wax only slightly increased the amount of UV-B light in the needles. Similar measurements were performed on broadleaf species (e.g., Bornman and Vogelmann, 1991; Day et al., 1992; Cen and Bornman, 1993; Alenius et al., 1995; Ruhland and Day, 1996; Karabourniotis and Bornman, 1999; Karabourniotis et al., 1999a; Liakoura et al., 2003; Qi et al., 2003c, 2010). Bornman and Vogelmann (1991) grew rapeseed (*Brassica campestris*), Abyssinian mustard (*Brassica carinata*), and alfalfa (*Medicago sativa*) under UV-B radiation: they observed an increase in leaf thickness, larger in rapeseed than in Abyssinian mustard or in alfafa, a decrease in chlorophyll content per unit leaf area, but variable change in UV-B transmission. Day et al. (1992) confirmed that difference in UV-B screening effectiveness results

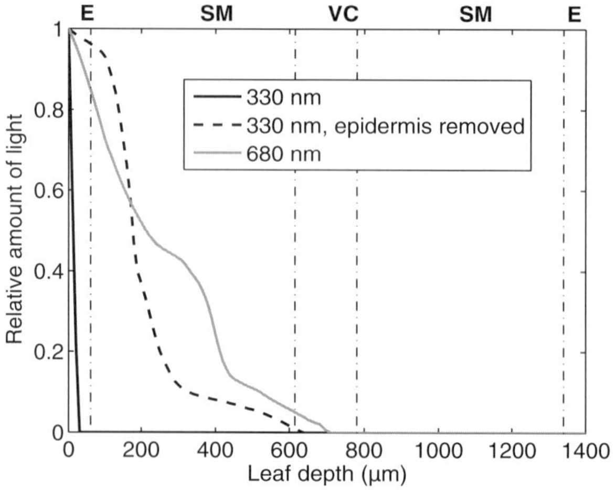

Figure 5.20 Typical amount of UV-B (300 nm) radiation, before and after a strip of epidermis (~ 50 μm wide) was removed, and red (680 nm) radiation as a function of depth in a one-year-old conifer white spruce (*Picea engelmannii*) needle. The vertical bars denote the thickness of various tissue layers: E (cuticle, epidermis, and hypodermis), SM (spongy mesophyll), VC (vascular cylinder or bundle) (redrawn from Day et al. (1992)).

more from different optical characteristics of leaf epidermis than from leaf thickness (Figure 5.20). Cen and Bornman (1993) have found a good correlation between the penetration of UV radiation in the dorsiventral leaves of rapeseed (*Brassica napus*) and changes in UV-screening pigments, including flavonoids. This confirmed the earlier work by Flint et al. (1985), who showed that low UV-B treatment resulted in the stimulation of flavonoid synthesis, therefore in decreased epidermal transmittance. With high UV-B treatment, transmittance values were closer to control levels, which suggested interaction with chlorophyll pigment synthesis. Sullivan et al. (1996) report very low UV-B transmission (0.5–1.0%) through the needle epidermis in loblolly pine (*Pinus taeda*) but subsequent needle thickening, reduced biomass, and slower development after exposure.

Which molecules absorb UV radiation? A comprehensive review of UV radiation screening pigments is given by Cockell and Knowland (1999). Flavonoids and related phenolic compounds, some xanthophylls, cuticular waxes and other lipids, and anthocyanins are responsible for the UV absorption in plant leaves (Solovchenko, 2010). Many of these compounds are contained within the vacuole of the epidermal cells, but they can also occur within specific organelles such as chloroplasts. Since they are easily soluble in methanol or other solvents, their absorbance spectra are provided by several papers. Lee and Lowry (1980) showed that young leaves produced on rapidly flushing shoots had markedly lower reflectance in the UV-B region than did mature leaves and contained higher levels of anthocyanins and total phenols. Kolb and Pfündel (2005) found species-specific differences and non-linear relationships between epidermal and phenolic UV absorbance. Qi et al. (2003a) report a field comparative study of dynamics and temporal changes of UV-B absorbing-compound concentration in 35 sun-exposed broadleaf trees over a growing season. Finally, Barnes et al. (2008) show that transmission of UV-A radiation into the leaf varies diurnally in great mullein (*Verbascum thapsus*), evening primrose (*Oenothera stricta*), and faba bean (*Vicia faba*), being least at mid-day, which they conclude is under active control, although the underlying mechanism is not understood.

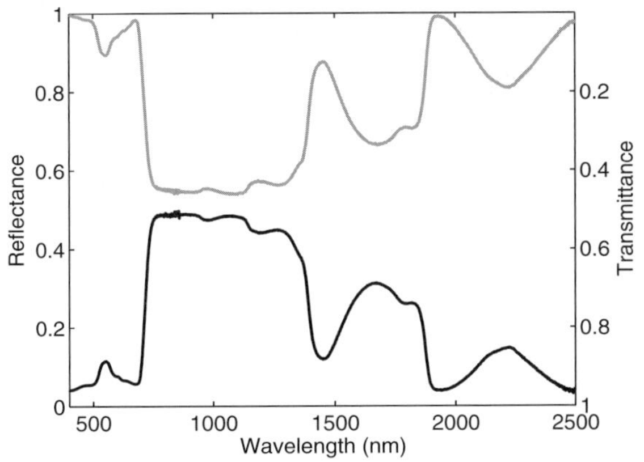

Figure 5.21 Reflectance (black line, left scale) and transmittance (gray line, right scale) spectra of fresh-green sunflower (*Helianthus annuus*) leaf. VIS = visible, NIR = near infrared, SWIR = shortwave infrared.

5.2.3 *Visible/Near Infrared/Shortwave Infrared*

The spectral domain that begins at 400 nm and ends at 3000 nm, but most often at 2500 nm, corresponds to 90% of the sunlight (by total energy). Figure 5.21 displays the typical form of reflectance and transmittance spectra measured on a sunflower leaf (*Helianthus annuus*) leaf from 400 nm to 2500 nm. This region is classically divided into three sub-domains where plant leaves display remarkable optical properties (e.g., Billings and Morris, 1951; Gates et al., 1965b; Knipling, 1970): the visible (400–800 nm) characterized by a strong absorption of light by the photosynthetic pigments in a green leaf; the near infrared (800–1100 nm) where absorption is limited to the biochemicals composing the dry plant matter of leaves but where multiple scattering within the leaf, related to the fraction of air spaces, that is to the internal structure, drives the reflectance and transmittance levels; and the shortwave-infrared (1100–2500 nm) which is also a zone of strong absorption, primarily by water in fresh leaves and secondarily by dry matter when the leaf wilts. In general, the curves for green leaves all have similar shapes, varying principally in magnitude. The internal absorption and scattering mechanisms account for the similarity in the shape of the reflectance and transmittance spectra. Note that leaf-to-leaf variability is usually higher for transmission than reflection. Hundreds of papers already detail the variation in spectral properties related to leaf biochemical composition and structure, which depends on many factors like the plant species, the phenological age of the leaves, the developmental or microclimate position of the leaf on the plant, whether it is stressed or not and if so, what kind of stressor, etc. All of these observations and experimental measurements are a prerequisite for any attempt to extract biophysical information, as seen in Chapter 6.

5.2.3.1 *Visible*

The visible, also called the PAR (photosynthetically active radiation) region, is characterized by a strong absorption of light by the photosynthetic pigments in a green leaf (chlorophylls, carotenoids, including xanthophylls), and the non-photosynthetic anthocyanin pigments, with two main absorption peaks in the blue and red, and a green minimum in-between. The reflectance spectrum displays low values; they are usually below 5% at 400 nm (blue) and rise gradually to a peak of about 12 to

15% at 550 nm (green) – this peak accounts for the green color of plants perceived by the human eye – and then declines to about 5% in the neighborhood of 675 nm (red). From this depression, they ascend steeply along the boundary between the red and the infrared (red-edge) to reach an asymptotic plateau of about 40 to 60% at 780 nm (near infrared). The rather sharp shoulder of the reflectance curve at 750 nm signals the long wavelength end of the absorption of visible light by chlorophylls. The spectral transmittance is of the same order of magnitude as the reflectance, except in the blue and the red where the values are close to zero. Horler et al (1983c) were among the first to show that the wavelength position of the red-edge, measured as the maximum slope from the first derivative, was closely correlated with the leaf chlorophyll concentration. Note that the absorption spectra for photosynthesis of leaves for light incident normally on the front surface are significantly different from the absorption spectra of equivalent suspensions of leaf extract.

5.2.3.2 Near Infrared

Willstätter and Stoll (1918) proposed the first theory to explain diffuse reflectance from a leaf on the basis of critical reflection of visible light at the spongy mesophyll cell wall–air interfaces. Indeed optical properties in the near infrared are mainly driven by multiple reflections and refractions within the leaf, due to differences in refractive index between air ($n = 1.0$) and hydrated cellulose walls ($n = 1.4$). Thus the shape, size, and three-dimensional arrangement of the cells in the mesophyll, as well as the fraction of air space, definitely play an important role in the way radiation is scattered. Knipling (1970) suggests that the important parameter in determining the level of reflectance, and mechanically of transmittance, is the number or total area of the air–wall interfaces more than the volume of air space. In this regard the palisade mesophyll of a leaf may be as important as the spongy mesophyll in the internal scattering of radiation. Sinclair et al. (1973) found unexpectedly high near-infrared reflectance from the palisade, attributed to diffuse reflectance from cell walls, particularly from the ventral side. After several reflections, radiation is eventually reflected back into the palisade region and emerges through the epidermal cells as diffuse reflectance. If the cell dimensions are generally too large to create Mie scattering, the chloroplast and grana that are about the same dimension as the wavelength of radiation are likely to do it (Gates et al., 1965b; Allen et al., 1970b). The Mie theory predicts a slight wavelength dependence of the scattering cross-section, which may explain the negative slope (falling from left to right) of the reflectance curve. This tilted plateau extends all the way to 1100 nm, with a small but consistent depression at 970 nm, which is certainly the first water absorption peak. The near-infrared region is characterized by relatively little or no internal absorption; it excludes pigments that are known to be completely transparent beyond 800 nm and is limited to the biochemicals composing the dry matter of leaves. Note that the interaction of visible and shortwave-infrared radiation with the leaf structure is not very different from the interaction of near-infrared radiation, as suggested by the high levels of reflectance and transmittance in white and dehydrated leaves (Knipling, 1970; Maas and Dunlap, 1989). Merzlyak et al. (2002) report even higher levels of reflectance and transmittance in the 780–900 nm region, resulting in only about 1% absorptance. They attribute previous numerous reports of absorption in the 5–20% range as due to measurement error from incompletely collecting all transmitted light. In a review of the diffuse and specular characteristics of leaf reflectance, Grant (1987) shows that the diffuse components of leaf reflectance originate primarily from the leaf's interior through multiple scattering but the specular components of reflectance arise at the leaf surface.

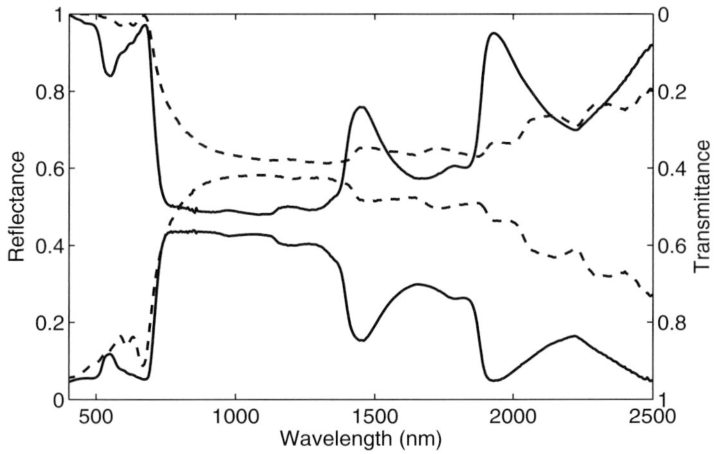

Figure 5.22 Reflectance (left scale) and transmittance (right scale) spectra of fresh (solid line) and dry (dashed line) Carolina poplar (*Populus canadensis*) leaf.

5.2.3.3 *Shortwave Infrared*

At wavelengths longer than 1100 nm, the reflectance and transmittance starts to fall off. Radiation is primarily absorbed by water at about 1200 nm, 1400 nm, 1900 nm, and 2500 nm in fresh leaves, and secondarily by dry matter in dry leaves, as effectively illustrated by the spectrum in a dehydrated Carolina (*Populus canadensis*) poplar leaf (Figure 5.22). The reflectance of a dried leaf is usually greater than that of the fresh leaf at all wavelengths.

Since they contain the same constituents, green leaves display relative qualitative constancy in their optical properties. Quantitative variation in these properties is associated with variations in their pigment, water and dry matter content, and with the anatomical structure of the leaves. These variations as a function of leaf biochemical composition and structure have been described in hundreds of papers that we will try to summarize in the Chapter 6 and Chapter 7.

5.2.3.4 *Fluorescence Emission*

Solar induced fluorescence emission is a complex signal that is superimposed on leaf reflectance and transmittance that reveals much about the physiological state of plants. As pointed out in the introduction, there is an extensive literature on this subject including dedicated books; therefore, we will only provide here some parts of the question. There are two fundamentally different but still complementary types of fluorescence: the blue-green fluorescence (400–630 nm) and the red-far red fluorescence (630–800 nm). Figure 5.23 displays a typical UV-A-induced fluorescence spectrum superimposed on the absorption spectrum of a methanolic extract. The magnitude of fluorescence emission is variable; it depends on leaf species, age, structural and chemical properties, and also on environmental factors. Changes in fluorescence emission are frequently expressed as ratios of two-band combinations of blue, green, red, and far-red regions of spectrum as indicators of plant vigor (Kim et al., 2001).

Blue-green fluorescence has a heterogeneous origin. Many potential fluorophores including polyphenols and hydroxycinnamic acids can contribute to photon emission in this region of the spectrum

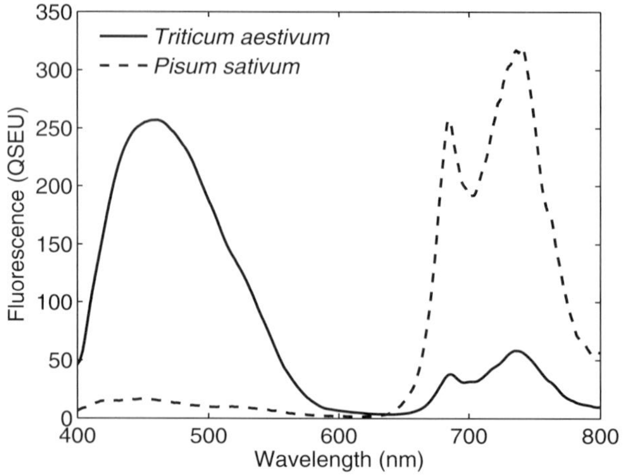

Figure 5.23 Fluorescence spectra of the adaxial side of wheat (*Triticum aestivum*) and pea (*Pisum sativum*) leaves excited at 355 nm. QSEU means quinine sulfate equivalent unit (by courtesy of Zoran Cerovic).

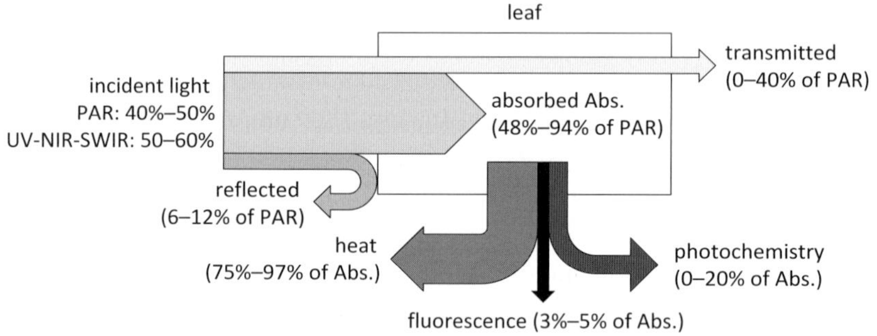

Figure 5.24 Different processes occurring to the energy reaching leaf surface (redrawn from Vidaver et al. (1991)).

(e.g., Broglia, 1993; Cerovic et al., 1999). It is well-known that a cellulosic matrix, for instance, white paper, fluoresces in this region. A maximum and a shoulder can be observed at approximately 440 nm and 530 nm, respectively. However, the emission spectra of all these molecular components are very wide, therefore it is difficult to identify a fluorophore by its emission spectrum alone.

Red (685–690nm) and far-red (~ 730–740nm) fluorescence, also called chlorophyll fluorescence (ChlF), has been more studied than blue-green fluorescence. Depending on the amount of incident radiation and the physiological state of the plant, variable proportions of the energy absorbed by photosynthetic pigments may be converted into a stable chemical form through photosynthesis. Excess energy may be dissipated and lost as heat, and some may be emitted as fluorescence (Figure 5.24).

These three processes occur in competition, such that any increase in the efficiency of one will result in a decrease in the yield of the other two. Hence, by measuring the yield of chlorophyll fluorescence, information about changes in the efficiency of photochemistry and heat dissipation can be gained (Maxwell and Johnson, 2000). Chlorophyll *a* which is associated with the two photochemical complexes, the photosystem I (PSI) and the photosystem II (PSII), is the only leaf fluorophore that emits photons in this region of the spectrum. ChlF can be induced by sunlight or laser excitation.

A typical chlorophyll fluorescence emission spectrum displays two maxima in the red region at approximately 685 nm (RF or F685) and in the red region at approximately 730 nm (FRF or F730). PSI contributes to the F730 emission band only, while PSII contributes to both F685 and F730. The ratio of the 685 nm to 730 nm fluorescence bands, $F685/F730$, has been particularly investigated, because at room temperature it is strongly dependent on chlorophyll concentration (Gitelson et al., 1998, 1999), and it can be used as a plant stress indicator (Lichtenthaler et al., 1998; Campbell et al., 2007). However, it is somewhat difficult to interpret because it is also sensitive to changes in leaf photosynthetic activity (Agati et al., 1995, 1996) and to external conditions such as light intensity or air temperature (Agati, 1998; Agati et al., 2000). Adaxial and abaxial faces of leaves also show differences in their optical behavior, with the $F685/F730$ ratio higher for abaxial than for adaxial leaves (Cordon and Lagorio, 2007b). Using the newly developed VIRAF (Visible, Infra-red, Reflection, Absorption, Fluorescence) spectrometer, Buschmann and Lichtenthaler (1988c) were among the first to show that ChlF emission might alter the shape of the reflectance spectra, especially in the red-edge region. This was confirmed by measurements of leaf reflectance during short-term stress, typically a few hours (Buschmann and Nagel, 1993b; Buschmann et al., 1994; Gamon and Surfus, 1999). Changes of reflectance were primarily caused by ChlF due to a decrease of photosynthetic activity and not by a loss of pigments. The effect of fluorescence in apparent reflectance has been quantitatively demonstrated by Zarco-Tejada et al. (2000). They carried out an experiment where the leaf is illuminated by a white light, then they used a long-pass filter that blocks radiant flux at wavelengths shorter than 695 nm. Without the filter, chlorophyll absorbs light and then produces ChlF, which is not the case when the filter is in place (Figure 5.25).

One can also study the kinetics of ChlF at different wavelengths. Kautsky and Hirsch (1931) reported for the first time that the intensity of the red fluorescence emitted by chlorophyll *a* varied during exposure of intact leaves to saturating blue light after a short period of adaptation to darkness. This effect, named after the first author as the Kautsky curve, was studied extensively in the following

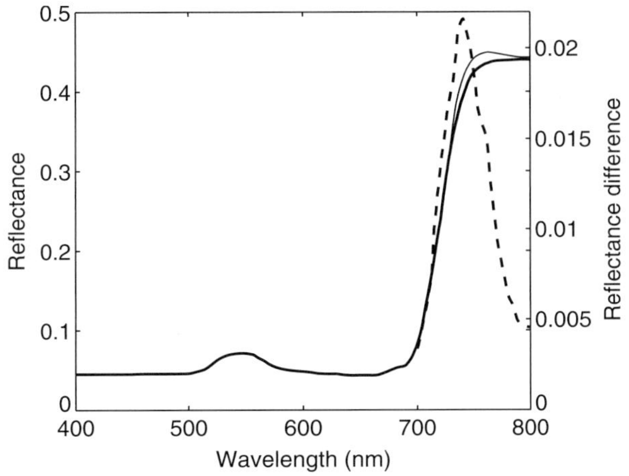

Figure 5.25 Reflectance spectrum of a dark-adapted sugar maple (*Acer saccharum*) leaf measured with no filter (thin line) and with a long-pass filter (thick line). The dashed line represents the difference (redrawn from Zarco-Tejada et al. (2000)).

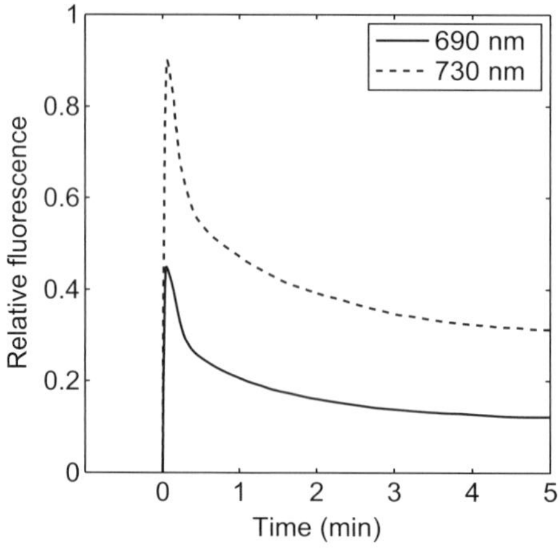

Figure 5.26 Time-resolved chlorophyll fluorescence kinetics (Kautsky effect: slow component) measured at 690 nm and 730 nm on an intact green leaf of tobacco (*Nicotiana tabacum*) using a portable field fluorometer (redrawn from Lichtenthaler and Rinderle, 1988).

decades and used as a non-invasive technique. As illustrated by Figure 5.26, the ChlF induction curve has a fast (less than a second) exponential phase (decrease of photochemical quenching), and a slow (few minutes duration) decay phase (increase of photochemical and non-photochemical quenching). Once again, the analysis of such a curve gave rise to many papers, but it is beyond the purpose of this book so it will not be developed here.

5.2.3.5 The Xanthophyll Cycle

As seen in Section 2.3.3, the xanthophyll cycle plays a key role in the protection of the photosynthetic apparatus. Under conditions of excess light, this non-photochemical quenching mechanism reduces the amount of energy that reaches the photosynthetic reaction centers. When dark-adapted leaves are exposed to bright light, violaxanthin is reversibly de-epoxidized to zeaxanthin via antheraxanthin. These are regulatory pigments closely linked to PSII light use efficiency. The conversion of violaxanthin to zeaxanthin can be studied with optical techniques because the absorption spectra of these pigments slightly shift in the green region (Yoshimura, 2001). This results in subtle changes in leaf optical properties that are more apparent in the difference spectrum. Bilger et al. (1989) studied differences in absorbance of cotton (*Gossypium hirsutum*) leaves. They identified a major peak at 505 nm, a minor one at 465 nm, a shoulder around 515 nm, and minor troughs at 455 and 480 nm. Gamon et al. (1990) measured the reflectance of sunflower (*Helianthus annuus*) leaves in dark state and light state. They showed changes in apparent reflectance at 531 nm (Figure 5.27). Note the double feature at 685 nm and 738 nm that is due to chlorophyll fluorescence.

To quantify the epoxidation state of the xanthophyll cycle pigments, Gamon et al. (1992, 1993) proposed a reflectance-based index, first called the physiological reflectance index and later renamed the photochemical reflectance index (PRI). It is analogous in formulation to the normalized difference vegetation index (NDVI) and incorporates narrow-band reflectance at 531 nm, the wavelength corresponding to the maximum xanthophyll cycle signal, and at 570 nm, a reference wavelength:

$$PRI = \frac{R_{531} - R_{570}}{R_{531} + R_{570}} \tag{5.7}$$

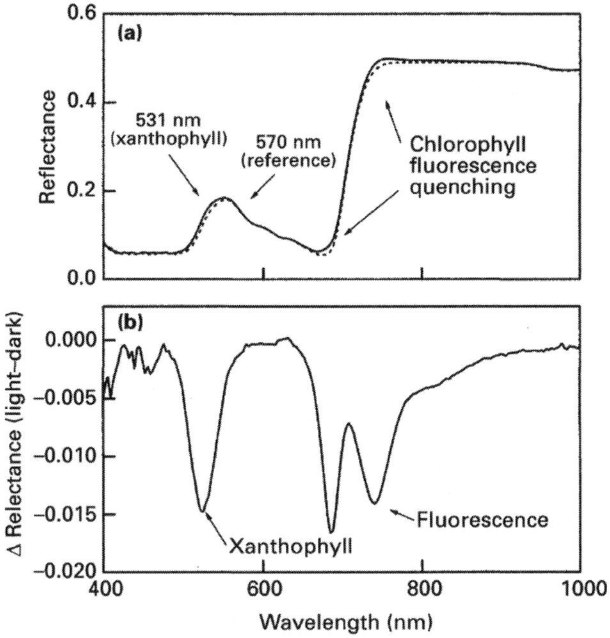

Figure 5.27 (a) Reflectance spectrum of a sunflower (*Helianthus annuus*) leaf in the dark state (solid line) and 10 min after exposure to white light (2000 μmol photons m^{-2} s^{-1}) using an halogen source (dotted line). (b) Difference spectrum (reflectance in dark state minus reflectance in light state) derived from the spectra in (a) (from Gamon and Surfus (1999), *New Phytologist*. Copyright © 1999 with permission from Blackwell).

This index is a measure of photosynthetic performance over a wide range of irradiances: Peñuelas et al. (1995a) found a linear relationship between PRI and the PSII light use efficiency assessed by $\Delta F/F_m{}'$ as

$$\Delta F/F_m{}' = (F_m{}' - F)/F_m{}' \tag{5.8}$$

where F and $F_m{}'$ are the leaf fluorescence measured under ambient light and under saturating light, respectively. However, the slope and intercept varied between species due to a diversity of leaf anatomy, morphology, and pigmentation; they also depend on environmental growth conditions. The same authors found a significant exponential relationship between the instantaneous light-use-efficiency coefficient, ε_i, and the PRI (Figure 5.28). ε_i is commonly used in studies of net primary productivity so that this relationship gave rise to many later articles, in particular in relation to plant stress (e.g., Gamon et al., 1997; Peñuelas et al., 1997; Peñuelas and Filella, 1998; Peñuelas et al., 2004; Filella et al., 2009; Garbulsky et al., 2011; Gamon and Berry, 2012; Kováč et al., 2013).

A large number of studies have used the PRI index for non-destructive monitoring of photosynthetic activity. Bisba et al. (1997) examined young and mature leaves of Oriental plane (*Platanus orientalis*) to determine whether the highly reflective and absorptive leaf hairs found on young leaves were due to a limited photodissipative capacity. Based on low presence of xanthophyll cycle pigments and no response to light at 531 nm, low concentrations of other carotenoid pigments and bleaching of chlorophyll pigments in young leaves compared to mature leaves, these authors

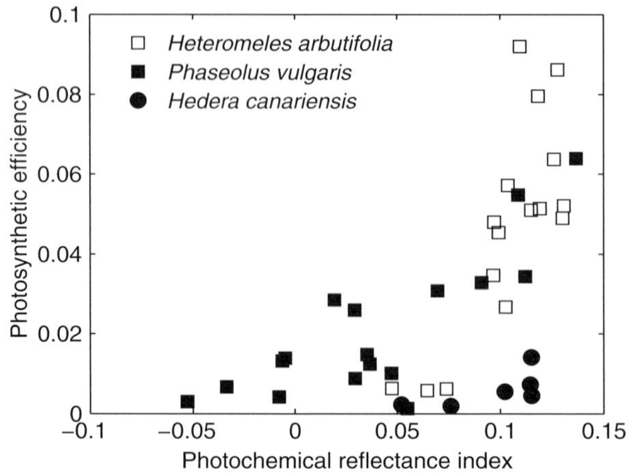

Figure 5.28 Relationship between photosynthetic efficiency (mol CO_2 mol^{-1} PPFD) and PRI, for randomly sampled leaves of three species in field conditions under PPFD ranging from shade to full sunlight. Each point represents a separate leaf (redrawn from Peñuelas et al. (1995)).

concluded that the transient leaf hairs provided some protection of the immature photosynthetic capacity in young leaves. Blanchfield et al. (2006) found that field-grown Pinot noir and Cabernet Sauvignon vines infected with phyloxera increased the concentration of photoprotective pigments as chlorophyll concentration decreased. The change in photoprotective pigments was detectable using the PRI index, which was a good measure of the total chlorophyll to carotenoid ratio. Inoue and Peñuelas (2006) found strong correlations between light use efficiency in soybeans grown under different soil moisture levels and an adjusted PRI. Thenot et al. (2002) found correlations between PRI and water stress in two semi-arid shrub species. Similarly, Ripullone et al. (2011) found strong correlations between PRI and chlorophyll fluorescence, and photosynthetic rates for a range of forest species water stress. *Quercus pubescens* was examined by Stylinski et al. (2000) at a naturally emitting CO_2 spring in Italy. They did not find any reduction in xanthophyll cycle pigments but they did find that the PRI well correlated with electron transport capacity. Gamon et al. (2001) developed a light use efficiency model based on absorbed photosynthetically active radiation (APAR) that is downregulated by the values of PRI in fertilized and unfertilized sunflower fields. They demonstrated that the PRI produced photosynthetic rates that agreed with the field measured rates when radiation efficiency declined.

5.2.3.6 *Infinite Reflectance*

The near-infrared reflectance of plant canopies increases as the leaf area index (LAI, defined as the ratio of total leaf area of plants to ground area covered by plants) increases, until a relatively constant value is reached called the infinite reflectance (R_∞). The effect of LAI can be simulated in the laboratory by stacking leaves in layers. In the equations of Allen and Richardson (1968), R_∞ is also directly related to the optical constant a (see Section 8.1.1). Spectral measurements have been made on a single leaf and on stacked leaves (e.g., Howard, 1966; Gausman et al., 1971a, 1976, 1981; Walter and Koch, 1981; Yoshimura et al., 1991b; Yamamoto et al., 1995; Yoshimura, 1998). The highest reflectances (lowest transmittances) and the largest inter-species differences are obtained at about 0.85 nm in the near-infrared reflectance plateau (Figure 5.29). Similarly, Westman and Price (1988)

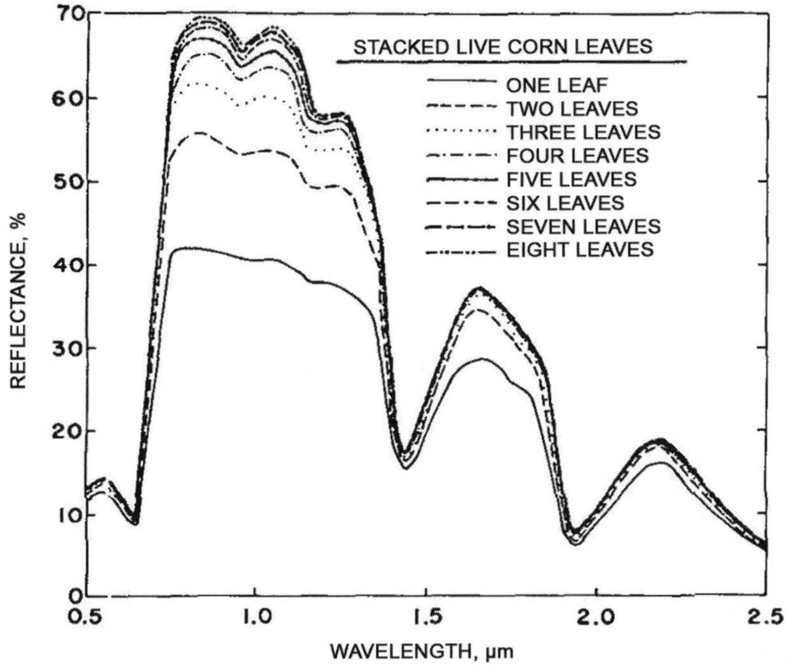

Figure 5.29 Spectrophotometrically measured reflectance of single and stacked maize leaves over the 0.5 to 2.5 µm waveband (from Gausman et al. (1976), *Agronomy Journal*. Copyright © 1976 with permission from the American Society of Agronomy).

report maximum reflectance for stacks of 7–8 conifer needles. Horler et al. (1983c) showed that the red-edge did not shift when fractional plant cover changed but did shift to longer wavelengths as leaf layers increased, attributing the response to light at wavelengths below 700 nm as being dominated by strong absorption by the first leaf in the stack and by scattering processes at longer wavelengths, and thus penetration of light to greater depths in the leaf stack.

5.2.4 Infrared

The first reflectance spectra of plant leaves in the infrared were published by Gates and Tantraporn (1952), who report that old leaves reflect more light between 3 µm and 25 µm than juvenile leaves, which have very different morphology to mature ones. However, Wong and Blevin (1967) and Salisbury (1986) found relatively little spectral variation with senescence in the 8–14 µm wavelength domain. Progressive changes in the thermal infrared transmission of virus-infected epidermal strips of cucumber cotyledons over the 1–15 µm range were found by Cochran et al. (1960), suggesting this technique could be used to monitor biochemical changes. Salisbury and Milton (1988), Elvidge (1988), and recently Fabre et al. (2011) measured the DHRF of leaves at different stages of drying in the 2.5–13.5 µm, 2.5–20 µm, and 2.5–15 µm regions, respectively. They all point out that the main effects occur at wavelengths shorter than 6 µm when a leaf is thoroughly dried, and that there is no significant influence of drying in the 8–14 µm region, where reflectance is more species dependent (Figure 5.30). Narayanan et al.

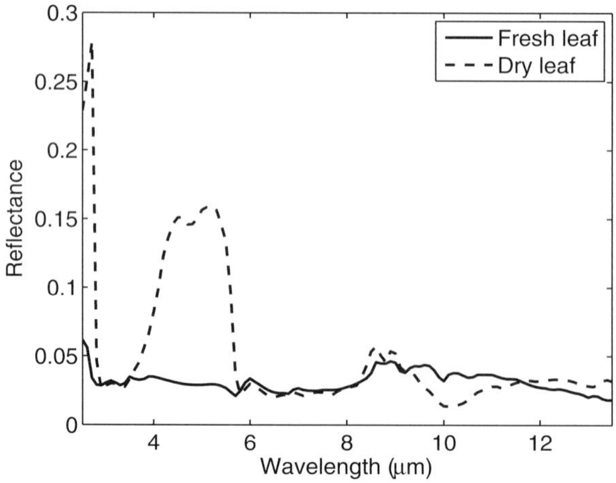

Figure 5.30 DHRF spectra of tap water and a black cherry (*Prunus serotina*) leaf before and after drying for 48 hours in a stream of dry air (redrawn from Salisbury and Milton (1988)).

(1991, 1992) confirm that leaf reflectance does not vary with water content to a noticeable degree in the 9–11 μm wavelength regions. They also show that reflectance starts to increase, or not, at water contents far beyond the wilting point. Gerber et al. (2011) added to these results by measuring both the reflectance and transmittance spectra of about 60 different leaves from 0.4 μm to 14.0 μm. They show that transmittance is also affected by leaf water content in the 3–5 μm atmospheric window. The most extensive reflectance spectrum of a plant leaf was published by Gausman and Hart (1974b) as between 2.5 μm and 40 μm. They collected healthy sweet orange (*Citrus sinensis*) leaves and others heavily coated with sooty-mold fungus, and compared their response showing that discrimination was still possible. The largest differences occurred between 3 μm and 3.5 μm – probably due to a difference in water content – and over the wavelength 6–40 μm interval. Unfortunately, the reflectance values are not absolute so that they are hardly comparable to more recent measurements.

Ribeiro da Luz (2006) and Ribeiro da Luz and Crowley (2007, 2010) recently investigated leaf optical properties in the thermal infrared by completing a comprehensive study between 8 μm and 14 μm. They show that reflectance is linked to leaf surface properties and that plant species identification can be achieved using their unique spectral signatures. Building on this work, in a series of papers, Ullah et al. (2012a, 2012b, 2012c, 2013, 2014) demonstrated both discriminating species with hyperspectral mid-infrared and thermal-infrared data and retrieval of leaf water content. Ullah et al. (2012a) measured leaves of 13 broadleaf species over the 2.5–6.0 μm and 8.0–14.0 μm regions with narrow spectral bands. They showed that separation of species pairs could be made with as few as two carefully selected bands and, in Ullah et al. (2012b), they demonstrated the use of genetic algorithms for identifying a subset of the bands suitable for discrimination and then using these bands in the spectral angle mapper to classify these species. Next, Ullah et al. (2012c) showed that the 2.5–6 μm region could be used to quantify leaf water content using wavelet analysis. Ullah et al. (2013a) found best prediction of leaf water content in the waveband ranges of 2.638–2.650 μm and 3.886–3.948 μm for an index they termed the mid-wave infrared normalized difference water index (MNDWI). Other combinations of mid-infrared bands yielded estimates of water content almost as good. Finally, Ullah et al. (2014) compared retrieval of leaf water content across the region from visible to thermal infrared wavelengths using PLSR models, and again found the mid-infrared bands at 2.65 μm, 3.89 μm, and 5.20 μm had the highest correlation to water content but bands at 1.60 μm and1.397 μm in the SWIR had predictions nearly as good. Villena et al. (2000) found that numerous narrow band absorptions and broad band absorptions could be identified with biopolymer groups in FTIR spectra from bush lily (*Clivia miniata*) cuticles. Hardin et al. (2013) report on a new technique using Fourier transform infrared attenuated total reflectance (FTIR-ATR) spectroscopy over the range from 650 to 4000 cm^{-1} to quantify *in situ* leaf cuticular waxes, whose chemistry was verified by gas chromatography.

As part of a study to better understand forest fire propagation, which is mainly controlled by radiative processes, Monod et al. (2009) and Acem et al. (2010) measured the optical properties of six Mediterranean broadleaf species (*Arbutus unedo, Genista, Juniperus oxycedrus, Quercus coccifera, Rosmarinus officinalis*) and a needle leaf species (*Pinus halepensis*). The energy equation involves an important variable, the extinction coefficient, which varies with plant type. These authors successfully calculate an equivalent plant medium built with individual leaves with known infrared radiative properties. In the aftermath of this work, Boulet et al. (2009, 2011) notice that most fire propagation simulators use basic radiation models, assuming that a plant leaf behaves as a black surface and that a flame is an ideal blackbody emitter. They compare the emissivity spectrum of kermes oak (*Quercus coccifera*) leaves to their flame emission spectrum and show that the reality is much more complex, and that these two assumptions are unlikely to be valid.

Before Monod et al. (2009) and Gerber et al. (2011), the only concomitant directional-hemispherical reflectance (DHRF) and transmittance (DHTF) spectra of plant leaves that we know of were published by Gates et al. (1965) in the 0.3–10 μm region and followed by Gruninger et al. (1992) in the 0.3–25 μm region. These authors show that leaves definitely cannot be regarded as an opaque medium, that is they transmit part of the incoming radiation in some spectral regions, so transmittance should be rigorously taken into account in the calculation of the emissivity, a key variable in leaf energy balance (see Section 12.1). From Kirchhoff's law, the emissivity is then given by (Clarke and Larkin, 1985; DeWitt and Nutter, 1988)

$$\varepsilon(\lambda) = 1 - \rho(\lambda) - \tau(\lambda) \tag{5.9}$$

However, due to the lack of transmittance data in the thermal infrared, some authors have considered leaves as opaque materials. They suppose that their transmittance is negligible and consequently relate their emissivity to reflectance only (e.g., Elvidge, 1988; Salisbury and Milton, 1988; Salisbury and D'Aria, 1992; Salisbury et al., 1994):

$$\varepsilon(\lambda) = 1 - \rho(\lambda) \tag{5.10}$$

In spite of this approximation, the datasets published by these authors form the first and, to date, the only open libraries of leaf emissivity. In most publications, the range from 0.94 to 0.99 is considered as an accepted order of magnitude for ε in the thermal infrared region (Idso et al., 1969). But one should keep in mind that the emissivity of a leaf equals its absorption, and therefore also depends on its transmittance. Let us take the example of a leaf whose reflectance and transmittance are 0.134 and 0.179 at 5 μm. We have $\varepsilon = 1 - 0.134 - 0.179 = 0.687$ and not $\varepsilon = 1 - 0.134 = 0.866$. Björn and Li (2011) tested the idea that silica accumulation in the epidermal tissue would reduce leaf temperatures by increasing infrared emissivity over the 2.5 to 22.3 μm range, which they concluded was unsupported by their measurements. Note that the emissivity of a single leaf is not representative of the emissivity of a vegetation canopy because of the multiple internal reflections resulting from the canopy architecture (Fuchs and Tanner, 1966). By analyzing integrated emissivity acquired from drying plant canopies or drawn from the ASTER spectral library (http://speclib.jpl.nasa.gov/), Olioso et al. (2007) emphasized the need for better knowledge of leaf optical properties in the infrared, especially as a function of water content. However, to date, the complexity of the phenomena involved discouraged the development of studies, which partly explains the scarcity of experimental measurements over this wavelength range.

5.2.5 Microwaves and Radio Frequencies

Leaves are among the most important sources of scattering in green plant canopies. In the microwave region, propagation of electromagnetic radiation through such a medium is controlled by their volume absorption coefficient κ_a and their volume scattering coefficient κ_s. They both depend on leaf size, shape, and orientation, three variables that can be parameterized by probability density functions; they also depend on canopy equivalent dielectric permittivity, which can be modeled in terms of the dielectric constants ϵ of the canopy constituents – a leaf can be regarded as a thin, non-magnetic lossy dielectric layer – and their individual volume fractions. As a consequence, several experiments have been conducted since the middle of the 1980s to measure the dielectric properties of plant leaves at different frequencies (Table 5.4). The three main domains that emerge are the radio waves (< 1 GHz), the microwaves (1–100 GHz), and terahertz waves (100–3000 GHz).

The result of these studies is that leaf relative permittivity ϵ is shown to mainly depend on water content, free ion concentration, and tissue structure. Hill et al. (1986) and Broadhurst et al. (1987) investigated the interaction of radio waves ranging from 10^{-2} Hz (Extremely Low Frequency) to 10^9 Hz (Ultra High Frequency) with the fleshy leaves of the succulent jade plant (*Crassula portulacea ovata*). Figure 5.31 shows an increase in ϵ' and ϵ'' curves with decreasing frequency. In order to study the effect of cell walls on the electrical properties, Broadhurst et al. (1987) also measured the liquid extracted from the leaf and observed no difference between 10^7 Hz and 10^9 Hz.

Hill et al. (1987) compared the dielectric properties of wilting *Rhoeo* leaves, measured over an 11-day period, and studied the effect of thickness by stacking up to eight common bean (*Phaseolus vulgaris*) leaves. They notice a decreasing response in both cases. Variation in the dielectric permittivity in this low-frequency range has been interpreted in terms of leaf ionic concentration. Saltas et al. (2007) show that tree leaves containing different amounts of heavy metals (Cd, Cu, Ni, Pb, Zn) display different dielectric properties. An interesting application is using the dielectric permittivity for environmental monitoring to identify spreading of pollution from industrial, domestic, or vehicle emissions.

Ulaby and Jedlicka (1984), El-Rayes and Ulaby (1987), and Chuah et al. (1997) extend the experimental observations to microwaves and especially investigate the effect of water content on the leaf dielectric properties. When water is tightly bound to the host material, its dielectric behavior is similar to that of ice, while free water is believed to exhibit the same dispersion properties as liquid water. Figure 5.32 displays the dielectric permittivity of maize (*Zea mays*) leaves, measured as a function of volume/tric moisture content. Sancho-Knapik et al. (2011a) present the prototype of

Table 5.4 *Measurement of leaf dielectric properties at different frequencies.*

Band	Frequency	Wavelength	Reference
RF	< 300 MHz	> 1 m	Du and Peake (1969), Baffioni et al. (1985), Hill et al. (1986, 1987), Broadhurst et al. (1987), El-Rayes and Ulaby (1987), Saltas et al. (2007), Afzal and Mousavi (2008)
P	0.3–1 GHz	0.3–1 m	
L	1–2 GHz	15–30 cm	Ulaby and Jedlicka (1984), El-Rayes and Ulaby (1987), Sancho-Knapik et al. (2011a)
S	2–4 GHz	7.5–15 cm	El-Rayes and Ulaby (1987)
C	4–8 GHz	3.75–7.5 cm	Ulaby and Jedlicka (1984), El-Rayes and Ulaby (1987), Mätzler and Sume (1989), Chuah et al. (1997), Chung (2007)
X	8–12 GHz	2.5–3.75 cm	Tan (1981), Ulaby and Jedlicka (1984), El-Rayes and Ulaby (1987), Harbinson and Woodward (1987), Sarabandi and Ulaby (1988), Mätzler and Sume (1989), Chuah et al. (1995, 1997)
Ku	12–18 GHz	1.67–2.5 cm	El-Rayes and Ulaby (1987), Chuah et al. (1997)
K	18–24 GHz	1.11–1.67 cm	El-Rayes and Ulaby (1987), Mätzler and Sume (1989), Lin et al. (2008)
Ka	24–40 GHz	0.75–1.11 cm	Mätzler and Sume (1989), Mätzler (1994), Hsieh (2003), Lin et al. (2008)
V	40–75 GHz	4.0–7.5 mm	
W	75–300 GHz	1–4.0 mm	Borel and McIntosh (1988), Mätzler and Sume (1989), Mätzler (1994), Hsieh (2003), Helhel et al. (2009)
T-ray	0.3–10 THz	30–1000 μm	Hu and Nuss (1995), Hadjiloucas et al. (1999, 2002, 2009), Schade et al. (2004, 2005), Hellicar et al. (2007), Jördens et al. (2009), Breitenstein et al. (2011), Zuo et al. (2011), Gente et al. (2013), Born et al. (2014a), Gente and Koch (2015)

a compact, low-budget, portable device that allows non-destructive estimates of the relative water content (RWC) of poplar leaves (*Populus euramericana*) by measuring their reflectivity at 1730 MHz (L-band). This technique seems to be very accurate, but it has been only validated on one species, so further experiments are required to study the effect of leaf structure on the signal. Sancho-Knapik et al. (2013) found best results using both the reflectivity in the S-band (2.4 GHz) and broadband ultrasonic spectroscopy for a reproducible, non-invasive technique following changing water relations in common grape vine (*Vitis vinifera*). The S-band data provided accurate estimates of changes in the dielectric permittivity and the non-contact, non-invasive broadband ultrasonic spectroscopy enables measurement of properties of the leaf turgor above the loss point, based on the frequency associated with the maximum transmittance and the maximum macroscopic elastic content of the leaf in the z-direction. Finally, since the permittivity of water is temperature dependent, the permittivity of the leaves is likely to be also temperature dependent. El-Rayes and Ulaby (1987) and Harbinson and Woodward (1987) studied the effects of repeated freeze-thaw cycles on different leaf species, between negative and room temperatures. A major result is that ϵ exhibits a hysteresis-like effect as a function of temperature, which may be due to frost damage of the leaf cells during water crystallization.

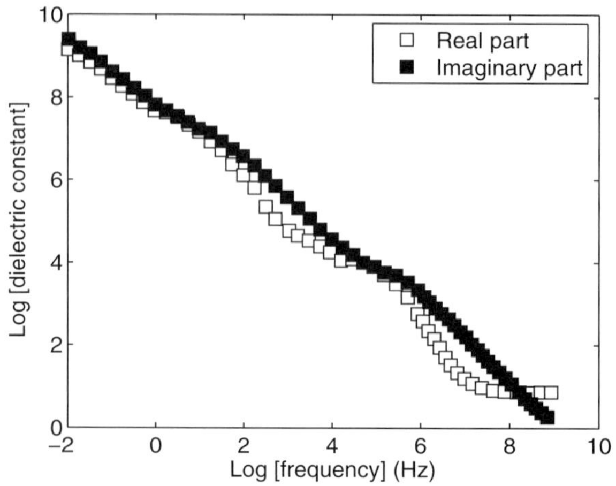

Figure 5.31 Real (ϵ', open symbols) and imaginary (ϵ'', filled symbols) parts of the room temperature complex relative permittivity vs. frequency for a jade plant (*Crassula ovata*) leaf. The data were obtained from three separate sets of equipment (10^{-3} to 10^4 Hz, 50 to 1.3 10^7 Hz, and 10^6 to 10^9 Hz) (redrawn from Broadhurst et al. (1987)).

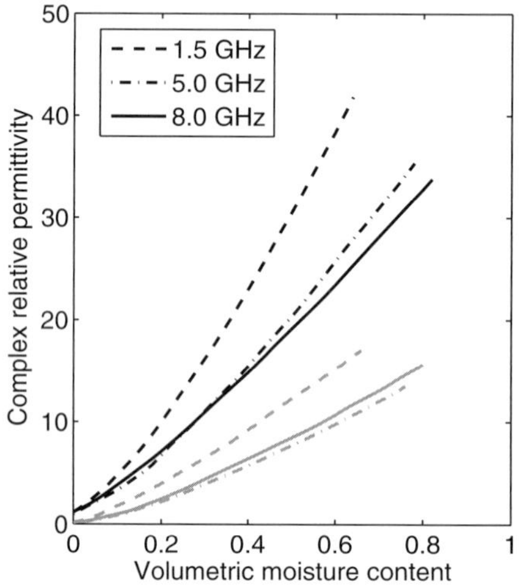

Figure 5.32 Real part (black lines) and imaginary part (gray lines) of the complex relative permittivity of maize (*Zea mays*) leaves measured at 1.5 (–2), 5.0 (3.5–6.5), and 8.0 (7.5–8.5) GHz, respectively, for different moisture contents and at a temperature of 23°C (redrawn from Ulaby and Jedlicka (1984)).

Terahertz radiation, which is strongly absorbed by water molecules, is another frequency for monitoring changes in leaf water content (e.g., Hadjiloucas et al., 1999, 2002; Jörden et al., 2009; Breitenstein et al., 2011; Zuo et al., 2011; Gente et al., 2013; Born et al., 2014a; Gente and Koch, 2015). One of the advantages is that it is a non-invasive and real-time technique; it has recently become possible to acquire a THz image of the entire blade, as seen in Figure 5.33. For instance, Schade et al. (2004, 2005) used THz synchrotron near-field images to show distinct structural

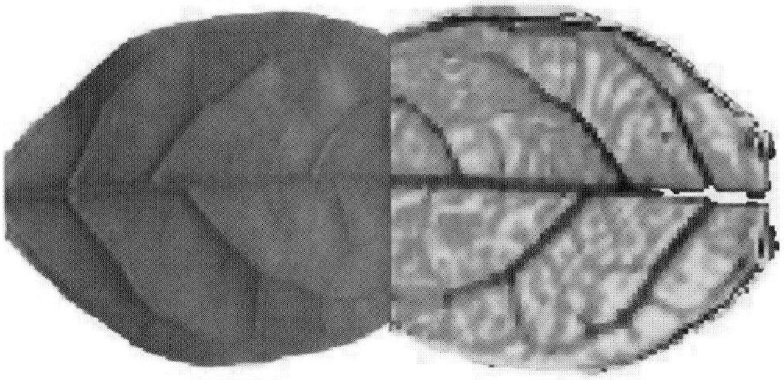

Figure 5.33 Images of leaf in optical domain (left) and at 615 GHz (right) (from Hellicar et al. (2007), in *Proc. 2nd International Conference on Wireless Broadband and Ultra Wideband Communications*. Copyright © 2007 with permission from the IEEE). (A black and white version of this figure will appear in some formats. For the color version, please refer to the plate section.)

differences between mesophytic Virginia creeper (*Parthenocissus* sp.) and xerophytic bay laurel (*Laurus nobilis*) leaves. Due to its unique features of penetration and resolution, it allows mapping of the spatial distribution of water during leaf drying (Hu and Nuss, 1995; Hellicar et al., 2007). Federici (2012) reviewed detection and mapping of water using terahertz imaging, including a review of applications for agriculture, such as changing diurnal water content in orchard foliage among many examples for other fields of study.

5.2.6 Ultra-Sounds

Although acoustic waves (or sound waves) are not electromagnetic waves, we will briefly describe here their interaction with plant leaves. Acoustic waves can travel through all kinds of materials including water. They generally travel a lot slower than electromagnetic waves, but the speed of each is dependent on the characteristics of the medium that it is traveling through so they can be used as a non-invasive method of investigation. There are only a few applications of this technique in plant science because measuring the acoustic characteristics of thin biological samples is generally difficult. Ultrasonic waves were first used by Torii et al. (1988) to determine the water content of tobacco (*Nicotiana tabacum*) leaves. Fukuhara (2002) accurately determined the phase velocity (speed of individual waves) and the attenuation coefficient (energy loss in the medium) of longitudinal ultrasonic waves transmitted by leaves suspended in water. They found good relationships between both acoustic characteristics and leaf thickness, one of the physical parameters that reflect their hardness, density, water potential, and photosynthetic activity (Figure 5.34). The phase velocity is smaller in leaves than in pure water (1497 m s^{-1}) because of their loosely packed parenchyma cells separated by intercellular air spaces.

Fukuhara et al. (2000, 2005, 2006) successfully applied this technique to tea (*Camellia sinensis*) leaves in order to provide information about plant maturation and predict the best times for harvesting, which greatly impacts the quality and the price of the tea. However, it also presents the drawback of using water as a coupling medium. Water tends to enter the leaves and to modify their acoustic

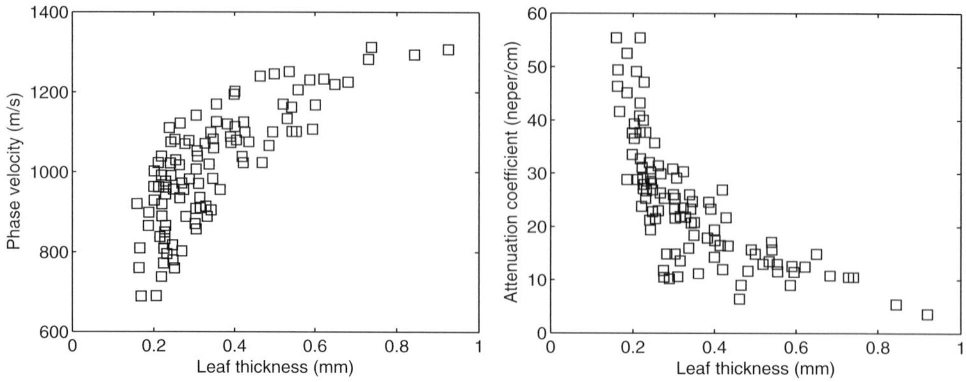

Figure 5.34 Thickness dependence of phase velocity (left) and attenuation coefficient (right) for 111 leaf species (redrawn from Fukuhara (2002)).

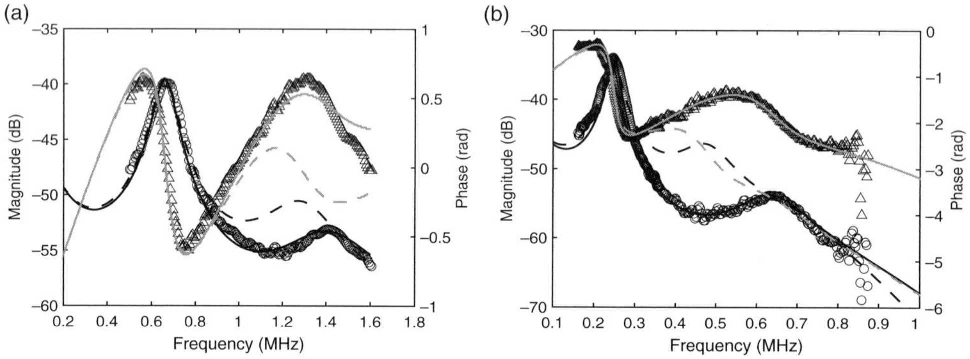

Figure 5.35 Magnitude and phase spectra of the transmission coefficient of (a) grape vine (*Vitis vinifera*) and (b) golden pothos (*Epipremnum aureum*) at normal incidence. O and Δ: measured magnitude and phase spectra, respectively. Dashed line: theoretical predictions of the one-layer model. Solid line: theoretical predictions of the bi-layer model (from Fariñas et al. (2013), *Applied Physics Letters*. Copyright © 2013 with permission from AIP Publishing LLC).

properties. To address this issue, new sensors, able to transmit and receive ultrasounds to and from the air at different frequencies, and advanced air-coupled ultrasonic spectroscopy techniques have been implemented. Transmittance of ultrasound through plant leaves reveal the presence of at least one resonant frequency where it is maximum. The magnitude and phase spectra of the transmission coefficient have been successfully reproduced using acoustic models (Figure 5.35); one-layer models mimicking an ideal homogeneous and isotropic monocot leaf (Álvarez-Arenas et al., 2009b), two-layer models (Fariñas et al., 2013), or four-layer models (Álvarez-Arenas et al., 2009a) mimicking heterogeneous and anisotropic dicot leaves (upper epidermis, palisade parenchyma, spongy mesophyll, and lower epidermis). This technique was recently refined by Sancho-Knapik et al. (2010, 2011b) and Fariñas et al. (2013, 2014a, 2014b) to determine variations of thickness, water content (relative water content, water potential), and dry matter in leaves.

The acoustical characteristics of leaves may have other applications than plant physiology. It is well-known that vegetation (leaves, but also branches, twigs, and wood) may reduce noise by absorbing sound vibrations and that broadleaf species partly lose their effectiveness in the winter when deciduous. Wilson and Dunton (2009) investigated the acoustic response of aquatic plants in the frequency band 0.5–2.5 kHz to improve the performance of sonar in mine hunting, which is severely reduced in seagrass meadows.

In an anecdotal way, a team of scientists made an amazing discovery (Simon et al., 2011). Deep in the rainforest of Cuba, they identified the rainforest vine (*Marcgravia evenia*), a rare plant that attracts nectar-feeding bats to its flower by manipulating sound. This plant evolved large, dish-shaped leaf above its red-and-pink buds to facilitate echolocation. This ingenious system does indeed help bats to navigate and find their flowers.

5.3 Leaf Color

Leaf color is a consequence of the many physical (structural scattering) and chemical (pigment absorption) phenomena that occur in the visible region. As an example, Zhang et al. (2009) found that the metallic luster in leaves of *Begonia rex* arose from light reflected from interior structure, suggesting it may originate from chlorophyll, cell arrangement, and air spaces. The wide variation in leaf color among some horticultural species and varieties has long been noted. One early study on commercial varieties of *Coleus* by Boye and Rife (1938) examined heredity of leaf color and pattern from selfed and backcross progeny and attributed the variation in these traits to a pair of heterozygous allelomorphs.

5.3.1 Variegation

Variegation is the appearance of white (or colors other than green) splotches at the leaf surface of plants mainly growing in forest understory environments. In a classic work based on a study of 55 species from 24 families, Hara (1957) identified two main mechanisms that account for foliar variegation, which may be produced by a lack of chloroplasts in certain tissues of the mesophyll, or by an air layer located just below the adaxial epidermis, which is the cause of back and forth light scattering. Many authors have assumed that variegation was due to chlorophyll deficiency alone (e.g., Fisher, 1986; Aluru et al., 2001), while only a few studies were in favor of its structural cause (e.g., Fooshee and Henny, 1990; Tsukaya et al., 2004; Sheue et al., 2012). For instance, Tsukaya et al. (2004) examined dropped tongue (*Schismatoglottis calyptrata*), a flowering plant in the *Araceae* family native to southeast Asia, Melanesia, Borneo, etc., and *Begonia*. They determined that the leaf areas with gray-green or light colored areas resulted from structural features where the connection between the palisade cells and the epidermis was loose and with air spaces between these tissues, while in green parts of the leaf, there was close contact between these cells. Sheue et al. (2012) studied the leaf anatomical structure for varieties of *Begonia* and found the color variation was controlled by air spaces between the tip of the palisade parenchyma cells and the epidermis. The distribution of chloroplasts and chlorenchyma cell shapes varied but the grana and stroma of chloroplasts were not different. Thus, it appears that both structure and pigments contribute to variation in leaf color. Why would plants in shaded environments reduce the absorption of light in portions of the leaf? One answer, developed in Section 12.5.1, may be that such leaves are perceived as not leaf-like or as already containing herbivorous insects (as leaf minors), thus avoiding being eaten by other insects (Lee, 2011).

5.3.2 Plant Names

Identifying a plant can be a tricky task, even for the experienced botanist, considering the huge number of species existing in the world. The numerous criteria of identification include leaf shape and texture but also, to a lesser extent, leaf color due to the subjective character of color. In his *Philosophia Botanica* of 1751, Swedish botanist Carl Linnaeus (1707–1778) warned his readers "not to put too much trust in colour" (Simpson, 2011). However, botanists sometimes gave names to plants with regard to their leaf color when it unambiguously represents a trait of the species. The rules that govern the naming of plants are complicated and were the source of much deliberation in the 19th and 20th centuries. Common names of plants typically apply to genera. Because they vary from one region to another, from one language to another, Latin has been adopted as the accepted language of botanical nomenclature (Gledhill, 1985).

In botanical nomenclature the scientific name of a plant is a combination of the generic name (genus) followed by a specific epithet (species) that is always used in conjunction with the genus. This name is followed by the name of the person or persons who first described a species, referred to as the scientific authority. These names are often shorted to an abbreviation. However, they can cause confusion and they are not obligatory, so we decided not to mention the scientific authority in this book. Both the genus and species names may be a reminder of leaf color, but the generic name is rarely associated with it. As an example, the genus *Tephrosia* (*Fabaceae*) is derived from the Greek word τεφρος (tephros) meaning "ash-colored" and referring to the grayish tint given to the leaves by their dense trichomes; the genus *Glaucium* (*Papaveraceae*) is derived from the Greek word γλαυκός (glaukos) meaning "gray-green." Most names expressing leaf color or surface property related to it are epithets (Table 5.5).

5.4 Light Gradients

A wide variety of spectral data are available for separate leaf tissues (Lin and Ehleringer, 1993). However, the light environment varies continuously across the leaf. The exponential decrease in the amount of light transmitted, which results from light attenuation by both absorption and scattering, can be predicted by theoretical models of light gradients across leaves (e.g., Seyfried and Fukshansky, 1983; Seyfried and Schafer, 1983; Kazarinova-Fukshansky et al., 1985; Terashima and Saeki, 1985; Fukshansky and Martinez von Remisowsky, 1992; Martinez von Remisowsky et al., 1992). However, leaves are more optically complex than is assumed by current models (e.g., Beer–Lambert law, Kubelka–Munk theory, or multilayer approach), so that it is difficult to estimate light quality or quantity at different depths. Vogelmann and Björn (1984) pioneered the experimental measurement of light gradients in plant tissues using a fiber optic driven through the cells by a motorized micromanipulator (see Section 4.7). Figure 5.36 shows the ratio of the amount of light inside an alfalfa (*Medicago sativa*) leaf to that outside (Vogelmann et al., 1989). Both green light (550 nm) and red light (680 nm) are rapidly attenuated as they travel through the lamina. Results at 450 nm would be similar to those obtained at 680 nm due to strong absorption of blue and red light by chlorophyll. The steeper light gradients at 550 nm are caused by multiple scattering of green light within the intercellular air spaces and/or cells, which increases the optical path length and therefore light absorption. This differential response induces changes in the composition of light available in the leaf.

Table 5.5 *The names of plants after leaf (or bract) color and surface aspect (Bailey, 1963; Gledhill, 1985; Eggli and Newton, 2004; Stearn, 2010).*

Character	Latin epithet	Meaning	Example
Black	ater	Black	*Vlokia ater*
	atratus	Blackened, dark	*Sedum atratum*
	atrofuscus	Black-brown	*Haworthia magnifica* var. *atrofusca*
	atropurpureus	Black-purple	*Crassula atropurpurea, Echeveria atropurpurea*
	atrovirens	Dark green	*Agave atrovirens*
Blue	caeruleus	Blue	*Molinia caerulea*
	caesius	Light blue	*Pachyphytum caesium, Tetragonia caesia*
Brown	ferrugineius	Light brown, with a little mixture of red	*Rhododendron ferrugineum*
	luridus	Dark yellow, dirty brown	*Agave lurida*
Green	glaucescens	Becoming blue-green	*Carpobrotus glaucescens, Synadenium glaucescens*
	glaucifolius	Comp. of Lat. *glaucescens*	*Trianthema glaucifolium*
	glaucophyllus	Bluish, blue-green	*Dasylirion glaucophyllum, Rosularia sempervivum* var. *glaucophylla*
	glaucus	Dull green, gray-green, blue-green	*Aloe glauca, Arctostaphylos glauca, Bulbine glauca, Festuca glauca, Gasteria glauca, Guioa semiglauca, Nicotiana glauca, Picea glauca, Quercus glauca, Salix glauca, Smilax glauca, Yucca glauca*
	olivaceus	Olive-green, olive-brown	*Echeveria olivacea, Lithops olivacea*
	pallidus	Pale green	*Delosperma pallidum, Sedum pallidum, Yucca pallida*
	viridissimus	The greenest	*Echeveria viridissima*
Gray	cinerarius	Ash-colored	*Senecio cineraria*
	cinereus	Ash-gray	*Eriogonum cinereum, Eucalyptus cinerea, Namibia cinerea*
	griseus	Gray	*Acer griseum, Aloe grisea, Begonia grisea, Echeveria grisea*
Red	cupreus	Coppery	*Alocasia cuprea*
	erythraeus	Red	*Sempervivum erythraeum*
	erythrocomos	Red leaf	*Rubus erythrocomos*
	erythrophyllus	Red leaf	*Aloe erythrophylla*
	erythrosorus	Having red spore cases	*Dryopteris erythrosora*
	purpuratus	Purple-red	*Plectranthus purpuratus*
	purpureus	Purple	*Setcreasea purpurea*
	rubens	Becoming red	*Crassula rubens, Sedum rubens*
	ruber	Pure red	*Acer rubrum, Quercus rubra, Synadenium compactum* var. *rubrum*
	rubineus	Ruby-red	*Crassula sieberiana* subsp. *rubinea*
	rubrotinctus	Red	*Sedum rubrotinctum*
	rubrovensus	Reddish striation	*Tylecodon rubrovenosus*

Table 5.5 (*cont.*)

Character	Latin epithet	Meaning	Example
	rubroviolaceus	Red-violet	*Aloe rubroviolacea*
	rufescens	Becoming reddish	*Anacampseros rufescens*
	sanguineous	Sanguine	*Rumex sanguineus*
Silver	argenteus	Silvery, silver-gray	*Antimima argentea, Elaeagnus argentea, Guettarda argentea, Lampranthus argenteus, Lupinus argenteus*
Violet	violaceus	Violet	*Opuntia violacea*
White	albescens	Becoming white	*Agave albescens*
	albicomus	White hair-like	*Mammillaria albicoma*
	albidior	Any kind of white a little soiled	*Agave mitis* var. *albidior*
	albidus	Comp. of Lat. *albidior*	*Cistus albidus, Machairophyllum albidum*
	albinota	White-spotted	*Rabiea albinota*
	albipunctus	Comp. of Lat. *albinota*	*Rabiea albipuncta*
	albivenius	White-veined	*Ipomoea albivenia*
	albomarginatus	White margin	*Agave albomarginata, Hosta undulata* var. *albomarginata*
	albovillosus	Densely hairy, villous	*Euphorbia albovillosa*
	albus	Dead white	*Populus alba, Salix alba*
	candicans	Whitish, hoary-white	*Quercus candicans*
	farinosus	Mealy, powdery	*Dudleya farinosa, Encelia farinosa, Sedum farinosum*
	leuconeura	White-veined	*Maranta leuconeura*
Yellow	luteoviridis	Yellow-green	*Gibbaeum luteoviride*
Brightness	lucidus	Bright, shining, clear	*Lindsaea lucida, Plectranthus lucidus, Sedum lucidum*
	lutatus	Muddy, dirty	*Astridia lutata*
	metallicus	Lustrous, metallic in appearance	*Aloe metallica, Echeveria gibbiflora* var. *metallica, Sansevieria metallica*
	micans	Gleaming, with slight metallic luster	*Drosanthemum micans*
	vitreus	Glassy, transparent	*Bulbine vitrea*
Pubescence or roughness	attonsa	Shaven	*Trichodiadema attonsum*
	ciliaris	Fringed, ciliate	*Aloe ciliaris*
	ciliatus	Ciliate	*Brownanthus ciliatus*
	ciliosus	Comp. of Lat. *ciliatus*	*Sedum radiatum* var. *ciliosum, Sempervivum ciliosum*
	glaber	Glabrous	*Rosularia glabra*
	glabriphyllus	Comp. of Lat. *glaber*	*Pelargonium glabriphyllum*
	hirsutissimus	Very hairy	*Operculicarya hirsutissima, Portulaca hirsutissima*

Table 5.5 (*cont.*)

Character	Latin epithet	Meaning	Example
	hirsutum	Roughly hairy	*Crassula hirsuta, Cucumis hirsutus, Raphionacme hirsuta*
	hirtifolius	Hairy	*Tylecodon hirtifolium*
	hirtipes	Comp. of Lat. *hirtifolius*	*Crassula hirtipes*
	hirtus	Comp. of Lat. *hirsutum*	*Pelargonium hirtum*
	malacophyllus	Softly fleshy or hairy	*Jatropha malacophylla, Orostachys malacophylla*
	pubescens	Pubescent, finely hairy, downy	*Agave pubescens, Betula pubescens, Polygonatum pubescens, Quercus pubescens*
	rugosifolius	Wrinkled	*Aloe rugosifolia*
	tomentellus	Felty matter	*Jatropha seineri* var. *tomentella, Pedilanthus tomentellus*
	tomentosus	Densely wooly, covered in matted hairs	*Carya tomentosa, Cerastium tomentosum, Diplazium tomentosum, Kalanchoe tomentosa, Pelargonium tomentosum, Populus tomentosa, Prunus tomentosa, Rhodomyrtus tomentosa, Stemodia tomentosa, Tilia tomentosa*
	velutinus	With a soft silky covering	*Astridia velutina, Crassula sericea* var. *velutina, Quercus velutina*
	velveteus	Velvet	*Impatiens morsei* "velvetea"
	viscatus	Sticky	*Aeonium lindleyi* var. *viscatum*
	viscidus	Viscid, glutinous	*Dudleya viscida*
	viscosus	Comp. of Lat. *viscatus*	*Haworthia viscosa*
Variegation	discolor	Of different colors	*Cissus discolor, Rhodiola discolor*
	fasciatus	Banded	*Haworthia fasciata, Sansevieria fasciata*
	guttatus	Spotted	*Begonia guttata*
	hieroglyphica	That resemble hieroglyphs	*Vriesea hieroglyphica*
	maculatus	Blotched, spotted	*Aloe maculata, Arum maculatum, Begonia maculata, Hypericum maculatum*
	maculosus	Spotted, mottled	*Agave maculosa, Sinocrassula indica* var. *maculosa*
	marginatus	Edged	*Dracaena marginata*
	marmoratus	Marbled	*Manfreda marmorata, Kalanchoe marmorata*
	punctatus	Dotted	*Aichryson punctatum*
	punctulatus	Small dots	*Ruschia punctulata*
	variegatus	Variegated	*Codiaeum variegatum, Vinca major variegata*
Vague	coloratus	Colored	*Agave colorata*
	concolor	Uniform in color	*Abies concolor*

The leaf epidermal layer is known to absorb harmful UV-A (~ 360 nm) and UV-B (~ 310 nm) radiation, acting as an effective filter that protects photosynthetically sensitive tissues from damage (see Section 5.2.2). Such an effect may be reinforced by the presence of trichomes or epicuticular waxes, which develop on the adaxial and abaxial side of certain leaf species. Therefore, leaf absorption profiles

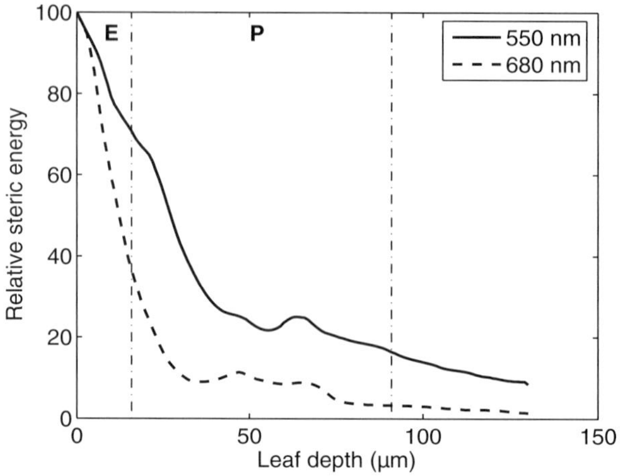

Figure 5.36 Distribution of transmitted light across an alfalfa (*Medicago sativa*) leaf. The boundaries of the upper epidermis (E), palisade (P), and spongy mesophyll (S) are shown at the top of the graph (redrawn from Vogelmann et al. (1989a)).

in the ultraviolet domain have been intensively studied on needles (Bornman and Vogelmann, 1988; DeLucia et al., 1992) and broadleaf species (Bornman and Vogelmann, 1991; Karabourniotis and Bornman, 1999a; Qi et al., 2002, 2003a). Qi et al. (2003c) showed that most of the UV light (98% at 310 nm and 96% at 360 nm) was attenuated within the first 5 μm of the upper epidermis.

Some authors noted that the palisade mesophyll might propagate light better than the spongy mesophyll (e.g., Terashima and Saeki, 1983; Knapp et al., 1988; Vogelmann, 1993; Vogelmann and Martin, 1993; Terashima et al., 2011). The allocation of light between the two mesophyll tissues is mainly due to differences in cell shape and in chlorophyll content. Cui et al. (1991) measured light gradients in leaves of spinach (*Spinacia oleracea*) grown under sunlit and shaded conditions. Although they have the same chlorophyll content, sunlit leaves are thicker than shaded leaves (631 μm vs. 456 μm). These authors showed that collimated light was propagated further into the tubular cells of the palisade of sunlit leaves in comparison to shaded leaves (120 μm vs. 76 μm at 450 nm). The particular shape of these cells seems to facilitate the penetration of light because of their central vacuole that presses the chloroplasts closely against the cell wall and serves as a transparent channel (Evans et al., 2004). Under diffuse light conditions, such as the sky on a cloudy day, or in the shade of plant canopies, internal light gradients are steeper in the palisade than when the leaves are irradiated with collimated light. It also shows that the transmission of light at 0° is much larger than at the other sampling orientations. Terashima et al. (2009) studied chloroplast density deeper into the leaf (lower palisade and spongy parenchyma) and found that absorption of green light at these depths had a much greater effect on photosynthesis than increased blue or red light. Brodersen and Vogelmann (2010) showed that depth of light penetration into the leaf of common snapdragon (*Antirrhinum majus*) varied with the incident light angle, wavelength, direct or diffuse light, and with the leaf's adaptation to prior light intensity. When the adaxial surface was irradiated with green light at 60° from the normal, the shape of the profile was changed and decreases faster with depth than when the light was normal to the surface.

These studies reveal the regulatory role of leaf anatomy on the distribution of light within the leaf, which optimizes absorption of light energy for photosynthesis (Vogelmann et al., 1996b; Vogelmann and Gorton, 2014). Nishio et al. (1993) studied the correlation between the light gradient and photosynthetic carbon fixation across leaves of spinach (*Spinacia oleracea*). Contrary to photosynthesis models that suggest maximal CO_2 fixation near the upper leaf surface, where light gradients are the steepest, they report that it occurs in the medial section of the leaf, at a depth of 100–250 μm in sunlit leaves and 80–120 μm in shaded leaves. This corresponds to the points at which approximately 90% of the blue and red lights are attenuated. This off-shifting of CO_2 fixation from the light gradient is at first sight surprising but it has been confirmed by other experiments (Takahashi et al., 1994; Sun et al., 1996, 1998; Vogelmann and Han, 2000; Vogelmann and Evans, 2002; Evans and Vogelmann, 2003) and the most recent models (Ustin et al., 2001). Anthocyanin in the mesophyll of red tawheowheo (*Quintinia serrata*) leaves restrict penetration of green light beyond the palisade parenchyma that reduces photosynthesis to the level of low light adapted plants (Gould et al., 2002). Fan et al. (2002, 2003) reported that although 20% of the transmitted red light in Japanese spindle (*Euonymous japonicus*) reached the spongy parenchyma, only a small fraction was absorbed for photosynthesis. Note that such techniques have also been used to study the distribution of chlorophyll fluorescence within leaves: when the adaxial side of an alfalfa (*Medicago sativa*) leaf is illuminated at 0° by blue light, the maximum fluorescence intensity measured in the two main peaks of fluorescence emission at approximately 685 nm and 730 nm occurs in the middle of the palisade mesophyll (Bornman et al., 1991; Koizumi et al., 1998).

Shortly after the first light profiles were published, a surprising optical phenomenon, predicted by models, was reported – the ability of plant tissues to trap light (e.g., Vogelmann and Björn, 1984, 1986; Mandoli and Briggs, 1984a, 1984b; Vogelmann, 1986, 1989; Pilarski and Rajba, 2004). Optical boundaries, which are created when the refractive index between two media changes very rapidly, control the penetration and distribution of light within leaves. These boundaries act as mirrors that can bounce photons back and forth between the different tissue layers, which can result in more light inside the leaf than outside. For instance, the amount of light immediately beneath the irradiated surface can exceed by several times that of incident light. The degree of light trapping depends upon the amount of absorption by the leaf. It is higher at 730 nm, where absorption is low, than at 660 nm, where absorption by chlorophyll is high. Until recently, the measurement of absorption profiles within plant leaves was more relevant to plant physiology than to remote sensing studies. The introduction of within-leaf chlorophyll profile gradients into leaf optical properties models may provide an improved estimation of this photosynthetic pigment by remote sensing techniques.

5.5 Near-Infrared and Fourier Transform Infrared Spectroscopy

5.5.1 Near-Infrared Spectroscopy

In the 1970s, near-infrared (reflectance) spectroscopy, commonly known as NIRS, offered a rapid, repeatable, accurate, and cost-effective alternative approach to conventional wet chemical analysis that was time-consuming and expensive. In the agricultural literature, the "near-infrared" region is taken in a broad sense, covering the wavelength range from 780 nm to 2500 nm and actually corresponds to the near infrared (760–1400 nm) and middle-infrared (1400–3000 nm), as defined in Section 4.1.1. Karl Norris is regarded as the father of modern near-infrared spectroscopic analysis. While working at the USDA Instrumentation Research Laboratory, Beltsville, USA, he developed the instruments and utilized computers to demonstrate that this region of the electromagnetic

Table 5.6 *Application of near-infrared spectroscopy to the analysis of leaf properties.*

Variable	Reference
Forage quality	Norris et al. (1976), Shenk et al. (1979, 1981), Himmelsbach et al. (1988), Smith and Flinn (1991), Garcia Ciudad et al. (2004)
Nutrients/heavy metals	Shenk et al. (1979, 1981), Clark et al. (1987), Smith et al. (1991), Gillon et al. (1999a), White et al. (2000), Kokaly (2001), Chen et al. (2002), Gislum et al. (2004), Petisco et al. (2005), Font et al. (2007), González-Martín et al. (2007), Guo et al. (2008), Yi et al. (2010b), van Maarschalkerweerd et al. (2013), Zhai et al. (2013)
Chemical content	Card et al. (1988), Wessman et al. (1988a), McLellan et al. (1991a, 1991b), Aber et al. (1994), Hattey et al. (1994), Martin and Aber (1994), Thygesen (1994), Bolster et al. (1996), Raymond and Schimleck (2002)
Litter decomposition	Joffre et al. (1992), Gillon et al. (1993), Lacaze and Joffre (1994), Gillon et al. (1999a, 1999b), Ono et al. (2003), Shepherd et al. (2003), Takahashi et al. (2004)
Intraspecific variability	Xu et al. (2009); Kokalj et al. (2011)
Interspecific diversity	Atkinson et al. (1997), Kudo et al. (2000), Lister et al. (2000), Castillo et al. (2008), Fan et al. (2010), Durgante et al. (2013)
Calorific value	Gillon et al. (1997)
Carbohydrate content	Hiukka (1998)
Phenolic compounds	Schulz et al. (1999), Soukupova et al. (2002)
Oil	Schimleck et al. (2003)

spectrum was very useful for quantitative analysis of agricultural samples. Many authors wrongly use the acronym of NIRS to describe any method of spectrum analysis involving reflectance/transmittance spectra of dry/fresh leaves. Here we will limit that definition to the analysis of reflected energy by dried, ground leaf samples or leaf extracts (see Section 4.5.4). The reflectance has two components, specular and diffuse; the specular component does not provide any information on the composition and concentration of unknown samples, so that only the diffuse component is used in the context of NIRS. Multivariate analyzes, such as stepwise linear regression analysis, partial least squares regression, or principal components analysis can be used to calibrate a prediction model. The independent variables are reflectances (or transformations of reflectances such as $\log(1/R)$, first and second derivatives) measured at a set of equally spaced wavelengths, and the dependent variable is leaf chemical composition or leaf functional properties (see Chapter 2). Once statistical models are developed, they can be used to separately predict several variables at the same time. According to Shenk et al. (1979), four criteria must be met for meaningful NIRS predictions: i) selection of calibration samples representative of the population to be predicted; ii) accurate laboratory analysis of the calibration samples; iii) choice of the correct mathematical treatment of the data for optimum information extraction; and iv) choice of wavelengths relevant to the total population of samples. The selection of wavelengths is not simple because of many factors including instrument design, data processing, wavelength selection algorithm, variables predicted, and calibration sample set. In addition, the number of wavelengths in an equation may differ from one variable to another.

There is a long list of variables that can be analyzed using NIRS (Table 5.6): forage quality (crude protein, neutral detergent fiber, acid detergent fiber, lignin, *in vitro* dry matter disappearance, dry matter intake, digestible energy intake); major (Ca, K, P); minor (Fe, Mn, Na, Zn) foliar nutrients (see Section 7.1.1); biochemical content (sugar, starch, protein, cellulose, lignin, total chlorophyll, nitrogen); litter decomposition (nitrogen, carbon, acid detergent fiber, acid detergent lignin, ash); interspecific variability, calorific value, carbohydrate content (starch, fructose, α-glucose, β-glucose, pinitol, sorbitol, inositol); polyphenol and alkaloid compounds (theogalline, gallic acid, caffeine, theobromine, epigallocatechin, epicatechin, epigallocatechin gallate, epicatechin gallate); and oil (1,8-cineole). Foley et al. (1998) reviewed the ecological applications of near-infrared reflectance spectroscopy.

Although, strictly speaking, NIRS does not concern the visible (400–760 nm) spectrum, it has been extended to this region for pigment extraction. Comparison of dried, ground samples and fresh leaves showed that if high determination coefficients were obtained for the first, they remained rather high for the second, although using different sets of wavelengths (e.g., Lacaze and Joffre, 1994; Martin and Aber, 1994; Jacquemoud et al., 1995b; White et al., 2000).

5.5.2 Fourier Transform Infrared Spectroscopy

This technique, commonly known as FTIR, is equivalent to NIRS but it operates in the thermal infrared (3–14 μm). It makes use of attenuated total reflectance (ATR) measurements that require that the sample be in contact with an internally reflecting element (see Section 4.2.4). It has been used to identify extracted constituents on plant leaves or to study intact leaves. For instance, it is specially adapted to investigate epicuticular waxes (Ribeiro da Luz, 2006; Bertoluzza et al., 1999; Hardin et al., 2011, 2013).

6

Variation Due to Leaf Structural, Chemical, and Physiological Traits

6.1 Structural Sources

6.1.1 Leaf Anatomy and Thickness

Changes in leaf internal structure may affect leaf optical properties to varying degrees in different parts of the solar spectrum, but they are most evident in the near infrared (NIR) where absorption by pigments is minimal. The mesophyll anatomy of leaves of terrestrial plants is highly variable. To illustrate the effect of such variability on leaf optical properties, Gausman et al. (1971c) selected 11 species displaying a wide range of internal structures and thicknesses (compact, dorsiventral, isolateral, and succulent). Gausman et al. (1973) completed this study two years later with 20 economically important agricultural crops. They showed that leaves from dicotyledonous species reflect greater amounts of radiation than leaves from monocotyledonous species. This is illustrated by the examples of a maize (*Zea mays*) leaf (monocot) and a potato (*Solanum tuberosum*) leaf (dicot). However, their anatomy is very different (see Section 2.1). The compact maize leaf with a continuous mesophyll anatomy has fewer intercellular air spaces than the porous dorsiventral potato leaf (Figure 6.1a–b). Therefore diffuse reflectance in the NIR (800–1000 nm) of maize is lowest (~ 43% vs. ~ 51%) and its transmittance highest (~ 47% vs. ~ 40%) (Figures 6.1c–d). This was confirmed by Asner et al. (1998) who showed that the mean leaf reflectance of the grass group was significantly higher than that of the woody dicot plant group throughout the visible (VIS) (400–700 nm) spectral region and consistently lower in the NIR (700–1400 nm). This is also the case for old leaves that are less compact and have more intercellular space than younger leaves.

The function of wall–air interfaces in reflectance of IR light is easily demonstrated. For instance, the internal structure of cotton leaves (*Gossypium hirsutum*) has been artificially altered by growing plants hydroponically with low-, medium-, and high-salinity (sodium chloride) substrate levels (Gausman et al., 1969a, 1969c), or by spraying plants with Cycocel ((2-chloroethyl) trimethylammonium (chloride)), a plant growth regulator (Gausman et al., 1970d). Cross-sections of the leaves in both experiments showed that treated samples were thicker than controls and that the fraction of intercellular spaces in the mesophyll increased. The reflectance measured between 750 nm and 1350 nm was 5–8% higher than that of control leaves. Conversely, their transmittance decreased by the same proportion. Thus the absorptance A calculated from the equation $A = 1 - R - T$, where R and T refer to the reflectance and transmittance respectively, changed very little, about 1–2%, over that wavelength interval. The role of wall–air interfaces is also supported by the observation that reflectance is reduced to one-half (and transmittance augmented) when intercellular spaces are infiltrated with water or mineral oil, which has a refractive index that approximates the refractive

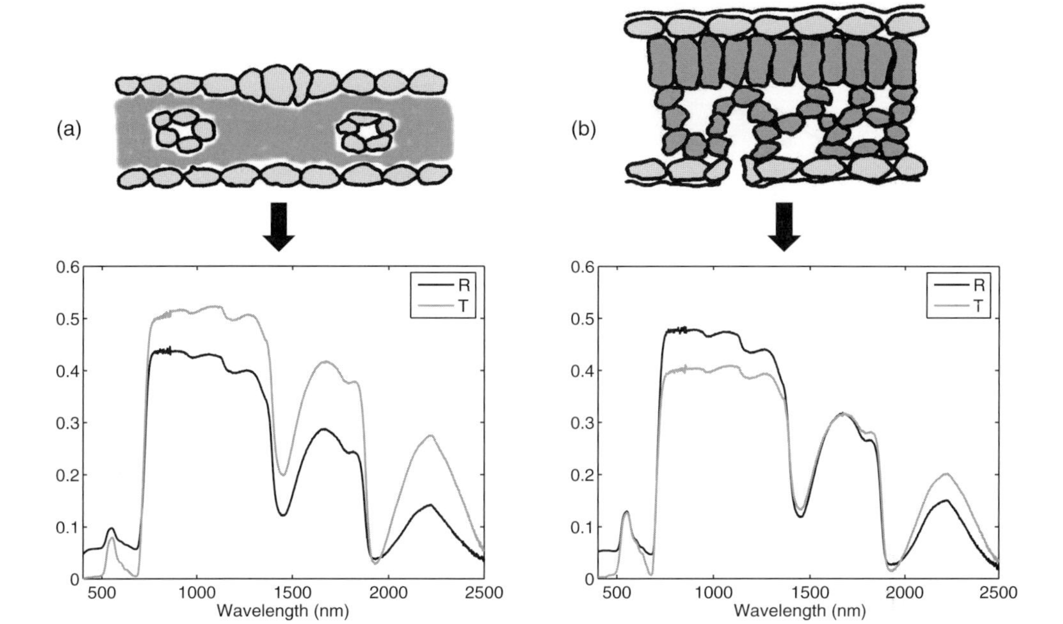

Figure 6.1 (Upper line) Leaf anatomy of typical: (a) maize (*Zea mays*) leaves; and (b) potato (*Solanum tuberosum*) leaves. (Lower line) Corresponding reflectance and transmittance spectra.

index of plant cell walls (Seybold, 1933b; Gausman, 1974; Buschmann and Nagel, 1993a). Peynado et al. (1979) studied the consequence of freezing on sour orange (*Citrus aurantium*) plants, which showed that ice crystallization in frozen leaves destroyed membrane semi-permeability and intracellular compartmentalization. Consequently, cellular water migrated to intercellular air spaces, reducing the number of cell wall–air interfaces, the amount of light scattering, and therefore leaf reflectance over the whole solar domain.

Statistical analyses have shown that the lamina thickness was another structural variable that could control leaf optical properties. Gausman et al. (1971b, 1973) obtained a positive (negative) correlation coefficient (R^2~0.60) between leaf thickness and reflectance (transmittance) at the 1.0 μm wavelength. This was confirmed by Knapp and Carter (1998), who studied the variability in leaf optical properties among 26 species representing a wide variety of growth forms in temperate zone habitats. They found that leaf thickness and specific leaf mass varied by ten-fold among the species sampled, with a high within-habitat variance. Strong relationships ($R^2 > 0.65$) were found between leaf thickness and both reflectance and transmittance, at the 0.85 μm wavelength (Figure 6.2).

Ourcival et al. (1999) measured the anatomical structure (leaf mass per area and tissue thickness) and the reflectance spectra of 87 leaves of the holm oak (*Quercus ilex*), an evergreen oak native to the Mediterranean region. They also found a strong correlation between leaf thickness, especially palisade parenchyma mesophyll thickness, and reflectance spectra. Using support vector machine (SVM) techniques, Qin et al. (2010) established some relationships between the leaf thickness and red-edge/near-infrared reflectance of several wine grape varieties, but showed cultivar dependence. Such relationships, which are not in accordance with the Lambert–Beer law, can be explained primarily in terms of probability that light will preferentially exit the leaf through the irradiated

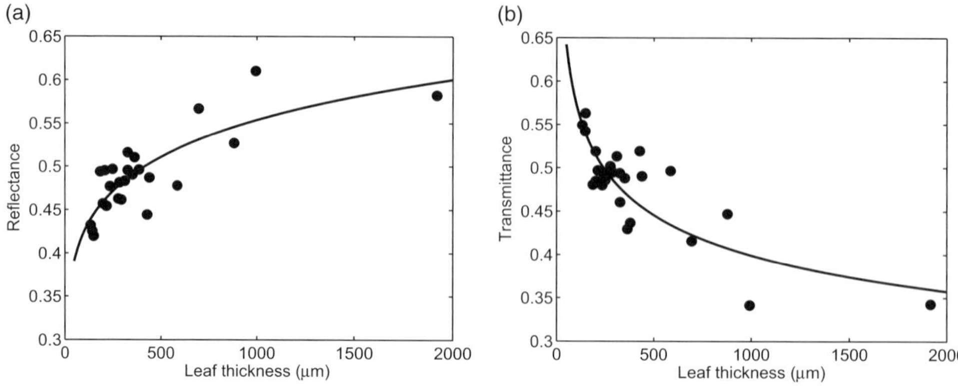

Figure 6.2 Relationship of 850 nm (a) reflectance and (b) transmittance with leaf thickness (ℓ). The regression equations are of the form $R = 0.2475\ \ell^{0.1166}$ and $T = 1.1955\ \ell^{-0.1588}$, respectively (redrawn from Knapp and Carter (1998)).

surface or through the opposite surface. The greater number of interfaces between cell walls and intercellular air in thicker leaves results in higher reflectance and lower transmittance. This is corroborated by other authors (e.g., Woolley, 1971; Baldini et al., 1997; Sui et al., 2010, 2012), although some studies provide a contradictory picture of this effect: for instance, Slaton et al. (2001) report the absence of a noticeable correlation between near-infrared reflectance (or transmittance) and leaf thickness, indicating that there is no apparent relation of leaf structure to leaf thickness. Light scattering is more closely related to internal structure (proportions of parenchyma) rather than to thickness *per se*. This was quantified by Castro-Esau et al. (2006b), who found that the NIR reflectance of seven tropical leaves was better correlated to the percentage of intercellular space than to the spongy mesophyll thickness.

In an intensely scattering environment, the primary effect of visible and shortwave infrared (SWIR) light scattering is to increase the light optical path length which, in turn, increases the probability of light absorption within leaf tissues (Figure 6.3). This lengthening of the optical path length, which explains the difference in optical properties between leaves and chloroplast suspensions, is referred to as the detour effect (e.g., Butler, 1964; Tsel'niker, 1975; Rühle and Wild, 1979; Garlaschi et al., 1989; Merzlyak et al., 2009). The detour effect leads both to increased absorption and to substantial flattening of the absorption spectrum, in comparison with the hypothetical spectrum that would be recorded in the absence of such an effect. It also leads to major deviations from the Beer–Lambert Law. Note that biominerals (Gal et al., 2012; Klančnik et al., 2014a) or sclereids (Karabourniotis et al., 1994; Karabourniotis, 1998) present in the leaves of some plant species can scatter the light within the leaf, causing a more equally distributed light flux.

6.1.2 The Sieve Effect

Absorbers in a chloroplast, a vacuole, or the cytoplasm are not distributed homogeneously throughout the leaf but are rather packaged. Therefore when light travels through a leaf, it may be intercepted and absorbed by foliar constituents or propagated without encountering them. This phenomenon, known as the sieve effect (also called package effect or flattening effect), reduces the probability of light absorption within leaf tissues. The sieve effect has been mainly applied to phenolic compound

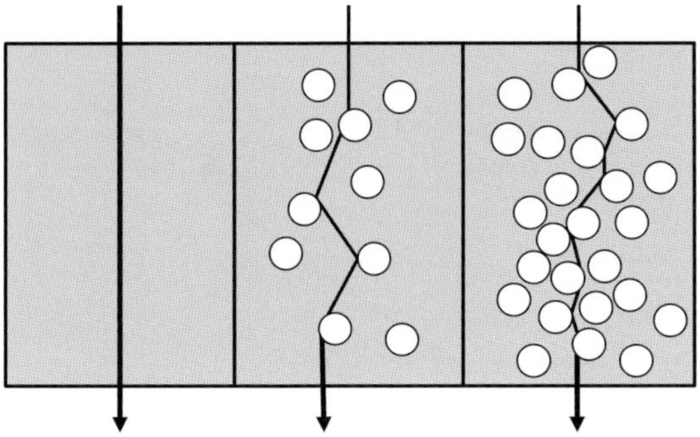

Figure 6.3 Light scattering increases path length of light through a leaf (redrawn from Evans et al. (2004)).

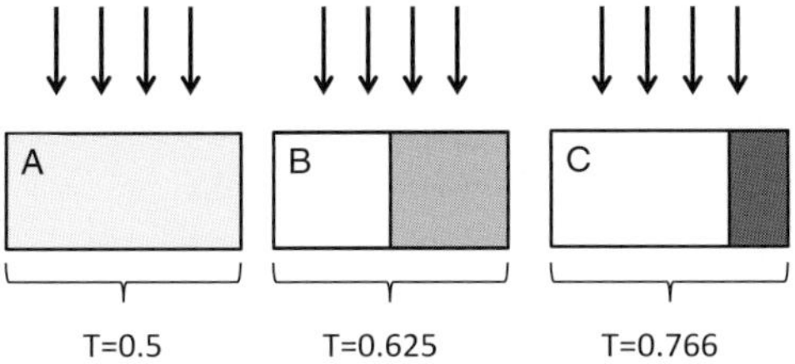

Figure 6.4 Simplified presentation of the sieve effect with chlorophyll in different proportions of the cell. Cells A to C contain the same amount of pigment (redrawn from Evans et al. (2004)).

absorption in the UV (Kolb and Pfündel, 2005), pigment absorption in the visible (Osborne and Raven, 1986; McClendon and Fukshansky, 1990a, 1990b; Evans et al., 2004), and water absorption in the shortwave infrared (Baranoski and Eng, 2007). Let us consider photosynthetic pigments, in which the location and density of chloroplasts largely determine the way light is absorbed within a leaf (Evans et al., 2004). The origin of the sieve effect is illustrated by Figure 6.4.

From this over-simplified scheme one can see that a spatial redistribution of the pigments leads to changes in measured transmittance, therefore absorptance, which is not explained by any changes in the properties of pigments. In case A, pigments are uniformly distributed throughout a non-scattering medium with an initial concentration of C_* (g cm^{-3}). In cases B and C, the entire amount of pigments is sequestered in a cell that occupies a fraction γ of the total volume, that is the fraction of air spaces equals $1 - \gamma$. Therefore the concentration changes into $C = C_*/\gamma$. The overall transmittance is given by

$$T_\gamma = (1 - \gamma)T_{air} + \gamma T_{medium} \tag{6.1}$$

where $T_{air} = 1$ and T_{medium} follows the Beer–Lambert Law (see Appendix F):

$$T_{medium} = \exp(-C\ k_a\ x) \tag{6.2}$$

with k_a the specific absorption coefficient of the pigment (cm^2 g^{-1}) and x the path length (cm). Thus, in case A, where $\gamma = 1$, Eq. 6.1 becomes

$$T_* = T_1 = \exp(-C_*\ k_a\ x) \tag{6.3}$$

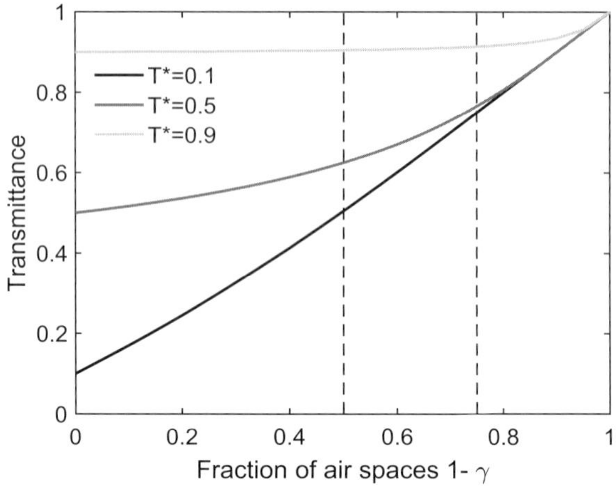

Figure 6.5 Model explaining the sieve effect on transmittance. The vertical lines correspond to two situations illustrated by Figure 6.4 ($\gamma = 0.5$ and $\gamma = 0.25$).

otherwise

$$T_\gamma = (1-\gamma)T_{air} + \gamma \exp\left(-\frac{C_*}{\gamma}\ k_a\ x\right) = (1-\gamma) + \gamma T_*^{1/\gamma} \tag{6.4}$$

Using this simple model, we can investigate the influence of the wavelength on the sieve effect (Figure 6.5). If we concentrate the pigments in one half of the cell only (case B, $\gamma = 0.5$) then the overall transmittance of the cell is $T_{0.5} = 0.505$ when $T_* = 0.1$, $T_{0.5} = 0.625$ when $T_* = 0.5$, and $T_{0.5} = 0.905$ when $T_* = 0.9$. If we concentrate them further into one-fourth of the cell (case C, $\gamma = 0.25$), then the overall transmittance of the cell is $T_{0.25} = 0.75$ when $T_* = 0.1$, $T_{0.25} = 0.766$ when $T_* = 0.5$, and $T_{0.25} = 0.9140$ when $T_* = 0.9$. This simulation shows that the increase of transmittance (loss of absorptance) by the sieve effect is large at wavelengths with strong absorption (low transmittance).

In Figure 6.5, the absorption coefficient of the medium $\sigma_a = C\ k_a$ is assumed to be constant along the whole depth, which of course is not true in a real leaf. Pigments in tissues are usually non-homogeneously distributed due to the layered structure, to cell type variety, and to the absence of pigments in some cells. Fukshansky (1978, 1981) proposed a statistical approach that takes into account such an effect. Finally, note that the sieve effect is not just a curiosity but rather something that can be used by a leaf to dynamically control the internal distribution of absorbed light energy (Vogelmann, 1993; Evans et al., 2004). It has been well documented that detour and sieve effects not only have opposite influences in the absorption profile of turbid materials, but they are also more pronounced in different regions of the light spectrum (e.g., Rühle and Wild, 1979; Fukshansky, 1981; Garlaschi et al., 1989; Terashima et al., 2009). The absorption increase caused by the detour effect favors the wavelengths where absorption is minimum (green, red-edge), while the absorption decrease caused by the sieve effect is more noticeable at the wavelengths where it is maximum (blue, red). In conclusion, for the same amount of pigments with the same specific absorption coefficient and a fixed thickness, the transmittance of the leaf can be different depending on the spatial distribution of pigments. Since chloroplasts can move within minutes in response to how much light they receive, this movement can influence leaf optical properties (see Section 6.1.4).

Some leaves display areas of transparent epidermis that is underlain by translucent water storage parenchyma (Krulik, 1980). These structures allow light to directly penetrate into the leaf's interior where it can be absorbed by chlorenchyma. This can occur naturally in "window-leaved" succulent desert plants or can be caused by insects, for instance, nocturnal caterpillars. The clear areas can take the form of scattered dots, lines, wide streaks, or even complete large portions of a leaf. Although strictly speaking such leaves do not cause a sieve effect as defined above, their transmittance could be calculated using the same formalism. These specialized structures do not seem to affect the health of the plant to any consistent degree. For instance, several genera of plants native to arid climates (e.g., South Africa and neighboring countries) produce succulent, cone-shaped leaves that may become mostly buried in sand blown by the wind. *Lithops* is a genus of fascinating succulent plants that are usually buried up to the leaf tips. These species have translucent portions of epidermal tissue known as "windows" that occur in a wide range of sizes and shapes among species. This achlorophyllous tissue allows incident radiant to penetrate into the succulent water-storage parenchyma (hydrenchyma) that fills the interior of the leaf, to secondly illuminate the surrounding chlorenchyma tissue from the inside of the leaf (Krulik, 1980; Simpson and Moore, 1984; Egbert et al., 2008). Figure 6.6 compares the transmittance spectra of columnar Peperomia (*Peperomia columella*) leaves measured in the window tissues and outside. It shows the high degree of transparency of the window cells. This additional radiant energy has been thought to optimize the absorption of light by window-leaved plants and to increase their photosynthetic rates without the need to increase leaf area and concomitant water loss. This theory was demolished by Egbert and Martin (2000) who showed that for three desert succulent species, photosynthesis was not enhanced by the windows. More recently, Martin et al. (2013) found that the enhanced light under high light conditions inhibited photosynthesis, but under cloudy, high rainfall conditions the windows maximized light absorption.

Some plants, such as lace plants (*Aponogeton madagascariensis*) or "Swiss cheese" plants (*Monstera deliciosa*), have perforated leaves with natural holes, referred to in botany as fenestration.

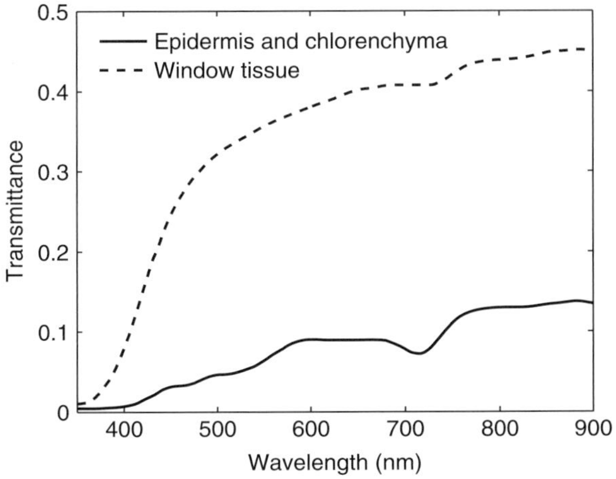

Figure 6.6 Transmittance spectra of columnar Peperomia (*Peperomia columella*) leaves (1.5 mm thick) with lamina containing epidermis and chlorenchyma and window tissue (redrawn from Krulik (1980)).

The size, shape, and quantity of the holes can vary greatly depending on the species and within a given species. What is the function of fenestration? Many theories have been suggested for such unusual leaf shape traits; some believe it reduces turbulence (see Section 12.1.3) while others believe the holes somehow camouflage the plants, hiding them from herbivores. It may be an adaptation to the plants' rainforest habitat. Muir (2013) used mathematical models to compare leaves with and without holes and showed an enhanced access to sunlight by the first. A fenestrated leaf can grow much larger while still maintaining the same amount of surface area. By spreading out its surface area over a larger region, it is actually more efficient at gathering what limited light is available.

6.1.3 Leaf Veins

As seen in Chapter 2, leaf venation shows enormous structural diversity (pattern, shape, size, thickness, density) across species of plants. Sack and Scoffoni (2013) reviewed the structure, function, development, evolution, ecology, and applications of leaf venation over time. For instance, Brodribb and Feild (2010) found that veins were a proxy for greater photosynthetic capacity, via greater gas exchange capacity. It could explain why *Angiosperms* rapidly diversified in the mid-Cretaceous period, about 100 million years ago, increasing in number and dominating other plant types such as *Pteridophytes* and *Gymnosperms*. In *Angiosperms*, the vascular system of the leaf is composed of a network of primary, secondary, and tertiary veins that may represent up to 25% of the leaf area (Figure 6.7). Vein density is generally measured as leaf vein length per unit leaf area or VLA (cm cm^{-2}). A range of software methods are in use to visualize and measure vein systems in cleared leaf images. However, Sack et al. (2014) warned against overestimating VLA when increasing image magnification for measurements.

In as much as the leaf sample is larger than the exit port of the integrating sphere or the entrance port of the leaf probe, it is generally recommended that the more or less translucent primary and secondary veins are avoided, because they are likely to be an important source of heterogeneity.

Figure 6.7 Close-up of a leaf showing the network of primary, secondary, and tertiary veins (Credit: temmuzcan, Copyright © 1999–2018 Getty Images).

However, the measured area needs to be large enough to average the properties of the mesh constituted by the leaf areolas. Indeed, veins can be the cause of the sieve effect, as seen in Section 6.1.2. They also protrude from the back of the leaf, or are centered, or protrude less from the front. And sometimes they appear as depressions on the leaf surface. Woolley (1971) pointed out that veins on the backs of the leaves tend to move the leaf surface away from the integrating sphere of the spectrophotometer, decreasing the reflectance of their adaxial side and the transmittance of their abaxial side. Reflectance images of various intact leaves were measured by Buschmann et al. (2012) in four selected bands (440, 550, 690, and 800 nm). The central vein and the major veins were clearly seen on the upper and/or lower sides at all wavelengths, depending on the species.

Monocotyledons are distinguished from dicotyledons in that most leaves have a parallel (or striate) venation system, with equally sized secondary veins that run lengthwise along the lamina between the midrib and leaf edge. Experiments performed by Woolley (1971) on maize (*Zea mays*), Combes et al. (2007) on fescue (*Festuca arundinacea*), and Comar et al. (2012) on durum wheat (*Triticum durum*), suggest that these leaves, for which the surface roughness is greater in the direction perpendicular to the central vein, reflect more diffuse light in a plane perpendicular to the veins than in a plane parallel to the veins. Taylor-Hell et al. (2005) pointed out that the simulation of light interaction with such leaves was simplified since their anisotropy is constant over the leaf. On the contrary, leaves of most dicots have netted (reticulate) venation that follows a branching pattern originating from the midrib, into successively smaller, secondary, and tertiary veins in a recursive fashion. In some leaves, even the very small veinlets are surrounded by bundle-sheath cells arranged into tightly packed sheaths, which separate the mesophyll into many small compartments. These extensions may guide light deeper into the mesophyll of thick sun-adapted leaves (Karabourniotis et al., 2000). As a consequence, the reflectance may vary from one spot to another, as shown by Ngo et al. (2013) on cabbage leaves.

6.1.4 Chloroplast Movement

Angiosperms have developed several protective mechanisms that are important to their survival. Some help them to mitigate damage to the photosynthetic apparatus under natural growth conditions, when light levels are higher than required for photosynthesis (Evans et al., 2004). One is blue light induced chloroplast movements in the cell, by receptors called phototropins (Christie, 2007). Phototropins activate changes in CHUP1 (Chloroplast Unusual Positioning 1), a protein associated with the actin filaments that form tracts along which chloroplasts circulate. They consequently are implicated in the spatial distribution of chloroplasts. Under low light intensity chloroplasts generally accumulate parallel to the leaf surface at the top and bottom of palisade cells (accumulation), while under high light intensity they arrange perpendicular to the leaf surface along the sides of cells (avoidance). The position of the chloroplasts in darkness is usually intermediate. Note that Palamaryuk and Guminetskii (1968) identified two other kinds of movement: i) rotation into a position in which the smallest cross-section of the chloroplast is normal to the incident beam; and ii) contraction-reduction of the volume of the chloroplast. Such movements have been hypothesized to optimize photosynthetic efficiency, for instance, by allowing light to penetrate deeper into leaves. The proportion of chloroplasts in face or profile position depends on species and on the growth conditions under which the leaves developed. This change is visible with the naked eye and a common demonstration is to attach an opaque mask to a low-light-adapted leaf and expose it to high

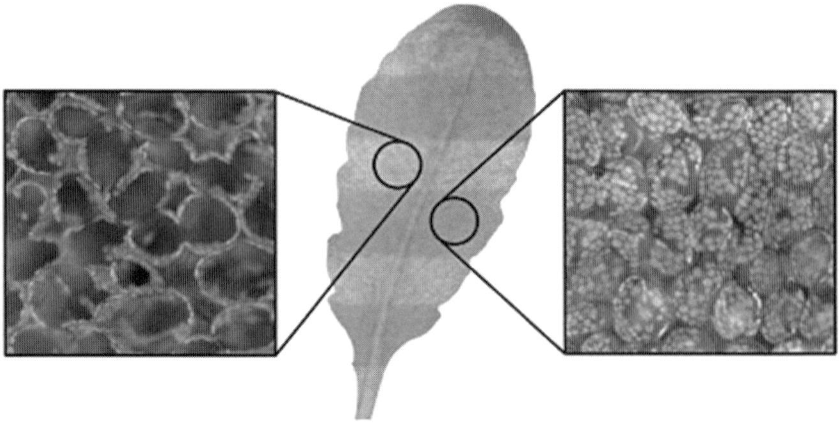

Figure 6.8 Visible patterns caused by arrangements of light-dependent chloroplasts in a thale cress (*Arabidopsis thaliana*) leaf (pale areas = chloroplasts in the high-light arrangement; darker areas = chloroplasts in the low-light arrangement) (from Davis et al. (2011), *Plant, Cell & Environment*. Copyright © 2011 with permission from John Wiley & Sons).

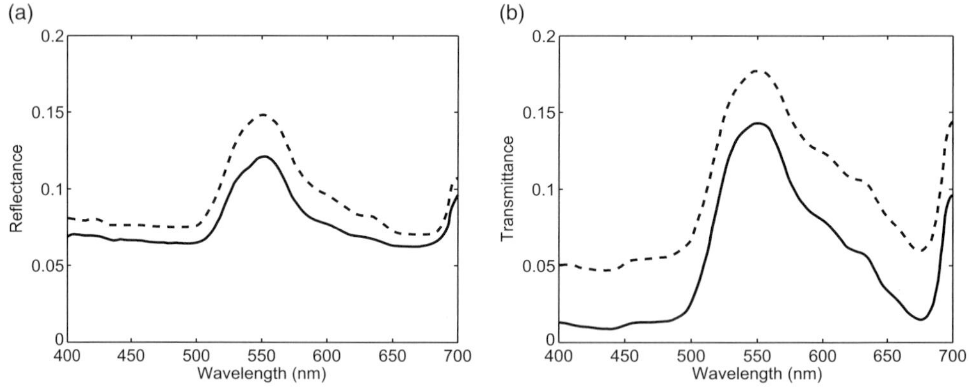

Figure 6.9 Reflectance (a) and transmittance (b) spectra for giant taro (*Alocasia macrorrhizos*) leaves with chloroplasts in the dark position (solid lines) and after movement to the high-light position (dashed lines) (redrawn from Gorton et al. (1999)).

light for an hour or so; when the mask is removed its image remains on the leaf, with the shaded areas darker green than the unshaded areas, as illustrated by Figure 6.8 (Seybold, 1956; Williams et al., 2003; Davis et al., 2011). This phenomenon is fully reversible as light fades. In return, leaf color may provide quantitative information on the kinetics of chloroplast movement.

The number and distribution of chloroplasts in the light path will alter leaf absorption because of the sieve effect (e.g., Rabinowitch, 1951; Zurzycki, 1961; Inoue and Shibata, 1973; Walczak and Gabrys, 1980; Hirata et al., 1983; Osborne and Raven, 1986; Gorton et al., 1999). The absorptance spectrum is increased under low light and decreased under high light. The reflectance and transmittance spectra follow the opposite trend as observed by Park et al. (1996), Gorton at al. (1999), and Kondo et al. (2004) on various plants (Figure 6.9). Davis et al. (2011) surveyed 24 plant species to examine how leaf anatomy influenced chloroplast movement and how the optical properties of leaves changed with chloroplast position. They showed light-dependent chloroplast movements for all species, but also high variability in leaf absorptance that was associated with cell size and shape. The narrower, more columnar cells found in sunlit leaves restrict the ability of chloroplasts to move while the broader, more spherical cells of shaded leaves allow greater chloroplast rearrangements. In low light conditions this would optimize light capture.

Chloroplast movement may also be a protective response to water stress. Zygielbaum et al. (2012) conducted experiments on maize (*Zea mays*) leaves and showed that the reflectance of stressed plants increased in the infrared, which is expected, but also in the visible. They discovered that although the amount of water decreased significantly during the early stages of stress, the amount of chlorophyll did not. The increase in visible light reflectance was, therefore, caused by chloroplast movement.

6.1.5 Upper and Lower Face

As reported by several authors (e.g., Shull, 1929; Moss and Loomis, 1952; Birkebak and Birkebak, 1964; Gates et al., 1965; Gausman et al., 1971c; Gupta and Woolley, 1971; WoolLey, 1971; Sinclair et al., 1973; Gauslaa, 1984; DeLucia et al., 1991; Stuckens et al., 2009), the bottom (or dorsal or abaxial) surfaces of most dorsiventral leaves usually reflect more light than the top (or ventral or adaxial) surfaces (Figure 6.10a). On average, the reflectance difference is about 6.6% but it can increase by up to 20% for some leaves. It indicates that the spongy parenchyma contribute more to light scattering than the palisade parenchyma of the leaf mesophyll. This distinction is manifest in the visible (400–700 nm) and shortwave infrared (1300–2500 nm) regions, for reflectance values of less than 0.40. It tends to disappear in the near-infrared region (700–1300 nm), where the reflectance is approximately equal for the two surfaces (Gates et al., 1965). An inversion may even occur (Woolley, 1971; Baldini et al., 1997; Stuckens et al., 2009). As for transmittance values, they are essentially alike in the visible and shortwave infrared regions, but in the reciprocal configuration, they can deviate by a few percent in the near-infrared region. An increase has been measured when the light falls on the abaxial side of the leaf (Figure 6.10b). This observation seems to be in conflict with the reciprocity theorem that states that it is impossible for any object to transmit more light in one direction than in the opposite direction (Woolley, 1971). However, recent studies show that this theorem, which is valid for ordinary isotropic materials, may not apply to one-dimensional stratified systems. Finally, absorptance tends to decrease when light falls on the abaxial side of the leaf (Figure 6.10c).

The behavior of compact leaves that have a non-differentiated mesophyll is somewhat different, as their optical properties depend very little on whether they are illuminated on the adaxial or abaxial side (Gausman et al., 1971c; Sinclair et al., 1973).

6.2 Chemical Sources

6.2.1 Pigments

As seen earlier, leaf pigments, primarily the chlorophylls but also the carotenoids and anthocyanins, are the cause of the high absorption of visible radiation (e.g., Rabideau et al., 1946; Shul'gin and Kleshnin, 1959; Gates et al., 1965). Variation in the relative concentrations of these pigments depend on many factors like the plant species, the phenological age of the leaves, the developmental or

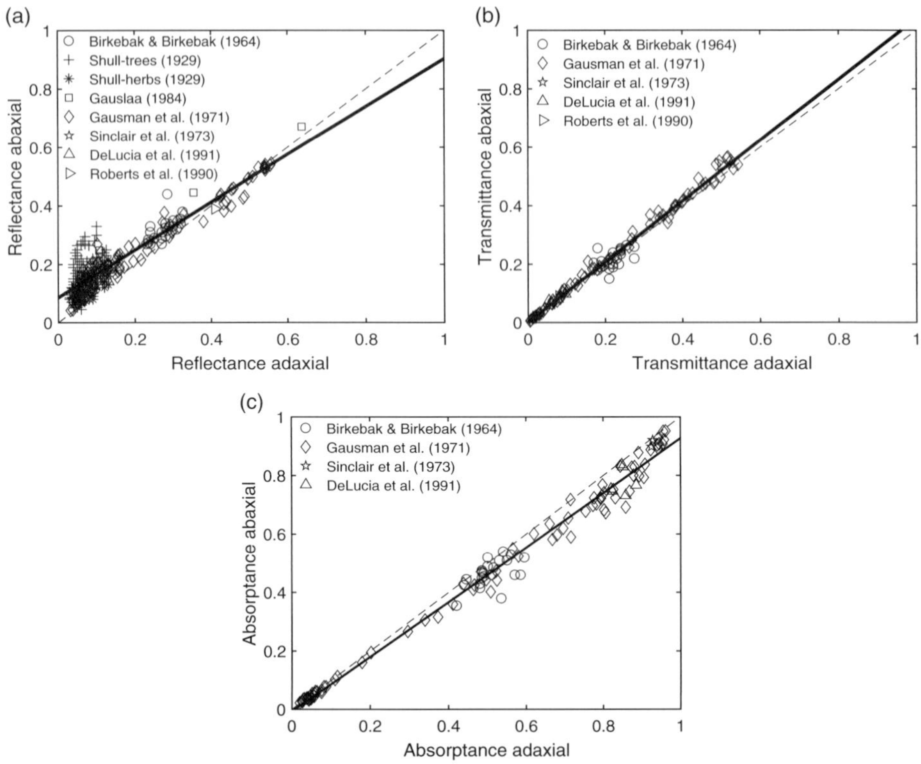

Figure 6.10 Relationship between adaxial and abaxial (a) reflectance ($R_{abaxial} = 0.8182 \times R_{adaxial} + 0.0851$)), (b) transmittance ($T_{abaxial} = 1.0393 \times T_{adaxial} - 0.0004$), and (c) absorptance ($A_{abaxial} = 0.9335 \times A_{adaxial} - 0.0065$), measured at various wavelengths: 430, 440, 460, 480, 500, 520, 540, 560, 580, 600, 620 640, 660, 680, 700 nm (Shull, 1929); 200–3500 nm (Birkebak and Birkebak, 1964); 550, 800, 1000, 1450, 1650, 1950, 2200 nm (Gausman et al., 1971c); 640–650 nm (Sinclair et al., 1973); 400–700, 700–1400, 400–1400 nm (Gauslaa, 1984); and 450, 675, 900 nm (Roberts et al., 1990a).

microclimate position of the leaf on the plant, whether it is stressed or not and if so, what kind of stressor, etc. They induce variation in the leaf optical properties. Jacquemoud and Baret (1990) grew maize seedlings in a greenhouse under different treatments and measured their leaf optical properties. As illustrated in Figure 6.11, the dark green soybean leaf containing high levels of chlorophyll and carotenoid has low reflectance both in the blue (400–500 nm) and in the red (640–690 nm) regions. One can notice a peak in the green at 550 m and a sharp increase in the near infrared beyond 700 nm. None of the important leaf pigments which absorb in the visible spectrum have absorption bands in the near-infrared region where they are generally transparent (Knipling, 1970). A decrease in chlorophyll content reduces the absorption of visible light. The reflectance of a light green leaf is therefore higher than that of a dark green leaf over the whole visible range. It should be noted that the absorption edge of the reflectance curves at 700 nm is shifted toward shorter wavelengths. The shift of the position of this edge toward shorter (blue shift) or longer (red shift) wavelengths may be a good measure of the amount of chlorophyll within the leaf (see Section 11.3). Etiolated leaves are pale yellow elongated leaves that can be obtained by growing seedlings in absolute darkness. They contain a small quantity of a green pigment known as protochlorophyll but not chlorophyll. It is difficult to

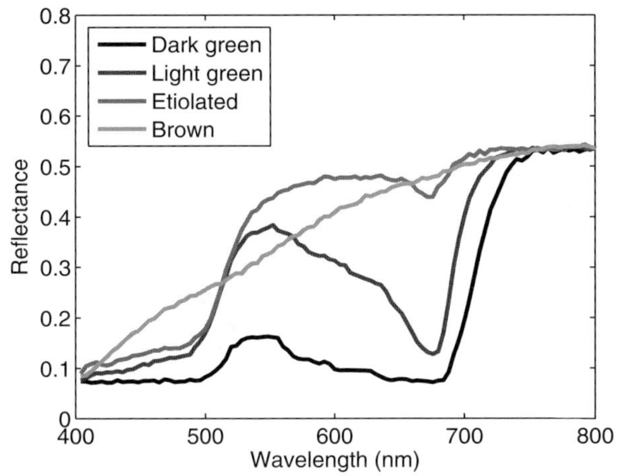

Figure 6.11 Variation of soybean (*Glycine max*) leaf reflectance as a function of pigment content (dark green, light green, etiolated, and brown leaves). The near-infrared reflectance values have been deliberately set to the same levels.

prevent chlorophyll formation prior to spectrophotometric measurements and experiments show that the conversion of protochlorophyll to chlorophyll by reduction under the influence of light is an extremely rapid process so that a small absorption peak is often found at 680 nm (Figure 6.11). The increase in leaf reflectance between 482 nm and 550 nm coincides with the *in vitro* decrease in absorptance associated with protochlorophyll and β-carotene (e.g., Semenenko, 1960a; Mandoli and Briggs, 1982; Maas and Dunlap, 1989). Finally, the reflectance spectrum of a brown leaf is very similar to that of a bare soil with a monotonic increase from the blue to the near infrared.

The variation in the reflectance spectra of leaves offers the opportunity to use measurements of reflected or transmitted radiation as a non-destructive method for quantifying pigments (e.g., Inada, 1985; Lichtenthaler et al., 1996; Merzlyak et al., 2003; Gitelson et al., 2006). For instance, Gitelson and Merzlyak performed a series of experiments on two deciduous species: chestnut (*Aesculus hippocustanum*) and maple (*Acer platanoides*), during autumn senescence (Gitelson and Merzlyak, 1994a, 1994b; Merzlyak and Gitelson, 1995; Gitelson et al., 2002). The maple leaves change color from dark green (sample 1) to completely yellow (sample 6) and they show a continuous, but wavelength dependent, variation in reflectance as a function of chlorophyll content (Figure 6.12). Within green leaves (samples 1 to 4), sensitivity to pigment variation is minimum from 400 nm to 500 nm and at 675 nm due to saturation effects, and maximum near 550 nm and 710 nm. With increasing amounts of chlorophyll in the leaves, the position of the maximum gradually shifts from 675 nm to 690 nm. Such results agree with model outputs (see Section 9.2.2). Because of the overlapping absorption features of the pigments, relating reflectance at a single wavelength is not straightforward and requires particular methods that are developed in Section 11.1.1. The same dataset was used a few years later to assess carotenoid content in plant leaves (Gitelson et al., 2002).

The participation of betalains and certain carotenoids in screening of light in the blue-green part of the spectrum has been reported. Curran et al. (1991) studied the effect of amaranthine, a red water-soluble pigment (betacyanin) located in the vacuoles of the edible amaranth (*Amaranthus tricolor*)

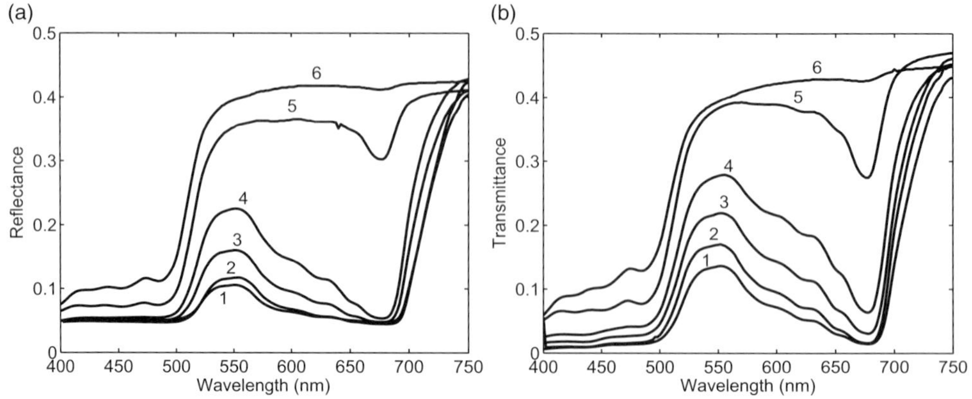

Figure 6.12 Reflectance (a) and transmittance (b) spectra of Norway maple (*Acer platanoides*) leaves containing different concentration of pigments: 1) ((Chl) = 1.30 μg/cm^2 and (Car) = 2.92 μg/cm^2); 2) ((Chl) = 1.89 μg/cm^2 and (Car) = 4.92 μg/cm^2); 3) ((Chl) = 9.75 μg/cm^2 and (Car) = 4.40 μg/cm^2); 4) ((Chl) = 18.92 μg/cm^2 and (Car) = 4.42 μg/cm^2); 5) ((Chl) = 33.30 μg/cm^2 and (Car) = 7.06 μg/cm^2); and 6) ((Chl) = 35.78 μg/cm^2 and (Car) = 8.75 μg/cm^2) (redrawn from Gitelson and Merzlyak (1994b)).

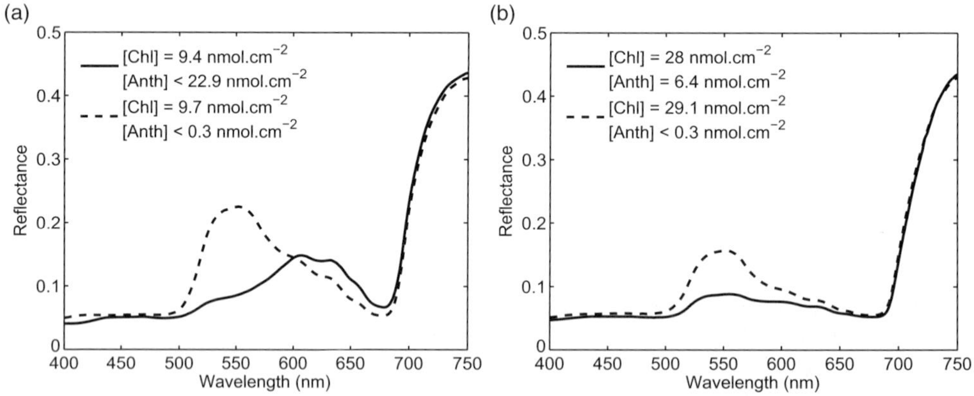

Figure 6.13 Reflectance spectra of Norway maple (*Acer platanoides*) leaves with different anthocyanin content and almost the same chlorophyll content: (a) low chlorophyll content; and (b) high chlorophyll content (redrawn from Gitelson et al. (2001a)).

leaves, on their reflectance spectra. They found that the relationship between the red-edge position and the chlorophyll content was affected by this pigment. Hormaetxe et al. (2005) tested the potential photoprotective role of red (retro-) carotenoids as light filters in common box (*Buxus sempervirens*) leaves but the results of their experiment did not support this hypothesis. Anthocyanins, like chlorophylls and carotenoids, absorb light in particular bands. Gitelson et al. (2001) compared the reflectance spectra of leaves with equal chlorophyll content but different anthocyanin content; hence the difference was solely attributed to anthocyanin. Figure 6.13 shows a marked difference near the green peak at 550 nm, the magnitude of which increases with anthocyanin content. These authors used these spectral features to design a new index called ARI (anthocyanin reflectance index) for anthocyanin content estimation from measured reflectance (see Section 11.1.1).

Finally, albino leaves are totally devoid of chlorophylls, carotenoids, and other colored pigments, therefore they represent extreme examples that are very useful to study leaf optical properties. They can be obtained by mutation (Smith et al., 1957) or by imbibing seeds for 24 hours in an aqueous solution of fluridone (1-methyl-3-phenyl-5-(3-(trifluoromethyl) phenyl)-4(1 H)-pyridinone), an inhibitor of carotenoid synthesis used as an aquatic herbicide to control invasive plants (Maas and Dunlap, 1989; Jacquemoud and Baret, 1990). Variegated leaves occur rarely in nature but are extremely common among indoor and outdoor ornamental oddities. They contain colored and white parts (Billings and Morris, 1951). In absence of pigments, leaf reflectance over most of the visible region is the same as the maximum leaf reflectance observed in the near infrared. Then it decreases below 450 nm or so due to absorption by non-pigment biochemical constituents such as phenolic compounds, nucleic acids, and proteins.

To give a full picture of pigment effects, black leaves may appear during the creation of an abscission zone or where leaves are attacked by a pathogen or parasite. They may also occur naturally in nature, though extremely rarely (Młodzińska, 2009). An example of a plant with very dark leaves is black mondo grass (*Ophiopogon planiscapus* "Nigrescens"), which accumulates an unusually high amount of chlorophyll *a* and *b*, and anthocyanins in superficial layers of the mesophyll. Hatier and Gould (2007) found that the adaxial surface of such leaves reflected approximately 4%, transmitted approximately 1%, and absorbed approximately 95% incident light across the 400–700 nm region (Figure 6.14). Absorption profiles of transmission of red and blue light were similar in green and black leaves, but green light was restricted to the uppermost palisade mesophyll layers in black leaves

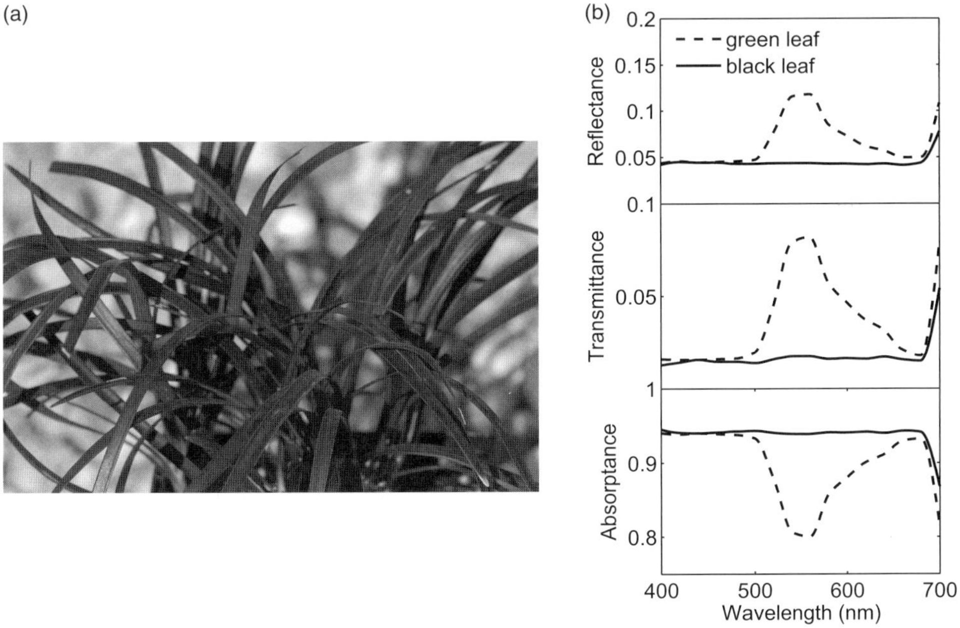

Figure 6.14 (a) Black mondo grass (*Ophiopogon planiscapus* "Nigrescens") leaves (© David Beaulieu, Landscaping Writer for *The Spruce*). (b) Reflectance, transmittance, and absorptance spectra of adaxial surfaces of green and black phenotypes of black mondo grass leaves (redrawn from Hatier and Gould (2007)). (A black and white version of this figure will appear in some formats. For the color version, please refer to the plate section.)

while it transmitted deeper in green leaves. The efficiency of black leaves to capture light for photosynthesis is therefore low and they grow only slowly. The description of leaf color variability would be incomplete without mentioning the gray leaves of *Corydalis benecincta*, an endemic alpine plant of China, which tend to be the same color as the rocks the plant grows on, which protects them against herbivores (Niu and Su, 2014; Niu et al., 2014).

6.2.2 Water

Water stress impacts plant productivity both in natural and crop plant communities. It results in partial or full stomatal closure, thus reducing stomatal conductance to limit water loss. This active regulation of gas exchange causes a decrease in CO_2 availability, which also becomes a limiting factor for photosynthesis. Continued absorption of photosynthetically active radiation induces an excited state in the chlorophyll reaction centers and release of free radicals which can damage pigments and proteins. The mechanisms of defense are highly variable from one species to another. Some strategies reduce photon absorption by heliotropic orientation of the leaves or by wilting or curling, in others the chloroplasts are reoriented in the palisade parenchyma (see Section 6.1.4), while others increase the dissipation of excess energy through the xanthophyll cycle (see Section 2.3.3), which temporarily dissipates light energy. In the longer term, plant adaptations include a reduction in chlorophyll content, leaf abscission, and leaf thickening. All of these mechanisms reduce photosynthetic efficiency. Measurement of leaf water status is usually expressed as the leaf water potential (Ψ), leaf water content (LWC), equivalent water thickness (EWT), or relative water content (RWC) (see Section 2.3.1). Many authors have established relationships between one of these quantities, mostly RWC, and leaf optical properties (e.g., Carlson et al., 1971; Hunt et al., 1987; Bowman, 1989; Carter, 1991; Cohen, 1991; Inoue et al., 1993; Linke et al., 2008). Some studied the natural variation of leaf water status, while others performed measurements on detached leaves that were allowed to air-dry at room temperature (25–30°C). It is also possible to hasten the drying process by arranging the leaves under a strong light source, a halogen lamp for instance, or by oven-drying them. Foley et al. (2006) followed changes in reflectance properties with time after leaves have been clipped from the plant. They found no common time limit and all regions of the spectrum changed at variable rates. They recommend management of water loss, for example, by wrapping petioles in moist paper enclosed in plastic bags. Note that water loss has contrasting effects on the mesophyll and the veins (Buriol et al., 1984a; Seelig et al., 2008a) and that it is somewhat difficult to obtain a progressive one. Figure 6.15 shows the typical pattern of spectral change in peanut (*Arachis hypogaea*) leaves during rapid dehydration.

An increase in reflectance (decrease in transmittance) is observed at all wavelengths with decreasing leaf moisture content from fully turgid to the dry state. The spectral response is highest in the shortwave infrared between 1300 nm and 2500 nm where strong water absorption bands are located. Changes also occur in the near infrared between 700 nm and 1300 nm due to water absorption, like in the SWIR, but also due to the arrangement of interfaces between cell walls and air that may change from the loss of cell turgor pressure. Coefficients of variation at each wavelength can be calculated for estimating relative sensitivity to a change in leaf water content. However, laboratory investigations have shown inconsistent results for the maximum value because these are not linear changes in the shape of the spectra. For instance, at high water contents the major water absorption bands near 1450 nm, 1940 nm, and 2500 nm become saturated so the regions of intermediate absorption near 1650 nm and 2200 nm and the weak absorption bands near 970 nm and 1200 nm may be more sensitive to differences in leaf water content

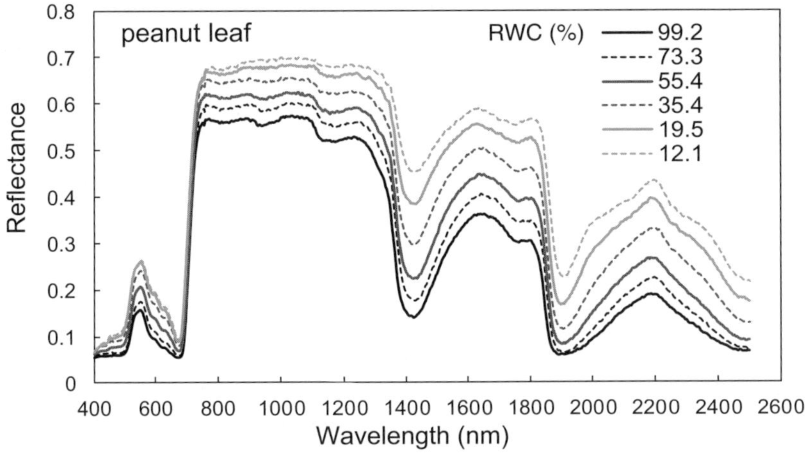

Figure 6.15 Spectral reflectance of peanut (*Arachis hypogaea*) leaves as a function of relative water content during dehydration (by courtesy of Yoshio Inoue).

(Figure 6.15). As for the visible range between 400 nm and 700 nm, it is generally little affected by leaf water content unless a concomitant change in anatomy or decrease in pigment content is observed (Inoue et al., 1993; Zygielbaum et al., 2009). The effects that cannot be explained solely by the absorption properties of water (chemistry, internal structure) are called secondary effects. Thus, sensitivity to *RWT* is greatest in sunlit versus shaded leaves and for adaxial versus abaxial surfaces (Carter, 1991; Carter and McCain, 1993). These secondary effects, principally variations in leaf structure including cell size, shape, etc., contribute to the difficulty in relating radiometric measurements to water content (Danson et al., 1992; Datt, 1999b). The first authors showed that the first derivative of the reflectance spectrum at selected wavelengths was insensitive to these anatomical and morphological differences.

6.3 Physiological Sources

6.3.1 Sunlit and Shaded Leaves

The access to light by plant leaves is fundamental to many biological processes (see Section 12.2). Forests, woodlands, and shrub formations are very important ecosystems that provide foundations for life on Earth through various ecological functions. All canopy layers (overstory, understory, and ground vegetation) are involved in forest dynamics. Plants are usually adapted to grow in direct sunlight or shady conditions. In response to the light intensity, they build sunlit and shaded leaves that are adapted to their natural environment. The difference between the two groups of leaves is primarily based on a combination of morphological, anatomical, ultrastructural, physiological, and biochemical features. Plants from various biomes growing in the shaded forest understory display similar traits. Although these characteristics may differ between individual plants or species, sunlit leaves, in general, are smaller and thicker with less intercellular air spaces than shaded leaves. Their palisade/spongy mesophyll ratio is higher and their specific leaf area (SLA) is lower. They contain less chlorophyll per leaf area, and they display higher chlorophyll *a*/*b* ratios. As a result, leaves exposed to sunlight reflect (absorb) more (less) visible light than the leaves in the shade, while their transmittance remains unchanged (e.g., Gausman, 1984; Lee and Graham, 1986; Lee et al., 1990; Poorter

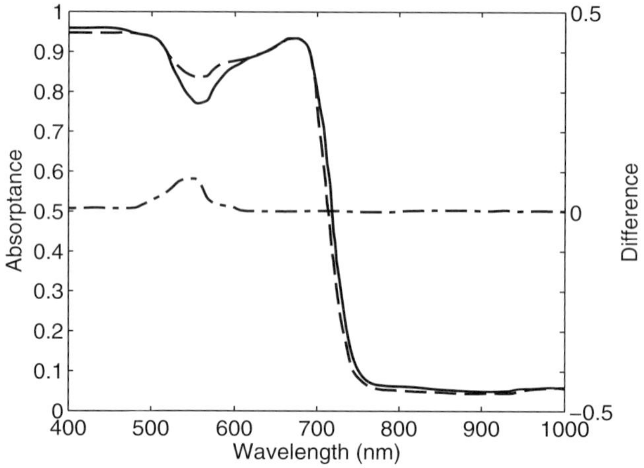

Figure 6.16 Mean leaf absorptance of 12 sunlit species (solid line) and 13 extremely shaded species (dashed line). Similarity between the two samples is shown by the difference spectrum (dash-dot line) (redrawn from Lee and Graham (1986)).

et al., 1995; Lei et al., 1996; Knapp and Carter, 1998; Souza and Valio, 2003). These differences are especially perceptible at 550 nm (Figure 6.16). The higher proportion of spongy mesophyll in shaded leaves may increase internal light scattering and thus affect light absorption, but also gaseous diffusion. Such leaves are consequently more efficient in harvesting sunlight at low light levels (DeLucia et al., 1991). Note that, although statistically significant, differences in leaf optical properties of sunlit and shaded leaves are generally not marked and measurements are sometimes inconsistent between species (Buschmann et al., 1990; Lei et al., 1996). Acevedo and Ataroff (2012) studied tree species of an Andean cloud forest in Venezuela, grouped according to vertical position in canopy, and found significant differences in leaf traits and spectra. Sunlit leaves are also exposed to higher levels of UV radiation than shaded leaves. The higher levels of soluble phenolic compounds in sunlit leaves compared with shaded leaves provide higher UV screening effectiveness (Caldwell et al., 1983; Lovelock et al., 1992; Liakoura et al., 2003; Neitzke and Therburg, 2003). In addition to the position of the leaves on the tree, Neitzke and Therburg (2003) showed an increase in UV-B radiation absorption by European beech (*Fagus sylvatica*) leaves along an elevation gradient, lined with an age effect over the course of the year.

The penetration of light into sunlit and shaded leaves has been also investigated. Cui et al. (1991) measured light gradients in spinach (*Spinacia oleracea*) leaves grown under sunny and shady conditions. Visible light travels further into sunlit leaves because of their elongated palisade cells. In the ultraviolet, the opposite is true. Liakoura et al. (2003) showed that UV absorbing compounds were found in higher concentrations in sunlit leaves of the Greek strawberry tree (*Arbutus andrachne*), a representative evergreen sclerophyllous plant of the Mediterranean region, than in shaded ones of the same species. Light gradients were consequently steeper in sunlit leaves.

6.3.2 Leaf Age and Development

The evolution of the spectral properties of a leaf during the growing season has been investigated by many authors. In the literature, one classically distinguishes three vegetative growth stages in the life of a leaf: the juvenile stage, maturity, and senescence. The juvenile stage starts with leaf emergence and finishes when it is fully expanded. Maturity represents the most important stage because the photosynthetic rate reaches its peak during this period. The onset of senescence can be defined as the time at which apparent photosynthesis begins to decline irreversibly due to chlorophyll breakdown. As leaves become more mature, both their mesophyll structure and biochemical content change. Juvenile leaves have small immature protoplasmic cells that are arranged in compact mesophyll layers. Their low water content is explained by the little water storage capacity of their vacuoles. Very young leaves exhibit primarily protochlorophyll that is rapidly replaced by chlorophyll. In contrast, the vacuolated cells of mature dicot leaves are loosely packed together, with large air spaces between them. They are consequently less compact than younger ones. The chlorophyll and water contents reach their maximum value. With further aging, the mesophyll structure of the leaf begins to deteriorate: the lateral walls of the palisade cells, followed later by those of the spongy cells, collapse; chlorophyll, starch, and protein decline; the preferential degradation of chlorophyll over carotenoids results in leaf yellowing and eventually the development of a bright yellow color; afterwards water-soluble anthocyanins may be produced giving the leaves their orange to red colors. The leaves drop in moisture concurrent with changes in cell geometry. Depending on the species, significant spectral fluctuations are produced in the visible, near, and shortwave infrared during the three growth stages.

Some researchers focused on changes in optical properties that occur during the transition from juvenile to adult phases, due to morphological and chemical changes. Shull (1929) observed a decrease in reflectance in the VIS associated with an increase in chlorophyll content as the plant grows, and an increase in reflectance at the end of the summer. He also mentioned a notable effect of carotenoids and anthocyanins on reflectance. The evolution of the spectral properties of a leaf of white oak (*Quercus alba*) during the growing season was investigated by Gates et al. (1965), who noted several interesting features (Figure 6.17). Initially the reflectance of the young leaf on 17 April 1964 exhibited primarily protochlorophyll absorption. The near-infrared reflectance is surprisingly high given that immature leaves have a compact mesophyll with small cells. As the leaf develops the red reflectance at 680 nm decreases and at first the green reflectance at 550 nm

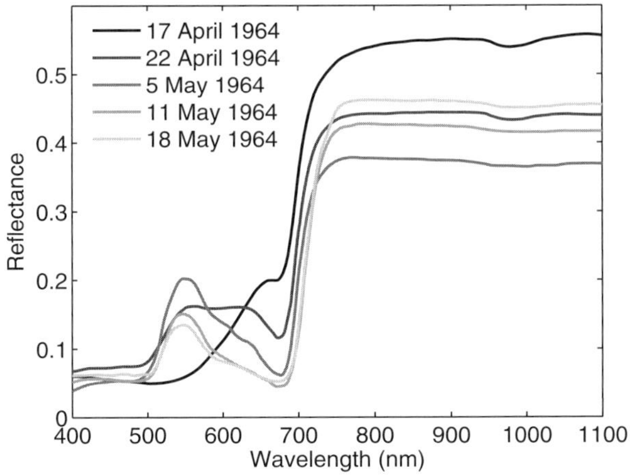

Figure 6.17 Evolution of the spectral reflectance of white oak (*Quercus alba*) leaf during the growing season (redrawn from Gates et al. (1965)).

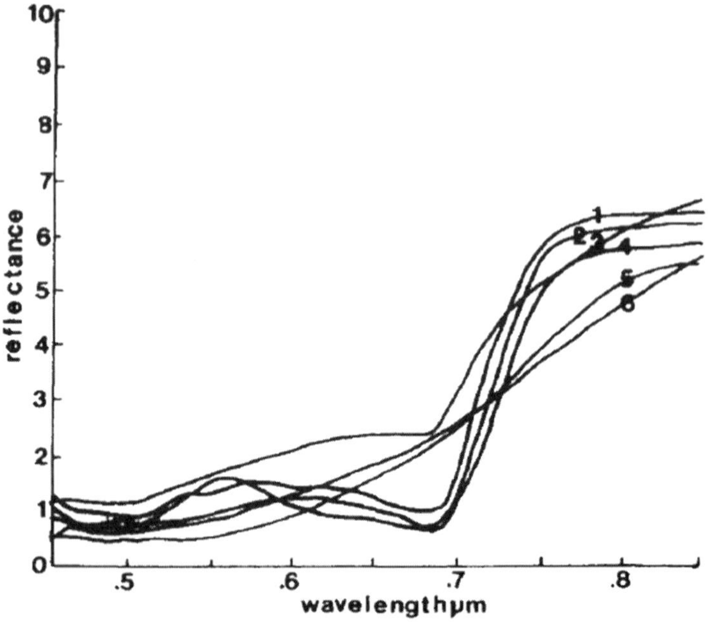

Figure 6.18 Characteristic reflectance spectra from senescing leaves of pin oak (*Quercus palustris*) with progressive symptomatic expressions: 1) green leaves; 2) green chlorotic leaves; 3) scarlet leaves with color dominated by chlorophyll; 4) scarlet leaves with color dominated by anthocyanins; 5) red leaves dominated by anthocyanins and carotenoids; and 6) bronze leaves (from Boyer et al. (1988), *Remote Sensing of Environment*. Copyright © 1988 with permission from Elsevier).

increases. After 5 May 1964, the latter decreases due to chlorophyll accumulation in the leaf and the near-infrared reflectance beyond 740 nm increases. Changes in leaf morphology explain why "older" leaves are brighter in the near infrared than "younger" ones. Similar results were obtained with other natural and cultivated plant species (e.g., Gausman et al., 1969b, 1970c, 1971b, 1972, 1974b; Brunner and Eller, 1977; Phan et al., 1979; Seyfried and Schafer, 1983; Carter et al., 1989; Walter-Shea et al., 1991; Middleton and Walter-Shea, 1995; Schultz, 1996; Léon and Downey, 2006).

Other researchers specifically studied foliar senescence. Boyer et al. (1988) measured visible and near-infrared reflectance and transmittance spectra of pin oaks (*Quercus palustris*) at the approach of autumn. In parallel, they performed anatomical studies but no pigment analysis, appraising the significant stages in the progression of leaf color visually. The transition from green leaves to red leaves produces profound changes in the spectral response (Figure 6.18). The reflectance spectrum is dominated by anthocyanins that mask the peak at 550 nm but in turn produce a large shoulder up to 650 nm. Alterations of internal geometry changes the visible wavelengths, as well as drying out which may indirectly affect chlorophyll content. Similar results were obtained by other authors (e.g., Sinclair et al., 1971; Merzlyak et al., 1997, 1999; Yoshimura, 1998; Castro and Sánchez-Azofeifa, 2008; Christie et al., 2013). Senescence may express itself in various ways, depending on the plant species. For instance, Matile et al. (1992) found that the structure and water content of gingko (*Gingko biloba*) leaves were practically unchanged throughout the senescence period. As a consequence, only the visible range was impacted by the overall loss of leaf pigments. They also brought to light the presence of a fluorescent chemical compound predominantly located in the adaxial epidermis that contributes to the golden luminous appearance of the autumnal foliage.

Only a few studies cover the whole vegetative season and the three growth stages (e.g., Tageyeva et al., 1960, 1961; Olson and Good, 1962; Trigui, 1983; Tanner and Eller, 1986; Miller et al., 1991; Masoni et al., 1994; Stone et al., 2005; Zhang et al., 2007; Fondom et al., 2009; Dillen et al., 2012; Mõttus et al., 2014). Of particular interest is the work of Roberts et al. (1998b), Woodall et al. (1998) and Liu et al. (2009a), who studied tropical and sub-tropical broadleaf species and the manner in which spectra change as leaves age in their natural environment. The most significant findings of Roberts et al. (1998b) are leaf flushing in the dry season and a consistent increase in absorptance at 830 nm with increasing leaf age, probably due to fixation of epiphylls on the leaf surface.

Finally, the effect of perennial plant aging has been poorly studied. Messier and Bellefleur (1987) evaluated the variation in spectral properties of yellow birch (*Betula alleghaniensis*) leaves collected on stands of different ages. They found that the shaded leaves of young trees (1–5 years old) absorbed significantly less radiation in the PAR than the shaded leaves of older trees (6–40 years old and more). The reason cited by these authors was that young trees produced more late leaves than older trees. Such late leaves, also called secondary leaves, have different structure and chemical content than primary leaves. The variation in leaf shape on a same plant is known as heterophylly. Middleton et al. (1997) also compared the optical properties of mature versus young trembling aspen (*Populus tremuloides*) and Jack pine (*Pinus banksiana*) stands. They did not find any noticeable difference.

6.4 Intraspecific Variation

General traits in both needles and broad leaves developed in similar ways over the course of evolution. For example, the waxy cuticle on the epidermis is an extremely important adaptation that has been conserved over long periods of geological time. Other selected traits can vary widely in terms of structure, composition, and function across space and time. Intraspecificity (literally within species), or being intraspecific, describes such variations and other issues occurring within individuals of a single species. The adaptation of a leaf to particular environmental settings is essential to life and evolution. Intraspecific variation exists both at any one place and between different places. If we measure the optical properties of a number of leaves belonging to one plant species at any one locality, they may differ – variation within a population – often showing a normal distribution. If we do the same with leaves belonging to one species, from different places, they may also differ – variation between populations, or geographical variation (Ridley, 2003). The maintenance of intraspecific diversity (i.e., genetic diversity) is an important component of adaptive evolution, driving the ability of plants to colonize habitats of wide ecological amplitudes and to tolerate environmental change. Different varieties of plants may react differently to environmental and growing conditions. Determination of plant varieties also became an important issue in the context of the expansion of international trade in the past half century. It has been based on morphological characters for a very long time, but new methods like leaf spectroscopy may provide another solution. Let us give some examples.

Viticulture is a traditional and important activity in many countries, which is directly impacted by anticipated climate change (see Section 6.6). Thousands of common grape vine (*Vitis vinifera*) varieties are available worldwide. Their identification and classification, a science called ampelography, have been traditionally done by visual comparison of the shape and color of the vine leaves and grape berries, more recently by wet chemistry methods based on isoenzymes or DNA analysis. Though obtained from a limited number of cultivars, and under controlled laboratory conditions, leaf spectroscopy has emerged as a promising method of grapevine variety identification (e.g., Renzullo et al., 2006b; Diago et al., 2013) or clone identification within a given genotype (Fernandes et al., 2015). It has also been applied to various crops such as cannabis (Walthall et al., 2006), strawberry (Deák et al., 2007), tomato (Xie et al., 2007; Xu et al., 2009), switchgrass (Foster et al., 2012), barley (Sun et al., 2012, 2013), and sugarcane (Guo et al., 2014), using various statistical methods, but with mixed results. For instance, cannabis (*Cannabis sativa*) is a flowering plant native to Central Asia that has spread worldwide because of its many uses. It is called marijuana when leaves contain tetrahydrocannabinol (THC), its principal psychoactive constituent. Walthall et al. (2006) cultivated multiple varieties of cannabis with various THC contents and measured their hemispherical leaf reflectance. They observed similar signatures with no notable differences between drug and fiber plants. On a different register, Deák et al. (2007) measured the leaf color of six strawberry (*Fragaria vesca*) cultivars growing in three different areas during the flowering stage. They found only small variations under optimal growth conditions. The success in separating cultivars may depend on the season when leaf spectral measurements are performed. Detection of genetically modified (or transgenic) plants, which have attracted a large amount of media attention in recent years, is still a burning question. As genotypes diverge genetically, so do spectral properties.

As mentioned in the Section 6.3, the occurrence of heterophylly in a species is a source of variation in leaf optical properties. Within-species variation has been investigated across varied natural environments, especially in tropical and sub-tropical regions. For instance, polymorphism, a discontinuous genetic variation resulting in the occurrence of several different forms or types of individuals among the members of a single species, is common in nature. Martin et al. (2007) and Martin and Asner (2009) found genetic adaptation in ohia lehua (*Metrosideros polymorpha*), a dominant flowering evergreen tree species endemic of Hawaii. One major source of variation for leaf traits of plants growing at different elevations was SLA, which was related to several variables such as leaf thickness, density, nitrogen, and pigment content. The potential of remote sensing to identify the gender of dioecious tree species has been explored by Arroyo-Mora et al. (2008) on two neotropical tree species, *Hyeronima alchorneoides* and *Virola koschnyi*. They applied several spectral indices and statistical tests to reflectance spectra and were able to ascertain which were the males and females. They found that females had higher photosynthetic pigments and nitrogen than males; they consequently photosynthesize at a higher rate to cover fruit production. Such studies of other dioecious species would ensure that male-to-female ratios are maintained when reforesting or during selective logging. Klančnik et al. (2012, 2014a) studied various aquatic plants growing in contrasting environmental conditions, for example, natant and emerged leaves, along water-land gradients and observed heterogeneous leaf anatomical and chemical properties within species, which affected their reflectance.

6.5 Interspecific Diversity

Most research focuses on interspecific variation, that is, the difference between species. Interspecificity (literally between species), or being interspecific, describes issues between individuals of separate species. Although intraspecific variability may be higher than interspecific variability, reflectance and transmittance spectra, which contain information regarding differences in leaf traits, have been employed in species discrimination for a long time. Are foliar optical properties able to discern fine-scale measurements of diversity at multiple levels? The answer is not simple since there is so much variation in these properties among individuals of the same species that discriminating among different species may not be possible in all situations. Spectral reflectance measurements in the optical domain have shown that although nearly all healthy leaves have a generally similar spectral response, differences in pigments such as chlorophyll, as well as differences in leaf structure, water content, and biochemical composition, can yield spectral signatures that are effective for discriminating among species (Ustin and Gamon, 2010). In contrast to visible and shortwave infrared data, thermal infrared spectra of broadleaf plants showed considerable spectral diversity, suggesting that such data eventually could be utilized to map vegetation composition (Ribeiro da Luz and Crowley, 2007; Ullah et al., 2012b). Another question arises concerning the relationships between variation in foliar optical properties and ecosystem functioning. Because of the cost and time required for characterizing heterogeneity within an ecosystem using classical methods, an increasing number of studies have paid attention to interspecific variation at both the leaf and canopy levels. We distinguish here between plants from the same family or genus and plants from different classes.

6.5.1 Inter-Family or Inter-Genus

Once again, Harold Gausman from USDA in Weslaco, TX, pioneered the discrimination of plant species using remote sensing. He and his colleagues measured the reflectance spectra of three citrus species (*Citrus aurantium, Citrus paradisi*, and *Citrus sinensis*) and showed differences in the visible that they attributed to differences in leaf chlorophyll content (Gausman et al., 1977). The genus *Eucalyptus* (*Myrtaceae*) is widely distributed in the southern hemisphere, mostly in Australia and Tasmania where they are native. To date, over 700 species of *Eucalyptus* are known. As they radiated across the country, they changed morphologically and physiologically to adapt to the various soils and climates. Due to their major commercial value, eucalyptus trees are now planted by both small farmers and large conglomerates in Asia, Australia, Europe, and the United States for fuel wood, lumber, biomass, essential oils, etc. Several methods have been applied to the reflectance spectra of eucalyptus leaves to determine whether there were spectral differences among the different species and, if the answer was yes, to identify the main regions (e.g., Coops et al., 2002; Magnussen et al., 2004; Kumar, 2007a; Kumar et al., 2010). Results indicated that, even though spectra were similar, there were statistically significant differences. Castillo et al. (2008) successfully discriminated between Tasmanian bluegum (*Eucalyptus globulus*) and shining gum (*Eucalyptus nitens*) species by applying different NIRS techniques. Similarly, Fan et al. (2010) studied the feasibility discriminating *Ephedra* plants of three different species with diffuse reflectance FT-NIRS. These plants are economically important in China because they contain secondary metabolites such as alkaloids, essential oils, tannins, flavonoids, and organic acids.

6.5.2 Inter-Class

A wide variety of studies have used laboratory spectra to discriminate between species in a wide range of ecosystems. The delineation and classification of ecologically homogeneous units is complex and is often performed by state-based agencies and many different classification schemes have been used so far. They include the Holdridge life zones classification system, a global

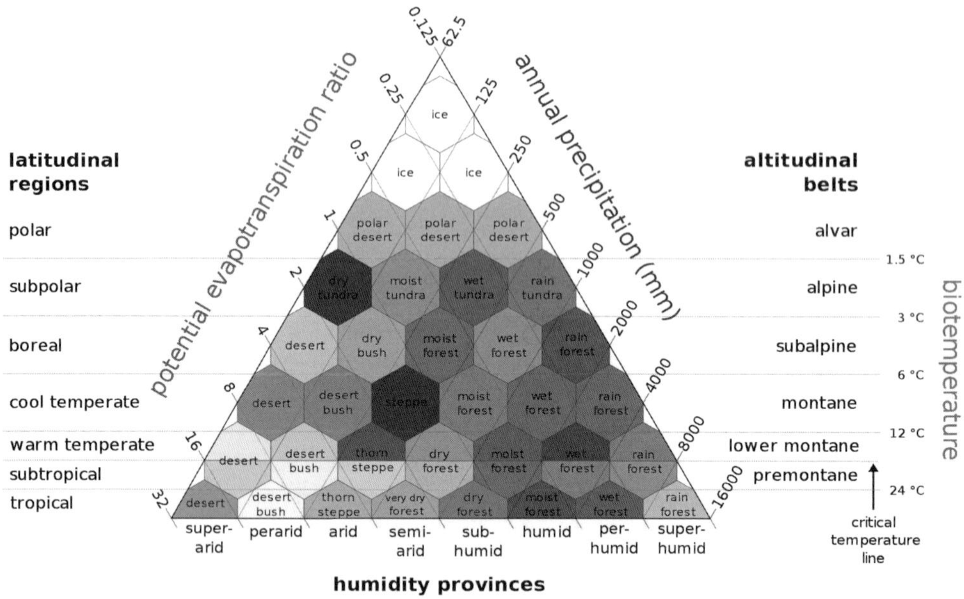

Figure 6.19 Holdridge life zones classification scheme (Holdridge, 1967) (https://en.wikipedia.org/wiki/Holdridge_life_zones). (A black and white version of this figure will appear in some formats. For the color version, please refer to the plate section.)

bioclimatic model based on limited empirical data. It consists of a triangle of hexagonal-shaped cells that relate the distribution of natural vegetation associations to two climate indices, average biotemperature and average annual precipitation. The potential evapotranspiration ratio depends on these two primary variables (Figure 6.19).

The diversity of vascular plants is very unevenly distributed across the Earth (Barthlott et al., 2005). For instance, some regions of the Siberian taiga only have one tree species per hectare, when there may be more than 450 tree species per hectare in a lowland forest of Ecuador. Interspecific variability, which contributes to biodiversity (a contraction of "biological diversity"), has been studied through leaf optical properties in many regions of the Earth: arid zones (e.g., Pearman, 1966; Sinclair and Thomas, 1970; Eller et al., 1983; Lewis, 2002), tundra (e.g., Kutyreva et al., 1960; Gauslaa, 1984), savannah (e.g., Schmidt and Skidmore, 2001), temperate and boreal forest (e.g., Serbin et al., 2014), temperate maritime climate (e.g., Roelofsen et al., 2014), warm temperate forest (e.g., Van Wittenberghe et al., 2014), subtropical forest (e.g., Fung et al., 1998, 2003; van Deventer et al., 2015), tropical heath forest (e.g., Cao, 2000), tropical mangrove (e.g., Wang and Sousa, 2009; Panigrahy et al., 2012), tropical mountain cloud forest (e.g., Velázquez-Rosas et al., 2010), tropical dry forest (e.g., Avalos et al., 1999; Sanches and Válio, 2006), and tropical rain forest (e.g., Lee and Graham, 1986; Cochrane, 2000; Poorter et al., 2000; Clark et al., 2005; Göttlicher et al., 2011; Durgante et al., 2013). Comparison between spectra involves advanced parametric and non-parametric classification approaches (see Chapter 11), but results generally do not distinguish one in particular in terms of performance.

Tropical forests are vast, mostly unexplored places. At the beginning of the 2000s, this ecosystem displays the highest biodiversity in the world and has been the center of attention. Two research teams initiated long-term research on the uniqueness of the spectral signatures of tropical tree species.

Gerardo Arturo Sánchez-Azofeifa and his colleagues from the University of Alberta studied the biodiversity loss and habitat fragmentation of tropical secondary dry forests. In particular, they studied the increasing dominance of lianas that take advantage of existing trunks and other lianas in disturbed habitats to climb up to the top of the canopy. The presence of up to 27 species of lianas has been documented in one single crown. Lianas rapidly form a carpet of leafy vegetation that shades the plants beneath and cause higher tree mortality. This layer on top of tree crowns has important implications for remote sensing studies, and could lead to misinterpretation of photosynthetic activity using vegetation indices; moreover, little or no differences between spectral signatures of lianas versus trees may limit the potential for tree classification. The separability these two groups has been explored using leaf spectral signatures collected during the dry season. The two life-forms were distinct at a dry forest site (Parque Natural Metropolitano, Panama) but not at a rainforest site (Fort Sherman, Panama) (Castro-Esau et al., 2004, 2006b; Kalacska et al., 2007; Sánchez-Azofeifa et al., 2009). Interspecies variability for dozens of tropical tree species has been also addressed at the leaf scale, based on leaf spectral reflectance, structural properties, and functional traits (Calvo-Alvarado et al., 2008; Rivard et al., 2008b; Alvarez-Añorve et al., 2012; Hesketh and Sánchez-Azofeifa, 2012, 2013; Ball et al., 2015). In particular, Hesketh and Sánchez-Azofeifa (2012) explored the nature and extent of seasonal spectral variation at both wet and dry tropical forest sites, to improve the accuracy of unsupervised classification.

Ecologist Gregory Asner and his colleagues from the Carnegie Institution for Science took an interest in the development of techniques for the analysis and interpretation of the biodiversity in tropical rain forests. They introduced the concept of spectranomics, aimed at providing a basis to potentially estimate biodiversity over large spatial scales using advanced airborne and satellite mapping techniques. Strong relationships have been identified between leaf phylogeny, chemistry, and spectroscopic properties (Asner, 2008, 2014; Asner and Martin, 2008b, 2009, 2011; Asner et al., 2009, 2011a, 2011b, 2012, 2014; Doughty et al., 2011; Féret and Asner, 2011; Feilhauer et al., 2015). Several important scientific research findings result from the spectranomics approach. For instance, in a study of 162 canopy species across a wide climatic gradient in Australia, Asner et al. (2009) found that variation in leaf chemical signatures, which are strongly related to leaf reflectance, varied far more in response to taxonomy and species richness than to changes in climate. Asner et al. (2011b) examined the separability of leaves growing on two natural untreated soils in lowland Amazonia, Inceptisols of fluvial origin (high fertility) and nitrogen-impoverished Ultisols (low fertility). They measured 21 leaf traits and spectral signatures among 600 coexisting trees and found that nearly 90% of the chemical variation could be attributed to taxonomic affiliation. They showed that taxa found on Inceptisols have a deep evolutionary history, whereas those found on Ultisols have undergone trait evolution at the species level.

Species identification has been also applied to crops (e.g., Farrar and Mapunda, 1977; Gausman et al., 1978a; Noble and Crowe, 2001b), wetlands (Khanna et al., 2011; Klančnik et al., 2015), medicinal plants (Kokalj et al., 2011), etc. The spread of non-native species into new habitats is of major global environmental concern. Leaf spectroscopy has been used to distinguish between native and invasive plant species, with the aim of mapping and monitoring the spread of the latter. Because of its narcotic potential cannabis may be considered as an illegal invasive plant. The "emerald green" color of the leaves may result from a combination of diffuse green reflectance from the interior of the leaves and specular reflectance of blue sky light from the surface of the leaves. Daughtry and Walthall (1998) compared the reflectance and transmittance spectra of leaves of cannabis with those of various plants occurring where marijuana is grown in the eastern United States, but they did not find significant distinctive spectral features. Burkholder et al. (2011) investigated the separability of tree of heaven (*Ailanthus altissima*) from four native species using leaf spectral reflectance measurements. The long-leaved wattle (*Acacia longifolia*), an invasive shrub native to southeastern Australia, has a negative impact on vegetation and functioning in Portuguese dune ecosystems. Lehmann et al. (2015) found that the low tannin concentration of this plant compared to all investigated neighboring species could favor its spectral identification in four visible infrared regions. While invasive plants occur in natural habitats, garden and agricultural weeds impact crop production, either in commercial settings or in home, fruit, and vegetable gardens. They may be native and non-native plants. There is a demand for systems which can discriminate between crop and weed plants, so enabling targeted application of remedial chemical treatments. Some studies reported differences in leaf reflectance among weed species and between weed and agronomically important crop species (e.g., Gausman et al., 1981; Walter and Koch, 1981; Franz et al., 1991; Favier et al., 1998; Vrindts and De Baerdemaeker, 1998; Zwiggelaar, 1998; Smith and Blackshaw, 2002, 2003; Henry et al., 2004; Noble and Crowe, 2005). However, these differences attributed to variation in mesophyll cell arrangement or pigment concentration do not seem to be specific enough to be used as an aid in identifying *in situ* leaves of various types of weed seedlings (Santos et al., 2012).

6.6 Climate Change

There is an abundance of evidence that the Earth's climate has changed in recent decades and that this change has had an influence on natural resources (Parmesan, 2006). Among the biological systems affected, phenology is one of the prominent bio-indicators of climate change, since the timing of recurring phenological events of deciduous plants is altered by changed weather conditions. However, the causes of the timing of these events in autumn, such as leaf coloring or leaf fall, are not totally understood, although theoretical knowledge exists about the triggering factors for leaf coloring and leaf fall of deciduous tree species (Estrella and Menzel, 2006; Doi and Takahashi, 2008; Archetti et al., 2013). Other expectations for climate change include warmer night temperatures that will extend the autumn period of leaf senescence and slow the degradation of chlorophyll pigments, thus minimizing the expression of vibrant reds, oranges, and yellows from carotenoid pigments. Besides changes in temperatures affecting leaf development, changing temperatures will have major impacts on photosynthesis, respiration, and evapotranspiration rates that differentially affect various species. Less predictable are changes in precipitation and biogeochemical cycling that can affect nutrient availability, thus altering physiological processes. However, such leaf changes should be observable by changes in leaf optical properties. Finally, characterizing biodiversity and understanding its consequences on ecosystem functioning are also important challenges in the context of environmental change. Some plant species like lianas are considered a significant fingerprint of the effects of global environmental change (Sánchez-Azofeifa et al., 2009).

7

Variations Due to Leaf Abiotic and Biotic Factors

The term "stress" was defined by Jackson (1986) as any disturbance that adversely influences plant growth. Various types of stress can be caused by abiotic (water deficit, nutrient deficiency, salinity, heavy metal, herbicide, air pollution, etc.) and biotic (bacteria, fungi, viruses, insects, etc.) factors. They may induce changes in leaf anatomy, chemistry, and physiology, which will result in changes in leaf optical properties (e.g., Carter, 1993). The challenge of remote sensing is to detect and diagnose stress as early as possible, before symptoms appear. Several approaches have been implemented with varying degrees of success. The main problem is that symptoms of stress caused by one factor are often confused with those caused by another factor. And when stress is caused by more than one factor, spectroscopic data may not provide enough information to identify these factors.

7.1 Abiotic Factors

7.1.1 Edaphic Factors: Mineral Nutrients

The difference between rich, fertile soil, and poor, infertile soil is essentially the mineral composition. To grow, plants need about 13 mineral nutrients in an available and balanced form: macronutrients (nitrogen, phosphorus, potassium, calcium, magnesium, and sulfur) and micronutrients (iron, copper, manganese, zinc, boron, molybdenum, and chlorine) that are essential but needed in small amounts. Some macronutrients may interact in the soil with micronutrients or bind with potentially toxic elements making them unavailable to plants. Greenhouse and field studies have shown that a deficiency or an excess in some of these minerals in the soil or other growing medium may cause visual symptoms that can be detected by studying changes in optical properties (Table 7.1). Once exposed, plants often have difficulty recovering. Such symptoms may be a powerful diagnostic tool for evaluating the nutrient status of plants (Taiz and Zeiger, 2010, http://5e.plantphys.net/). However, the use of such properties as a tool in nutrient deficiency diagnostics requires further understanding. Such experiments are difficult to set up because the first signs of deficiency or excess may appear several weeks, sometimes even several months, after the installation of the experiment. Attempts to correlate spectral changes directly with metal concentrations in plant tissues have not been successful so far, which suggests only indirect effects. In particular, the lack of common protocols has made it difficult for researchers to compare and validate their results. Some use fresh leaves, others dry compacted powders, some measure the directional-hemispherical reflectance factor, while others use a leaf clip. One should also keep in mind that a given individual symptom is seldom sufficient to make a definitive diagnosis of a plant nutrient status. Many of the classic deficiency symptoms such as tip

Table 7.1 *Estimation of leaf macronutrients using remote sensing techniques.*

		Reference	
Nutriment	Symbol	Near-infrared spectroscopy	Leaf optical properties
Primary macronutrients			
Nitrogen	N	McLellan et al. (1991a, 1991b), Aber et al. (1994), Martin and Aber (1994), Bolster et al. (1996), Gillon et al. (1999a), White et al. (2000), Kokaly (2001), Gislum et al. (2004), Petisco et al. (2005)	Judkins and Wander (1950), Thomas and Oerther (1972), Al-Abbas et al. (1974), Gausman et al. (1979), Younes et al. (1974), Nelson et al. (1986), MacNeil et al. (1987), Massantini et al. (1992), Blackmer et al. (1994), McMurtrey III et al. (1994), Peñuelas et al. (1994), Schepers et al. (1996), Li et al. (1999), Liew et al. (1999), Ponzoni and Gonçalves (1999), Moran et al. (2000), Tarpley et al. (2000), Graeff et al. (2001), Johnson (2001), Buscaglia and Varco (2002), Read et al. (2002), Bortolot and Wynne (2003), Graeff and Claupein (2003), Middleton et al. (2003), Yang et al. (2003), Fridgen and Varco (2004), Baltzer and Thomas (2005), Coops and Stone (2005), Jørgensen et al. (2007), Wang et al. (2007), Ulissi et al. (2011), Zhai et al. (2013)
Phosphorus	P	Shenk et al. (1979, 1981), Clark et al. (1987), Gillon et al. (1999a), Chen et al. (2002), Petisco et al. (2005)	Al-Abbas et al. (1974), Younes et al. (1974), Milton et al. (1991), Ponzoni and Gonçalves (1999), Graeff et al. (2001), Li et al. (2006), Jørgensen et al. (2007), Wang et al. (2007), Zhang and Li (2008), Liu et al. (2011b), Zhai et al. (2013)
Potassium	K	Shenk et al. (1981), Clark et al. (1987), Guo et al. (2008)	Al-Abbas et al. (1974), Younes et al. (1974), Wang et al. (1997), Ponzoni and Gonçalves (1999), Fridgen and Varco (2004), Wang and Bai (2005), Jørgensen et al. (2007), Yi et al. (2010a), Zhai et al. (2013)
Secondary macronutrients			
Calcium	Ca	Shenk et al. (1981), Clark et al. (1987), Petisco et al. (2005)	Al-Abbas et al. (1974), Li et al. (2005a,b), Liew et al. (1999)
Magnesium	Mg	Clark et al. (1987), Smith et al. (1991)	Al-Abbas et al. (1974), Masoni et al. (1996), Graeff et al. (2001)
Sulfur	S		Al-Abbas et al. (1974), Masoni et al. (1996)
Micronutrients			
Iron	Fe	Clark et al. (1987), Basayigit et al. (2014)	Mariotti et al. (1996), Masoni et al. (1996), Adams et al. (2000a, 2000b), Graeff et al. (2001), Li et al. (2006), Chi et al. (2009a)
Copper	Cu	Clark et al. (1987), van Maarschalkerweerd et al. (2013)	Howard et al. (1971), Schwaller et al. (1981), Adams et al. (2000a, 2000b), Schellekens et al. (2005), Liu et al. (2008a), Yang et al. (2009)
Manganese	Mn	Clark et al. (1987)	Deroin and Deroin (1990), Adams et al. (1993, 1999), Masoni et al. (1996), Adams et al. (2000a, 2000b)
Zinc	Zn	Clark et al. (1987), Yi et al. (2010b)	Adams et al. (2000a, 2000b), Su et al. (2007a)

burn, chlorosis, and necrosis are characteristically associated with more than one mineral deficiency and also with other stresses that by themselves are not diagnostic for any specific nutrient stress.

7.1.1.1 Nitrogen (N)

The green color of plant leaves is generally associated with chlorophyll, which is positively correlated with nitrogen concentration. This is why leaf chlorophyll content is often used as an indicator of soil nitrogen, biomass, and crop productivity potential. Plants have a high requirement for N, more than any other mineral nutrient, which is found in chlorophyll, proteins, enzymes like RuBisCO that stores 50–60% of the soluble protein in plant leaves, amino acids, nucleic acids, and energy transfer molecules like adenosine triphosphate (ATP) (Evans, 1989). As a consequence, deficiency will reduce the amount of proteins and associated chlorophylls (Khamis et al., 1990). Because N is often the most limiting nutrient after water, detection of N status has been a major priority for crop monitoring as well as in understanding the nutrient relations of natural systems. Absorption features in near infrared, shortwave infrared, and thermal infrared that have been related to particular foliar chemical concentrations. Chapter 3 indicates that many wavelengths in the shortwave infrared (SWIR) are sensitive to N bonds with H. However, most authors have focused on the use of visible bands to predict N, taking advantage of the relationship to chlorophyll and carotenoid pigments. Early studies showed correlations between N and reflectance in the visible spectrum on fresh leaves for a variety of herbaceous and woody crops. Judkins and Wander (1950) used a reflectometer to calibrate the nitrogen concentration in leaves of apple (*Malus domestica*), peach (*Prunus persica*), and grape vine (*Vitis vinifera*) and showed that accurate estimates of N concentration could be retrieved. Thomas and Oerther (1972) extended this relationship by measuring nitrogen concentration in sweet pepper (*Capsicum annum*) from spectrophotometry by regression between concentration and reflectance. Al-Abbas et al. (1974) examined the reflectance, transmittance, and absorptance of maize (*Zea mays*) leaves at 30 wavelengths for N concentrations and found significant correlations from analysis of variance (ANOVA) for bands in the infrared wavelengths. Younes et al. (1974) found increases in maize leaf reflectance across the visible between 13 and 35 days after emergence and a decline in the near-infrared reflectance after the 28th day. Following these pioneering studies, research expanded to examine native woody species (e.g., Gausman et al., 1979; Nelson et al., 1986). These authors and subsequently others found strong relationships with leaf N in the visible bands around 550 nm and changes in broadening or narrowing the red-edge position (e.g., MacNeil et al., 1987; Blackmer et al., 1994; Moran et al., 2000; Tarpley et al., 2000; Buscaglia and Varco, 2002; Read et al., 2002; Middleton et al., 2003). Li et al. (1999) focused on the near-infrared reflectance of maize leaves between 650 nm and 1150 nm to predict N, calculating the first derivative and then applying partial least squares (PLS) regression, and principal components regression (PCR), where they found strong linear correlations from the multivariate regression for different growth stages. Johnson (2001) found strong correlations both for fresh leaves and modeled leaves (using LEAFMOD) at 1675 nm, 2075 nm, and 2160 nm. More recent studies have used PLS regression (Fridgen and Varco, 2004; Jørgensen et al., 2007; Ulissi et al., 2011), although recently Zhai et al. (2013) found stronger results with support vector machine (SVM) regression methods. Studies on dry ground leaves from multiple species have found stronger correlations with infrared bands than visible bands, as might be expected when pigments have degraded. Kokaly (2001) found strong correlations between 2054 nm and 2172 nm using continuum removal analysis. Several studies looked at interactions of other nutrients with the correlations between N and reflectance. Fridgen and Varco (2004) concluded that detection required sufficient K, while Ponzoni and Gonçalves (1999) concluded that the changes they observed in reflectance were due to K deficiency and not N. Thus interpretations of causal relationships between pigment concentrations, N, and leaf reflectance are complicated by age, other environmental resources, and stress. Yang et al. (2003) have used a leaf color chart instead of a chlorophyll meter (SPAD) to estimate the leaf nitrogen status of rice (*Oryza sativa*) for timing of fertilizer applications. This application is developed in Chapter 12.

7.1.1.2 Phosphorus (P)

Phosphorus is a key macronutrient absorbed by plants predominantly in the monovalent orthophosphate form $H_2PO_4^-$. It is the second most limiting nutrient, essential at all phases of plant growth. For instance, in reforested soils where deficiency of phosphorus is common, a significant loss of productivity is generally observed. The reason for this is that phosphorus is an essential component of genetic material so that a deprivation will adversely affect cell division and expansion (Li et al., 2006). It also hinders metabolic processes leading to sugar buildup, which is associated with anthocyanin formation. Phosphorus is not a component of chlorophyll but a deficiency may lead to high chlorophyll content when nitrogen is present in abundant supply; hence leaves of some P-deficient plants display a dark green or reddish-purple color. Milton et al. (1991) tested for the first time the hypothesis that phosphorus deficiency could cause morphological and spectral changes in hydroponically grown soybean (*Glycine max*) plants. Only slight variations occurred in the green and the red-edge regions. Also, P-deficient plants did not display the red-shift generally observed as leaves mature. Other authors confirmed the potential of the visible and near-infrared reflectance of leaves to estimate phosphorus content (e.g., Ponzoni and Gonçalves, 1999; Graeff et al., 2001; Li et al., 2006; Wang et al., 2007; Zhang and Li, 2008; Liu et al., 2011b; Zhai et al., 2013). Phosphorus concentration has also been determined on forage samples by NIRS (e.g., Shenk et al., 1979, 1981; Gillon et al., 1999a; Chen et al., 2002; Petisco et al., 2005; Zhai et al., 2013).

7.1.1.3 Potassium (K)

Plants absorb potassium ions K^+, which can be readily leached from soil. Potassium is known to influence leaf color. Low levels may reduce chlorophyll concentration, therefore photosynthesis, and vice versa. In contrast to nitrogen deficiency, chlorosis is irreversible in potassium deficiency, even if potassium is given to the plants. Because it is very mobile within the plant, visible symptoms only develop on young leaves in the case of extreme deficiency. The symptoms preferentially affect the edge and interveinal spaces of these leaves, which become necrotic and turn brown as the deficiency progresses, while yield and quality decline. Potassium anomalies have been predicted using partial least-squares regression or other advanced statistical methods (e.g., Fridgen and Varco, 2004; Yi et al., 2010a; Zhai et al., 2013).

7.1.1.4 Calcium (Ca)

Desert soils and water generally have plenty of calcium, so deficiency problems are rare. Excessive calcium can limit the availability of other nutrients. Calcium plays a fundamental physiological role in leaf structure and signaling. Calcium ions Ca^{2+} are a key component in plant cells: they maintain the osmotic potential within vacuoles; they act as a stabilizing element in membranes; they strengthen cell walls; and they act as a secondary messenger for a multitude of signals. A deficiency in calcium

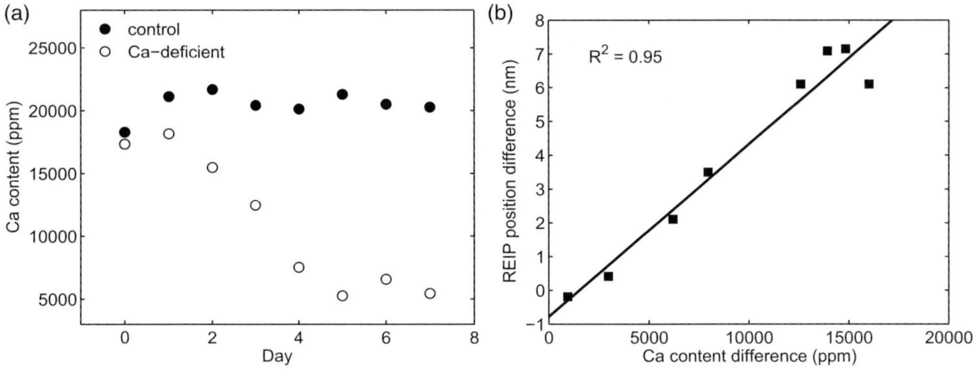

Figure 7.1 (a) Calcium content in control Ca-deficient plants as a function of time. (b) Relationship between Δ (REIP) and Δ(Ca) (redrawn from Li et al. (2005a)).

usually results in smaller plants with dead necrotic spots around the base of the leaves. Due to the very low mobility of this nutrient, new leaves (top of plant) are distorted or irregularly shaped. However, very slow growing plants with a deficient supply of calcium may re-translocate sufficient calcium from older leaves to maintain growth with only a marginal chlorosis of the leaves. While plants are generally subjected to prolonged periods of stress when spectral data are acquired, some authors have attempted to detect plant stress before its consequences are noticeable or irreversible. For this they measured variations in leaf reflectance (Liew et al., 1999), leaf color (Li et al., 2005b) or position of the red-edge inflection point (Li et al., 2005a), as illustrated in Figure 7.1.

7.1.1.5 Magnesium (Mg)

Plants absorb magnesium like potassium, as an ion which can be readily leached from soil. Magnesium is an essential macronutrient that plays an important role in photosynthesis because it forms the central atom of chlorophyll. Therefore, without sufficient amounts of magnesium, plants begin to degrade the chlorophyll in the old leaves. Mg-deficient leaves show advanced interveinal chlorosis, giving the leaves a marbled appearance. In its advanced form, magnesium deficiency may superficially resemble potassium deficiency. Due to the mobility of magnesium in the phloem, the plant will first break down chlorophyll in older leaves, which turn yellow at the edge leaving a green arrowhead shape in the center of the blade, and transport the magnesium to younger leaves which have greater photosynthetic needs. Early in the growing season, magnesium intake by grazing ruminants may be insufficient to meet their metabolic demand, especially during lactation, resulting in hypomagnesemic tetany. NIRS calibration equations have been developed to determine the magnesium concentration in forage samples: crested wheatgrass (*Agropyron cristatum*), tall fescue (*Festuca arundinacea*), and alfalfa (*Medicago sativa*) samples (Clark et al., 1987), and ryegrass (*Lolium perenne*) samples (Smith et al., 1991). At the leaf scale, the reflectance and transmittance of plants growing under Mg-deficient conditions are higher in the visible than those of normal plants leaf, and their absorptance lower (Al-Abbas et al., 1974; Masoni et al., 1996).

7.1.1.6 Sulfur (S)

Plants absorb sulfur in the form of sulfate. The visual symptoms of sulfur deficiency may appear as a yellowing of the foliage, similar to that produced by N deficiency. However, unlike

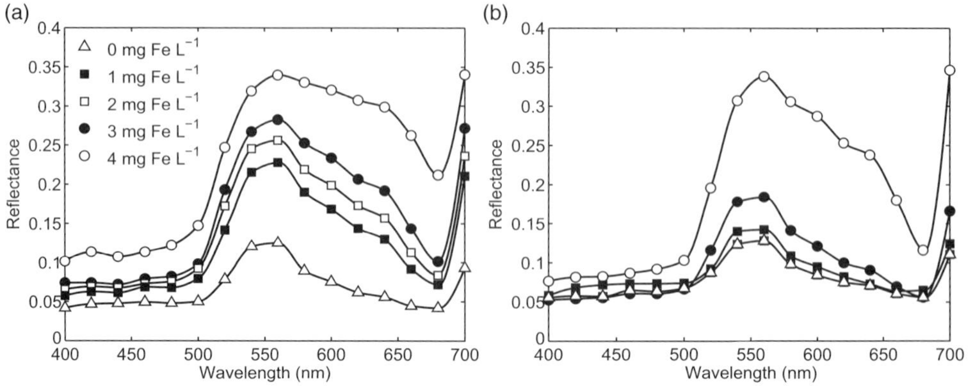

Figure 7.2 Measured spectra of reflectance of the third maize (*Zea mays*) leaf (a), and eighth sunflower (*Helianthus annuus*) leaf (b) grown with five iron concentrations in nutrient solution (redrawn from Mariotti et al. (1996)).

nitrogen, sulfur does not readily move about in the plant so that younger leaves in the upper canopy turn yellow first, sometimes followed by older leaves. With advanced deficiency, brown lesions and/or necrotic spots may develop along the petiole, and the leaves tend to become more erect and to curl downward.

7.1.1.7 Iron (Fe)

Iron is taken up by the root of plants as either Fe^{2+} (ferrous cations) or Fe^{3+} (ferric cations). It is involved in photosynthesis, respiration, chlorophyll formation, and many enzymatic reactions so it is an essential element for plants, and agricultural production is lowered whenever it is deficient. The most common symptom for iron deficiency is chlorosis that primarily affects young leaves and then evolves into an overall chlorosis, ending as a totally bleached leaf. The leaves may recover upon application of iron before they turn completely white and the bleached areas develop necrotic spots. Like sulfur, iron has low mobility, which explains why Fe-deficiency symptoms appear first on the youngest leaves. Mariotti et al. (1996) studied the effects of iron deficiency-induced modifications in the spectral properties of maize (*Zea mays*) and sunflower (*Helianthus annuus*) leaves. They showed an increase in reflectance and transmittance (therefore a decrease in absorptance) at all wavelengths. The greatest variations were observed at 555 nm while they were negligible at 430 nm. Such variations were attributed to changes in leaf chlorophyll concentration, which is well correlated with leaf iron concentration. Thus, the latter is indirectly responsible for modifications in leaf spectral properties (Figure 7.2). Similar results have been obtained by other authors (Li et al., 2006; Chi et al., 2009a; Basayigit et al., 2014).

7.1.1.8 Copper (Cu)

Plants absorb copper ions Cu^{2+}. Copper is an essential element for plant growth and it is a component of many enzymes. Copper deficiency in plants may be difficult to diagnose and may not be recognized by the plant producer until it has reached a stage where correction is no longer possible. Symptoms include chlorosis, necrosis, leaf distortion, and terminal dieback. Wilting can also occur in the young leaves, indicating structural weaknesses due to reduced lignification of the xylem elements. This is known as "white tip disease" in cereals. Howard et al. (1971) compared the visible and near-infrared reflectance spectra of ponderosa pine (*Pinus ponderosa*) needles taken from trees growing in a copper-rich area to the spectra of trees growing in a background area of low copper content. The highest correlation (R = 0.51) between copper content and spectral reflectance occurred at 810 nm. Schwaller et al. (1981, 1983) conducted investigations to provide data on sugar maple (*Acer saccharum*) plant response to copper: they showed an increased leaf reflectance over the entire range of investigated wavelengths (475–1650 nm). Liu et al. (2008a) explored the spectral response of Chinese cabbage (*Brassica rapa* subsp. *chinensis*), a vegetable widely cultivated in China, to copper content. They showed an increase of reflectance in the green, a decrease in the near infrared, and a blue shift of the red-edge.

7.1.1.9 Manganese (Mn)

Plants absorb manganese as an ion through their foliage as well as their roots. The early stages of the chlorosis induced by manganese deficiency are somewhat similar to iron deficiency, although the pattern is not as distinct as with iron. They begin with a light chlorosis of the young leaves and netted veins of the mature leaves, especially when they are viewed through transmitted light. As the stress increases, the leaves take on a gray metallic sheen and develop dark freckled and necrotic areas along the veins. A purplish luster may also develop on the upper surface of the leaves. Deroin and Deroin (1990) performed one of the most extensive studies on the effect of manganese on European beech (*Fagus sylvatica*) leaf optical properties. The depletion of manganese in the substratum induced phytomorphological changes (leaf thickness, tissue volume) that in turn induced variation in leaf near-infrared reflectance. Adams et al. (1993, 1999, 2000a, 2000b) measured the reflectance and fluorescence of Mn-deficient soybean (*Glycine max*) leaves. They showed a significant relation to manganese content, with a strong effect of leaf age and nodal position, and developed a yellowness index sensitive to manganese content.

7.1.1.10 Zinc (Zn)

Plants absorb zinc as an ion through their foliage as well as their roots. In the early stages of zinc deficiency, yellowing occurs between the veins of the new leaves and pitting develops in the interveinal upper surfaces of the mature leaves. As the deficiency progresses, these symptoms develop into an intense interveinal necrosis but the main veins remain green, as in the symptoms of recovering iron deficiency. In many plants, especially trees, terminal leaves become very small and may be rosetted. Su et al. (2007a) compared the reflectance spectra of Zn-treated and control of barley (*Hordeum vulgare*) leaves: the latter displayed structural changes such as a decrease in intercellular space, palisade, and epidermal cell size, and biochemical changes such as a decrease in chlorophyll content. This produced a decrease in absorption at around 680 nm and a blue shift of the red-edge.

7.1.1.11 Multi-Element Analysis

Some studies just do not focus on one or a few elements, but on many of them to determine their relationships (e.g., Jørgensen et al., 2007; Menesatti et al., 2010; Pacumbaba and Beyl, 2011; Ordóñez et al., 2013; Basayigit et al., 2014).

7.1.2 Edaphic Factors: Salt

All soil contains salts, many of which are essential nutrients for plants. The ions responsible for salination are Na^+, K^+, Ca^{2+}, Mg^{2+}, and Cl^-. Some of these have been investigated in Section 7.1.1. Salinity occurs when soluble salts are present in excess amounts in soil and water. Saline soils found in arid or semi-arid areas generally arise from natural processes like weathering of mineral rocks. In these areas, there is often inadequate rainfall or drainage to move the salt down through the soil so it can leach away from plant roots. Saline soils are also a scourge to agriculture. When marginal land is cleared of native vegetation and replaced with shallow-rooting crop plants, this causes the underground water table to rise, moving salts up to the soil surface. Irrigation water is also often saline in these areas, which adds to the level of salts that the plant must tolerate. The effect of soluble salt concentration on plant yield has been studied for numerous species. Once inside the cell, salt can cause severe stress to the plant. In particular, it inhibits metabolic processes such as protein synthesis and photosynthesis (Vieira Santos, 2004). And when it rises to toxic levels in older leaves, it can reduce plant growth and cause plant death. Thomas et al. (1967) have observed a significant increase (decrease) in visible-near-infrared reflectance (transmittance) from single leaves of cotton (*Gossypium hirsutum*) exposed to saline conditions, attributed to an increased leaf thickness and volume of the intercellular spaces. Similar results have been obtained by Carter and Myers (1963) with grapefruit (*Citrus* × *paradisii*) leaves and Hernández et al. (2014) with melon (*Cucumis melo*) leaves, while Gausman et al. (1969a, 1970a) with cotton (*Gossypium hirsutum*) leaves and Garriga et al. (2014) with strawberry (*Fragaria chiloensis*) leaves conversely report a decreased in reflectance and an increase in transmittance. To complicate matters further, Krezhova et al. (2009a, 2009b, 2011) measured the effect of salinity on leaf spectral reflectance of soybean (*Glycine max*) leaves and observed a decrease in reflectance in the visible, a red shift, and no significant changes in the near infrared. Lorenzen et al. (1990) grew two salt marsh plants, sea aster (*Tripolium pannonicum*) and sea purslane (*Halimione portulacoides*) under different salinity levels and hypoxic conditions and determined their leaf absorptance in four wavebands. They showed a variation in net photosynthesis with increased salinity but no detectable effect of salinity on leaf absorptance. This result was confirmed by Tilley et al. (2007) on other plants collected in coastal freshwater wetlands.

7.1.3 Edaphic Factors: Heavy Metals and Soil Pollution

Heavy metals are natural components of the Earth's crust, which become concentrated in soils either naturally or as a result of human activities. The term originally refers to the noxious effects of cadmium (Cd, 112.4 amu), mercury (Hg, 200.6 amu), and lead (Pb, 207.2 amu), all of which have a relatively high atomic mass unit compared to iron (Fe, 55.9 amu) (note: amu stands for atomic mass unit). It has since been applied to any other metallic chemical element that is toxic or poisonous at low concentrations: arsenic (As, 74.9 amu), chromium (Cr, 52.0 amu), cobalt (Co, 57.9 amu), nickel (Ni, 57.7 amu), copper (Cu, 63.5 amu), zinc (Zn, 65.4 amu), selenium (Se, 79.0 amu), thallium (Tl, 204.4 amu), or silver (Ag, 107.9 amu). These elements, some of which have been investigated in Section 7.1.1, because they are essential to plant growth, have the special feature that they are not degraded or destroyed. Due to the complexity of natural ecosystems, studies aiming to correlate the spectral response of plants with geochemical anomalies in the soil, especially those soils associated with mineral deposits, may be somewhat confusing and even inconsistent. Clijsters and Van Assche (1985) reviewed the *in vitro* action of heavy metals (nickel, thallium, cadmium, cobalt, chromium, and zinc) on the photosynthetic apparatus and compared them with the effects observed *in vivo*. They showed that these elements are phytotoxic and strongly decrease photosynthesis under sub-lethal concentrations. However, this effect is moderate *in vivo* unless the toxicity threshold exceeds the point where plant metabolism is affected. A reduction in photosynthesis can be due to a decrease of the chlorophyll content or changes in the ratio

Chl *a:b*. Carotenoids seem to be less sensitive than chlorophylls to the accumulation of heavy metals in plants. The spectral response of plants subjected to anomalous concentrations of heavy metals in growing media has been studied since the early 1980s. It often happens that differences in vegetation types are related to differences in soil (Cannon, 1971). Plant species whose presence is determined by soil geochemistry are often referred to as indicators. But there are many situations where we wish to detect changes in soil geochemistry that may not produce any obvious change in the species composition or density of the plant cover. Horler et al. (1980) grew garden pea (*Pisum sativum*), soybean (*Glycine max*), and sunflower (*Helianthus annuus*) plants in culture solutions containing various concentrations of Cd, Cu, Pb, or Zn. The loss of chlorophyll caused by the high concentration of heavy metals in the plant root zone induced an increase in leaf reflectance in the visible. Conversely, in the near and shortwave infrared, these authors found a decrease in reflectance with increasing metal concentration, which they attributed to an ill-defined change in internal leaf structure or, most certainly, to an increase in water content in these hydroponically grown plants. In apparent conflict with these results, Schwaller et al. (1981, 1983) with sugar maple (*Acer saccharum*) and Labovitz et al. (1983) with white oak (*Quercus alba*) leaves, found that the reflectance increased in the infrared with increasing metal concentration. Horler et al. (1983a) investigated the biogeochemical nature of the spectral shift of the red-edge inflexion point centered near 680 nm: they reported the same 5–20 nm shift to shorter wavelength that was seen in airborne radiometer data. In order to determine which metallic elements would produce this shift, Milton et al. (1988, 1989a) and Milton and Mouat (1989b) tested the effect of different metallic elements (selenium, arsenic, copper acetate, copper sulfate, zinc acetate, zinc sulfate, cobalt acetate, and nickel sulfate) on the spectral reflectance of plants grown in a greenhouse at the United States Geological Survey in Reston, Virginia. The presence or absence of spectral changes was attributed to the mobility of the metal cations, which affects their availability to plants. Only the addition of arsenic, cobalt acetate, and nickel sulfate resulted in a significant increase in reflectance between 500 nm and 650 nm and a spectral shift of the red-edge to shorter wavelength (Figure 7.3). Selenium, in contrast, produced a shift of the red-edge to longer wavelengths and the other elements produced only few spectral changes. These findings, suggesting little spectral variations in reflectance between mineralized and background sites, have been confirmed by other authors, who concluded that detection from airborne spectrometers might be difficult (Singhroy et al., 1989, 2000; Mendelssohn et al., 2001). However, compared to previously published data, the most recent studies advocate for a correlation between spectral characteristics and heavy metal contamination: arsenic (e.g., Font et al., 2004, 2005; Sridhar et al., 2007b; Slonecker et al., 2009; Bandaru et al., 2010), cadmium and vanadium (Rosso et al., 2005; Cheng et al., 2008), cesium (Su et al., 2007), chromium (Sridhar et al., 2007b), lead (Liu et al., 2010b), mercury (Dunagan et al., 2007; Yao et al., 2012), and cadmium and zinc (Sridhar et al., 2007a; Liu et al., 2010a, 2011a). Note that the latter did not observe any effect of strontium accumulation on leaf structural and spectral characteristics. There has been renewed interest for this topic in recent years; several papers have explored the concept of vegetation stress as bio-indicators of soil contamination using visible-infrared reflectance spectroscopy (e.g., Font et al., 2002, 2007; Kupková et al., 2012; Rathod et al., 2013; Del'Arco Sanches et al., 2014; Van Deventer and Cho, 2014; Xue et al., 2014).

Gold is not, strictly speaking, a heavy metal, in spite of its high atomic mass unit (Au, 196.9 amu). Currently, the metal is found in outcrops, where the ore appears at the surface, or it is detected through exploratory drilling. Vegetation could offer an alternative method to find untapped gold deposits. Plants absorb and transport gold particles within their leaves at trace amounts (Shacklette et al., 1970). Gold is

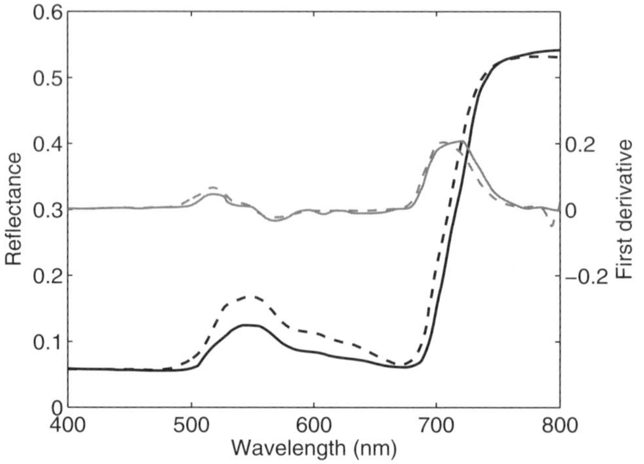

Figure 7.3 Laboratory reflectance spectra (left axis) and first derivative for reflectance spectra (right axis) of greenhouse-grown blue plantain lily (*Hosta ventricosa*) plants doped with $NiSO_4$ (dashed line) compared with average curve for control plants (solid line) (redrawn from Milton et al. (1988)).

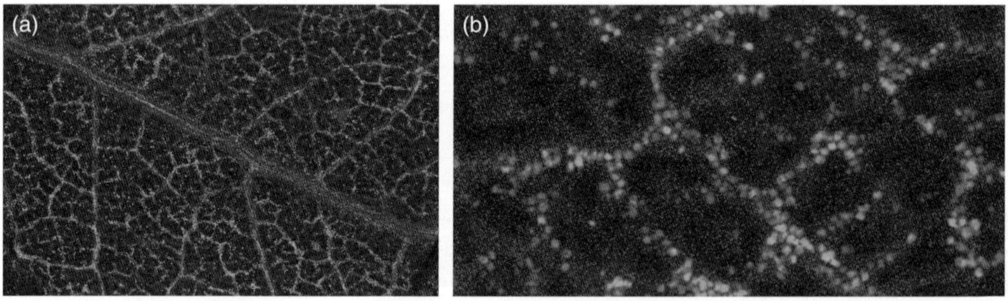

Figure 7.4 (a) Synchrotron image showing distribution of Cu (blue), Sr (oxalate crystals, green), and Au (red) in natural *Eucalyptus* leaf specimen. (b) Detailed view showing isolated particulate Au grains (red) and two of which are attached to the surface of Ca oxalate crystals (green) (from Lintern et al. (2013), Nature *Communications*, Copyright © 2013 with permission from Macmillan Publishers Limited). (A black and white version of this figure will appear in some formats. For the color version, please refer to the plate section.)

not toxic although it may inhibit plant growth in the form of Au^{3+} ions or Au nanoparticles. Xu and Ma (2004) studied plants growing in the Hetai gold deposit region of South China. Poisoned leaves are smaller and rougher, with yellow spots on the surface, and they may show signs of wilting. The cell and chloroplast structure of pine needles is deformed and even broken. Their reflectance was 10–30% higher than that measured for control needles. However, there is no unequivocal evidence that this increase in reflectance was due to gold actually absorbed by plants and that currently measured Au was not a result of dust contamination of samples. Using X-ray imaging detectors, Lintern et al. (2013) have detected particles of gold in the branches and leaves of eucalyptus trees growing in the Kalgoorlie region of Western Australia. They could probe in remarkable detail gold particles in the leaves, which would have been untraceable by other techniques (Figure 7.4).

Petroleum hydrocarbons, the occurrence of which is due to both natural and anthropogenic sources, are the most common chemicals involved in soil contamination and subsequently a major source of environmental degradation. In regions of oil exploration and exploitation, hydrocarbon spillage regularly affects crop and natural vegetation growing across a range of ecosystems. For instance, plants growing in salt marshes, estuaries, and mangrove forests may be directly exposed to marine pollution by petroleum oils that is transported by tidal flooding and can be deposited in the sediment. Because of the difficulty in accessing wetlands on the ground, remote sensing may be the proper tool to monitoring them. To evaluate the possibility of detecting spectral changes in vegetation induced by petroleum oils, Rosso et al. (2005) grew Virginia glasswort (*Salicornia virginica*) in the presence of light and heavy crude oil at sublethal and lethal concentrations. They observed significant decreases in growth and photosynthesis caused by the toxicity of some oil components, especially those that have a high molecular weight and lack mobility. Such effects were sufficient to produce significant changes in reflectance. Emengini and Ugbelase (2013) and Emengini et al. (2013) recently extended this study to the thermal infrared. They showed that leaf temperature was unsuitable for detecting oil pollution in maize crops, but that the application of spectral indices could enhance the early detection of oil pollution.

7.1.4 Herbicides

Upon exposure to herbicides that exceed the appropriate injury thresholds, plants may develop phytotoxicity symptoms including abnormalities in growth and chlorosis, for example, leaf yellowing or bleaching. The efficacy of a particular herbicide has been long appraised visually. Such estimations are subjective, that is, they may vary between observers, and they may not be established by standard biochemical and biometric techniques at very low concentrations. Some studies have investigated the potential of spectroscopy to evaluate the efficacy, or damage, of herbicides commonly used in agriculture (Table 7.2). The latter can be added to the nutrition medium in increasing concentration or spread on the leaves at different phenological stages. The comparison between control and treated plants showed that herbicides generally produce an increase in visible reflectance, but effects are not consistent across species for the near-infrared region. Mohammed et al. (2000) showed that current-year red pine (*Pinus resinosa*) needles were more sensitive to herbicide than older needles, especially to imazapyr, where a pronounced increase in reflectance was observed in the visible. Krezhova et al. (2007a,b) found statistically significant differences between the spectral reflectance characteristics of garden pea (*Pisum sativum*) leaves treated with fluridone, acifluorfen, and glyphosate compared to control ones. Depending on the herbicide and its concentration, changes are observed preferentially in the visible or in the near infrared. Recently, Zhao et al. (2014a, 2014b) investigated the early detection of leaf injury caused by glyphosate treatments on soybean (*Glycine max*) and cotton (*Gossypium hirsutum*) by PROSPECT inversion (see Section 9.3).

7.1.5 Air Pollution

Understanding the effects of air pollutants on natural ecosystems and predicting the possible consequences of various levels of exposure are of major biological concern. Air pollution problems are primarily concentrated near urban and industrial areas but they may affect remote, non-urban regions, after wind has carried air pollutants many kilometers from their sources. They mostly have a negative impact on plants (Emberson, 2003). Visual observations have been the first approach employed to evaluate injury caused by air pollution on foliage, which can range from mild chlorosis to severe necrosis and the death of much or all of the exposed tissues. Note that it is often impossible to tell

Table 7.2 *Effect of herbicides on leaf optical properties.*

Herbicide	Plant species	Reference
2,4-D		
Atrazine	*Pisum sativum*	Krezhova et al. (2005)
Acifluorfen		Krezhova et al. (2007a)
Fluridone		
	Pinus banksiana	Mohammed et al. (2000)
Glyphosate	*Pisum sativum*	Krezhova et al. (2007b)
	Glycine max, Gossypium hirsutum	Zhao et al. (2014a, 2014b)
Imazapyr	*Pinus banksiana*	Mohammed et al. (2000)
Mepiquat chloride	*Gossypium hirsutum*	Richardson and Gausman (1982)
Oxyethylene docosanol	*Nicotiana tabacum*	Tanaka (1968)
Triclopyr	*Pinus banksiana*	Mohammed et al. (2000)

whether such symptoms are caused by air pollutants or normal senescence. Moreover, directly correlating specific gases to spectral responses may be problematic as plants respond to different stress types in basically the same way, following a generalized senescent response. Several experiments have been carried out in open-top or closed fumigation chambers, at either the laboratory or the field level. These consisted of exposing seedlings to reactive air pollutants, for brief (a few hours/days/weeks) or long (a few months/years) periods, at more or less realistic levels, and in interpreting possible physical/physiological changes in terms of spectral features (Table 7.3). The treatments involved one or more pollutants, for example, ozone plus acid mist, ozone plus sulfur dioxide or carbon dioxide, or sulfur dioxide plus carbon dioxide. Despite species-specific responses, results are generally consistent.

7.1.5.1 Ozone (O_3)

While stratospheric ozone shields us from ultraviolet radiation (see Section 5.2.2), in the troposphere this highly reactive molecule is considered as a dynamic secondary pollutant, the concentration of which varies from day to day, season to season, and year to year. It enters leaves through stomata during normal gas exchange and causes oxidative stress. The most common symptoms associated with ozone exposure include flecks (tiny light-tan irregular spots < 1 mm diameter), stipples (small darkly pigmented areas $\sim$ 2–4 mm diameter), bronzing, and reddening. Additionally, ozone-exposed leaves have lower water content and different internal structure than do controlled leaves. This damage, which is not always visible on leaves, in turn decreases photosynthesis, plant growth, biomass accumulation, and crop yield (e.g., Knudson et al., 1977; Heath, 1994; Bagard et al., 2008). Therefore the effects of ozone on leaf spectroscopy has been the first and most widely studied (Meroni et al., 2009). Runeckles and Resh (1975) showed that chronic injury, which is confined to the palisade layer, was spectroscopically evident from 450–750 nm where reflectance increased dramatically. Ozone has also been observed to induce moisture stress and changes in anatomy and chloroplast membrane structure and distribution in certain species by interfering with stomatal behavior. Gausman et al. (1978b) found that ozone-exposed cantaloupe (*Cucumis melo*) leaves had higher reflectance in the SWIR due to a decrease of water content from 90.3% for control to 82.6% for very severely damaged leaves. Prompted by the reported decline in the health of forests in Europe and

Table 7.3 *Main atmospheric pollutants affecting leaf optical properties.*

Pollutant	Main source of emission	Main leaf effects	Reference	Plant species
Gas: ozone (O_3)	Secondary pollutant formed by NOx and hydrocarbons	Entry via stomatal conductance; oxidizing potential causes chlorosis and necrosis; loss and disassociation of the thylakoids in the chloroplast; reduced growth to mortality; forest decline	Runeckles and Resh (1975)	*Phaseolus vulgaris*
			Gausman et al. (1978b)	*Cucumis melo*
			Omasa et al. (1980a)	*Helianthus annuus*
			Schutt et al. (1984b)	*Lycopersicon esculentum*
			Westman and Price (1988)	*Pinus jeffreyi, Sequoiadendron giganteum*
			Ustin and Curtiss (1990)	*Pinus contorta, Pinus ponderosa, Pseudotsuga menziesii*
			Essery and Morse (1992)	*Picea abies*
			Carter et al. (1992, 1995)	*Liriodendron tulipifera, Pinus strobus, Pinus taeda*
			Rudorff et al. (1995)	*Glycine max*
			Kraft et al. (1996)	*Trifolium repens, Triticum aestivum, Zea mays*
			Meinander et al. (1996)	*Pinus sylvestris*
			Gäb et al. (2006)	*Fagus sylvatica*
			Meroni et al. (2008, 2009)	*Populus deltoides*
			Di Vittorio and Biging (2009)	*Pinus ponderosa*
			Ainsworth et al. (2014)	*Glycine max*
Gas: sulfur dioxide (SO_2)	Energy production; industry; commercial and domestic heating	Visible foliar damage; reduced growth; fatal for many lichens and ferns; forest decline, especially broadleaf types	Omasa et al. (1980a, 1980b)	*Helianthus annuus*
			Schutt et al. (1984b)	*Lycopersicon esculentum*
			Gemmell and Colls (1992)	*Vicia faba*
			Rudorff et al. (1995)	*Glycine max*
			Lovchikova et al. (1997)	*Acer saccharum, Picea abies*
			Jenkins et al. (2012)	*Gunnera insignis*
			Rogge et al. (2012)	*Brassica napus, Populus deltoides, Populus trichocarpa, Triticum aestivum*

Table 7.3 (*cont.*)

Pollutant	Main source of emission	Main leaf effects	Reference	Plant species
			Liu et al. (2006)	*Acmena acuminatissima, Castanopsis fissa, Cryptocarya concinna, Pinus massoniana, Schima superba*
Gas: carbon dioxide (CO_2)			Carter et al. (1995)	*Liriodendron tulipifera, Pinus strobus*
			Meinander et al. (1996)	*Pinus sylvestris*
			Carter et al. (2000)	*Acer saccharum*
			Kakani et al. (2004)	*Gossypium hirsutum*
			Thomas (2005)	*Adenanthera pavonina, Intsia palembanica, Sindora echinocalyx*
			Noomen and Skidmore (2009)	*Zea mays*
			Jenkins et al. (2012)	*Gunnera insignis*
			Bellante et al. (2014)	*Medicago sativa*
Gas: methane (CH_4)			Smith et al. (2004a,b)	*Hordeum vulgare, Phaseolus vulgaris, Vicia faba*
			Smith et al. (2005)	*Brassica napus*
			Noomen et al. (2006)	*Zea mays*
Gas: nitrogen oxides (NOx)	Energy production; transport	Nutrient enrichment under low concentration; leaf necrosis from increased acidity; reduced plant growth; increased sensitivity to secondary stress	Lovchikova et al. (1997)	*Acer saccharum, Picea abies*
Hydrometeors: acid mist/rain			Westman and Price (1988)	*Pinus jeffreyi, Sequoiadendron giganteum*
			Carter et al. (1992)	*Pinus taeda*
			Essery and Morse (1992)	*Picea abies*
			Williams and Ashenden (1992)	*Trifolium repens*

Table 7.3 (*cont.*)

Pollutant	Main source of emission	Main leaf effects	Reference	Plant species
			Zhumar and Zaitseva (2003a, 2003b)	*Pinus* sp.
			Song et al. (2010)	*Altingia chinensis, Cinnamomum camphora, Cyclobalanopsis myrsinae, Ilex rotunda, Machilus pauhoi, Manglietia yuyuanensis, Quercus myrsinaefolia*
			Jin et al. (2013)	
Aerosol: dust			Eller and Brunner (1975), Eller (1977b, 1977c)	*Rhododendron catawbiense*
			Eller and Willi (1977a)	*Hedera helix*
			Sieghardt (1990)	*Erysimum sylvestre*
			Wang et al. (2012)	
			Luo et al. (2013)	*Euonymus japonicus*
			Xu and Yu (2013)	*Ficus microcarpa*
			Wu and Wang (2014)	
			Yan et al. (2014)	
Aerosol: volcanic ash	Volcanoes		Yamamoto et al. (1993)	*Glycine max*
Aerosol: urban pollution	Transport; energy production; industry; domestic heating	Variable depending on chemistry; injury from wet or dry deposition; reduced plant growth; increased sensitivity to secondary stress	Wang et al. (2008)	*Ficus microcarpa*
			Khavanin Zadeh et al. (2012, 2013, 2014)	*Carpinus betulus, Tilia* s.p.
			Zamblé Fidèle et al. (2014)	*Ficus polita*
Aerosol: radionuclides			Davids and Tyler (2003)	*Betula pendula, Pinus sylvestris*
Negative ions			Jovanic and Jovanic (2001)	*Zea mays*

North America in the 1990s, some experiments were set up to assess the impacts of O_3 on various species of common conifers (e.g., Ustin and Curtiss, 1990; Essery and Morse, 1992; Carter et al., 1992, 1995; Meinander et al., 1996; Di Vittorio and Biging, 2009). They are, strictly speaking, difficult to compare because the reflectance spectra were obtained over needle bundles, branches, or seedling canopies at a short distance (~ 1 m) from the plants. Overall, following exposure to ozone, the spectral reflectance displays similar patterns at all wavelengths. The increase in reflectance occurring across the visible and the blue shift of the red-edge are indicative of chlorosis;. Responses are much less consistent in the near infrared and the shortwave infrared than in the visible since reflectance may increase, remain unchanged, or decrease. Further experiments performed on plants of agricultural or forestry interest employed image processing methods (Omasa et al., 1980a; Kraft et al., 1996), NIR spectroscopy (Gäb et al., 2006), and active or passive fluorescence, and related spectral indices such as the photochemical reflectance index (e.g., Schreiber et al., 1978; Rudorff et al., 1995; Meroni et al., 2008; Ainsworth et al., 2014). They confirmed that the intensity of reflectance response to ozone depends on exposure duration, concentration, and genotype.

7.1.5.2 Sulfur Dioxide (SO_2)

Sulfur dioxide is an atmospheric contaminant primarily produced by the combustion of sulfur-containing fossil fuels, therefore concentrated in the vicinity of coal-fired power plants and industrial areas. It can also be related to volcanic degassing activity in the natural environment. Like ozone, sulfur dioxide is phytotoxic, penetrating into leaves primarily in gaseous form through the stomata and can produce acute injury in the form of foliar necrosis, even after relatively short duration exposure. Injury is indicated by bleached or pigmented (tan to reddish brown or dark brown, depending on species) necrotic interveinal areas on broad-leaved plants and chlorotic spots and bands or brown tips on needles of conifers. The consequences of SO_2 toxicity are reduced growth, cell death, plant organ death, and whole plant death (Malhotra, 1977; Shimazaki et al., 1980). However, the effects on plants depend on a number of factors including gas concentration, duration of exposure, exposure environment, and plant species. Omasa et al. (1980a, 1980b) measured changes in leaf temperature pattern during SO_2 fumigation with a thermal camera. They showed that these patterns were comparable to those of visible injury occurring on the same leaf (Figure 7.5).

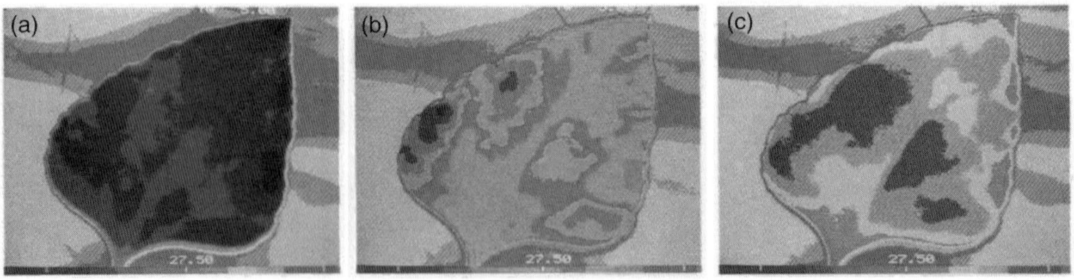

Figure 7.5 Changes in thermal pattern of sunflower (*Helianthus annuus* L.) leaf during SO_2 fumigation: (a) 0 min; (b) 20 min; and (c) 80 min (from Omasa et al. (1980b), *Research Report from the National Institute for Environmental Studies*. Copyright © 1984 with permission from National Institute for Environmental Studies, Japan). (A black and white version of this figure will appear in some formats. For the color version, please refer to the plate section.)

Figure 1.1 Illustrations of leaf structures: (left) from the *De historia Stirpium*; and (right) from the *Kitab-i hasha'ish* depicting bifacial leaf color.

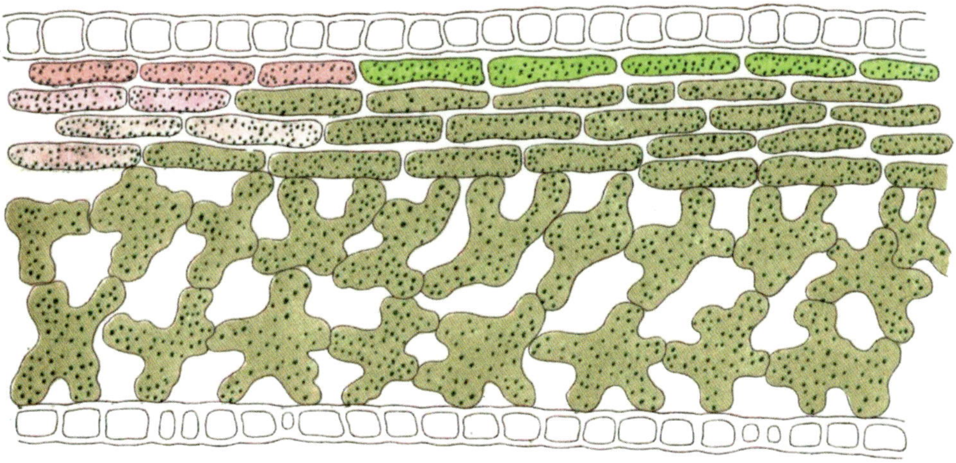

Figure 1.4 Cross-section of a European dogs tooth (*Erythronium dens-canis*) leaf. The epidermal cells are colorless; the upper mesophyll is made of elongated cylindrical cells that contain "granular" and "gelatinous" chlorophyll; the cytoplasm is either red or colorless, as it contains erythrophyll, the red coloring matter of leaves, or not; the white areas are due to the interposition of a thin air layer between the epidermis and the upper mesophyll (Morren, 1858a).

Figure 1.5 Decomposition of light by a solution of an English laurel green leaf showing several main absorption bands (from Brewster (1834), *Transactions of the Royal Society of Edinburgh*. Copyright © 1834 with permission from Cambridge University Press).

Figure 2.2 A few examples of plant types.

Figure 2.29 Variation of Japanese cherry (*Prunus serrulata*) leaf color during leaf development.

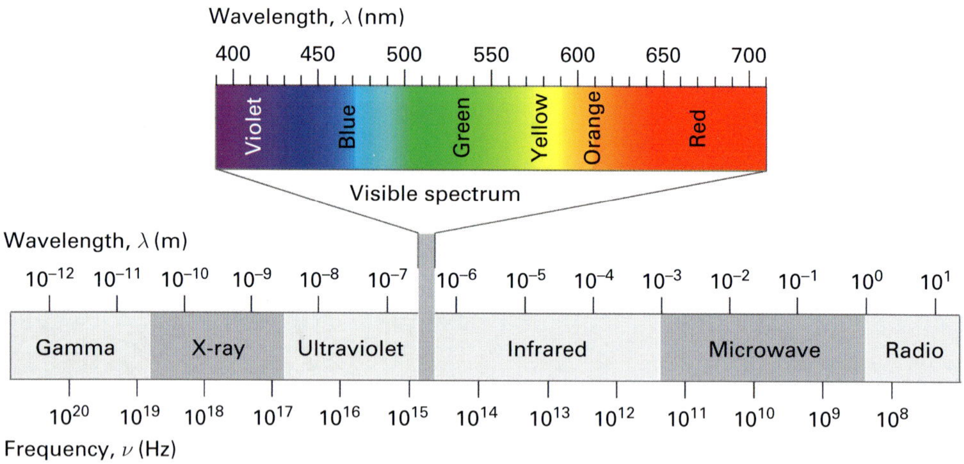

Figure 4.2 Electromagnetic spectrum (by courtesy of Averill and Eldredge, 2011).

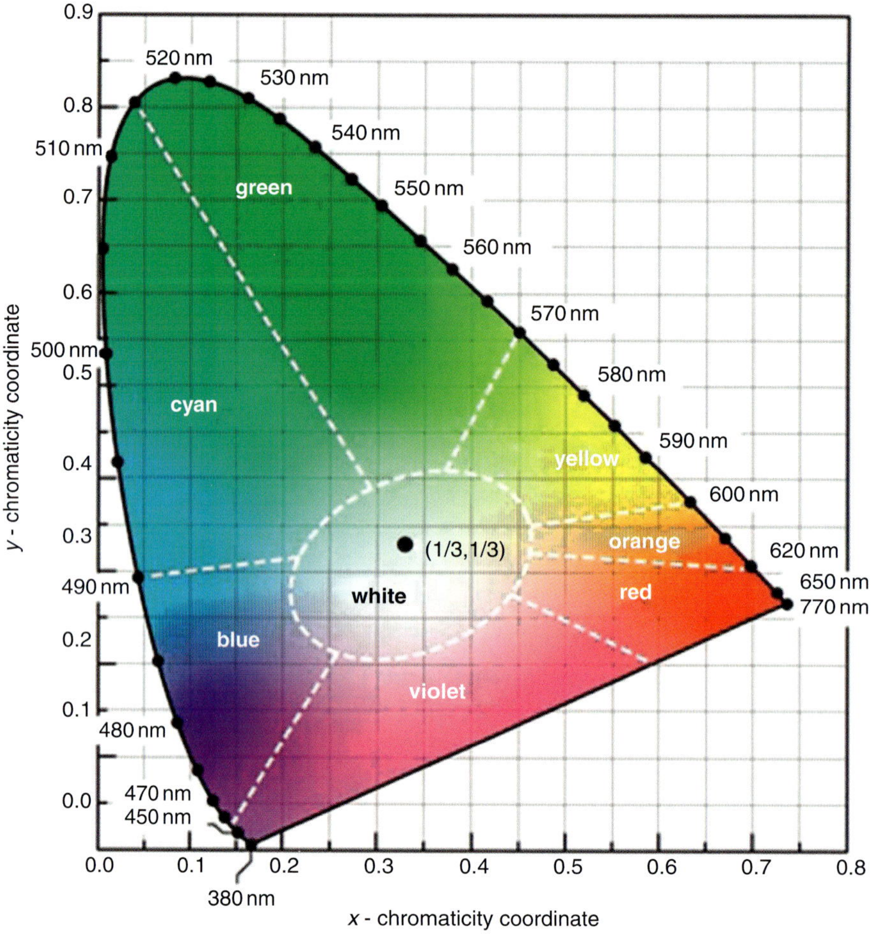

Figure 4.10 CIE 1931 (x, y) chromaticity color diagram relating coordinates and wavelengths (from Schubert (2006) Copyright © with permission from Cambridge University Press).

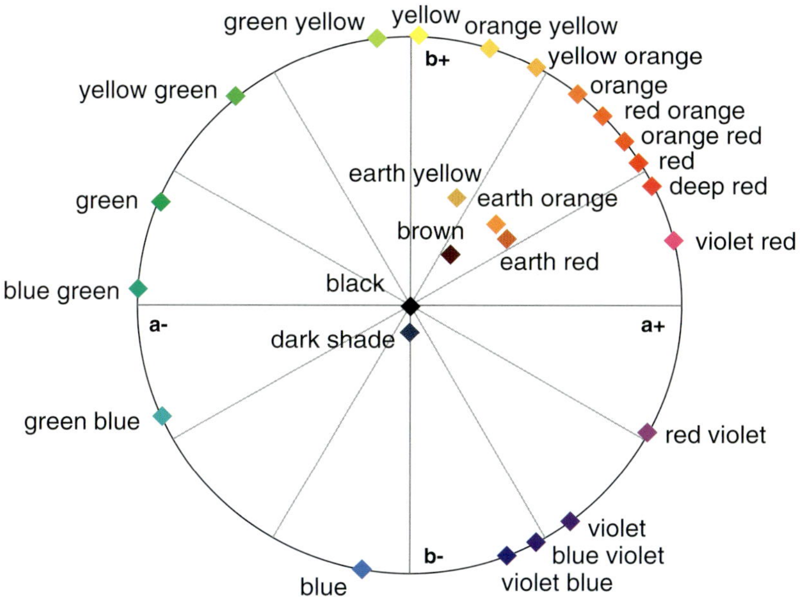

Figure 4.12 The CIELAB a*b* plane.

Figure 5.1 Iridescent leaves of: (a) octopus tree (*Schefflera actinophylla*); (b) peacock fern (*Selaginella willdenowii*); (c) peacock begonia (*Begonia pavonina*); (d) sedge (*Mapania caudata*); (e) *Diplazium cordifolium* from Borneo; and (f) Persian shield (*Strobilanthes dyeriana*) (from *Vertical Garden* by Patrick Blanc, www.verticalgardenpatrickblanc.com/inspiration/leaf-colors).

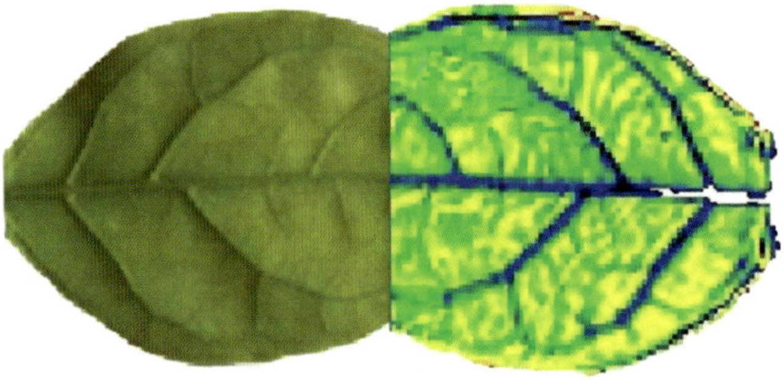

Figure 5.33 Images of leaf in optical domain (left) and at 615 GHz (right) (from Hellicar et al. (2007), in *Proc. 2nd International Conference on Wireless Broadband and Ultra Wideband Communications*. Copyright © 2007 with permission from the IEEE).

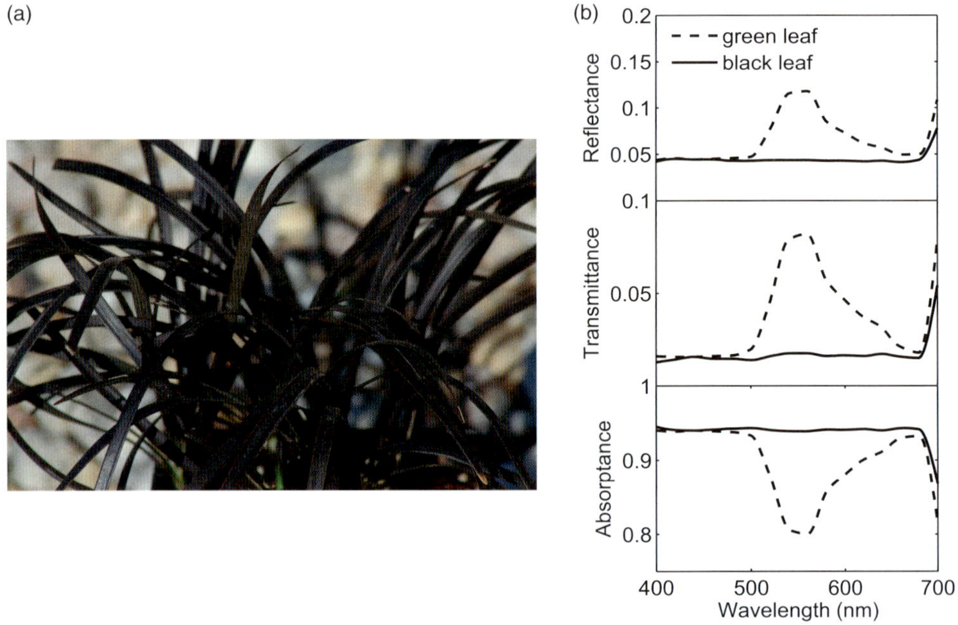

Figure 6.14 (a) Black mondo grass (*Ophiopogon planiscapus* "Nigrescens") leaves (© David Beaulieu, Landscaping Writer for *The Spruce*). (b) Reflectance, transmittance, and absorptance spectra of adaxial surfaces of green and black phenotypes of black mondo grass leaves (redrawn from Hatier and Gould (2007)).

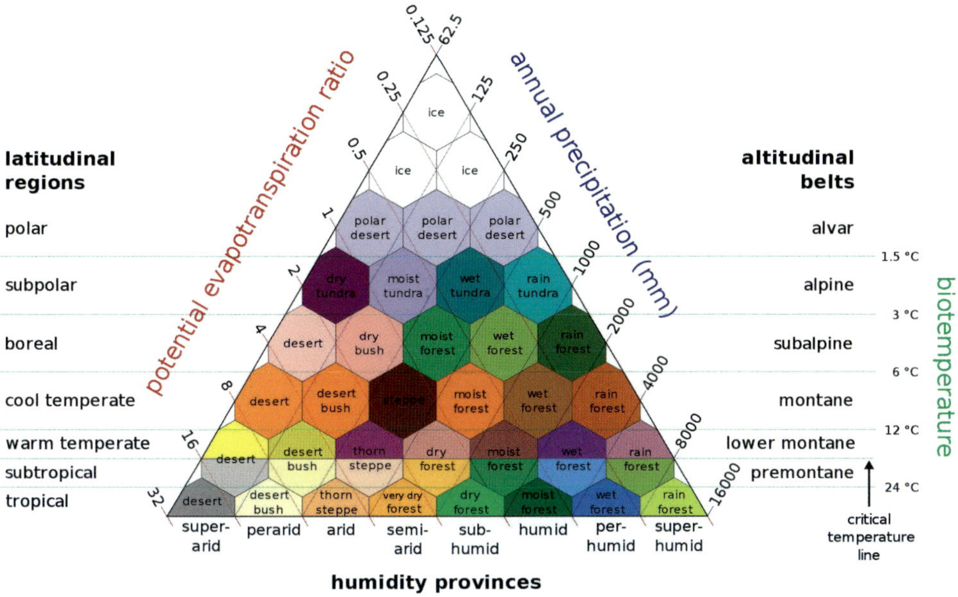

Figure 6.19 Holdridge life zones classification scheme (Holdridge, 1967) (https://en.wikipedia.org/wiki/Holdridge_life_zones).

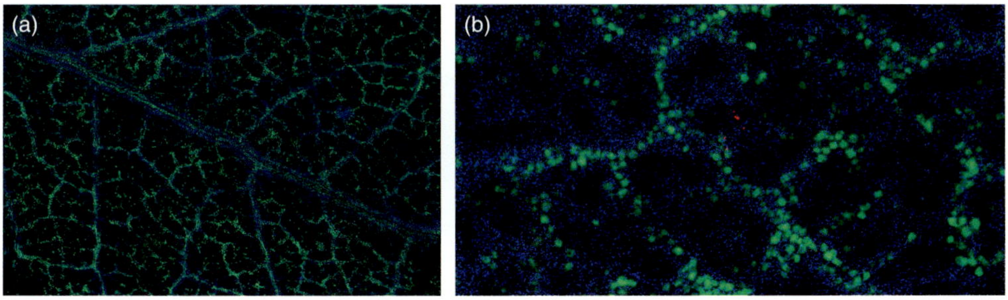

Figure 7.4 (a) Synchrotron image showing distribution of Cu (blue), Sr (oxalate crystals, green), and Au (red) in natural *Eucalyptus* leaf specimen. (b) Detailed view showing isolated particulate Au grains (red) and two of which are attached to the surface of Ca oxalate crystals (green) (from Lintern et al. (2013), Nature *Communications*, Copyright © 2013 with permission from Macmillan Publishers Limited).

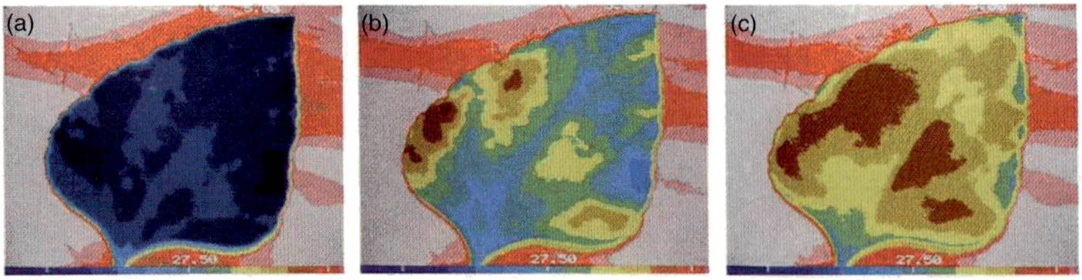

Figure 7.5 Changes in thermal pattern of sunflower (*Helianthus annuus* L.) leaf during SO_2 fumigation: (a) 0 min; (b) 20 min; and (c) 80 min (from Omasa et al. (1980b), *Research Report from the National Institute for Environmental Studies*. Copyright © 1984 with permission from National Institute for Environmental Studies, Japan).

Figure 7.7 Consequences of some viruses on leaf color: (a) maize dwarf mosaic virus; (b) grapevine leafroll-associated virus; (c) tomato mosaic virus; and (d) cucumber mosaic virus.

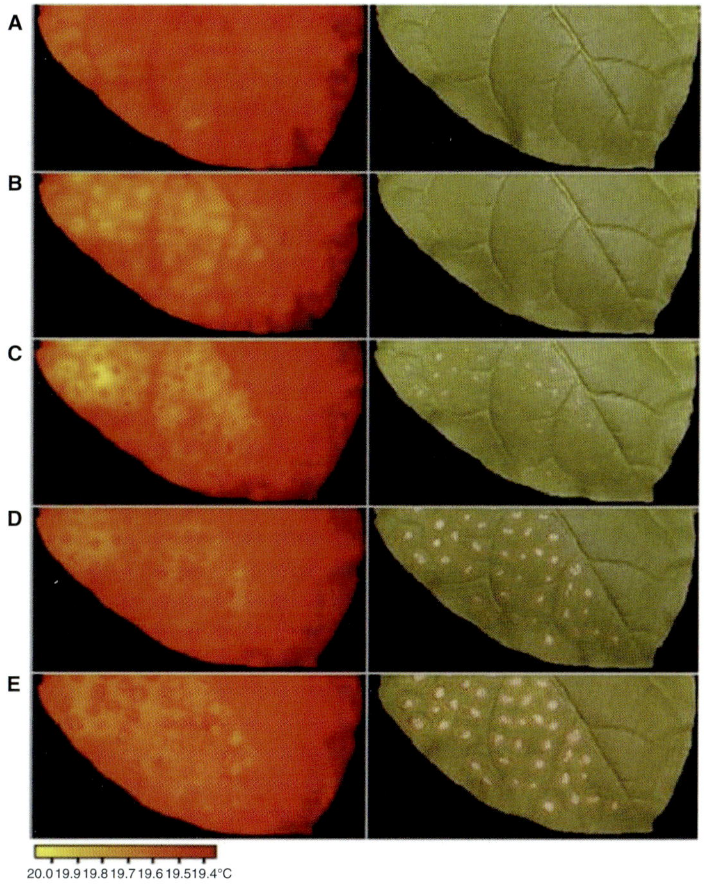

Figure 7.8 Images of a tobacco (*Nicotiana tabacum*) leaf infected by tobacco mosaic virus at 21°C. The images were taken at various times: 26 hours (A), 39 hours (B), 3 days (C), 5 days (D), and 8 days (E). (left) thermal infrared images and (right) visible images (from Chaerle et al. (1999), *Nature Biotechnology*. Copyright © 1999 with permission from Nature Publishing Group).

Figure 7.9 Consequences of Huanglongbing on leaf color: (a) sweet orange (*Citrus* × *sinensis*) foliage from São Paulo, Brazil demonstrating asymmetrical "blotchy mottle" symptom relative to the mid vein; (b) South Florida pomelo (*Citrus maxima*) tree displaying "zinc-pattern-deficiency" with inter-veinal chlorotic symptoms (from Gottwald et al. (2007), *Plant Health Progress*. Copyright © 2007 with permission from The Plant Management Network).

Figure 7.10 Consequences of some fungi on leaf color: (a) southern corn leaf blight; (b) wheat yellow rust; (c) bean chocolate spot; and (d) sugar beet powdery mildew.

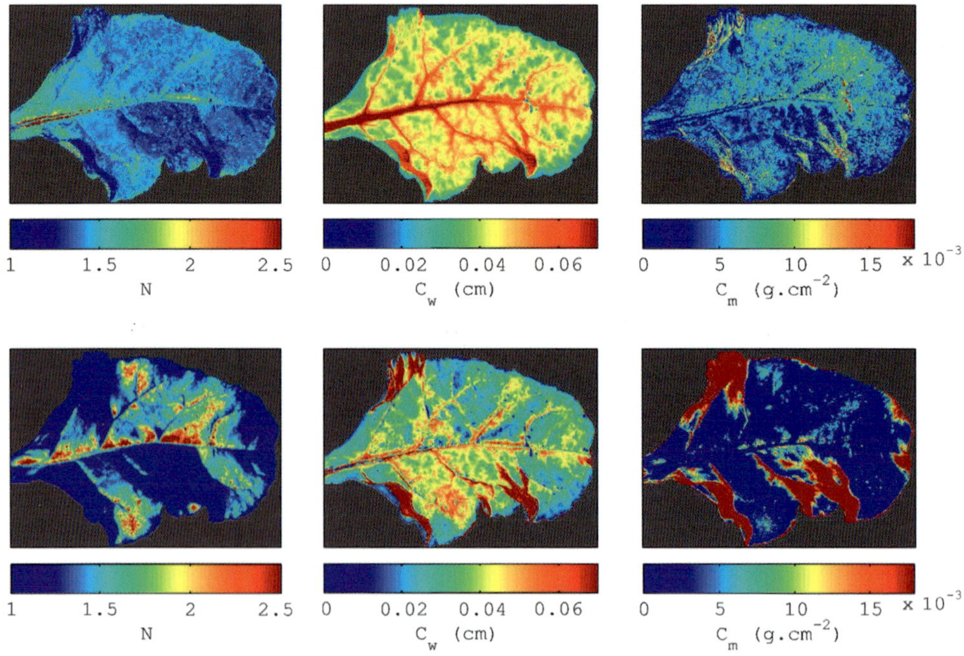

Figure 9.16 Leaf characteristics determined by inversion of the PROCOSINE model (upper figures) and the PROSPECT model (lower figures) in the SWIR (from Jay et al. (2016), *Remote Sensing of Environment*. Copyright © 2016 with permission from Elsevier).

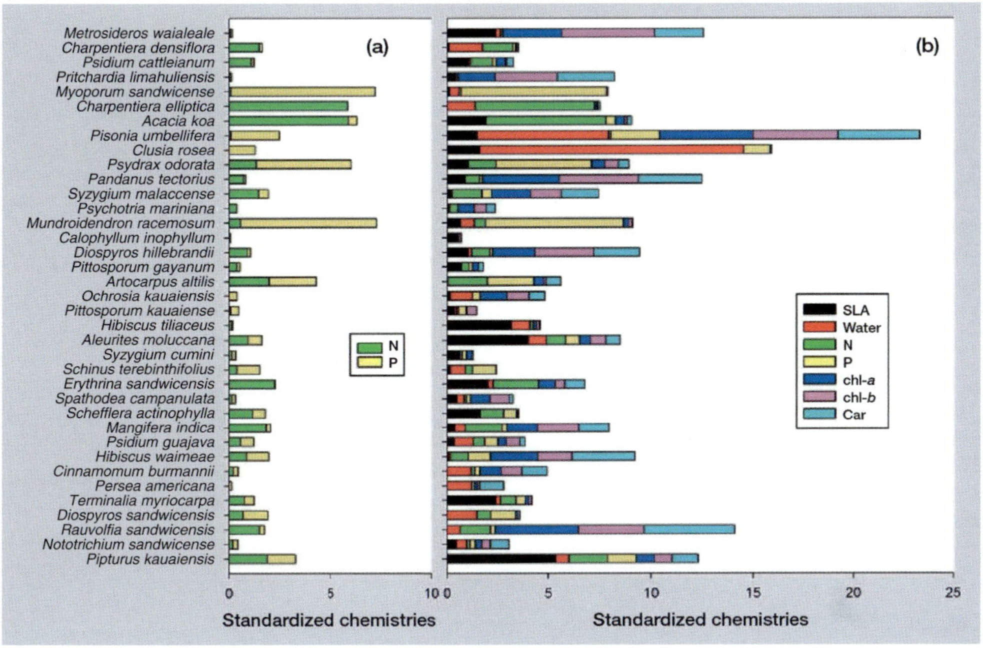

Figure 11.15 Chemical fingerprints of canopy species found in a rainforest site on Kauai, Hawaii, based on: (a) leaf nitrogen (N) and phosphorus (P) concentrations; and (b) seven leaf properties: specific leaf area (SLA), water content (water), nitrogen (N), phosphorus (P), chlorophyll *a* (chl-a), chlorophyll *b* (chl-b), and total carotenoids (Car) (from Asner and Martin (2009), *Frontiers in Ecology and the Environment*. Copyright © 2009 with permission from the Ecological Society of America).

Figure 12.24 The leaves from four populations of *Corydalis hemidicentra* covering the main distribution area of this species: from the left to the right, Pujinlangba (PJ), Caishichang (CSC), Tianbaoshan (TBS), and Huangguanfeng (HGF) (from Niu et al. (2017), *Proceedings of the Royal Society of London. Series B.* Copyright © 2017 with permission from The Royal Society).

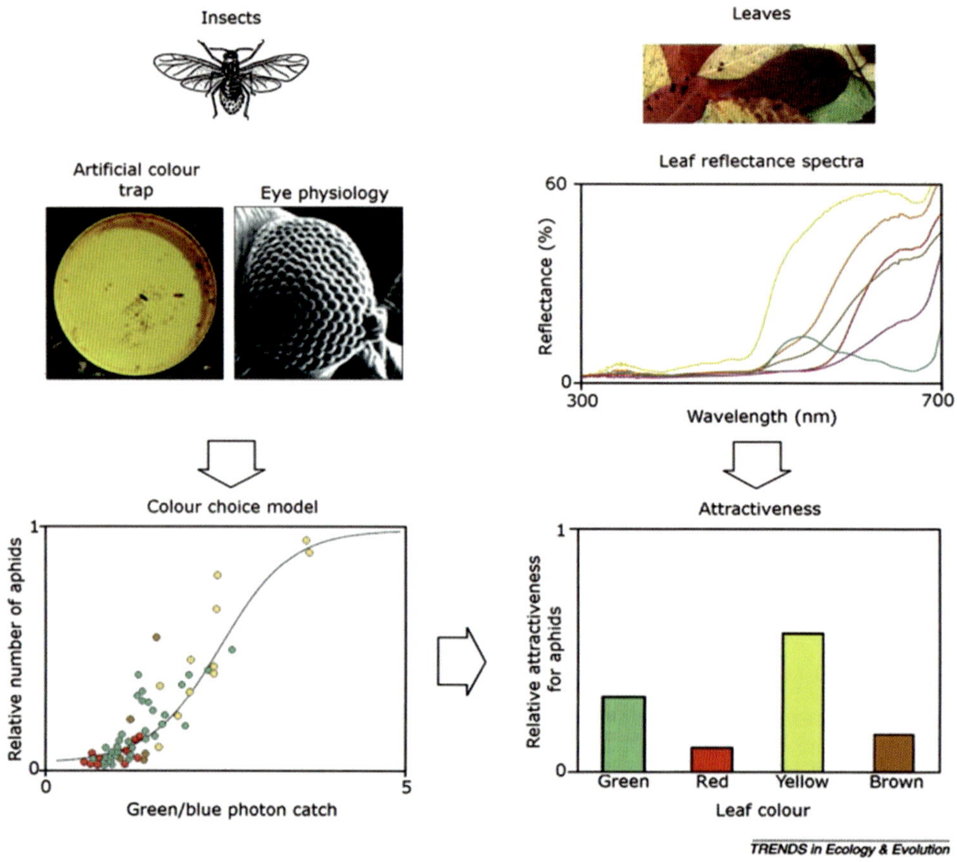

Figure 12.17 Measuring color preference in insects. An objective color measure (the leaf reflectance spectra) of leaves is integrated with a color choice model for aphids (derived from artificial color trap measurements and the physiology of aphid color vision) to obtain a measure of the attractiveness for aphids of leaf colors according to color names based on human vision (from Archetti et al. (2009c), *Trends in Ecology & Evolution*. Copyright © 2009 with permission from Elsevier).

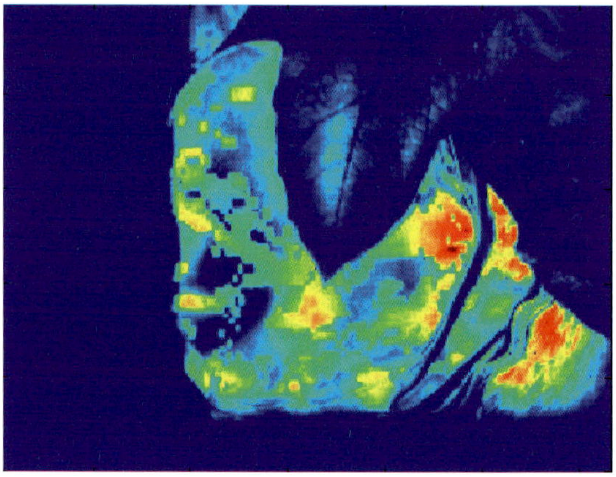

Figure 12.25 SWIR image showing reflectance contrast between foliage and non-foliage objects (from Leong (2008), *POINTER*. Copyright © 2008 with permission from the Singapore Government).

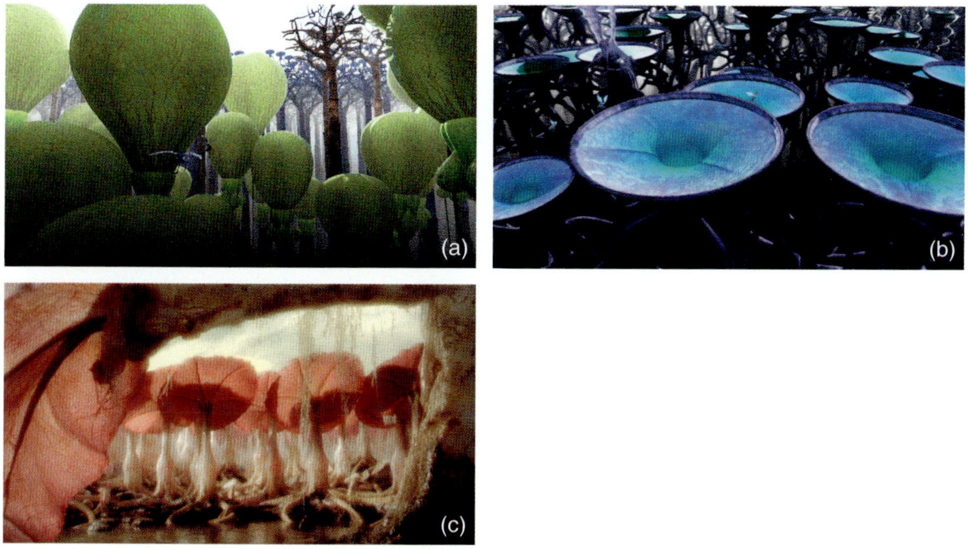

Figure 12.29 Computer artwork depicting a scientist's impression of alien plant forms called: (a) Balloon plants, 5 m in diameter and 80 m height; (b) Pagoda plants; and (c) Stinger fans (©2005 Big Wave Productions/Science Source. All rights reserved).

Figure 12.30 Martian landscape as imagined by Tikhov: a low blue vegetation covering a red soil. A sandstorm, frequent on this planet, is rising on the horizon (from Tikhov (1960)).

Figure 12.34 Virtual autumn colors of a sugar maple (*Acer saccharum*) tree (from Mochizuki et al. (2007), *The Journal of the Society for Art and Science* © 2007 with permission from the Society for Art and Science).

Figure 12.35 (a) Leaf degradation of rose tree affected by color changes, insect attacks, and geometric shrinking (from Peyrat et al. (2008), *The Visual Computer* © 2009 with permission from Springer). (b) Autumn leaves stacking on the grass (from Desbenoit et al. (2006) © 2006 with permission from *The Eurographics Association*). (c) Simulated shape of medium wilting cucumber (*Cucumis sativus*) leaf (from Lu et al. (2009), *International Journal of Computer Games Technology* © 2009 with permission from Hindawi Publishing Corporation).

Figure 12.37 *"Autumn Leaves" 2008*, inkjet printed computer drawing on paper, 35 × 45 1/2" (© David Hockney, www.davidhockney.co/).

Figure 12.38 Eucalyptus, oil on canvas, 100.2 cm × 120.5 cm (G. Balla, 1914).

Gemmell and Colls (1992) exposed faba bean (*Vicia faba*) seedlings to 15 levels of sulfur dioxide during ten days and measured their leaf optical properties between 400 nm and 900 nm. They first studied the effect of SO_2 on reflectance and transmittance using statistical methods, at ten wavelengths selected as being representative of a particular region of the spectrum. They showed that, for all experiments, the near-infrared reflectance (transmittance) of the exposed plants was lower (higher) than that of the controls, They attributed this to an increased number of dead mesophyll cells. However, due to the increase of the number of air–cell wall interfaces, this would typically have the opposite effect. These authors give another explanation that seems more reliable; cell damage would induce localized conversion of polyphenol oxidase to opaque pigments, increasing absorptance. The effect of sulfur dioxide on the leaf in the visible region of the spectrum is more classical with an increase of reflectance and transmittance as a result of SO_2-induced reductions in leaf chlorophyll content. They also investigated the first derivatives of the spectra. Similar results were obtained by Lovchikova et al. (1997) and Liu et al. (2006). Finally, note the original experiment conducted by Jenkins et al. (2012) on poorman's umbrella (*Gunnera insignis*) leaves along an elevation gradient corresponding to SO_2 gradients on vegetation along the south flank of Volcán Turrialba, Costa Rica.

7.1.5.3 Carbon Dioxide (CO_2)

Atmospheric carbon dioxide has risen from a preindustrial partial pressure close to 270 ppmv to about 412 ppmv at present, and is projected to double sometime after 2080. Although CO_2 enrichment is considered to cause little plant stress, some responses including decreased leaf pigmentation, changed anatomic structure, and increased cuticular wax deposition have been reported in the literature. They should alter leaf optical properties in the 400–700 nm waveband accordingly. Carter et al. (2000) and Thomas (2005) tested this hypothesis on sugar maple (*Acer saccharum*) and three tropical tree species (*Adenanthera pavonina, Intsia palembanica*, and *Sindora echinocalyx*), respectively. The first authors observed a nonsignificant increase in leaf reflectance under elevated CO_2, while the second author found a significant effect, large enough, if widespread, to have implications for global net radiative forcing. Plant physiological responses to various atmospheric stress agents may also be counterbalanced by higher carbon dioxide concentrations as a result of increased photosynthesis. As a consequence, the optical properties of damaged leaves may be modified in the presence of elevated carbon dioxide concentration in the atmosphere. A review of the literature offers conflicting results regarding the combined effects of CO_2 and ozone (Carter et al., 1995; Meinander et al., 1996), CO_2 and air temperature (Carter et al., 2000), CO_2 and ultraviolet-B radiation (Kakani et al., 2004), and CO_2 and sulfur dioxide (Jenkins et al., 2012). More recently, with the advent of CO_2 sequestration, spectroscopy has been applied to monitor carbon dioxide influx into soils and the resulting plant stress (Noomen and Skidmore, 2009; Male et al., 2010; Bellante et al., 2014).

7.1.5.4 Methane (CH_4)

Vegetation changes have been reported in the vicinity of gas pipelines buried in the soil. Aerial contact of plants with methane does not produce harmful effects. Plant stress, which is manifest as decreases in growth and chlorosis of the leaves, is mainly due to subsurface gas leaks. Even though the mechanisms have not been totally identified, it may be a response to the displacement of soil-oxygen from the soil, which then inhibits root function. Smith et al. (2004a) detected methane-induced stress in common barley (*Hordeum vulgare*) and faba bean (*Vicia faba*) plants, at least 7 days before visible symptoms were observed. For that they determined the first derivative of the reflectance spectrum and calculated the ratio at 725 nm to that at 702 nm within the red-edge. Since other stresses may cause similar responses, they also studied the effects of soil-oxygen displacement using argon, nitrogen, and water logging (Smith et al., 2004b), or the effects of dilute herbicide and extreme shade (Smith et al., 2005). Noomen et al. (2006) tested continuum removal techniques in pigment and water absorption regions (see Section 11.2) and the photochemical reflectance index to test whether natural gases, namely methane and ethane (C_2H_6), affects vegetation reflectance.

7.1.5.5 Acid Mist

Acid rain is a worldwide and complex environmental problem which affected the health of forests in Europe and eastern North America in the 1970s. Sulfur dioxide and nitrogen oxides released by the burning of fossil fuels such as oil and coal both combine with water in the atmosphere to create acid rain. It does not usually kill trees directly. Instead, it is more likely to weaken the trees by damaging their leaves. The effects of wet acid deposition on vegetation are species-specific and not well defined. Several authors have reported visible foliar injury on some plant species resulting from exposure to highly acid rainfall treatments (pH ~ 2–4). To model the influence of acid rains on needles, Zhumar and Zaitseva (2003a, 2003b) applied 0.1, 1, 5, and 10% solutions of sulfuric acid to them during half an hour or one hour and measured their reflectance from 470 nm to 800 nm. They found that the increase in sulfuric acid concentration resulted in a decrease in chlorophyll content and, surprisingly, in a decrease of needle reflectance. This was due to the filling of the intercellular air spaces by the acid, as shown using the LIBERTY model (see Section 8.1.2). Significant plant growth reduction, however, may occur in other species without the presence of visible leaf lesions. Such changes may also have an indirect effect on plants by increasing their susceptibility to other stresses which have the potential to cause acute injury to plants. For that reason simulation of acid mist on leaf optical properties has often been studied in combination with pollutant gases that can be found in rural environments, even at low concentrations: O_3 (Westman and Price, 1988; Carter et al., 1992; Essery and Morse, 1992), and $O_3 + SO_2 + NO_2$ (Williams and Ashenden, 1992). The main spectral changes in polluted plants are observed in the red-edge and near-infrared regions, but results may be contradictory according to the papers and there is no clear explanation for such responses to acidic precipitation. More recently, southern China has become one of the most serious acid rain polluted regions in the world, where it constitutes a potential threat to the health of the local forest (Li et al., 1996; Xie et al., 2009; Song et al., 2010; Jin et al., 2013). Actually, acid mist, which consists of very fine droplets of water in air, can be considered as aerosols.

7.1.5.6 Aerosols

Atmospheric aerosols consist of fine liquid or solid particles suspended in a gas. They may originate from many sources and have important environmental effects. Research on the effects of aerosol pollution on plants has not received the same level of attention as that given to gas pollutants such as O_3 or SO_2. An increase in quarrying, open-cast mining, and road traffic suggest that dust deposition onto vegetation may be increasing, affecting plant growth, leaf morphology, and photosynthetic efficiency (Farmer 1993; Mandre and Tuulmets, 1997). Eller and his colleagues (Eller and Brunner, 1975; Eller 1977b, 1977c; Eller and Willi, 1977a) reported an increase in the absorptance of leaves covered by road dust compared to clean leaves, especially between 750 nm and 1350 nm. It induced an increase in leaf temperature of the order of 2–4°C, a result observed by other authors. Since temperature is one of the major parameters controlling photosynthesis and respiration, such leaves were more likely to be susceptible to water stress. Research conducted later confirmed these results (Sieghardt, 1990; Sharifi et al., 1997). Once again, the increase in pollution in China has initiated studies in the effect of dust on leaf reflectance (e.g., Wang et al., 2012; Luo et al., 2013; Xu and Yu, 2013; Wu and Wang, 2014; Yan et al., 2014).

The development of industry and traffic in many cities has contributed to the increase in air pollution, which runs parallel with the amount of total suspension particles in the atmosphere. Urban atmospheres are typically a complex mixture of air pollutants in gaseous, liquid, as well as solid phases. The accurate monitoring of air quality is an important issue for urban management. However, it requires expensive equipment which is often only placed in a small number of sites. The bio-indicator approach uses the chemical and physical changes in organisms to directly assess the effects of air pollution on the environment. Wang et al. (2008) collected leaves of Chinese banyan (*Ficus microcarpa*) trees in Guangzhou, China, a city located in one of the world's most important industrial areas. They compared the spectral reflectance of polluted leaves with leaves after the surface had been cleaned using dry cotton. The reflectance of polluted leaves was 3–5.5% higher than that of clean leaves in the visible, while it was 10–15% lower in the near infrared. Khavanin Zadeh et al. (2012, 2013, 2014) and Zamblé Fidèle et al. (2014) recently evaluated a simple and cheap non-destructive biomonitoring method to assess urban habitat quality in Gent (Belgium) and Abidjan (Ivory Coast), respectively.

Volcanic eruptions often release atmospheric ash in great quantities. The largest particles constitute an inhomogeneous optically thick cloud that is quickly deposited; the smallest spread out and remain in the atmosphere for a long time. Hirano et al. (1992) conducted a study to infer the effects of volcanic white and black ash accumulating on a leaf surface on its reflectance, temperature, and then photosynthesis. They showed that volcanic ash with dark color and small particle size reduced photosynthesis through shading, increasing temperature and plugging stomata. Yamamoto et al. (1993) applied different amounts of white and black ash to soybean (*Glycine max*) leaves and measured their reflectance spectra. The reflectance increased in the blue and the red, and decreased in the near infrared (Figure 7.6).

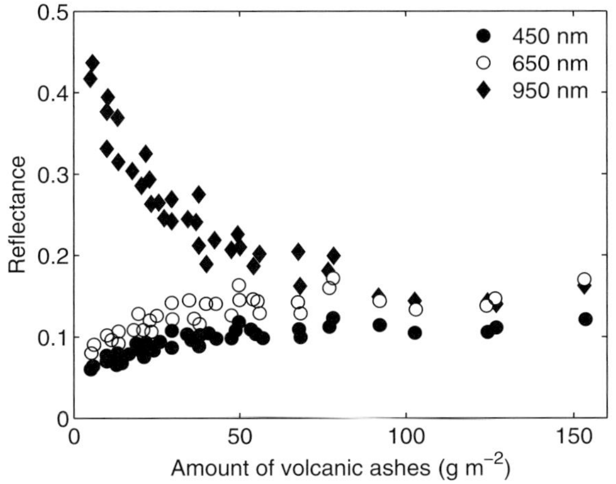

Figure 7.6 Relationship between amounts of white volcanic ashes piled on soybean (*Vicia faba*) plant leaves and reflectance (450 nm, 650 nm, and 950 nm) (redrawn from Yamamoto et al. (1993)).

Although they are not strictly speaking aerosols, suspended particle of various kinds can be sprayed on the foliage of crops to limit environmental stresses (water and heat) and to reduce disease or insect damage, a role that can be played by pubescence and cuticular wax. This technique is called particle film technology. The microscopic layer of mineral particles on the surface of the leaf allows water vapor and carbon dioxide to pass through the film, but changes their optical properties. Abou-Khaled et al. (1970) applied a uniform white coating of kaolin, a chemically inert fine white clay, to the upper surface of some leaves and observed significant changes in their spectral properties. The decrease in absorption due to increased leaf whiteness induced a decrease in leaf temperature and transpiration (see Section 12.1.4). The photosynthetic rate of individual leaves is generally decreased under low light intensities and increased under high light intensities.

7.2 Biotic Factors

Plants are very sensitive to insect attack, but also to viral and bacterial infections. The number and diversity of these biological agents are considerable, as well as their effects on plants. As for many abiotic factors, they are directly responsible for chlorosis and physiological reactions. Many studies connecting the pigment content to biotic aggressions have been conducted for several years. Because of climate change and the proliferation of invasive plants, many species are currently impacted by organisms formerly confined in other climates or other ecosystems.

7.2.1 Plant Diseases: Viruses, Bacteria, and Fungi

Plant diseases may be caused by viruses, bacteria, fungi, or oomycetes (or oomycota). They play an important role in limiting fruit and vegetable production, causing considerable seasonal losses. As a consequence, increasing production costs have generated considerable interest in rapid disease detection and monitoring systems to help improve disease and pest control (Agrios, 2004). Leaf spectroscopy is one of the techniques that are used for detecting plant disease (Sankaran et al., 2010b). So far, about 70 papers have investigated its potential to follow physiological, biochemical, and anatomical changes which occur in tissues as a result of such infections (Table 7.4). The earliest research was conducted to ascertain if such diseases could be detected from aerial photographs a few days before visual symptoms become apparent, which has had variable success (e.g., Colwell, 1956). Recently fluorescence imagery and digital infrared thermography proved to be better suited for the detection of disease-induced changes in plant leaves, long before visual symptoms appear (e.g., Oerke and Steiner, 2010). However, we need to bear in mind that diseases remain difficult to diagnose because they display the same set of symptoms as plants that are stressed by other causes. Measurements of absolute temperatures are likewise inappropriate for disease assessment, especially under varying environmental conditions.

7.2.1.1 Viruses

Viruses on occasion can be beneficial, but for the most part, they are harmful to plants. They can persist for many years, before they appear as a problem. First, plant foliage may appear yellow, or as mosaic patches of yellow, light green, or white (Figure 7.7). In addition, the plants are often stunted or misshapen. For instance, the leaves may be rolled, or swollen or puckered, or they may be abnormally

Table 7.4 *Main plant diseases that gave rise to spectroscopic studies. For fungus diseases, (A) stands for* Ascomycetes *and (B) for* Basidiomycetes.

Plant	Type	Disease	Acronym	Latin name	Reference
		Herbaceous crops			
Barley	Fungus	Powdery mildew (A)		*Blumeria graminis*	Lorenzen and Jensen (1989)
		Brown rust (B)		*Puccinia hordei*	Teng and Close (1977)
Bean	Fungus	Chocolate spot (A)		*Botrytis fabae*	Malthus and Madeira (1993)
		Bean rust (B)		*Uromyces phaseoli*	Runeckles and Resh (1975)
Celery	Fungus	Sclerotinia rot (A)		*Sclerotinia sclerotiorum*	Huang and Apan (2006), Machado et al. (2015)
Cotton	Fungus	Verticillium wilt (A)		*Verticillium dahliae*	Chen et al. (2012a)
Cucumber	Virus	Cucumber mosaic virus	CMV	*Cucumovirus*	Cochran et al. (1960)
	Oomycetes	Downy mildew		*Pseudoperonospora cubensis*	Lindenthal et al. (2005), Oerke et al. (2006)
Eggplant	Fungus	Grey mold (A)		*Botrytis cinerea*	Wu et al. (2008)
Maize	Virus	Maize dwarf mosaic virus	MDMV	*Potyvirus*	Ausmus and Hilty (1972)
	Fungus	Southern corn leaf blight (A)	SCLB	*Cochliobolus heterostrophus*	Ausmus and Hilty (1972), Safir et al. (1972)
Oat	Virus	Barley yellow dwarf virus	BYDV	*Luteovirus*	Colwell (1956)
Potato	Oomycetes	Late blight		*Phytophthora infestans*	Taoutaou et al. (2010, 2012)
Rice	Fungus	Rice blast (A)		*Magnaporthe grisea*	Kobayashi et al. (2003), Zhang et al. (2011)
		Brown spot (A)		*Cochliobolus miyabeanus*	Liu et al. (2007, 2008c)
Soybean	Virus	Mung bean yellow mosaic virus	MYMV	*Begomovirus*	Gazala et al. (2013)
Sugar beet	Virus	Beet necrotic yellow vein virus	BNYVV	*Benyvirus*	Sandwald (1981), Steddom et al. (2003)
	Fungus	Powdery mildew (A)		*Erysiphe betae*	Mahlein et al. (2010, 2012, 2013), Rumpf et al. (2010)

Table 7.4 (*cont.*)

Plant	Type	Disease	Acronym	Latin name	Reference
		Cercospora leaf spot (A)		*Cercospora beticola*	Chaerle et al. (2004), Mahlein et al. (2010, 2012, 2013), Rumpf et al. (2010), Bauer et al. (2011)
		Sugar beet rust (B)		Uromyces *beticola*	
		Sugar beet root rot (B)		*Rhizoctonia solani*	Hillnhütter et al. (2012)
Sugarcane	Virus	Fiji leaf gall	FDV	*Fijivirus*	Purcell et al. (2009)
		Sugarcane yellow leaf virus	ScYLV	*Polerovirus*	Grisham et al. (2010)
	Fungus	Brown spot (A)		*Cercospora longipes*	Patil and Bodhe (2011)
Tobacco	Virus	Tobacco mosaic virus	TMV	*Tobamovirus*	Polischuk et al. (1997), Chaerle et al. (1999, 2002, 2004), Lenk et al. (2007)
	Oomycetes	Black shank		*Phytophthora nicotianae*	Yusuf and He (2011)
Tomato	Virus	Tomato mosaic virus	ToMV	*Tobamovirus*	Krezhova and Yanev (2007); Krezhova et al. (2010)
	Bacteria	Bacterial spot of tomato		*Xanthomonas perforans*	Jones et al. (2010)
Watermelon	Virus	Watermelon mosaic virus	WMV	*Potyvirus*	Cardenas et al. (1972), Blazquez and Edwards (1986)
	Fungus	Fusarium wilt (A)		*Fusarium oxysporum*	Blazquez and Edwards (1986)
	Oomycetes	Downy mildew		*Pseudoperonospora cubensis*	
Wheat	Virus	Wheat streak mosaic virus	WSMV	*Tritimovirus*	Savenkov et al. (2004)
	Fungus	Powdery mildew (A)		*Blumeria graminis*	Graeff et al. (2006), Tian et al. (2011, 2012b), Zhang et al. (2012a, 2012b), Yuan et al. (2014a)
		Take-all (A)		*Gaeumannomyces graminis* var. *graminis*	Graeff et al. (2006)
		Spot blotch (A)		*Cochliobolus sativus*	Tian et al. (2012b)

Table 7.4 (*cont.*)

Plant	Type	Disease	Acronym	Latin name	Reference
		Yellow rust (B)		*Puccinia striiformis* var. *striiformis*	Tian et al. (2011, 2012b), Zhang et al. (2014), Zhao et al. (2012, 2014c), Yuan et al. (2013, 2014a)
		Black rust (B)		*Puccinia graminis*	Colwell (1956)
		Brown rust (B)		*Puccinia triticina*	Colwell (1956), Tian et al. (2011, 2012b), Ashourloo et al. (2014)
			Woody crops		
Apple	Fungus	Apple scab (A)		*Venturia inaequalis*	Delalieux et al. (2007, 2009), Oerke et al. (2011)
Avocado	Fungus	Laurel wilt		*Raffaelea lauricola*	Sankaran et al. (2012)
Citrus	Bacteria	Citrus canker	CC	*Xanthomonas axonopodis*	Edwards et al. (1975)
		Huanglongbing	HLB	*Candidatus Liberibacter asiaticus*	Hawkins et al. (2010), Sankaran et al. (2010a, 2011), Windham et al. (2011)
	Fungus	Citrus melanose (A)		*Phomopsis citri*	Pydipati et al. (2005, 2006)
		Citrus scab (A)		*Elsinoë fawcettii*	
		Citrus greasy spot (A)		*Mycosphaerella citri*	
		Sooty mold (A)		*Capnodium citri*	Hart and Myers (1968), Gausman and Hart (1974b)
	Oomycetes	Citrus foot rot		*Phytophthora nicotianae*	Gausman et al. (1970b)
Grapefruit	Oomycetes	Foot rot		*Phytophthora nicotianae*	Gausman et al. (1970b)
Grapevine	Virus	Grapevine leafroll-associated virus	GRLaV	*Closterovirus*	Naidu et al. (2009)
	Oomycetes	Grapevine downy mildew		*Plasmopara viticola*	Calcante et al. (2012)

Table 7.4 (*cont.*)

Plant	Type	Disease	Acronym	Latin name	Reference
		Forest trees			
Eucalyptus	Fungus	Mycosphaerella leaf disease (A)	MLD	*Teratosphaeria* sp.	Pietrzykowski et al. (2006), Barry et al. (2011)
Hevea rubber tree	Fungus	Corynespora leaf fall disease	CLFD	*Corynespora cassiicola*	Jusoff et al. (2010)
Horse chestnut	Fungus	Leaf blotch (A)		*Phyllosticta paviae*	Bertoluzza et al. (1999)
Humid forests	Epiphylls				Toomey et al. (2009)
Monterey pine	Fungus	Sphaeropsis blight (A)		*Sphaeropsis sapinea*	Coops and Stone (2005)
	Fungus	Dothistroma needle blight (A)		*Dothistroma septosporum*	Stone et al. (2003)
Slash pine	Fungus	Ectomycorrhizal fungus (B)		*Pisolithus arrhizus*	Cibula and Carter (1992)
		Dothistroma needle blight (A)		*Dothistroma septosporum*	Stone et al. (2003)

Figure 7.7 Consequences of some viruses on leaf color: (a) maize dwarf mosaic virus; (b) grapevine leafroll-associated virus; (c) tomato mosaic virus; and (d) cucumber mosaic virus. (A black and white version of this figure will appear in some formats. For the color version, please refer to the plate section.)

narrow. Unlike bacteria and fungi, viruses are not spread by water or wind. Instead, they must physically enter the plant. One of the most common vectors for viruses is insects. Insects feed on infected plants and then transmit the viruses into healthy plants when they feed again. Other pathways include plant propagation, contact by humans, and from infected seed.

Cochran et al. (1960) studied progressive changes in infrared absorption, from 1.0 μm to 15.0 μm, of the lower epidermis of cucumber cotyledons induced by the cucumber mosaic virus. They found that the greatest difference occurred in the 1.0–2.0 μm range and differences increased with increasing incubation periods. Ausmus and Hilty (1972) noticed that maize dwarf mosaic virus-infected maize leaves possessed similar reflectance in the visible (0.4–0.75 μm) but significantly lowered reflectance than healthy leaves in the near infrared (0.8–2.60 μm). Cardenas et al. (1972) also found that cucumber leaves inoculated with watermelon mosaic virus had lower reflectance over the entire solar spectrum, even though the eye could not discern visual differences between them and control leaves. Sandwald (1981) did not observe significant differences in the reflectance of sugar beet (*Beta vulgaris*) leaves, displaying rhizomania disease compared to controls. Recently, spectral indices or advanced statistical analyses proved to be helpful in the differentiation of healthy and diseased leaves for several kinds of pathologies and species (e.g., Polischuk et al., 1997; Steddom et al., 2003; Naidu et al., 2009; Purcell et al., 2009; Grisham et al., 2010; Gazala et al., 2013).

Salicylic acid may be produced by tobacco (*Nicotiana tabacum*) leaves in reaction to an infection by tobacco mosaic virus, as a resistance signal, inducing localized stomatal closure and increase in tissue temperature. Chaerle et al. (1999, 2002, 2004) could co-localize thermal and visual lesions using infrared and video images, respectively, and showed that the first lesions were 0.3–0.4°C warmer than the surrounding tissue and were detectable about 8 hours before the initial appearance of necrotic lesions (Figure 7.8). Lenk et al. (2007) likewise combined fluorescence and multispectral images and showed that early symptoms of virus infection could be seen in fluorescence images, especially in the blue (UVF440) and green (UVF550), about 20 hours before the human eye could detect changes or is detectable in a true-color image.

7.2.1.2 Bacteria

There are millions of bacteria, the majority of which are beneficial to plants. However, some can cause diseases in plants. They are most active in warm and humid environments. Among all symptoms of bacterial infection, the most common is leaf spot. The bacteria that attack the plants produce a toxic chemical that kills the surrounding plant cells. The plant then reacts defensively by killing off the surrounding plant cells, thereby isolating the infected cells. Bacteria can spread in several ways, including insects, splashing water, contact with other diseased plants, or from tools. They enter plants through tiny openings either through damage or cuts, but also through natural openings in the plant itself. One disease, citrus greening disease, also known as Huanglongbing (HLB), has focused the attention of the remote sensing community in recent years. It was discovered in July 2004 in Brazil and since then it has seriously affected citrus production in a number of countries, especially in Asia and Africa. Wherever the disease has appeared, citrus production has been compromised with the loss of millions of trees. The early symptoms of HLB on leaves are vein yellowing and an asymmetrical chlorosis referred to as "blotchy mottle" (Figure 7.9a). Leaves may be small and upright with a variety of chlorotic patterns that often resemble mineral deficiencies such as those of zinc, iron, and manganese (Figure 7.9b). Leaf samples can be collected, dried, ground, and analyzed in a matter of minutes by spectroscopy to detect citrus greening disease. Hawkins et al. (2010) and Sankaran et al. (2010a) used

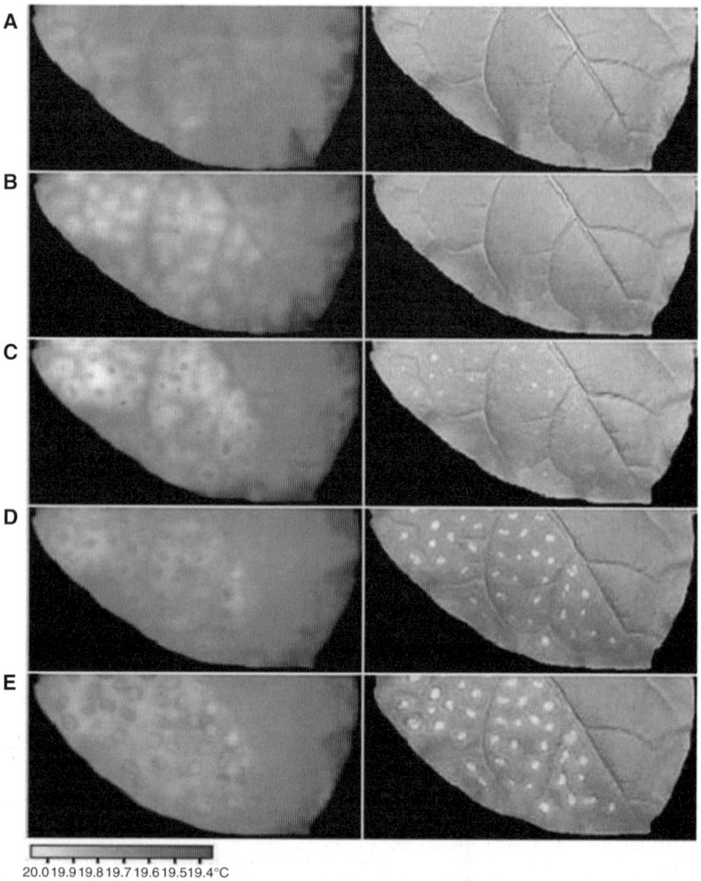

Figure 7.8 Images of a tobacco (*Nicotiana tabacum*) leaf infected by tobacco mosaic virus at 21°C. The images were taken at various times: 26 hours (A), 39 hours (B), 3 days (C), 5 days (D), and 8 days (E). (left) thermal infrared images and (right) visible images (from Chaerle et al. (1999), *Nature Biotechnology*. Copyright © 1999 with permission from Nature Publishing Group). (A black and white version of this figure will appear in some formats. For the color version, please refer to the plate section.)

Fourier transform infrared-attenuated total reflection (FT-IR-ATR) spectroscopy as a useful, rapid, and inexpensive tool in the early identification of plants containing citrus greening disease. Near-infrared reflectance spectroscopy (NIRS) techniques have also been successfully applied to spectra collected in field on fresh leaves (Sankaran et al., 2011) and in the lab on dry ground leaves (Windham et al., 2011). The success of these techniques largely depends on the quality of the calibration.

7.2.1.3 Fungi and Oomycetes

Unlike bacteria, there are thousands of fungi that are harmful to plants. They are present in the soil and above ground, and symptoms of fungal attack can appear above and below ground. Fungal spores are very small and light, and can travel great distances through the air to infect other plants or trees. They are also spread by water, animals and insects, and people. The majority of phytopathogenic fungi belong to the *Ascomycetes* and the *Basidiomycetes*. Plants can display leaf spots, mildews, rusts, wilts, molds, etc. (Figure 7.10).

Figure 7.9 Consequences of Huanglongbing on leaf color: (a) sweet orange (*Citrus* × *sinensis*) foliage from São Paulo, Brazil demonstrating asymmetrical "blotchy mottle" symptom relative to the mid vein; (b) South Florida pomelo (*Citrus maxima*) tree displaying "zinc-pattern-deficiency" with inter-veinal chlorotic symptoms (from Gottwald et al. (2007), *Plant Health Progress*. Copyright © 2007 with permission from The Plant Management Network). (A black and white version of this figure will appear in some formats. For the color version, please refer to the plate section.)

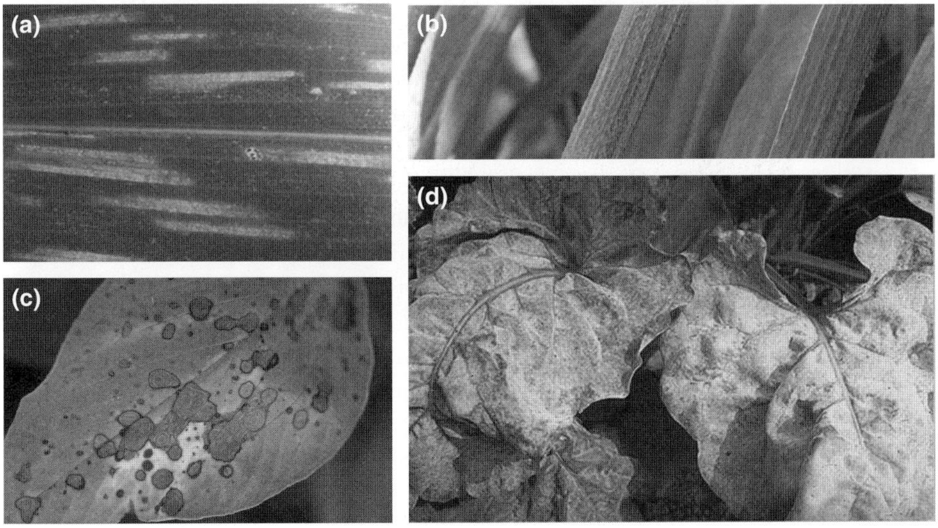

Figure 7.10 Consequences of some fungi on leaf color: (a) southern corn leaf blight; (b) wheat yellow rust; (c) bean chocolate spot; and (d) sugar beet powdery mildew. (A black and white version of this figure will appear in some formats. For the color version, please refer to the plate section.)

Leaf spots are round blemishes found on the leaves infected by parasitic fungi. They are brown or black, sometimes with a yellow halo, usually uniform in size. The proportion of the leaf infected controls their optical properties (e.g., Malthus and Madeira, 1993; Pydipati et al., 2005, 2006; Liu et al., 2007, 2008c; Mahlein et al., 2010, 2012, 2013; Rumpf et al., 2010; Patil and Bodhe, 2011), as illustrated in Figure 7.11.

Mildew is easily recognized on foliage, usually on the upper leaf surface, with white to gray powdery patches. Disease severity depends on many factors, including cultural practices, variation in

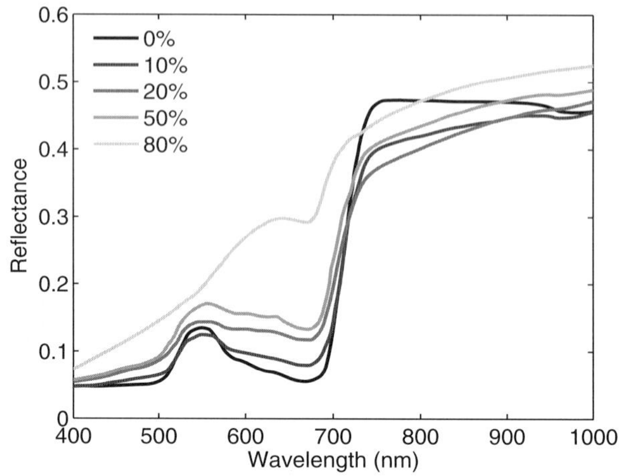

Figure 7.11 Spectral reflectance measured under controlled conditions of sugar beet (*Beta vulgaris*) leaves affected with *Cercospora* leaf spot at different disease severities (redrawn from Mahlein et al. (2010)).

weather conditions, level of cultivar susceptibility, and regional and in-field location. Mildew may occur during all stages of growth. The fungi result in loss of yield and therefore have received increasing attention over the past few years. Leaves of severely infected plants turn brown and can be monitored by spectroscopy (e.g., Blazquez and Edwards, 1986; Lorenzen and Jensen, 1989; Graeff et al., 2006; Rumpf et al., 2010; Tian et al., 2011; Mahlein et al., 2010, 2012, 2013; Calcante et al., 2012; Zhang et al., 2012a, 2012b; Yuan et al., 2014a) or by thermal infrared imagery. For instance, the oomycetes *Pseudoperonospora cubensis* causing downy mildew of cucumber induces an increase in overall leaf temperature at early stages of infection (Lindenthal et al., 2005; Oerke et al., 2006).

Mold or rot – it is difficult for a non-specialist to tell the difference between mold and rot – is a widespread fungal disease that can infect all plant organs. Gray mold, also known as Botrytis blight, is a common form caused by the pathogenic fungus *Botrytis cinerea*. Infections are identified as grayish-colored soft, dusty spots on leaves. As the infection progresses, the color changes to brown and can cover most of the leaf, causing it to wilt. Under especially humid conditions, grayish filaments may appear on the leaves that contain spores. To support a study on infrared aerial color photography for detection of populations of brown soft scale in citrus groves, Hart and Myers (1968) measured the diffuse reflectance of citrus leaves having varying amounts of sooty mold. Mature leaves with no trace of sooty mold had reflectance values of 58% and 53% at 770 nm and 1300 nm (the extremes of the region recorded on the film used in their study), respectively (Figure 7.12). Leaves heavily coated with sooty mold had reflectance values of 9% and 23% at respective wavelengths. A light coating of mold reduced reflectance by 14% at 770 nm and 2% at 1300 nm; a medium coating reduced reflectance 29% and 10%, respectively. Gausman and Hart (1974b) examined the effect of the sooty mold of citrus (*Capnodium citri*) and found consistently large reflectance differences across the 2.5–40 μm range between infected and non-infected sweet orange (*Citrus* × *sinensis*) leaves. Partial least squares regression analysis has been used to study the effect of *Sclerotinia sclerotiorum*, a plant pathogenic fungus that causes a disease called white mold, on celery (*Apium graveolens*) (Huang and Apan, 2006) or common bean (*Phaseolus vulgaris*) leaves

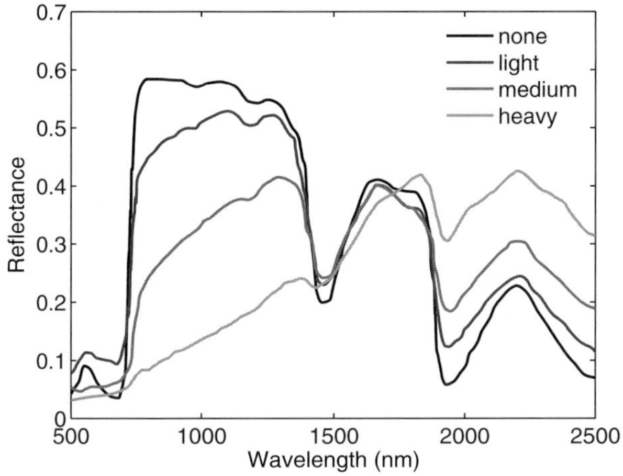

Figure 7.12 Effect of sooty mold on reflectance (percent) of sweet orange (*Citrus sinensis*) leaves coated with varying amounts of the fungus (redrawn from Hart and Myers (1968)).

(Machado et al., 2015), or the effect of *Botrytis cinerea* on eggplants (*Solanum melongena*) leaves (Wu et al., 2008).

There are more than 5000 known species of rust on plants. It is most often found on mature plants where symptoms appear primarily on the surfaces of lower leaves. Early on, white, slightly raised spots can be observed on the undersides of leaves. After a short period of time, these spots become covered with rust-colored spore masses. Later, leaf pustules may turn yellow-green and eventually black. Severe infestations will deform and yellow leaves and cause leaf drop. Colwell (1956), while working primarily with cereal rusts, suggested using infrared film for recording any disease that interfered with the internal reflection of light within leaves. Keegan et al. (1956), with Colwell's assistance in collecting leaf specimens, did extensive research on effects of black rust (*Puccinia graminis*) and brown rust (*Puccinia triticina*) on leaf reflectance of wheat (*Triticum aestivum*) in the 0.4–1.08 μm range. They compared spectral reflectance of healthy leaves with those with rusted foliage. Results indicated that plants infected with wheat rust were better reflectors of red wavelengths than healthy plants. The maximum difference between the reflectivity of healthy and rusted foliage occurred between 0.6 μm and 0.7 μm; the second peak occurred between 0.75 μm and 0.9 μm. Other authors have quantified leaf rust infection by spectroscopy spectral indices or statistical methods (e.g., Runeckles and Resh, 1975; Teng and Close, 1977; Mahlein et al., 2010, 2012, 2013; Rumpf et al., 2010; Bauer et al., 2011; Tian et al., 2011; Zhao et al., 2012, 2014c; Yuan et al., 2013, 2014a; Ashourloo et al., 2014; Zhang et al., 2014).

Vascular wilt diseases are caused by parasitic fungi (or bacteria) infecting hundreds of eudicot species, including many important crop plants. Most common are species of *Fusarium* and *Verticillium* fungi. The wilt enters the xylem tissue and impedes the flow of water through the plant. In addition to wilting, other symptoms include chlorosis, yellowing, and necrosis following the rapid drying of the leaves. An early paper by Blazquez and Edwards (1986) demonstrated that wilt symptoms were observable across the visible and SWIR spectrum in watermelon (*Citrullus vulgaris*) leaves and used reflectance differences between ratios at 500–640 nm and 660–900 nm to quantify the severity of the infection. More recently, Chen et al. (2012a) also found differences across the spectrum in cotton (*Gossypium hirsutum*) leaves and used a first difference analysis, particularly at the red-edge, to quantify the infection. Sankaran et al. (2012) used linear discriminant analysis, quadratic discriminant analysis (QDA), Naïve-Bayes classifier, and bagged decision trees (BDT) on avocado (*Persea Americana*) leaves and found all had high accuracy for prediction of symptomatic plants. But none of these studies were able to identify infected but asymptomatic plants.

Blight is defined by symptoms of rapid chlorosis and death of infected plant tissues (leaves, branches, flowers) and is caused by many different pathogens. Safir et al. (1972) found increased red reflectance in maize (*Zea mays*) leaves infected with southern corn leaf blight caused by *Helminthosporium maydis* after 40 hours of infection, when disease lesions became visible to the eye. More recently, Malthus and Madeira (1993) measured leaf reflectance of faba beans (*Vicia faba*) that were infected with *Botrytis fabae*. They found increases in chlorophyll *b* content that were proportional to the extent of the leaf surface area infected but no change in the chlorophyll *a* content. They found no changes in leaf reflectance before visible symptoms were seen, although the early spectral changes were described as a "flattening" of the spectrum in the visible region and a decrease in the near infrared that they attributed to collapse of the leaf cells. Stone et al. (2003) for Dothistroma needle blight, caused by *Dothistroma septosporum*, a necrotrophic fungal pathogen, and Coops and Stone (2005) for *Sphaeropsis sapinea*, a common and widely distributed pine fungus, which causes crown wilt, cankers, and shoot blight, found that the most pronounced spectral feature was a flattening of the visible and near-infrared spectral region and a lower slope of red-edge, as observed on Monterey Pine (*Pinus radiata*) needles as chlorosis developed. *Phytophthora infestans* is the agent of late blight (or potato blight). It is the most important pathogen on potato crop that caused the well-known great Irish famine in the 1840s. By using FTIR techniques, Taoutaou et al. (2010, 2012) measured the total composition of healthy and diseased potato (*Solanum tuberosum*) leaves infested by late blight. They found the spectral signature of key molecules involved in plant defense in the range 650–1800 cm^{-1}. Then they used these signatures as specific markers for resistance/susceptibility in several potato genotypes.

Scab disease affects trees and shrubs and is caused by fungi in the genus *Venturia* and related asexual forms in the genera *Fusicladium* and *Fusicladosporium*. Infection occurs in wet weather in spring and summer. Infection can cause sooty, dark spots on the leaves, premature leaf loss, aborted flowers, and shriveled fruit. Pydipati et al. (2005, 2006) converted their data into hue and saturation levels first and then applied a neural net to identify leaf scab disease in citrus (*Citrus* × *sinensis*). Other authors have quantified the effects of the *Ascomycetes* fungus *Venturia inaequalis* on apple leaves by using spectral indices (Delalieux et al., 2009) or thermal infrared imagery (Oerke et al., 2011).

Other examples of the impact on spectral properties of fungi in plants are provided hereafter. The foot rot disease in citrus is caused by the fungi *Phytophthora citrophthora* and *Phytophthora parasitica* that produce a gummy extrudate and wood rot at or near grafts on citrus trees. Leaves from infected trees were chlorotic and the loss of pigments was observable across the spectrum between 400 nm and 750 nm and as white leaves (compared to red in healthy leaves) in color infrared film (Gausman et al., 1970b). Pietrzykowski et al. (2006) and Barry et al. (2011) examined the *Ascomycetes Teratosphaeria* spp. (formerly known as *Mycosphaerella* spp.) that causes chlorosis and necrosis in Tasmanian bluegum (*Eucalyptus globulus*), and found wavelengths between 679 nm and 695 nm were most sensitive to detecting necrosis in the presence of chlorosis and reddening of the foliage. Finally, in contrast to these studies, Cibula and Carter (1992) examined the differences in reflectance between slash pine (*Pinus elliotii*) inoculated with the beneficial fungal ectomycorrhizae and trees without this root symbiont. They identified wavelengths of maximum sensitivity to the ectomycorrhizae in the red-edge region and they found increased reflectance at 700 nm, causing a blue shift in the plants that were not inoculated compared to those with the ectomycorrhizae. The latter are formed with the roots of about 2% of plant species, mostly woody temperate trees and shrubs (Smith and Read, 2008). The infection of horse chestnut (*Aesculus hippocastanum*) leaves by the fungus *Guignardia aesculi* that causes leaf blotch disease on this species in the United States and Europe have been investigated by Bertoluzza et al. (1999) using FTIR spectroscopy. Compared to sound leaves, the spectra of diseased leaves showed a weakening of the aromatic domain bands (1650–1500 cm^{-1}) and intensity inversion of two bands of this domain (1650 cm^{-1} and 1610 cm^{-1}) that could be a spectroscopic maker of that disease.

7.2.2 Plant Diseases: Insects and Other Animals

Insects can be beneficial to plants at one stage and harmful at another, for example, at the larval stage, or harmful to one species but beneficial to others. Only a small fraction is considered plant pests. The mechanism for how insects identify the host plant is uncertain. Chemical odors and visual characteristics like color are often cited (Bernays and Chapman, 1994). As developed in Section 12.5.1, insects may use visual clues to identify host plants; for instance, Bálint et al. (2013) investigated the UV-A (340–400 nm) and visible (400–650 nm) spectrum finding that reflectance in these wavelengths were

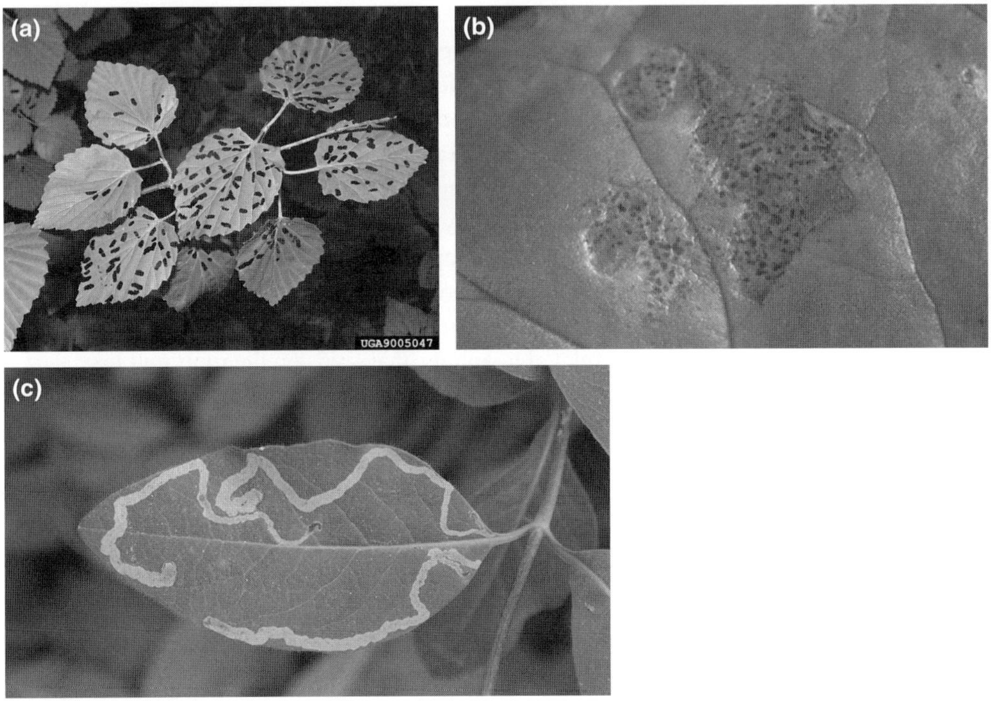

Figure 7.13 Consequences of some insect on leaf morphology and color: (a) damage of leaf beetle (chewing insect) on *Viburnum* sp. (photo courtesy of Paul Weston, Cornell University); (b) damage of *Frankliniella occidentalis* (sucking insect) on lettuce (*Lactuca sativa*) (photo courtesy of INRA); and (c) damage of diptera (leaf miner) on honeysuckle (*Lonicera periclymenum*) (photo courtesy of Krzysztof Ziarnek).

negatively associated with thrip infestation. Harmful insects generally injure leaves directly from feeding on foliage or the phloem (Figure 7.13), or they can indirectly injure them by introducing a bacteria, virus, or fungal infection as described above.

There are hundreds of phytophagous insect pests that feed on leaves, either as adults or larvae. Table 7.5 sums up works relating the extent and severity of insect damage to leaf optical properties. The type of damage is related to the type of mouthparts of the insect. Chewing insects (grasshoppers, caterpillars, and beetles) are the cause of holes or notches in foliage, leaf skeletonizing (removal of tissue between the leaf veins), or leaf defoliation to the extreme. Locusts provide a good example of a particularly destructive insect when they form migrating swarms, eating everything in their path. Rasping insects (e.g., thrips) scrape the surface of foliage, removing the waxy cuticle and destroying the cell walls, and then they suck up the spilled contents from the damaged cells.

Sucking insects (e.g., aphids, scales, leafhoppers, and true bugs) feed by sucking phloem or xylem sap from plant tissues. This type of feeding can cause spotting or stippling of foliage, and leaf curling. The California pine aphid *Essigella californica* feeds on needles of old Monterey pine (*Pinus radiata*) trees, causing small isolated zones of yellowing along the needle length that can quickly extend with high insect density. Coops and Stone (2005) showed that symptoms such as changes in the chlorophyll content, cellular integrity, and moisture content could be detected by VIS-NIR spectroscopy. Spider mites feed on the underside of leaves, piercing the chloroplast-containing cells. The decrease in leaf chlorophyll content can be spectrally detected in the visible and near-infrared regions and the damage quantified using vegetation indices (MacNeil et al., 1974, 1987; Reisig and Godfrey, 2007; Herrmann et al., 2012). The psyllid *Glycaspis baileyi* is endemic to Australia. It forms large populations on *Eucalyptus* trees. The female lays several hundred eggs on succulent young leaves. Resulting nymphs insert stylets into leaf, feed, and secrete wax filaments called lerps, which are composed of wax and solidified honeydew. They significantly influence leaf reflectance depending on the species composition and density of insects present. Stone et al. (2001, 2003) also showed an increase in levels of anthocyanins in insect-damaged leaves of mature foliage, as well as reduced rates of photosynthesis. Aphids may cause severe damage to cereal crops, leading to yield loss. Damage occurs during feeding when the aphids inject phytotoxic substances into leaves and remove assimilates from leaf vascular tissues. Wavelengths sensitive to aphid damage have been identified by Riedell and Blackmer (1999) and Yuan et al. (2014a). Finally, grape phylloxera is a small, pale yellow sap-sucking insect, related to aphids, that feeds on the roots and leaves of grape vine (*Vitis vinifera*). Current detection methods rely on time-consuming ground surveying, which involves detailed examination of root systems. Virus-infected vines show decreased chlorophyll – the leaves become chlorotic – increased carotenoid/chlorophyll content and increased non-photochemical quenching that can be measured by spectroscopy (e.g., Baldy et al., 1996; Renzullo et al., 2005; Blanchfield et al., 2006; Powell et al., 2006).

Oviposition damage is due to insects (e.g., moths, sawflies, and flies) that lay eggs into leaf tissue. The larva, also called leaf miners, live in and eat the tissues between the upper and lower surfaces of leaves, creating distinctive tunnel patterns visible as translucent lines or blotches on leaves. Linear, serpentine, digitate, and tentiform mines can be found on various types of plant leaves. Yamazaki (2010) remarks that the diverse morphology of leaf mines has fascinated naturalists for a long time; that ecologists have used them to examine the population dynamics of leaf miners; and that evolutionary biologists have investigated their evolution based on leaf fossils. Leaf mines are also often structured so as to improve the inner microclimate. In order to study alterations in microclimate when an herbivore feeds on host

Table 7.5 *Main plant diseases caused by insects that gave rise to spectroscopic studies.*

Order	Name	Latin name	Plant species	Reference
Coleoptera	Mountain pine beetle	*Dendroctonus ponderosae*	Lodgepole pine	Ahern (1988), Cheng et al. (2010)
Diptera	Leaf miner	*Agromyzidae*	Tomato	Xu et al. (2005, 2007)
		Phytomyza ranunculi	*Ranunculus japonicus*	Yamazaki (2010)
Hemiptera	Phylloxera	*Daktulosphaira vitifoliae*	Grapevine	Baldy et al. (1996), Renzullo et al. (2005), Blanchfield et al. (2006), Powell et al. (2006)
	California pine aphid	*Essigella californica*	Monterey pine	Coops and Stone (2005)
	Sirex woodwasp	*Sirex noctilio*	Patula pine	Mutanga et al. (2007), Ismail et al. (2008)
	Greenfly	*Aphis gossypii*	Cotton	Reisig and Godfrey (2007)
	Russian wheat aphid	*Diuraphis noxia*	Wheat	Riedell and Blackmer (1999)
	Greenbug	*Schizaphis graminum*		
	English grain aphid	*Sitobion avenae*		Yuan et al. (2014a)
	Psyllid	*Glycaspis baileyi*	Eucalyptus	Stone et al. (2001, 2005)
	Citrus blackfly	*Aleurocanthus woglumi*	Citrus	Gausman and Hart (1974a), Hart et al. (1976)
	Brown soft scale	*Coccus hesperidum*		Hart and Myers (1968), Gausman and Hart (1974a)
	Citrus mealybug	*Planococcus citri*		Gausman and Hart (1974a)
	Leaf miner	*Profenusa japonica*	Japanese rose	Yamazaki (2010)
Lepidoptera	Cedar tip moth	*Hypsipyla robusta*	Red cedar	Cunningham and Floyd (2004)
	Spotted tentiform leaf miner	*Phyllonorycter blancardella*	Apple	Pincebourde and Casas (2006)
	Rice leaf roller	*Cnaphalocrocis medinalis*	Rice	Li et al. (2009), Liu et al. (2012)
	Moth	*Lyonetia euryella*	Sakaki	Yamazaki (2010)
		Liocrobyla desmodiella	Tick-trefoil	
Thysanoptera	Onion thrips	*Thrips tabaci*	Cabbage	Bálint et al. (2013)
Thripidae	Sugarcane thrip	*Fulmekiola serrata*	Sugarcane	Abdel-Rahman et al. (2010)
Trombidiformes	Two-spotted spider mite	*Tetranychus urticae*	Greenhouse pepper	Herrmann et al. (2012)
	Spider mite	*Tetranychus* spp.	Cotton	Reisig and Godfrey (2007)
	European red mite	*Panonychus ulmi*	Apple	MacNeil et al. (1974)
	White apple leafhopper	*Typhlocyba pomaria*		MacNeil et al. (1987)

Table 7.6 *Main plant diseases caused by nematodes that gave rise to spectroscopic studies.*

Nematoda	Reniform nematode	*Rotylenchulus reniformis*	Cotton	Gausman et al. (1975a)
	Rice white tip	*Aphelenchoides besseyi*	Rice	Li et al. (2009)
	Sugar beet nematode	*Heterodera schachtii*	Sugar beet	Hillnhütter et al. (2012)

plant, Pincebourde and Casas (2006) measured the transmittance spectra of feeding windows and green patches in apple (*Malus communis*) leaves infested by a moth, the spotted tentiform leaf miner. As expected, feeding windows that form translucent patches due to the removal of chlorophyllian tissues transmit much more radiation than unfed areas. Xu et al. (2005, 2007) studied the infestation of tomato (*Lycopersicon esculentum*) leaves by leaf miner flies and showed that leaf reflectance significantly decreased in the NIR (800–1100 nm), while it increased in the SWIR (1450–1900 nm).

Note that when some insects or mites feed on or lay eggs in plants, they may inject a chemical into the plant that causes it to grow abnormally and produce a gall.

Some insects have indirect effects on leaf optical properties. More than 20 species of bark beetles (family *Scolytidae*) are economically and ecologically important forest pests that can kill large numbers of trees during epidemics, particularly old trees in stand replacing events, or trees weakened by drought or air pollution (Rose and Lindquist, 1994). Bark beetles lay eggs in the phloem layer and the growing larva can kill or injure the tree by blocking the flow of phloem to the roots. Ahern (1988) reported changes in leaf reflectance in lodgepole pine (*Pinus contorta*) following mountain bark beetle attack and found the green peak, red-edge and near infrared were useful to identify different levels of beetle infestation. Similar examples are provided by the wood-boring pest *Sirex noctilio*, a species of horntail that attacks all commercial pine species in South Africa (Mutanga et al., 2007; Ismail et al., 2008). Secondarily, insects can also transmit diseases in plants by introducing other pathogens to the plant tissues. Most of the plant diseases transmitted by insects involve plant viruses, but there are examples from all plant pathogen groups (e.g., fungi, bacteria, and nematodes).

Nematodes are multicellular animals that can infest plants and change their optical properties (Table 7.6). For instance, Gausman et al. (1975a) showed that leaves of cotton (*Gossypium hirsutum*) plants stressed by *Rotylenchulus reniformis*, a nematode largely distributed in tropical, subtropical, and warm temperate zones, have less reflectance than leaves of non-stressed plants over the entire 0.5 μm to 2.5 μm waveband. The nematode *Aphelenchoides besseyi* is most often associated with a disease in which the leaves of the rice (*Oryza sativa*) plant turn white in the meristematic regions followed by necrosis. Li et al. (2009) could identify infected plant using principal component analysis. Finally, Hillnhütter et al. (2012) used a hyperspectral imaging system (400–1000 nm) to detect symptoms caused by the nematode *Heterodera schachtii* on sugar beet (*Beta vulgaris*) plants.

8

Comprehensive Reviews of Leaf Optical Properties Models

A model is a simplified mathematical representation of a phenomenon that simulates its functioning. Science has long used, explicitly or implicitly, approaches that associate models with pure theory and experimentation. These three lines of research are inseparable. We expect a theory to explain the experimental results and predict new results; from an experiment we generally expect it to verify the validity of existing theories and to collect new data. With the exception of a few cases where experimentation is not possible, the confrontation of a model with experimental data is needed to validate it. Sometimes, however, this comparison is misleading. Coincidentally, a false theory may be consistent with experimental results, and conversely a theory may be correct but in disagreement with experience. Thus, there are "good" and "bad" experiments. In the following we will assume that the experimental data have been calibrated, checked, and any systematic errors removed (see Chapter 4). We can consider two main approaches:

- *The statistical approach*: experimental data are collected of various objects, and we try to obtain a correlation between the objects properties and the measurements (see Section 11.1).
- *The analytical approach*: we try to understand the mechanisms of the observed phenomena and to incorporate this knowledge into a model.

Physical models are essential to understand how electromagnetic radiation interacts with leaf elements, but also to directly relate observed optical properties to leaf biophysical attributes. Their complexity depends on what we want to use the model for. In the case of a model inversion, an essential first step is to validate it and the choice of the model to do this is governed by a number of rules. The French mathematician René Thom wrote that if we use a sufficient number of parameters or polynomial functions of higher degree, we can always construct a mathematical model describing any phenomenology (Thom, 1980). However, this is not the fundamental problem; it is the construction of a model that does not require many parameters. Thus, there is a conflict between the strict adherence to the empirical data and the number of parameters. If there are many parameters, the fit is good but the model complicated; if there are few, the model is simple but the fit is bad. The best models are those that balance, with some compromises, a few parameters, and a good fit. As seen in Section 4.2, when a leaf intercepts a light beam, it can be reflected, transmitted, or absorbed. The total reflected fraction is the sum of two terms – a fraction R_d is caused by multiple scattering within the leaf tissues and a fraction R_s is reflected at the leaf surface:

$$R = R_d + R_s \tag{8.1}$$

where the subscripts d and s stand for diffuse and surface – or for specular, since most leaves display such optical behavior – respectively. In the case of very thin leaves, part of the incident radiation may travel through the leaf without being affected, so that the total transmitted light should also be rigorously split into two fractions. In practice, only the diffuse transmittance is considered in models. As pointed out by Wang et al. (2005), leaf optical properties are particularly difficult to simulate due to their intricate underlying structure and their complex and subtle interaction with light.

8.1 Different Approaches for Leaf Diffuse Optical Properties

The propagation of light through plant leaves is governed primarily by absorption and scattering interactions. While experimental measurements of leaf optical properties were progressing, deterministic approaches using diverse representations of light interactions with plant leaves were also developed. These models are distinguished by the underlying physics and by the complexity of the leaf. The simplest consider the blade as a single scattering and absorbing layer, while in the most complicated, all cells are described in detail by their shape, size, position, and biochemical content. Whatever the approach, these models have improved our understanding of the interactions of light with plant leaves. Baranoski and Rokne (2004), Ustin et al. (2004), Jacquemoud (2004), and Jacquemoud et al. (2009a) extensively reviewed computer-based leaf models which, from the late 1960s to the present, have improved our understanding of the interaction of light with plant leaves. These can be categorized into different classes, arranged in order of increasing complexity (e.g., Hecht, 1976; Philips-Invernizzi et al., 2001). Each model listed in Table 8.1 can provide a faithful simulation of leaf optical properties.

8.1.1 Plate Models

A model in which the scattering medium is approximated by plane-parallel layers has been used by numerous workers as a basis for reflectance theory. The plate model developed by Allen et al. (1969) considered a compact plant leaf as a semi-transparent plate with plane-parallel surfaces and initially assumed that the incident light was partially isotropic. That requirement is equivalent to the assumption that the surfaces are rough. The ray with unit intensity incident at the surface of such a leaf undergoes multiple reflections and transmissions, as shown in Figure 8.1.

This concept was first developed in 1833 by the British astronomer George Biddell Airy (Yeh, 1988). Consider a beam incident on the plate, which is partially reflected and partially transmitted at the first interface. The transmitted part is subsequently reflected back and forth between the two interfaces. The expression for the total reflectance of the plate R can be derived by summing the amplitudes of successive reflections and refractions. We obtain

$$\begin{aligned} R &= r_{12} + t_{12}\,\tau\,r_{21}\,\tau\,t_{21} + t_{12}\,\tau\,r_{21}\,\tau\,r_{21}\,\tau\,r_{21}\,\tau\,t_{21} + t_{12}\,\tau\,r_{21}\,\tau\,r_{21}\,\tau\,r_{21}\,\tau\,r_{21}\,\tau\,r_{21}\,\tau t_{21} + \ldots \\ &= r_{12} + t_{12}\,t_{21}\,r_{21}\,\tau^2\left(1 + r_{21}^2\,\tau^2 + r_{21}^4\,\tau^4 + \ldots\right) \\ &= r_{12} + \frac{t_{12}\,t_{21}\,r_{21}\,\tau^2}{1 - r_{21}^2\,\tau^2} \end{aligned} \tag{8.2}$$

where r_{ij} and t_{ij} are the average reflectivity and transmissivity, respectively, from medium i into medium j, and τ is the fraction of light transmitted through the medium in isotropic light. The total

Table 8.1 *Categorization of leaf optical properties models. (*) Only the articles corresponding to the development of the* PROSPECT *model are cited here; those which apply will be detailed in Chapter 9.*

Type	Name	References
Beer–Lambert		Baret et al. (1988)
Plate models		Allen et al. (1969), Allen et al. (1970a), Kaufmann and Hartmann (1988)
	PROSPECT (*)	Jacquemoud and Baret (1990), Fourty et al. (1996), Jacquemoud et al. (1996), Baret and Fourty (1997), Jacquemoud et al. (2000)
	PROSPECT-5 (*)	Féret et al. (2008)
	PROSPECT-D	Féret et al. (2017)
	PROSPECT-VISIR	Gerber et al. (2011)
	DLM	Stuckens et al. (2009)
	QSPECT	Ma et al. (2007)
N-flux models	FRT	Allen and Richardson (1968), Fukshansky et al. (1991), Yamada and Fujimura (1991), Martinez von Remisowsky et al. (1992), Conel et al. (1993a,b), Richter and Fukshansky (1996a)
Compact spherical particle models	LIBERTY	Dawson et al. (1995, 1998a)
	LIBERTY3P	Di Vittorio (2009), MacArthur and Malthus (2012)
Radiative transfer theory		Hemenger (1977), Ma et al. (1990), Hanrahan and Krueger (1993), Berdnik and Loiko (1999), Berdnik and Mukhamed'yarov (2001), Donner and Jensen (2005), Habel et al. (2007), Habel (2009, 2010), Krekov et al. (2009), Zakharov et al. (2009, 2010)
	LEAFMOD	Ganapol et al. (1998), Wang et al. (2005)
Stochastic models	LFMOD1	Tucker and Garratt (1977)
	SLOP	Maier et al. (1999)
Ray tracing models	FSM	Allen et al. (1973), Brakke and Smith (1987), Kumar and Silva (1973)
	ABM, ABM-U, ABM-B	Baranoski and Rokne (1997, 1999, 2001)
	RAYTRAN	Govaerts et al. (1996), Ustin et al. (2001)

transmittance of the plate T is also determined by summing the components of transmissions to infinity:

$$\begin{aligned} T &= t_{12}\ \tau\ t_{21} + t_{12}\ \tau\ r_{21}\ \tau\ r_{21}\ \tau\ t_{21} + t_{12}\ \tau\ r_{21}\ \tau\ r_{21}\ \tau\ r_{21}\ \tau\ r_{21}\ \tau\ t_{21} + \dots \\ &= t_{12}\ t_{21}\ \tau\big(1 + r_{21}^2\ \tau^2 + r_{21}^4\ \tau^4 + \dots\big) \\ &= \frac{t_{12}\ t_{21}\ \tau}{1 - r_{21}^2\ \tau^2} \end{aligned} \tag{8.3}$$

Therefore the total absorptance of the plate A is given by $A = 1 - R - T$. r_{12} is calculated from Fresnel's equations for an incidence angle θ and a refractive index of the medium n (see Appendix E). t_{12} is easily deduced via the relation $r_{12} = 1 - t_{12}$. Remembering that $t_{21} = t_{12}/n^2$ and

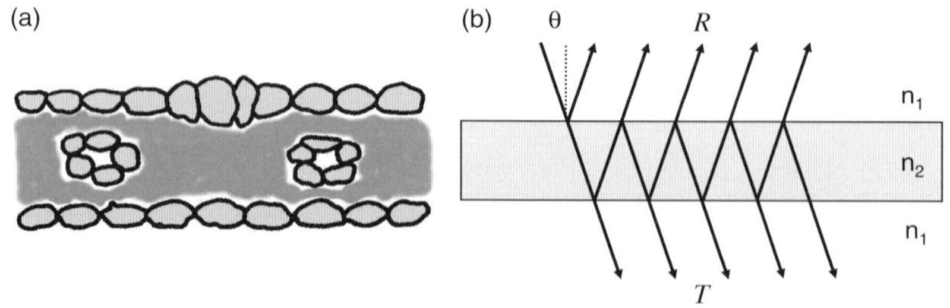

Figure 8.1 Schematic representation of a monocot leaf (a) and diagram of a ray traced through a single plate (b).

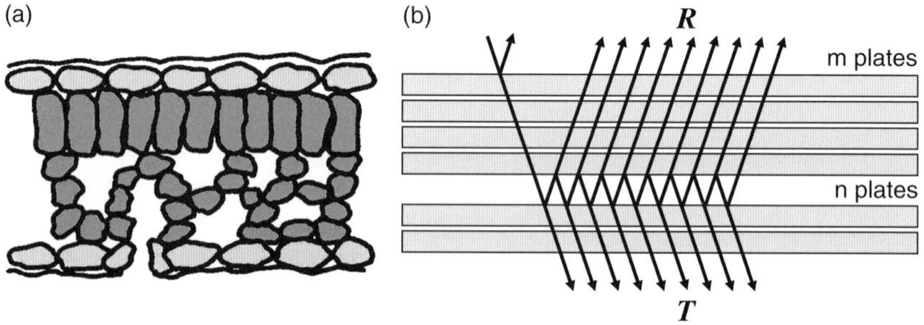

Figure 8.2 Schematic representation of a dicot leaf (a) and multiple reflections produced by a set of $N = m + n$ plates (b).

that $r_{21} = 1 - t_{21}$ avoids unnecessary calculations. τ is related to the absorption coefficient of the plate k through the Beer–Lambert law. Consequently, the reflectance and the transmittance of the plate only depend on its index of refraction n and its absorption coefficient k. This model was successful in reproducing the reflectance spectrum of a compact maize (*Zea mays*) leaf characterized by few air–cell wall interfaces (Allen et al., 1969). However, plant leaves are not compact but represent a wide range of anatomical structures which are species dependent (see Section 1.1). Since plant leaves cannot be described as a unique compact layer, the plate model detailed above is not suited for simulating leaf optical properties.

Allen et al. (1970a) followed by Breece and Holmes (1971) rapidly extended the plate model to non-compact leaves by introducing the generalized plate model which simply consists of stacking elementary plates. This new model accounts for the development of intercellular spaces in the leaf mesophyll. The leaf is conceptually subdivided into N uniform compact plates separated by $N - 1$ air spaces. Such a system was solved for reflectance and transmittance many years ago by the Irish mathematician George Gabriel Stokes in his glass plate theory (Stokes, 1862), which contemplated a system of $N = m + n$ glass plates, and obtained the reflectance $R(m + n)$ and the transmittance $T(m + n)$ of this system in terms of the reflectances and transmittances of the two subsets consisting of m and n plates (Figure 8.2). The radiation flux which strikes the m plates from above is partially reflected and partially transmitted. The fraction $T(m)$ falls on the n plates, the fraction $T(m)R(n)$ is reflected, and the fraction $T(m)T(n)$ transmitted. The fraction $T(m)R(n)$ falls from beneath the m plates, the new fraction $T(m)R(n)T(m)$ is transmitted, and the fraction $T(m)R(n)R(m)$ reflected, etc.

By summing all the reflected and transmitted fractions, we obtain two geometrical series:

$$\begin{aligned} R(m+n) &= R(m) + T(m)R(n)T(m) + T(m)R(n)R(m)R(n)T(m) + \dots \\ &= R(m) + \frac{T(m)^2 R(n)}{1 - R(m)R(n)} \end{aligned} \tag{8.4}$$

and

$$\begin{aligned} T(m+n) &= T(m)T(n) + T(m)R(n)R(m)T(n) + \dots \\ &= \frac{T(m)T(n)}{1 - R(m)R(n)} \end{aligned} \tag{8.5}$$

Note that if we set $m = 1$ and $n = \infty$, Eq. 8.4 becomes

$$R_\infty = R + \frac{T^2 R_\infty}{1 - RR_\infty} \tag{8.6}$$

This gives us

$$R_\infty = \frac{1 + R^2 - T^2 \pm \sqrt{(T^2 - R^2 - 1)^2 - 4R^2}}{2R} \tag{8.7}$$

To solve his system, Stokes first proceeded to show that the following expression was constant:

$$\frac{1 + R(m)^2 - T(m)^2}{R(m)} = \frac{1 + R(n)^2 - T(n)^2}{R(n)} = 2\cos\alpha \tag{8.8}$$

For convenience, he denoted this invariant by $2\cos\alpha$ and solved Eq. 8.8 for T^2 to obtain a simple expression for the transmittance in terms of the reflectance and, with it, to eliminate T in Eq. 8.4. There are various ways to solve this system of recursive series (e.g., Olf, 1988), the solution of which can be put in a simple form:

$$\frac{R(N)}{b^N - b^{-N}} = \frac{T(N)}{a - a^{-1}} = \frac{1}{ab^N - a^{-1}b^{-N}} \tag{8.9}$$

where the terms $a = \alpha = \beta^{-1}$ and $b = \sqrt{k}$ depend on the reflectance and transmittance of one layer $R(1)$ and $T(1)$ through the relationships:

$$k = \frac{\beta\left(\alpha - R(1)\right)}{\alpha\left(\beta - R(1)\right)} \tag{8.10}$$

$$\alpha = \frac{R^2(1) - T^2(1) + 1 + \sqrt{\Delta}}{2R(1)} \tag{8.11}$$

$$\beta = \frac{R^2(1) - T^2(1) + 1 - \sqrt{\Delta}}{2R(1)} \tag{8.12}$$

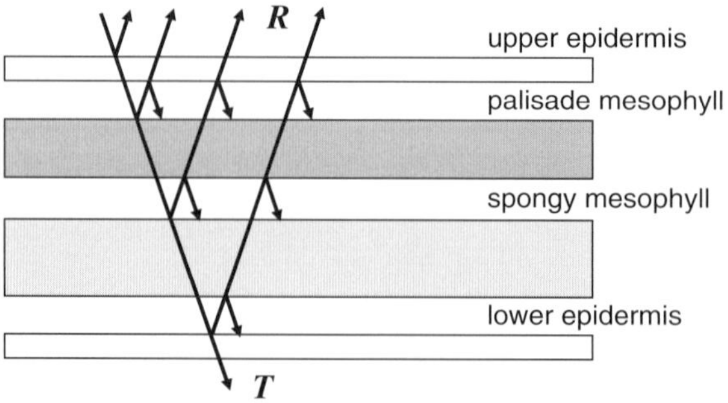

Figure 8.3 Schematic representation of light propagation in a dorsiventral leaf.

$$\Delta = \left(R(1) + T(1) + 1\right)\left(R(1) + T(1) - 1\right)\left(R(1) - T(1) + 1\right)\left(R(1) - T(1) - 1\right) \quad (8.13)$$

One can note that $\alpha\beta = 1$. This discrete approach has given rise to an abundant literature (e.g., Benford, 1923; Gronwall, 1926; Benford, 1946; Tuckerman, 1947; Olf, 1988; Dahm and Dahm, 1999). It can be extended to a continuous distribution where N does not need to be an integer. Now in widespread use in the remote sensing community, the PROSPECT (Leaf Optical Properties Spectra) model was designed this way (Jacquemoud and Baret, 1990). It was among the first radiative transfer codes to accurately simulate the hemispherical reflectance and transmittance of various plant leaves and conditions (monocots, dicots of healthy or senescent leaves) over the solar spectrum from 400 nm to 2500 nm. Several versions of PROSPECT have been made widely available to the research community. They will be described in detail in Chapter 9.

The differences in optical properties of the upper side and lower side of dorsiventral (also called bifacial or asymmetric) leaves have been described in Section 6.1.5. Ma et al. (2007) with QSPECT (Q is for "quatre," four in French, and SPECT is derived from PROSPECT) and Stuckens et al. (2009) with DLM (Dorsiventral Leaf Model) distinguished four layers: upper epidermis, palisade mesophyll, spongy mesophyll, and lower epidermis. In QSPECT, the optical properties of each of the four layers are calculated using PROSPECT with different values for biochemical content and structure, and multiple scattering may occur between each layer (Figure 8.3). In DLM, the mesophyll layers may be tightly connected or separated by a thin layer of air, and two adjacent layers are separated by air spaces for only a fraction of their surface area f_{air} and are connected for the remainder. f_{air} determines the scattering behavior within the leaf and is equivalent to the structure parameter N in PROSPECT.

8.1.2 Compact Spherical Particle Models

None of these models were adapted to needle-shaped leaves, which cannot be treated as discrete plane parallel layers. As part of his doctoral thesis, Terence Dawson adapted Melamed's formulation for light interaction with suspended powders and designed the LIBERTY model (Leaf Incorporating Biochemistry Exhibiting Reflectance and Transmittance Yields) specifically to calculate both dried and fresh slash pine (*Pinus elliottii*) needle optical properties (Dawson et al., 1995, 1998a). Consider discrete particles making a uniform compact layer and reflecting light diffusely according to the laws of geometrical optics. The calculation of the reflectance and transmittance of this layer was carried

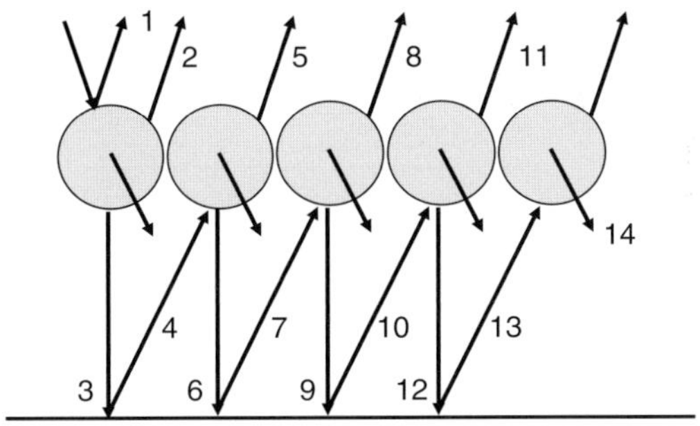

Figure 8.4 Schematic representation of the scattering of light by suspended powders as viewed by Karvaly (1970).

out by Melamed (1963) for spherical particles that are assumed to be large compared to the wavelength. It involves internal reflections within particles and scattering of light between particles in the same relative plane (Figure 8.4; Karvaly, 1970).

8.1.2.1 Interaction of Radiation with One Cell

Application of Lambert's cosine law over the whole volume of an idealized spherical particle yields the following expression for single-pass transmittance:

$$M = \frac{2}{(kd)^2}\left(1 - (kd + 1)\exp(-kd)\right) \tag{8.14}$$

where $k = k(\lambda)$ is the particle absorption coefficient at the wavelength λ and d is the particle diameter. The transmittance of this particle including multiple internal reflections is written as

$$\tau = \frac{(1 - \overline{m}_i)M}{1 - \overline{m}_i M} \tag{8.15}$$

where $\overline{m}_i$ is the average internal reflection coefficient of the particle arising from reflection of photons at the sphere–vacuum interface. It is the Fresnel reflection coefficient (or reflectivity) integrated over all angles up to the critical angle $\alpha_c = \sin^{-1}(1/n)$, with n being the refractive index of the medium.

8.1.2.2 Interaction of Radiation with a Layer of Cells

Consider spheres of equal diameter and arranged in a compact hexagonal form. The reflectance of an infinite number of spherical cell layers, R_∞, including the correction provided by Mandelis et al. (1990), is given by an infinite series summation that converges to

$$R_\infty = 2x\overline{m}_e + x(1 - 2x\overline{m}_e)\tau\left\{\frac{(1 - x)(1 - \overline{m}_e)\tau R_\infty + x(1 - \overline{m}_e R_\infty)}{(1 - \overline{m}_e R_\infty) - (1 - x)(1 - \overline{m}_e)\tau R_\infty}\right\} \tag{8.16}$$

where x is the fraction of radiation scattered in the upward direction, expressed as a fraction of 4π steradians, and $\overline{m}_e$ is the average external reflection coefficient of the particle arising from reflection of photons at the vacuum-sphere interface. The reflectance R_∞ is obtained by solving the quadratic

Eq. 8.16 where the negative root of the discriminant only is retained for physically meaningful values ($R_\infty \leq 1$). Mandelis et al. (1990) propose the following expression for x:

$$x = \frac{x_u}{1 - \left(1 - x_u\left(1 - \exp(-kd)\right)\right)T} \tag{8.17}$$

where x_u represents the probability for diffuse scattering in the upward direction ($0.1<x_u<0.4$). The analytical expressions for the reflection coefficients $\overline{m}_i$ and $\overline{m}_e$ as a function of the refractive index n can be found in Appendix E. Examination of Eq. 8.16 shows that, in the absence of underlying leaf material, the reflectance of a single layer of leaf cells R is equal to

$$R = 2x\overline{m}_e + x(1 - 2x\overline{m}_e)\tau \tag{8.18}$$

Benford (1946) derived the transmittance of a single layer T in terms of the reflectance of a single layer R and the reflectance of infinite layers R_∞ by solving Eq. 8.6:

$$T = \sqrt{\frac{(R_\infty - R)(1 - R_\infty R)}{R_\infty}} \tag{8.19}$$

In summary, by treating the leaf as an aggregation of cells, with multiple radiation scattering between cells, output reflectance and transmittance spectra are a function of three structural parameters (average cell diameter in μm, intercellular air space, needle thickness) and the combined absorption coefficients of leaf biochemicals (chlorophyll content in mg m^{-2}, water content in g m^{-2}, lignin and cellulose content in g m^{-2}, and nitrogen content in g m^{-2}). Di Vittorio (2009) and MacArthur and Malthus (2012) enhanced the LIBERTY model by considering the chlorophylls *a* and *b* separately, by introducing total carotenoid and anthocyanin content, by recalibrating the specific absorption coefficients, and by simplifying some variables. To date, LIBERTY remains the only model used for this purpose. It is simple to use and has been coupled with several vegetation canopy or ecosystem models (see Section 12.4).

Barton (2001) performed a theoretical sensitivity analysis of LIBERTY to test the influence of heterogeneity in chlorophyll distribution on leaf reflectance. He considered a chlorotic conifer leaf displaying a fraction of yellow mesophyll f_1 with a chlorophyll content C_1 and a fraction of green mesophyll f_2 with a chlorophyll content C_2, where $f_1 + f_2 = 1$. Therefore, the mean leaf chlorophyll content is $C = f_1 \times C_1 + f_2 \times C_2$ and its mean reflectance is $C = f_1 \times R_1 + f_2 \times R_2$. One main result of this work was that small yellow areas had a disproportionately large influence on overall leaf reflectance. For instance, a chlorotic leaf with $f_1 = 0.25$ and C_1 has the same reflectance (400–700 nm) as a homogeneous leaf with $f_1 = 1$ and $C_2 = 0.6 \times C_1$. Such simulations have been used to design spectral indices sensitive to pigments (Coops and Stone, 2005) or water (Wang et al., 2009a) content. Recalibration and inversion of the model have also been performed on different kinds of needles to estimate their chemical content: European larch (*Larix deciduas*) (Kobayashi et al., 2002), Norway spruce (*Picea abies*) (Schlerf et al., 2003), and jack pine (*Pinus banksiana*, Moorthy et al., 2003b, 2008). LIBERTY has also been used to study damage caused by sulfuric acid in pine needles (Zhumar and Zaitseva, 2003a, 2003b), by nitrogen deficiency and two infectious agents in Monterey pine (*Pinus radiata*; Coops and Stone, 2005), and by water stress in Dahurian larch (*Larix gmelinii*) (Wang et al., 2009a). Hernández-Clemente et al. (2011) also used it to study photosynthesis under stressful environments in Scots pine

(*Pinus sylvestris*) and Austrian pine (*Pinus nigra*). Finally, although they are based on different criteria, LIBERTY and PROSPECT have been compared by some authors (Moorthy et al., 2003b, 2008; Di Vittorio, 2009; Wang and Ju, 2017).

8.1.3 N-Flux Models

These models derive from the Kubelka–Munk (KM) theory, which considers the leaf as a slab of absorbing (absorption coefficient k) and diffusing (scattering coefficient s) material (Kubelka and Munk, 1931; see Appendix G). Infinite lateral extension and the absence of boundary reflections at the top and the bottom of the slab are first assumed in order to eliminate edge effects. The theory provides simple analytical formulae for the diffuse reflectance and transmittance in terms of k, s, and the thickness d of the medium. We consider hereafter the special case where k and s are independent of depth in the material but the theory can be extended to the case where they are both functions of depth with the same functional dependence (Mandelis and Grossman, 1992). The radiation field inside the slab consists of diffuse fluxes propagating in opposite directions. In the two-flux model, the intensities of the fluxes traveling in the forward $I(x)$ and backward $J(x)$ directions at depth x along the spatial axis and at any wavelength λ are coupled according to a system of two first-order differential equations for ascending and descending fluxes:

$$\begin{cases} \dfrac{dI(x)}{dx} = -(k+s)\ I(x) + s\ J(x) \\ \dfrac{dJ(x)}{dx} = -s\ I(x) + (k+s)\ J(x) \end{cases} \tag{8.20}$$

Note that if $s = 0$ (there is no scattering), the differential equations degenerate into the Beer–Lambert law (see Appendix F). Here the incident flux is designed $I_0 = I(0)$ and considered unity. The reflectance R and transmittance T of the slab are given by the boundary conditions $R = J(0)$ and $T = I(d)$:

$$\frac{J(0)}{b^d - b^{-d}} = \frac{I(d)}{a - a^{-1}} = \frac{1}{ab^d - a^{-1}b^{-d}} \tag{8.21}$$

where a and b can be written as a function of k and s. The similarity between this expression and the solution of the generalized plate model in Eq. 8.9 is worthwhile. In the more rigorous four-flux model, radiation within the medium consists of two collimated and two diffuse fluxes propagating in opposite directions (Figure 8.5).

The flow of total fluxes can be written in differential form as

$$\begin{cases} \dfrac{dI_c}{dx} = -(k+s)\ I_c + s\ J_c \\ \dfrac{dJ_c}{dx} = -s\ I_c + (k+s)\ J_c \\ \dfrac{dI_d}{dx} = -2\Big(k + s(1-f)\Big)\ I_d + 2s(1-f)\ J_d + sfI_c \\ \dfrac{dI_d}{dx} = 2\Big(k + s(1-f)\Big)\ J_d - 2s(1-f)I_d - s(1-f)I_c \end{cases} \tag{8.22}$$

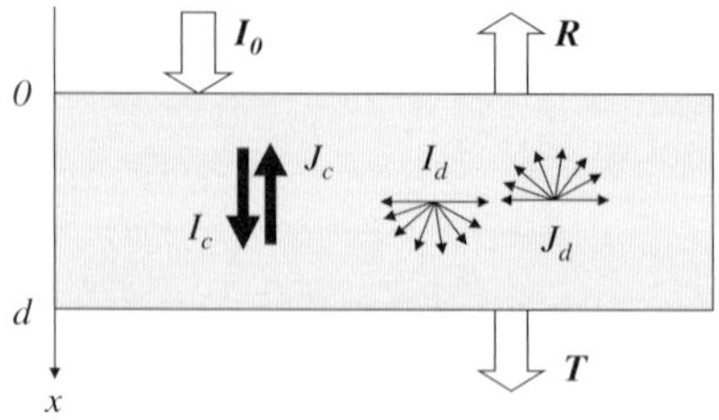

Figure 8.5 Four-flux Kubelka–Munk model.

where I_c and J_c are the upward and downward collimated radiant fluxes, I_d and J_d are the corresponding diffuse fluxes, and f is the asymmetry factor that describes the anisotropy of scattering. From the solutions of the previous system, we obtain collimated-collimated, collimated-diffuse, and diffuse-diffuse components of reflectance and transmittance. According to Vargas and Niklasson (1997), the four-flux model compares well with numerical solutions of the transfer radiative equation or with highly accurate Monte Carlo simulations. This is not surprising since the N-flux equations are a simplification of the radiative transfer theory. Despite their simplicity, two-flux models (e.g., Allen and Richardson, 1968; Cordon and Lagorio, 2007a; Terashima et al., 2009) and four-flux models (Fukshansky et al., 1991; Martinez von Remisowsky et al., 1992; Richter and Fukshansky, 1996a) have been successfully used in the forward mode to calculate the optical parameters of plant leaves. Yamada and Fujimura (1991) proposed a more sophisticated version in which the leaf was divided into four parallel layers: the upper cuticle, the palisade parenchyma, the spongy mesophyll, and the lower cuticle. The KM theory was applied with different parameters in each layer, and solutions were coupled with suitable boundary conditions to provide the leaf reflectance and transmittance as a function of the scattering and absorption coefficients. The leaf reflectance and transmittance were determined in the visible region and interpreted in terms of chlorophyll content and the four-layer model became a nondestructive method for the measurement of photosynthetic pigments by inversion. This approach was validated on leaves of both monocots – rice (*Oryza sativa*) and dragontree (*Dracaena draco*) – and dicots – hortensia (*Hydrangea* sp.) and chinquapin (*Chrysolepis* Hjelmqvist). But Yamada and Fujimura went further, reproducing the difference between reflectance from above and below, which is especially significant in dicotyledonous leaves. Terashima et al. (2009) applied the KM theory to reconstruct the light environment and absorption profile of camellia (*Camellia japonica*) leaves to study leaf photosynthesis. They showed a good agreement between the modeled and measured light absorption gradients, the latter being derived from experimental data. The full leaf biochemistry was introduced by Conel et al. (1993a, 1993b), who used a two-flux model to study the influence of water, protein, cellulose, lignin, and starch on leaf middle infrared reflectance. Unfortunately, they did not validate their model. A very simple model, directly derived from the expression of the reflectance, was used to estimate the chlorophyll content of wheat leaves (Andrieu et al., 1988). Finally, note that the parameter N in the generalized plate model, which describes the leaf internal structure, plays a role similar to that of the scattering coefficients in the KM model. One of the disadvantages of such approaches is that consideration of multiple scattering by mesophyll objects, the size of which are comparable to the wavelength of incident radiation (cells, organelles, air bubbles, and others), is impossible to characterize.

8.1.4 Radiative Transfer Equation

Propagation of light in plant leaves can, in principle, be described using fundamental electromagnetic theory. In this case, the leaf is considered a random medium with a spatially varying permittivity ε_r, and variations in the electric field can be described using the Maxwell theory. However, such a treatment would be troublesome because of the complexity of the problem and a lack of precise knowledge of ε_r. Instead the problem can be simplified by considering only the flow of energy through the medium. Although the radiative transfer equation was developed for three-dimensional media, the propagation of light along one direction will be considered hereafter. Let $I(x, \mu)$ be the intensity of monochromatic radiation at a position x in the interior of the leaf, traveling in a small volume element $dx \times ds$, in the direction θ. We design $\mu = \cos\theta$ and $\mu' = \cos\theta'$, which are the axial cosines of the angle of light propagation before and after scattering, respectively. On the one hand, radiation is lost by absorption and diffusion as it propagates into the medium:

$$\mu \frac{dI(x, \mu)}{dx} = -\sigma_e(x) I(x, \mu) \tag{8.23}$$

where $\sigma_e(x)$ is the extinction coefficient of the medium defined as $\sigma_e(x) = \sigma_k(x) + \sigma_s(x)$. Equation 8.23 is nothing but the famous Beer–Lambert law (see Appendix F). On the other hand, $I(x)$ is intensified by two processes: coherent single scattering emerging from other directions and thermal emission of the medium, at the local thermodynamic equilibrium. The medium includes inhomogeneities in the form of small particles that scatter the radiation beam in all directions. The scattering is called coherent because the scattered radiation has the same wavelength as the incident radiation. The distribution of scattered radiation over all the directions is characterized by a phase function $P(\mu, \mu')$ that is normalized to 1:

$$\frac{1}{4\pi} \int_{4\pi} P(\mu, \mu') d\Omega' = 1 \tag{8.24}$$

$P(\mu, \mu')$ is the probability that the radiation falling in the direction μ of incidence will be scattered in the direction μ' of observation. Consequently, the multiple scattering source function that corresponds to radiation falling on a volume element of the medium from all directions is equal to

$$\frac{1}{4\pi} \sigma_s(x) \int_{4\pi} P(\mu, \mu') I(x, \mu) d\Omega' \tag{8.25}$$

The thermal emission of this small volume element is expressed in terms of the Planck function for the intensity of radiation of an ideal blackbody:

$$\sigma_e(x)\ M\Big(T(x)\Big) \tag{8.26}$$

where $M\Big(T(x)\Big)$ is given in Appendix C. Then the source function $J(x, \mu)$ is the sum of these two processes:

$$J(x, \mu) = \frac{1}{4\pi} \sigma_s(x) \int_{4\pi} P(\mu, \mu')\ I(x, \mu) d\Omega' + \sigma_e(x)\ M\Big(T(x)\Big) \tag{8.27}$$

The radiative transfer equation that expresses the infinitesimal variation of flux dI in the infinitesimal slab at position x is an integro-differential transport equation:

$$\mu \frac{dI(x,\mu)}{dx} = -\sigma_e(x)\ I(x,\mu)\ + \frac{1}{4\pi}\sigma_s(x)\int_{4\pi} P(\mu,\mu')\ I(x,\mu)d\Omega' + \sigma_e(x)\ M\Big(T(x)\Big) \tag{8.28}$$

In the solar domain from 400 nm to 2500 nm, there is no thermal emission but intense scattering, so that Eq. 8.28 simplifies to

$$\mu \frac{dI(x,\mu)}{dx} = -\sigma_e(x)\ I(x,\mu) + \frac{1}{4\pi}\sigma_s(x)\int_{4\pi} P(\mu,\mu')I(x,\mu)d\Omega' \tag{8.29}$$

There are very few cases in which such integro-differential equations can be directly solved. The most famous solution was derived for the case of isotropic scattering by Chandrasekhar (1960). Iterative approaches for numerically calculating transport quantities are generally required. Solving Eq. 8.29 with boundary conditions provides the diffuse reflectance and transmittance of the medium. Compared with canopy level, only a few models directly use the radiative transfer equation at leaf level. The mathematical complexity of the algorithms, and the poor information we have on leaf internal structure and biochemical distribution, lead to strong simplifications which make such an approach less efficient as compared to more robust formulations. As noted by Krekov et al. (2009), the mesophyll can be considered as a polydisperse structure of quasi-spherical cells in air or as a polydisperse system of air bubbles in liquid. Because of the high water content of fresh leaves, Ma et al. (1990) modeled the leaf as a slab of water with an irregular surface containing randomly distributed spherical scatterers, the size distribution of which is Gaussian (Figure 8.6).

Radiation transfer theory and Kirchhoff rough surface scattering theory was applied to obtain the reflectance and transmittance of the leaf. The model was compared with bidirectional measurements made on maize (*Zea mays*), potato (*Solanum tuberosum*), and cherry laurel (*Prunus laurocerasus*) leaves using an He-Ne laser operating at 632.8 nm, for normal illumination. Different configurations of polarization were tested with success on these leaves. In LEAFMOD (Leaf Experimental Absorptivity Feasibility MODel), the leaf is compared to a homogeneous mixture of biochemicals which scatter and absorb light (Ganapol et al., 1998). Absorption σ_k and scattering σ_s coefficients were calculated using reflectance and transmittance spectra simulated with PROSPECT, and reflectance spectra of stacks of hollyleaf cherry (*Prunus lyonii*) leaves measured at different water stages. Following the exploratory study of Hanrahan and Krueger (1993) and Franzke and Deussen (2003) for computer graphics

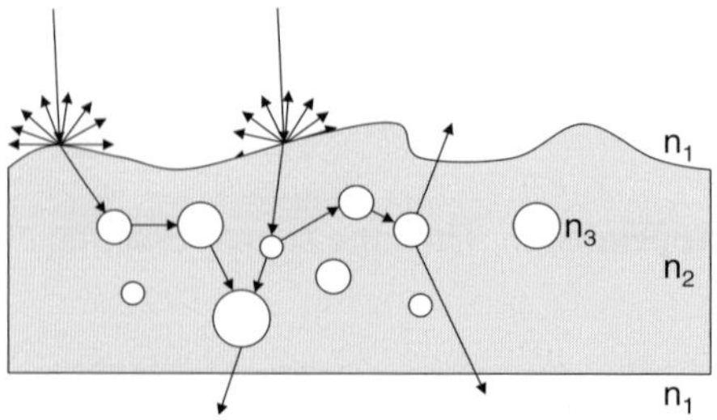

Figure 8.6 Geometry of the optical scattering model.

applications, Wang et al. (2005) also used LEAFMOD to calculate subsurface scattering inside leaf tissues. In order to better represent the differentiated mesophyll of dicots (see Section 2.1), Hemenger (1977) developed a two-layer model that accurately accounts for the non-uniformity of the spongy mesophyll. These authors showed an interesting and important property of leaves, that they absorb green light much more effectively than extracted pigments. Berdnik and Mukhamed'yarov (2001) also proposed a transport model considering two homogeneous layers differing in their concentrations of absorbing dyes and scattering particles. The first layer, that is, the palisade parenchyma, was made of small spherical particles (average radius of 2 μm) similar in volume to chloroplasts and having a refractive index of 1.42 (see Section 3.4). The layer below, that is, the spongy mesophyll, was made of large spherical particles full of air (average radius of 9.5 μm) and then having a refractive index of 1. The refractive index of the surrounding medium in both layers was set to 1.36 as water, and the size distribution of particles was described by a lognormal distribution. The novelty of this paper lies in the introduction of the Mie theory to calculate the scattering cross-section of the particles, the size of which is comparable to the wavelength of incident radiation. The transport equation in the one-dimensional approximation is solved by an original method of δ anisotropy separation in the zeroth azimuthal harmonic of the scattering indicatrix (Berdnik and Loiko, 1999). Experimental values of reflectance and transmittance obtained on Chinese holly (*Ilex cornuta*) leaves showed good agreement with the model, both for illumination from above and below. As mentioned above, the Mie theory can be used to approximate tissue scattering at the cellular level by assuming cells are homogeneous spheres of a single size. While it has been successful in describing light propagation in bulk tissue, it cannot describe the light interaction with complex structures such as individual cells of arbitrary shape containing multiple organelles. To do this Dunn and Richards-Kortum (1996) used the finite-difference time-domain (FDTD) technique, a powerful computational method that is a full vector, three-dimensional solution of Maxwell's equations. Zakharov et al. (2009) developed a three-dimensional mathematical model of interactions of optical radiation with plant tissue based on the Monte Carlo method. It is difficult to use this method with plant leaves because considerable calculation times are required due to mesophyll structural complexity. The multiple scattering in plant tissues can be taken into account by using the P_L approximation in which the radiation intensity is approximated using a spherical harmonic expansion (Donner and Jensen, 2005; Zakharov et al., 2010). In order to achieve real-time performance and faithfully represent translucency into leaves, light transport models through a leaf based on an analytical diffusion dipole approximation (Jensen et al., 2001; Jensen and Buhler, 2002) have been implemented (Habel et al., 2007; Habel, 2009, 2010). Indeed, while subsurface scattering has negligible impact on the visual appearance of the leaf when it is oriented in the direction of the Sun, it is the dominant factor when it is oriented in the opposite direction.

8.1.5 Stochastic Approach

Tucker and Garratt (1977) proposed an original model, LFMOD1, where the radiation transfer is simulated by a Markov chain, which is a stochastic process. As seen in Figure 8.7, they partitioned a boxelder maple (*Acer negundo*) leaf into two independent tissues (palisade parenchyma and spongy mesophyll) and defined eight types of photon states (incident solar radiation, specularly reflected radiation, diffuse reflected radiation, diffuse transmitted radiation, absorbed, and scattered radiation in each tissue). The random variable photon state can only take discrete values, like "absorbed in the palisade parenchyma" or "scattered in the spongy mesophyll."

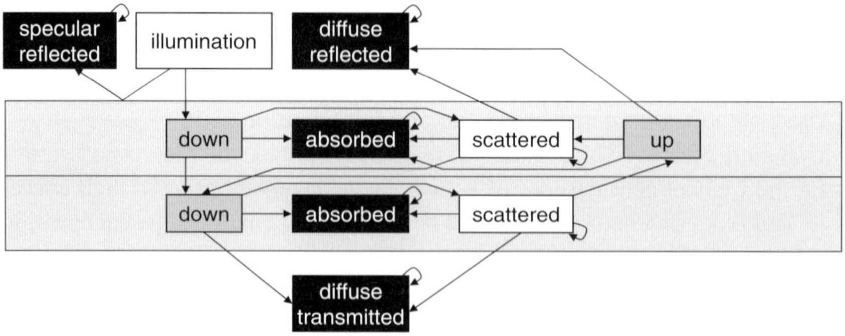

Figure 8.7 Representation of a typical dicotyledon leaf in LFMOD1. The boxes figure the different photon states. The black ones show the absorbing, the white ones show the non-absorbing, and the gray ones show the illumination states. The arrows show possible transitions (redrawn from Tucker and Garrett (1977)).

The system is then totally described by ten compartments connected by arrows that correspond to single-step transition probabilities from one photon state i to another state j between two compartments:

$$p_{ij} = \Pr(X_1 = j \mid X_0 = i) \tag{8.30}$$

The transition probabilities p_{ij} are most conveniently represented in a square array $\mathbf{P}$ of dimension 10×10 called transition matrix, in which each row sums to one. They are set on the basis of the optical properties of the leaf material. The probabilities for the occurrence of these states are described by the elements of a column vector $\mathbf{p}_i$ called the state vector. Starting with an initial state vector $\mathbf{p}_0$ representing the incident radiation, the next state vector is obtained as $\mathbf{p}_1 = \mathbf{P}\,\mathbf{p}_0$ and after n steps we obtain

$$\mathbf{p}_n = \mathbf{P}^n\,\mathbf{p}_0 \tag{8.31}$$

In the limit, the steady state that yields both the reflectance and transmittance is defined to be

$$\mathbf{p} = \lim_{n \to \infty} \mathbf{P}^n\,\mathbf{p}_0 \tag{8.32}$$

Tucker (1980) used the LFMOD1 model to predict leaf reflectance spectra as a function of the leaf equivalent water thickness. The SLOP (Stochastic model for Leaf Optical Properties) model (Lüdeker and Günther, 1990; Maier et al., 1999; Maier, 2000) is an improved version of the stochastic model, which differs in that the leaf is partitioned into four different tissues instead of two used in the original work of Tucker and Garratt (1977). It implies introduction of additional photon states (Figure 8.8).

8.1.6 Ray Tracing Models

Previous leaf optical properties models can impressively simulate the reflectance and transmittance over the solar spectrum, but they are not sophisticated enough to describe, for instance, the absorption profile of light inside the leaf or the bidirectional reflectance at the blade surface. Among various approaches, only ray tracing techniques can account for the complexity of internal leaf structure as it appears in a photomicrograph. They require a detailed description of individual cells and their unique arrangement

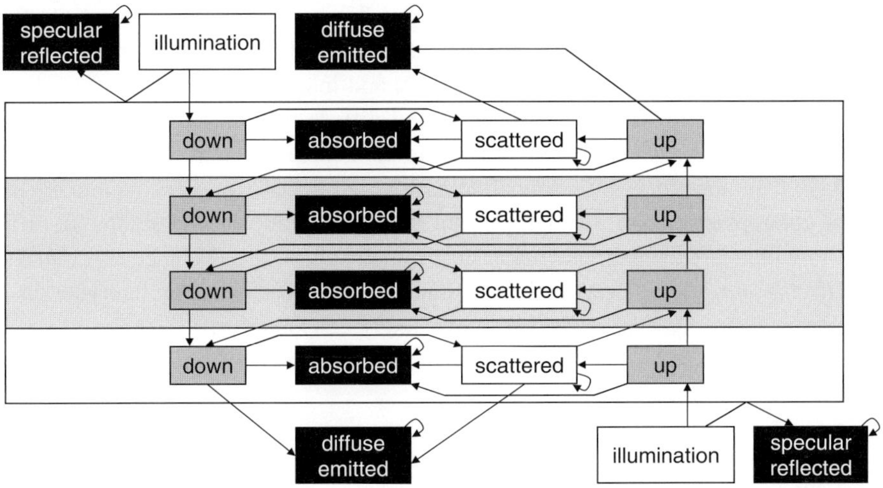

Figure 8.8 Representation of a typical dicotyledon leaf in SLOP. The boxes calculate the different photon states. The black ones are absorbing, the white ones are non-absorbing, and the gray ones are the illumination states. The arrows show possible transitions (redrawn from Maier et al. (1999)).

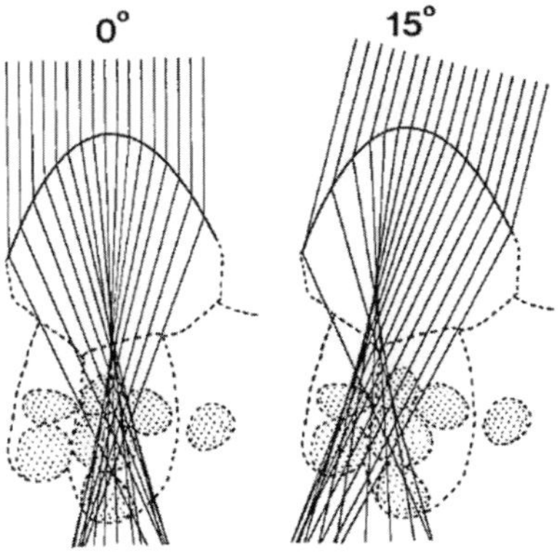

Figure 8.9 Ray diagram showing light refraction by leaf epidermal cells of Queen Anthurium (*Anthurium warocqueanum*) at zenith angles of 0° and 15° (from Bone et al. (1985), *Applied Optics*. Copyright © 1985 with permission from The Optical Society of America).

inside tissues. The optical constants of leaf materials (cell walls, cytoplasm, pigments, air cavities, etc.) also have to be defined. Using the laws of reflection, refraction, and absorption, it is then possible to simulate the propagation of individual photons incident on the leaf surface. Once a sufficient number of rays have been simulated, statistically valid estimates of the radiation transfer in a leaf may be deduced. The technique has been applied with a number of variants. The first studies were performed at the cell level: Senn (1908), Haberlandt (1914), and more recently Gabrys-Mizera (1976) and Bone et al. (1985) constructed geometrical models of light transmission through the cross-sections of plant cells of various shapes, in particular epidermal cells, the shape of which might influence the path of the incident beams (Figure 8.9).

Research efforts have also been directed toward understanding the interaction of light with two-dimensional models of entire leaves. Willstätter and Stoll (1918) are often cited as the first authors who proposed a theory of leaf optics on the basis of critical reflection of light at cell wall–air interfaces. Allen et al. (1973) built an albino boxelder maple (*Acer negundo*) leaf consisting of two medial cell walls characterized by their indexes of refraction ($n = 1.52$) and intercellular space air ($n = 1.00$). A 100 circular arcs model was used to test the specular and diffuse nature of the reflection. Simulations performed at 1 μm where absorption is weak led to an underestimation of the reflectance and an overestimation of the transmittance. Shortly afterward, by adding two more media into the model (cell sap, $n = 1.36$; chloroplasts, $n = 1.42$), Kumar and Silva (1973) increased the internal diffusion and could predict to within 2% the experimentally determined values of reflectance and transmittance of a soybean (*Glycine max*) leaf. However, they found that the cell wall–air interface caused more deviation of the incident ray than any other single interface (Figure 8.10). They did not, however, investigate how their four-media optical model would diffuse or distribute the transmitted light. Brakke

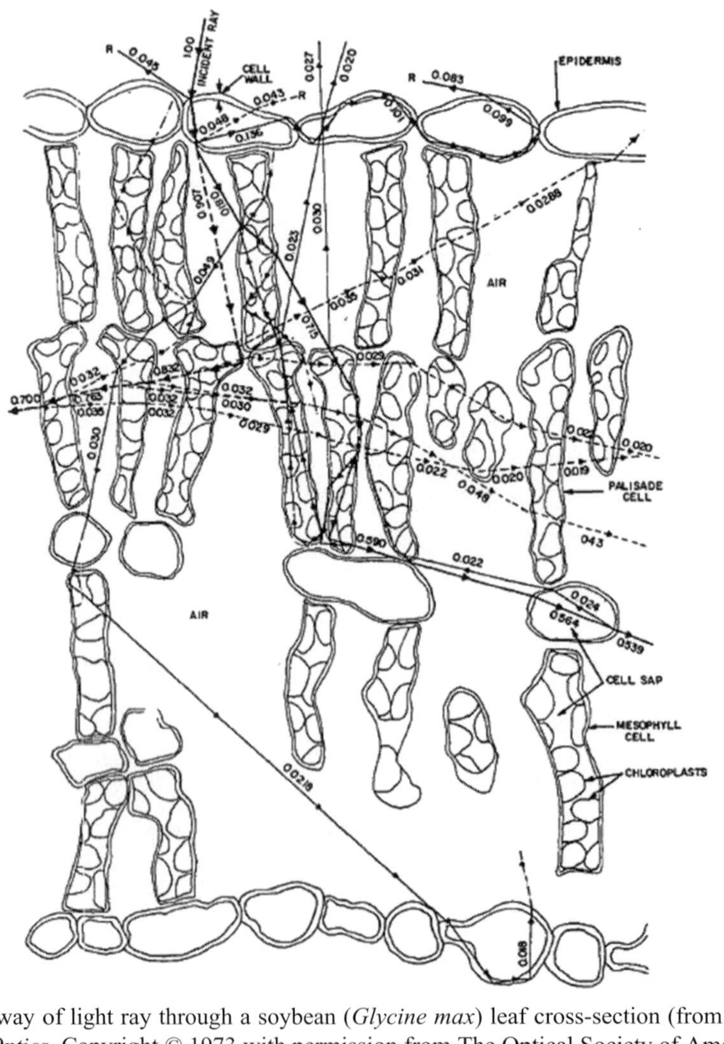

Figure 8.10 Pathway of light ray through a soybean (*Glycine max*) leaf cross-section (from Kumar and Silva (1973), *Applied Optics*. Copyright © 1973 with permission from The Optical Society of America).

and Smith (1987) incorporated more realistic representations of leaf anatomy by digitizing the geometrical structure of the maple leaf of Allen et al. (1973); Kumar and Silva (1973), and Nicoletti and Adams (2000) did the same with soybean (*Glycine max*) and rice (*Oryza sativa*) leaves, respectively. The first authors also attempted to model the red pine (*Pinus resinosa*) needle described in Westman and Price (1988) the year after. A needle-shaped leaf, which displays increased geometry compared to a lamina-shaped one, required about 600 arcs.

Scattering mainly occurs at the cell wall air boundaries. Mie scattering at the chloroplasts is negligible as they have almost the same refractive index as the surrounding cytoplasm (e.g., Fukshansky et al., 1993). Since the cell sizes are much larger than the wavelength in the modeled spectral region, scattering is treated independently of the wavelength. Regardless of the approach, the absorption phenomena that characterize leaf optical properties outside the near-infrared plateau have been ignored. Moreover, in all these models, leaves were always described as two-dimensional objects, although the three-dimensional structure of these organs is very important to their physiological function. For this reason, Govaerts et al. (1996) and Jacquemoud et al. (1997) used a three-dimensional ray tracing model, RAYTRAN (Govaerts and Verstraete, 1998), on a virtual three-dimensional dorsiventral leaf, to characterize the light environment, including absorption, scattering, and transmission, within and between cells. Cells of variable size, cell wall thicknesses, chemistry, and air spaces were modeled (Figure 8.11) and implications for absorption profiles, light harvesting, and photosynthesis were successfully investigated (Ustin et al., 2001).

Xiao et al. (2016) drew on the work of Govaerts et al. (1996) to develop three-dimensional leaf models that can be used to evaluate the internal light environment of a leaf and its implications for photosynthesis. Kallel (2018) recently introduced light polarization in Monte Carlo ray tracing simulations of leaf bidirectional reflectance distribution function (BRDF). For applications in computer graphics, Baranoski

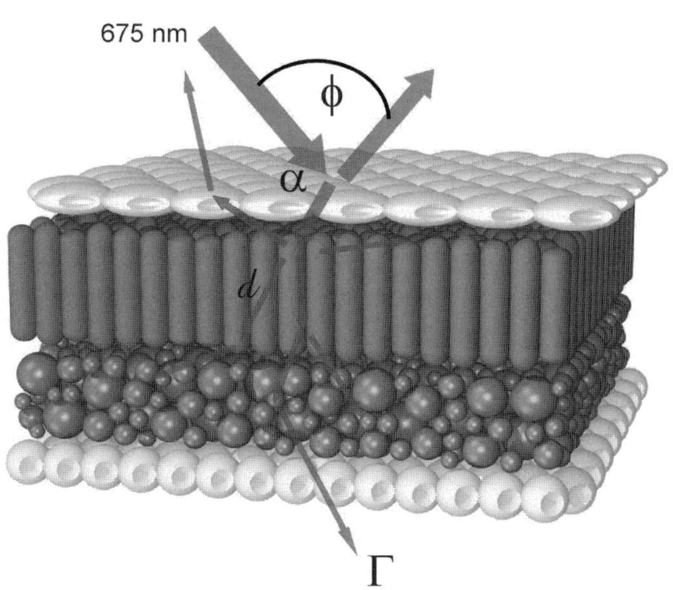

Figure 8.11 Perspective view of a virtual bifacial dicotyledon leaf illuminated at 675 nm. The cross-section of the target leaf is 300 mm × 300 mm × 170 mm (from Ustin et al. (2001), *Plant Cell & Environment*. Copyright © 2001 with permission from Blackwell Publishing).

and Rokne (1997, 1999) proposed a model called ABM (Algorithmic BDF Model), which takes into account biologically meaningful parameters that affect light scattering and absorption in leaves. Such an approach was new in this research community. Instead of modeling cells individually as in RAYTRAN, the propagation of light within the leaf tissues is simulated as a stochastic process where the transition probabilities are driven by refraction at each interface (Fresnel reflection coefficients) and absorption in each tissue (Beer–Lambert law). These authors successfully simulated the BRDF, Bidirectional transmittance distribution function (BTDF), Directional-hemispherical reflectance factor (DHRF), and Directional-hemispherical transmittance factor (DHTF) of a soybean (*Glycine max*) leaf but ABM seemed computationally expensive. They later introduced the FSM (foliar scattering model) model, which gains efficiency over ABM by pre-computing reflectance and transmittance values and applying a simplified scattering model (Baranoski and Rokne, 2001). Both the ABM and FSM models are based on Monte Carlo ray tracing. These models have been recently adapted to better account for leaf complexity. Baranoski (2006) developed the ABM-U and ABM-B to investigate the interaction of light with unifacial and bifacial leaves that are characteristic of monocots and dicots, respectively. Finally, Baranoski and Eng (2007) introduced the sieve effect that affects the optical properties of heterogeneous media.

8.1.7 Summary

Other approaches have been tested like the domain wall model of Brito and Freire (2006), in which light neutral particles play the role of electromagnetic radiation and the domain wall with internal structures plays the role of a plant leaf, but results are unconvincing. Contrary to accepted dogma, much more work is required before we will completely understand leaf diffuse optical properties. This knowledge is nevertheless crucial to develop more accurate relationships between these properties and important leaf functional characteristics, or to improve models which are directly used to interpret remote sensing data when coupled with canopy reflectance models. More understanding of the relationships between structure and function at the leaf level are clearly needed to drive model improvements. Better optical characterization of more biochemical compounds is also needed to expand the range of biochemistry that can be detected. To give an example, the separation of all leaf photosynthetic pigments (chlorophyll *a*, *b*, carotenoids, and xanthophylls) is still an issue and yet this information would significantly improve understanding of the biological controls on photosynthesis. Additionally, other aspects of leaf optics like fluorescence are not developed here but provide critical information about photosynthetic function. Only a few models are formulated to specifically include the chlorophyll fluorescence signal. Progress on the next generation of optical models requires improvements in understanding detailed cell and leaf anatomy for leaves having adaptations to different environmental conditions.

8.2 Different Approaches for Leaf Fluorescence

As seen in Section 5.2, steady-state chlorophyll fluorescence (ChlF) spectroscopy has proved to be a promising tool to effectively assess photosynthetic rates and to monitor stress conditions of plants. However, it is an intricate response of environmental and physiological factors so that its application in plant physiology may be limited as a diagnostic tool, because the measured spectrum is distorted by the absorption and scattering properties of the tissue. Attempts to retrieve the intrinsic fluorescence, that is, to remove these effects, have been made by using radiative transfer models or

Table 8.2 *Categorization of leaf ChlF models.*

Type	Name	Fluorescence
Beer–Lambert law		Stahl et al. (1989), Agati et al. (1993), Nauš et al. (1993, 1994), Ounis et al. (2001)
Plate models	FLUORMODLEAF	Pedrós et al. (2010), Zhao et al. (2015)
	FLUSPECT	Verhoef (2011), Verrelst et al. (2015)
N-flux models	FRT	Fukshansky and Kazarinova (1980), Rosema et al. (1991), Zarco-Tejada et al. (2000), Ramos and Lagorio (2004)
Stochastic models	SLOP	Maier (2000)
Monte Carlo algorithms		Sušila and Nauš (2007), Zakharov et al. (2009, 2010)

approximate solutions of these models that describe light propagation in a turbid medium. Note that most fluorescence models have been developed for other applications than plant physiology, like medical diagnosis of human diseases (e.g., Richards-Kortum et al., 1989; Wu et al., 1993; Gardner et al., 1996; Welch et al., 1997) or quality color control for the paper industry (e.g., Emmel, 2003), where fluorescence spectroscopy is widely used. Some of them have been adapted to fluorescence of plant tissues (Table 8.2). It is important to note that the correctness of these models for simulating reflectance, transmittance, and upward and downward fluorescence, in the same package is crucial.

8.2.1 Beer–Lambert Attenuation Law

Although it is not suited to simulate reflectance, the Beer–Lambert attenuation law may provide reasonable light profiles within the blade, especially when the optical properties are dominated by absorption (e.g., Vogelmann and Evans, 2002; Evans and Vogelmann, 2006). It has also been used to calculate leaf fluorescence in the case of a compact leaf (e.g., Stahl et al., 1989; Agati et al., 1993; Nauš et al., 1993, 1994; Ounis et al., 2001). Consider a semi-infinite plate with defined thickness and homogeneously distributed chlorophyll content, illuminated by incident light I_0. We define an infinitesimal vertical layer dx inside the plate, located at the distance x from the plate surface and receiving the flux $I(x)$ (Figure 8.12).

The intensity of light at this distance x is given by (see Appendix F)

$$I(\lambda_{ex}, x) = I_0(\lambda_{ex}, 0)\exp\Big(-\alpha(\lambda_{ex})x\Big) \tag{8.33}$$

where $I_0(\lambda_{ex}, 0)$ is the intensity of light at the leaf surface and $\alpha(\lambda_{ex})$ is the absorption coefficient (m^{-1}) of the leaf at the excitation wavelength. The fluorescence emission $df(\lambda_{ex}, \lambda_{em}, x)$ in the infinitesimal layer dx is proportional to the amount of photons absorbed $da(\lambda_{ex}, x)$ in this layer. It can be written as

$$df(\lambda_{ex}, \lambda_{em}, x) = \phi(\lambda_{ex}, \lambda_{em})da(\lambda_{ex}, x) \tag{8.34}$$

where λ_{ex} and λ_{em} are the excitation and emission wavelengths, respectively, and $\phi(\lambda_{ex}, \lambda_{em})$ is a source function. The latter, also called spectral fluorescence yield (SFY) of the elementary layer, is defined as the probability that one photon absorbed by the medium at λ_{ex} be emitted at λ_{em}. In the

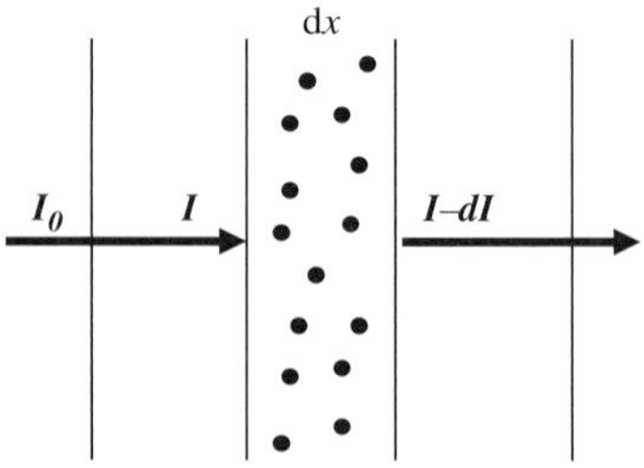

Figure 8.12 Absorption and emission in an infinitely thin fluorescent layer.

case of a single fluorophore, it can be written as the product of: i) the excitation efficiency spectrum $\xi(\lambda_{ex})$; ii) the fluorescence quantum yield Q, ratio of the number of photons emitted to the number of photons absorbed; and iii) the fluorescence emission spectrum $\eta(\lambda_{em})$ normalized to unity so that $\int_0^{\infty} \eta(\lambda_{em}) d\lambda_{em} = 1$.

$$\phi(\lambda_{ex}, \lambda_{em}) = \xi(\lambda_{ex}) Q \, \eta(\lambda_{em}) \tag{8.35}$$

The difficulty of this formalism lies in the definition of three terms. The downward fluorescence signal that can be detected is obtained as the integral of $df(\lambda_{ex}, \lambda_{em}, x)$ over the whole plate. If the reabsorption of light emitted by chlorophyll fluorescence is theoretically accounted for in this approach, the influence of light scattering is only considered by Agati et al. (1993).

8.2.2 Plate Models

In the scope of the FluorMOD project, funded by the European Space Agency in preparation for the FLEX (FLuorescence EXplorer) mission, an integrated leaf-canopy fluorescence model was developed (Miller et al., 2005). The leaf-level model called FLUORMODLEAF, which simulates forward and backward chlorophyll *a* fluorescence spectra of leaf exposed to normally incident excitation light, is based on the PROSPECT-5 model (Pedrós et al., 2010). Besides the five input parameters of PROSPECT, it uses the relative absorption cross-section of photosystem I (PSI) and photosystem II (PSII) and fluorescence lifetimes of PSI and PSII as inputs (Pedrós et al., 2008). Like PROSPECT, it assumes that a leaf is modeled by a stack of N homogeneous absorbing plates with rough surfaces, except that the intra-plate medium is fluorescent. Scattering is accounted for by internal reflections at the interface between the medium and the surrounding air. The way Pedrós et al. (2010) calculate the fluorescence is elegant and powerful. Consider an elementary layer of thickness D (by convention $D = 1$) including an infinitesimal layer of thickness dx, parallel to the surface, at position x (Figure 8.13).

Solving the system of linear equations relating all the fluxes allows calculating the absorbed flux da along dx in the first instance, and then the infinitesimal upward df_u and downward df_d fluorescence fluxes. The upward and downward fluorescence signals $F_u(1)$ and $F_d(1)$ of the whole plate are then obtained as the integral of df_u and df_d from $x = 0$ to $x = 1$. Consider a stack of N fluorescent plates, the upward and downward fluorescence signals of which are $F_u(N)$ and $F_d(N)$. If a new plate is added at the bottom of the stack, $F_u(N+1)$ and $F_d(N+1)$ can be recursively computed. Zhao et al. (2015) implemented the EFAST (Extended Fourier amplitude sensitivity test) method to perform

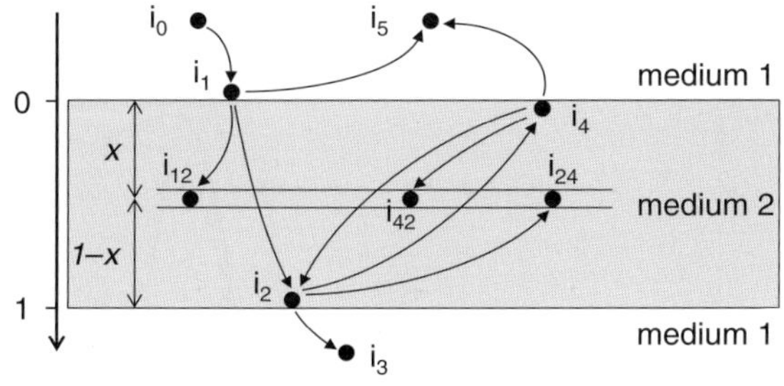

Figure 8.13 Flux network showing the excitation fluxes at position x.

a sensitivity analysis of FLUORMODLEAF. They showed that the chlorophyll content and the fluorescence lifetimes of PSI and PSII were the most sensitive parameters of the model; they could also retrieve the fluorescence parameters of wheat (*Triticum* sp.), soybean (*Triticum* sp.), cotton (*Gossypium hirsutum*), and peanut (*Arachis hypogaea*) leaves by inversion of the model using a Bayesian approach.

8.2.3 N-Flux Models

Because the optical properties of most media are dominated both by scattering and absorption, Allen (1964) modified the KM equations to include fluorescence. Radiation within the medium consists of two diffuse fluxes propagating in opposite directions. The flow of total diffuse flux across a horizontal slab of thickness dz at any wavelength λ can be written in differential form as

$$\begin{cases} \dfrac{d(I+F^{+})}{dz} = -(k+s)(I+F^{+}) + s\,(J+F^{-}) + \dfrac{1}{2}\,P \\ -\dfrac{d(J+F^{-})}{dz} = s\,(I+F^{+}) - (k+s)\,(J+F^{-}) + \dfrac{1}{2}\,P \end{cases} \tag{8.36}$$

where s and k are the linear scattering and absorption coefficients for diffuse light (m^{-1}), respectively, I and J are the upward and downward radiant fluxes at depth z, respectively, F^{+} and F^{-} are the upward and downward fluorescence fluxes at depth z, respectively, and P is the fluorescence emission flux (assumed to be isotropic) at depth z. The latter assumes that all absorbed photons of I and J in the photosynthetic active radiation (PAR) region contribute to the excitation of the photosystems and is defined as integration over the wavelength range 400–700 nm:

$$P(z) = \xi\,\eta \int_{400}^{700} k(I+J)d\lambda \tag{8.37}$$

where ξ is the photon fluorescence efficiency, that is, the fraction of absorbed upward and downward flux that contributes to fluorescence excitation, and η is the fluorescence emission spectral distribution function. Fukshansky and Kazarinova (1980), Bonham (1986), Shakespeare and Shakespeare (2003), and Ramos and Lagorio (2004) solved the extended KM equations by successive approximations in an analytical way. Rosema et al. (1991) and later Zarco-Tejada et al. (2000) preferred

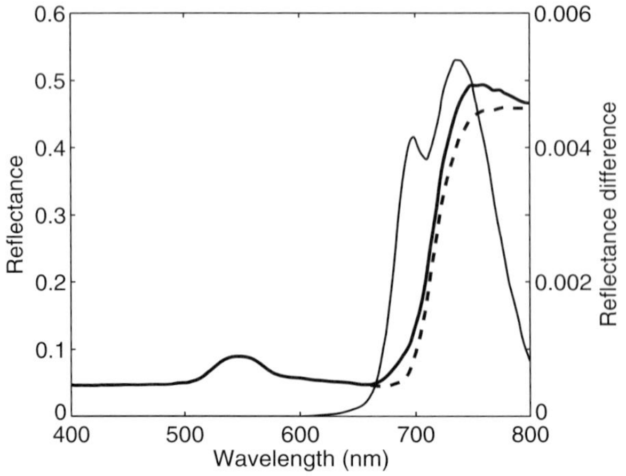

Figure 8.14 Leaf reflectance with fluorescence (plain thick line) and without fluorescence (dashed thick line) simulated by the FRT (Fluorescence-Reflectance-Transmittance) model ($\phi = 0.085$, $C_{ab} = 50$ µg cm^{-2}, leaf thickness = 0.075 mm). The fine dark line indicates the difference between the two spectra (redrawn from Zarco-Tejada et al. (2000)).

a numerical way called the adding-doubling method. Finally, Emmel and Hersch (1998) and Emmel (2000) proposed a matrix method. Of course, all the methods should provide the same results. Since plant leaves cannot be considered as homogeneous layers due to the existence of differentiated tissues (epidermis, palisade, and spongy mesophyll), the matrix formulation provided by Yamada and Fujimura (1991) was used by Zarco-Tejada et al. (2000) to improve simulations. These authors assumed the leaf to be represented as a stack of three layers – top epidermal layer, compact inner layer containing the chloroplasts and cellular material, and lower epidermal layer – and calculated its apparent reflectance with the superimposed effects of fluorescence (Figure 8.14).

Two-flux approximation may fail if scattering is strongly anisotropic and weak. The FLUSPECT model is an extension of the PROSPECT simulating leaf ChlF with fluorescence quantum efficiencies of photosystem I (PSI) and photosystem II (PSII) as inputs (Verhoef, 2011; Verrelst et al., 2015; Vilfan et al., 2016, 2018). In this model, PROSPECT is used to calculate the reflectance ρ and the transmittance τ of a single plate, from which the absorption k and backscattering s coefficients of the KM model are derived. Then the authors apply the efficient adding-doubling method to solve Eq. 8.36 and derive fluorescence emission in forward and backward directions.

8.2.4 Stochastic Approach

Little information is available on the implementation of chlorophyll fluorescence in SLOPE (Maier, 2000). The radiative transfer has been modeled in a wavelength sequential manner, starting with the shortest, because chlorophyll fluorescence causes an energy transfer from shorter to longer wavelengths. After the absorption in one wavelength channel has been calculated, the number of absorbed photons in each layer by each pigment is multiplied by a fluorescence quantum yield specific to this pigment. These fluorescence photons are then distributed over the longer wavelengths according to a given fluorescence spectrum. Figure 8.15 shows the apparent reflectance and transmittance spectra

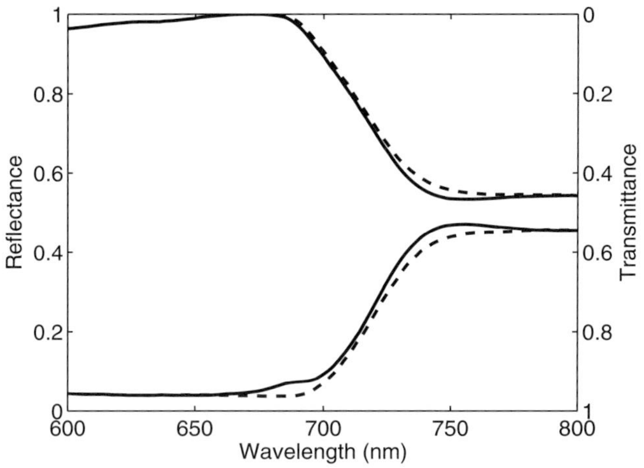

Figure 8.15 Reflectance and transmittance spectra of a linden (*Tilia* sp.) leaf with fluorescence (plain thick line) and without fluorescence (dashed thick line) simulated with SLOPE (redrawn from Maier (2000)).

of a linden (*Tilia* sp.) leaf. As can be seen, the reflectance is strongly modified in the red region, and relatively less in the red-edge up to the near-infrared plateau. In the red, about 50% of the radiance is due to fluorescence, which seems to be overestimated. The position of the red-edge is shifted 3 nm to shorter wavelengths. In the near-infrared plateau, fluorescence amounts only to a few percent of the signal, which agrees with experiments.

8.2.5 Monte Carlo Algorithms

As seen earlier, the pathway of light in an absorbing and scattering medium can be described by the radiative transfer equation or by using ray tracing models. A Monte Carlo photon transport model has been implemented by Sušila and Nauš (2007) for simulating leaf reflectance, transmittance, and fluorescence of a compact leaf represented by a semi-infinite slab. The approach of Zakharov et al. (2009) is different, as they considered four layers (upper and lower epidermis, palisadem and spongy parenchyma) and used a Monte Carlo algorithm for solving numerically the radiative transfer equation, the scattering phase function of which is a Henyey–Greenstein function accounting for the anisotropic nature of scattering in leaf tissue. To reduce the number of iterations needed to obtain a reasonable calculation error for the light fluxes and save calculation time, Zakharov et al. (2010) proposed to use a different method called P_L approximation, in which the radiation intensity is expanded in terms of spherical harmonics.

8.3 Different Approaches for Leaf Surface Reflectance Properties

As seen in Chapter 5, a large variety of plant leaves definitely do not have Lambertian surfaces and display a strong specular reflection, resulting from the presence of wax on their surface. If leaf directional properties do not have much influence on vegetation reflectance as measured from a plane or a satellite, collecting data from a short distance on a plant canopy using a camera or a small footprint LiDAR may require having information about leaf BRDF to avoid misinterpreting them in terms of leaf chemistry. Few studies have been performed on the modeling of leaf BRDF because of the lack of experimental measurements and because most canopy reflectance models assume that leaves are basically Lambertian. Moldau (1967) first proposed a mathematical model where the surface component was calculated based on Fresnel reflection from a rough surface but, to our knowledge, his work published in Russian has not been translated, and therefore, has not spread into the global scientific community. Woolley (1971) already imagined the leaf surface as composed of specular facets but he did not develop the corresponding model from this concept. Computer graphics researchers, who have intensively worked on surface scattering of many inorganic objects during the last five decades, have taken an interest in plant leaves only recently. They first ran their models on such semi-transparent biological objects without raising the question of their validity. The latest studies, however, make extensive use of information on plant tissues.

8.3.1 Parametric Equations

Brakke et al. (1989) measured the reflectance factor of northern red oak (*Quercus rubra*) and red maple (*Acer rubrum*) leaves in the principal plane and wished to find a mathematical expression that could reproduce its shape. Their main concern was with the physical meaning of the retrieved parameters. After an unsuccessful attempt with polynomials, they divided the leaf scattering indicatrix into two sections, one on either side of the peak value (Figure 8.16) and an exponentiated cosine function of the form:

$$R(\theta_r) = R_m \cos^b\left(\frac{\theta_r - \theta_m}{\omega}\right) \tag{8.38}$$

Equation 8.38 was fitted to each section. In polar coordinates, R is the fractional reflectance in the direction θ_r from the normal of the leaf, R_m is the maximum reflectance occurring at θ_m, and ω is an angle of transformation so that scattering in the plane of the leaf will be zero. A separate transformation was required for the data of the other side. Note that continuity at θ_m was ensured by this empirical equation.

Greiner et al. (2007) measured the scattering properties of sugar maple (*Acer saccharum*) and eastern cottonwood (*Populus deltoides*) leaves. They used a Lambertian distribution to model diffuse

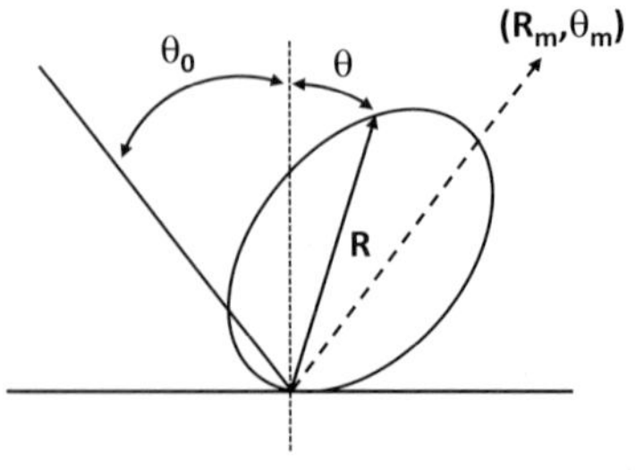

Figure 8.16 Schematic for leaf scattering: θ_0 is the zenith angle of the incident ray, θ is the view zenith angle, R is the fractional reflectance, and the subscript m indicates the point of maximum reflectance (redrawn from Brakke et al. (1989)).

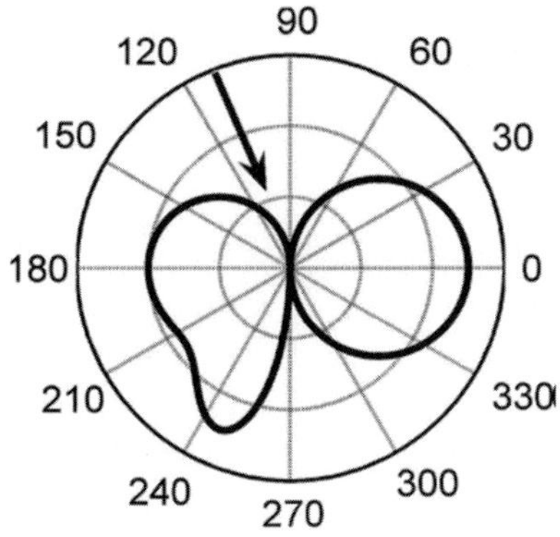

Figure 8.17 Simulation of the BSDF of a sugar maple leaf (*Acer saccharum*) illuminated at an incidence angle $\theta_i = 110°$ shown by the arrow. The lobe on the left is due to reflection and the lobe on the right is due to transmission (from Greiner et al. (2009), *Applied Optics*. Copyright © 2009 with permission from the Optical Society).

scattering and a reversed Rayleigh probability distribution to model specular reflection. The normalized sum of the two components was then expressed as

$$R(\theta_i, \theta_v) = \left(1 - F_s(\theta_i)\right) \cos\theta_v + F_s(\theta_i)\frac{x}{\sigma^2}\exp\left(-\frac{x^2}{2\sigma^2}\right) \tag{8.39}$$

where $x = \frac{\pi}{2} - \theta_v$, $\sigma = \frac{\pi}{2} - \theta_m$ and F_s is the fractional specular reflection. θ_i, θ_v, and θ_m are the incident, viewing, and maximum signal angles, respectively. Equation 8.39 is also valid for transmittance, where the bidirectional scattering distribution function (BSDF) is assumed to be perfectly Lambertian and therefore the specular reflection fraction is set to zero. Figure 8.17 shows the fit of Eq. 8.39 to the BSDF of a sugar maple leaf (*Acer saccarum*) leaf.

In the middle of the 1990s, Rahman et al. (1993) designed The RPV (Rahman-Pinty-Verstraete) semi-empirical parametric model to represent surface anisotropy patterns with three input parameters (ρ_0, k, Θ) only. The model splits a bidirectional reflectance factor (BRF) field into an amplitude component $(0 \leq \rho_0 \leq 1)$ and an associated angular field describing the anisotropic behavior of the surface. The latter is the product of three functions accounting for both the illumination θ_i and viewing θ_v directions. The BRF of a surface is written as

$$\rho_s(\theta_i, \theta_v) = \rho_0 \times M(\theta_i, \theta_v, k) \times F(g, \Theta) \times H(G, \rho_c) \tag{8.40}$$

The angular function M is a modified Minnaert function that allows representing the basic shape of the angular field through the parameter k. Typically, k is close to 1.0 for a quasi-Lambertian surface, lower than 1.0 when a bowl-shaped pattern dominates and greater than 1.0 when a bell-shaped pattern dominates. The other angular functions add more flexibility to the anisotropy classes described above. The function F is based on the Henyey–Greenstein function, and the parameter Θ controls the relative amount of forward $(\Theta \geq 0)$ and backward $(\Theta \leq 0)$ scattering, depending on the sign. The function H allows modeling the hot spot, which depends on a geometrical factor G and a factor ρ_c. Biliouris et al. (2009) inverted the RPV model on BRF measurements performed on 60 European beech (*Fagus sylvatica*) leaves. They found a good fit of the data $(0.71 < R < 0.99)$ for all wavelengths between 400 nm and 2500 nm and all light/sensor angular positions. However, interpretation of the retrieved parameters is still an issue.

8.3.2 Physical Models

The challenge is to provide analytical models of reflection by a rough surface which successfully predicts the experimental finding, in particular off-specular peaks that are observed on plant leaves. Such models have been first designed for inorganic materials but, until recently, none of them took into account subsurface scattering which is important for plant tissues. Because surface roughness of a leaf is large compared to the wavelength of the incident light, we can apply the Kirchhoff rough surface scattering theory. One approach to modeling leaf BRDF is to separate the diffuse $(BRDF_{diff})$ and specular $(BRDF_{spec})$ components:

$$BRDF = BRDF_{diff} + BRDF_{spec} \tag{8.41}$$

8.3.2.1 Diffuse Component

The diffuse component characterizes absorption and scattering processes within the leaf volume, thus it is assumed to be Lambertian and to strongly depend on the wavelength such that $BRDF_{diff}$ may be written as

$$BRDF_{diff}(\lambda, \theta_i, \theta_v, \varphi_v) = \frac{k_L(\lambda)}{\pi} \tag{8.42}$$

where $1/\pi$ is the BRDF of a perfect Lambertian scatterer and $k_L(\lambda)$ is the Lambert coefficient. Note that the diffuse component does not vary with the incidence angle.

8.3.2.2 Specular Component

There are several ways to model the shape of the specular component. To assess the effect of leaf BRDF on canopy BRF, Reyna and Badhwar (1985) introduced a leaf specular reflectance factor ρ in the SAIL model, the mathematical expression of which is not however explicitly given in their paper. Since no value of ρ was currently available, they treated it as a parameter to be determined. For an optically smooth leaf surface, Vanderbilt and Grant (1985) propose to use the Fresnel equations, distinguishing between the parallel component $F_{||}$ and the perpendicular $F_{\perp}$ component:

$$F(\theta_i, n) = \frac{F_{||}(\theta_i, n) + F_{\perp}(\theta_i, n)}{2} \tag{8.43}$$

If the angle of incidence θ_i is not null, then

$$F_{||}(\theta_i, n) = \frac{\tan^2(\theta_i - \theta_t)}{\tan^2(\theta_i + \theta_t)} \text{ and } F_{\perp}(\theta_i, n) = \frac{\sin^2(\theta_i - \theta_t)}{\sin^2(\theta_i + \theta_t)} \tag{8.44}$$

where the angle of refraction θ_t is computed with Snell's law $\theta_t = \arcsin(\sin \theta_i / n)$ and n is the refractive index. Reflection takes place in the plane defined by the leaf normal and the incident ray, and the reflection angle θ_r is equal to the angle θ_i. Note that in the limiting case $\theta_i = 0$ (the incident ray goes along the normal). Equation 8.44 does not work and then

$$F(0, n) = \left(\frac{n-1}{n+1}\right)^2 \tag{8.45}$$

When θ_i increases, F tends toward 1, which is not realistic for plant leaves because their cuticular wax is rarely optically smooth. To account for the reduction in the amount of specularly reflected light, Vanderbilt and Grant (1985) introduced an attenuation factor $K(\theta_i)$ that was assumed to depend on the angle of incidence θ_i and to be identical for all foliage surfaces. Nilson and Kuusk (1989) are the first to propose an analytical expression for K that was later called the leaf hair index (Kuusk, 1995a):

$$K(\theta_i) = \exp\left(\frac{-2\ k\ \tan\theta_i}{\pi}\right) \tag{8.46}$$

where k is a parameter characteristic of leaf surface. The approaches of Ross and Marshak (1989) and Myneni and Ross (1991) are slightly different, as they also used the Fresnel equations but without the attenuation factor, introducing sophistication into the simulation of the specular reflection. While the previous models considered that all reflected energy concentrated in a direction, they simulated a fluctuation in the orientation of leaves to scatter the specular energy in a wider set of directions.

The microfacet theory is a recurrent approach to model the specular component of rough surfaces. Consider a surface of area S illuminated by a parallel light beam in the direction (θ_i, φ_i) and viewed by a sensor in the direction (θ_v, φ_v), both measured from the surface normal. In the following we will assume that $\varphi_i = 0°$ (principal plane), so that only the relative azimuth angle φ_v will be considered. The calculation of the specular component is based on the Torrance–Sparrow model (e.g., Torrance and Sparrow, 1967; Cook and Torrance, 1981). A recurrent approach to model the shape of this component assumes that, on a small scale, the leaf surface is composed of N_f oriented mirror-like microfacets of equal area S_f that form V-cavities (Figure 8.18). These microfacets are identified by the direction of their normal $\omega = (\alpha, \phi)$ and by the statistical distribution of their slopes $D(\alpha, \phi)$ that specifies the roughness of the surface. We define θ_a the angle of incidence between the normal to the facet and the illumination direction. The angles θ_a and α are derived from basic geometry as $\cos 2\theta_a = \cos\theta_i \cos\theta_v + \sin\theta_i \sin\theta_v \cos\varphi_v$ and $\cos\alpha = \frac{\cos\theta_i + \cos\theta_v}{2\cos\theta_a}$.

Given the direction $\omega = (\alpha, \phi)$ and the infinitesimal solid angle $d\omega = \sin\alpha\ d\alpha\ d\phi$ centered in this direction, the probability for the normal of a microfacet to be inside the solid angle $d\omega$ is $D(\alpha, \phi)\ d\omega$. The area of the plane surface is linked to the facet area by the following expression:

$$S = \iint_{2\pi} S_f N_f D(\alpha, \phi) \cos\alpha d\omega = S_f N_f \iint_{2\pi} D(\alpha, \phi)\ \cos\alpha d\omega = S_f N_f D_{\text{int}} \tag{8.47}$$

Therefore

$$D_{\text{int}} = \frac{S}{S_f N_f} \tag{8.48}$$

Only facets producing specular reflection between the illumination direction and the viewing direction are accounted for. The BRDF of the specular component is decomposed into a Fresnel factor F,

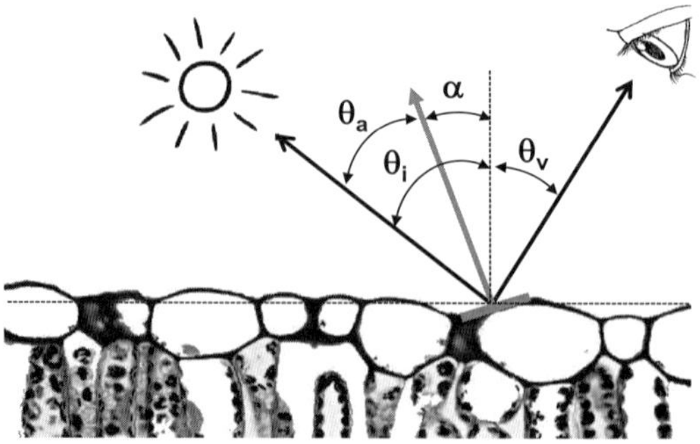

Figure 8.18 Notations for the main directions and angles in the principal plane.

the microfacet slope distribution function D, and a geometrical attenuation (masking and shadowing) factor G:

$$BRDF_{spec}(\lambda, \theta_i, \theta_v, \varphi_v) = \frac{1}{4\cos\theta_i\,\cos\theta_v}\frac{D(\alpha, \phi)}{D_{\text{int}}}F\Big(n(\lambda), \theta_a\Big)G(\theta_i, \theta_v, \varphi_v) \tag{8.49}$$

Equation 8.49 is the same expression as Cook–Torrance BRDF, except that we have a factor of 4 in the denominator instead of π.

The Fresnel factor $F\Big(n(\lambda), \theta_a\Big) \in [0, 1]$ describes how light is reflected by a single facet at the boundary between air and leaf surface material. It depends on the refractive index $n(\lambda)$ of leaf material and the angle of incidence θ_a between the normal to the facet and the illumination direction. In the solar domain, the cuticle which coats the leaf epidermal cells is assumed to be a low absorbing medium compared to the leaf interior. Thus the absorption of the cuticle can be neglected and its refractive index assumed real. The Fresnel factor is calculated for a plane dielectric boundary between two semi-infinite media and unpolarized light as (see Appendix E)

$$F\Big(n(\lambda), \theta_a\Big) = \frac{1}{2}\left(\frac{g - \cos\theta_a}{g + \cos\theta_a}\right)^2\left(1 + \left(\frac{\cos\theta_a(g + \cos\theta_a) - 1}{\cos\theta_a(g - \cos\theta_a) + 1}\right)^2\right) \tag{8.50}$$

with $g^2 = n(\lambda)^2 + \cos^2\theta_a - 1$. When cuticle absorption cannot be neglected, like in the thermal infrared, this factor is expressed as a function of the real and imaginary parts of the complex refractive index of the medium.

The microfacet slope distribution function $D(\alpha, \phi) \in [0, \infty[$, normalized to unity, represents the fraction of the facets that are oriented in the direction $\omega = (\alpha, \phi)$. Various formulations can be found in the optics literature (e.g., Beckmann, Blinn-Phong, GGX/Trowbridge-Reitz). The widely used Beckmann distribution function is a zero-mean Gaussian distribution, which depends on the angle α and a roughness parameter σ that is the root-mean-square slope of leaf surface (e.g., Cook and Torrance, 1981; Oren and Nayar, 1995):

$$D(\alpha, \sigma) = \frac{D_{\text{int}}}{\pi\sigma^2\cos^4\alpha}\exp\left(-\frac{\tan^2\alpha}{\sigma^2}\right) \tag{8.51}$$

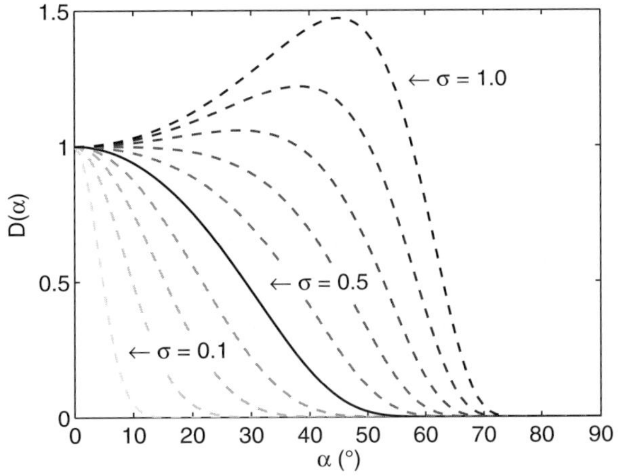

Figure 8.19 Adjusted probability density function of facet orientation for different values of the roughness parameter σ varying from 0.1 to 1.

When the surface is rough (large values of σ), orientations of microfacets are very dispersed; when the surface is smooth (small values of σ), the microfacet normals are closer to the average normal. The variation of D as a function of α for different values of the roughness parameter σ is illustrated by Figure 8.19.

In order to take into account the azimuthal anisotropy that characterizes the leaf surface of many monocots, Comar et al. (2014) decomposed the roughness parameter into two perpendicular components in the so-called anisotropic Cook–Torrance (ACT) model:

$$D(\alpha, \phi, \sigma_x, \sigma_y) = \frac{D_{\text{int}}}{\pi \sigma_x \sigma_y \cos^4 \alpha} \exp\left(-\tan^2 \alpha \left(\frac{\cos^2 \phi}{\sigma_x^2} + \frac{\sin^2 \phi}{\sigma_y^2}\right)\right) \tag{8.52}$$

The geometrical attenuation factor $G \in [0, 1]$ (sometimes noted *GAF*) is a multiplicative factor lying between zero and one, which accounts for the shadowing and masking of one facet by another. Once again, several formulations can be considered (e.g., Neumann, Cook–Torrance, Kelemen). Ideally it should depend on roughness and micro-facet distribution, but for simplicity's sake, the one below that only depends on the illumination and viewing directions can be used (Blinn, 1977):

$$G(\theta_i, \theta_v, \varphi_v) = \min\left(1, \frac{2\cos\alpha \cos\theta_i}{\cos\theta_a}, \frac{2\cos\alpha\cos\theta_v}{\cos\theta_a}\right) \tag{8.53}$$

8.3.2.3 BRDF model

For isotropic surface roughness, the final expression for leaf *BRDF* is

$$BRDF_{spec}\left(k_L(\lambda), n(\lambda), \sigma\right) = \frac{k_L(\lambda)}{\pi} + \frac{1}{4\cos\theta_i \cos\theta_v} \times \frac{e^{-\tan^2\alpha/\sigma^2}}{\pi\sigma^2 \cos^4\alpha} \times F\left(n(\lambda), \theta_a\right) G(\theta_i, \theta_v, \varphi_v) \tag{8.54}$$

with θ_i, θ_v, φ_v, θ_a, and α the angles defined above. Several articles are based on this approach (Bousquet et al., 2005; Donner and Jensen, 2005; Wang et al., 2005; Chelle, 2006 with the Simple yet Realistic BRDF Model; Lao et al., 2006, 2009; Habel et al., 2007; Xie et al., 2010, 2017; Uto and Kosugi, 2012, 2013a, 2013b; Comar et al., 2014; Jay et al., 2016). One should keep in mind that this approach is a simplification of real leaf surfaces since other features like trichomes on hairy leaves, for instance, may involve other optical phenomena like diffraction (Fuhrer et al., 2004).

8.3.2.4 Inversion of a Leaf BRDF Model

Bousquet et al. (2005) inverted a leaf BRDF model over directional reflectances measured experimentally on the adaxial and abaxial sides of 11 leaf species (Combes et al., 2007). A full acquisition corresponded to 4 illumination directions θ_i (8°, 21°, 41°, and 60°), 98 viewing directions θ_v, and 400 wavelengths λ ranging between 480 nm and 880 nm. Because data acquired at viewing zenith angles higher than 70° were unreliable, only 65 viewing directions of the 98 measured were used, which represented 260 reflectance measurements at each wavelength. The three input parameters (k_L, n, σ) were retrieved at each wavelength separately by minimizing the merit function χ^2 defined as

$$\chi^2(\lambda) = \sum_{\theta_i,\theta_v,\varphi_v} (BRDF_{\text{measured}} - BRDF_{\text{modeled}})^2 \tag{8.55}$$

The optimization was constrained by fixing the lower and upper parameter bounds. Results are presented in Figure 8.20 for cherry laurel (*Prunus laurocerasus*), European beech (*Fagus sylvatica*), and common hazel (*Corylus avellana*).

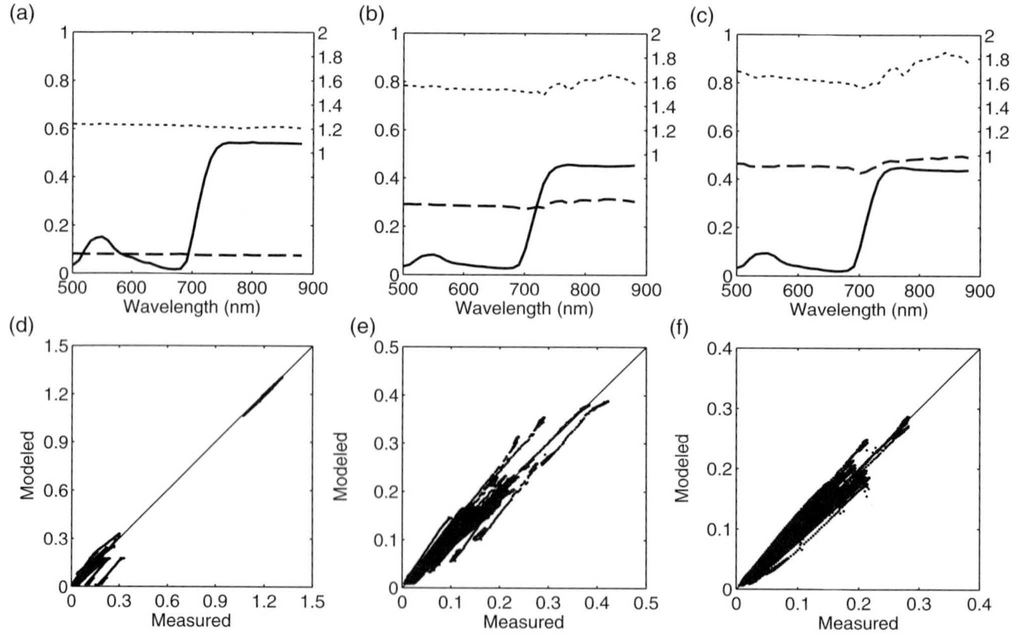

Figure 8.20 Three-parameter model inversion at each wavelength for cherry laurel (*Prunus laurocerasus*) (a), European beech (*Fagus sylvatica*) (b), and common hazel (*Corylus avellana*) (c). Parameter values as a function of wavelength: line for k_L, dashed for σ on the left scale, dotted for n on the right scale; and corresponding measured vs. modeled BRDF (d–f) (redrafted from Bousquet et al. (2005)).

Table 8.3 *Mean leaf surface characteristics estimated in the 480–880 nm range.*

Leaf	Refractive index n	Roughness parameter σ
Laurel	1.22	0.078
Beech	1.56	0.29
Hazel	1.68	0.46

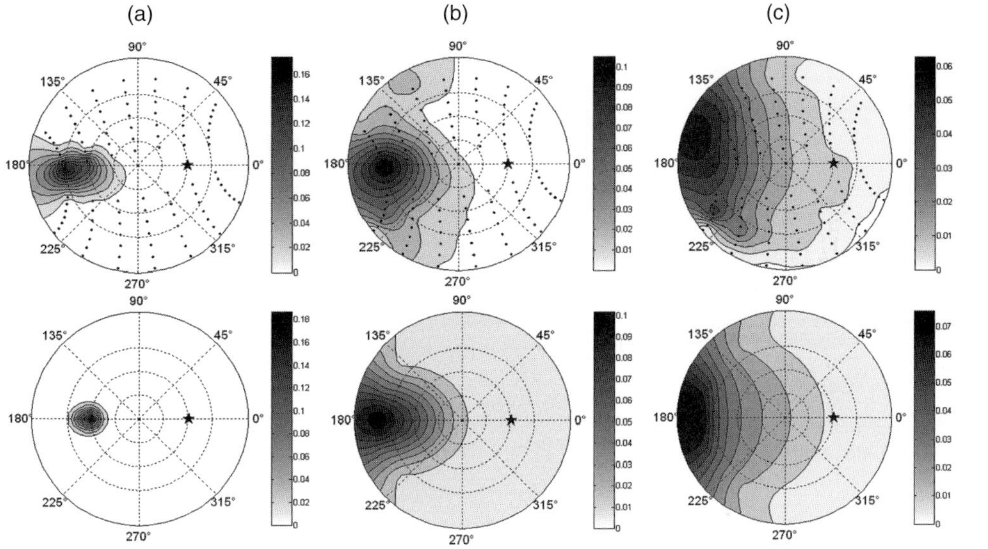

Figure 8.21 Measured (top) and modeled (bottom) BRDF for $\theta_i = 41°$ at wavelengths of minimum reflection (scale: 0–0.1 sr^{-1}): (a) cherry laurel at 662 nm, (b) European beech at 669 nm, and (c) common hazel at 664 nm (redrawn from Bousquet et al. (2005)).

Measured versus modeled BRDF values agree well. Compared to k_L, the shape of which mimics the reflectance of a plant leaf, the parameters n and σ vary slowly over the entire spectrum. The spectral invariance of the specular component is consequently a good approximation, especially in the visible. The mean values of the retrieved cuticle refractive index and roughness parameter from 480 nm to 880 nm are listed in Table 8.3. The roughness parameter is minimum for laurel, which has a smooth thick cuticle, maximum for hazel that has pubescent faces, and intermediate for beech which has an intermediate surface roughness structure.

Figure 8.21 shows the measured and modeled BRDF at wavelengths of minimum reflection. They reveal very similar directional shapes. This proves the ability of the model to fit the directional variations of the reflected light that did not penetrate into the leaf.

The retrieved model parameters can be used to estimate the diffuse and specular contributions to the DHRF at each wavelength, that is, to evaluate the amount of light that did not penetrate into the leaf:

$$DHRF(\theta_s, \lambda) = DHRF_{diff}(\theta_i, \lambda) + DHRF_{spec}(\theta_i, \lambda) \tag{8.56}$$

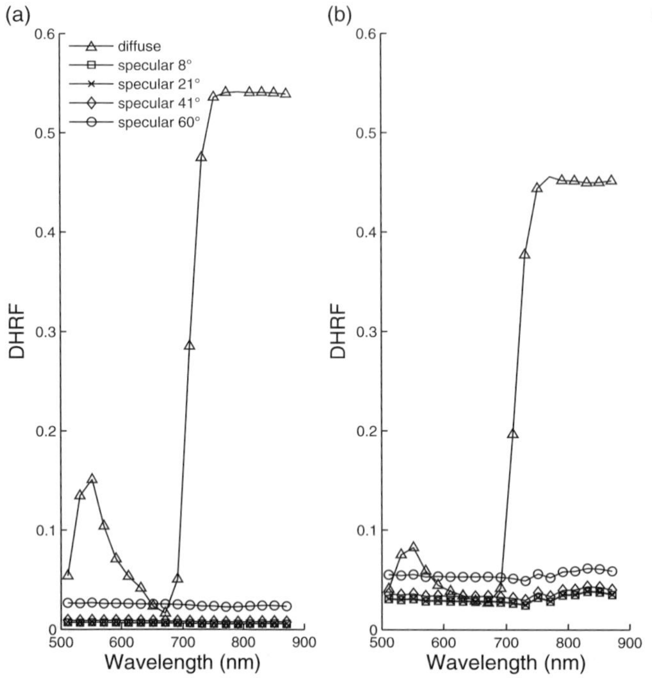

Figure 8.22 Estimated diffuse and specular contributions to the DHRF for cherry laurel (a), European beech (b), and common hazel (c) (redrawn from Bousquet et al. (2005)).

with

$$DHRF_{diff}(\lambda,\theta_i) = \iint_{2\pi} BRDF_{diff}(\lambda,\theta_i,\theta_v,\varphi_v)\cos\theta_v \sin\,\theta_v d\theta_v d\varphi_v = k_L(\lambda) \tag{8.57}$$

and

$$DHRF_{spec}(\lambda,\theta_i) = \iint_{2\pi} BRDF_{spec}\left(n(\lambda),\sigma(\lambda),\theta_i,\theta_v,\varphi_v\right)\cos\theta_v \sin\,\theta_v d\theta_v d\varphi_v \tag{8.58}$$

$DHRF_{spec}$ can be numerically computed. Figure 8.22 shows both contributions for the four incidence angles, θ_i. The specular contribution varies very slowly with the wavelength, regardless of the leaf type and the incidence angle. It is almost the same at incidence angles θ_i equal to 8°, 21°, and 41° and double at $\theta_i = 60°$. The specular contribution can be higher than the diffuse one for wavelengths of strong absorption.

8.3.3 Ray Tracing Models

Govaerts et al. (1996) pioneered the simulation of the optical properties of a standard dicot leaf modeled as a three-dimensional object with the RAYTRAN ray tracing model (see Chapter 10). They approximated the epidermal cells by oblate ellipsoids and reported very encouraging results in terms of leaf BRDF and BTDF. Bousquet (2007) consolidated this first attempt using a real European beech

(*Fagus sylvatica*) leaf collected in the outdoors and analyzed in the laboratory. He measured its biochemical content and three-dimensional structure to fix the parameters of the model. Such an approach requires a detailed description of leaf anatomy that is somewhat difficult to assess. In the ABM of Baranoski and Rokne (1997), surface reflectance is accounted for by perturbing the reflected rays using a warping function, which is a probability density function based on the exponential cosine distribution defined in Eq. 8.59. The perturbation is performed through angular displacements, α_e and β_e, where α_e corresponds to the polar angle with respect to the ideal reflection or ideal transmission direction and β_e corresponds to the azimuth angle around the ideal reflection or ideal transmission direction. These angles are given by

$$\alpha_e = \arccos\left((1 - \xi_1)^{\frac{1}{ob+1}}\right) \tag{8.59}$$

and

$$\beta_e = 2\pi\xi_2 \tag{8.60}$$

where ξ_1 and ξ_2 are computer-generated random numbers uniformly distributed over the interval $[0, 1]$ and *ob* is the oblateness of the epidermal cells (Figure 8.23).

Fuhrer et al. (2004) generated a plant skeleton using an L-system and distributed individual hairs across the surface of each leaf. They studied the trichome distribution on thale cress (*Arabidopsis thaliana*) leaves, as well as their length and shape, and they modeled hairs as curved generalized cylinders (Figure 8.24). A Monte Carlo ray-tracer was then used to produce images of several hairy plant species.

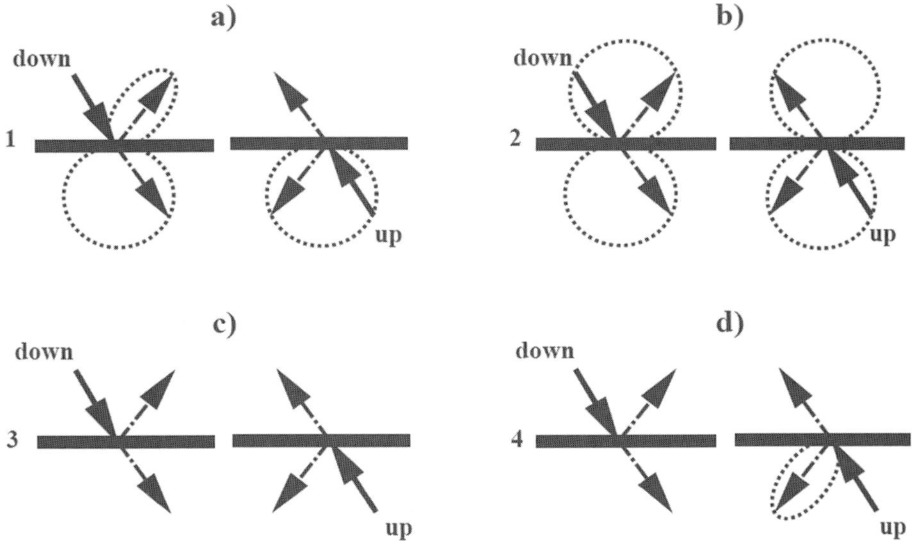

Figure 8.23 Perturbations performed by the ABM on the rays distributions at the four interfaces in the upward and downward directions of propagation considering the adaxial surface on the top and the abaxial surface on the bottom (from Baranoski and Rokne (1997). Copyright © 1997 with permission from the Eurographics Association).

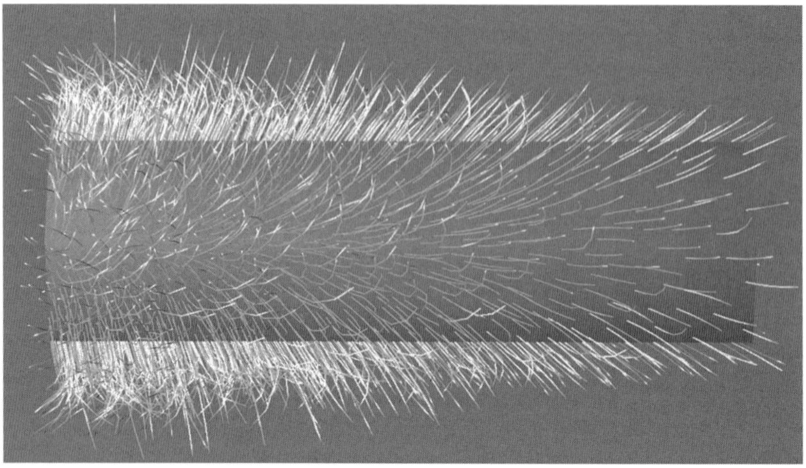

Figure 8.24 Model of leaf hair: distribution of different types and shapes, curly hairs on the left end and straight hairs on the right end (from Fuhrer et al. (2004). Copyright © 2004 with permission from the Association for Computing Machinery).

8.4 Terahertz, Microwaves, and Radio Waves Scattering Models

Leaves are the main constituents of plant canopies, so knowledge of their scattering and extinction properties is very important in radar remote sensing. Several models of backscattering cross-sections of leaves have been proposed in the literature. At high microwave frequencies (e.g., X band at υ = 8–12 GHz or λ = 2.5–3.75 cm), a typical leaf is electrically thin with lateral dimensions comparable to or larger than the wavelength. If the blade thickness is smaller than $\lambda/50$, the leaf can be considered as an infinitesimally thin resistive sheet (Figure 8.25a, Senior et al., 1987). The leaf can also be represented as a slab of finite thickness τ consisting of a thin dielectric layer with a relative permittivity $\widetilde{\epsilon} = \epsilon' - i\epsilon''$ (Figure 8.25b, Sarabandi and Ulaby, 1988; Oh and Hong, 2007).

In both models, the resistivity R of the leaf expressed in ohms per square ($\Omega/\square$) equals

$$R = \frac{i\, Z_0}{k\ \tau(\widetilde{\epsilon} - 1)} \tag{8.61}$$

with $k = 2\pi/\lambda$ and Z_0 the impedance of free space (376.730 Ω). The relative permittivity, and therefore the resistivity, is mainly controlled by the leaf gravimetric moisture content. When $R = 0$. the sheet appears perfectly conducting and when $R = \infty$, it ceases to exist. The electromagnetic scattering properties of a rectangular resistive sheet of finite dimensions have been calculated for different incident angles and polarizations by developing expressions for the electric and magnetic fields and applying the appropriate boundary conditions. Leaf reflectivity and transmissivity have been also derived using a similar flat sheet model by Mätzler and Sume (1989).

At millimeter wave frequencies (e.g., V band at υ = 40–75 GHz or λ = 4.0–7.5 mm or W band at υ = 65–300 GHz or λ = 1–4 mm), however, the blade thickness is a significant fraction of a wavelength. Sarabandi et al. (1990) showed that it was necessary to take into account the internal structure of the leaf. The latter can be modeled as two dielectric layers, which have different water content and therefore different dielectric constants. To compute the scattering at these higher frequencies, these

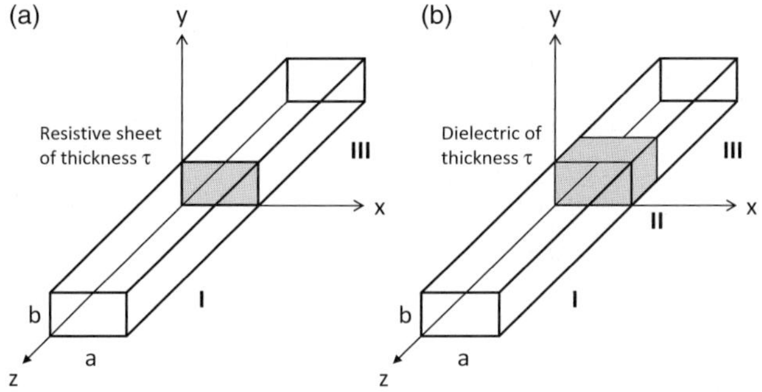

Figure 8.25 Rectangular waveguide with a dielectric sheet (a), or slab (b) of dimensions $a \times b$ and thickness τ at $z = 0$ (redrawn from Sarabandi and Ulaby (1988)).

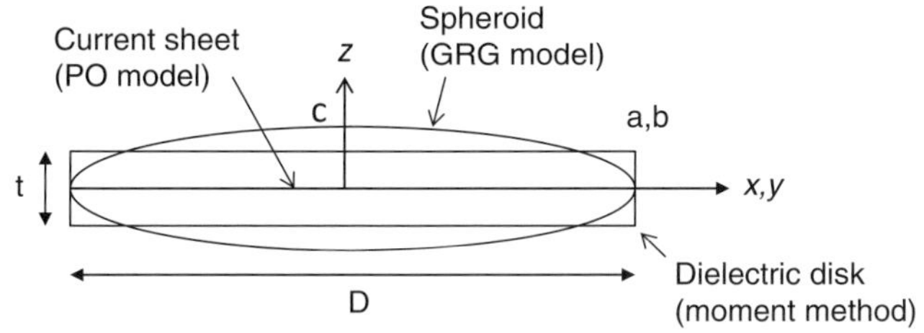

Figure 8.26 Cross-sections of a dielectric slab for three different methods (redrawn from Oh and Hong (2007)).

authors examined two types of physical optics approximations referred to as the Volume Integral Physical Optics (VIPO) method and the Surface Current Physical Optics (SCPO) method. The VIPO approximation is more complicated but it provides actual dielectric profiles.

At low frequencies (e.g., L band at $\upsilon = 1$–2 GHz or $\lambda = 15$–30 cm; P band at $\upsilon = 0.3$–1 GHz or $\lambda = 30$–100 cm), the leaf disk is small compared to the wavelength. The Generalized Rayleigh-Gans (GRG) approximation can be used to compute the scattering cross-section of leaves with flat surfaces (Karam et al., 1988; Karam and Fung, 1989; Pan and Narayanan, 2002; Oh and Hong, 2007). This approximation uses the internal fields of a thin Rayleigh cylinder to modify the terms of the Born (Rayleigh-Gans) approximation (Figure 8.26).

At terahertz frequencies ($\upsilon = 0.3$–10 THz or $\lambda = 30$–1000 μm), it is important to identify the main components of the leaf that actually interact with the electromagnetic wave, such as water, air, and dry matter (Figure 8.27).

The permittivity of such a medium can be described by an effective medium theory (EMT). The Landau, Lifshitz, Looyenga (LLL) model is commonly used to obtain the effective dielectric function of complex heterogeneous mixtures consisting of irregular-shaped particles (see Section 3.1.4). Therefore, it is well-suited to calculate the effective permittivity of a plant leaf. The third-order extension of the LLL model can be written in the form:

$$\sqrt[3]{\epsilon(\upsilon)} = \xi_{\mathrm{w}} \sqrt[3]{\epsilon_{\mathrm{w}}(\upsilon)} + \xi_{dm} \sqrt[3]{\epsilon_{dm}(\upsilon)} + \xi_{\mathrm{a}} \sqrt[3]{\epsilon_{\mathrm{a}}(\upsilon)} \tag{8.62}$$

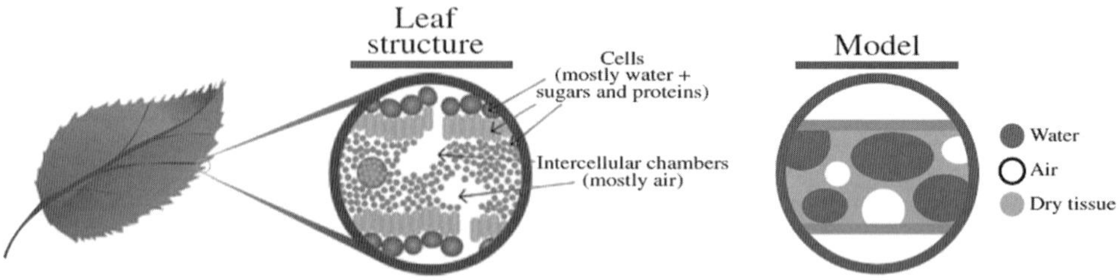

Figure 8.27 Cross-section of a plant leaf mainly composed mainly of air cavities, water, and dry matter (from Gente et al. (2013). Copyright © 2013 with permission from Springer).

where ξ and ϵ are the concentration and the permittivity of the components, respectively. Subscripts *w*, *dm*, and *a* refer to water, dry matter, and air, respectively. The effective refractive index and absorption coefficient can then be calculated at different frequencies for leaves at different water contents (Jördens et al., 2009; Gente et al., 2013; Gente and Koch, 2015; Born et al., 2014b).

9

Modeling Leaf Optical Properties: PROSPECT

As seen in Chapter 8, radiative transfer models have advanced our understanding of light interception by plant leaves throughout the history of remote sensing. They describe absorption and scattering, the two main physical processes involved during the interaction of light with matter. The open-source PROSPECT leaf optical properties model (Jacquemoud and Baret, 1990; Lu et al., 2013) has been the most extensively used radiative transfer model over the past 25 years. PROSPECT originally simulated the directional-hemispherical reflectance and transmittance of leaves at fine wavelength resolution for the 400 nm to 2500 nm region of the solar spectrum. Since then it has been coupled with a surface bidirectional reflectance distribution function (BRDF) model; adapted to take into account chlorophyll *a* fluorescence; extended to the short-wave infrared domain; and several chemical constituents have been added to the original ones over the years.

9.1 The PROSPECT Model

9.1.1 Optical Properties of One Plate: the Plate Model

PROSPECT is based on the Allen et al. (1969, 1970a) representation of the leaf as one or several absorbing plates with rough surfaces giving rise to isotropic diffusion. Let us consider a single plate (see Figure 8.1), where we showed that its reflectance can be written as (see Eq. 8.2)

$$R = r_{12} + \frac{t_{12}\ t_{21}\ r_{21}\ \tau^2}{1 - r_{21}^2\ \tau^2} \tag{9.1}$$

and its transmittance as (see Eq. 8.3)

$$T = \frac{t_{12}\ t_{21}\ \tau}{1 - r_{21}^2\ \tau^2} \tag{9.2}$$

where r_{ij} and t_{ij} are the average reflectivity and transmissivity, respectively, from medium i into medium j, and τ is the fraction of light transmitted through the medium. r_{ij} and t_{ij} are calculated from Fresnel's equations. In the visible, near infrared, and middle infrared, there is no appreciable absorption at the interface so that the reflectivity is related to the transmissivity by the simple relation $r_{ij} = 1 - t_{ij}$. Because the results for the transmissivity are somewhat simpler to interpret, we will use it in the following description. When measuring the directional-hemispherical reflectance and transmittance (see Section 4.5), the incident radiation is perpendicular to the leaf blade. However, the leaf surface is not flat so it intercepts light rays in more than one direction. The surface micromorphology of leaves displays at least three scales of roughness: hairs and veins, epidermal cells, and epicuticular wax. In PROSPECT, leaf surface roughness is assumed to be caused by the size, shape, and organization of epidermal cells. Consequently, it is intuitively mimicked by assuming the incident radiation in PROSPECT lights up a horizontal plane supposed to be perfectly flat at all angles between 0 and α (degree) such that

$$t_{12} = t_{av}(\alpha, 1, n) \tag{9.3}$$

where $t_{av}(\alpha, 1, n)$ is detailed in Appendix E. Due to multiple scattering, light is assumed to be diffuse inside the plate so that r_{21} and t_{21} are averaged values integrated over the entire hemisphere ($\alpha = 90°$). One can show that

$$t_{21} = t_{av}(90, n, 1) = n^{-2} t_{av}(90, 1, n) \tag{9.4}$$

Jacquemoud and Baret (1990) set the angle α to 60°, a value empirically determined from a limited dataset. Although long used, this angle seems to be overestimated in measurements gathered in recent years. While modeling leaf BRDF, Bousquet et al. (2005) physically linked the probability density function of facet orientations D (Cook and Torrance, 1981) to the surface roughness parameter σ and to the angle α, which is formed by the direction of the leaf normal and the direction of a facet normal (see Section 8.3.2). The versions 4 and 5 of PROSPECT published by Féret et al. (2008) suggest using $\alpha = 40°$, which appears to be a more realistic value. Note that this angle should logically change from one leaf to the next.

The transmission τ for isotropic light passing through the plate at a distance x (m) is calculated by a simple Beer–Lambert law integrated over the entire hemisphere and is given by (see Appendix F)

$$\tau(x) = (1 - kx)\exp(-kx) + k^2 x^2 \Gamma(0, kx) \tag{9.5}$$

where $\Gamma(a, x)$ is the "upper" incomplete gamma function and k (m^{-1}) is the absorption coefficient of the plate. In PROSPECT, the leaf thickness is not explicitly taken into account but is characterized as the number of elementary plates. In consequence, $x = 1$ in Eq. 9.5 that simplifies to

$$\tau = (1 - k)\exp(-k) + k^2 \Gamma(0, k) \tag{9.6}$$

The absorption coefficient k, which varies as a function of the wavelength λ, contains contributions from the n individual leaf biochemical constituents:

$$k(\lambda) = \sum_{i=1}^{n} C_i \times k_i(\lambda) \tag{9.7}$$

where C_i is the content of the *ith* constituent in the elementary layer (units of mass per unit leaf area) and $k_i(\lambda)$ is the corresponding specific absorption coefficient (units of leaf volume per unit mass).

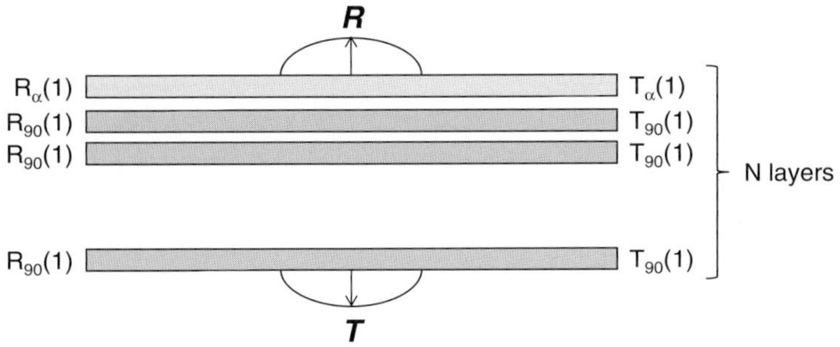

Figure 9.1 Schematic representation of PROSPECT.

9.1.2 Optical Properties of N Plates: The Stokes System

The generalized plate model, or Stokes system, consists of stacking elementary plates. In order to take into account the anisotropic structure of the incident beam ($\alpha = 40°$) impinging on the top of the stack of plates, it was slightly modified by separating the first plate from the $N - 1$ others (Figure 9.1).

Consider 1 layer overlaying $N - 1$ layers (Figure 9.2a), the reflectance and the transmittance are $R_\alpha(1)$ and $T_\alpha(1)$, respectively, for the first plate, and $R_{90}(N-1)$ and $T_{90}(N-1)$, respectively, for the $N - 1$ other plates because the light flux is assumed to be isotropic inside the leaf. The total reflectance and transmittance are expressed as

$$\begin{cases} R_\alpha(N) = R_\alpha(1) + T_\alpha(1)R_{90}(N-1)T_{90}(1) + T_\alpha(1)R_{90}(N-1)R_{90}(1)R_{90}(N-1)T_{90}(1) + \ldots \\ \qquad = R_\alpha(1) + \dfrac{T_\alpha(1)T_{90}(1)R_{90}(N-1)}{1 - R_{90}(1)R_{90}(N-1)} \\ T_\alpha(N) = T_\alpha(1)T_{90}(N-1) + T_\alpha(1)R_{90}(N-1)R_{90}(1)T_{90}(N-1) + \ldots \\ \qquad = \dfrac{T_\alpha(1)T_{90}(N-1)}{1 - R_{90}(1)R_{90}(N-1)} \end{cases} \tag{9.8}$$

If we now consider $N - 1$ layers overlaying 1 layer (Figure 9.2b), the reflectance and the transmittance are $R_\alpha(N-1)$ and $T_\alpha(N-1)$, respectively, for the first plates and $R_{90}(1)$ and $T_{90}(1)$, respectively, for the others:

$$\begin{cases} R_\alpha(N) = R_\alpha(N-1) + T_\alpha(N-1)R_{90}(1)T_{90}(N-1) + T_\alpha(N-1)R_{90}(1)R_{90}(N-1)R_{90}(1)T_{90}(N-1) + \ldots \\ \qquad = R_\alpha(N-1) + \dfrac{T_\alpha(N-1)T_{90}(N-1)R_{90}(1)}{1 - R_{90}(1)R_{90}(N-1)} \\ T_\alpha(N) = T_\alpha(N-1)T_{90}(1) + T_\alpha(N-1)R_{90}(1)R_{90}(N-1)T_{90}(1) + \ldots \\ \qquad = \dfrac{T_\alpha(N-1)T_{90}(1)}{1 - R_{90}(1)R_{90}(N-1)} \end{cases} \tag{9.9}$$

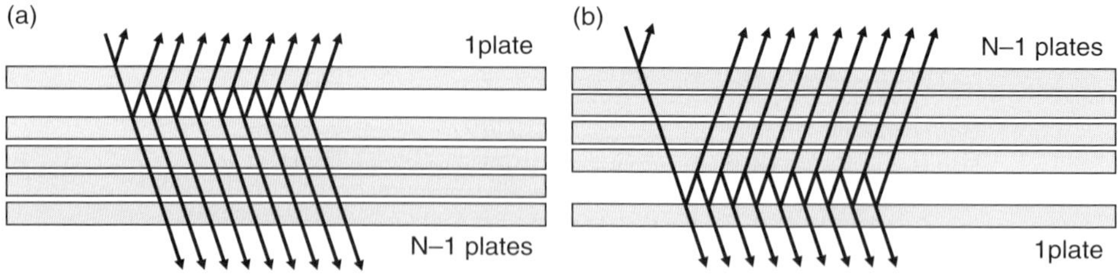

Figure 9.2 Stack of 1 plate over $N-1$ plates (a) and $N-1$ plates over 1 plate (b).

The solution of this system of recursive series has been given by Jacquemoud (1992):

$$R_\alpha(N) = \frac{R_\alpha(1)\left(ab^{N-1} - a^{-1}b^{1-N}\right) + \left(T_\alpha(1)T_{90}(1) - R_\alpha(1)R_{90}(1)\right)\left(b^{N-1} - b^{1-N}\right)}{ab^{N-1} - a^{-1}b^{1-N} - R_{90}(1)(b^{N-1} - b^{1-N})} \tag{9.10}$$

and

$$T_\alpha(N) = \frac{T_\alpha(1)\left(a - a^{-1}\right)}{ab^{N-1} - a^{-1}b^{1-N} - R_{90}(1)(b^{N-1} - b^{1-N})} \tag{9.11}$$

where

$$a = \frac{1 + R_{90}(1)^2 - T_{90}(1)^2 + \sqrt{\delta}}{2R_{90}(1)} \tag{9.12}$$

$$b = \sqrt{\frac{\beta\left(a - R_{90}(1)\right)}{a\left(\beta - R_{90}(1)\right)}} \tag{9.13}$$

$$\beta = \frac{1 + R_{90}(1)^2 - T_{90}(1)^2 - \sqrt{\delta}}{2R_{90}(1)} \tag{9.14}$$

$$\delta = \left(R_{90}(1)^2 - T_{90}(1)^2 - 1\right)^2 - 4R_{90}(1)^2 \tag{9.15}$$

9.1.3 Evolution of the PROSPECT *Model*

Originally the PROSPECT model used three input parameters (Jacquemoud and Baret, 1990): the structure parameter N (number of compact layers specifying the average number of air–cell walls interfaces within the mesophyll); the total chlorophyll content C_{ab} (µg cm^{-2}); and the equivalent water thickness C_w (cm≡g cm^{-2}). Since then, several different versions of PROSPECT have been developed, including new leaf biochemical constituents (mostly pigments and cell wall molecules), new capabilities (dorsiventral leaf, directional effects, fluorescence), new wavelengths (mid-wave infrared), and new calibration procedures (refractive index and specific absorption coefficients). The evolution of PROSPECT over time is summarized in Figure 9.3.

Motivated by the existence of specific absorption bands in the shortwave infrared for chemical bonds in cellulose and lignin (~ carbon) or proteins (~ nitrogen), that are potentially measurable by

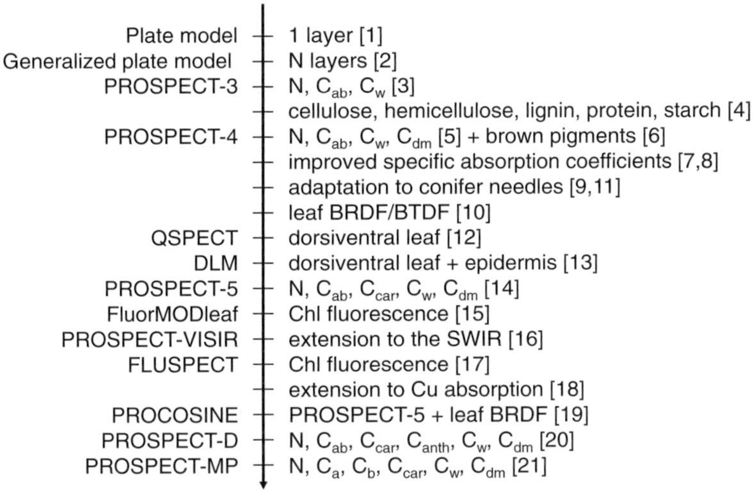

Figure 9.3 Evolution of the PROSPECT model: [1] Allen et al. (1969), [2] Allen et al. (1970a), [3] Jacquemoud and Baret (1990), [4] Fourty et al. (1996), [5] Baret and Fourty (1997), [6] Baret (personal communication), [7] Jacquemoud et al. (2000), [8] le Maire et al. (2004), [9] Zarco-Tejada et al. (2004b), [10] Bousquet et al. (2005), [11] Malenovský et al. (2006), [12] Ma et al. (2007), [13] Stuckens et al. (2009), [14] Féret et al. (2008), [15] Pedrøs et al. (2010), [16] Gerber et al. (2011), [17] Verhoef (2011), [18] Zhu et al. (2014), [19] Jay et al. (2016), [20] Féret et al. (2017), and [21] Zhang et al. (2017).

remote sensing (Wessman et al., 1989; Curran, 1989), scientists introduced them into the model. During the summer of 1993, an experiment at the Joint Research Centre (Ispra, Italy) built a database called LOPEX (Leaf Optical Properties Experiment) associating visible and infrared spectra of dry and fresh vegetation elements (leaves, conifer needles, stems, etc.) with physical measurements (thickness, water content, specific leaf area) and biochemical analyses (chlorophyll *a+b*, proteins, cellulose, lignin, etc.) (Hosgood et al., 1995). LOPEX was used to introduce the full leaf biochemistry into PROSPECT (Fourty et al., 1996; Jacquemoud et al., 1996; Fourty and Baret, 1998). A limit to the resolution of biochemical constituents arose in the inversion of the model when it was discovered that protein content could not be retrieved, or with great difficulty, in fresh leaves because the presence of strong water absorption features masked the protein absorptions and in dry leaves, and because cellulose and lignin were inconsistently identified and quantified as separate constituents. As a consequence, the model was simplified to the point that it now considers the dry matter content C_m (g cm^{-2}) as a whole instead of treating the leaf biochemical constituents individually (Baret and Fourty, 1997; Jacquemoud et al., 2000). That simplification, nevertheless, provides one interesting measurement: indeed C_m is equivalent to specific leaf area or *SLA* (cm^2 g^{-1}), a widely used key variable in plant ecology because it is easily measured and is correlated with plant growth, light interception, gas exchange, and photosynthesis (e.g., Dijkstra, 1989; Meziane and Shipley, 1999; Shipley and Vu, 2002). For instance, after a net uptake of 1 kg of carbon, wheat (*Triticum aestivum*) will produce 20 m^2 of leaf area (allocation rate of 65% and $SLA = 30$ m^{2}kg^{-1}), while saltbush (*Atriplex* sp.) will produce only 2 m^2 (allocation rate of 30% and $SLA = 7$ m^{2}kg^{-1}). C_m can also be accurately retrieved by model inversion. In short, the four input parameters of PROSPECT-4 were the leaf structure parameter, the total chlorophyll content, the equivalent water thickness, and the dry matter content. In order to obtain smooth derivatives of leaf

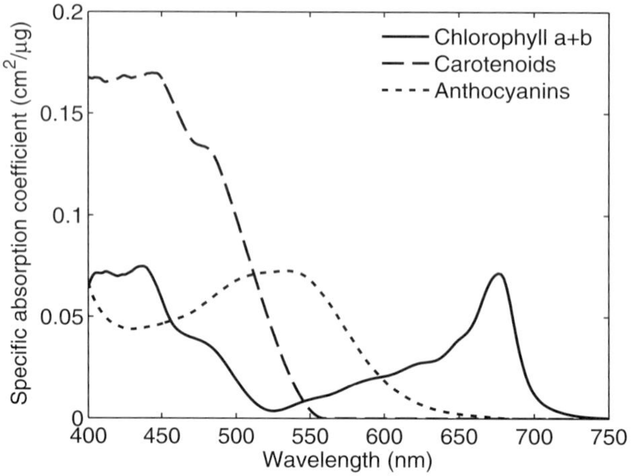

Figure 9.4 Specific absorption coefficients of chlorophylls, carotenoids, and anthocyanins in PROSPECT-D.

spectra, le Maire et al. (2004) decided to recalculate the specific absorption coefficients at 1 nm resolution. They took this opportunity to reshape the specific absorption of chlorophyll in the red-edge region, the shoulder of which was too sharp. Barry et al. (2009) found PROSPECT-4 improved chlorophyll estimates in eucalyptus leaves to about 5 μg cm^{-2} compared to PROSPECT-3. The subsequent version of the model called PROSPECT-5 constitutes a real advance with the separation of the foliar pigments into the total chlorophyll and the total carotenoid content (Féret et al., 2008). In addition to the fact that the new specific absorption coefficients of these pigments agree well with *in vitro* coefficients, the retrieval of chlorophyll content has been improved while very encouraging results were obtained with carotenoids. The reconstruction of reflectance and transmittance spectra over the solar domain was also excellent with small errors. Recently, anthocyanins have been added to chlorophylls and carotenoids in a new version of the model called PROSPECT-D where the suffix 'D' stands for dynamic (Féret et al., 2017). Figure 9.4 shows the specific absorption coefficient $k_i(\lambda)$ as a function of wavelength for pigments. PROSPECT-D seems to outperform all the previous versions.

PROSPECT has been adapted to simulate the optical properties of non-flat targets, such as conifer needles (e.g., Zarco-Tejada et al., 2004b; Moorthy et al., 2003a, 2003b; Malenovský et al., 2006; Moorthy et al., 2008). These efforts led to a recalibration of the specific absorption coefficients of leaf constituents to improve comparisons between simulated and measured optical properties. A version suited to semiarid species has also been designed (Ma et al., 2012; Li and Wang, 2013). The reassessment of the refractive index and specific absorption coefficients of PROSPECT has actually given rise to several studies, either to improve the simulations or to adapt the model to a particular leaf species. For instance, Chen and Weng (2012) applied the Kramers–Konig relations (see Section 3.1.3) to generate a complex refractive index varying with the leaf mass composition and absorption properties, therefore introducing spectral details that improved the model outputs. A specific absorption coefficient for copper ions was added and tested (Qu et al., 2014; Zhu et al., 2014). Wang et al. (2015b) have recalibrated the specific absorption coefficients of protein and carbon-based compounds (cellulose + lignin) using the above-mentioned LOPEX database and increased their retrieval accuracy both on fresh and dry leaves. In order to directly derive leaf nitrogen from nondestructive field measurements, Yang et al. (2015) replaced the specific absorption coefficient of chlorophyll with that of nitrogen in the so-called N-PROSPECT.

PROSPECT was also adapted to account for leaf surface directional reflectance (Bousquet et al., 2005; Jay et al., 2016), as detailed in Section 8.3.2, and chlorophyll *a* fluorescence emission (Pedrós et al., 2010), as detailed in Section 8.2.

9.2 Direct Mode: Sensitivity Analysis

To quantify the sources of variation in leaf optical properties, sensitivity analyses can be performed. Ideally, this would be achieved by using very large datasets together with the measured biophysical parameters. Such datasets are rare, however, and where they do exist, they are relevant to specific ecosystem types and for restricted period of time. In the direct mode, sensitivity analysis of the model output is a crucial step in model verification and validation, which ensures that the response of the computational model to its input parameters conforms to theory (Saltelli et al., 2000). This technique is used to determine how different values of an independent variable will impact a particular dependent variable under a given set of assumptions. The taxonomy of sensitivity analysis methods is complex but one can divide them into two broad categories: local methods and global methods. The local sensitivity analysis methods examine the response of model output by varying model parameters one at a time around a local neighborhood of their central values. The global sensitivity analysis methods examine the global response (i.e., response averaged over variation of all parameters) of model output by exploring a finite region.

9.2.1 First-Order Simulations

Considering a first-order formalism that is common to all versions of PROSPECT as amended from time to time, three "basic" parameters are required to run the model: the leaf structure parameter N; the absorption coefficient k; and the refractive index of leaf material n. The absorption coefficient varies between 0 and a value that depends of the chemical composition of leaves, so it is a little hard to use. We prefer to do simulations with the transmission of the elementary layer τ which ranges from 0 (infinite absorption) to 1 (no absorption). Figure 9.5 shows the variation of leaf reflectance and transmittance as a function of τ for three values of N and $n = 1.45$. For $\tau = 0$ ($k = \infty$), the reflectance equals 0.0352 and the transmittance equals zero regardless of the number of layers N. It means that there is a direct link between the reflectance at high absorption wavelengths and the refractive index of leaf material. For $\tau = 1$ ($k = 0$), the reflectance of a very compact leaf ($N = 1$) equals 0.3833 and it transmittance 0.6167. However, in spite of the controversy caused by Merzlyak et al. (2002), it is generally accepted that this situation does not really exist; the absorptance in the near infrared is not null. Consider a very compact leaf ($N = 1$) with a very small amount of water and dry matter; the transmission of this elementary layer τ is between 0.98 and 0.99, its reflectance between 0.3635 and 0.3732, and its transmittance between 0.5934 and 0.6049. These values are in good agreement with leaf spectra measured in the solar and thermal infrared domains.

The effect of the refractive index on leaf reflectance and transmittance is illustrated by Figure 9.6. For your information, the two extreme values in the visible are $n = 1.33$ for water and

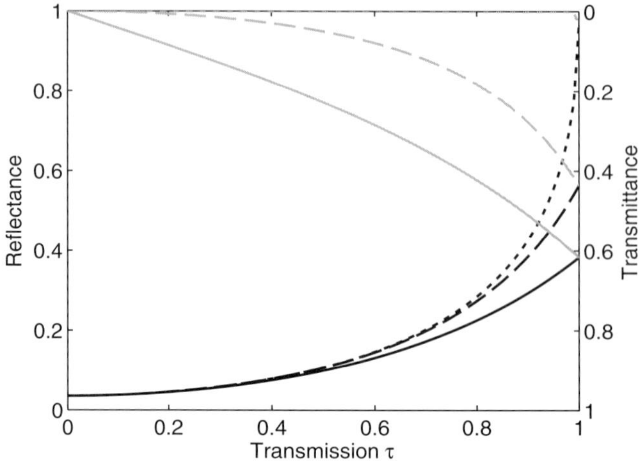

Figure 9.5 Reflectance (black, left scale) and transmittance (gray, right scale) of a leaf made of $N = 1$ layer (solid line), $N = 2$ layers (dash line), and $N = 100$ layers (dotted line), as a function of the transmission τ. In this simulation the refractive index is fixed to $n = 1.45$.

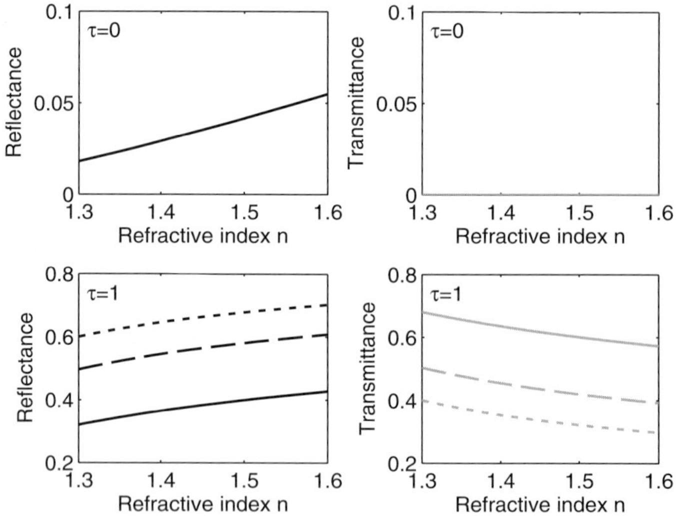

Figure 9.6 Reflectance (black, left plots) and transmittance (gray, right plots) of a leaf made of $N = 1$ layer (solid line), $N = 2$ layers (dash line), and $N = 3$ layers (dotted line), as a function of the refractive index n. In this simulation, the transmission of the elementary layer is fixed to $\tau = 0$ (upper plots) and $\tau = 1$ (lower plots).

$n = 1.56$ for crystalline cellulose (see Section 3.6). For $\tau = 0$ ($k = \infty$), the reflectance is between 0.0180 and 0.0550 and the transmittance equals zero regardless of the number of layers N. For $\tau = 1$ ($k = 0$), the value depends on the number of layers. As for the most compact leaf ($N = 1$), the reflectance is between 0.3202 and 0.4276 and the transmittance between 0.5724 and 0.6798, but the most likely values are 0.3833 and 0.6167, as seen above. These simulations show that leaf

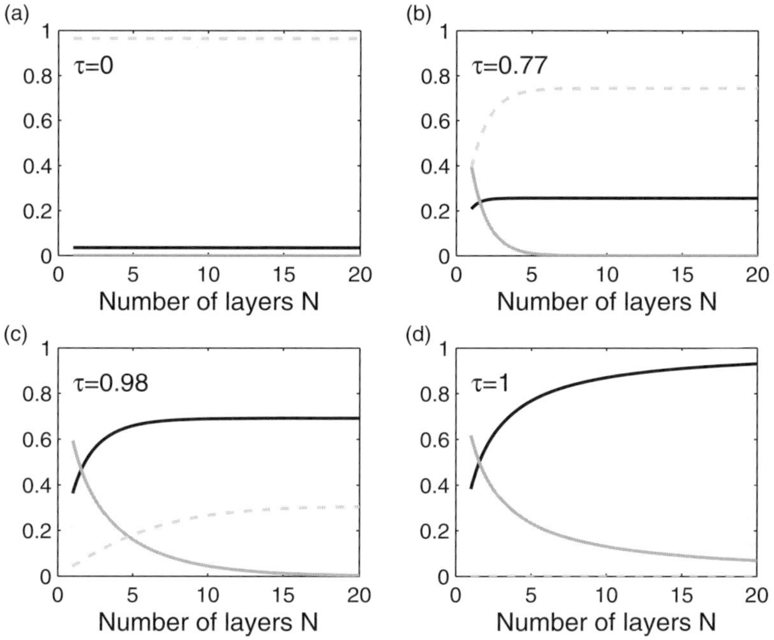

Figure 9.7 Effect of the number of layers N on leaf reflectance (black, solid line), transmittance (gray, solid line), and absorptance (light gray, dash line) for different values of transmission of the elementary layer: (a) $\tau = 0$; (b) $\tau = 0.77$; (c) $\tau = 0.98$; and (d) $\tau = 1$. In this simulation the refractive index is fixed to $n = 1.45$.

optical properties are sensitive to the refractive index, an optical constant that is unfortunately imperfectly known.

The last simulation shows the effect of the leaf structure parameter N on leaf optical properties. The equations used in PROSPECT allow studying the effect of an infinite number of plates starting from a single plate, a classical problem in optics. Figure 9.7 shows the variation of leaf reflectance, transmittance, and absorptance after the addition of a certain number of plates. It is seen that they reach an asymptote as more plates are added. In the case of infinite absorption ($\tau = 0$), the reflectance, the transmittance, and the absorptance are constant and equal to 0.0352, 0, and 0.9648, respectively (Figure 9.7a). This result confirms the earlier comments. In the case of no absorption ($\tau = 1$), the reflectance increases from 0.3833 to 1, the transmittance decreases from 0.6167 to 0, and the absorptance is constant and equal to 0 when N varies from 1 to infinity (Figure 9.7d). Such asymptotic values are actually obtained for $N = 100$ layers. The case of $\tau = 0.77$ is interesting because the transmittance and absorptance of the elementary layer both approximate 0.39, when the reflectance is 0.2086 (Figure 9.7b). Asymptotic values are reached for $N = 5$ layers for reflectance (0.2566) and $N = 10$ for transmittance (0.0001) and absorptance (0.7433). The transmission of a "realistic" compact layer case is 0.98 in the near infrared (Figure 9.7c). The reflectance, transmittance, and absorptance of this layer are 0.3635, 0.5934, and 0.0431; 0.5203, 0.3962, and 0.0834 for $N = 2$; 0.5955, 0.2845, and 0.1200 for $N = 3$; and 0.6919, 0, and 0.3081 for an infinite number of layers. Such asymptotic values are actually obtained for $N = 40$ layers.

9.2.2 One-Factor-at-a-Time Sensitivity Analysis

The one-factor-at-a-time (OFAT) or one-at-a-time (OAT) sensitivity analysis is the most widely used engineering technique and is also called differential sensitivity analysis. It consists of varying the value for one parameter (or factor) at a time, while all other input parameters remain constant, and studying the impact of each parameter on the spectra. It is easy to implement but may be computationally expensive. Consider a model containing p independent input parameters $X = (X_1, X_2, \ldots, X_p) \in \mathbb{R}^p$ and one dependent variable, the model output $Y = f(X) \in \mathbb{R}$. The variance of Y is calculated using the general error propagation formula, i.e.:

$$V(Y) = \sum_{i=1}^{p} \left(\frac{\partial Y}{\partial X_i} \right)^2 V(X_i) \tag{9.16}$$

The variance in Y is utilized as a measure of uncertainty in model predictions, while the variance in X_i, weighted by the first-order partial of Y with respect to X_i, provides a measure of model sensitivity to X_i. Let us consider PROSPECT-D, the latest version of the model including anthocyanins (Féret et al., 2017). Simulated spectra in Figure 9.8 show that N and C_m affect the entire wavelength range between 400 nm and 2500 nm; C_{ab}, C_{car}, and C_{ant} only affect the wavelength range from 500 nm to 800 nm; and C_w only affect the wavelength range from 800 nm to 2500 nm, respectively.

Such a method can be found in other studies (e.g., Jacquemoud et al., 1996; Ceccato et al., 2001; Ramalingam et al., 2001; Zarco-Tejada et al., 2003b; Shi et al., 2005, 2007; Wang et al., 2011b; Hunt and Daughtry, 2014; Conejo et al., 2015; Croft et al., 2015). This mode of model evaluation has serious limitations, first, because the choice of the initial parameter set is often arbitrary and decided without rules and second, because it does not provide insight into how the interactions between parameters influence the response of the model. In many real-world systems, factors rarely act independently.

9.2.3 Global Sensitivity Analysis

To overcome the limitations of local methods, one special class of sensitivity analysis techniques, referred to as global sensitivity analysis, was developed in the late 1980s (Saltelli et al., 2000). In this technique, the model parameters (factors) are varied simultaneously in model runs, rather than sequentially. The full range of model parameter variability is therefore exhaustively searched rather than just a local area around a mean value. In doing so, results of such a global sensitivity analysis provide information on the interactions of the parameters in driving model output variability. Although global methods themselves can be subdivided into finer categories, one class is based on a decomposition of the variance of the output into terms corresponding to the different inputs and their interactions. The variance-based methods of global sensitivity indices are based on Sobol' sensitivity indices (Sobol', 1993), which became very popular among practitioners due to the easiness of its interpretation (Appendix H). The global sensitivity analysis techniques differ in their algorithms for exploring multi-dimensional spaces: full factorial design, quasi-Monte Carlo methods, Sobol' sequences (or LP_T sequences), or some form of stratified sampling, such as the Latin hypercube sampling. Global sensitivity analyses of

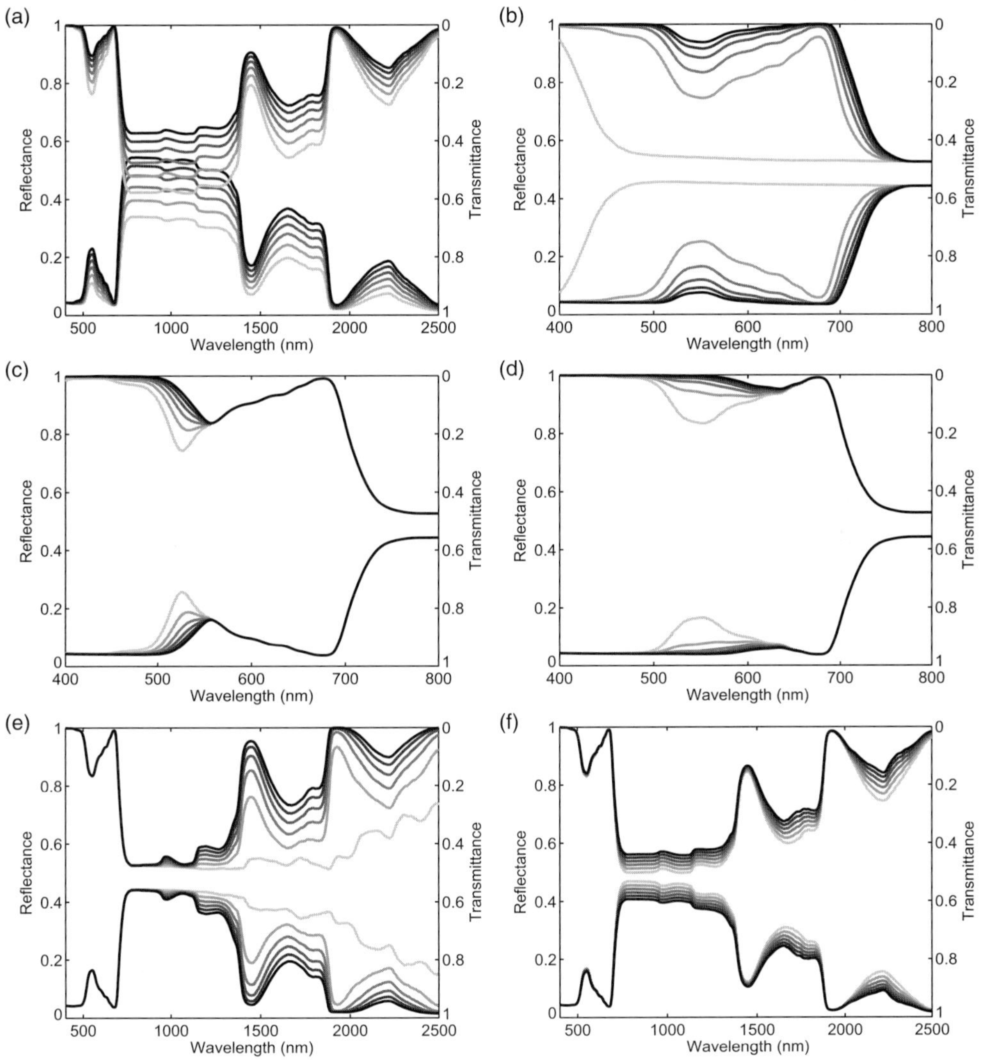

Figure 9.8 Prediction of leaf reflectance (lower lines) and transmittance (upper lines) for different values of PROSPECT-D input parameters. Variations around a standard leaf defined by ($N = 1.5$, $C_{ab} = 32$ μg cm^{-2}, $C_{car} = 8$ μg cm^{-2}, $C_{ant} = 0$ μg cm^{-2}, $C_w = 0.016$ cm, and $C_m = 0.009$ g cm^{-2}) are considered with: (a) the leaf structure parameter (N= 1, 1.25, 1.5, 1.75, 2, 2.25); (b) the leaf chlorophyll content (C_{ab}= 0, 16, 32, 48, 64, 80 μg cm^{-2}); (c) the leaf carotenoid content (C_{car}= 0, 4, 8, 12, 16, 20 μg cm^{-2}); (d) the leaf anthocyanin content (C_{ant}= 0, 8, 16, 24, 32, 40 μg cm^{-2}); (e) the leaf equivalent water thickness (C_w= 0, 0.008, 0.016, 0.024, 0.032, 0.040 cm); and (f) the leaf dry matter content (C_m= 0.003, 0.006, 0.009, 0.012, 0.015, 0.018 g cm^{-2}). The carotenoid content in subplot (b) is not constant but equal to the chlorophyll content divided by four to keep the simulation realistic.

various versions of PROSPECT have been performed in different wavelength domains using different approaches (Table 9.1). Although they globally provide similar results, the method used for performing the calculations and the range of variation for each parameter may have an influence on them.

Table 9.1 *Global sensitivity analysis of* PROSPECT *using different sampling methods. Range of variation for each input parameter: leaf structure parameter N*, chlorophyll content C_{ab} (µg cm^{-2}), carotenoid content C_{car} (µg cm^{-2}), equivalent water thickness C_w (cm), dry matter content C_m (g cm^{-2}), protein content C_{pro} (g cm^{-2}), and cellulose + lignin content $C_{cel+lig}$ (g cm^{-2}). The values listed in the table are rounded. NC is the number of combinations. * Gamma (instead of normal) distributions are used.

	N	C_{ab}	C_{car}	C_w	C_m	C_{pro}	$C_{cel+lig}$	NC	Method
Ceccato et al. (2001)	1–4	×	×	0.0001–0.07	0.002–0.040	×	×	9987	eFAST
Bowyer and Danson (2004)	×	×	×	0.0078–0.02	0.009–0.018	×	×	5000	eFAST
Pavan et al. (2004)	1–3	5–95	×	0.0025–0.05	0.003–0.019	×	×	1331	DOES
Shi et al. (2008)	1–4	0–60	×	0.0001–0.07	0.001–0.040	×	×	4996	eFAST
Li and Wang (2011)	1–3	10–60	×	0.0026–0.02	0.003–0.011	×	×	10000	Sobol
Xiao et al. (2014a, 2014b)	1–2	10–120	5–30	0.005–0.04	0.002–0.014	×	×	2000	eFAST
Wang et al. (2015 a, 2015b)	1–3	×	×	0.001–0.05	×	0.0001–0.002	0.0001–0.005	?	FAST
Jay et al. (2016)	1.52 ±0.23	32.8 ±18.87	8.51 ±3.92	0.0122± 0.0061	0.0078± 0.0036	×	×	5000	eFAST

Table 9.2 *Values of the input parameters used for the computations conducted with* PROSPECT-D.

Parameter	Unit	Column of the $L_{343}7^{57}$ design	Range of variation	Levels
N	unitless	1	1 – 2.25	1.0625 | 1.25 | 1.4375 | 1.625 | 1.8125 | 2.00 | 2.1875
C_{ab}	μg cm^{-2}	2	0 – 80	4 | 16 | 28 | 40 | 52 | 64 | 76
C_{car}	μg cm^{-2}	9	0 – 20	1 | 4 | 7 | 10 | 13 | 16 | 19
C_{ant}	μg cm^{-2}	17	0 – 40	2 | 8 | 14 | 20 | 26 | 32 | 38
C_w	g cm^{-2} or cm	26	0 – 0.04	0.002 | 0.008 | 0.014 | 0.020 | 0.026 | 0.032 | 0.038
C_m	g cm^{-2}	32	0.0030 – 0.0180	0.0037 | 0.0060 | 0.0083 | 0.0105 | 0.0127 | 0.0150 | 0.0172

9.2.3.1 Design of Experiments for Simulation (DOES)

Also called experimental design, this method is a structured approach for conducting experiments. It allows quantifying the relative contributions of the input parameters of a model on their outputs, and then organizes them into a hierarchy (e.g., Benoist et al., 1994). Hyper Graeco-Latin squares, an extension of the basic Latin hypercube design, are efficient fractional factorial designs where all factors have the same number of levels. As an orthogonal subset of the complete table (full factorial design), they greatly reduce the number of simulations. They were used for the first time in radiative transfer models by Bacour et al. (2002b) with the PROSAIL model, and later by Pavan et al. (2004) with PROSPECT-4. The experiment table used hereafter with PROSPECT-D is named $L_{343}7^{57}$: it is designed to study up to 57 factors with 7 levels in $N_e = 343$ experiments. Because this plan is not complete – as seen above, a complete plan would include 7^{57} experiments – columns of some actions are aliases of others, that is, they are linearly combined and therefore unsuitable to conduct a sensitivity study. For those reasons, only six independent parameters are considered with this table. The values of the parameters are summarized in Table 9.2: the lowest (highest) values correspond to the lower (upper) bounds of the ranges of variation, increased (decreased) by 5%.

Let us assume that the level m of a parameter X_i appears N_m times in the experiment table. The effect $E_{X_i,m}(\lambda)$ can be then expressed as

$$E_{X_i,m}(\lambda) = \frac{1}{N_m} \sum_{N_m} Y_{X_i,m}(\lambda) - \overline{Y}(\lambda) \tag{9.17}$$

where $\overline{Y}(\lambda)$ is the averaged leaf reflectance (or transmittance) over the N_e numerical experiments, in the waveband λ. A positive (or negative) effect indicates that, for the corresponding level of the parameter, the reflectance increases (or decreases) by this percentage, in relation to the mean reflectance. The assessment of the effect of leaf structure parameter N, chlorophyll content C_{ab}, and water content C_w on leaf reflectance is illustrated in Figure 9.9. For instance, when C_{ab} is equal to 4 μg cm^{-2} (the lowest value), the computed reflectance is about 118% higher than the mean reflectance at 700 nm, 210% at 670 nm, and 35% at 550 nm, and there is no effect at 800 nm. When C_{ab} is equal to 76 μg cm^{-2} (the highest value), the reflectance is about 56% lower than the mean reflectance at 700 nm, 46% at 670 nm, and 22% at 550 nm, while there is no effect at 800 nm. Therefore, the slope of the curve representing the parameter effect is instructive about the sensitivity

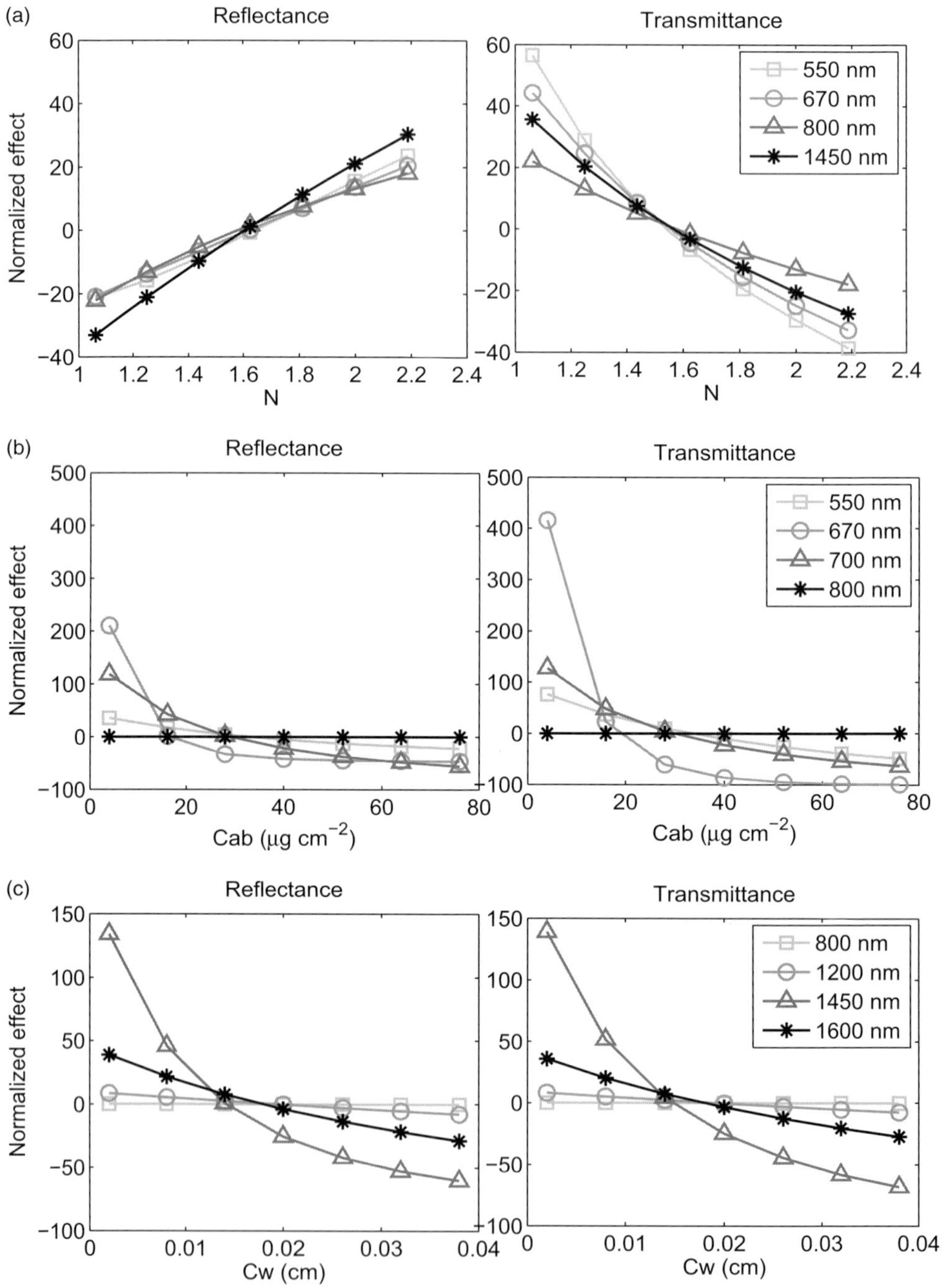

Figure 9.9 Effect of the leaf: (a) structure parameter N; (b) chlorophyll content C_{ab}; and (c) water content C_w on leaf (left) reflectance and (right) transmittance at four wavelengths.

of that parameter. A positive (negative) slope means that increasing the value of the given parameter leads to an increase (decrease) the reflectance. Also, the steeper the slope, the more sensitive the reflectance is to variations in the considered parameter. Figure 9.9a clearly show a quasi-linear positive (negative) effect of N on reflectance (less on transmittance) at all wavelengths. Figures 9.9b show an exponential effect of C_{ab} at wavelengths where chlorophyll displays a high absorption coefficient (670 nm and 700 nm) and a quasi-linear effect at wavelengths where it displays a lower absorption coefficient (550 nm). At 670 nm saturation occurs for chlorophyll contents greater than 40 μg cm^{-2}. Finally, Figure 9.9 shows an exponential effect of C_w at wavelengths where water displays a high absorption coefficient (1450 nm) and a quasi-linear effect at wavelengths where it displays a lower absorption coefficient (1200 nm and 1600 nm). The curves obtained for leaf transmittance are very similar. From a general point of view, transmittance is more sensitive to leaf parameters than reflectance. These results agree well with the simulated spectra in Figure 9.8.

A way to rank the most influential parameters into a hierarchy is achieved by considering the relative contribution of each factor to the leaf reflectance or transmittance. For each wavelength, such a contribution index characterizes the contribution of each parameter to the output's variance. It is evaluated by

$$C_{X_i}(\lambda) = \frac{1}{N_l}\sum_{N_l} E_{X_i,m}(\lambda) \Big/ \frac{1}{N_e}\sum_{N_e}\left(Y(\lambda) - \overline{Y}(\lambda)\right)^2 \tag{9.18}$$

where N_l is the number of levels taken by each parameter ($N_l = 7$). Figure 9.10 illustrates the contribution to the leaf reflectance, expressed in percent, of each input parameter with respect to the wavelength. In the visible, C_{ab} has the greatest influence (~90% in the red and ~20% in the blue), followed by C_{ant} (~60% in the green) and, equally, by C_{car} (~10% in the blue) and N (~10% in the green). Note that the sum of the first-order contributions ranges from 30% in the blue to 100% in the near infrared. Low values in the visible are mainly due to interaction effects between pigments. Xiao et al. (2014) obtained similar results using the eFAST method. In the near-infrared, N is found to have

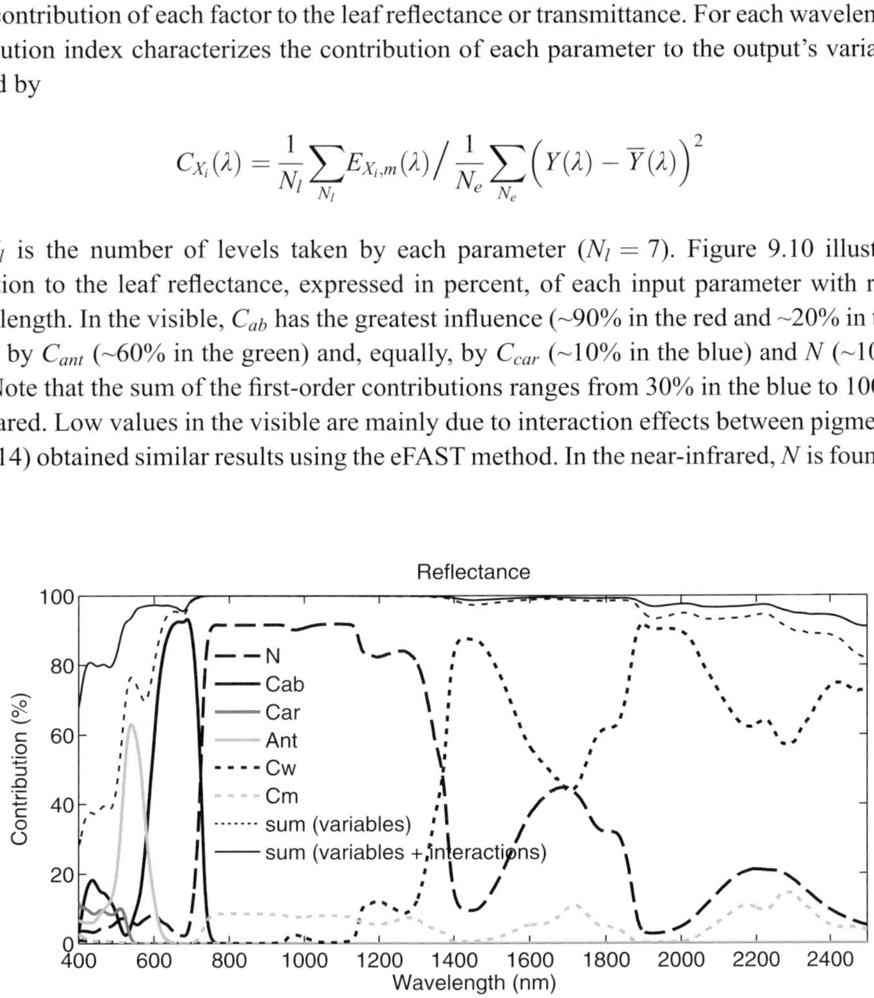

Figure 9.10 Contribution as a percentage of the six input parameters of PROSPECT-D to leaf reflectance spectrum. The interactions considered in this simulation only concern pigments and are detailed in Figure 9.11.

the greatest influence (~90%), while C_m accounts for the remaining part (~10%); there is no interaction between these two parameters in this domain. After 950 nm, C_w has great influence, accounting for about 10%, 85%, and 90% of contributions in the absorption peaks of water at 1200 nm, 1450 nm, and 1950 nm, respectively. N and C_m may also significantly affect the reflectance spectra outside these absorption peaks, although the contribution of the dry matter content never exceeds 10%. It may affect the retrieval of dry matter.

Interactions between parameters occur when the effect of parameter A on a response depends on the level of parameter B. The combined change in two parameters may produce a greater effect than the sum of effects expected from either variable alone. Such information may be helpful in inversion because if the parameters are not independent, the inverse problem may be ill-posed. The interaction effect between parameter A (level m) and parameter B (level n) is calculated as follows:

$$E_{A_m,B_n}(\lambda) = \overline{Y}_{A_m,B_n}(\lambda) - \overline{Y}(\lambda) - E_{A_m}(\lambda) - E_{B_n} \tag{9.19}$$

and their contribution is calculated by

$$C_{A_m,B_n}(\lambda) = \frac{1}{N_A N_B} \sum_{N_A} \sum_{N_B} E^2_{A_m,B_n} \Big/ \frac{1}{N_e} \sum_{N_e} \left(Y(\lambda) - \overline{Y}(\lambda) \right)^2 \tag{9.20}$$

The results presented in Figure 9.11 show that the interactions between pigments may have a greater impact on the reflectance than some parameters alone. This is typically the case of the carotenoid content. The sum of all second-order contributions is about 40% from 400 nm to 500 nm.

9.2.3.2 *Fourier Amplitude Sensitivity Test (FAST)*

FAST is one of the most popular global sensitivity analysis techniques (Xu and Gertner, 2011). In this method, a probability density function (PDF) and a range of variation are specified for each of the model input parameters. Then a Monte Carlo procedure is used to produce adequate parameter sets.

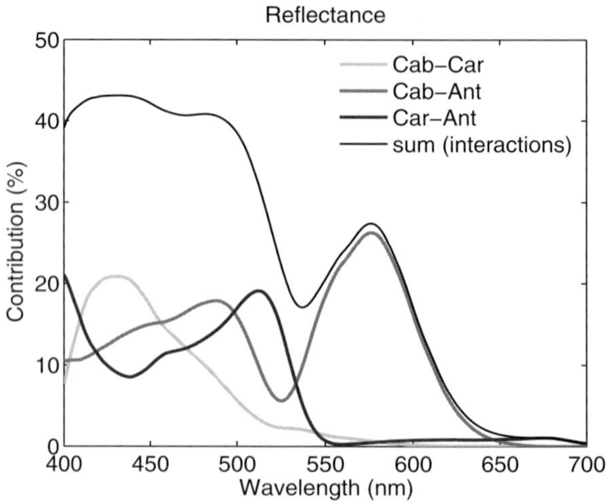

Figure 9.11 Contribution as a percentage of the interactions of the three pigments of PROSPECT-D to leaf reflectance spectrum.

Table 9.3 *First-order and total-order indices calculated on* PROSPECT-4 *(Ceccato et al., 2001).*

	First-order index		Total-order index	
	ρ_{820}	ρ_{1600}	ρ_{820}	ρ_{1600}
N	0.744	0.395	0.744	0.411
C_w	0	0.350	0	0.364
C_m	0.256	0.216	0.256	0.225

A periodic sampling approach is used to generate a search curve, which approximately scans the entire parameter space of the model inputs. The periodic sample of each parameter is assigned with a characteristic frequency (i.e., a distinct integer), which is used to determine the parameter's contribution to the variance of a model output based on a Fourier transformation (e.g., Saltelli and Bolado, 1998; Saltelli et al., 1999, 2000). Wang et al. (2015b) applied the FAST method to estimate the effects of PROSPECT parameters on the reflectance of fresh and dry leaves, from 1100 nm to 2500 nm, using the ranges of input parameters presented in Table 9.2. They first performed a new calibration of an early version of the model that decomposed the dry matter content into protein (C_{pro}) and cellulose + lignin ($C_{cel+lig}$) content (Fourty et al., 1996; Jacquemoud et al., 1996). For fresh leaves, they showed a large contribution of C_w, followed by N, while the influence of C_{pro} and $C_{cel+lig}$ was very low (< 5%). Compared to fresh leaves, the influence of C_{pro} and $C_{cel+lig}$ on leaf spectra for dry leaves is greater in wavelengths longer than 1500 nm.

Whereas FAST can only determine first-order Sobol' indices for all the parameters of a model (Appendix H), a more efficient sampling procedure called eFAST (Extended Fourier Amplitude Sensitivity Test) can be used to estimate total-order indices. Global sensitivity analyses of PROSPECT-4 have been performed using eFAST to quantify both the effect and interactions of model parameters (Ceccato et al., 2001; Bowyer and Danson, 2004; Shi et al., 2008; Xiao et al., 2014; Jay et al., 2016). If there is no *a priori* information to guide the selection of the statistical distributions, uniform distributions are generally chosen for each input parameter on their range of variation (Table 9.2). Ceccato et al. (2001) were the first to perform a global sensitivity analysis of PROSPECT; they selected different distributions for N and C_m (uniform) and for C_w (logarithmic), and they calculated first-order and total-order indices for these three parameters at 820 nm and 1600 nm (Table 9.3). The sum of the first-order indices equals 100% at 820 nm and 96.1% at 1600 nm. It means that there is no or little interaction between the parameters. At 820 nm, the total-order indices show that N has the greatest effect on leaf reflectance (74.4%) and that C_m accounts for the remaining part of the uncertainty of the output (25.6%). At 1600 nm, N still has the greatest effect on reflectance (41.1%), whereas C_w and C_m respectively account for 36.4% and 22.5% of the variance.

Bowyer and Danson (2004) extend the sensitivity analysis of PROSPECT-4 to the shortwave infrared (SWIR) spectrum, but they do not consider the leaf structure parameter so that their results cannot be compared to other studies. The work performed on PROSPECT-4 (Shi and Sun, 2007; Shi et al., 2008; Li and Wang, 2011) and on PROSPECT-5 (Xiao et al., 2014) show similar results over the solar spectrum. It can be seen in Figure 9.12 that the most sensitive region to variation in C_{ab} is unsurprisingly located in the visible; that N has the greatest effects in the near infrared, but also in the visible and shortwave infrared; that the effect of C_w is maximum in the shortwave infrared, especially

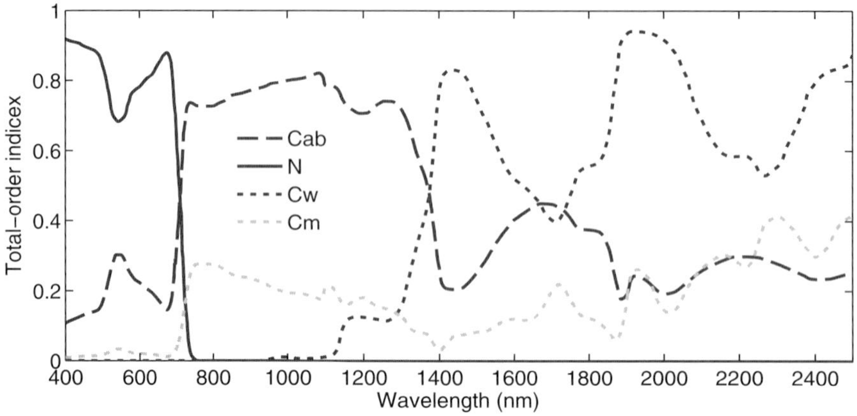

Figure 9.12 Total-order indices of PROSPECT-4 parameter calculated using the eFAST method (redrawn from Shi et al. (2008)).

near the three main water absorption features at around 1420 nm, 1940 nm, and 2480 nm; and that the effect of C_m is much lower, which means the retrieval of dry matter by model inversion may be more difficult.

9.2.4 Calibration of Spectral Indices and Other Methods

The development of spectral indices or other retrieval methods for analysis of leaf chemical and structural properties has long relied on experimental datasets. There is a large variety of published spectral indices for leaf pigment, water, or dry matter retrieval (see Section 11.1), but inter-comparison studies intended to find the "best" index in terms of reliability have underscored several problems. In particular, reaching consensus is difficult because the calibration (training) and validation datasets have long been limited to one or few closely related species, and they do not represent the whole range of possible conditions encountered in nature. Therefore methods based on small-sized datasets may lead to specialized indices that perform poorly when applied to a wide variety of plant leaves and conditions. The best way to design efficient indices with broad applicability is to use large datasets containing thousands of leaves spanning a large range of chemical composition and distribution, phenological stage, and leaf anatomy. Unfortunately, it is difficult and expensive to obtain such large datasets. Moreover, measurement errors associated with both parameters of interest and the radiometric data may propagate into uncertainties and biases in the indices.

Synthetic leaf reflectance and/or transmittance spectra generated using PROSPECT, and their corresponding input parameters, can be used to develop statistical relationships between leaf optical and chemical or structural properties, which are applied to experimental data without any readjustment. It requires that the model be faithful enough to offer a good representation of reality in leaves, otherwise model uncertainties may dominate all other sources of uncertainties for some leaves. For example, Le Maire et al. (2004) tested more than 60 published chlorophyll indices on a simulated database containing about 11 583 spectra to determine those that performed best over a wide range of species. The range for each input parameter was derived from experimental observations. These authors found that the early version of PROSPECT that they used did not perform well due to the rough specific absorption coefficients for chlorophyll and dry matter. Wang et al. (2012) came to a similar conclusion with spectral indices dedicated to the estimation of chlorophyll, water, and dry matter content, and Wang et al. (2016) with spectral indices dedicated to the estimation of carotenoids. Hunt and Daughtry (2014) successfully used PROSPECT to predict the relationship between chlorophyll content and the SPAD-502 value (see Section 4.6). They found that changing the leaf structure parameter resulted in different relationships, which were similar to the measured range of variation from calibration equations found in the literature, and suggested that tracing back to the transmittance signal in the red and the near infrared may improve the estimation of leaf chlorophyll. The addition of carotenoids in PROSPECT-5, as well as the new calibration of the specific absorption coefficients, has provided substantial improvements. Several methods used in remote sensing to retrieve vegetation chemical composition (see Chapter 11) have been trained on synthetic datasets and validated against observations; multiple stepwise regression analysis (Liu et al., 2014; Casa et al., 2015); partial least squares regression (Féret et al., 2011); spectral indices (Wang et al., 2016; Zhou et al; 2017); or continuous wavelet analysis (Blackburn and Ferwerda, 2008; Cheng et al., 2012, 2014; Fang and Ju, 2015; Li et al., 2016). Yang et al. (2015) calibrated a version of the model called N-PROSPECT and including leaf nitrogen density (*LND* in μg cm^{-2}) as an input parameter instead of chlorophyll content, then they applied a partial least squares regression to determine *LND* from leaf reflectance. In most studies, the datasets were usually generated by combining discrete values of the input parameters of the model (e.g., full factorial design, Latin hypercube design) or by assigning uniform distributions to them. It produces unrealistic combinations, which may affect the performance of the spectral indices or other retrieval methods for analysis of leaf chemical constituents. Féret et al. (2011) proposed another strategy, where they generated leaf spectra by randomly selecting leaf chemical and structural properties based on the multivariate normal distributions and covariance matrix produced from experimental datasets collected in various ecosystems. This approach provides robust predictions and a new perspective into understanding leaf spectral responses to leaf traits across a wide range of plant species.

Lewis and Disney (2007) developed a new approximation to the PROSPECT model by applying the spectral invariant approach to leaf internal scattering, that is, relating leaf biochemical constituents to leaf single scattering albedo ω, in a similar manner to that used to describe canopy-level scattering. They varied the refractive index n over the range 1.27 to 1.52 and the leaf structure parameter N over the range 1.0 to 2.5. They showed that if N highly affects leaf reflectance and transmittance in the near infrared, it has a small impact on ω, confirming global sensitivity analyses of the PROSAIL model. The impact of N was somewhat higher, particularly for low absorption coefficients.

9.3 Model Inversion

PROSPECT has been used in different ways to retrieve leaf characteristics. In the forward (or direct) problem, the input parameters of the model are given, and we want to calculate leaf reflectance and/or transmittance. Conversely, the inverse problem consists of extracting the biophysical properties of the leaf from experimental measurements (Figure 9.13). Prior information on leaf traits also may be included in the retrieval process to improve performance. The inversion of PROSPECT has given rise to many articles. The success in using this model can be explained by the relatively small number of input variables required, combined with good computer efficiency. Several retrieval techniques can

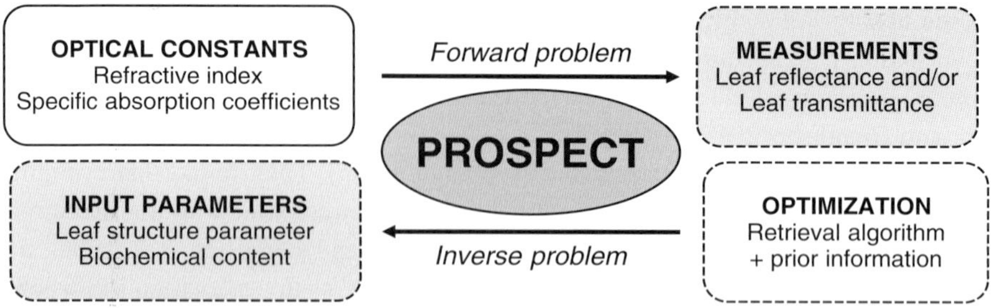

Figure 9.13 Forward and inverse problems.

be used, but they can be split into two main approaches, depending on whether emphasis is put on radiometric data or on the parameters of interest (Baret and Buis, 2008).

9.3.1 Approaches Based on the Parameters Space: Artificial Neural Network

In this approach, a parametric model representing the inverse model is adjusted over a learning dataset, which can be built using experimental measurements or generated using PROSPECT. It basically consists of fitting a surface response, that is, a relationship between the parameters of interest (explanatory variables) and leaf reflectance and/or transmittance at several wavelengths (response variables). Once calibrated, the parametric model is run to compute these parameters from the observed radiometric data. Although often considered as a "black box," artificial neural network (ANN) has been exploited to adjust such surface responses (see Section 12.7). ANN was first used to invert PROSPECT in the frame of the RAMIS project (see Section 4.6). It was viewed as an efficient way to retrieve leaf chemical constituents using broadband reflectances or transmittances, depending on the version of the prototype (de Rosny et al., 1995; Conejo et al., 2010, 2015). To overcome two difficulties with regard to the insufficient number of leaf spectra and the large number of spectral bands, with high correlations between adjacent bands, Shi and Sun (2007) simulated reflectance spectra using PROSPECT and performed a global sensitivity analysis to reduce the spectral dimensionality of the problem. They trained an ANN on four bands and band combinations (spectral indices) sensitive to chlorophyll and leaf water. Results of the estimation of these two parameters on rice (*Oryza sativa*) leaves showed very satisfactory accuracy.

9.3.2 Approaches Based on the Radiometric Space

Look up tables (LUT) and iterative methods for optimization are, among other minimization techniques, the most widely used to infer leaf characteristics using PROSPECT. No proper calibration step is required in this approach, which consists in finding the best match between measured and simulated radiometric values. The goodness of the fit is quantified by a cost function J (also called merit function) that may explicitly account for measurements and model uncertainties. This function is the sum over the N observational configurations (wavelengths and/or direction) of squared differences between measured (R_{meas} and T_{meas}) and simulated (R_{mod} and T_{mod}) values, weighed by the uncertainties (σ^2) associated with both the radiometric values and model ambiguities. When available, prior information may be used to assess the cost function, the most general form of which can be written as

$$J = \sum_{i=1}^{N} \frac{\left(R_{\text{meas},i} - R_{\text{mod},i}\right)^2}{\sigma_{R,i}{}^2} + \sum_{i=1}^{N} \frac{\left(T_{\text{meas},i} - T_{\text{mod},i}\right)^2}{\sigma_{T,i}{}^2} + \sum_{j=1}^{k} \frac{\left(v_j - v_j^{\text{prior}}\right)^2}{\sigma_{\text{meas},j}{}^2} \tag{9.21}$$

where v_j is the estimated value of the simultaneously retrieved parameter and v_j^{prior} is the measured prior value of the parameter j. When no prior information is available and when uncertainties associated to each configuration used are assumed independent and Gaussian, we can define an unconstrained cost function.

LUTs are conceptually simple, although their implementation may be tricky. The first step consists in generating a large table of radiometric values (broadband or spectral values) by running the model in forward mode with known input parameters. The maximum and minimum values of the parameters are set based on prior information. The parameter space can be sampled systematically by crossing all possible combinations of the parameters in fixed steps, or randomly with values uniformly distributed in the specified range. To avoid very large tables that could be difficult to manipulate, a limited number of input parameters are generally chosen. In the inversion step, for a measured spectrum with unknown parameters, the closest modeled radiometric values are searched in the table according to a cost function (Eq. 9.21) and their average parameters are used as inversion results. The uncertainties associated with the radiometric values may be accounted for in the distribution of the solution. The implementation of an LUT in algorithmic operational chains is very efficient because the radiative transfer model is run off-line. The first attempt to invert PROSPECT using an LUT was made by de Rosny et al. (1995), who varied the leaf structure parameter, the chlorophyll and carotenoid contents, and built a three-dimensional table containing 10^3 simulations. They checked the validity of the method on synthetic data. LUTs generated by PROSPECT have been used to retrieve leaf chemical content (e.g., Romero et al., 2007, 2012; Miao et al., 2011; Ali et al., 2015, 2016; Buddenbaum and Hill, 2015).

The inverse problem can be solved by applying iterative methods for optimization, which consist in finding the best values of the input parameters of PROSPECT that minimize the cost function J. As seen earlier, that function may explicitly take into account measurements and model uncertainties and prior information on the parameters. Note also that if only reflectance values are available, J can be calculated with the first term of Eq. 9.21. Because of the high nonlinearity of the model, the inverse problem is numerically solved using adapted algorithms (gradient, quasi-Newton, Levenberg–Marquardt, simplex, etc.). Although the algorithms may converge toward a local minimum of the cost function, PROSPECT does not seem to be much affected by the so-called ill-posed problem, such that the solution is generally unique and depends continuously on the data. Jacquemoud and Baret (1990) made the first attempt to estimate leaf chlorophyll and water content from reflectance and transmittance spectra using PROSPECT. They showed a good global fit between measured and modeled spectra (Figure 9.14) and a good retrieval of leaf chemical content.

Since then, inversions of the successive versions of PROSPECT have been performed over thousands of leaf species gathered in various ecosystems to retrieve all the input parameters, or a selection (Table 9.4). Problems of estimation can be improved using specific merit functions and specifying the wavelengths for each parameter compared to the standard version that optimizes with

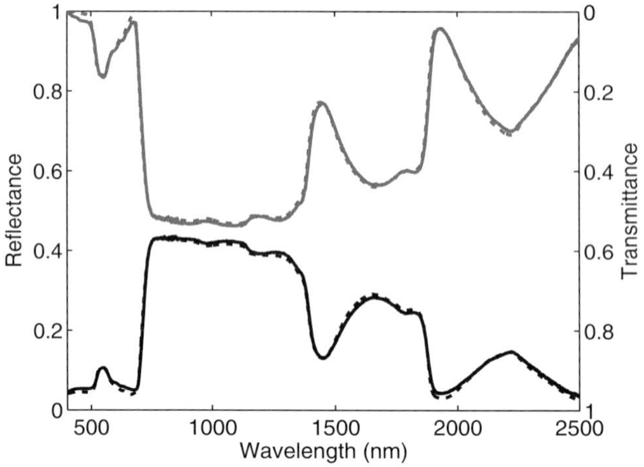

Figure 9.14 Graphical representation of PROSPECT-5 model output, showing measured tulip tree (*Liriodendron tulipifera*) leaf reflectance (black dots) and transmittance (gray dots) spectra, model fitted reflectance (black line), transmittance (gray line). The retrieved values for this leaf are $N = 1.28$, $C_{ab} = 37.4\ \mu g.cm^{-2}$, $C_{car} = 7.14\ \mu g.cm^{-2}$ $C_w = 0.0117$ cm, and $C_m = 0.0050\ g.cm^{-2}$.

one merit function (Li and Wang, 2011). Ma et al. (2012) found that for plants with polymorphic leaves, inversions of PROSPECT improved retrieval of biochemicals if each leaf type was separately calibrated. Using simulations, Riaño et al. (2005b) showed that in wet leaves, water masks the spectral absorption of dry matter and they proposed using a species-specific value of dry matter that improved inversion results. Generally, the results clearly demonstrate the ability of the model to estimate the basic leaf biophysical properties with a very good accuracy. The direct comparison between the values of the parameters retrieved from the inversion and those available from laboratory measurements is a *sine qua non* condition to truly validate PROSPECT. Indeed, a good fit of the spectral measurements by the model does not necessarily imply that the functional dependencies between the radiometric data and the leaf parameters are correctly established.

9.4 Link of PROSPECT with a Leaf BRDF Model

As seen in Section 8.3.2, Bousquet et al. (2005) derived a simple and general expression of leaf BRDF, which has been resumed and improved by other authors (e.g., Chelle, 2006; Stuckens et al., 2009; Comar et al., 2014; Jay et al., 2015, 2016). That model requires three input parameters: the refractive index of leaf material n, the roughness parameter σ, and the Lambert coefficient k_L. It was found that n and σ were almost wavelength independent in the VIS-NIR, so that one could replace them by their mean values; and that the spectrum of k_L displayed a general shape similar to that of leaf directional-hemispherical reflectance as measured in an integrating sphere, such as simulated by PROSPECT. As well as BRDF, leaf directional-hemispherical reflectance factor (DHRF) can be split into two components, a diffuse component $DHRF_{diff}$ and a specular component $DHRF_{spec}$:

$$DHRF(\lambda, \theta_i) = DHRF_{diff}(\lambda, \theta_i) + DHRF_{spec}(\lambda, \theta_i) \tag{9.22}$$

Table 9.4 *Retrieval of leaf biochemical constituents by inversion of* PROSPECT. *The star (*) in the second column means that the model has been recalibrated. Several inversion algorithms based on iterative methods for optimization have been used: QN = Quasi-Newton, S = Simplex, GRG = generalized reduced gradient, CG = Conjugate Gradient, LM = Levenberg–Marquardt, and SA = Simulated Annealing. Inversions were performed on leaf reflectance (R) and/or transmittance (T).*

Version of the model	Reference	Wavelength range (nm)	Inversion method	RT	Input parameter	n	R^2	RMSE
PROSPECT	Jacquemoud and Baret (1990)	400–2500	?	RT	C_{ab}	52	0.98	3.67 μg cm^{-2}
					C_w	38	0.96	0.0029 cm
PROSPECT-2	Jacquemoud et al. (1996)	400–2500	QN	RT	C_{ab}	63	0.68	×
					C_w	63	0.95	×
					C_{pro}	63	0.10	×
					$C_{cel+lig}$	63	0.39	×
	Botha et al. (2006)	400–2500	S	RT	$C_{ab,11}$	195	0.14	6.45 μg cm^{-2}
					$C_{ab,12}$	194	0.66	3.48 μg cm^{-2}
					$C_{ab,21}$	83	0.43	5.41 μg cm^{-2}
					$C_{ab,22}$	112	0.64	3.72 μg cm^{-2}
					$C_{pro,11}$	195	0.01	0.0058 mg cm^{-2}
					$C_{pro,12}$	194	0.17	0.0010 mg cm^{-2}
					$C_{pro,21}$	83	0.25	0.0019 mg cm^{-2}
					$C_{pro,22}$	112	0.07	0.0011 mg cm^{-2}
	Wang et al. (2015b)	400–2500	?	RT	C_{pro}	63	0.66	0.00020 g cm^{-2}
					$C_{cel+lig}$	63	0.83	0.00039 g cm^{-2}
PROSPECT-SWIR	Baret and Fourty (1997)	1300–2400	S	RT	C_w	58	×	0.0025 cm
					C_m	58	×	0.0016 g cm^{-2}
PROSPECT-3	Jacquemoud et al. (2000)	400–2500	S	RT	C_{ab}	63	0.67	9.1 μg cm^{-2}
					C_w	63	0.95	0.0018 cm
					C_m	63	0.65	0.0022 g cm^{-2}
	Newnham and Burt (2001)	400–2500	GRG	RT	C_{ab}	28	0.78	×
					C_w	28	0.93	×
	Li et al. (2003)	400–900	CG+S	RT	C_{ab}	×	0.93	1.99 μg cm^{-2}
	Riaño et al. (2005b)	400–2500	?	RT	C_w	245	0.94	×
					C_m	245	0.38	×
					FMC	245	0.33	×
	Renzullo et al. (2006a)	400–1600	LM	R	C_{ab}	96	0.70	10.63 μg cm^{-2}
	Barry et al. (2009)	400–1600	?	RT	C_{ab}	112	0.84	11.57 μg cm^{-2}
	Buddenbaum et al. (2011)	400–2500	CG	R	$C_{ab,1}$	×	0.78	×
					$C_{ab,2}$	×	0.54	×
					C_w	236	0.75	×
	Colombo et al. (2008)	400–1600	QN	RT	C_w	144	0.65	×
					C_m	144	0.01	×

Table 9.4 (*cont.*)

Version of the model	Reference	Wavelength range (nm)	Inversion method	RT	Input parameter	n	R^2	RMSE
PROSPECT-4	Féret et al. (2008)	400–2500	S	RT	$C_{ab,1}$	64	×	28.92 μg cm^{-2}
					$C_{ab,2}$	106	×	7.83 μg cm^{-2}
					$C_{ab,3}$	276	×	5.85 μg cm^{-2}
					$C_{ab,4}$	41	×	14.25 μg cm^{-2}
					$C_{w,1}$	580	×	0.0017 cm
					$C_{w,3}$	276	×	0.0020 cm
					$C_{w,4}$	41	×	0.0057 cm
					$C_{m,1}$	580	×	0.0035 g cm^{-2}
					$C_{m,3}$	276	×	0.0026 g cm^{-2}
					$C_{m,4}$	41	×	0.0049 g cm^{-2}
	Barry et al. (2009)	400–1600	?	RT	C_{ab}	112	0.86	6.57 μg cm^{-2}
	Li and Wang (2011)	400–2500	CG	R	C_{ab}	350	0.56	8.11 μg cm^{-2}
					C_w	350	0.63	0.0012 cm
					C_m	350	0.84	0.0008 g cm^{-2}
	Li and Wang (2013)	400–2500	CG	R	C_{ab}	120	0.53	9.14 μg cm^{-2}
	* Ma et al. (2012)	400–2500	?	R	C_{ab}	100	0.75	9.46 μg cm^{-2}
					C_w	100	0.84	0.0047 cm
					C_m	100	0.77	0.0038 g cm^{-2}
	Zhao et al. (2014a)	400–2500	SA	R	$C_{ab,1}$	?	0.87	1.23 μg cm^{-2}
					$C_{ab,2}$	?	0.89	0.91 μg cm^{-2}
					$C_{w,1}$	?	0.76	0.0005 cm
					$C_{w,2}$	?	0.71	0.0124 cm
					$C_{m,1}$	?	0.58	0.0042 g cm^{-2}
					$C_{m,2}$	?	0.51	0.0003 g cm^{-2}
	* Wang et al. (2015a)	400–2500	S	RT	C_{ab}	47	0.54	7.72 μg cm^{-2}
					C_w	47	0.66	0.0014 cm
					C_m	47	0.64	0.0022 g cm^{-2}
	* Zhang and Wang (2015)	400–800		R	C_{ab}	68	0.47	2.88 μg cm^{-2}
PROSPECT-5	Féret et al. (2008)	400–2500	S	RT	$C_{ab,1}$	64	×	32.35 μg cm^{-2}
					$C_{ab,2}$	106	×	7.06 μg cm^{-2}
					$C_{ab,3}$	276	×	5.17 μg cm^{-2}
					$C_{ab,4}$	41	×	12.57 μg cm^{-2}
					$C_{car,1}$	64	×	5.35 μg cm^{-2}
					$C_{car,2}$	106	×	3.22 μg cm^{-2}
					$C_{car,3}$	276	×	4.22 μg cm^{-2}
					$C_{car,4}$	41	×	3.08 μg cm^{-2}
PROSPECT-D	Féret et al. (2017)	400–2500	S	RT	C_{ab}	?	×	4.76 μg cm^{-2}
					C_{car}	?	×	2.23 μg cm^{-2}
					C_{ant}	?	×	1.51 μg cm^{-2}

PROSPECT simulates leaf hemispherical reflectance and/or transmittance, accounting for both specular and diffuse reflections, for illumination direction normal to the leaf blade:

$$R_{prospect} = DHRF(\lambda, 0°) \quad (9.23)$$

Evaluation of Eq. 9.22 for $\theta_i = 0°$ gives

$$R_{prospect} = DHRF_{diff}(\lambda, 0°) + DHRF_{spec}(\lambda, 0°) \quad (9.24)$$

Assuming that $DHRF_{diff}$ is equal to $k_L(\lambda)$ for all illumination angles and that $DHRF_{spec}$ is wavelength independent, Eq. 9.24 may be written as

$$R_{prospect} = k_L(\lambda) + DHRF_{spec}(0°) \quad (9.25)$$

One can extract $k_L(\lambda)$ from Eq. 9.25 and substitute it into Eq. 8.42, leading to a new expression for the leaf BRDF model, the parameters of which are now wavelength independent:

$$BRDF(\lambda, \theta_i, \theta_v, \varphi_v) = \frac{R_{prospect} - DHRF_{spec}(n, \sigma, 0°)}{\pi} + BRDF_{spec}(n, \sigma, \theta_i, \theta_v, \varphi_v) \quad (9.26)$$

Bousquet et al. (2005) successfully tested the ability of this model to fit the measured BRDF, as shown in Figure 9.15.

However, this formulation is not compatible with plate models and does not obey energy conservation and reciprocity laws. Moreover, the diffuse component is constant, when it should

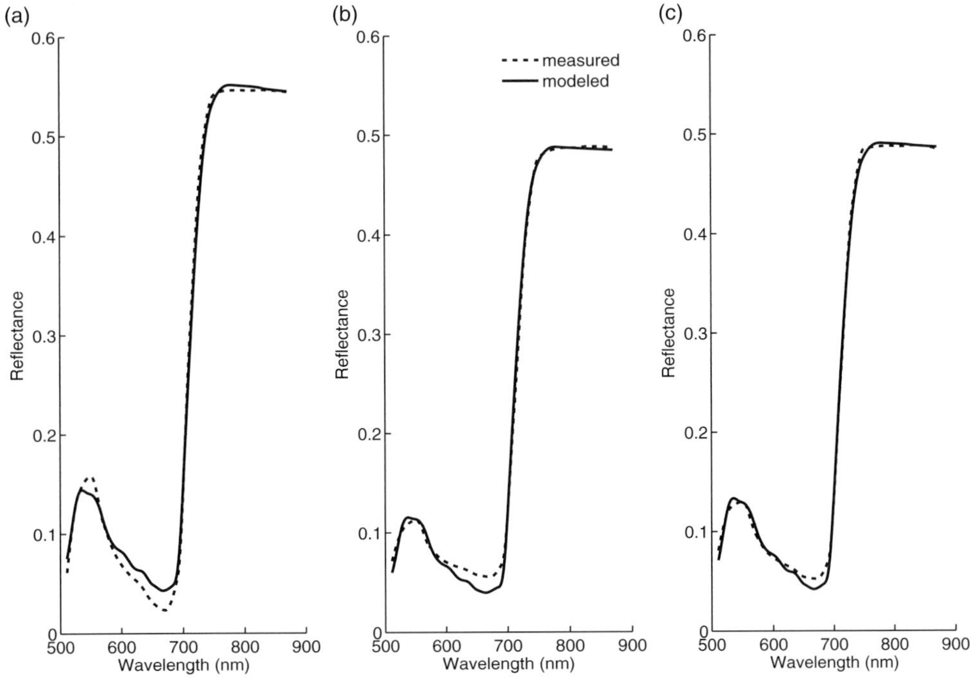

Figure 9.15 $DHRF(\lambda, 0)$ simulated with the BRDF model (dashed line) and fitted with the PROSPECT-5 model (solid line) for laurel (a), beech (b), and hazel (c).

decrease as the incident angle goes down. Stuckens et al. (2009) and Jay et al. (2016) considered the fraction of light reflected by the first wax layer and that does not reach the bottom layer, and extended Eq. 9.22 to

$$BRDF(\lambda,\theta_i,\theta_v,\varphi_v) = \left(1 - DHRF_{spec}(\lambda,\theta_i)\right)BRDF_{diff}(\lambda,\theta_i) + BRDF_{spec}(\lambda,\theta_i,\theta_v,\varphi_v) \qquad (9.27)$$

By integrating Eq. 9.27 over a hemisphere, one obtains

$$DHRF(\lambda,\theta_i) = \left(1 - DHRF_{spec}(\lambda,\theta_i)\right)k_L(\lambda) + DHRF_{spec}(\lambda,\theta_i) \qquad (9.28)$$

This equation provides the basis to relate leaf directional-hemispherical reflectance (DHR) to leaf BRDF through the Lambert coefficient:

$$k_L(\lambda) = \frac{DHRF(\lambda,\theta_i) - DHRF_{spec}(\lambda,\theta_i)}{1 - DHRF_{spec}(\lambda,\theta_i)} \qquad (9.29)$$

Assuming that $\theta_i = 0°$ and that $DHRF(\lambda, 0°) = R_{prospect}$

$$k_L(\lambda) = \frac{R_{prospect} - DHRF_{spec}(\lambda,0°)}{1 - DHRF_{spec}(\lambda,0°)} \qquad (9.30)$$

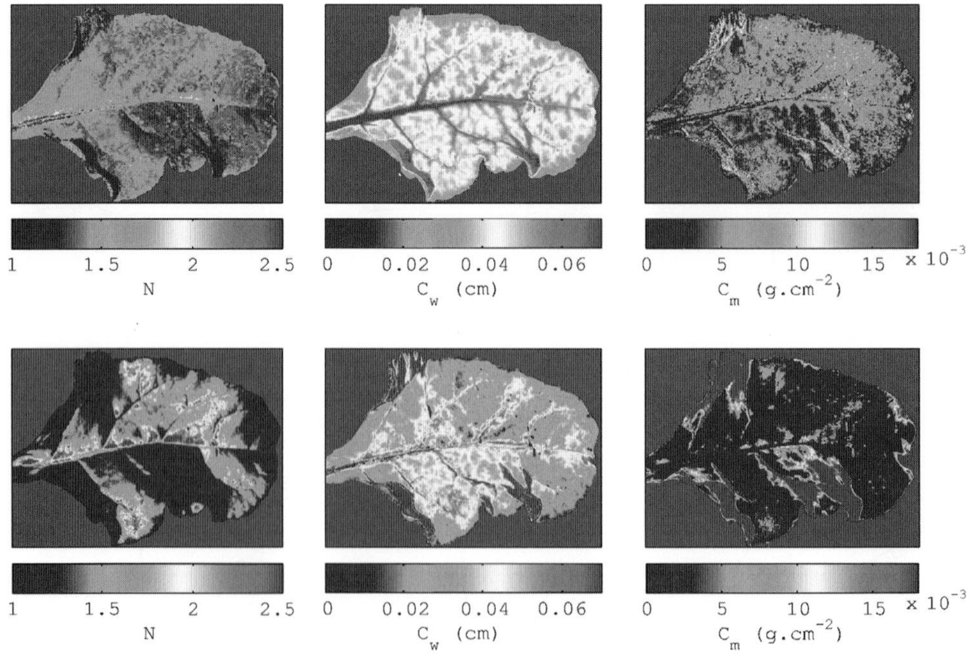

Figure 9.16 Leaf characteristics determined by inversion of the PROCOSINE model (upper figures) and the PROSPECT model (lower figures) in the SWIR (from Jay et al. (2016), *Remote Sensing of Environment*. Copyright © 2016 with permission from Elsevier). (A black and white version of this figure will appear in some formats. For the color version, please refer to the plate section.)

then

$$BRDF(\lambda, \theta_i, \theta_v, \varphi_v) = \frac{1}{\pi}\frac{1 - DHRF_{spec}(\lambda, \theta_i)}{1 - DHRF_{spec}(\lambda, 0°)}\left(R_{prospect} - DHRF_{spec}(\lambda, 0°)\right) + BRDF_{spec}(\lambda, \theta_i, \theta_v, \varphi_v) \tag{9.31}$$

Equation 9.31 is the heart of the PROCOSINE model which couples the COSINE (ClOse-range Spectral ImagiNg of lEaves) model with PROSPECT for a robust retrieval of foliar content at the millimeter scale using hyperspectral imagery (Jay et al., 2016). Note that if $DHRF_{spec}(\lambda, \theta_i) = DHRF_{spec}(\lambda, 0°)$ (the diffuse component is constant), Eq. 9.31 simplifies in Eq. 9.26. Figure 9.16 shows that, compared to PROSPECT, PROCOSINE makes improvements to leaf chemical mapping.

10

Modeling Three-Dimensional Leaf Optical Properties: RAYTRAN

Great progress has been made over the last two decades in the simulation of photon transport within vegetation canopies using radiosity or ray tracing models. At the leaf scale, similarly, it is possible to track a single photon from cell to cell and to derive the optical properties of the entire blade by following the paths of hundreds, thousands, or even millions of photons (see Section 8.2.5). Ray tracing techniques require a detailed description of leaf geometrical properties, as well as knowledge of the mechanisms involved in the scattering and absorption of light at different levels of organization from organelle to leaf and at different wavelengths. While modeling leaf anatomy in two dimensions is relatively easy, the generation of a three-dimensional leaf structure remains a challenging problem. One major impediment is the general lack of information on the detailed structure of leaf tissues (Parkhurst, 1982). First, in real tissues, cells present a great diversity of shape, size, and arrangement. They are generally associated together in leaf tissues, which are agglomerations of neighboring cells in close contact (Gibson and Ashby, 1997). Second, compared to other cellular solids, plant cells do not completely fill the available space. Intercellular air spaces may occupy a significant volume fraction, which varies with plant species, leaf tissue, as well as with environmental conditions (e.g., sun-illuminated or shaded leaves, hydrophytic or xerophytic leaves). Only three-dimensional models provide a meaningful representation of the spatial structure of the anatomy of a leaf. Then various assumptions must be made about the shape and size of the cells and their spatial arrangement in tissues, if efficient numerical computations are to be performed repetitively. In particular, cells will be represented through simple geometrical objects, and tissues by juxtapositions of such objects. Although the relationships between the shape and function of objects or living systems has been exhaustively investigated in the past (Thompson, 1992), these questions have experienced renewed interest (e.g., Prusinkiewicz and Lindenmayer, 1990; Guyon and Troadec, 1994; Hildebrandt and Tromba, 1995).

10.1 Three-Dimensional Structure of Plant Leaves

Building an artificial leaf is a three-step process that involves tissue observation, reconstruction, and modeling. The three-dimensional structure of plant leaves is an important factor for light interception, CO_2 diffusion, water transport, biomechanics, and other leaf functions. For instance, the surface area of mesophyll cell walls exposed to intercellular spaces is one of the most important parameters related to photosynthesis. Many synthetic images of natural objects are strikingly realistic and some of them are so realistic that they approach the visual appearance of real pictures. However, representations can be misleading since visualization algorithms are generally designed to extrapolate surface properties. For purposes other than pure visualization, the three-dimensional representation of leaf internal structure requires validation in terms of physiological processes (biochemical content) and solid mechanics (arrangement of the cells). In contrast to the avalanche of publications related to plant biochemistry and metabolism, ultrastructural studies on plant cells or tissues have been largely ignored in recent years and it has remained a qualitative field. The reasons are as much historical as technical. With recent progress in genetics and analytical biochemistry, entire traditional disciplines of botany, such as plant anatomy, are considered too descriptive and not sufficiently innovative to warrant research support, and so are now in jeopardy. As a result, the biosynthesis, structure, and function of many molecules in the cell are well known, while the macroscopic description of leaf cells and tissues remains undescribed. Nonetheless, it is well known that natural selection has led to the proliferation of a variety of internal leaf structures whose influence on the physiology, growth, and development of plant leaves seems evident.

10.1.1 Characterization of Leaf Anatomy: A Problem of Observation

An instrument with two lenses for seeing much-enlarged images of minute objects was invented early in the 17th century and received the name *microscopium* in 1625, but its potential as a tool for natural-history investigations remained unappreciated for another quarter-century (Stearn, 1967). Among the five pioneers of biological microscopy were Marcello Malpighi, Antoni van Leeuwenhoek, Robert Hooke, Jan Swammerdam, and Nehemiah Grew, the latter of whom explored the identification of three-dimensional structures in plants and made some of the first attempts at reconstructing vascular elements in a plant leaf by introducing depth and perspective to his illustrations (Grew, 1682). The three-dimensional anatomical description of leaf cellular structure is a non-trivial problem (e.g., Parkhurst, 1982; Moreno et al., 2006; Brodersen and Roddy, 2016). Ideally, the leaf anatomy should be characterized without destroying the cells and tissues. However, due to the small size of the studied objects, the heterogeneity of leaf internal structure, and technological limitations, direct observations using nondestructive methods still fail to reach the same resolution as those derived from destructive methods. As seen earlier, the leaf internal structure is a very complex medium, which contributes to this difficulty.

10.1.1.1 Destructive Methods

Destructive methods have long been practiced in botanical research to study the internal leaf structure at both cell and tissue levels. Serial cross-sections perpendicular to the blade are generally performed in two stages: the leaf sample is first fixed, dehydrated, and embedded in paraffin or epoxy resin before it is cut using a microtome. With light or low power microscopy, one uses sections as thin as possible, but of finite thickness (5–10 μm); in high power microscopy, the slices are called ultrathin (about 10× finer). In a second stage, the section is prepared for microscopy. At each stage some systematic errors may be introduced, for instance, compression of the sample can occur during sectioning or preparation, which would result in a loss of valuable anatomical and physiological information. As developed later in this chapter, some three-dimensional information can be deduced from a single slice but serial cross-sections are generally necessary to fully access the third dimension.

10.1.1.2 Nondestructive Methods

Different techniques such as magnetic resonance imaging (Walter et al., 1989; Veres et al., 1993), confocal and multiphoton laser scanning microscopy (Lhotáková et al., 2008; Wuyts et al., 2010), fluorescence microscopy (White et al., 1996), optical coherence microscopy (Hettinger et al., 2000), or high-resolution X-ray computed tomography (Stuppy et al., 2003; Blonder et al., 2012) have been tested on leaves and other plant organs like roots, stems, fruits, seeds, buds, etc.

As an example, observations of plant tissues using magnetic resonance imaging (MRI) have been attempted on leaves (e.g., Brown et al., 1988; Veres et al., 1991; Chudek and Hunter, 1997). The sample is placed in a strong homogeneous electromagnetic field and a radio-frequency field. The magnetic field causes the protons, chiefly those associated with water, to precess synchronously about the field vector at the Larmor frequency. Application of a 90° radio-frequency pulse shifts the protons away from the equilibrium position. In this process, the protons absorb energy that is later reemitted. The reemitted signal can be spatially localized to create a digital image of the spatial distribution of water in plant tissue (Walter et al., 1989). This nondestructive technique provides plant anatomists and physiologists with a tool for truly *in vivo* microscopy. However, obtaining quality images of plant leaves with MRI is technically difficult due to the laminar shape of this organ and its small size which requires a special device – small radio-frequency coils – to obtain a good signal to noise ratio (SNR). Moreover, the different components of leaves have different magnetic susceptibilities so that in the end only a few protons are available to provide signals for imaging. In consequence, the attempts to extract slices from leaves have been largely unsuccessful. In the future, however, improvements in SNR should allow distinguishing between two different tissues or two different cells.

Another example is given by synchrotron X-ray computed microtomography that offers several advantages, including a wide energy range, a high flux, and small source size and beam divergence, for scanning samples at high resolution within a short period without requiring extensive sample preparation (Verboven et al., 2015; Watté et al., 2015). By rotating a sample positioned in an X-ray beam, hundreds of digital two-dimensional images can be reconstructed to produce a three-dimensional model. This technique has been recently employed to obtain the model of a tomato (*Lycopersicon esculentum*) leaf at a resolution of 0.75 μm. As seen in Figure 10.1, the high lateral coherence of synchrotron radiation X-rays provides sufficient contrast to visualize individual cells in a dense tissue, therefore to produce statistics useful in plant anatomy (size, shape, and distribution), but the contrast in the laminography images is insufficient to distinguish cell organelles such as the disc-shaped chloroplasts and the vacuoles.

These nondestructive methods give information about tissues, their extent, the way they are interconnected, and their air spaces but, until recently, none of them really provides the single-cell resolution required for the reconstruction and analysis of the cellular organization of leaf tissues.

10.1.2 Three-Dimensional Reconstruction

Different techniques for three-dimensional reconstruction, all based on destructive methods, have existed for nearly 80 years. They may be classified into three groups: stereology, reconstructed media, and computer modeling.

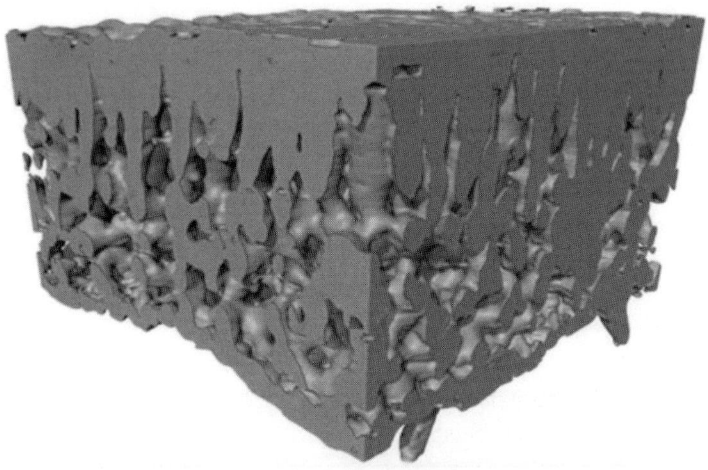

Figure 10.1 Synchrotron computed laminography image of a section of a tomato (*Lycopersicon esculentum*) leaf (MeBioS @ KU Leuven) .

10.1.2.1 Stereological Methods

Stereology is the three-dimensional interpretation of two-dimensional images (e.g., Elias, 1971; Toth, 1982; Kubinova et al., 2004). When a thin section is examined, only two-dimensional structures are seen. Spheres become circles, planes become lines, and lines become points. We can generalize rules as a principle: "a section through an n-dimensional object is, in general, an (n–1)-dimension figure. Conversely, an n-dimensional figure in a section, results in general, from cutting an (n+1)-dimensional object." The second statement is the basis of stereology. Although many structural parameters of cells and tissues can be quantified using this technique, it appears that botanists have made little use of it (Parkhurst, 1982). It is unfortunate because stereology seems suitable to study leaf tissues which are highly non-isotropic media. Fundamental properties such as volume fractions, surface areas, or number of particles could be determined. For instance, the amount of cell surface area in contact with air space is a physiologically important variable whose effects on photosynthetic rates have been investigated (Gibeaut and Thomson, 1989). However, stereology provides only partial information and cannot be used to reconstruct a cellular tissue in three dimensions. The next two methods require serial cross-sections.

10.1.2.2 Digital Methods

The problem is here to create three-dimensional models from a set of digital images of serial cross-sections (Kubinova et al., 2004). For a long time, stacking transparent slides sequentially was the only way to imagine the three-dimensional structure of plant organs (Banchoff, 1996). In the 1980s, the full view of such a structure was made possible with the advent of computer graphics systems (Delozier et al., 1987). Today, two types of methods are classically used: surface and volume reconstruction (Geiger, 1993). In surface reconstruction, the problem consists of building a surface between contours in adjacent cross-sections. A polyhedral representation with triangular faces is well adapted to ray-tracing methods (Watté et al., 2015). In volume reconstruction, such as the Delaunay reconstruction, volume elements consist of tetrahedra that are adapted to the object's shape (Figure 10.2).

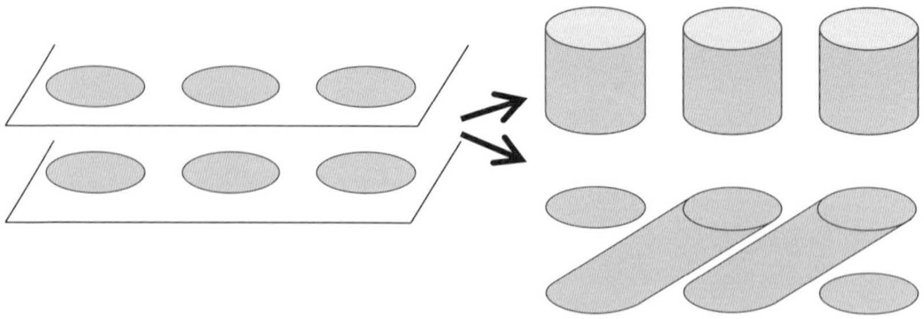

Figure 10.2 A set of six contours, pairwise superimposed in two horizontal, parallel planes shows two possible connections (redrawn from Geiger (1993)).

Figure 10.3 A three-dimensional reconstructed porous medium (from Adler et al. (1992), *Water Resources Research*. Copyright © 1992 with permission from John Wiley & Sons).

10.1.2.3 Statistical Methods

So far these methods have been used in Earth Science to statistically characterize porous media (Adler, 1992; Adler et al., 1992). The three-dimensional randomly generated medium is a statistical representation with the average properties of the leaf (Figure 10.3) in contrast to the variability typical of a real leaf. The porosity and the correlation function of the pore space measured on leaf cross-sections are used to derive the geometrical properties of the simulated material. Such a method should adapt to botanical characterization of the spongy mesophyll with its complex network of cells and air spaces. However, these models do not accurately represent the leaf tissue structure as do the following methods.

10.1.3 Two-Dimensional Modeling

The modeling of leaf anatomical properties in two dimensions is much simpler than in three. Gabryś-Mizera (1976) constructed a geometric model of cells of funaria moss (*Funaria hygrometrica*) and

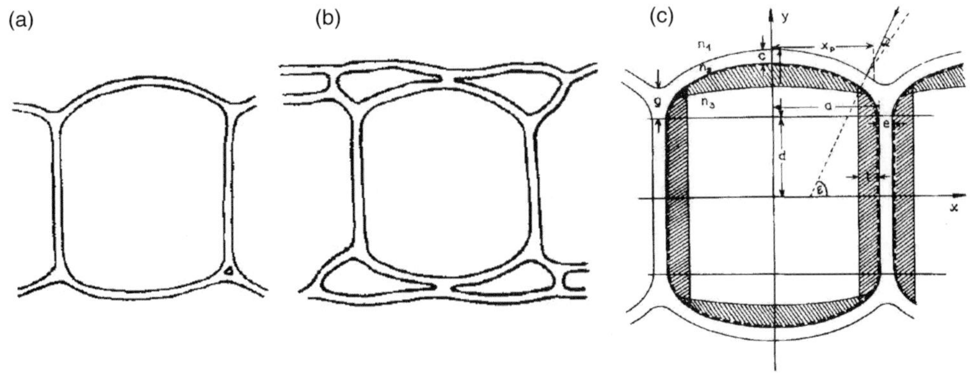

Figure 10.4 Cross-sections of typical cells of: (a) funaria moss (*Funaria hygrometrica*); (b) star duckweed (*Lemna trisulca*) leaves; and (c) the geometrical model used for the calculation of light energy distribution on the walls of chloroplast containing cells (from Gabryś-Mizera (1976), *Photochemistry and Photobiology*. Copyright © 1976 with permission from The American Society for Photobiology).

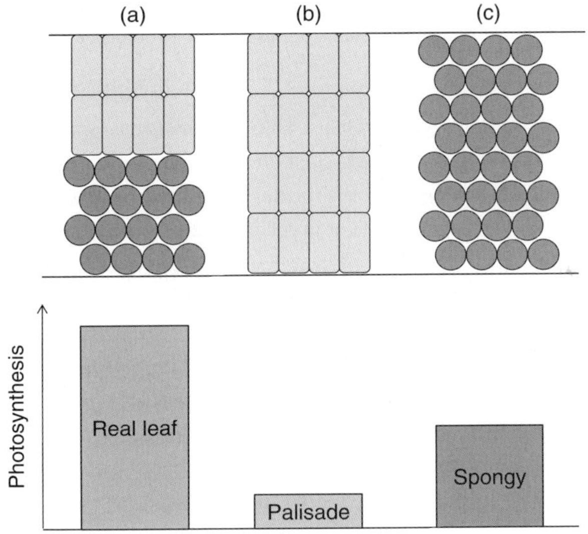

Figure 10.5 Relative photosynthetic rates calculated for three tissue distributions: (a) real leaf; (b) all palisade; and (c) all spongy cells (redrawn from Parkhurst (1986)).

star duckweed (*Lemna trisulca*) leaves on the basis of photographs of their cross-sections to study the intensity distribution of light falling on the internal surface of cell walls and the phototactic movements of chloroplasts. The shape of the cell was simplified as a cylindrical form (Figure 10.4).

In order to understand why natural selection led to the differentiation of leaf mesophyll into palisade and spongy parenchyma, two structurally very different tissues as seen in Section 2.1, Parkhurst (1986) designed a mathematical two-dimensional model for CO_2 diffusion inside leaves and applied it to three leaves including a real structure and two extreme anatomical structures representing all palisade or all spongy mesophyll cells (Figure 10.5). It turned out that the leaf with the structure found in nature was the most efficient in terms of photosynthetic activity, because the combination of the two cell types optimizes the capture of light energy and diffusion of gases into the mesophyll.

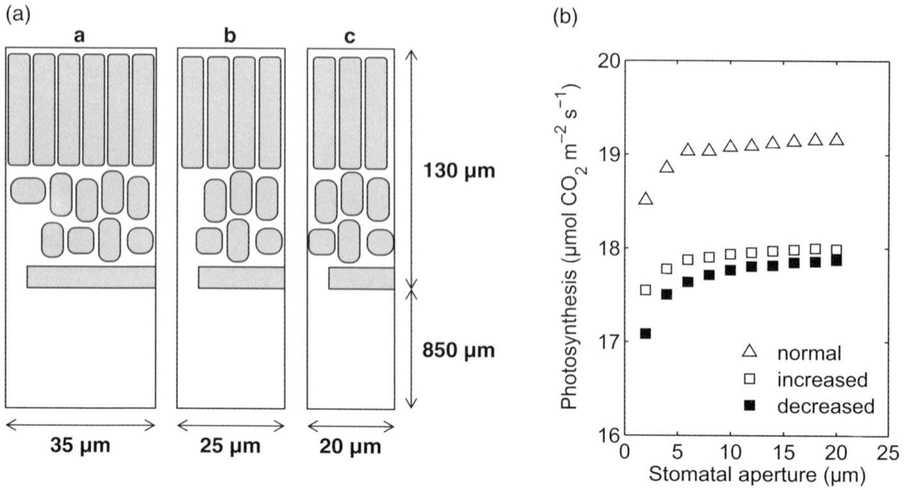

Figure 10.6 (left) schematization of hypostomatous leaf internal structure for: a) 25% decreased (225 stomata mm^{-2}); b) normal (300 stomata mm^{-2}); and c) 25% increased (375 stomata mm^{-2}) stomatal density. (right) effect of stomatal density on photosynthesis (redrawn from Pachepsky and Acock (1996)).

Such an approach was used by Pachepsky and Acock (1996, 1998), who measured the geometric parameters of leaf anatomy and implemented a two-dimensional spatial grid made of polygonal structures covering the domain of a leaf cross-section. They validated their 2DLEAF model for tomato (*Lycopersicon esculentum*) leaves. This model simulates: i) transport of CO_2 and water vapor in the intercellular spaces and in the boundary layer adjacent to a leaf; ii) fluxes of CO_2 across cell surfaces due to assimilation; and iii) fluxes of water vapor across cell surfaces due to the difference between cellular and intercellular water vapor pressure. In particular, they studied the effect of stomatal density, showing the impact of this variable on photosynthesis (Figure 10.6).

These mesophyll cells are definitely not realistic, especially when the precision becomes crucial to the model performance. Abera et al. (2016) recently proposed a cell growth algorithm for virtual leaf tissue generation based on the biomechanics of plant cells in tissues. The algorithm can account for typical distinctions in epidermal layers, palisade mesophyll layer, and spongy mesophyll layer which have characteristic differences in the shape of cells, arrangement of cells, and void fractions present in each layer (Figure 10.7). Moreover, the geometrical properties of the virtual leaf tissues and real leaf tissues are very similar.

10.1.4 Three-Dimensional modeling

In the first half of the last century, the cell shapes of parenchyma tissue were first modeled as polyhedra, two of which are the rhombic dodecahedron (10 parallelogram faces) and the tetrakaidekahedron (8 hexagonal faces and 6 square faces). These two polyhedra that conveniently fill space when assembled, as seen in Figure 10.8, have produced a rich body of literature (e.g., Lewis, 1923; Hulbary, 1944; Macior and Matzke, 1951). However, although these forms are closely approached by real parenchyma cells or even attained in certain simple and homogeneous plant tissues, they still do not represent the cell shapes in leaf tissues that are not composed of regularly packed identical units, but are instead formed from cells of different sizes and shapes with differing numbers of faces and edges.

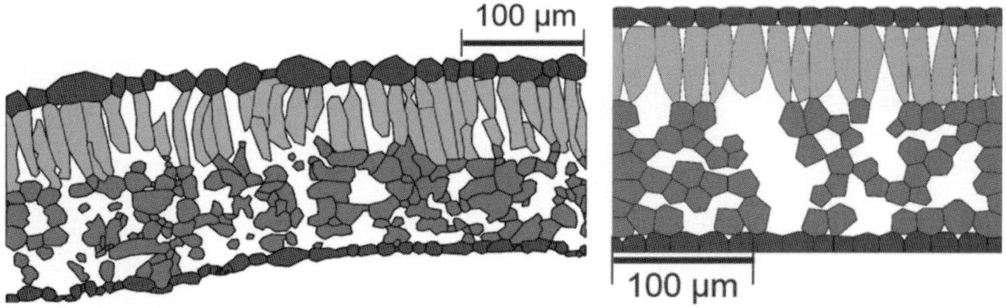

Figure 10.7 Visual comparison of real tomato (*Lycopersicon esculentum*) leaf tissues (left) and virtual tissues (right) (from Abera et al. (2016), *Acta Horticulturae*. Copyright © 2016 with permission from the International Society for Horticultural Science).

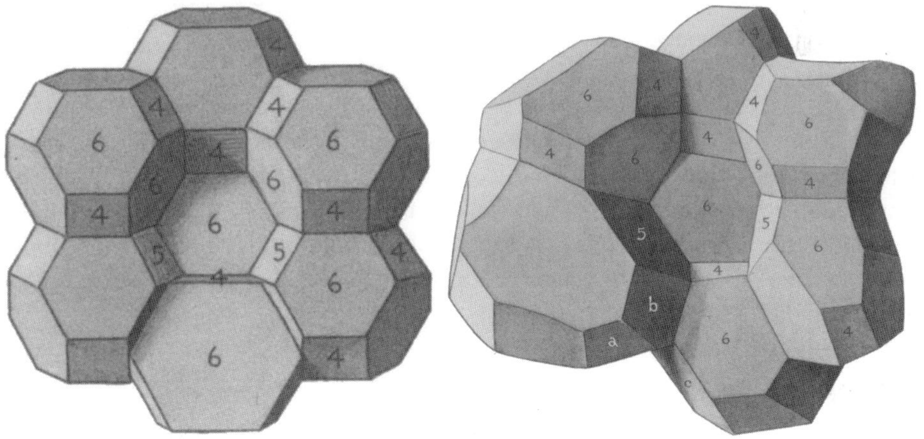

Figure 10.8 A group of orthic tetrakaidecahedra (left) and of 7 cells showing 18 surfaces (right) (from Lewis (1923), *Proceedings of the American Academy of Arts and Sciences*. Copyright © 1923 with permission from American Academy of Arts & Sciences).

Lewis (1923) attempted to formulate explicit rules for division orientations in cell layers, but in the absence of a suitable formalism he could not organize them into a general framework. The first models of cell division assumed more simple cells like cubic (Korn, 1974) or hexagonal (Lück et al., 1988) cells that provided an approximate representation of the epidermis of various plant tissues (rhizophore, apex, root, etc.), allowing the authors to understand and simulate its growth (Figure 10.9). But, once again, these shapes were definitely not realistic and adapted to plant leaves, especially when the precision becomes crucial to the model performance.

There is a lack of models for plant tissue development, including the differentiation and division of cells in three dimensions, which follow precise genetically controlled rules. Indeed, the arrangement of cells in space is a consequence of mitosis. Although the mechanics of morphogenesis is still an issue, simple tissues have been generated by computers in two dimensions (Abbott and Lindenmayer, 1981) and three dimensions (Lindenmayer, 1984). In order to produce more realistic scenarios, variability has to be introduced. At this stage, all the difficulty rests in the introduction of alternative

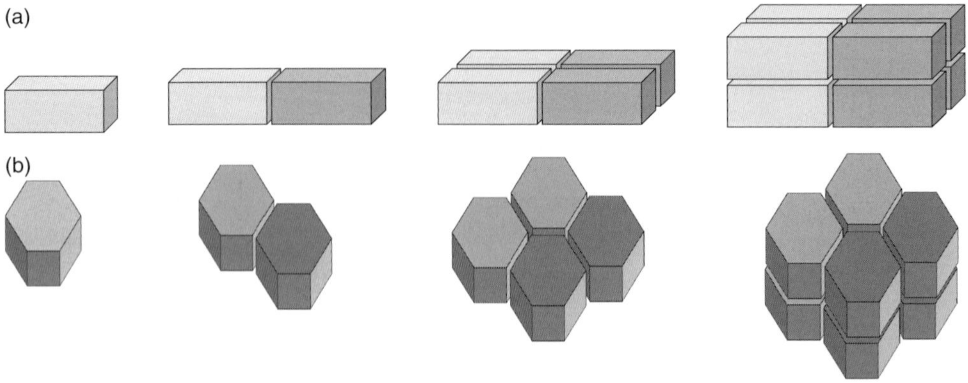

Figure 10.9 (a) Development of a block of cells by successive divisions of an initial cubicle (redrawn from Korn (1974)). (b) Generating system producing an alternating S and Z tetrads (redrawn from Lück et al. (1988)).

growth processes that happen at random. In plain language, how to integrate the misfires of nature? Better knowledge of the statistical properties of leaf anatomy would be helpful to improve the construction of our virtual leaves or tissues, but because of the experimental limitations mentioned above, we are obliged to make many assumptions and simplifications. As a consequence, we will set this question aside and focus on the purpose of three-dimensional representation of plant leaves. As seen above, it should be possible to simulate actual cells by defining polyhedra with many faces varying in number from 4 to 50 or more (Romberger et al., 2004). Multiplying the number of faces actually tends to smooth the cell shape and makes it converge to a sphere-like volume. This is the main approach to date.

Working on this assumption, Govaerts et al. (1996) and later Xiao et al. (2016), considered cells to be primitive solid objects (spheres, ellipsoids, cones, cylinders, triangular prisms, etc.) that can be carved out and assembled using Constructive Solid Geometry (CSG). A primitive is typically described by a set of parameters including the coordinates of its center point, a length, width, and radius value. Then primitives can be combined into compound objects using Boolean operators like union, intersection, and difference to define more complex objects (Mortenson, 1985). Generating cells is the first step, but binding them together is a second challenge. Section 10.2 details the construction of a typical dicot leaf. Similarly, Aalto et al. (1999) constructed a three-dimensional model of a conifer needle to study the diffusion of CO_2 through stomatal pores and further into intercellular air spaces. The mesophyll was treated as a porous object and the vicinity of one stoma is described by two media, air and leaf mesophyll. Such a model cannot be used for purposes other than gas diffusion, which is mainly concerned with surface properties. To perform computations on a system as complex as a tissue with cells bonded to other cells, a number of assumptions are required about the shape and size of the cells, and the connections between cells. This requires making many decisions that are necessary to create an abstract and simplified version of reality. Aalto and Juurola (2002) have built a one-year-old silver birch (*Betula pendula*) leaf using the GAMBIT® software designed for computational fluid dynamics (Figure 10.10a). Spongy and palisade mesophyll cells are represented by spheres and cylinders, respectively, and they both contain chloroplasts (darker dots). A capillary tube extending through the cuticle of the abaxial face controls the stomatal opening and gas exchange with the atmosphere. Ivanova et al. (2006) used SIMAGIS® MesoPlant™

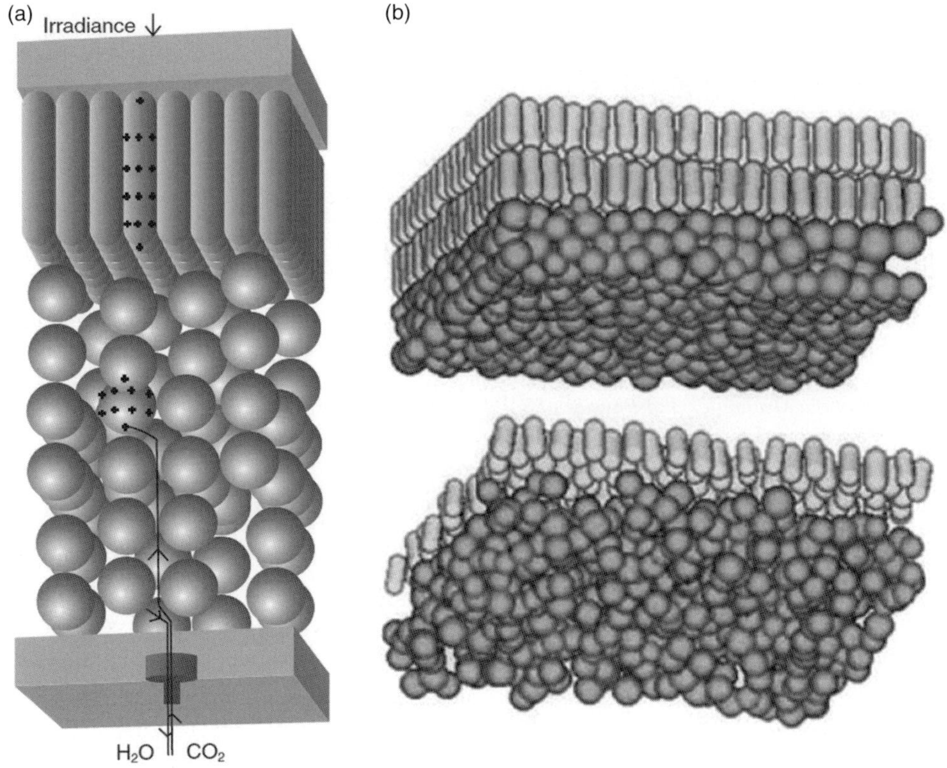

Figure 10.10 Three-dimensional model of (a) a silver birch (*Betula pendula*) leaf (from Aalto and Juurola (2002), *Plant, Cell & Environment*. Copyright © 2002 with permission from John Wiley & Sons, Inc.) and (b) of a sun (top) and a shade (bottom) fireweed (*Chamerion angustifolium*) leaf (from Ivanova et al. (2006), *Russian Journal of Plant Physiology*. Copyright © 2006 with permission from Springer).

(http://smartimtech.com/mesoplant.htm) to analyze cross-sections of fireweed (*Chamaerion angustifolium*) leaves and S3D SpheroPolyhedra™ to build their three-dimensional model (Figure 10.10b, http://smartimtech.com/modeling/polyhedrons.htm). Some random structure is introduced in the three-dimensional figures of fixed shapes and sizes.

As seen in Figure 10.11, there are no connections between the spongy mesophyll cells. That is not very realistic compared to the complexity of the spongy tissue (Figure 10.11a). One could draw inspiration from engineering cellular solids, in particular solid foams that describe materials with a highly dispersive solid phase arranged into cells that can be either open or closed (Figure 10.11b and 10.11c).

As seen in Section 10.1.1, the three-dimensional microstructure of a tomato leaf has been obtained by synchrotron X-ray computed tomography. A mesh was automatically generated from the voxel-based images representing this complex structure, distinguishing between several subcellular structures. Over this mesh, a ray tracing algorithm was applied to predict the optical properties of the leaf (Watté et al., 2015). Furthermore, a reaction-diffusion model was solved to simulate its photosynthetic properties (Berghuijs et al., 2016, 2017; Ho et al., 2016), which enables exploration of the functional implications of anatomical variations in leaves.

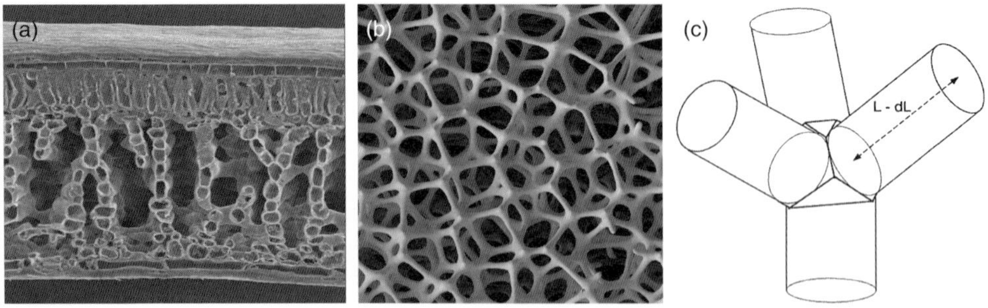

Figure 10.11 (a) Cross-section of a leaf showing the epidermis, palisade mesophyll, and spongy mesophyll (www.kasiazieminska.com/). (b) Micrograph of foam (https://n-e-r-v-o-u-s.com/). (c) Foam solid volume consisting of one truncated tetrahedron and four truncated cylindrical members (from Wang and Cuitiño (2000), *Journal of the Mechanics and Physics of Solids*. Copyright © 2000 with permission from Elsevier).

10.2 Construction of a Three-Dimensional Leaf Model

The internal leaf structure and the optical properties of each of the media constituting the cell elements must be carefully simulated if the reflectance and transmittance of a leaf is to be accurately estimated. The work of Govaerts et al. (1996), which remains relevant 20 years later, is described below.

10.2.1 Schematic Representation of a Leaf Cell

Although plant leaves may present numerous anatomical structures, and leaf cells may vary largely in shape and size according to the foliar tissue type (protective tissue such as the cuticle, conductive tissue such as veins, parenchyma such as ground tissue, etc.), basic cell structures are relatively uniform because of common cell functions (Mohr and Schopfer, 1995). Cells are surrounded by a wall and an interior plasma membrane containing the cytoplasm with several organelles (nucleus, ribosomes, mitochondria, chloroplasts, amyloplasts, endoplasmic reticulum, etc.), and a central vacuole which may occupy up to 90% of the cell volume (see Section 2.1). Chloroplasts, variable in number, are located within the cytoplasm of all photosynthesizing cells (Figure 10.12).

Each of these elements scatters or absorbs the light penetrating into the plant cell. At all scales where the size of the particles is much larger than the wavelength, the difference in refractive indexes between two different media creates optical boundaries. When the particle size is less than or about equal to the wavelength, Rayleigh and Mie scattering may occur. While the dimension of plant cells with respect to wavelengths in the solar domain is too large to induce such phenomena (Gates et al., 1965; Sinclair et al., 1973), the cytoplasmic organelles, such as chloroplasts or starch granules, which have sizes comparable to optical wavelengths, and large molecules like proteins, do scatter light. As seen in Chapter 3, absorption results essentially from electronic transitions and vibrations by polyatomic molecules. Electronic transitions mainly occur in the porphyrin rings found in chlorophyll; vibrational absorptions in poly-atomic molecules involve another category of chemical compounds: water, which fills the vacuole, and cellulose, hemicellulose, and lignin, which are mainly located in the cell walls where they act to strengthen and protect plant structures.

A typical cell may be defined as a set of concentric objects, as "Russian doll" structures, filled with three different media: cell wall material (cellulose, hemicellulose, and lignin), chlorophyll, and water

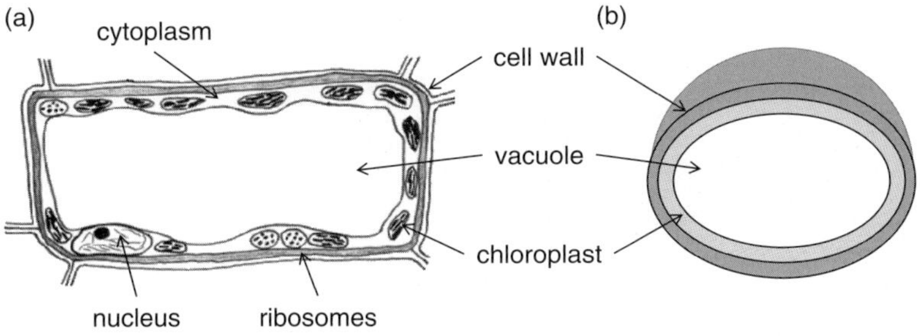

Figure 10.12 (a) Schematic representation of a mesophyll leaf cell showing the complexity of the different organelle membranes. (b) For modeling purposes, the internal cell structure is represented as composed of solutions bounded by two homogeneous membranes.

(Figure 10.12). These media are assumed homogeneous, that is, their physical properties are independent of the direction. Each cell constituent is therefore characterized by a volume, a refractive index n, and a specific absorption coefficient k to describe the partitioning of light among the reflected, transmitted, or absorbed fractions. The selection of adequate n and k spectra is not necessarily the easiest task. For water, the complex refractive index of pure liquid water can be used in the model; for cell walls, the refractive index of leaf material and the specific absorption coefficient of dry matter taken from PROSPECT (see Chapter 9) represent a good compromise; for photosynthetic pigments (primarily chlorophyll), the specific absorption coefficient also taken from PROSPECT can be used. As for the refractive index, the values determined *in vitro* by Dobek et al. (1994) for example, or if nothing else those published for water, can be used. Indeed, Richter and Fukshansky (1994) suggested that the pigment particles immersed in the cytoplasm and adjacent to the plasma membranes had practically the same refractive index as their microenvironments. In that respect they called them soft particles. Although pigment molecules attached to the thylakoid membranes have highly clumped distributions, Govaerts et al. (1996) considered a simplified distribution to be homogeneous within the layer, as shown in Figure 10.12. The simplified assumption of a homogeneous chlorophyll solution within the plasma membrane may reveal some limitations if finer detail about light absorption by chloroplasts is desired to investigate photosynthetic mechanisms. This simplified representation of a leaf cell still permits to take into account the basic cell functions. Note that the number of media may be changed.

10.2.2 Schematic Representation of Leaf Tissues

As seen in Chapter 8, a model is an abstract and simplified version of reality. The internal structure of plant leaves varies from species to species, therefore a number of assumptions about the shape and size of the cells have to be made (Figure 10.13). Their density (number of cells per unit area) and connections are another problem that should be addressed. Indeed, cells are not isolated in real tissues but are bound to neighboring cells in multiple directions by their shape and by plasmodesmata, the cytoplasmic bridges that allow exchange of chemical signals between cells. They also may be separated by intercellular air spaces. The three models described below, called DICOT1, DICOT2, and DICOT3, are based on the representation of typical dicotyledon leaves, with the palisade and spongy mesophyll tissues between two layers of epidermal cells. The requirements to build these

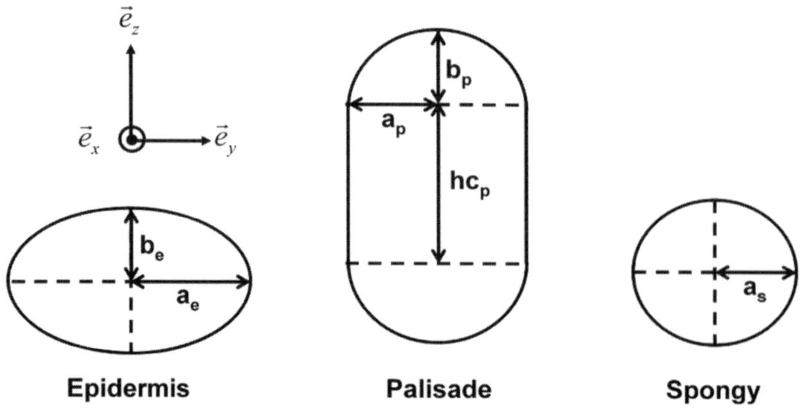

Figure 10.13 Leaf cells (by courtesy of Laurent Bousquet).

models are first to define the cell shape, size, and spatial arrangement in different tissues, and second, to derive the equations that calculate the volume of each medium (Appendix I). These equations are then used to compute the total amount of the various media in each tissue.

10.2.2.1 The Epidermis

The epidermis is made up of a single layer of colorless cells, which entirely cover both faces of the leaf and have few if any chloroplasts except in stomatal guard cells. Therefore the cells only have two media, cell walls and water. Both the upper and lower epidermises are modeled as layers of compact ellipsoidal cells, the oblateness of which can vary so as to control the leaf surface roughness (Figure 10.13a and Appendix I). Normally, no intercellular spaces are present in this particular tissue, which governs gas exchange between the leaf and its environment through stomatal pores. These stomatal openings are ignored here. As seen in Figure I.1 of Appendix I, the epidermal cells fit together like pieces of a jigsaw puzzle (Niklas, 1992). Although very different arrangements can be created by varying the way cells are located in space, a simple but realistic epidermal layer has been defined based on photographs (Martin et al., 1989).

10.2.2.2 The Palisade Parenchyma Cells

As seen in Section 2.1.2, the palisade parenchyma is made of narrow, elongated cells oriented perpendicularly to the leaf surface and usually arranged in one or two layers subjacent to the adaxial epidermis. They contain the largest concentration of chloroplasts. Although compactly arranged, they have little mutual contact due to the long, intercellular voids along their anticlinal walls, as shown in paradermal sections of palisade parenchyma (Martin et al., 1989; Niklas, 1992). The palisade cells have been modeled as cylinders terminated by two identical spherical caps (Figure 10.13b and Appendix I).

10.2.2.3 The Spongy Mesophyll Cells

The spongy mesophyll often appears as a network of cells whose spatial arrangement does not seem to display any regular organization. In contrast to the other two tissues, the cells have variable shapes and sizes, and the intercellular air spaces, which facilitate the circulation of gases inside the leaf, may occupy up to 50% of the tissue volume. Therefore, it is almost impossible to describe them by individual volumes. However, the porosity of typical spongy tissue is isotropic and nondirectional (Parkhurst, 1986). Therefore, the spongy mesophyll structure may be modeled by spheres or ellipsoids of different sizes, located at random, such that the occupied volume corresponds to observed values (Figure 10.13c and Appendix I). Although this representation is approximate, it permits a description of the fraction of air spaces and the isotropy in that tissue compatible with values found in the literature. It appears that this simplification underestimates the surface area of these cells, the highly lobed shape of which enhances gas exchange. Nonetheless, such a statistical approach is appropriate to describe an irregular tissue which does not comply with simple rules. Furthermore, this approximation is not expected to strongly affect the scattering of light in that layer.

10.2.3 Construction of Two Typical Dicot Leaves: DICOT1 *and* DICOT2

DICOT1 was the first model built by Govaerts et al. (1996) to simulate photon transport in a three-dimensional artificial leaf. The dimensions of the leaf cells were determined by observations under a microscope and published in the literature. For instance, typical dimensions are 15 μm × 15 μm × 60 μm for palisade cells and 18 μm × 15 μm × 20 μm for spongy mesophyll and epidermal cells (e.g., Gates et al., 1965; Ma et al., 1990). These characteristics were selected to ensure realistic cell density, and thereby air-space volume. The thickness of the cell membranes is fixed for the cell wall and the remaining volume is assigned to water and chlorophyll in such a way that these constituents occur in the correct concentrations as described in the literature. For a given set of parameters, a program automatically computes the position of each cell and generates the whole leaf, as shown in Figure 10.14. Details about the construction of DICOT1 are provided in Appendix I.

The geometric construction of DICOT2 defined in Ustin et al. (2001) is similar to that of DICOT1, with the exception of the epidermis thickness. The upper epidermis is 17.5 μm thick and irregular, while the lower epidermis is 25 μm thick and regular (Table 10.1). Note that the thickness of the

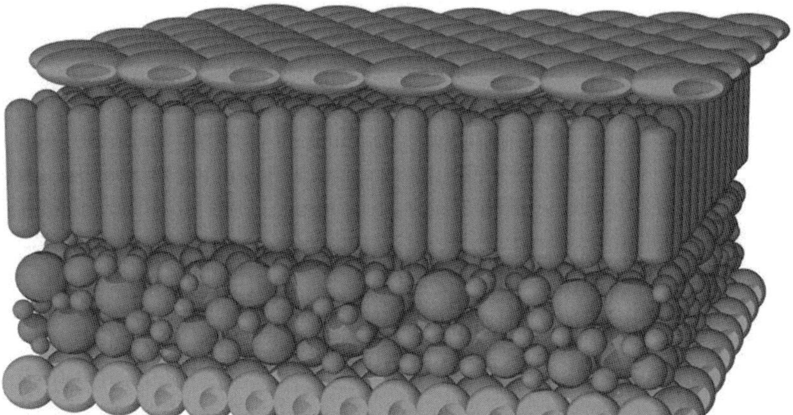

Figure 10.14 Perspective view of DICOT1. The size of the represented target is 300 μm × 300 μm × 170 μm (by courtesy of Yves Goverts (Govaerts, 1996)).

Table 10.1 *Thickness of the tissues defining the* DICOT1, DICOT2, *and* DICOT3 *virtual leaves.*

Leaf model	Upper epidermis	Palisade parenchyma	Spongy mesophyll	Lower epidermis
DICOT1	17.5 μm	65 μm	50 μm	17.5 μm
DICOT2	17.5 μm	65 μm	50 μm	25 μm
DICOT3	10 μm	40 μm	60 μm	10 μm

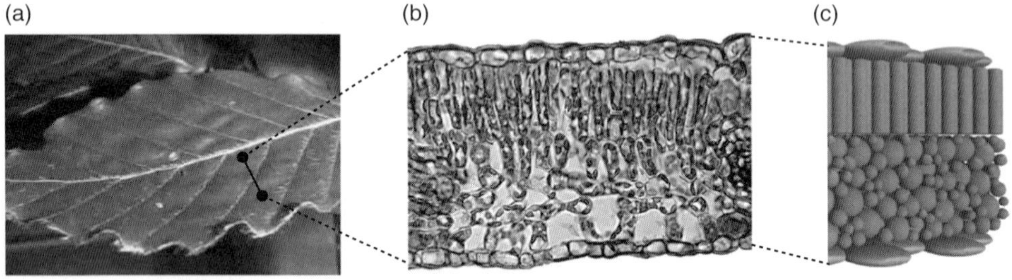

Figure 10.15 The European beech (*Fagus sylvatica*) leaf used to build the DICOT3 model. Photo of the whole leaf (a), cross-section (height 120 μm, width 200 μm, thickness 8 μm, stained by astrablue 0.5% aq. and Ziehl-Neelsen carbol-fuchsin 10% solution) (b), and DICOT3 model (c) (Bousquet, 2007).

whole leaf may exceed the sum of the thicknesses of each tissue, according to the layers of air spaces that separate them.

10.2.4 Construction of a European Beech Leaf: DICOT3

When DICOT1 and DICOT2 were the three-dimensional models of a conceptual dicot leaf, DICOT3 depicted in Figure 10.15 is directly inspired by a European beech (*Fagus sylvatica*) leaf picked in July 2005 in the campus of the University of Paris-Saclay, France (Bousquet, 2007). European beech is a dominant deciduous tree species native to temperate Europe. As previously, the construction of the leaf model requires specifying the cell size and thickness of the main tissues (Table 10.1). These variables were determined in the French National Museum of Natural History by histological examination of photos taken through an optical microscope. Fresh material has been fixed and stored in a mixture of acetic acid (5%), formaldehyde (5%), and ethanol (90%). After dehydration and wax embedding, 8 μm transverse sections were made through the limb using a slide microtome and then stained. The fraction of air spaces in the spongy mesophyll was also roughly estimated from these photos at 43%. Note that this leaf model does not have a cuticle, whereas the European beech leaf displays a cuticle of about 1 μm thick. Thus the cell walls of the upper epidermis are the first medium to be hit by light.

In comparison with DICOT1 and DICOT2, improvements have been made in the construction of DICOT3. The choice of the correct density for each leaf constituent is crucial to derive correct volumes per unit leaf area. For instance, for a given chlorophyll content (mass per unit leaf area), an overvaluation of chlorophyll density tends to minimize the volume occupied by pigments, which impacts the absorption efficiency of the chlorophyll layer. Fukshansky et al. (1993), who measured the mass and sizes of various chloroplasts, report a chlorophyll density of 4.7×10^{-2} g cm^{-3}. As for water and cell walls, density is less critical and is taken from the literature. The comparison of the biochemical content of the leaf model to experimental measurements shows that chlorophyll and dry matter contents are of the same order of magnitude, whereas water content is undervalued by about 50% in the model compared to the real leaf, in spite of realistic tissue thicknesses. This means that the leaf model is not compact enough and contains too many air spaces. Since water absorption is weak in the visible (VIS) and near-infrared domains, it only makes an impact in the shortwave infrared (SWIR) domain. The distribution of photosynthetic pigments in the parenchyma is then critical because of its effects on absorption profiles. Although chlorophyll gradients may vary from one species to another or according to environmental conditions (e.g., sunlit and shaded leaves), a high content is generally reached in the palisade mesophyll. Then the pigment content gradually decreases with depth in the leaf (Cui et al., 1991; Pilarski, 1999). Fukshansky et al. (1993) report a distribution of two-thirds in the palisade mesophyll and one-third in the spongy mesophyll. It was set to 64% in the palisade mesophyll and 36% in the spongy mesophyll in DICOT2.

10.3 The RAYTRAN Model

The ray tracing code RAYTRAN is a radiative transfer model designed to investigate light propagation in complex terrestrial environments (Govaerts and Verstraete, 1998). It was designed as a virtual laboratory, where scenes can be described explicitly in three dimensions at spatial scales relevant to simulate actual radiometric measurements. RAYTRAN relies on three main hypotheses: i) incident radiation can be simulated with a finite number of photons which do not interact between themselves, and can be either collimated, to represent direct illumination, or distributed angularly, to represent diffuse light, or both; ii) the medium can be described with geometrical primitives; and iii) optical properties of the elements can be defined with probability distribution functions. Monte Carlo procedures are used to generate incident photons, to define the type of interaction when collisions occur, and to compute the scattering angles; the propagation of light is described in terms of geometric optics. RAYTRAN has been mainly used to simulate the bidirectional reflectance distribution function (BRDF) of vegetation canopies. It has also been used to compute radiation transfer in virtual plant leaves (Govaerts et al., 1996; Ustin et al., 2001).

The program is divided into several parts. The initialization of the model involves the construction of the target (see Section 8.1.6) and the choice of an illumination model. The model allows the specification of the wavelength of the incident radiation, and later interactions with objects can be made dependent on this wavelength. Photons are generated in the forward mode, that is, from the light source to the target, and tracked from interaction to interaction throughout the leaf cell structure, until they are absorbed or escape from the leaf. The simulation of the type of interaction is performed by generating a random number u uniformly distributed in the interval $[0, 1]$:

- if $u<R$, rays are specularly reflected at the interface between two dielectrics according to Fresnel's equations (see Appendix E);
- if $R<u<R+T$, rays are refracted in the direction determined by Snell's law;
- if $u>R+T$, rays are absorbed by the medium by Beer–Lambert law (see Appendix F).

Given the incident direction Ω_1 and the normal Ω_L to the leaf surface, the direction of reflection Ω_2^R may be expressed as (Glassner, 1989)

$$\Omega_2^R = \Omega_1 - 2(\Omega_L\Omega_1)\Omega_L \tag{10.1}$$

and the direction of transmission (refraction) Ω_2^T is given by

$$\Omega_2^T = n_{21}\Omega_1 + \Omega_L\left(n_{21}(-\Omega_L\Omega_1) - \left(1 + n_{21}^2(\Omega_L\Omega_1)^2 - 1\right)^{1/2}\right) \tag{10.2}$$

where $n_{21} = n_1/n_2$, n_1 is the index of refraction of the medium into which the ray is initially propagating, and n_2 is the index of refraction of the intercepted medium. The absorption coefficient k of each medium mentioned in Chapter 3 is used to estimate the ray free path d_a when traveling in the medium:

$$d_a = \frac{1}{k}\ln u \tag{10.3}$$

If d_m is the maximum distance the ray can cover in the medium, it will be absorbed when $d_a < \mathrm{d_m}$. Ray path statistics are accumulated to compute the bidirectional reflectance and transmittance factors, and also the light extinction profile inside the leaf. The hemispheres above and below the scene are divided into m equal area elementary surfaces S_l. RAYTRAN simply counts how many rays escape through each surface. The bidirectional reflectance factor f_l of the elementary surface S_l is calculated as follows (Ross and Marshak, 1989):

$$f_l = \pi\frac{N_l}{N\Delta\Omega_l}, \quad l = 1\ldots m, \tag{10.4}$$

where N_l is the number of photons that cross the surface S_l, N is the total number of generated photons, and $\Delta\Omega_l$ is the projected solid angle corresponding to the elementary surface S_l.

10.4 Radiative Transfer Simulations

Although the three leaf models described above (DICOT1, DICOT2, and DICOT3) are simplified compared to actual leaves, they nevertheless allow the representation of a fair degree of complexity. Computation of the radiation regime in such complex media is not particularly difficult with RAYTRAN.

10.4.1 Bidirectional Reflectance and Transmittance

The bidirectional reflectance and transmittance of plant leaves has been simulated in the near infrared (NIR) at illumination zenith angles ranging from 0° to 60°. In this experiment, surface roughness was controlled by changing the oblateness of the epidermal cells: $o_e = 0.7$ for a rough leaf, $o_e = 0.4$ for a moderately rough leaf, and $o_e = 0.2$ for a smooth leaf. Figure 10.16 shows that the BRDF and BTDF of the leaves with a rough epidermis are almost Lambertian. Even though such behavior may be observed in leaves covered by hair (see Section 5.1.3), most leaves exhibit a specular reflection peak. As the epidermis becomes smoother, the reflectance becomes more specular, especially for high illumination zenith angles. At large zenith angles, the transmittance is

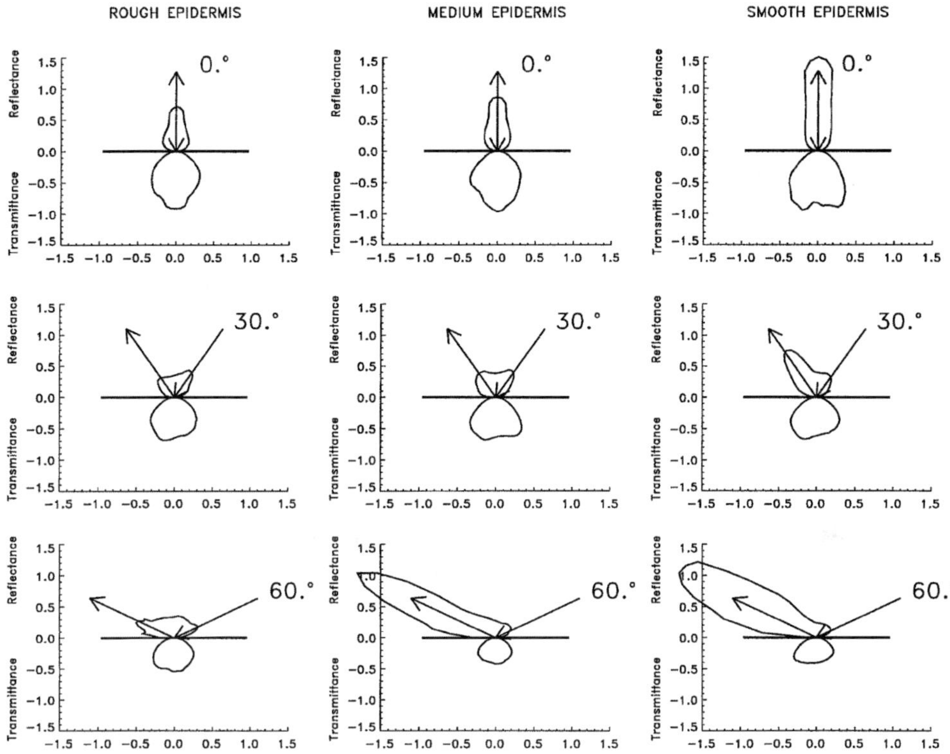

Figure 10.16 Bidirectional reflectance and transmittance of DICOT1 in the principal plane, in polar coordinates, for illumination angles of 0°, 30°, and 60°. The roughness parameter o_e of the epidermis cells was set to 0.7 (rough), 0.4 (moderately rough), and 0.2 (smooth) (from Govaerts et al. (1996), *Applied Optics*. Copyright © 1996 with permission from the Optical Society of America).

also affected by the illumination direction, as reported elsewhere. Note also that the specular component exhibits an off-specular peak as predicted by theory.

More realistic simulations of leaf bidirectional reflectance would require a careful description of the roughness of the epidermis. For instance, it would be useful to understand why various leaves adopt a particular roughness, and how the structure and therefore the leaf optical properties, evolve in space and time. As seen in Section 5.1.5, convex epidermis cells of some leaves tend to focus light on the subjacent palisade cells, but they also tend to reflect light isotropically. Such leaves are often found in the lower part of the canopy, where direct solar radiation does not easily penetrate. They exhibit a larger absorptance than those with a smoother epidermis, and have rather Lambertian bidirectional reflectance. Conversely, leaves with a smooth epidermis exhibit a more specular reflectance and absorb less radiation; they are expected to be found in the upper part of the canopy, where direct sunlight is predominant. Through this mechanism, leaves may protect themselves against excessive direct solar radiation and transmit light to the lower layers of the canopy.

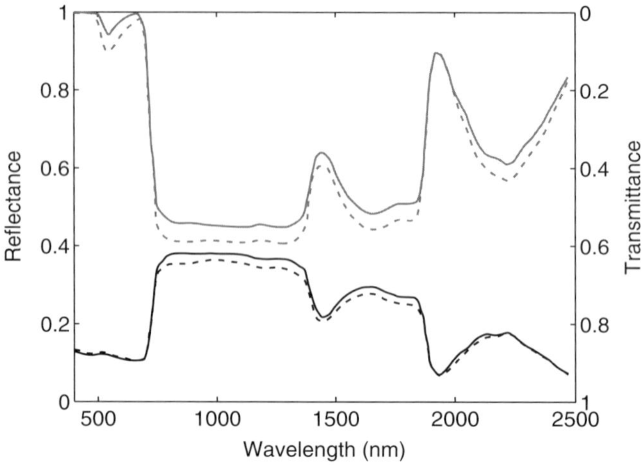

Figure 10.17 Leaf bi-hemispherical reflectance and transmittance of DICOT1 with two different oblatenesses of epidermal cells: $o_e = 0.7$ (solid line) and $o_e = 0.2$ (dashed line) (from Govaerts et al. (1996), *Applied Optics*. Copyright © 1996 with permission from the Optical Society of America).

10.4.2 Bi-Hemispherical Reflectance and Transmittance Spectra

If the adaxial face of the leaves is illuminated by an isotropic point light source with a conical angular aperture of 90°, one can calculate the leaf bi-hemispherical reflectance R_{bh} and transmittance T_{bh}. This is typically a situation occurring under cloudy conditions. The absorptance A_{bh} is simply derived through the relationship $A_{bh} = 1 - R_{bh} - T_{bh}$. The spectra displayed in Figure 10.17 have been generated on DICOT1 with two different epidermal cell oblatenesses. These spectra are typical of a plant leaf, with a strong absorption in the VIS and SWIR attributed to pigment and water absorption, respectively, and a plateau in the NIR attributed to leaf internal structure. However, in detail, one can observe some difference with laboratory measurements. For instance, the reflectance spectra simulated by RAYTRAN in the blue and in the red display unexpected absorption peaks by pigments, when the transmittance spectra are more familiar. Several factors may be involved, such as the strong assumption of a homogeneous chloroplast membrane, but increasing the absorption by the chlorophylls does not improve the output of the model. It is more likely that leaf surface reflectance is overestimated. Indeed, leaves covered by a thick wax layer display very similar spectral features. In the NIR, the reflectance and transmittance levels are more characteristic of monocots than dicots, where R_{bh} that is smaller than 0.4 is underestimated and T_{bh} that is greater than 0.6 is overestimated. Note that the reflectance (transmittance) would be even smaller (greater) if the simulated spectra were directional-hemispherical instead of bi-hemispherical (Gorton et al., 2010). By correcting the absorption coefficients of leaf constituents, in particular chlorophyll, and by improving leaf anatomical structure, Bousquet (2007) could increase the directional-hemispherical reflectance and decreased the transmittance in the NIR, but could not totally fix the flattening of the green peak observed in previous simulations. For 5° to 65° incidence angles, this author found that the absorptance was almost constant while the reflectance increased and the transmittance decreased, as observed by Brandt and Tageyeva (1967).

10.4.3 Vertical Light Attenuation

Light absorption profiles through leaves have been modeled using Beer–Lambert law (Terashima and Saeki, 1985); others have adopted more complicated approaches (e.g., Seyfried and Fukshansky, 1983; Fukshansky and von Remisowsky, 1992; Richter and Fukshansky, 1996b). In RAYTRAN, virtual detectors can be equally positioned along a vertical axis to determine the relative downward light flux $\Phi_\downarrow(z)$ and the relative upward light flux $\Phi_\uparrow(z)$ perpendicular to the leaf. Each time a ray reaches the upper (lower) surface of the sensor, the downward (upward) flux counter is incremented by 1. Rays collected in this manner are then divided by the total number of emitted rays. Therefore, the relative net flux is calculated for each sensor as the difference between the downward and upward fluxes, $\Phi_n(z) = \Phi_\downarrow(z) - \Phi_\uparrow(z)$. In this scheme, the incident light is $\Phi_\downarrow(0) = 1$, the leaf reflectance is $R = \Phi_\uparrow(0)$, and the leaf transmittance is $T = \Phi_\downarrow(l)$, with l the thickness of the blade. Therefore the leaf absorptance can be calculated as

$$A = 1 - R - T = 1 - \Phi_\uparrow(0) - \Phi_\downarrow(l) \tag{10.5}$$

Since the transmittance of the leaves at 675 nm is negligible – at this wavelength, they are considered as optically thick objects – their absorptance is the fraction of light not reflected, i.e.:

$$A = 1 - \Phi_\uparrow(0) \tag{10.6}$$

By definition, $\Phi_n(0) = \Phi_\downarrow(0) - \Phi_\uparrow(0) \Leftrightarrow \Phi_\uparrow(0) = 1 - \Phi_n(0)$, then

$$A = \Phi_n(0) \tag{10.7}$$

The sampling area is assumed to be surrounded by identical patches in all directions, so that lateral boundary conditions are periodic and the model actually simulates a leaf of infinite extent. Figure 8.11 illustrates how light penetrates the leaf at 675 nm. The predicted vertical distributions of light are in close agreement with microprobe experimental results reported in Section 5.4, including small variations in net flux and upwelling fluxes at the transition between palisade and spongy mesophyll in the profiles measured using a fiber-optic probe (Vogelmann et al., 1989). Eighty percent or more of the light is exponentially absorbed in the palisade parenchyma. Note that the relative downward flux in the upper epidermis can exceed 100% of the incident flux at the leaf surface because the same ray may be scattered several times inside a cell of this tissue, and hence be counted more than once by a detector. This concentration of light has been reported for various leaves (e.g., Vogelmann and Björn, 1984; Vogelmann et al., 1989, 1991; Vogelmann, 1993); these authors found that light fluxes within leaves could exceed by three to four times the incident light in the near-infrared spectral region.

In nature, sunlight normally strikes the adaxial surface of leaves and only small amounts of transmitted light reach the chloroplasts near the lower surface. To investigate how bifacial leaf anatomy and biochemistry affect light propagation in different orientations, Ustin et al. (2001) simulated light profiles with RAYTRAN. Simulations were based on variations in the orientation of the DICOT2 leaf model. The leaves were tilted vertically and horizontally so that either their upper side or their lower side faced the Sun, which was set at a zenith angle of 25°. It resulted in the four arrangements illustrated by Figure 10.18: horizontal leaf with light incident on palisade parenchyma (#1); vertical leaf with light incident on palisade parenchyma (#2); horizontal leaf with light incident on spongy mesophyll (#3); and vertical leaf with light incident on spongy mesophyll (#4). The

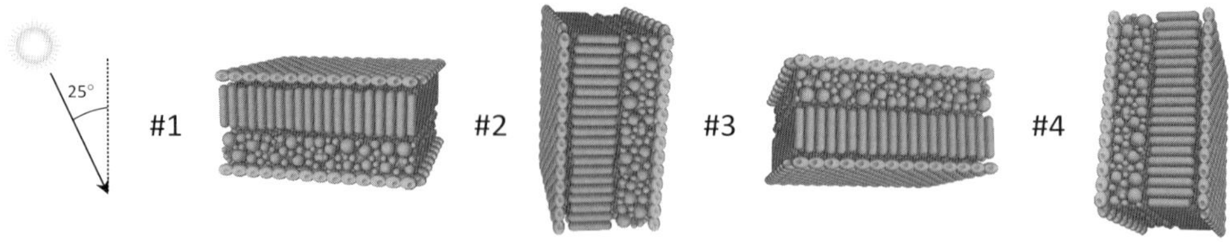

Figure 10.18 Four orientations selected to simulate the absorption profiles of the DICOT1 leaf.

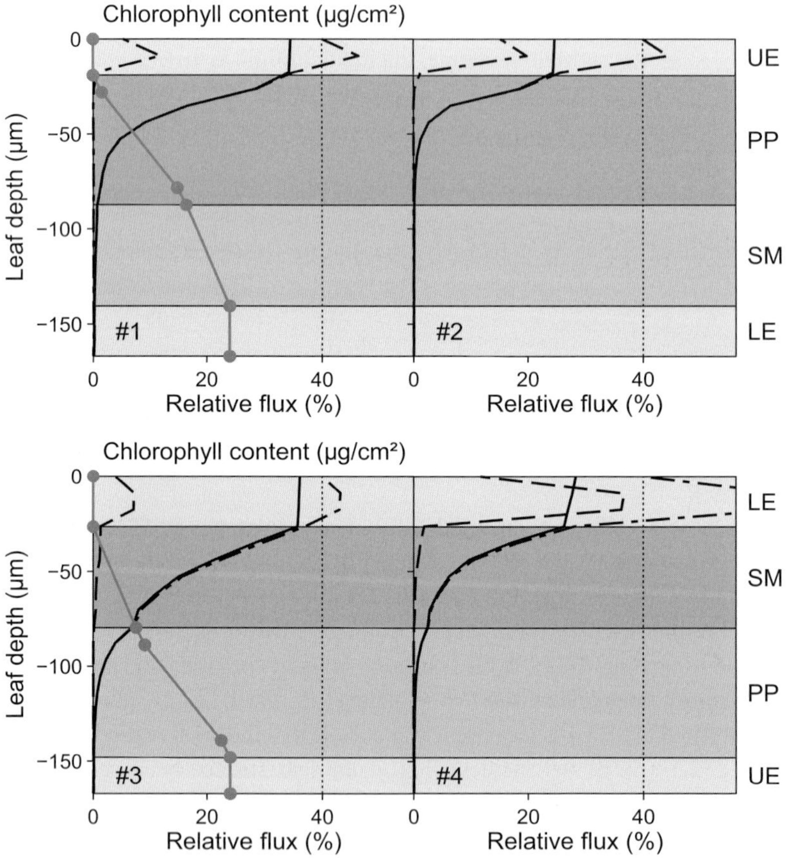

Figure 10.19 Normal distribution of light (relative downward flux $\Phi_{\downarrow}(z)$, dashed line; relative upward flux $\Phi_{\uparrow}(z)$, dash-dot line; relative net flux $\Phi_n(z)$, solid line) at 675 nm inside a horizontal dorsiventral leaf with light incident on the palisade parenchyma (#1 and #2) and light incident on the spongy mesophyll (#3 and #4), for illumination zenith angles of 25° (#1 and #3) and 65° (#2 and #4). UE = upper epidermis, PP = palisade parenchyma, SM = spongy mesophyll, and LE = lower epidermis. Gray discs indicate the cumulative chlorophyll concentration, expressed in μg cm^{-2}.

positions #2 and #4 can be also viewed as horizontal leaves illuminated by the Sun in late afternoon, at a zenith angle of 65°. Basically, the position #3 only occurs on windy days.

Light gradients simulated at 675 nm perpendicular to the leaf surface are shown in Figure 10.19, which illustrates how cell distribution (palisade parenchyma vs. spongy mesophyll) affects the

Table 10.2 *Reflectance (R), transmittance (T), and absorptance (A) derived from the light fluxes simulated on the* DICOT2 *and* DICOT3 *leaves.*

	DICOT2				DICOT3			
	675 nm				675 nm		740 nm	
	25° (#1)	65° (#2)	115° (#3)	155° (#4)	5°	65°	5°	65°
R	0.135	0.382	0.098	0.292	0.150	0.350	0.500	0.650
T	0	0	0	0	0	0	0.500	0.300
A	0.865	0.618	0.902	0.708	0.850	0.650	0	0.050

Table 10.3 *Reflectance (R), transmittance (T), and absorptance (A) of isolated tissues of the* DICOT3 *leaf displayed in Figure 10.21, at 674 nm and 740 nm, for directional (dir) or diffuse (dif) incident light (Bousquet, 2007).*

	Upper epidermis				Palisade parenchyma				Spongy mesophyll			
	675 nm		740 nm		675 nm		740 nm		675 nm		740 nm	
	dir	dif	dir	dif	dir	dif	dir	dif	dir	dif	dir	dif
R	0.126	0.218	0.127	0.219	0.053	0.077	0.132	0.241	0.020	0.036	0.336	0.429
T	0.871	0.778	0.871	0.778	0.470	0.156	0.856	0.737	0	0	0.661	0.568
A	0.003	0.004	0.002	0.003	0.477	0.767	0.012	0.022	0.980	0.964	0.003	0.003

distribution of upwelling and downwelling photons in photosynthetically active radiation wavelengths. Net flux profiles vary strongly with leaf orientation and incident light zenith angle.

The optical properties of the leaf are gathered in Table 10.2. Light is definitely more efficiently absorbed in horizontal leaves (#1 and #3) than it vertical leaves (#2 and #4). As seen earlier, rotating the leaf from horizontal to vertical is equivalent to increasing the illumination zenith angle, that is, the fraction of specular reflectance. Consequently, the effectiveness of light absorption is less in vertically oriented leaves. It has already been shown experimentally (Brakke, 1994) and with simulations (Govaerts et al., 1996). With few exceptions, vertical leaves are actually symmetrical, with palisade cells beneath both the adaxial and abaxial epidermis (Poulson and DeLucia, 1993). The attenuation of light occurs rapidly when the palisade parenchyma is illuminated. Nearly 100% of the rays are absorbed in this tissue, that is, within the initial 90 μm of the adaxial leaf surface. These results are consistent with observations that palisade cells aid in redistributing collimated light within the leaf for maximum efficiency of photosynthesis. Because the 45% air space in spongy mesophyll greatly increases the photon path length, this tissue does not attenuate light as fast as the palisade parenchyma when illuminated, but the total absorptance is higher for profile #3 than for #1, and is higher for #4 than for #2. This confirms experimental observations by Yates (1981) and DeLucia et al. (1991). Bousquet (2007), who performed similar simulations on the horizontal DICOT3 leaf at 5° and 65° illumination angles, and at 675 nm and 740 nm, found values of the same order of magnitude (Table 10.3).

Because the parenchyma tissues receive light that has been diffused by the epidermis, Bousquet (2007) ran simulations on leaf tissues taken separately using both collimated (5° angle of incidence) and diffuse light (Figure 10.20).

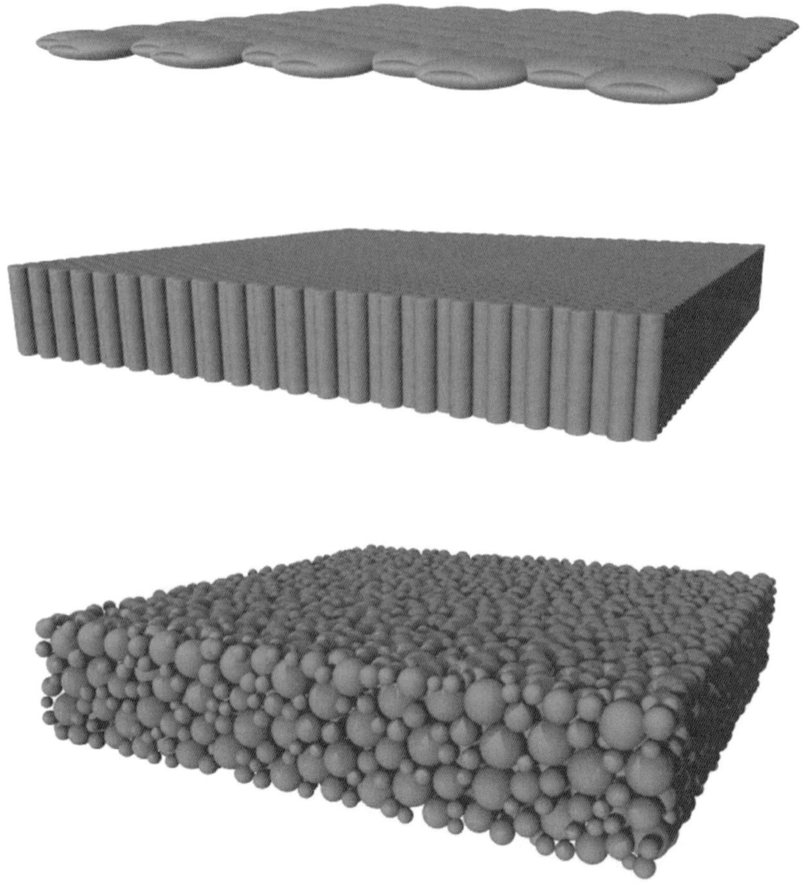

Figure 10.20 Perspective view of the three tissues composing the DICOT3 leaf model (Bousquet, 2007).

The results of the simulations are gathered in Figure 10.21 and Table 10.3. The reflectance of the epidermis alone equals approximately 13%, whatever the wavelength. In a real leaf, epidermal cells actually make a unique compact layer that is in contact with the underlying palisade mesophyll so their overall reflectance would be lower. Scattering mainly occurs in the palisade and spongy parenchyma. At 740 nm, where absorption is low, the palisade behaves like the epidermis although its cells are much more elongated. This implies that scattering mostly occurs in the cell caps but not so much in the cylinders. In the spongy mesophyll, the downward and upward fluxes vary continuously, showing that scattering occurs everywhere in this tissue.

Brodersen and Vogelmann (2010) performed direct measurements of the propagation of direct and diffuse light into an entire leaf. They found no significant differences in the shapes of the light absorption profiles, contrary to the results shown in Figure 10.21 for separate tissues. Regarding the effect of absorption at 674 nm, the palisade parenchyma and spongy mesophyll have different behaviors. The net flux decreases exponentially in the spongy parenchyma, whereas its decline is smoother in the palisade parenchyma. This effect is linked to the cylindrical structure of the palisade cells, where some rays penetrate the inner part of the cell, which is filled with water, and then are

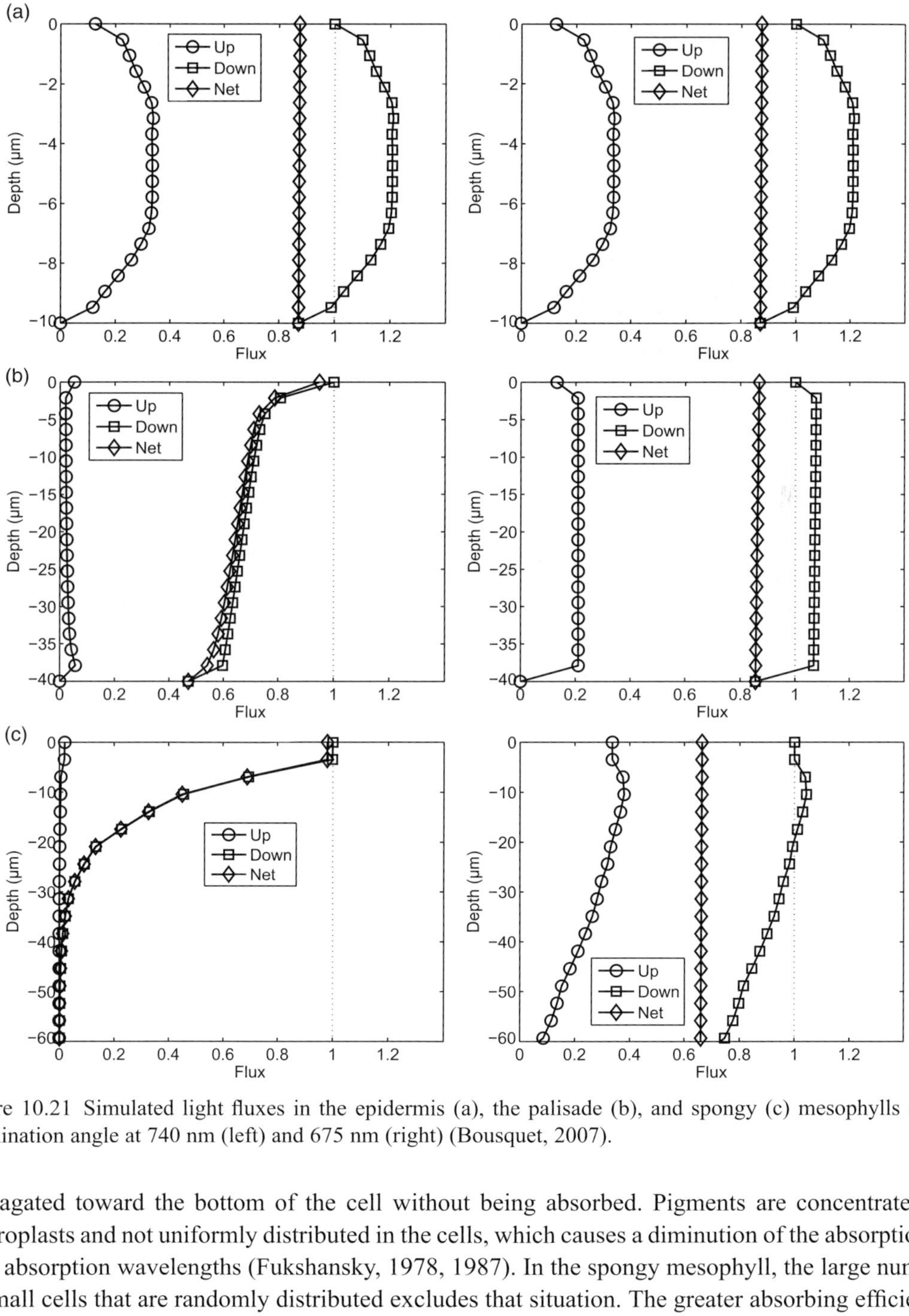

Figure 10.21 Simulated light fluxes in the epidermis (a), the palisade (b), and spongy (c) mesophylls at 5° illumination angle at 740 nm (left) and 675 nm (right) (Bousquet, 2007).

propagated toward the bottom of the cell without being absorbed. Pigments are concentrated in chloroplasts and not uniformly distributed in the cells, which causes a diminution of the absorption at high absorption wavelengths (Fukshansky, 1978, 1987). In the spongy mesophyll, the large number of small cells that are randomly distributed excludes that situation. The greater absorbing efficiency of the spongy has been noted by Vogelmann and Evans (2002).

Although chlorophyll is more concentrated in the palisade mesophyll that contains 64% of the chloroplasts of the leaf, the spongy parenchyma is the most absorbing tissue. More than 95% of the incident light is absorbed whatever the illumination condition. Figure 10.21 also shows a faster decrease of the net downward flux in the spongy mesophyll than in the palisade mesophyll. The chlorophyll content and the anatomic structure are the two factors driving light absorption. The first determines the fraction of light that is absorbed while going through the tissue, and the second controls the length of the ray path through the tissue. Because of its numerous spherical cells, the spongy mesophyll is the most scattering tissue. Light diffusion is thus essential to absorption efficiency. When setting a specific absorption coefficient for a leaf model, one should also take into account this effect.

10.5 Coupling RAYTRAN with a Photosynthesis Model

The photosynthetic response of dorsiventral leaves to light absorption put in contrasting positions has been measured experimentally to study the effect of irradiating adaxial versus abaxial surfaces (e.g., Moss, 1964; Syvertsen and Cunningham, 1979; Terashima, 1986; Evans et al., 1993; Ögren and Evans, 1993; Terashima et al., 2009) and horizontal versus vertical surfaces (e.g., Yates, 1981; DeLucia et al., 1991). The absorption profiles of Figure 10.19 can be used as inputs of a simple photosynthesis model. Among all the equations proposed in the literature to simulate the response of photosynthesis to light intensity, Ustin et al. (2001) chose the non-quadratic hyperbola defined by Prioul and Chartier (1977). This equation is generally used at the scale of an entire leaf but Terashima and Saeki (1985) applied it to the sections of a Japanese camellia (*Camellia japonica*) leaf segmented into ten vertical layers. The whole leaf photosynthesis P is then expressed as the sum of the gross photosynthetic rate of each layer p_j. Note that a similar approach was proposed by Fukshanski and Martinez von Remisowsky (1992) on southern catalpa (*Catalpa bignonioides*) and umbrella-tree (*Magnolia tripetala*) leaves with another photosynthesis model. For layer j, the form of the non-quadratic hyperbola model is

$$\theta_j^2 p_j^2 - \left(\phi_j A_j + p_{\max,j}\right) p_j + \phi_j A_j p_{\max,j} = 0 \tag{10.8}$$

where p_j is the gross photosynthetic rate (μmol CO_2 m^{-2} s^{-1}), θ_j is the convexity index ($0 \leq \theta_j \leq 1$), ϕ_j is the quantum yield for CO_2 fixation (μmol CO_2 μmol^{-1} photons), A_j is the amount of absorbed light (μmol photons m^{-2} s^{-1}), and $p_{\max,j}$ is the gross photosynthetic rate at light saturation (μmol CO_2 m^{-2} s^{-1}). When expressed in μmol CO_2 mg^{-1} chlorophyll s^{-1}, the latter must be multiplied by the chlorophyll content $C_{ab,j}$ (mg chlorophyll m^{-2}). p_j is then the smallest of the two solutions of Eq. 10.8. To run the photosynthesis model, the gradients within the leaf of the terms described above are required. The DICOT2 leaf has been segmented into 19 layers, each having the same thickness: two layers for the upper epidermis, eight for the palisade mesophyll, six for the spongy mesophyll, and three for the lower epidermis. Figure 10.22 shows the gradient in absorption A_j and Figure 10.23 the gradient in chlorophyll $C_{ab,j}$ deduced from Figure 10.19.

The total absorptance reported in Table 10.2 equals 0.865, 0.618, 0.902, and 0.708 for profiles #1, #2, #3, and #4, respectively. Decreasing the incidence angle from 65° to 25° decreases the absorptance by 24.4% when the adaxial surface is illuminated and by 19.4% when the abaxial surface is illuminated. In the latter case, the pattern of light absorption is slightly inverted at the transition from spongy mesophyll to palisade parenchyma, probably because of the higher absorption coefficient in this tissue.

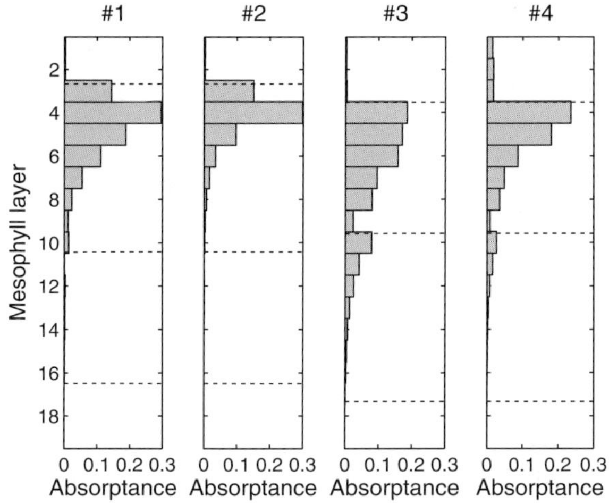

Figure 10.22 Gradients in light absorption (A_j) within the leaf when illuminated at 675 nm on the adaxial surface (#1 and #2) and on the abaxial surface (#3 and #4). The dashed lines represent the boundaries between the leaf tissues (redrawn from Ustin et al. (2001)).

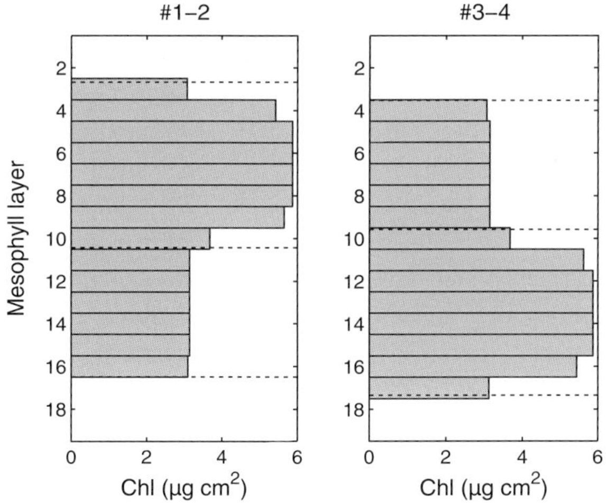

Figure 10.23 Gradients in chlorophyll concentration $C_{ab,j}$ within the leaf. The dashed lines represent the boundaries between the leaf tissues (redrawn from Ustin et al. (2001)).

Some assumptions are necessary to calculate the gross photosynthetic rates of the layers. For instance, although ϕ_j and $p_{\max,j}$ depend on many physiological parameters, such as the internal temperature and the CO_2 and O_2 concentrations at the intercellular spaces, they can be considered as constants (Terashima and Saeki, 1985): $\phi_j = \phi = 0.05$μmol CO_2 μmol^{-1} photons for all the layers, $p_{\max,j} = p_{\max,1} = 0.0125$μmol CO_2 mg^{-1} chlorophyll s^{-1} for the palisade parenchyma, and $p_{\max,j} = p_{\max,2} = 0.01$ μmol CO_2 mg^{-1} chlorophyll s^{-1} for the spongy mesophyll. This difference accounts for the higher electron transport activity in the palisade parenchyma (Terashima and Inoue, 1984). The convexity index θ_j is representative of the convexity of the hyperbola; although it has been reported to

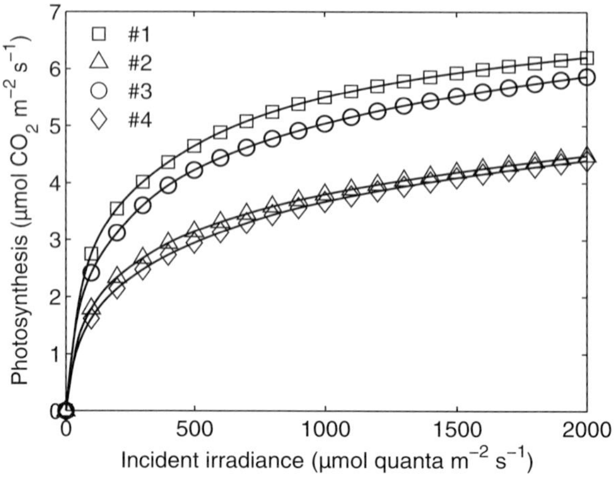

Figure 10.24 The response of gross photosynthetic rate to irradiance incident on the adaxial (#1 and #2) and abaxial (#3 and #4) leaf surfaces, oriented horizontally (#1 and #3) and vertically (#2 and #4) (redrawn from Ustin et al. (2001)).

be correlated with the light gradient inside the leaf and the chlorophyll content, in this study it was assumed to be constant and was set to $\theta_j = 0.95$, a common value in C3 plants. The response curves shown in Figure 10.24 describe the relation between photosynthesis and illumination. Photosynthesis increases with increasing light until all the chloroplasts are saturated. These curves are consistent with experimental data or simulations performed on dorsiventral leaves; the asymptotic gross photosynthetic rates are also comparable to experimental data published in the literature, which are very variables. The low levels may be due to the high values of θ_j in the simulation. Figure 10.24 also shows that photosynthesis is less when leaves are lit from below (#3) rather than from above (#1). The difference between the two curves can be explained by the difference in optical properties between the palisade parenchyma and the spongy mesophyll. Considering that vertical leaves receive 47% less total incident irradiance than horizontal leaves for purely geometrical reasons (Yates, 1981), and that their orientation increases their total reflectance (decreases absorptance) by 20–25% as seen earlier, their gross photosynthesis should be reduced. This is confirmed by the results shown in Figure 10.24. Once again, incident light on the adaxial surface seems to be advantageous, although the difference is small in this case.

As pointed out by Jeje and Zimmermann (1983), Nishio et al. (1993), and Sun et al. (1996, 1998), the profiles of gross photosynthetic rate, which are related to carbon fixation within leaves, do not follow within-leaf light gradients closely. For adaxial illumination, one can see a maximum in the middle of the palisade parenchyma that is in agreement with experiments (Figure 10.25). Since no hypotheses have been made on the Rubisco distribution, our simulations show that the latter is not the only factor controlling carbon fixation. Moreover, an increase in light irradiance from 400 to 2000 µmol photons $m^{-2}\ s^{-1}$ leads to saturation in the first half of this tissue, while abaxial illumination induces saturation in the entire spongy mesophyll. Ustin et al. (2001) found that orientations #1 and #2 were better adapted to increasing light intensities than #3 and #4. Figure 10.25 also shows that CO_2 fixation is very low in the spongy mesophyll (almost nil for vertical leaves) when the adaxial surface

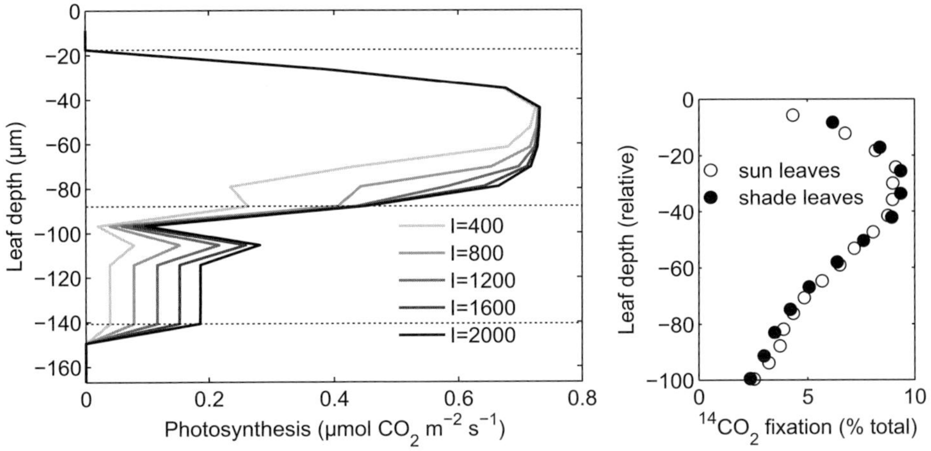

Figure 10.25 (Left) Profiles of gross photosynthetic rate within the leaf for different incident irradiation intensities on the adaxial surface (#1). The dashed lines represent the boundaries between the leaf tissues. (Right) Normalized CO_2 fixation across sun and shade spinach (*Spinacia oleracea*) leaves (redrawn from Sun et al. (1996)).

is illuminated. However, the palisade parenchyma is still important for light absorption when the abaxial surface is illuminated. This pattern is consistent with the results of physiological models; for instance, Parkhurst (1986) and Pachepsky and Acock (1998) found significantly higher photosynthetic rates in simulated leaves containing only spongy mesophyll cells than in leaves containing only palisade parenchyma.

11

Extraction of Leaf Traits

Leaf directional-hemispherical or bidirectional reflectance and transmittance spectra can directly feed canopy reflectance models as input parameters, but the measurement of these properties is not an end in itself. A large number of spectral analysis methods have been proposed to detect plant biochemistry, ranging from simple band ratios to inversion of radiative transfer methods of varying complexity. The estimation of leaf biophysical parameters is often developed in parallel with the estimation of canopy characteristics, using the same methods as detailed below. The first studies linking leaf reflectance with its chemical content date from nearly one century ago. They have long remained qualitative due to the poor precision of the spectrophotometers to measure the leaf optical properties and the wet chemistry methods to measure the leaf chemical constituents. Despite recent developments of new algorithms, sensors, and platforms, Majeke et al. (2008) illustrated the need for further advances in remote sensing and image analysis and strongly suggested the need for more development to standardize and improve the methods.

11.1 Combinations of Narrow Bands

The combination of reflectance into spectral indices is a classical approach in remote sensing. It developed from the limited spectral information in early multispectral instruments and the limited computational capabilities in past decades. Such indices have continued to be actively used because of their simplicity and because they often yield good results for individual datasets. Since differences in scattering properties among leaves produce additive offsets and multiplicative effects in reflectance spectra, these methods typically rely on identifying two or three wavelengths, rarely more, where the biochemical of interest causes an absorption in one or more bands and another where it does not absorb light, allowing the methods to detect relative absorption differences between samples. Like vegetation indices at the canopy level, a simple non-linear mathematical relationship f is established between the biochemical component of interest and the leaf optical properties:

$$C = f\Big(R(\lambda_1), \ldots, R(\lambda_n)\Big) = f(R_{\lambda_1}, \ldots, R_{\lambda_n}) \tag{11.1}$$

where $R(\lambda_i)$ refers to reflectance at a specific spectral band or wavelength λ_i and C stands for a biochemical of interest (e.g., chlorophyll, carotenoids, anthocyanins, and water). Typically, only one biochemical is analyzed at a time, thus possible interactions have been ignored. The function f can be linear, a polynomial of higher degree, or an exponential model, but most published spectral indices are quasi-linearly correlated to leaf biochemical constituents. Despite problems in identifying specific wavebands for individual constituents, the earliest attempts to use spectroscopy focused on estimating the concentration (g g^{-1}) of total photosynthetic pigments or water using simple ratios between red and near-infrared (NIR) or near-infrared and shortwave-infrared (SWIR) bands. As interest in narrow-band spectrometry increased, other ratio based indices or simple transforms of band combinations were developed as practical methods of analysis, as did the use of first and second derivatives, which were applied across the full spectrum. Derivative analyses, discussed in Section 11.3, have been primarily focused on the red-edge, which is the long wavelength edge of the chlorophyll absorption, around 700–750 nm. Relationships similar to Eq. 11.1 have been defined:

$$C = f\left(D(\lambda_1), \ldots, D(\lambda_n)\right) = f(D_{\lambda_1}, \ldots, D_{\lambda_n}) \tag{11.2}$$

where $D(\lambda_i)$ henceforth refers to the first or second derivative of reflectance at a specific spectral band or wavelength λ_i. Traditional techniques for evaluating the performance of these spectral indices are regression-based statistics, such as the coefficient of determination (R^2), the mean squared error (MSE), or the root mean square error (RMSE). To measure the sensitivity of a vegetation index to biophysical parameters, Ji and Peters (2007) even developed a "sensitivity function."

11.1.1 Photosynthetic Pigments

Benedict and Swidler (1961) are the first to propose a logarithmic relation between leaf reflectance at 625 nm and total chlorophyll content (g m^{-2}). Thomas and Gausman (1977) determined the correlations between chlorophyll and carotenoids in eight types of crops and then applied multiple linear regression to link pigment content to reflectance at three wavelengths (450 nm, 550 nm, and 670 nm). They highlighted the main issues encountered in estimating pigments by remote sensing, namely the strong correlations between their contents and the lack of robustness of such a model when applied to different types of plants. The principle of a spectral index sensitive to chlorophyll is illustrated in Figure 11.1. For a given anatomical structure, the reflectance R_1 measured at any wavelength in the visible (VIS) varies more or less with C_{ab}. In contrast, the near-infrared reflectance R_2 is insensitive to leaf pigments but highly sensitive to leaf structure. Thus the natural variation in leaf anatomies prevents us from using only R_1 to compare the chlorophyll content of two leaves. Simple ratios or normalized difference indices, like the Normalized Difference Vegetation Index (NDVI), are the most widely applied at the canopy scale, and allow users to minimize this problem. Using narrow wavebands at the edge of the chlorophyll absorption feature (e.g., 705 nm) rather than at the middle of the red region, as illustrated in Figure 11.1, generally improves sensitivity to a wider range of chlorophyll concentrations (Gamon and Surfus, 1999). But variation in leaf thickness or structure is known to affect the performance of chlorophyll indices (Serrano, 2008).

There are a large number of narrow-band spectral indices targeted at plant pigments. The ability of ratio-based indices and first and second derivatives of the reflectance and the natural logarithm of one divided by the reflectance to predict concentrations of chlorophyll (Table 11.1), carotenoids (Table 11.2), and anthocyanins (Table 11.3) has been reviewed by dozens of authors (e.g., Le Maire et al., 2004; Blackburn, 2007a; Rodríguez-Pérez et al., 2007; Ustin et al., 2009; Xue and Yang, 2009; Ju et al., 2010; Main et al., 2011; Kochubey and Kazantsev, 2012; Neuwirthová et al., 2017; Sonobe

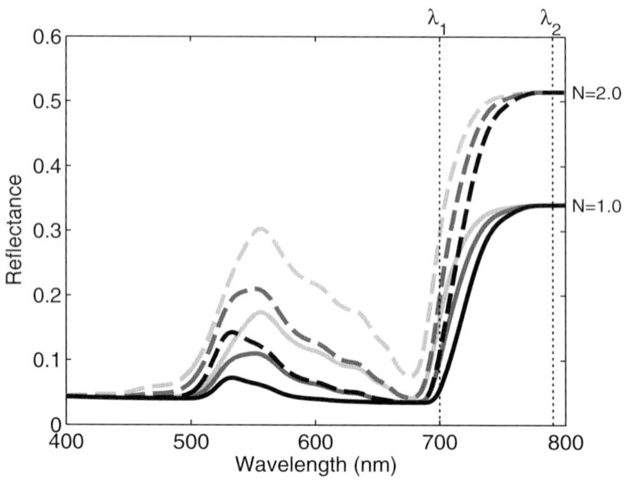

Figure 11.1 Principle of a spectral index using PROSPECT-D. The lines and dotted lines represent two distinct foliar structures, and the gray levels correspond to various chlorophyll contents (light gray = less pigment).

and Wang, 2017a,b). These lists are not exhaustive but they cover the commonly used methods for leaf analysis, while also including more recent examples. Strong linear or non-linear relationships, typically either power or exponential fits, can be developed but maximum correlations are produced at somewhat different wavelengths that depend on the sample set of leaves, the index used, and whether the pigment is expressed on an area (content) or mass (concentration) basis (Grossman et al., 1996). Finally, there are a few examples in the literature where chlorophyll *a* and chlorophyll *b* have been separately assessed from reflectance data using empirical models. Among the first, Chappelle et al. (1992) and Blackburn (1999) used band ratios to quantify the concentration of chlorophyll *b*.

Detection of individual accessory pigments has been given even less attention by the remote sensing community, despite their importance in light capture, photosystem protection, and in various growth and development functions. This lack is due to the difficulty in resolving the overlapping absorptions of individual pigments and their high degree of autocorrelation.

This survey of published indices, however, reveals that there is no consensus on wavelengths or methods, mainly because these studies are often obtained from a specific, limited, experimental dataset. Intercomparison studies are intended to find the most reliable methods and identify discrepancies, as seen in Figure 11.2 (e.g., Richardson et al., 2002; Sims and Gamon, 2002; Gitelson et al., 2003; Féret et al., 2011).

Designing a "universal" spectral index is challenging because of the limited size of the calibration and validation datasets that are used. Indices may behave differently because of variations in pigment distribution, phenological stage, leaf structure, and plant species. As a consequence, methods based on small-sized datasets often lead to specialized indices that apply poorly to the vast majority of leaves. Synthetic datasets containing simulated leaf reflectance and transmittance spectra help to overcome this obstacle (e.g., Danson and Bowyers, 2004; le Maire et al., 2004, 2008). This approach was improved by Féret et al. (2011) who developed new optimized spectral indices based on a "realistic" synthetic database of leaf reflectance spectra. They first compiled a comprehensive experimental dataset including 1442 leaf samples across 185 plant genera, characteristic of the

Table 11.1 *Spectral indices developed as chlorophyll indicators, where* R_λ *and* D_λ *represent the leaf reflectance and the first derivative respectively at the indicated wavelength* λ*, and listed by first year of publication. The subscripts* B, G, R, NIR, sw, mw, *and* lw *stand for blue, green, red, near infrared, short wavelength, medium wavelength, and long wavelength, respectively.*

Index	Short	Formula	Reference
Simple Ratio	**SR**	R_{NIR}/R_B, R_{NIR}/R_G, R_{NIR}/R_R, R_B/R_G, R_B/R_R, R_R/R_G	Aoki et al. (1980, 1986), Carter (1994), Gitelson and Merzlyak (1994b, 1996, 1997), Gitelson et al. (1996), Datt (1998), Maccioni et al. (2001), Sims and Gamon (2002)
Red-edge Inflexion Point	**REIP**	$R''(\lambda_i) = 0$	Horler et al. (1983c), Curran et al. (1995), Lichtenthaler et al. (1996)
Derivative index	**DI**	D_{715}/D_{702}, D_{715}/D_{sw}, D_{725}/D_{702}, D_{lw}/D_{sw}, D_{lw}/D_{702}, D_{725}/D_{sw}, D_{mw}/D_{sw}, D_{mw}/D_{702}	Kochubey et al. (1987, 1988), Smith et al. (2004), Kochubey and Kazantsev (2007, 2012)
Ratio Analysis of Reflectance Spectra	**RARS**	$RARS_a = R_{675}/R_{700}$ $RARS_b = R_{675}/(R_{650} \times R_{700})$	Chappelle et al. (1992), Blackburn (1999)
Chlorophyll Normalized Difference Index	**Chl NDI**	$Chl\ NDI = (R_{750} - R_{705})/(R_{750} + R_{705})$	Gitelson and Merzlyak (1994a, 1994b), Gamon and Surfus (1999), Richardson et al. (2002), Sims and Gamon (2002)
Summed Reflectance Index	**SRI**	$S_1 = \int_{700}^{750} (R_\lambda/R_{555} - 1)d\lambda$ $S_2 = \int_{700}^{750} (R_\lambda/R_{705} - 1)d\lambda$	Gitelson and Merzlyak (1994a, 1994b)
Logarithm of Reciprocal Reflectance	**LRR**	$\log R_{737}^{-1}$	Yoder and Pettigrew-Crosby (1995)
Normalized Difference Vegetation Index	**NDVI**	$NDVI = (R_{NIR} - R_R)/(R_{NIR} + R_R)$	Bacci et al. (1996), Datt (1998)
Green NDVI	**gNDVI**	$gNDVI = (R_{750} - R_G)/(R_{750} + R_G)$	Gitelson et al. (1996), Datt (1998, 1999a)
Pigment Specific Simple Ratio	**PSSR**	$PSSR_a = R_{800}/R_{675}$ $PSSR_b = R_{800}/R_{650}$	Blackburn (1998a 1998b, 1999), Sims and Gamon (2002)
Pigment Specific Normalized Difference	**PSND**	$PSND_a = (R_{800} - R_{675})/(R_{800} + R_{675})$ $PSND_b = (R_{800} - R_{650})/(R_{800} + R_{650})$	Blackburn (1998a,1998b)
Eucalyptus Pigment Indices	**EPI**	$Chl_{a,b} = \alpha_{a,b} \times \left(R_{672}/(R_{550} \times R_{708})\right)^{\beta_{a,b}}$	Datt (1998)
Yellowness Index	**YI**	$YI = -10(R_{580} - 2R_{624} + R_{668})/\Delta\lambda^2$ $(R_{850} - R_{710})/(R_{850} - R_{680})$	Adams et al. (1999) Datt (1999a)

Table 11.1 (*cont.*)

Index	Short	Formula	Reference
Reciprocal Reflectance	**RR**	R_{700}^{-1}	Gitelson et al. (1999)
Modified Simple Ratio	**mSR**	$mSR = (R_{750} - R_{445})/(R_{705} - R_{445})$	Sims and Gamon (2002)
Modified Normalized Difference Index	**mNDI**	$mNDI = (R_{750} - R_{705})/(R_{750} + R_{705} - 2R_{445})$	Sims and Gamon (2002)
Derivative Chlorophyll Index	**DCI**	D_{705}/D_{722}	Zarco-Tejada et al. (2002)
Double Difference	**DD**	$DD = (R_{749} - R_{720}) - (R_{701} - R_{672})$	Le Maire et al. (2004)
Red Edge Symmetry	**RES**	$RES = (R_{718} - R_{675})/(R_{755} - R_{675})$	Ju et al. (2010)

Table 11.2 *Spectral indices developed as carotenoid indicators, where* R_λ *represents the leaf reflectance at the indicated wavelength* λ*, and listed by first year of publication.*

Index	Short	Formula	Reference
Ratio Analysis of Reflectance Spectra	**RARS**	$RARS_c = R_{760}/R_{500}$	Chappelle et al. (1992)
Structure Insensitive Pigment Index	**SIPI**	$PSND_c = (R_{800} - R_{445})/(R_{800} - R_{680})$	Peñuelas et al. (1995), Sims and Gamon (2002)
Pigment Specific Simple Ratio	**PSSR**	$PSSR_c = R_{800}/R_{500}$	Blackburn (1998a,1998b)
Eucalyptus Pigment Index	**EPI**	$Car = \alpha \times \left(R_{672}/(R_{550} \times R_{708})\right)^{\beta}$	Datt (1998)
Photosynthetic Reflectance Index	**PRI**	$PRI = (R_{531} - R_{570})/(R_{531} + R_{570})$	Gamon and Surfus (1999)
Carotenoid Reflectance Index	**CRI**	$CRI_{550} = R_{510}^{-1} - R_{550}^{-1}$ $CRI_{700} = R_{510}^{-1} - R_{700}^{-1}$	Gitelson et al. (2002)
Modified Carotenoid Reflectance Index	**mCRI**	$mCRI_G = \left(R_{510-520}^{-1} - R_{560-570}^{-1}\right) \times R_{NIR}$ $mCRI_{RE} = \left(R_{510-520}^{-1} - R_{690-710}^{-1}\right) \times R_{NIR}$	Gitelson et al. (2006)
CARotenoid Index	**CARI**	$CARI = R_{720}/R_{521} - 1$	Zhou et al. (2017)

Table 11.3 *Spectral indices developed as anthocyanins indicators, where* R_λ *represents the leaf reflectance at the indicated wavelength* λ, *and listed by first year of publication.*

Index	Short	Formula	Reference
Red:Green Ratio	**RGR**	$RGR = R_R/R_G$	Gamon and Surfus (1999), Sims and Gamon (2002)
Anthocyanin Reflectance Index	**ARI**	$ARI = R_{550}^{-1} - R_{700}^{-1}$	Gitelson et al. (2001)
Modified Anthocyanin Reflectance Index	**mARI**	$mARI = \left(R_{530-570}^{-1} - R_{690-710}^{-1}\right) \times R_{NIR}$	Gitelson et al. (2006)

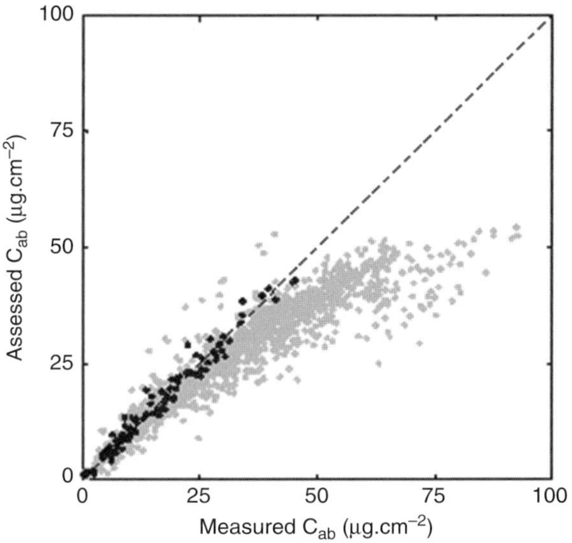

Figure 11.2 Assessment of chlorophyll content using a relationship linking $NDVI_{red\text{-}edge}$ to measured C_{ab}, adjusted on birch leaves (black dots, $N = 98$), and applied to a much larger experimental dataset (gray dots, $N = 1442$) (from Féret et al. (2011), *Remote Sensing of Environment*. Copyright © 2010 with permission from Elsevier).

Earth's dominant vegetation types and derived its statistical properties, such as the probability density function of all biochemical constituents and their cross-correlations. The synthetic dataset computed using the PROSPECT-5 model is assumed to be representative of the natural variability in leaf chemical constituents.

Although pigment indices are correlated with potential photosynthetic activity, they overestimate photosynthesis under conditions where the capacity to absorb energy exceeds the capacity to fix CO_2 into carbohydrate, as happens when stomatal closure limits CO_2 uptake, for example, during drought conditions and when light intensity is greater than the capacity for CO_2 reduction. A number of approaches have been taken from estimating the fraction of absorbed photosynthetic radiation available for photosynthesis to measurements of photosynthetic regulatory mechanisms (see Section 12.2). Researchers (e.g., Gamon et al., 1992; Peñuelas et al., 1995; Gamon and Qiu, 1999; Gamon et al., 2001; Gamon and Berry, 2012; Sonobe and Wang, 2016) investigated the use of a normalized ratio termed the Photochemical Reflectance Index (PRI), that is sensitive to changing xanthophyll cycle pigments. The PRI uses R_{531} as an indicator of xanthophyll cycle activity and R_{730}, or any band with no absorption, as a reference wavelength. As seen in Section 2.3.3, when incident light energy exceeds that needed for photosynthesis, plants protect their photosynthetic machinery by dissipating the excess energy through the xanthophyll cycle (Demmig-Adams, 1990). Violaxanthin undergoes a reversible bond configuration to the zeaxanthin form under high light, thus reducing light absorption in the green spectral region, in particular at 531 nm, and preventing oxidation of the photosynthetic reaction centers. Stylinski et al. (2000) showed that the PRI was well correlated to photosynthetic electron transport capacity in pubescent oak (*Quercus pubescens*) exposed to long-term elevated CO_2 enrichment. Takayama et al. (2008) showed an almost immediate decrease in PRI following petiole cutting, indicating a response to increased heat dissipation, supporting its role in monitoring photosynthetic functionality. Subsequently, Stylinski et al. (2002) tracked seasonal changes in carotenoid pigments, carotenoid to chlorophyll ratio, and monitored photosynthetic activity of mature evergreen semiarid shrubs using the PRI. More recently, Gitelson et al. (2017a) evaluated seasonal changes at the leaf level in PRI and found strong correlations with carotenoid to chlorophyll ratios. Further evaluation at the stand level (Gitelson et al., 2017b) showed strong correlations with green leaf area index (LAI), the one-sided leaf area per unit ground surface area) and stand chlorophyll content, but it was insensitive to light use efficency (LUE), the ratio of net primary productivity to absorbed photosynthetically active radiation, and not clearly related to carotenoid to chlorophyll ratios. Other authors found that PRI was sensitive to changes in leaf water content (LWC) and suitable for monitoring water stress (e.g., Thenot et al., 2002; Inoue and Peñuelas, 2006; Ripullone et al., 2011; Hmimina et al., 2014).

11.1.2 Water and Dry Matter

The same kind of ratio approach has been used to detect leaf water, taking advantage of either narrow-band water absorption features in the near infrared or broader absorption features in the shortwave infrared (Table 11.4). Spectral indices have been used to retrieve leaf water potential (Ψ), foliage moisture content (FMC), leaf water content (LWC), equivalent water thickness (EWT), or the relative water content (RWT), the main parameters that describe the water status of plants (see Section 2.3.1). They are based on the theoretical absorption depth produced by different amounts of water at wavelengths in the NIR. As for pigments, comparison of several indices has been proposed (e.g., Inoue et al., 1993; Rodríguez-Pérez et al., 2007; Colombo et al., 2008; Seelig et al., 2008a, 2008b, 2008c). Many indices are affected by large changes in water content, while others are better suited to detect water stress. Wu et al. (2012) applied the method developed by Ji and Peters (2007) to investigate the strength of three spectral indices: Moisture Stress Index (MSI); Water Index (WI); and Normalized Difference Water Index (NDWI) or variant) to LWC. They found similar trends, such as a monotone decreasing sensitivity as FMC or EWT increase, but a more tortuous relationship with LWC.

Few indices have been published for the retrieval of dry matter (Table 11.5). The problem of these indices is that their sensitivity to leaf mass per area (LMA) or leaf dry matter content ($LDMC$) (see Section 2.4) is reduced considerably in fresh green leaves because of the presence of liquid water, which has an absorption coefficient almost equal to dry materials in the SWIR.

Table 11.4 *Spectral indices developed as water indicators, where* R_λ *represents the leaf reflectance at the indicated wavelength* λ*, and listed by first year of publication.*

Index	Short	Formula	Water	Reference
Leaf Water Content Index	**LWCI**	$\frac{-\ln(1-(R_{820}-R_{1600}))}{-\ln(1-(R_{820}-R^{FT}_{1600}))}$	**RWC**	Hunt et al. (1987, 1989), Maki et al. (2004)
Moisture Stress Index	**MSI**	R_{1600}/R_{820}	**EWT**	Hunt et al. (1987), Datt (1999b), Ceccato et al. (2001), Danson and Bowyer (2004), Wu et al. (2012), Zhang et al. (2012c)
			FMC	Danson and Bowyer (2004), Wu et al. (2012), Zhang et al. (2012c)
			LWC	Wu et al. (2012)
Simple Ratio	**SR**	R_{520}/R_{720}, R_{860}/R_{1240}, R_{900}/R_{970}, R_{1128}/R_{1152}, R_{1187}/R_{1096}, R_{1200}/R_{1430}, R_{1300}/R_{1450}, R_{1650}/R_{1430}	**EWT**	Aoki et al. (1988), Inoue et al. (1993), Peñuelas and Inoue (1999), Zarco-Tejada et al. (2003b), Foley et al. (2006), Seelig et al. (2008a, 2008b, 2008c), Zygielbaum et al. (2009), Zhang et al. (2012c), Mobasheri and Fatemi (2013)
		R_{900}/R_{970}, R_{1529}/R_{1416}, R_{1599}/R_{819}	**RWC**	Peñuelas and Inoue (1999), Schlemmer et al. (2005), Kim et al. (2015)
		R_{550}/R_{850}, R_{680}/R_{850}, R_{1200}/R_{850}, R_{1450}/R_{850}, R_{1689}/R_{1657}	ψ	Fujino et al. (2002), Kakani et al. (2007)
		R_{860}/R_{1240}	**FMC**	Zhang et al. (2012c)
		R_{1397}/R_{1614}, R_{3814}/R_{5219}, R_{6730}/R_{11310}	**LWC**	Ullah et al. (2013, 2014)
First Derivative	**FD**	$R'(\lambda)$	**RWC**	Danson et al. (1992), Pu et al. (2004), Sun et al. (2008)
			LWC	Kumar (2007b)
Water Index	**WI**	R_{900}/R_{970}	**FMC**	Peñuelas et al. (1993), Dawson et al. (1998b), Danson and Bowyer (2004), Wu et al. (2012), Zhang et al. (2012c), Neto et al. (2017)
			EWT	Datt (1999b), Danson and Bowyer (2004), Wu et al. (2012), Zhang et al. (2012c)
			RWC	Peñuelas and Inoue (1999)
			LWC	Wu et al. (2012)

Table 11.4 *(cont.)*

Index	Short	Formula	Water	Reference
Ratio of TM Bands	**TM5 / TM7**	R_{1650}/R_{2218}	**EWT**	Datt (1999b), Danson and Bowyer (2004), Zhang et al. (2012c)
			FMC	Danson and Bowyer (2004), Zhang et al. (2012c)
Normalized Difference Vegetation Index	**NDVI**	$(R_{800}-R_{680})/(R_{800}+R_{680})$	**EWT**	Datt (1999b), Peñuelas and Inoue (1999)
		$(R_{900}-R_{680})/(R_{900}+R_{680})$ $(R_{857}-R_{1241})/(R_{857}+R_{1241})$	**RWC**	Peñuelas and Inoue (1999), Kim et al. (2015)
Structure Insensitive Pigment Index	**SIPI**	$(R_{800}-R_{445})/(R_{800}-R_{680})$	**EWT**	Peñuelas and Inoue (1999)
			RWC	Peñuelas and Inoue (1999)
Three Band Ratio Index	**TBRI**	$TBRI_{975}=2\times R_{975}/(R_{930}+R_{1100})$ $TBRI_{1200}=2\times R_{1200}/(R_{1100}+R_{1275})$	**RWC**	Pu et al. (2002)
Normalized Difference Water Index	**NDWI**	$(R_{860}-R_{\lambda})/(R_{860}+R_{\lambda})$ with $\lambda=1240,1640,2130$	**EWT**	Datt (1999b), Danson and Bowyer (2004), Maki et al. (2004), Wu et al. (2012), Zhang et al. (2012c), Mobasheri and Fatemi (2013)
			FMC	Dawson et al. (1998b), Danson and Bowyer (2004), Maki et al. (2004), Wu et al. (2012), Zhang et al. (2012c)
			LWC	Wu et al. (2012), Ullah et al. (2013, 2014)
Global Vegetation Moisture Index	**GVMI**	$\dfrac{(R_{NIR}+0.1)-(R_{SWIR}+0.02)}{(R_{NIR}+0.1)+(R_{SWIR}+0.02)}$	**FMC**	Danson and Bowyer (2004)
			EWT	Danson and Bowyer (2004)
Difference Water Index	**DWI**	$R_{1396}-R_{1615}$, $R_{2591}-R_{4220}$, $R_{6730}-R_{11460}$	**LWC**	Ullah et al. (2013, 2014)
Water Balance Index	**WABI**	$(R_{1490}-R_{531})/(R_{1490}+R_{531})$ $(R_{1500}-R_{538})/(R_{1500}+R_{538})$ $(R_{1485}-R_{550})/(R_{1485}+R_{550})$	ψ	Rapaport et al. (2015)

11.2 Absorption Band Depth

To identify minerals in reflectance spectra of Earth and planetary surfaces, Clark and Roush (1984) proposed a method termed the continuum removal, or baseline normalization, commonly used in laboratory infrared spectroscopy and which has been applied to mineral, soil, plant, and leaf reflectance spectra. In this method it is possible to remove absorption features of no interest and

Table 11.5 *Spectral indices developed as dry matter indicators, where* R_λ *represents the leaf reflectance at the indicated wavelength* λ*, and listed by first year of publication.*

Index	Short	Formula	Dry matter	Reference
Normalized Difference	**ND$_{LMA}$**	$(R_{1340} - R_{1710})/(R_{1340} + R_{1710})$	**LMA**	Le Maire et al. (2008)
Normalized Dry Matter Index	**NDMI**	$(R_{1649} - R_{1722})/(R_{1649} + R_{1722})$	**LMA**	Wang et al. (2011a, 2011b)
Moisture Stress Index	**MSI**	R_{1600}/R_{820}	**LMA**	Zhang et al. (2012c)
Normalized Difference Water Index	**NDWI**	$(R_{860} - R_\lambda)/(R_{860} + R_\lambda)$ with $\lambda = 1240, 1640, 2130$	**LMA**	Zhang et al. (2012c)
Ratio of TM Bands	**TM5/TM7**	R_{1650}/R_{2218}	**LMA**	Zhang et al. (2012c)
Simple Ratio	**SR**	R_{860}/R_{1240}	**LMA**	Zhang et al. (2012c)
Water Index	**WI**	R_{900}/R_{970}	**LMA**	Zhang et al. (2012c)

thus to isolate individual absorption features of interest. The first step consists of determining the convex hull (or convex envelope) of a set of *n* points in the Euclidean plane representing the top of a spectrum. Note that it is the smallest convex set containing the spectrum. Linear segments or Gaussian band shapes can then be used to approximate the continuum that connects local spectral maxima, which represents the background reflectance in the absence of the absorbing material (Figure 11.3). Each local absorption feature is characterized manually according to each specific application by identifying two local maxima or shoulders, which surround an absorption band center (Table 11.6).

The continuum line is established by joining the two endpoints across the absorption feature of interest and is equal to

$$R_c(\lambda) = \frac{R_2 - R_1}{\lambda_2 - \lambda_1}(\lambda - \lambda_1) + R_1 \tag{11.3}$$

where R_1 is the reflectance at λ_1 and R_2 the reflectance at λ_2. The continuum line provides a normalized baseline to compare band depths that are independent of the local brightness of the reflectance spectrum. Once the continuum line is established, the continuum-removed spectrum $R'(\lambda)$ is calculated by dividing the original reflectance values, $R(\lambda)$, for each point in the absorption feature, by the corresponding values of the continuum line, $R_c(\lambda)$, at the corresponding wavelength:

$$R'(\lambda) = \frac{R(\lambda)}{R_c(\lambda)} \tag{11.4}$$

$R'(\lambda)$ is equal to a percent reflectance of 1.0 for wavelength regions where the spectra have the same reflectance, and less than 1.0 where absorption features occur. As an example, the water

Table 11.6 *The location of the main absorption features.*

Foliar constituent	First maximum λ_1 (nm)	Second maximum λ_2 (nm)	Band center λ_0 (nm)	Reference
Chlorophyll	400	550		Noomen et al. (2006)
Chlorophyll	408	518	484	Curran et al. (2001)
Chlorophyll	550	750	670	Noomen et al. (2006), Shi et al. (2006)
Chlorophyll	588	750	672	Curran et al. (2001)
Water	920	1120	975	Pu et al. (2003)
Water	1070	1320	1200	Pu et al. (2003)
Water	1116	1284	1190	Curran et al. (2001)
Water	1370	1570	1450	Noomen et al. (2006)
Dry matter	1634	1786	1708	Curran et al. (2001)
Water	1650	1850	1750	Tian et al. (2001), Pu et al. (2003)
Dry matter	1652	1778	1728	Kokaly and Clark (1999)
Dry matter	2006	2196	2188	Curran et al. (2001)
Water	1870	2170	1940	Noomen et al. (2006)
Nitrogen	2010	2222	2100	Kokaly (2001)
Dry matter	2030	2218	2106	Kokaly and Clark (1999)
Dry matter	2222	2378	2306	Curran et al. (2001)
Dry matter	2238	2366	2304	Kokaly and Clark (1999)

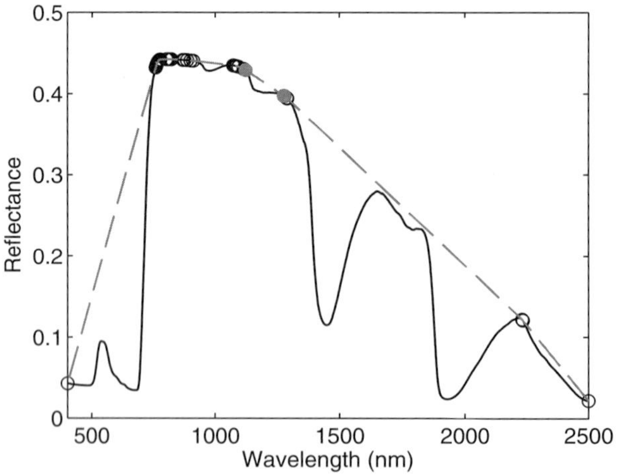

Figure 11.3 Example reflectance spectrum of a leaf simulated using PROSPECT-D and showing common absorption features and connecting line segments.

absorption feature at 1162 nm is shown in Figure 11.4a with a continuum line connecting its shoulders. Figure 11.4a presents the same feature rescaled as a fraction of the continuum reflectance. Some characteristics under the continuum can be computed, as seen in Figure 11.4b: the wavelength position of the minimum continuum removed reflectance λ_0; the absorption depth D_0; the width of the

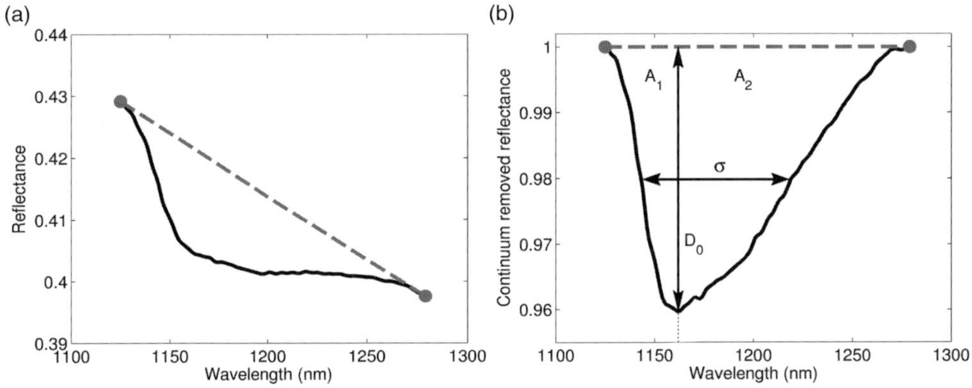

Figure 11.4 Continuum analysis for the absorption feature at 1162 nm of the leaf reflectance spectrum simulated in Figure 11.3: (a) measured reflectance spectrum between the two continuum endpoints of the feature at 1125 nm and 1279 nm and continuum line; (b) continuum-removed spectrum (R') showing some absorption features: position (λ_0), absorption depth (D_0), and area ($A = A_1 + A_2$).

absorption feature σ (full wavelength width at half depth); the area on the left of the absorption center A_1; the area on the right of the absorption center A_2; the total area $A = A_1 + A_2$; and the asymmetry of the feature that can reveal if it is composed of one or more overlapping absorption features (ratio of A_1 to A_2).

The band depth of each point in the absorption feature, $D(\lambda)$, is computed from the continuum-removed reflectance values, $R'(\lambda)$, by

$$D(\lambda) = 1 - R'(\lambda) \tag{11.5}$$

Kokaly and Clark (1999) propose a normalization procedure to minimize the influence of external factors in the area of the absorption feature. The normalized band depths within the continuum-removed absorption band, D_{NO}, are calculated by dividing the band depth of each channel, $D(\lambda)$, by the band depth at the band center, D_0:

$$D_{NO}(\lambda) = \frac{D(\lambda)}{D(\lambda_0)} = \frac{D(\lambda)}{D_0} \tag{11.6}$$

where the band center actually corresponds to the minimum of the continuum-removed absorption feature (Table 11.6). These authors also investigate normalization against the area of the absorption feature (A):

$$D_{NA}(\lambda) = \frac{D(\lambda)}{A} \tag{11.7}$$

Stepwise regression is then used to identify optimal combinations of band depths for providing accurate estimates of foliar constituents. The continuum removal technique works best to define the wavelength intervals, strength, and asymmetry of an absorption feature but it can also be used to discriminate different concentrations of materials, if a standard is available to calibrate the non-linear response of reflectance with increasing biochemical concentrations. This method was tested and validated to retrieve leaf pigments (e.g., Curran et al., 2001; Schlerf et al., 2003; Shi et al., 2006; Chen et al., 2007b; Kumar, 2007a; Noomen et al., 2006; Sanches et al., 2014), water (e.g., Curran et al., 2001; Tian et al., 2001; Pu et al., 2003; Rodríguez-Pérez et al., 2007), nitrogen (Kokaly, 2001) and other biochemical constituents (e.g., Kokaly and Clark, 1999; Curran et al., 2001; Kokaly and Skidmore, 2015). For example, Kokaly (2001) investigated the broadening of the 2.1 μm absorption by increasing nitrogen concentrations in leaves. This feature is present in both fresh and dry leaves and it has been proposed that it is related to the amide bonds in proteins but this has not been proven. Kováč et al. (2012, 2013) tracked dynamic changes in the xanthophyll cycle of Norway spruce (*Picea abies*) needles using a spectral index called ANMB (area under curve normalized to maximal band depth) and defined in the 507–556 nm wavelength range. Rivard et al. (2008a) has identified two major problems with continuum removal method. First, the appropriate shape of the continuum is not always known and different researchers have used a constant, linear, or curved baseline fit either locally or to the entire spectrum. Second, continuum removal consists of ratioing the continuum and the spectrum, a process which is non-linear and which results in continuum-removed spectra that, in theory, cannot serve as inputs for linear models, such as spectral unmixing.

11.3 Spectral Shifts

As mentioned in Section 11.1, sharp variations in leaf optical properties are common from the violet to the infrared regions of the electromagnetic spectrum. These fairly wide features are mathematically defined by the wavelength of maximum slope where the spectrum changes from being concave upward to concave downward or, in other words, by the wavelength at which the sign of the curvature changes:

$$\frac{\partial R(\lambda)}{\partial \lambda} \text{ maximum or } \frac{\partial^2 R(\lambda)}{\partial \lambda^2} = 0 \tag{11.8}$$

The sharp increase in leaf reflectance (or transmittance) from the red (670 nm) to the near infrared (800 nm), which is referred to as the red-edge, is the most famous. It corresponds to the long wavelength edge of the chlorophyll absorption feature and was first reported by Gates et al. (1965). It is generally characterized by the red-edge position (REP) or by the inflection point of the red-edge λ_i. The REP is centered on 700 nm but this position is different between species. Moreover, as spectral resolution in the visible spectrum and signal/noise in spectrometers have increased, several studies have identified the existence of two or more peaks in first-derivative spectra between 690 nm and 720 nm (e.g., Horler et al., 1983b, 1983c; Boochs et al., 1990; Gitelson et al., 1996; Ju et al., 2010). The first main peak position has been attributed to the absorption spectrum of chlorophyll; the second and consecutive peaks has been attributed to a possible variation in the chlorophyll *a*/*b* ratio, a hypothesis that has been invalidated, or a contribution from Chl fluorescence (Blackburn, 2007a). As seen in Figure 11.5, calculating a maximum becomes challenging. Its computation, reproducibility, and reliability are still a question at issue. Various numerical methods have been developed to approximate the REP (e.g., Baranoski and Rokne, 2005; Cho and Skidmore, 2006).

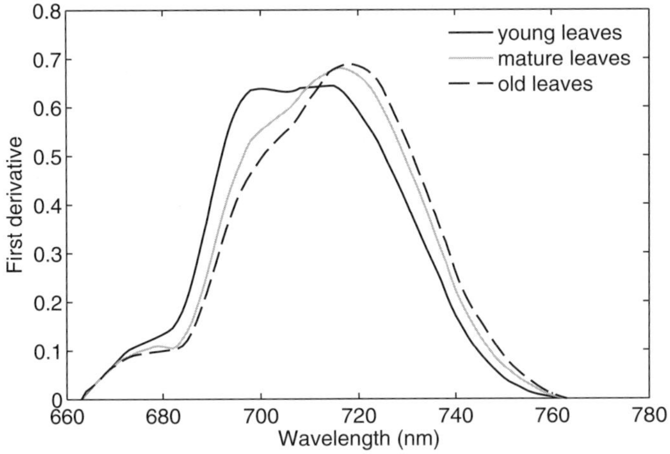

Figure 11.5 Red-edge first derivative curves of maize leaves at different ages (redrawn from Cho and Skidmore (2006)).

11.3.1 Methods for Determination of the REP

11.3.1.1 Spectrum Derivation

The easiest way to determine the REP consists of calculating the first derivative of the reflectance (or transmittance) spectrum (Horler et al., 1983b; Curran et al., 1991). One can use the forward difference formula:

$$R'(\lambda) = \lim_{\Delta\lambda \to 0} \frac{R(\lambda_{k+1}) - R(\lambda_k)}{\Delta\lambda} \tag{11.9}$$

or the backward difference formula:

$$R'(\lambda) = \lim_{\Delta\lambda \to 0} \frac{R(\lambda_k) - R(\lambda_{k-1})}{\Delta\lambda} \tag{11.10}$$

or alternatively the central difference formula:

$$R'(\lambda) = \lim_{\Delta\lambda \to 0} \frac{R(\lambda_{k+1}) - R(\lambda_{k-1})}{2\Delta\lambda} \tag{11.11}$$

The second way is to calculate the second derivative:

$$R''(\lambda) = \lim_{\Delta\lambda \to 0} \frac{R(\lambda_{k+1}) - 2R(\lambda_k) + R(\lambda_{k-1})}{(\Delta\lambda)^2} \tag{11.12}$$

with $R(\lambda_{k+1})$ the reflectance at the $k+1$ waveband, $R(\lambda_k)$ the reflectance at the k waveband, and $R(\lambda_{k-1})$ the reflectance at the $k-1$ waveband. Preliminary smoothing may be required due to the sensitivity of derivatives to noise (Demetriades-Shah et al., 1990). Moreover, because continuity and spectral resolution of the reflectance spectrum may be inadequate, a number of replacement techniques have been proposed.

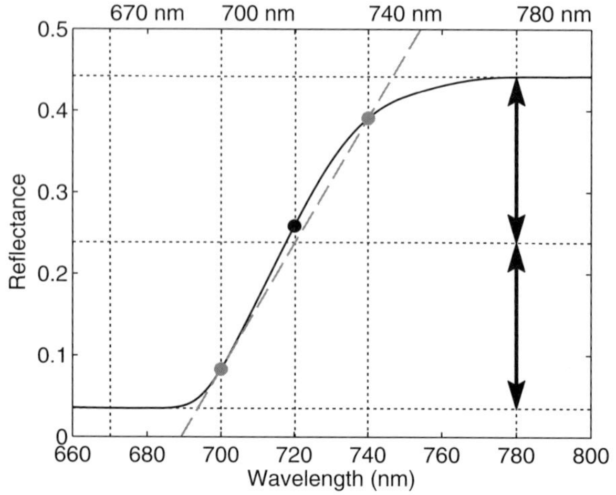

Figure 11.6 Practical determination of the red-edge inflection point λ_i using the method described in Guyot et al. (1992). The reflectance spectrum is the same as in Figure 11.3.

11.3.1.2 Linear Four-Point Interpolation

Guyot et al. (1992) show that the REP can be determined using a limited number of narrow spectral bands. They assimilate the reflectance curve at the red-edge to a straight line and the inflexion point to a center of symmetry (Figure 11.6).

λ_i is calculated in two steps: first the reflectance at the inflection point R_i is equal to the arithmetic mean of R_1 and R_4, and then the reflectance values measured in the red ($\lambda_1 = 670$ nm) and near infrared ($\lambda_4 = 780$ nm), respectively:

$$R_i = \frac{R_1 + R_4}{2} \tag{11.13}$$

On the basis of earlier modeling work, Guyot et al. (1992) report that λ_i can vary between $\lambda_2 = 700$ nm and $\lambda_3 = 740$ nm. The equation for a straight line joining the two reflectance points at these two wavelengths is equal to

$$R(\lambda) = \frac{R_3 - R_2}{\lambda_3 - \lambda_2}(\lambda - \lambda_2) + R_2 \tag{11.14}$$

which can be rewritten as

$$\lambda = \frac{\lambda_3 - \lambda_2}{R_3 - R_2}\left(R(\lambda) - R_2\right) + \lambda_2 \tag{11.15}$$

then the inflexion point of the red-edge is calculated by

$$\lambda_i = \lambda_2 + (\lambda_3 - \lambda_2)\frac{R_i - R_2}{R_3 - R_2} \tag{11.16}$$

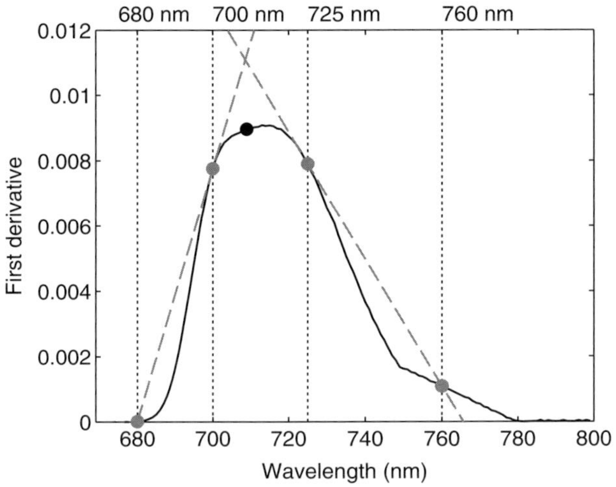

Figure 11.7 Schematic representation of the linear extrapolation technique for extracting the REP described in Cho and Skidmore (2006). The reflectance spectrum is the same as in Figure 11.3.

11.3.1.3 Linear Extrapolation

Cho and Skidmore (2006) developed a linear extrapolation technique for calculating λ_i that does not depend on characterizing a peak in the first derivative. It consists of calculating the intersection point of two straight lines fitted to the flanks of the red-edge feature in the first derivative (Figure 11.7).

The equation of the first line may be determined using two points located in the far-red (680–700 nm):

$$R'(\lambda) = a_1\lambda + b_1 \tag{11.17}$$

and the equation of the second line may be determined using two points located in the near-infrared (725–760 nm):

$$R'(\lambda) = a_2\lambda + b_2 \tag{11.18}$$

with a_1 and a_2 the slopes and b_1 and b_2 the intercepts of the straight lines. At the intersection, the two lines have equal derivatives, therefore the REP is simply given by

$$\lambda_i = \frac{b_2 - b_1}{a_1 - a_2} \tag{11.19}$$

The authors demonstrated interest in this approach when applied to the wavebands of imaging spectrometers that may have a "low" spectral resolution, since only four coordinate points are required to calculate the REP: the waveband combinations are 683.30 nm, 702.04 nm, 731.21 nm, and 759.97 nm for AVIRIS; 680 nm, 695 nm, 725 nm, and 756 nm for HyMap; and 681.20 nm, 701.55 nm, 732 nm, and 762.60 nm for Hyperion. However, no information is provided when full resolution spectra ($\Delta\lambda = 1$ nm) measured with field or laboratory spectro-radiometers are used to calculate the REP. Figure 11.7 actually shows an underestimation of the REP ($\lambda_i = 708.6$ nm) compared to the real value obtained by calculating the wavelength position of the maximum of the first derivative ($\lambda_i = 713$ nm).

11.3.1.4 Lagrangian Technique

The idea is to construct a Lagrange interpolating polynomial of degree n, which passes through $n+1$ points. These points may be the reflectances, or their derivatives, calculated by using Eq. 11.20 and Eq. 11.21. For instance, Dawson and Curran (1998) fitted a parabola ($n = 3$ points) to the first derivative values of the red-edge reflectance curve:

$$R'(\lambda) = R'_1 \frac{(\lambda - \lambda_2)(\lambda - \lambda_3)}{(\lambda_1 - \lambda_2)(\lambda_1 - \lambda_3)} + R'_2 \frac{(\lambda - \lambda_1)(\lambda - \lambda_3)}{(\lambda_2 - \lambda_1)(\lambda_2 - \lambda_3)} + R'_3 \frac{(\lambda - \lambda_1)(\lambda - \lambda_2)}{(\lambda_3 - \lambda_1)(\lambda_3 - \lambda_2)} \tag{11.20}$$

and then performed a second derivative to determine the REP at the local maximum by solving Eq. 11.21:

$$R''(\lambda) = R'_1 \frac{2\lambda - \lambda_2 - \lambda_3}{(\lambda_1 - \lambda_2)(\lambda_1 - \lambda_3)} + R'_2 \frac{2\lambda - \lambda_1 - \lambda_3}{(\lambda_2 - \lambda_1)(\lambda_2 - \lambda_3)} + R'_3 \frac{2\lambda - \lambda_1 - \lambda_2}{(\lambda_3 - \lambda_1)(\lambda_3 - \lambda_2)} \tag{11.21}$$

One of the advantages of this approach is that the three band positions λ_1, λ_2, and λ_3 need not be equally spaced. However, although it is suitable for coarsely sampled spectra, it may be not capable of resolving the destabilizing effect of the double peak feature when determining the REP.

11.3.1.5 Polynomial Fitting

In this method, the reflectance spectrum within the spectral range from 680 nm to 740 nm is fitted by a polynomial of degree p in a least-squares sense:

$$R(\lambda) = \sum_{i=0}^{p} a_i \lambda^i \tag{11.22}$$

The REP of each measured reflectance spectrum is subsequently defined to be the maximum of the first derivative of this polynomial. Hoque and Hutzler (1992) used a sixth-order polynomial on beech (*Fagus sylvatica*) leaves and Cho and Skidmore (2006) a fifth-order polynomial on maize (*Zea mays*) leaves.

11.3.1.6 Rational Function Fitting

In this approach, Baranoski and Rokne (2005) assume that the red-edge can be modeled by a function of the form:

$$R(\lambda) = \frac{a + b\lambda + c\lambda^2}{1 + d\lambda + e\lambda^2} \tag{11.23}$$

where a, b, c, d, and e are four variables to be determined, one after another, by simple linear calculations.

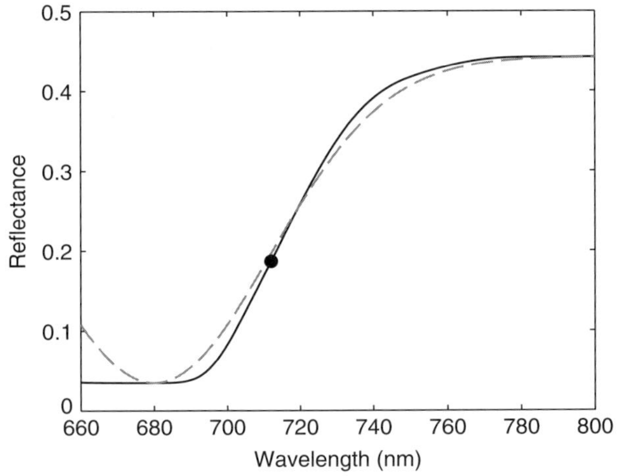

Figure 11.8 Illustration of red-edge of plant leaf reflectance (black line) and associated inverted Gaussian model (gray dashed line). The reflectance spectrum is the same as in Figure 11.3.

11.3.1.7 Inverted Gaussian Fitting

Miller et al. (1990, 1991) suggested that the spectral shape of the red-edge could be approximated by one half of an inverted Gaussian function (Figure 11.8).

This model represents the red-edge by the equation:

$$R(\lambda) = R_S - (R_S - R_0)\exp\left(-\frac{(\lambda_0 - \lambda)^2}{2\sigma^2}\right) \tag{11.24}$$

with λ the wavelength, R_S the near-infrared shoulder reflectance, R_0 the minimum reflectance corresponding to the chlorophyll absorption well, λ_0 the wavelength corresponding to R_0, and σ the Gaussian function deviation parameter. Accordingly, an expression may be found for the position of the inflection point of the red-edge λ_i by deriving twice Eq. 11.24 with respect to λ and equating the expression to zero:

$$\lambda_i = \lambda_0 + \sigma \tag{11.25}$$

The model is fitted to the reflectance data using standard numerical methods to generate parameters that represent them over the entire red-edge spectral region. The REP ($\lambda_i = 711.8$ nm) found in Figure 11.8 is close to the real value obtained by calculating the wavelength position of the maximum of the first derivative ($\lambda_i = 713$ nm). However, Miller et al. (1990) have noticed that inappropriate constraints, for example, wrong initial guesses for the model parameters, could lead to significant errors. Moreover, since non-linear fitting methods may be computationally intensive, they suggest an alternative linearization method developed by Bonham-Carter (1988) that permits the determination of two of the four inverted Gaussian model parameters. They calculate the parameters R_S (mean reflectance in the 670–685 nm region) and R_0 (mean reflectance in the 780–795 nm region) and rearrange Eq. 11.24 to give

$$\ln\left(\frac{R_S - R(\lambda)}{R_S - R_0}\right) = -\frac{(\lambda_0 - \lambda)^2}{2\sigma^2} \tag{11.26}$$

which may be written as a simple straight line:

$$R^* = \alpha\lambda + \beta \tag{11.27}$$

where $R^* = \sqrt{\ln\left(\frac{R_S - R(\lambda)}{R_S - R_0}\right)}$, $\alpha = \frac{1}{\sqrt{2}\sigma}$, and $\beta = -\lambda_0\alpha$. The expressions of the slope α and intercept β can be obtained by a least squares fit and then the two unknown parameters λ_0 and σ are easily determined.

11.3.1.8 Surface Response Model

Baret et al. (1992) formally derived the PROSPECT model with respect to the wavelength to determine the position of several inflexion points in the solar domain from leaf reflectance or transmittance. In the red-edge, they observe a strong dependence of the REP on chlorophyll content, but also on the mesophyll internal structure through the N parameter. The results presented in Figure 11.9 are somewhat different compared to what they published. The simulation of reflectance spectra with PROSPECT-D, like those presented in Figure 11.1, shows a red shift of the REP when C_{ab} increases, and a blue shift of the REP when N increases. Transmittance spectra would display a red shift of the REP when both C_{ab} and N increase. The values of REP have been determined using the inverted Gaussian method described above. As for the reflectance, they range from 690 nm to a maximum of 722 nm for $N = 1$ and from 690 nm to a maximum of 713 nm for $N = 2.5$ (Figure 11.9a). As for the transmittance, they range from 690 nm to a maximum of 711 nm for $N = 1$ and from 690 nm to a maximum of 716 nm for $N = 2.5$ (Figure 11.9b). The effect of N is inverse, contrary to Baret et al. (1992).

Since C_{ab} and N explain most of the variance of leaf optical properties in the red-edge, two narrow wavebands should theoretically be sufficient to invert PROSPECT from reflectance or transmittance data and retrieve these variables and then to compute λ_i. Consequently, a one-to-one function F should relate λ_i to the reflectance (or transmittance) at the 672 nm (maximum absorption wavelength) and 780 nm (near-infrared plateau):

$$\lambda_i = F\Big(R(672), R(780)\Big) \tag{11.28}$$

λ_i can be very accurately fitted (RMSE = 0.73 nm) by a polynomial surface of the fourth order that represents the function F.

11.3.1.9 Comparison of Methods

Some studies compared the performance of a number of these methods (e.g., Baret et al., 1992; Dawson and Curran, 1998; Baranoski and Rokne, 2005; Cho and Skidmore, 2006). For given chlorophyll content, they report a wide range of variation of the REP. The sampling characteristics

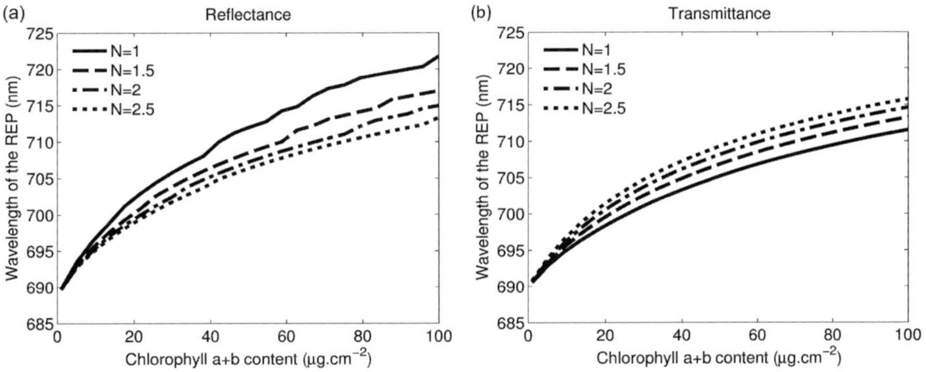

Figure 11.9 Sensitivity analysis of λ_i to chlorophyll concentration C_{ab} and leaf structure parameter N using PROSPECT-D. (a) Reflectance; (b) Transmittance.

of the sensor, the choice of the wavelengths, and the fact that the method is supervised or not, may considerably influence the results. For instance, Dawson and Curran (1998) pointed out that the linear method is faster than the Lagrangian technique and the inverted Gaussian fitting, but that it overestimates the REP by approximately 10 nm. These differences cast some doubt on the actual location of the REP.

11.3.2 Sources of Variation

In the 1980s, the availability of imaging spectrometers with sufficient spectral resolution to define the red-edge allowed it to become the subject of intense interest in geobotany as a means to detect underlying geological conditions. At leaf level, the spectral features observed in the red-edge are first and foremost related to chlorophyll content and, to a lesser extent, to mesophyll internal structure (e.g., Gates et al., 1965; Horler et al., 1983b, 1983c; Belanger, 1990; Curran et al., 1991, 1995; Guyot et al., 1992; Moss and Rock, 1991; Vogelmann et al., 1993; Gitelson et al., 1996; Lichtenthaler et al., 1996; Richardson et al., 2002). Stacking leaves leads to the same result because it places more chlorophyll within the field-of-view of the sensor. An increase (decrease) in chlorophyll content shifts the inflexion point λ_i toward longer (shorter) wavelengths and increases (decreases) the absolute value of the first and second derivatives. These are known respectively as the blue-shift and the red-shift. Different relationships have been found between C_{ab} and λ_i, with strong correlation coefficients and significance measures in both monocot and dicot plants (Figure 11.10):

$$C_{ab} = f(\lambda_i) \tag{11.29}$$

The chlorophyll *a*/*b* ratio is suspected to have minor effect on the REP compared to the total chlorophyll content (Curran et al., 1995). Other pigments that also absorb radiation in the region of the red-edge may disrupt this relationship. Let us take an actual instance with amaranth (*Amaranthus tricolor*) leaves, which contain the pigment amaranthin. Curran et al. (1991) show a near-linear relationship between the REP and chlorophyll concentration for amaranthin concentrations below 0.075 mg/g, but no relationship at higher concentrations.

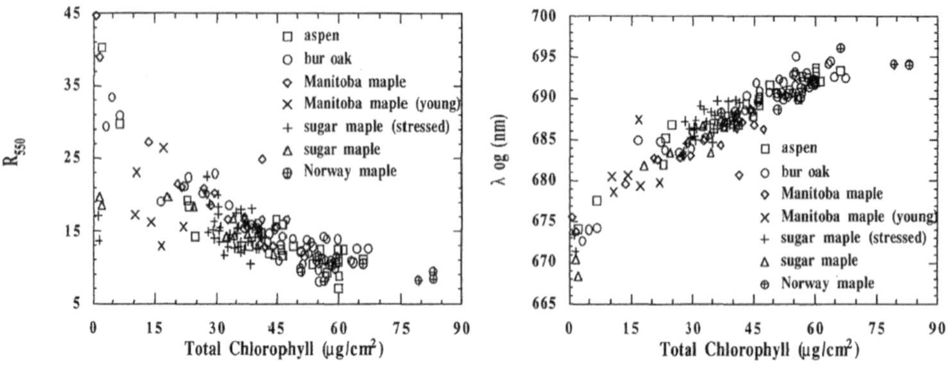

Figure 11.10 Relationship between red-edge parameters and total chlorophyll (from Belanger (1990)).

Some studies trace the position of the red-edge through phenological development, showing that it shifts to longer wavelengths as plants mature (e.g., Horler et al., 1983b, 1983c; Schutt et al., 1984a; Miller et al., 1991; Belanger et al., 1995; Cipar et al., 2008). Since it is also a good measure of the amount of chlorophyll within the leaf, it is often presented as a sensitive indicator of foliar stress (e.g., Ustin and Curtiss, 1990; Wessman, 1990; Belanger, 1990; Curran et al., 1990, 1991; Wickham et al., 1993; Gitelson et al., 1996; Hongo et al., 1998; Nishida et al., 2000; Smith et al., 2004a,b; Li et al., 2005a, 2005b, 2005c; Xie et al., 2009) or foliar nitrogen content (e.g., Moran et al., 2000; Middleton et al., 2002). Pollution or any other type of environmental stressors impacts the physiological or physio-chemical properties of plants. The result is often an overall reduction in chlorophyll content due to metabolic processes (e.g., photosynthesis, respiration). This reduction in chlorophyll leads to a blue-shift and is detectable by comparison of the REPs of suspected stressed vegetation to non-stressed controls. Buschmann and Lichtenthaler (1988) note that fluorescence emission may influence the reflectance in the red-edge, while Zhumar (1998) shows that the mesophyll structure can also move the REP. The double peak mentioned earlier is shown by Zarco-Tejada et al. (2003a) to be due to steady-state fluorescence emission, while le Maire et al. (2004) demonstrated that the specific absorption coefficient of chlorophyll alone can explain it. This question is still at issue, although Dobrowski et al. (2005) showed that the fluorescence signal tracks the onset of both heat and drought stress.

Note that similar techniques have been applied to the inflection points associated with water status in several regions of the SWIR. Inoue et al. (1993) found that the first derivatives at 1121 nm and 2008 nm were closely related to leaf LWC and RWC in maize (*Zea mays*), peanut (*Arachis hypogaea*), soybean (*Glycine max*), and wheat (*Triticum aestivum*). The location of the second inflection point shifted over a range from 1880 nm to 2080 nm, according to the degree of dehydration of the leaves.

11.4 Statistical Approach

This approach consists of adjusting a statistical model connecting explanatory variables (leaf optical properties at any measured wavelength) to response variables (leaf biochemical properties) by using a training database. To ensure an easy use and interpretation, relationships are generally limited to the combinations of a reduced number of narrow spectral bands and they estimate only one variable at a time. The choice of wavelengths or the leaf biochemical constituents may be predetermined or not. Several statistical techniques of near-infrared reflectance spectroscopy (NIRS) originally used to relate the reflectance spectra of dried ground plant material to the concentrations of substances, such as protein, cellulose, oil, and fiber (e.g., Williams and Norris, 1987; Burns and Ciurczak, 2007), have been applied to single, fresh, and dry leaves. They proved to be as accurate as the wet chemical techniques and, above all, much faster.

11.4.1 Multiple Linear Regression Model

A multiple linear regression model establishes a regression equation relating one variable, that is, the concentration of a leaf biochemical constituent C_i, to another variable, such as leaf reflectance $R_i(\lambda_j)$ (or a function of reflectance, or transmittance, or their derivatives) at a few preselected wavelengths λ_j. It is written as

$$C_i = \beta_0 + \sum_{j=1}^{m} \beta_j R_i(\lambda_j) + \varepsilon \tag{11.30}$$

with β_0 the intercept, $\beta_1, \ldots, \beta_m$ the regression coefficients, also called the slopes, and ε the error term in the model. The word "linear" in "multiple linear regression" refers to the fact that the model is linear in the regression coefficients. The number of wavelengths m is generally less than three. Suppose now that n leaves have been sampled in the frame of a laboratory or field experiment. There are n measured values of concentration and n associated measured reflectance spectra, such as n values for each of m different wavelengths. Equation 11.30 can be expressed compactly and conveniently using matrix notation:

$$Y = X\beta + \varepsilon \tag{11.31}$$

where Y is an $n \times 1$ column vector of concentrations, X is a $n \times (m+1)$ matrix containing the reflectance values, β is the $(m+1) \times 1$ column vector of the unknown regression coefficients, and ε is a $n \times 1$ column vector of error values. The simplest method used to determine β, called the ordinary least squares estimation, consists of minimizing the sum of squared residuals. It leads to

$$\hat{\beta} = (X^t X)^{-1} X^t Y \tag{11.32}$$

where X^t is the transpose of the matrix X. A restricted number of samples (calibration set) are used to determine the coefficients β_j and Eq. 11.32 is tested on a larger independent set (validation set) to estimate C. In parallel, a linear correlation coefficient can be computed between the biochemical constituent C and reflectance at each wavelength in the range 400–2500 nm, as seen in Figure 11.11.

This method has been used to estimate leaf pigments (e.g., Aoki et al., 1980; Yoder and Daley, 1990; Gitelson and Merzlyak, 1996; Datt, 1998, 1999c; Daughtry et al., 2000; Carter and Spiering, 2002; Chen et al., 2007b) and water content (e.g., Carlson et al., 1971; Ripple, 1986; Bowman, 1989; Danson et al., 1992; Inoue et al., 1993; Yamamoto et al., 1994; Yu et al., 2000). Nevertheless, when $p = 1$, it is limited by the fact that absorption phenomena are generally not linear but rather

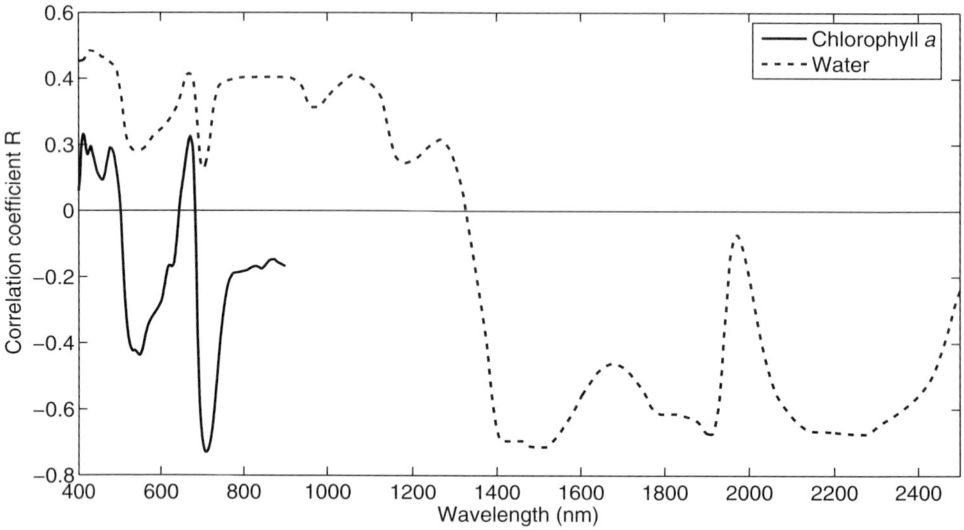

Figure 11.11 Correlation coefficient spectrum between leaf reflectance and leaf chlorophyll *a* (redrawn from Datt (1999c)) or water (redrawn from Danson et al. (1992)).

exponential or hyperbolic, depending on the wavelength. Ordoñez et al. (2010, 2013) applied an extension of the classical multivariable model called the functional linear regression model, where the matrix X and vector β in Eq. 11.31 are replaced by the function:

$$Y = \lambda + \int_0^T x(t)\gamma(t)dt + \varepsilon \tag{11.33}$$

where λ is a constant that represents the model translation term, $x(t)$ is the vector of functional inputs, $\gamma(t)$ represents the functional coefficients of the regression, and ε is the error occurring in the approximation process. These authors could predict leaf moisture and dry mass, and the concentrations of nitrogen, phosphorous, potassium, calcium, iron, and magnesium of vine leaves from hyperspectral reflectance data with correlation coefficients between 0.77 and 0.96.

11.4.2 Stepwise Linear Regression Analysis

In stepwise linear regression analysis (SLRA), the number and position of wavelengths λ_i are selected iteratively by the procedure. To avoid overfitting, this number should not exceed the number of samples and a minimum of 10 samples is usually required. The choice of the wavelengths is determined one at a time by an automatic procedure where the first step is to choose the one which is the best statistical predictor. The second independent variable to be added to the regression equation is that which provides the best fit in conjunction with the first variable. Further variables are then added in this recursive fashion, adding at each step the optimum variable, given the other variables already in the equation. Goodness of fit may be measured by various criteria: the coefficient of multiple determination, R^2, which measures the proportion of variation in the dependent variables accounted for by the other variables in the system; and the standard error of prediction (SEP) between the predicted and measured values. In this method, each variable is studied independently, and no consideration is given to variable interaction. Stepwise linear regressions were first applied to ground and dried leaf samples of nonagricultural plants (e.g., Card et al., 1988; Peterson et al., 1988; Wessman et al., 1988a; McLellan et al., 1991a, 1991b; Gillon et al., 1993, 1999a; Bolster et al., 1996; Kokaly and Clark, 1999). Note that $\log 1/R$ (a measure of absorbance, see Section 4.5.5) and its first and second derivatives have been often used as dependent variables to study powder samples. Since calibration equations derived from dry plant material cannot be applied directly to fresh leaves, new wavelengths have to be selected for application to remotely sensed data (e.g., Curran et al., 1992; Martin and Aber, 1994; Jacquemoud et al., 1995b; Yoder and Pettigrew-Crosby, 1995; Grossman et al., 1996; Fourty and Baret, 1998; Gillon et al., 2002, 2004; Chen et al., 2007b; Yi et al., 2007). Excellent results have been obtained with this method on both fresh and dry material.

However, major points of criticism have been reviewed by several authors: stepwise variable selection yields R-squared values that are badly biased to be too high; the selected wavelengths are often inconsistent between datasets, depending on whether the regression is performed on reflectance or transmittance, on fresh or dry leaves; whether the chemistry is expressed on a weight ($g\ g^{-1}$) or area basis ($g\ m^{-2}$); using the reflectance R, $\log(1/R)$ or the nth derivative $d^nR/d\lambda^n$ also results in different wavelengths; and finally some of the wavelengths that are selected in the stepwise regression may not correspond to any absorption maxima of the chemical of interest (Curran, 1989), as illustrated by Figure 11.12 for protein estimation (Jacquemoud et al., 1995b). The selected wavelengths in the visible part of the spectrum are non-attributable to absorption features of the chemical of interest and are probably associated with chlorophyll or other related pigments. This may be due to a mix of factors, including the correlation between proteins and pigments, limitations of the stepwise procedure, the instrument noise, and the lack of fine absorption features (Curran et al., 1992). As most of the stretching modes (harmonics and overtones) of proteins occur in the SWIR region, it would be interesting to limit analysis to just this region.

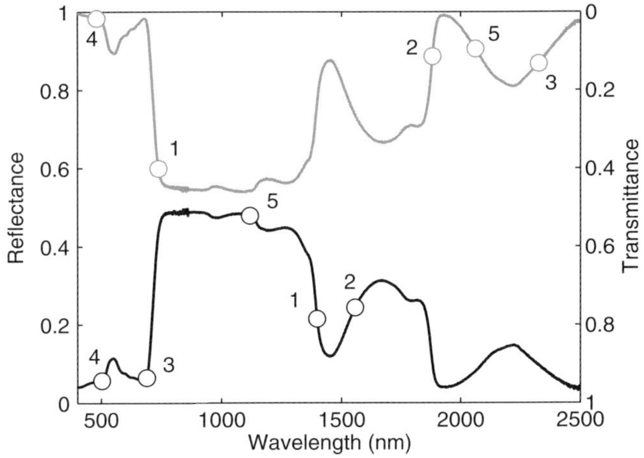

Figure 11.12 Reflectance and transmittance spectra of a typical fresh leaf illustrated by Figure 5.22. Wavelengths selected by the multiple stepwise regression analysis on protein are represented by circles in the order of the selection (redrawn from Jacquemoud et al. (1995b)).

11.4.3 Principal Component Analysis

Principal component analysis (PCA) is an unsupervised statistical method, widely used in remote sensing to explore multispectral or hyperspectral data. The main purpose of PCA is to reduce the dimensionality of a dataset consisting of a large number of variables that may be correlated. It uses an orthogonal transformation to convert a set of independent variables into a set of uncorrelated variables, called principal components. These are constructed as linear combinations of the original variables. Imagine that we have a set of p reflectance (or transmittance) spectra of leaf samples, measured in n wavelengths (variables). We arrange them in a $p \times n$ matrix M and wish to represent each set of measurements using n' components, with n' much less than n. The PCA identifies n' principal components, being defined as the n'-vectors $u_j, j = 1 \ldots n'$, corresponding to the directions where the measurements exhibit the maximum variance. The first component u_1 corresponds to the direction of maximum variance, the second one u_2 to the direction of maximum variance uncorrelated to u_1, and so on. Usually, the first few components capture most of the variation in the entire original dataset; however, their interpretation may be problematic. Bell and Baranoski (2004) used it to reduce the dimensionality of plant leaf spectral databases. The components derived from PCA can then be correlated with sample characteristics such as species or varieties (e.g., Gillon et al., 1999a; Xu et al., 2009; Kokalj et al., 2011; van Deventer et al., 2015), growth environment (Richardson et al., 2003b), and variation in leaf anatomy or chemical composition due to plant stress or disease (e.g., Liu et al., 2007, 2012; Wu et al., 2008; Li et al., 2009; Sankaran et al., 2012; van Maarschalkerweerd et al., 2013; Thomas et al., 2017) of the samples. The application of PCA has been reported in many studies as an effective data compression technique, especially for hyperspectral remote sensing data because it can solve the high dependence and autocorrelation in adjacent wavebands. A method that estimated Lambert coefficients based on a dichromatic model-PCA method was developed by Uto and Sosugi (2013a). This method utilized the shading distribution as determined from a leaf-scale hyperspectral image. The principal axis of the diffuse class when plotted as the first and second PCA axes is the chlorophyll content. The method works independent of the sky illumination, whether clear or overcast, thus this could be a practical approach for measurements in areas of persistent cloud cover.

11.4.4 Partial Least Square Regression

Multivariate regression models built on hyperspectral data suffer from overfitting and collinearity. The partial least square regression (PLS or PLSR) is a combination of multiple linear regression and PCA that is used to address more sophisticated data analysis problems (Wold et al., 1984, 2001). It is a robust technique since the biochemical constituents do not need to be treated independently of the explanatory variables, that is the spectral bands may be highly correlated. It uses all the spectral information, unlike the stepwise regression method which uses only a limited number of wavelengths. The main purpose of PLS is to build a linear model:

$$Y = XB + E \tag{11.34}$$

where Y is a $n \times p$ biochemical constituent matrix (n samples by p constituents), X is a $n \times m$ spectral reflectance matrix (n samples by m bands), B is a $m \times p$ regression coefficient matrix, and E is an error matrix which has the same dimensions as Y. Usually, the variables in X and Y are centered by subtracting their means and scaled by dividing by their standard deviations. The method produces factor scores that are linear combinations of the original predictor variables, so that there is no correlation between the factor score variables used in the predictive regression model. The $n \times k$ factor score matrix T is first computed:

$$T = XW \tag{11.35}$$

for an appropriate $m \times k$ weight matrix W. Note that in a principal components regression (PCR), the weight matrix W is made of the eigenvectors of X^tX and reflects the covariance structure between the predictors variables, while in the PLS regression it reflects the covariance structure between the predictor X and response Y variables. A linear regression model is then performed to produce a $k \times p$ matrix of regression coefficients Q, the weights for Y, such that

$$Y = TQ + E \tag{11.36}$$

Once Q is computed, the above regression model is equivalent to

$$Y = XB + E \tag{11.37}$$

where $B = WQ$, which can be used as a predictive regression model.

PLSR has been widely used in chemometrics to analyze the composition of forage, food, etc. Other researchers have started to determine whether foliar chemistry or traits can be quantified using PLS regressions or other learning methods (Table 11.7).

Gillon et al. (2002, 2004) used PLS to determine foliage moisture content (FMC) in seven Mediterranean evergreen species from dried, ground leaf reflectance spectra (Figure 11.13). Li et al. (2007, 2008) investigated the performance of genetic algorithms with partial least squares (GA-PLS) modeling to retrieve live FMC and its components, equivalent water thickness (EWT), and dry matter content (DMC), from fresh leaf reflectance spectra. Antunez et al. (2008) used PLS to correlate leaf water potential (ψ) with reflectance in sweet cherry (*Prunus avium*) leaves and found good correlation coefficients and SEP across the visible bands in the ranges of 490–560 nm, 560–600 nm, 600–630 nm, and 630–700 nm. Ullah et al. (2014) used PLS and found that highest correlations and lowest RMSE for LWC in six species was in the midwave infrared (2.6–6.0 μm), while the SWIR had nearly as good a prediction. Zhang et al. (2012c) wanted to use the visible-near-infrared reflectance to predict LWC in the presence of the strong pigment absorptions. The best model they obtained from PLSR used a first difference spectrum with Savitzky–Golay smoothing, followed by subsetting the spectrum into 60 band segments using Backward Interval PLS, that was developed by Nørgaard et al. (2000).

Stepwise regression calibrations and partial least squares were used by Ourcival et al. (1999) on evergreen oak (*Quercus ilex*) leaves collected from mesic to xeric sites over different seasons and from different parts of the tree canopies. They found good relations for both anatomical characteristics (leaf mass per area, leaf thickness) and biochemical concentrations (nitrogen). Coops et al. (2002) sampled leaves of 9 species of *Eucalyptus* at 14 sites along climate and nutrient gradients in New South Wales and examined whether spectral decomposition using factor analysis could identify the most important environmental variables. Only a few factors explained the variance in the spectra but these included chlorophyll, nitrogen, proteins, and water. Jørgensen et al. (2007) tried to discriminate between deficiencies in nitrogen (N), phosphorus (P), and potassium (K), and non-

Table 11.7 *Application of PLSR to estimate leaf traits: leaf water content (LWC), fuel moisture content (FMC), equivalent water thickness (EWT), dry matter content (DM) or leaf mass per area (LMA), stem or leaf water potential (ψ), nitrogen (N).*

Variable	Water	Sample	Spectrum type	Reference
LWC	Fresh	Leaf	R, log(1/R)	Min et al. (2004), Jiang et al. (2005a), Zhang et al. (2012c), Ullah et al. (2014)
FMC	Dry	Powder	R	Gillon et al. (2002, 2004)
	Fresh	Leaf	R, R', log(1/R)	Min et al. (2004), Li et al. (2007, 2008)
EWT	Fresh	Leaf	R, R', log(1/R)	Li et al. (2007, 2008), Feng et al. (2016)
DM or LMA	Fresh	Leaf	R', log(1/R)	Ourcival et al. (1999), Li et al. (2007, 2008)
ψ	Fresh	Leaf	R	Antunez et al. (2008), Santos and Kaye (2009), Rapaport et al. (2015)
N	Dry	Powder	R	Hiukka (1998), Gislum et al. (2004)
	Fresh	Leaf	R, log(1/R)	Min and Lee (2003, 2005), Min et al. (2004), Cohen et al. (2010), Hellmann et al. (2015)
Polyphenols	Dry	Powder	R, log(1/R)	Chen et al. (2006, 2008)
C_{ab}	Fresh	Leaf	log(1/R)	Jiang et al. (2005b, Jiang and Lu, 2006)
Bacteria	Fresh	Leaf	log(1/R)	Jones et al. (2010)
Fungi	Fresh	Leaf	R	Liu et al. (2007), Pu et al. (2008), Yuan et al. (2014a, 2014.b), Machado et al. (2015)
Nutrients	Dry	Powder	R, log(1/R)	Yi et al. (2010b), van Maarschalkerweerd et al. (2013), Galvez-Sola et al. (2015)
	Fresh	Leaf	R	Jørgensen et al. (2007), Yi et al. (2010a), Zhai et al. (2013)
Species classification	Dry	Powder	log(1/R)	Richardson et al. (2003b)
	Fresh	Leaf	R	Castro-Esau et al. (2004), Asner and Martin (2009), Xu et al. (2009), Diago et al. (2013), Fernandes et al. (2015), van Deventer et al. (2015), Cavender-Bares et al. (2016)
Carbohydrates	Dry	Powder	R	Hiukka (1998), Serbin et al. (2014)
		Leaf	R	Serbin et al. (2014)
Oil	Dry	Leaf	R	Gudi et al. (2015)
Seed yield	Fresh	Leaf	R	Zhang and He (2013)
Leaf traits	Fresh	Leaf	R, T	Coops et al. (2002), Asner et al. (2009), Yang et al. (2016a), Chavana-Bryant et al. (2017), Wu et al. (2017)
Transpiration	Fresh	Leaf	R, R'	Wang and Jin (2015)
Caffeine	Dry	Leaf	R	Khanchi et al. (2007)

deficient control conditions (C) in spring barley (*Hordeum vulgare*) plants using leaf spectral reflectance. To avoid overfitting the PLS, Asner et al. (2009) applied a variant called Prediction Residual Error Sum of Squares (PRESS) to estimate the contents of the following molecules or elements in leaves: chlorophyll *a*, chlorophyll *b*, carotenoids, anthocyanins, nitrogen, phosphorus, water and, specific leaf area (SLA) (Figure 11.14). Except for the anthocyanins, the results obtained

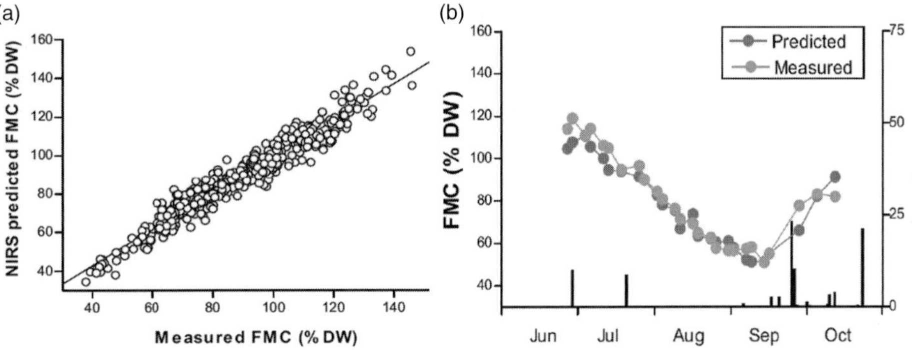

Figure 11.13 NIRS prediction of FMC for all species combined (a) and giant heather (*Erica arborea*) leaves (b) (from Gillon et al. (2002)).

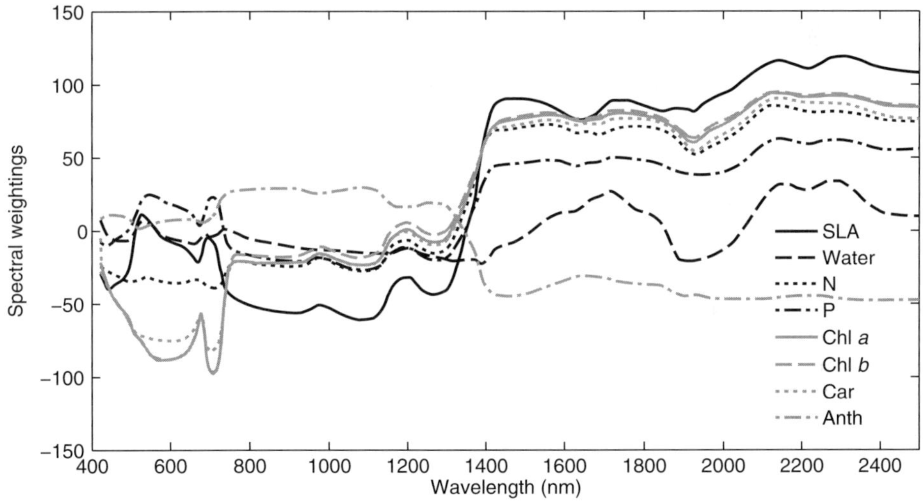

Figure 11.14 Partial least squares (PLS) reflectance weighting factors for each leaf chemical constituents. Wavelengths of maximum importance in determining leaf properties are those with spectral weightings that diverge from zero (redrawn from Asner et al. (2009)).

are very encouraging. The statistical methods contribute to decreasing the volume of explanatory variables while limiting to some wavelengths. Recently, Zhai et al. (2013) used PLSR and support vector machine regression (SVMR) to characterize the nitrogen (N), phosphorous (P), and potassium (K) contents of diverse plant species in China's Yixing region and obtained cross-validation correlations of about 70% each for each chemical element using SVMR and 0.66, 0.64, and 0.54 respectively using PLSR. Zhang and He (2013) used PLSR to predict seed yield in oilseed rape (*Brassica napus*) leaves and got the strongest correlations with seed weights at the wavelengths 543 nm, 686 nm, 718 mn, 741 nm, 824, and 994 nm, with a correlation prediction of 0.71 and a root mean square error of prediction (RMSEP) of 23.72. Pu et al. (2008) extended this to examine whether the state of health could be determined from leaves from California live oak (*Quercus agrifolia*) that were being monitored for signs of sudden oak death syndrome, caused by the fungal pathogen, *Phytophthora ramorum*. They used two classification algorithms, penalized discriminant analysis (PDA), a version of Fisher's linear discriminant analysis (LDA), and cross correlogram spectral matching (CCSM), to discriminate between healthy and infected leaves. They did find small differences in LWC between infected and healthy leaves at wavelengths between 700 nm and 1100 nm but did not find strong correlations for wavelengths that separated healthy leaves from infected ones.

Other researchers have focused on detection of different species, varieties, and even clones from spectroscopy using PLS and related algorithms. Castro-Esau et al. (2004) showed that they could get good classifications of liana versus trees from tropical forest sites in Panama, with the best results from the dry versus the wet forest site. They used PCA and various supervised testing various parametric and nonparametric classifiers. While they measured chlorophyll *a* and *b* concentrations and found it was lowest in the lianas, they speculated other structural parameters might have been more important, especially differences in the phenologies of the two groups, which were most pronounced in the dry site. In the scope of the Spectranomics project (http://spectranomics.stanford.edu/), Asner and Martin (2009) studied the chemical uniqueness and diversity among plant species in the Hawaiian rainforests because the spectroscopy seems to be driven primarily by the chemical composition of the foliage. The main idea was to develop chemical "fingerprints" of species (Figure 11.15).

Diago et al. (2013) explored whether leaf spectroscopy over the 380 nm to 1028 nm region could be used to consistently identify three common wine grape (*Vitis vinifera*) varieties (Tempranillo,

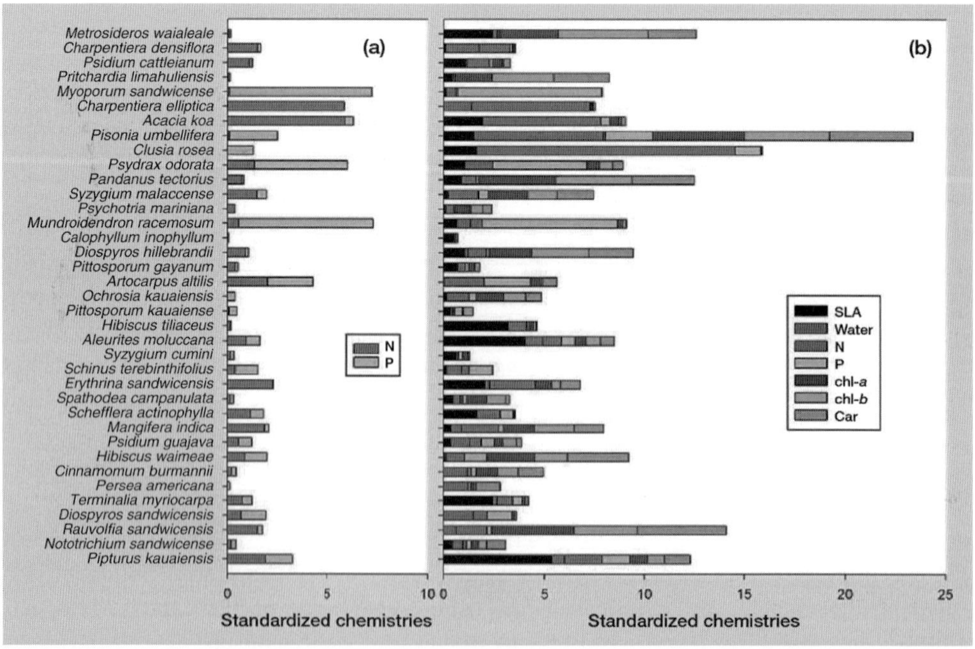

Figure 11.15 Chemical fingerprints of canopy species found in a rainforest site on Kauai, Hawaii, based on: (a) leaf nitrogen (N) and phosphorus (P) concentrations; and (b) seven leaf properties: specific leaf area (SLA), water content (water), nitrogen (N), phosphorus (P), chlorophyll *a* (chl-a), chlorophyll *b* (chl-b), and total carotenoids (Car) (from Asner and Martin (2009), *Frontiers in Ecology and the Environment*. Copyright © 2009 with permission from the Ecological Society of America). (A black and white version of this figure will appear in some formats. For the color version, please refer to the plate section.)

Grenache, and Cabernet Sauvignon). They found using cross-validation on PLS regressions produced accuracies that exceeded 92%. Fernandes et al. (2015) examined whether leaves of different clones of Cabernet Sauvignon could be identified using leaf spectroscopy over the 634 nm to 759 nm region with PLS regression and Monte Carlo cross-validation, and found accuracies of 98.2–100%.

11.4.5 Bayesian Model Averaging

The continued inconsistent responses between researchers regarding predictive wavelengths and highly variable accuracies led to an interest in methods that incorporate model probability and uncertainty in the predictions. One such method that has become widely used in a variety of fields, including environmental applications, is Bayesian modeling. Bayesian model averaging (BMA) takes into account the uncertainty in model selection. It quantifies the merits of many candidate models and synthesizes a weighted average model that produces improved predictions. BMA was used by Zhao et al. (2013) on 27 leaf spectral-chemical datasets that represented more than 80 global species to estimate nitrogen, hydrogen, carbon, cellulose, lignin, chlorophyll (*a* or *b*), carotenoid, polar and non-polar extractives, leaf mass per area, and equivalent water thickness. BMA significantly reduced overfitting ($r^2 = 0.80$) with realistic error estimates compared to partial least square regression (PLSR) and stepwise linear regression analysis (SLRA).

11.5 Wavelet Transform

To analyze periodic functions, Fourier analysis, which consists of breaking a signal into sine and cosine waves of various frequencies, is ideal. Leaf reflectance spectra are not stationary and require other signal processing methods. The wavelet transform, which is not restricted to periodic functions, has emerged as a promising tool in laboratory spectroscopy to decompose spectra and to detect local absorption features. It can be implemented as continuous wavelets transform (CWT) or discrete wavelets transform (DWT). Only CWT is described in this section. Wavelets, $\psi_{a,b}$, are a class of mathematical functions that are produced by scaling (contraction) and shifting (translation) a mother wavelet, ψ. In spectrum analysis, they are defined as analytical functions of wavelength:

$$\psi_{a,b}(\lambda) = \frac{1}{\sqrt{a}}\psi\left(\frac{\lambda - b}{a}\right) \tag{11.38}$$

where a and b are two positive real numbers called the scaling factor (the wavelet's width) and the shifting factor (the wavelet's position), respectively. The mother wavelets are confined in a finite interval, that is, they do not extend from $-\infty$ to $+\infty$ like a sine and a cosine function, and they satisfy some mathematical requirements like the condition of zero mean:

$$\int_{-\infty}^{+\infty} \psi(\lambda)d\lambda = 0 \tag{11.39}$$

Selection of the wavelet shape is one of the important decisions on the user's part (Figure 11.16).

In the wavelet transform, the leaf reflectance spectrum $R(\lambda)$ is convolved with the wavelet $\psi_{a,b}$ to extract the wavelet coefficients, $W_R(a, b)$, which are given by

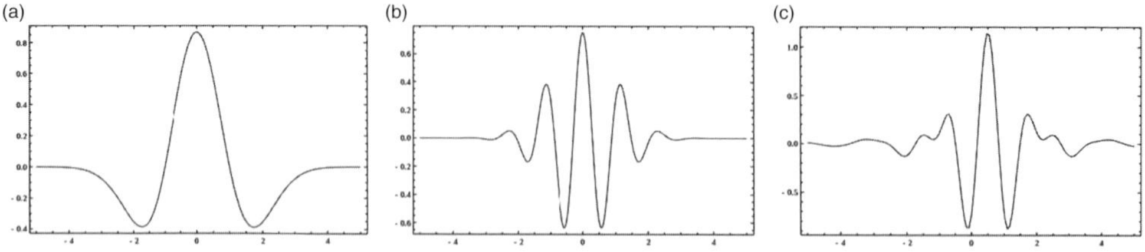

Figure 11.16 Three examples of mother wavelets: (a) Mexican hat; (b) Morlet; and (c) Meyer.

$$W_R(a,b) = \langle R, \psi_{a,b} \rangle = \int_{-\infty}^{+\infty} R(\lambda)\psi_{a,b}(\lambda)d\lambda \tag{11.40}$$

They represent the wavelet power which measures the correlation between the scaled and shifted mother wavelets and the signal at different scales for different sections of the signal. They constitute an $m \times n$ matrix, also called a scalogram, where m is the maximum scale available and n is the number of wavelengths. Since wavelet decomposition at all possible scales would be computationally expensive and lead to a large data volume, the dimensions of the scalogram is reduced by decomposing the reflectance spectra at dyadic scales $a = 2^j$ (or levels j). These scales are proportional to the effective length of the wavelet compressed or stretched at that scale. Low-scale components of the scalogram capture absorption features of fine spectral details and high-scale components emulate the overall continuum of the reflectance spectrum. This is illustrated in Figure 11.17, where the wavelength is displayed along the horizontal direction and the scale along the vertical axis. To summarize, wavelet transform of a spectrum yields a vector of wavelet coefficients that are assigned to different frequencies. Each band expands over the complete wavelength domain and responds to a certain frequency range of the spectrum. By selecting the appropriate wavelet coefficients, a spectral model can be established between the coefficients and biochemical contents. Hence, this technique has proved its potential to capture much more of the information contained within high-resolution spectra than previous approaches.

Leon et al. (2003) published an early paper using the Haar wavelet transform to extract features from cotton (*Gossypium hirsutum*) and soybean (*Glycine max*) reflectance spectra because of the similarity between the spectral curve and the so-called Haar wavelet, which is considered one of the simplest possible. Soon after, Blackburn (2004) used wavelets to decompose hyperspectral data to estimate chlorophyll a and b and carotenoid pigment concentrations and found that they outperformed narrow-band spectral indices and stepwise regressions. This result was confirmed by several subsequent studies performed on both reflectance and first derivative spectra (e.g., Blackburn, 2007b; Reum and Zhang, 2007; Blackburn and Ferwerda, 2008; Huang and Blackburn, 2011; Liao et al., 2013). Wavelet transforms were also used to predict water content from leaf spectra for a wide range of plant species, in the solar (e.g., Cheng et al., 2011, 2012; Fang and Ju, 2015; Li et al., 2016) and the thermal infrared (Ullah et al., 2012c, 2012d) domains. The two datasets (LOPEX and PANAMA) used by Cheng et al. (2012) showed consistent selection of wavelet features at three scales. Ullah et al. (2012c) used wavelet analysis on reflectance spectra from nine plant species at different levels of water content spectra from the 2.5 μm to 14 μm and obtained good estimates of LWC from six wavelet features in the mid infrared (2.5–6.0 μm). Cheng et al. (2014) and Li et al. (2016) used several datasets to simulate a larger database using PROSPECT to derive leaf mass per area (LMA). They

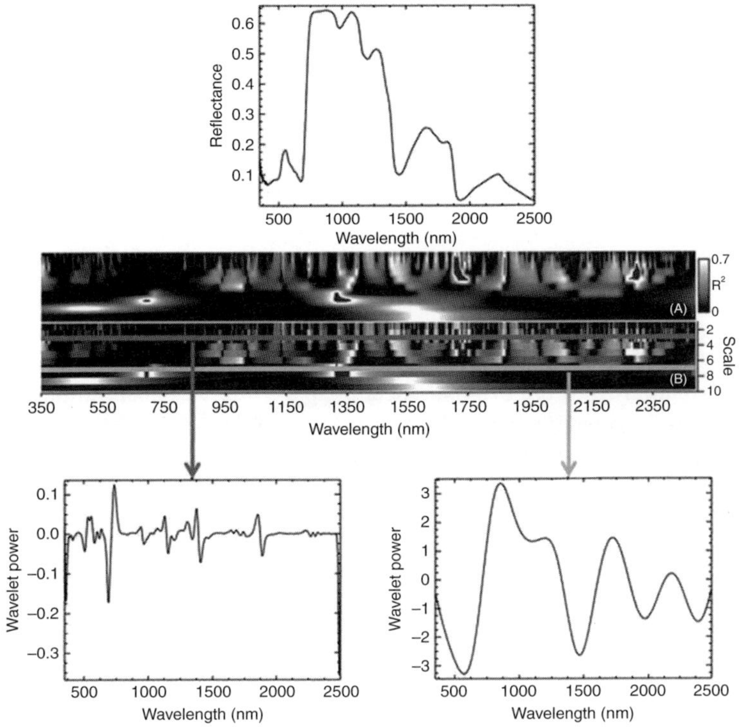

Figure 11.17 Schematic representation of the continuous wavelet transform workflow. (a) scale interpolated correlation scalogram; (b) scale (3 and 7) replicated scalogram. The dark areas encompass features with the highest R^2 (top 1%) between water content and wavelet power (from Cheng et al. (2010), *Remote Sensing of Environment*, © 2010 with permission from Elsevier).

identified five wavelet features to estimate LMA, and the wavelet analysis was the only model tested that consistently produced good predictive capability. Vegetation stress caused by biotic or abiotic factors has also been studied using wavelet transforms (Kempeneers et al., 2005; Cheng et al., 2010; Liu et al., 2010a, 2011b; Zhang et al., 2012b, 2014). Finally, Zhang et al. (2008) used a wavelet transform to remove noise and reduce the dimension of a tobacco (*Nicotiana tabacum*) dataset, prior to application of statistical methods.

11.6 Spectral Mixture Analysis

Spectral mixture analysis (SMA) is a method that allows simultaneous determination of the contributions of all biochemical components to leaf optical properties. It assumes that a foliar spectrum is the sum of the absorption features of each chemical weighted by its concentration, so that it reduces the spectral information $R(\lambda)$ into independent sources of variability, termed the endmembers. This method was derived from analytical chemistry and later extended to studies of environmental chemistry. It was first developed for remote sensing geological exploration (Adams et al., 1986), although it was previously reported by Adams and Adams (1984) as a method to remove the spectral characteristics of plants from Landsat data and reduce confusion in detecting the geological substrate. It was first applied to differentiating semiarid shrub communities and quantifying their fractional cover by Ustin et al. (1986) and expanded in Smith et al. (1990) for low canopy cover semiarid ecosystems. Roberts identified both green and dry plant material as endmembers in an early effort to map plant constituents (Roberts et al., 1990b). At the leaf level, one considers the specific absorption coefficients $k_i(\lambda)$ of chlorophyll, water, protein, cellulose, lignin, etc. as endmembers and the coefficients C_i are the concentrations to be retrieved (e.g., Goetz et al., 1990; Aber et al., 1994; Hlavka et al., 1997):

$$R(\lambda) = (a + b\lambda)\exp\left(-\sum_{i=1}^{n} k_i(\lambda)C_i\right) \tag{11.41}$$

The unknown concentrations C_i are those producing the best fit of $R(\lambda)$. Conceptually, the method produces a transformation from the original spectral space into a concentration space for the endmembers and could be considered statistically analogous to a canonical PCA, where the endmembers represent the statistically distinct spectral classes but which are not necessarily orthogonal, as would be the case for a PCA. Because endmembers are not required to be orthogonal axes for the specific dataset, they can be used to compare different sets of data. A further advantage over PCA is that the user knows what the endmembers represent, while the exact composition of standard PCA axes is less certain. A disadvantage is that the spectral materials must be distinct to avoid misidentification and cannot be simulated from mixtures of the other endmembers. This results in only a small number of endmembers that can be identified in any dataset, even using narrow band spectroscopy data. Roberts and his colleagues (Roberts et al., 1998a; Dennison and Roberts 2003a, 2003b; Dennison et al., 2004) developed the concept into the multiple endmember mixture model (MESMA) to differentiate between different plant species based on spectral properties. This differs from SMA in that multiple libraries of each type of land cover component are developed and the algorithm determines the best fit endmembers on a pixel-by-pixel basis (in contrast to analyzing the entire image with one set of endmembers).

Smith et al. (1994) developed a concept for defining a multidimensional spectral projection in which the object of interest was variable while the variability of other materials was hidden in this projection, which they termed foreground-background analysis (FBA). This concept was further developed for quantifying leaf chemistry, by weighing vectors which simultaneously extract concentrations of carbon, cellulose, and nitrogen and water content. Defined by Pinzón et al. (1998) as the hierarchical foreground-background analysis (HFBA), it is a more general and sophisticated spectral mixing technique equivalent to a single neuron in a neural net.

11.7 Artificial Neural Networks

Artificial neural networks (ANNs) are data driven self-adaptive non-linear systems. They are not related to physical models but they are useful in the modeling of complex relations where non-linear effects are known to be involved. They can identify and learn correlated patterns between input datasets and corresponding target values. They resemble the human brain in two respects: the network acquires knowledge through a learning process, and the interconnection strengths known as synaptic weights are used to store the knowledge. ANNs work like biological neurons assembled in layers. Each neuron receives information from at least one neuron of the preceding layer and transmits it to the neurons of the next layer after transforming it. Each connection is evaluated by a

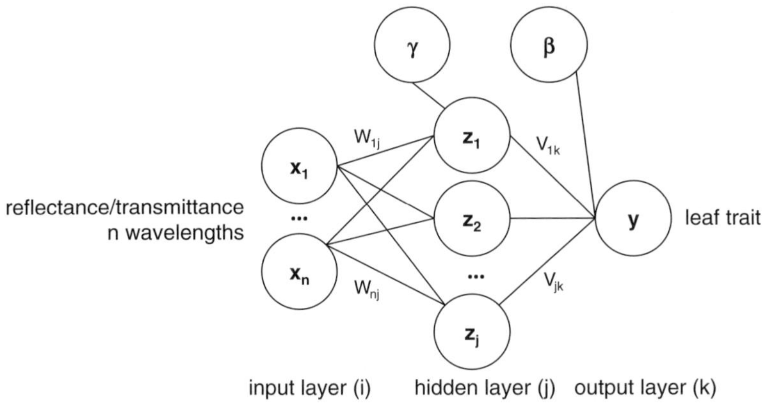

Figure 11.18 Typical feed-forward neural network composed of three layers. The input layer of the network is made of neurons made of the spectral reflectance (or transmittance or absorptance). The output layer of the network has one or several neurons representing a plant species, one or more chemical concentrations, or a disease condition.

real number, called the weight coefficient, which reflects the degree of importance of the given connection in the neural network. At the end, the output layer proceeds as the output of the entire network. Many different types of ANNs have been developed, the basic principles of which are very similar. The back propagation (BP) neural network algorithm is one of the most commonly used. It is also called a multi-layer feed-forward (MLF) neural network because of there is no signal propagation within a layer or to a previous layer. For example, Figure 11.18 displays a typical MLF neural network composed of three layers: the input layer, the hidden layer, and the output layer. The ANN model is defined by the number of neurons in these layers. The number of independent variables and dependent variables determines the number of nodes in the input and output layers respectively. There are a number of common transfer functions used at the hidden layers and the output layer: linear function, log-sigmoid function, tan-sigmoid function, etc.

ANNs proved attractive in early studies to analyze remote sensing data or to classify vegetation types when the training (or learning, or calibration) dataset is small relative to the validation dataset. They have been first applied on dry leaf powder or leaf extracts to discriminate between plant species (e.g., Hana et al., 1997) or to determine the chemical composition of plants (e.g., Khanchi et al., 2007; Zhang et al., 2008). For instance, Hana et al. (1997) used it to classify Burley versus flue-cured tobacco (*Nicotiana tabacum*) and identify native US Burley tobacco from a non-native one. The results of the calibration model and the true performance for classifying tobacco species were (100%, 100%), (99.38%, 99.39%), and (95.19%, 99.26%) for the quadratic classifier, back-propagation neural network, and linear network, respectively. The identification of native tobacco and its true performance were (100%, 100%) using a quadratic classifier, (89.12%, 88.46%) using a back-propagation network, and (80.68%, 79.62%) using a linear network. Li and He (2008) used ANNs on intact tea (*Camellia sinensis*) leaves to discriminate varieties based on their visible/near infrared reflectance spectral characteristics. Liu et al. (2016) compared ANNs and a Bayes algorithm to identify two invasive coastal plants (*Phragmites australis* and *Spartina alterniflora*) in the Yangtze estuary. They showed better predictions with the second method compared to the first. Although customarily applied to classification problems, ANNs have also been adopted for estimating foliar pigments (e.g., Chen et al., 2007b; Kira et al., 2015), water and or dry matter (Riaño et al., 2005a; Zakaluk and Ranjan, 2007; Chen et al., 2012b; Mirzaie et al., 2014), nitrogen (Chen et al., 2007a; Yi et al., 2007; Yao et al., 2011; Zhang et al., 2013; Lestari et al., 2015; Afandi et al., 2016), and other biochemicals (Dawson et al., 1998b) from reflectance/transmittance measurements. They generally produced more accurate leaf chemical concentrations when compared to those generated with spectral indices or traditional statistical approaches like single band correlations, PLS, and stepwise regression. One explanation is that they can handle non-linear relationships and non-Gaussian distributed data and that they are able to account for some of the spectral variation which can be attributed to the other biochemicals. On a different register, Ravi et al. (2008) used ANNs in a spectral mixture analysis that produced results consistent with the correct proportions of different leaf types. Finally, ANNs were investigated to detect plant stress due to biotic (e.g., Faruque et al., 2001; Pydipati et al., 2005; Wu et al., 2008; Li et al., 2009; Zhang et al., 2011) or abiotic (e.g., Liu et al., 2011b; Temizel et al., 2014) factors. When trained on data simulated by radiative transfer models, ANNs can be considered as an inversion method of these models. This is developed in Section 11.9.

11.8 Model Inversion

While experimental measurements of leaf optical properties were progressing, determinist models based on diverse representations of light interactions with a plant leaf were also developed (see Chapter 5). The inverse problem is critical when the radiometric signal has to be interpreted in terms of leaf biophysical characteristics. A general description of inverse theory, which is beyond the scope of this book, can be found in Tarantola (2005). In brief, we may represent the mathematical model of a physical system, for instance, PROSPECT, by the symbol **m** and the variables needed to completely describe the system by the symbol **x**. The physical observation, here shown as the reflectance **R** (or the transmittance **T** or both of them), is formally expressed by

$$\mathrm{R} = \mathrm{m}(\mathrm{x}) + \varepsilon \tag{11.42}$$

Equation 11.42 is a short notation for the set of equations $R_i = m_i(x_1, x_2, \ldots) + \varepsilon_i$ for $(i = 1, 2, \ldots)$. ε_i represents the residual errors between simulated and measured reflectance. If some variables are perfectly known, for example, the refractive index and specific absorption coefficients, they can be considered as fixed constants. If they are uncertain, for example, the leaf biophysical characteristics, they must enter the parameter set **x**. The inverse problem of a nonlinear model such as PROSPECT is based on the minimization of a cost (or misfit) function δ^2 that concurrently measures the discrepancies between: i) the observed and simulated reflectance; and ii) the variables to estimate and the associated prior information:

$$\delta^2 = \sum_{i=1}^{n} \left(\frac{R_i - m_i(x_1, x_2, \ldots, x_k)}{\sigma_{Ri}} \right)^2 + \sum_{j=1}^{k} \left(\frac{x_j - x_j^{prior}}{\sigma_{xj}} \right)^2 \tag{11.43}$$

where σ_{Ri} and σ_{xj} are respectively the diagonal elements of the error covariance matrices on observations and parameters. σ_{Ri} corresponds to the measurement/model uncertainties and σ_{xj} to the uncertainty of any prior information on the variables to estimate x_j^{prior}, such as obtained independent of the measurements. The distribution of σ_{Ri} and σ_{xj} is assumed to be Gaussian but, if statistics are not available, the cost function simplifies the increasing risk of converging on local minima due to the non-unicity of the inverse solution. Equation 11.43 does not incorporate boundary constraints or relationships between parameters. Two main categories of inversion approaches have been exploited: the first emphasizes the first part of the misfit function such that the solution corresponds to a minimum distance between reflectance measurements and model simulations, with respect to the *a priori* constraints. Iterative optimization, Markov Chain Monte Carlo methods, and look-up tables belong to this category. The other family of approaches emphasizes the second part of the misfit function, focusing on the space of canopy biophysical variables, where a parametric model is adjusted over the surface response between reflectance and the biophysical variables of interest. Spectral indices (see Section 11.1) are related to the biophysical variables through parametric models and neural networks, which belong to this second category.

11.8.1 Approaches Based on the Observation Space

Two iterative minimization techniques have been implemented: the simplex and the quasi-Newton algorithms mainly differ from downhill search algorithms, in their capacity to avoid being trapped in local minima, and in constraints on the range of variation of each unknown variable to be estimated. It is difficult to make general recommendations as to the choice of one algorithm versus the other because at the leaf level, they are both well-adapted to the problem being considered. Iterative minimization methods have proved to be efficient for case studies when a limited number of observations are to be processed. However, operational implementation of such methods using field sensors and satellite instruments might be facilitated by using pre-computed simulated databases. Look-up tables are a simple technique that consists of generating a training table with the model for a discrete set of input variables covering their prescribed range of variation. They offer the advantage of providing the global minimum where the variable space is sufficiently sampled, thus avoiding the tricky problem of local minima encountered with analyzing small datasets. Sensitivity analyses of the model can contribute to helping choose better sub-domains where reflectance is sensitive to a variable, when all of them are varying, and as in the case of iterative methods, prior information can be introduced.

11.8.2 Approaches Focusing on the Space of Leaf Variables

The use of spectral indices may be considered an inverse technique when radiative transfer models are used to calibrate parametric equations between these indices and some leaf characteristics. Machine learning techniques such as ANNs also belong to this type of approach and have lately become a popular method to operationally invert models because of their high computational speed, once they are calibrated. The machine learning models interconnect a set of inputs (leaf reflectance and/or transmittance) to a set of outputs (leaf biophysical variables), assuming that they are functionally related, during the so-called "learning phase." Their performances mainly depend on the characteristics of the training database, for which no explicit assumption about the physics of radiative transfer within plant leaves is mandatory, so that pairs of reflectance observations and corresponding biophysical variables can be used, in theory, to train a neural network. However,

Table 11.8 *Statistics of the inversion on RAMIS signal by neural networks.*

	C_{ab}		C_w		C_m	
	RMSE	R^2	RMSE	R^2	RMSE	R^2
Training	7.43	0.954	0.00232	0.666	0.00178	0.746
Validation	7.30	0.955	0.00289	0.603	0.00251	0.646

because such datasets that gather both radiometric and biophysical data are scarce and prone to significant uncertainties associated with laboratory measurements of leaf characteristics, the observations are conveniently replaced by model simulations. This approach has been successfully applied to relate leaf biochemistry to their reflectance and/or transmittance spectra measured (e.g., Chen et al., 2007a, 2007b) or simulated by PROSPECT (e.g., de Rosny et al., 1995; Le Maire et al., 2004; Pavan et al., 2004; Shi and Sun, 2007; Conejo et al., 2015) and LIBERTY (Dawson et al., 1998a; Schlerf et al., 2003). Calibrated versions of PROSPECT and LIBERTY were able to retrieve chlorophyll content in saxaul (*Haloxylon ammodendron*) leaves over the full range of values, with R^2 of 0.55 or 0.53, respectively (Li and Wang, 2013). In the frame of the RAMIS (*RAdiomètre portatif de Mesure In Situ*) project (see Section 4.6). Pavan et al. (2004) conducted inversions of PROSPECT-4 on five broadband transmittances with a feed-forward neural network where the neurons were arranged in two layers. The first was composed of 16 sigmoid neurons and the second of 4 linear neurons. The neural network was trained for 300 000 epochs. Extraction of the leaf biochemistry on both the training set and the control set produced very good results, as seen in Table 11.8. In particular, the estimation errors on chlorophyll *a+b* content decreased by 5 μg cm^{-2} when compared to the spectral indices. This is not surprising since the neural networks can take into account spectral variation attributed to the other leaf biochemicals or to leaf structure.

12
Applications of Leaf Optics

Applications of leaf spectroscopy have many different end uses. Leaf level information is crucial to quantify the state of physiological processes, for example, the energy budget and transpiration. It is used to monitor photosynthetic rates and respiration rates. It provides a basis to scale environmental processes from the molecule to the planet. Leaf spectroscopy is also used in remote sensing studies to calibrate processes and to provide ground truth data for interpretation, and in agriculture to indirectly calibrate foliar nutrients like nitrogen concentration. Color is also used by insects and animals for host selection (or avoidance) at wavelengths from the ultraviolet to the thermal infrared. In experimental settings, spectroscopy is often used as a non-destructive sampling method to follow a process over time. Finally, it can help guide research for the search for life on exoplanets to possible mechanisms for energy capturing processes.

12.1 Leaf Energy Budget

Campbell (2000) defines environmental physics as the study of the physical interactions of living organisms with their environment. As such it involves analyses of energy and mass exchange between plant leaves and their surroundings, as well as continuity principles governing the conservation of energy and mass. Studying the leaf energy budget is of significant interest to biometeorologists, ecologists, and agronomists who want to predict rates of evapotranspiration, photosynthesis, or respiration, and sensible or latent heat fluxes. Gates (1965, 1968, 1976, 1980a, 1980b) provided a theoretical basis for the leaf energy balance. The first law of thermodynamics, which is an expression of the principle of conservation of energy, states that energy can be converted from one form to another but cannot be created or destroyed. Although leaves are exposed to different energy fluxes, the main contributors to the energy budget are absorbed radiation from shortwave solar radiation to longwave thermal infrared radiation, originating from the surrounding atmosphere and the terrestrial environment (Figure 12.1).

For a leaf at equilibrium, all energy absorbed by the upper and lower surfaces must be balanced by a combination of emitted thermal infrared radiation, convected and conducted heat, and latent heat loss by transpiration (e.g., Knoerr and Gay, 1965; Idso et al., 1966; Wuenscher, 1970; Ye et al., 2013; Monson and Baldocchi, 2014):

$$Q_a = Q_r \pm C \pm LE + P \tag{12.1}$$

where Q_a is absorbed energy (W m^{-2}), both visible and infrared; Q_r is reradiated energy (W m^{-2}); C is energy lost or gained by convection (W m^{-2}); LE is energy lost as latent heat of vaporization of water

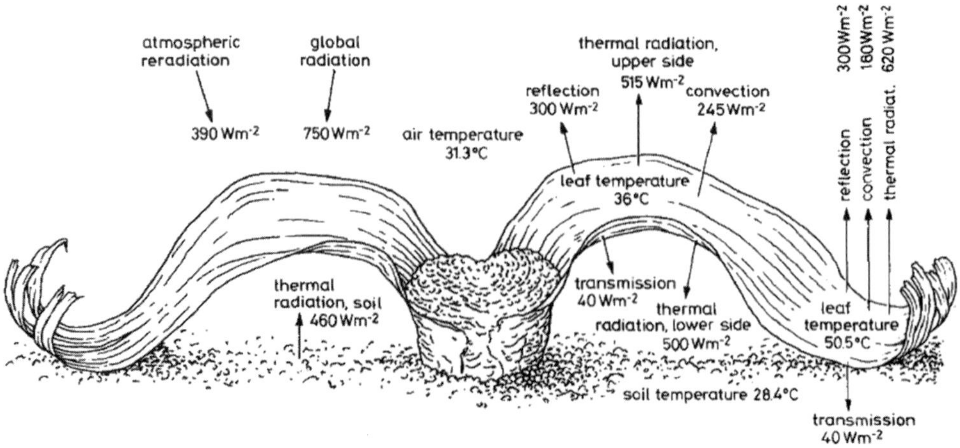

Figure 12.1 Energy balance of parts of two tree tumbo (*Welwitschia mirabilis*) leaves on March 30, 1977, in a coastal desert area of Namibia (from Schulze et al. (1980), *Oecologia*. Copyright © 1980 with permission from Springer).

in transpiration (W m^{-2}); and P is energy used in photosynthetic CO_2 fixation (W m^{-2}). Contrary to soil, snow, and water, which are thermally massive objects, there is no term for conduction of energy in Eq. 12.1. Energy flow toward the leaf is considered positive, and away from the leaf is negative. Energy balance occurs when energy gains and losses are in balance. Well hydrated leaves typically reach equilibrium quickly, often within minutes. The reason is because biological systems must function within limited temperature ranges that are suitable for their metabolic processes or they will die. Because plants are sessile organisms that are exposed to incident solar radiation, leaves express various mechanisms that keep energy absorption within optimal ranges, including avoidance (pubescence, size), mechanical (orientation, wilting, rolling), physiological (stomatal control, osmotic adjustment, senescence), and other adaptations. Leaves are cooled by emission of longwave radiation. Only a small fraction, usually less than 2% of the total absorbed radiation, is stored or used to convert chemical energy by the process of photosynthesis or to perform other metabolic processes, so we can neglect P in Eq. 12.1.

12.1.1 Absorbed Energy

The total solar radiation S_{total} received on the upper leaf surface consists of direct and diffuse radiation:

$$S_{total} = S_{direct} \times \cos\,\theta_i + S_{diffuse} \tag{12.2}$$

where S_{direct} and $S_{diffuse}$ are the incident direct and diffuse solar radiation from 0.3 μm to 4 μm (W m^{-2}) and $\cos\,\theta_i$ refers to the cosine of the angle between the leaf normal and the sunbeam direction. As seen in Section 4.1.4, the total radiated power of a blackbody is given by Planck's law and is a function of the fourth power of the absolute temperature of its surface. The atmospheric thermal radiation from 4 μm to 100 μm (W m^{-2}) is expressed as

$$M_{air} = \varepsilon_{air}\sigma T_{air}^4 \tag{12.3}$$

where ε_{air} is the mean air emissivity, σ is the Stefan-Boltzmann constant (5.67×10^{-8} W m^{-2} K^{-4}), and T_{air} is the air temperature (K). The energy absorbed by the upper plant leaf surface $Q_{a,u}$ can be written as

$$Q_{a,u} = A_{leaf} \times S_{total} + \varepsilon_{leaf} \times M_{air} \tag{12.4}$$

where $A_{leaf} = 1 - R_{leaf} - T_{leaf}$ is the mean leaf absorptance in the solar domain, and ε_{leaf} the mean leaf emissivity, that is., the mean leaf absorptance for infrared radiation. The total solar radiation is reflected by the underlying soil before hitting the lower leaf surface. The soil thermal radiation from 4 μm to 100 μm (W m^{-2}) is expressed as

$$M_{soil} = \varepsilon_{soil}\sigma T_{soil}^4 \tag{12.5}$$

where ε_{soil} is the mean soil emissivity, and T_{soil} the soil temperature (K). As a consequence, the energy absorbed by the lower leaf surface, $Q_{a,l}$ can be written as

$$Q_{a,l} = A_{leaf} \times \rho_{soil} \times S_{total} + \varepsilon_{leaf} \times M_{soil} \tag{12.6}$$

where ρ_{soil} is the soil reflectance in the solar domain. Then the total energy absorbed by the leaf equals

$$Q_a = Q_{a,u} + Q_{a,l} = A_{leaf} \times (1 + \rho_{soil}) \times S_{total} + \varepsilon_{leaf} \times (M_{air} + M_{soil}) \tag{12.7}$$

The heat load varies significantly based on the orientation of the leaf, with planophile (horizontally oriented) leaves absorbing more light than erectophile leaves. For a given species, ε_{leaf} does not vary much, although large differences in absorption occur depending on the phenological stage or state of health of the leaf.

12.1.2 Reradiated Energy

Reradiated energy is radiation resulting from the emission of previously absorbed radiation. For a leaf, Q_r symbolically equals

$$Q_r = 2\ \varepsilon_{leaf}\sigma T_{leaf}^4 \tag{12.8}$$

where ε_{leaf} is the mean leaf emissivity in the thermal infrared domain, and T_{leaf} the leaf surface temperature (K). The factor 2 in Eq. 12.8 is because each surface of the leaf reradiates energy.

12.1.3 Convective Heat Transfer

Sensible heat is convected by the air. Convection occurs whenever the temperature of a leaf surface is above or below the ambient air temperature. The energy flux in convection C is given by

$$C = 2\ h_c\ (T_{leaf} - T_{air}) \tag{12.9}$$

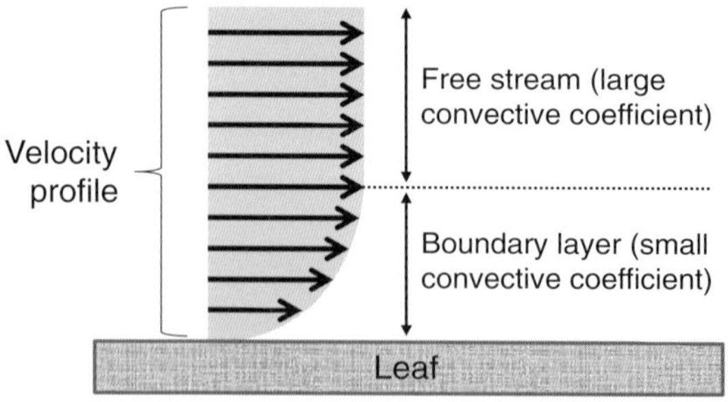

Figure 12.2 Leaf boundary layer.

where T_{leaf} and T_{air} are the temperature of leaf and air (K), and h_c is a convection coefficient (W m^{-2} K^{-1}). Leaf temperature is often different to the temperature of the air. During the day, when radiant energy from the Sun heats the leaf surface, the leaf receives energy and is warmed unless energy is reduced by latent heat exchange. At night, when solar radiation no longer strikes it, energy is lost from the leaf to the air. Moreover, a boundary layer develops as air flows across a leaf (Schuepp, 1993). The boundary layer is a thin air layer in the immediate vicinity of the leaf surface where transport can be either laminar or turbulent flow, or a mixture. It retards the transfer of heat, CO_2, O_2, H_2O, and other trace gases from the leaf to the surrounding air (Figure 12.2).

Mechanical engineers have proposed many expressions for h_c corresponding to different geometrical configurations occurring in nature. For leaves that look like flat plates, it is approximated by

$$h_c = \frac{\rho_{air} C_{p_{air}} \kappa_{air}}{k_{leaf}} \left(\frac{u}{d_{leaf}} \right)^{1/2} \tag{12.10}$$

where ρ_{air} is the density of air (kg m^{-3}), $C_{p_{air}}$ is the specific heat capacity of air (J kg^{-1} K^{-1}), κ_{air} is the thermal diffusivity of air (m^2 s^{-1}), k_{leaf} is a factor related to leaf morphology (shape, thickness, venation), u is the ambient wind speed (m s^{-1}), and d_{leaf} is the characteristic dimension of the leaf (m) in the direction of wind flow across the leaf. k_{leaf} has been determined experimentally in a wind tunnel. For air temperature in the boundary layer of 20° to 25°C, the value $k_{leaf} = 4 \times 10^{-3}$ m $s^{-0.5}$ is commonly used in the literature for flat leaves. d_{leaf} is the length or diameter of the leaf (the shape exerts a secondary influence). Based on hydrodynamic theory for laminar flow adjacent to a flat surface, the average boundary layer thickness (m) can be defined as

$$\delta_{bl} = k_{leaf} \left(d_{leaf}/u \right)^{1/2} \tag{12.11}$$

Table 12.1 presents the variation of δ_{bl} with typical leaf size and wind speed. As leaf size increases, the boundary layer thickness increases; as wind speed increases, it decreases.

Equation 12.10 can be rewritten as

$$h_c = \frac{\rho_{air} C_{p_{air}} \kappa_{air}}{\delta_{bl}} = \frac{\lambda_{air}}{\delta_{bl}} \tag{12.12}$$

Table 12.1 *Thickness of the boundary layer (mm) adjacent to a flat leaf for a variety of leaf size (m) and wind speed (m s^{-1}). Note that 1 m s^{-1} equals 3.6 km h^{-1} (adapted from Nobel, 2009).*

	u (m s^{-1})					
d_{leaf} (m)	0.1	0.5	1.0	5.0	10.0	15.0
0.005	0.89	0.40	0.28	0.13	0.09	0.07
0.01	1.26	0.57	0.40	0.18	0.13	0.10
0.05	2.83	1.26	0.89	0.40	0.28	0.23
0.10	4.00	1.79	1.26	0.57	0.40	0.33
0.50	8.94	4.00	2.83	1.26	0.89	0.73

with $\lambda_{air} = \rho_{air} C_{p_{air}} \kappa_{air}$ the thermal conductivity of air (W m^{-1} K^{-1}). One can define $r_{bl} = \delta_{bl}/\kappa_{air}$, the boundary layer resistance (s m^{-1}), to exchange of heat between leaf and air. It depends on leaf surface characteristics, in particular leaf pubescence (Meinzer and Goldstein, 1985). Then Eq. 12.12 becomes

$$h_c = \frac{\rho_{air} C_{p_{air}}}{r_{bl}} \tag{12.13}$$

Let us calculate the energy flux in convection C across the air boundary layer, assuming the following conditions: if $d_{leaf} = 0.10$ m and $u = 1.0$ m s^{-1}, Eq. 12.11 leads to $\delta bl = 4 \times 10^{-3}$ $(0.1/1.0)^{1/2} = 1.26 \times 10^{-3}$ m. If $T_{air} = 20°\text{C}$ and $P_0 = 1.013$ bar, then tables give $\rho_{air} = 1.205$ kg m^{-3}, $C_{p_{air}} = 1.00682 \times 10^3$ Jkg^{-1}K^{-1}, and $\kappa_{air} = 2.143 \times 10^{-5}$ m^{2}s^{-1}. The thermal conductivity of air is $\lambda_{air} = 0.026$ Wm^{-1}K^{-1}. As a consequence, if $T_{leaf} = 25°\text{C}$, then Eq. 12.9 and Eq. 12.12 lead to $C = 2\,\frac{0.026}{1.26 \times 10^{-3}}\,(25 - 20) = 206$ Wm^{-2}.

Leaf shape affects the boundary layer thickness, causing leaves with a narrow cross-section to have a thinner boundary layer than leaves with a wider cross-section (e.g., a round shape). For conifer needles that look like cylinders, the convection coefficient h_c is approximated by

$$h_c = \frac{C_{p_{air}} \rho_{air} \kappa_{air}}{k_{needle}} \left(\frac{u}{d_{needle}^2} \right)^{1/3} \tag{12.14}$$

where d_{needle} is the diameter of the needle (m).

In both cases the effect is modified by wind, where a long path length increases boundary layer and a short path length reduces it. Figure 12.3 shows a simulation of the effective boundary layer thickness as a function of wind speed and leaf size.

Note that leaf shredding and compound leaves, which are common in plants of arid regions, interrupt the developing boundary layer, as illustrated by Figure 12.4.

12.1.4 Latent Heat Flux

The amount of water vaporized in the process of transpiration can be computed with the help of diffusion theory. The air in the substomatal cavity is assumed to be saturated with moisture at the

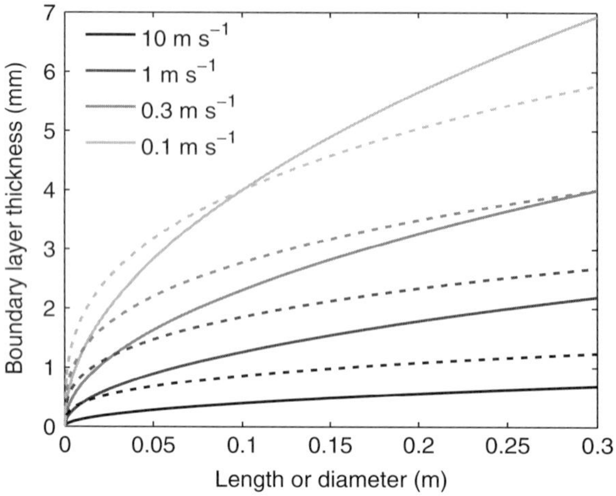

Figure 12.3 Mean thickness of the air boundary layer adjacent to a flat leaf (plain line) and a cylinder (dotted line) at various wind speeds. The length for the flat leaf represents the mean distance across it in the directions of the wind (redrawn from Nobel (2009)).

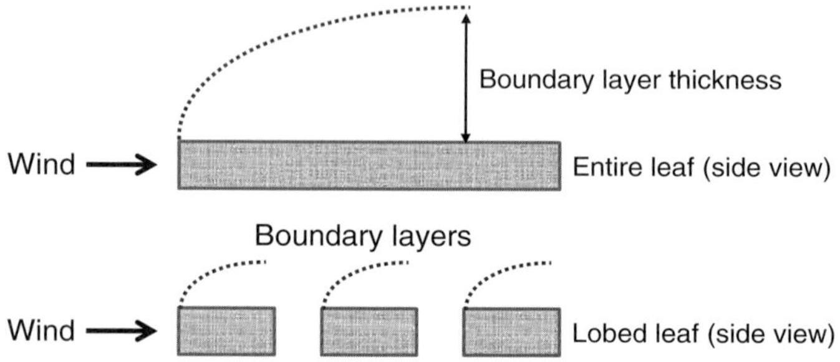

Figure 12.4 Effect of leaf path length on boundary layer thickness (redrawn from Ehleringer, http://ehleringer.net/).

temperature of the leaf. Thus, there is a water vapor concentration gradient between this region and the free air beyond the leaf. This gradient causes water vapor to diffuse up through the stomatal channel and across the boundary layer of air near the leaf surface. The latent heat flux LE may be expressed as

$$LE = 2\,\lambda \frac{e_{leaf} - e_{air}}{r_{tot}} \tag{12.15}$$

where λ is the latent heat of vaporization (J g^{-1}), e_{leaf} and e_{air} are the water vapor density of leaf and air (g m^{-3}), and r_{tot} is the total leaf resistance (stomatal and aerodynamic) (s m^{-1}). Note that leaf conductance is calculated by taking the reciprocal of the resistance. The term $e_{leaf} - e_{air}$ may be written as

$$e_{leaf} - e_{air} = s\left(T_{leaf} - T_{air}\right) \tag{12.16}$$

where s is the slope of the curve relating air temperature to saturation water–vapor pressure over ice or water.

12.1.5 Applications

Leaf energy budget is closely linked to leaf temperature, which affects many physiological processes (Schulze et al., 1980). For instance, photosynthetic enzymes that contribute to CO_2 fixation generally have optimal functioning at temperatures around 25–30°C and when temperatures exceed 35°C, they begin to denature. Buriol et al. (1984a, 1984b) developed a non-destructive method to determine leaf water content based upon observation of leaf temperature kinetics following changes in leaf energy balance. Assume that an additional radiative energy ΔQ_a is provided to the leaf, for instance, by switching on a spotlight of mean flux density 210 W m^{-2} for a few seconds, one can show that

$$\Delta Q_a = MC\left(\frac{\partial T}{\partial t}\right)_{t=0} \tag{12.17}$$

where MC is the leaf surface caloric capacity (J m^{-2} K^{-1}) and $(\partial T/\partial t)_{t=0}$ is the initial rate of leaf temperature variation under additional radiation (K s^{-1}). MC can be decomposed into two terms:

$$MC = C_w SH_w + C_m SH_m \tag{12.18}$$

where C_w and C_m are the leaf water and dry matter contents (kg m^{-2}), and SH_w and SH_m are the corresponding specific heats. $SH_w = 4180\ \ \text{J kg}^{-1}\ \text{K}^{-1}$ and $SH_m = 1250\ \ \text{J kg}^{-1}\ \text{K}^{-1}$. Then C_w, the unknown variable in Eq. 12.18, can be determined. Wuenscher (1970) followed by Parkhurst (1976) studied the effect of hair coating on the leaf energy budget in common mullein (*Vesbascum thapsus*) by comparing leaves with and without hair. Although Wuenscher did not notice any significant difference in their optical properties, this is not the case in other species as discussed in Section 5.1.3. Parkhurst suggested that the shaved leaf may have greater transmission, thus less energy absorption. The hairy leaves were warmer than the shaved ones, apparently primarily due to higher stomatal resistance. A denser hair layer increased the boundary layer thickness δbl and decreased convective heat loss. The leaf energy budget also provides useful information about photosynthesis and transpiration rates (e.g., Gates, 1964, 1966; Leuning, 1989; Inoue et al., 1990; Pieruschka et al., 2010). The last example deals with water loss from the leaf which is regulated by stomata (aperture size and density) and aerodynamic resistance. A method to measure transpiration of single hypostomatous leaves utilizes two leaves: one (leaf 1) remains undisturbed while the other (leaf 2) is prevented from transpiring by applying an impermeable film. According to Eq. 12.4 to Eq. 12.7, steady state energy balances for the two leaves can be written as

$$Q_a = Q_{r1} + C_1 + LE_1 = 2\ \varepsilon_{leaf}\sigma T^4_{leaf,1} + 2\ h_c\ \left(T_{leaf,1} - T_{air}\right) + LE_1 \tag{12.19}$$

and

$$Q_a = Q_{r2} + C_2 = 2\ \varepsilon_{leaf}\sigma T^4_{leaf,2} + 2\ h_c\ \left(T_{leaf,2} - T_{air}\right) \tag{12.20}$$

Subtracting Eq. 12.20 from Eq. 12.19 eliminates Q_a and T_{air} to yield

$$0 = 2\ \varepsilon_{leaf}\sigma\left(T^4_{leaf,1} - T^4_{leaf,2}\right) + 2\ h_c\ \left(T_{leaf,1} - T_{leaf,2}\right) + LE_1 \tag{12.21}$$

$T_{leaf,1} - T_{leaf,2}$ can be written as a small difference of temperatures ΔT. It implies that $T_{leaf,2} = T_{leaf,1} - \Delta T$ so that $T^4_{leaf,2} = T^4_{leaf,1} - 4T^3_{leaf,1}\Delta T + 6T^2_{leaf,1}\Delta T^2 - 4T_{leaf,1}\Delta T^3 + \Delta T^4$. For $\Delta T < 5$ K this equation can be reduced to $T^4_{leaf,2} \approx T^4_{leaf,1} - 4T^3_{leaf,1}\Delta T$, which leads to

$$T^4_{leaf,1} - T^4_{leaf,2} \approx 4T^3_{leaf,1}\left(T_{leaf,1} - T_{leaf,2}\right) \tag{12.22}$$

Combining Eq. 12.21 and Eq. 12.22 and solving for LE_1 yields

$$LE_1 = \left(8\ \varepsilon_{leaf}\sigma T^3_{leaf,1} + 2\ h_c\right)\left(T_{leaf,2} - T_{leaf,1}\right) \tag{12.23}$$

Other equations to calculate transpiration rates are provided by Kümmerlen et al. (1999), among others.

12.2 Photosynthesis – Leaf Carbon Budget

Each species is adapted to live in a particular set of environmental conditions. So plants show different adaptations for photosynthesis, in response to the environmental conditions under which they evolved (e.g., Smith et al., 1997; Evans et al., 2004). For example, in the desert, leaves control evaporation by reducing leaf size, the number of stomata, thickening their epicuticular wax, growing a thick layer of hairs, and so on. Leaf optical properties, which are an external manifestation of such processes, are consequently likely to change. Measurement of absorptance, the fraction of light absorbed by leaves given by $A = 1 - (R + T)$, where R is the reflectance and T is the transmittance, is important. As seen earlier, since most of the absorbed radiation is converted to heat, leaf absorptance over the total solar spectrum (0.3–4.0 μm) can be used to calculate leaf energy balance. Absorptance also indirectly affects photosynthesis through the effect on leaf temperature. In the visible (0.4–0.7 μm), leaf absorption by pigments: i) determines the proportion of incident radiation that can be used in photosynthesis; and ii) partly protects the photosynthetic apparatus against excess light stress. Gorbunova et al. (1960) studied the optical adaptability of plants along a latitudinal gradient in the Soviet Union, ranging from the 60th to the 73th parallel north. They showed higher photosynthetic rates and carbon fixation, apparently associated with the greater absorption of solar radiation by plants, from the south to the north. Ehleringer (1981) showed that the 0.3–4.0 μm absorptance and the 0.4–0.7 μm absorptance were tightly correlated, in agreement with other studies. He also pointed out that they could both be reliably calculated by knowing the leaf absorptance at 625 nm. Modifications of solar-radiation absorption by leaf surface structures, such as epicuticular waxes and epidermal hairs, definitely have an impact on whole-leaf photosynthesis, as shown on brittlebush (*Encelia farinosa*), a common desert shrub of northern Mexico and the southwestern United States (Ehleringer, 1986). Leaves of this plant display variable pubescence depending on climatic conditions. Pubescent white leaves have a temperature 6°C lower than nonpubescent green leaves, while their transpiration rate is lower and their photosynthetic rate is higher.

Note that the absorption spectrum should not be confused with the action spectrum, which reflects the spectral efficiency of photosynthetic activity to incident electromagnetic radiation. In an influential paper published to provide a standard definition of the term "photosynthetically active radiation" (PAR), which had been previously described empirically, McCree (1972) measured the action spectrum, absorptance, and quantum yield of CO_2 uptake in 22 species of crop plants. For all species

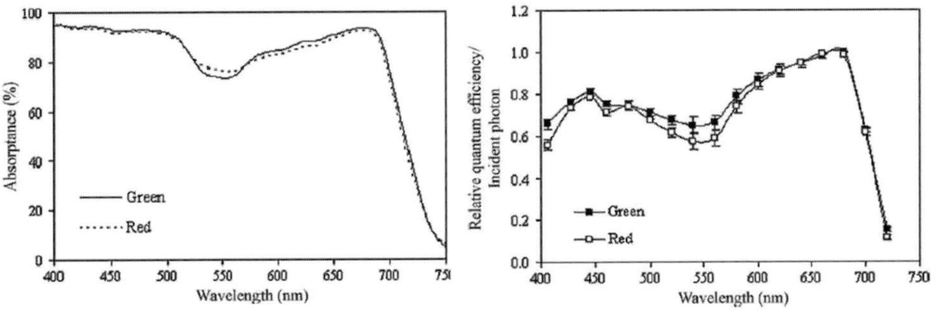

Figure 12.5 Absorptance (left) and action (right) spectra of green and reddish rose (*Rosa* sp.) leaves (from Paradiso et al. (2011), *Scientia Horticulturae*. Copyright © 2011 with permission from Elsevier).

and conditions the quantum yield curve displayed two broad maxima, centered at 440 nm and 620 nm, with a shoulder at 670 nm. Since then, comparisons have been reported for a considerable number of plant species (e.g., Inada, 1976). The leaves of higher plants generally show a divergence between the absorption and action spectra, especially in the blue and ultraviolet regions of the spectrum (Figure 12.5). This discrepancy is partly due to accessory pigments (e.g., Inada, 1980).

Leaf anatomy plays a major role in the distribution of light and CO_2 in the mesophyll (e.g., Terashima, 1989; Terashima and Hikosaka, 1995; Evans, 1999; Terashima et al., 2011). For efficient use of the light energy, light needs to be delivered to all the chloroplasts distributed along the cell surfaces throughout the leaf. Internal gradients of absorbed light (see Section 5.4) and of chlorophyll fluorescence emission (e.g., Vogelmann and Evans, 2002) have been measured experimentally on leaves illuminated on their adaxial or abaxial surfaces (Evans et al., 2004; Evans and Vogelmann, 2006). Such measurements using optical fibers are not as straightforward as they seem. Light gradients decrease exponentially across the leaf. Red or blue light is strongly absorbed by chloroplasts located near the illuminated surface of the leaf whereas green light, which is absorbed only weakly, penetrates further into the mesophyll and reaches chloroplasts deeper in the leaf. Terashima et al. (2009) noted that leaf photosynthesis was increased by this mechanism. Where is carbon fixed? Figure 12.6 shows that gradients in photosynthetic capacity (the maximum amount of carbon that can be fixed) are very different from the light gradients. Maximum carbon fixation and RuBisCO content occur in the middle of the leaf and coincide with one another, but they do not follow the exponential shape of light gradient. The contribution of the spongy mesophyll to carbon reduction was higher than expected: 60% in the palisade parenchyma versus 40% in the spongy mesophyll (Nishio et al., 1993; Vogelmann et al., 1996a, 1996b). It appears that there are factors other than light that determine the profiles of photosynthetic capacity within leaves.

The relationship between leaf optical properties and some photosynthetic parameters, for example, the maximum rate of RuBP (ribulose-1,5-bisphosphate) carboxylation, $V_{c\,max}$, and the maximum rate of electron transport, J_{max}, gave rise to several studies. Stylinski et al. (2000) found a close relationship between J_{max} and the narrow-band photochemical reflectance index (PRI), an optical indicator of the xanthophyll cycle pigments (see Section 12.3.3), in foliage of pubescent oak (*Quercus pubescens*) trees. Recent papers have explored the ability to relate variations in spectral optical properties to the key parameters $V_{c\,max}$ and J_{max}, as well as other leaf functional properties. Doughty et al. (2011) and Serbin et al. (2012) combined spectroscopic data with partial least square regression (PLSR) modeling to

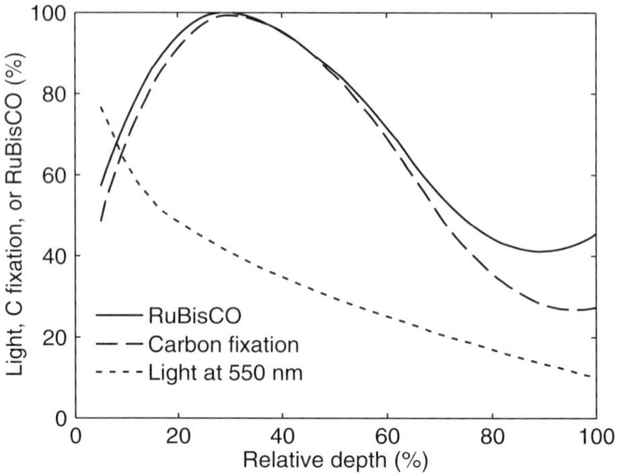

Figure 12.6 Measured gradients of carbon fixation, light at 550 nm and RuBisCO at depth in spinach (*Spinacia oleracea*) leaves (adapted from Vogelmann et al. (1996a, 1996b)).

estimate levels of those two photosynthetic parameters. Dillen et al. (2012) were the first to study leaf reflectance in relation to photosynthetic capacity in two co-occurring deciduous tree species, northern red oak (*Quercus rubra*) and white birch (*Betula papyrifera*), over a growing season.

Direct and diffuse light are not equivalent to leaves. With the eruption of Mt Pinatubo in 1991 and the subsequent increase in sulfate aerosols in the atmosphere, several authors had the opportunity to study the effects of diffuse light on plant growth and the carbon cycle. It seems that the directionality of light affects photosynthetic processes differently when scaling from the leaf to landscape. Increased rates of photosynthesis have been observed at the canopy level due to increased proportion of diffuse light. The reason is that two leaves receiving moderate light produce more photosynthesis than if one receives bright light while the other is in deep shade (Farquhar and Roderick, 2003). At the leaf level direct and diffuse light may be utilized differently. Leaves measured under diffuse light reflect more and transmit less light than under direct light (e.g., Brodersen and Vogelmann, 2007; Gorton et al., 2010). Their overall absorptance may be a few percent lower, as well as their photosynthesis (Brodersen et al., 2008). Finally, leaves that develop under direct high-light conditions use direct light more effectively than diffuse light, while those that develop under diffuse, low-light conditions use diffuse and direct light equally well.

12.3 Proximal Sensing

In many applications like agriculture, forestry, or global ecology, it is crucial to determine the *in situ* biochemical content of plant leaves, such as chlorophyll, water, and leaf mass per area (LMA). Recent applications in precision farming mobilized the scientific community to assess total chlorophyll content, which can be roughly related to nitrogen content, one of the key variables for measuring plant canopies in precision farming. This farming concept takes into account intra-field variability and ultimately aims at applying the appropriate amounts of inputs, for instance, fertilizers, to various locations within a field to achieve an economically profitable yield while preserving the environment by reducing unnecessary inputs (Moran et al., 2004). The monitoring of water content is also subject to special attention in agriculture because exposure to water deficit, which can be spectrally related to the content, limits productivity of crop plants. In natural vegetation, it is essential for drought assessment and prediction of wildfire susceptibility (Ceccato et al., 2001). The LMA (or dry matter content) is equivalent to the inverse of specific leaf area (SLA), a widely used key variable in plant ecology which is correlated with rates of plant growth, light interception, gas exchange, and photosynthesis (e.g., Meziane and Shipley, 2001; Wright et al., 2004; Kattge et al., 2011).

The development of the first spaceborne hyperspectral sensors, like EO-1 Hyperion and Compact High Resolution Imaging Spectrometer (CHRIS), have demonstrated the potential to monitor crops and better understand ecosystem functioning through measuring the radiance of targets in contiguous spectral bands. The next generation of spaceborne sensors, such as PRecursore IperSpettrale della Missione Applicativa (PRISMA) or Environmental Mapping and Analysis Program (EnMAP), are dedicated to monitoring crops and natural ecosystems. They will provide an opportunity to map variation in canopy biochemistry, for the moment restricted to chlorophyll and water content. In order to validate these new remote sensing products, several international field campaigns have been started (Justice et al., 2000), which require the characterization of a suite of leaf biochemicals among other variables.

While determination of water and dry matter content "only" requires a precision balance and an oven, which can be brought into the field, the extraction and measurement of photosynthetic pigments involves much more complicated and expensive wet chemistry methods (see Chapter 3.2). Quantifying other biochemicals (e.g., leaf nitrogen, cellulose, lignin) requires resources beyond those routinely available at field sites. Moreover, these measurements are almost always destructive and consequently, difficult to implement when the experimental site is distant from the laboratory. Sampling bias is another concern as it is nearly impossible to access leaves from the upper canopy of tall trees, not to mention the problem of collecting a statistically significant set of samples over large heterogeneous landscapes to understand the variance in these properties. As soon as cut, the leaf starts to dehydrate and to lose pigments by degradation, as well as other metabolic degradation products. All these factors limit sampling capacity, thus the representativeness of such measurements. The *in situ* measurement of these biochemical constituents can be performed by non-destructive techniques, generally optical ones for pigments (e.g., chlorophyll meters) or water, although these are usually tied to destructive measurements for calibration.

12.3.1 Pigments

Foliar color has been of great interest and value to resource managers and scientists as a visual indicator of plant health. This information is linked with growth rates, nutrient deficiencies, plant diseases, and other plant processes. Prior to the use of digital cameras, health assessments often relied on simple, visual scales to determine foliar color. At the end of the 19th century, Ville (1889) already noticed the relationship between the color of leaves and the richness of the soil. "The color considerably changes when the fertilizers are missing in the soil; its intensity increases or decreases, the color remains green or turns yellow, according to whether phosphorus, potassium or nitrogen are missing in the soil." Then he used a kind of color chart adapted to marijuana (*Cannabis sativa*) to estimate the amount of fertilizer to be applied. Grime (1961) used color standards established in 1912 by Robert Ridgway, an American ornithologist specializing in systematics, in order to assess the severity of leaf chlorosis. Since then several affordable guides and color charts depicting gradients of

Figure 12.7 4-panel Leaf Color Chart (see http://www.nitrogenparameters.com/about.html).

leaf color were developed to help standardize color/assessments. These charts are a tool for describing and analyzing color in terms of three attributes: hue, value (bright), and chroma (saturation), which are arranged in orderly scales of equal visual steps. Such charts include the Munsell® Plant Tissue Color Charts (Munsell Color, X-Rite, Inc.), the GLOBE Plant Color Guide (www.globe.gov/), and the Leaf Color Chart (LCC). The LCC has been jointly developed by the International Rice Research Institute (IRRI) and the Philippines Rice Research Institute (PhilRice) from a Japanese prototype named "Standard rice leaf color scale" (Furuya, 1986), for the purpose of measuring the required quantity of nitrogen to be applied to a rice field and thereby to get maximum productivity (Figure 12.7). LCCs are specific to the crops in a region, for instance, LCCs developed at the University of California, Davis, using California rice varieties, produce a spectrum of colors that do not match the rice leaves found in Asia (Witt et al., 2005). To date, about 300 LCCs are in use in the Sacramento Valley, as an integral part of the fertility management program for farm managers and private consultants.

Leaf color charts make it easy to estimate foliar nitrogen status and evaluate seasonal changes in leaf color, but because this approach is somewhat subjective, they lack quantitative rigor. More precise methods of foliar color based on reflectance and transmittance measurements have been developed since the 1950s. Handheld photometers like the SPAD-502 (Markwell and Blevins, 1999; Figure 12.8) or other optical devices are now commonly used in the scientific community to measure leaf chlorophyll content (e.g., Hardwick and Baker, 1973; Wallihan, 1973; MacNicol et al., 1976; Hardacre et al., 1984; Yadava, 1986; Dwyer et al., 1991; Madeira et al., 2000, 2003; Azia and Stewart, 2001; Uddling et al., 2007; Hunt and Daughtry, 2014), nitrogen content (e.g., Blackmer et al., 1993; Blackmer and Shepers, 1995; Schepers et al., 1997; Bonfil et al., 2005), or both biochemicals (e.g., van den Berg and Perkins, 2004; Wang et al., 2004; Han et al., 2009). Other leaf pigments have been investigated, like anthocyanins (e.g., Manetas et al., 1998; van den Berg and Perkins, 2005) and flavonoids (e.g., Goulas et al., 2004; Cerovic et al., 2012). More generally, they have been used to

Figure 12.8 SPAD-502.

evaluate plant physiological processes including leaf development (Thompson et al., 1996) and photosynthesis (e.g., Earl and Tollenaar, 1997; Torres Netto et al., 2002; Bauerle et al., 2004).

Note that, although they should yield similar exponential relationships to chlorophyll content, even different chlorophyll meters of the same model may give slightly different responses (Schaper and Chacko, 1991). Comparisons between instruments of different commercial brands have also been performed: SPAD-502 vs. custom-built instruments (Monje and Bugbee, 1992; Pavan et al., 2004); SPAD-502 vs. CCM-200 (Richardson et al., 2002; Cerovic et al., 2012); SPAD-502 vs. atLEAF+ (Zhu et al., 2012); SPAD-502 vs. Hydro N-Tester (Uddling et al., 2007); and TYS-A vs. CCM-200 (Lunagaria et al., 2015).

An advantage of chlorophyll meter optical devices is greater accuracy of measurements compared to color charts and, in some cases, portability for field applications. They now are often coupled with global positioning systems to provide spatial information. Disadvantages of these systems are that they have been sold at prohibitive prices for many agricultural users; they generally do not provide information on the concentrations of different pigments (or classes of pigments, e.g., chlorophyll *a* and *b* or carotenoids); and, moreover indices have no units and require a calibration that depends on the leaf species due to the wide range of anatomical structures in nature (e.g., Richardson et al., 2002; Marenco et al., 2009). As seen in Section 6.1.1, the blade structure is known to change the leaf optical properties, not only in the near-infrared where absorption is weak but also across the whole spectrum. These errors result in reduced accuracy when leaves of various species are analyzed by the chlorophyll meter. Irradiance and plant water status have been shown to affect SPAD 502 readings (Hoel and Solhaug, 1998; Martinez and Guiamet, 2004). In addition, measurements are laborious on many mature dicot leaves that have high chlorophyll concentration and high LMA (Yamamoto et al., 2002). This particularly affects species with thicker persistent leaves. For instance, the SPAD-502 measurement is restricted to detection of chlorophyll, although many users use the measurement to predict nitrogen concentration by assuming a linear relationship between these biochemicals. Because of these limitations, each sensor type has to be calibrated for the plant species of interest by destructive wet chemistry measurements. As so rightly pointed out by Markwell (2002), "the gold-standard for chlorophyll determination remains extraction of the pigments into a solvent, followed by spectrophotometric analysis."

Scientists made some attempt to better assess the spatial variability of leaf pigments/nitrogen or to detect plant stress remotely using digital analysis of images acquired in different color spaces (see Section 4.3): color photographs (Andrieu et al., 1992), video images (e.g., Spomer et al., 1988; Nakatani and Kawashima, 1994; Kawashima and Nakatani, 1998; Tian et al., 2010), scanners (Shibghatallah et al., 2013), digital cameras (e.g., Yuzhu et al., 2011; Lunadei et al., 2012; Yang et al., 2014), and hyperspectral cameras (Behmann et al., 2015). At the canopy level, Yang et al. (2014) found a mismatch between green chromatic leaf color and total pigment concentration in late spring and early summer, with greenness peaking 20 days earlier than chlorophyll concentration, although redness was significantly correlated with anthocyanin concentration in the fall. Nonetheless, others have found strong correlations between chromaticity and leaf chemistry. Yuzhu et al. (2011) report strong correlations between chromaticity and leaf nitrogen in pepper plants.

Last but not least, Bingen Cortazar and colleagues in the Department of Electrical Engineering at UCLA have employed Google Glass as a flexible photometer for measuring the chlorophyll content of leaves in the field (Cortazar et al., 2015). The glasses take two images of a leaf, which is placed in the illuminator device and illuminated with either red or white light-emitting diodes. The system has been validated using the SPAD-502. The entire apparatus is presented as an inexpensive mobile tool for rapid diagnosis of plant health using non-destructive measurements. The light box could be manufactured for less than \$30 (Figure 12.9a), although the Google Glass costs about \$1600. Other applications requiring a rear-camera-equipped iPhone, iPod Touch, or iPad are "on the go" methods. We can mention the FieldScout GreenIndex+ (Spectrum Technologies, Inc.) that has been developed to measure the chlorophyll content of plants and compute nitrogen application recommendations. It determines the Dark Green Color Index (DGCI) of plant leaves, a number between 0 and 1 related to the chlorophyll content. The application requires using a specific green and yellow color standard on a target board in order for the DGCI to be computed accurately (Figure 12.9b). A similar system based on Android and dedicated to rice leaf color analysis has been recently published (Intaravanne and Sumriddetchkajorn, 2015).

Figure 12.9 SPAD-502. Measurement of leaf chlorophyll concentration using (left) the Google Glass mobile system (source: http://dailybruin.com/2015/02/23/ucla-lab-develops-google-glass-app-for-rapid-diagnosis-of-plant-health/), and (right) the FieldScout GreenIndex+ (source: www.specmeters.com/greenindexsupport/).

Figure 12.10 The RAMIS instrument performing measurements in a greenhouse.

12.3.2 Water and Dry Matter

Monitoring water content in vegetation is subject to special attention in agriculture because it limits productivity of crop plants exposed to water stress. In natural ecosystems, it is essential for drought assessment and prediction of wildfire susceptibility. Measuring leaf water content using electromagnetic waves mainly relies on two techniques based upon observation of leaf temperature kinetics, which allows tracking changes in leaf energy balance (see Section 12.1), and on transmission measurements of terahertz radiation (see Section 8.4). Attempts by Bonfil et al. (2005) to link SPAD units to leaf water content failed. This was not surprising and emphasizes the need for measuring wavebands other than those for chlorophyll. Han et al. (2009) designed an instrument similar to the SPAD 502 with three LEDs at 650 nm, 940 nm, and 975 nm (water absorption peak) but no results are provided in their article. RAdiomètre portatif de Mesure In Situ (RAMIS) was the first portable field radiometer dedicated to simultaneous determination of total chlorophyll content, EWT, and LMA (Pavan et al., 2004; Conejo et al., 2010) (Figure 12.10). To date, there is no commercial portable device devoted to water content measurement.

12.3.3 Leaf Wetness

The French mathematician and essayist Yves-Marie André was somewhat poetic in his *Discours sur les merveilles de l'arc-en-ciel* (André, 1746). In translation, he wrote:

> Je me promène le matin dans une belle prairie couverte de rosée, & aussitôt que le soleil levant y vient répandre ses rayons, je vois sur l'herbe plus de brillants, que l'on n'en vit jamais paraître sur la tête des rois. Là des rubis, là des topazes ; ici des émeraudes, des saphirs, des diamants, des perles de toutes les couleurs.[1]

[1] I go for a morning walk in a nice meadow covered with dew, and as soon as the rising sun sheds its rays, I can see on the grass more brilliants, than has ever been seen to appear on the head of Kings. There rubies, there topazes; here emeralds, sapphires, diamonds, pearls of all colors. (Cited by Zuppiroli and Bussac, 2001).

Leaf wetness is a generic term to describe water on leaf surfaces. It can be caused by precipitation, irrigation, guttation, or dew. It is an important agrometeorological parameter for many agricultural and environmental applications including plant disease forecasting, deposition of atmospheric pollutants like ozone, effectiveness of pesticides, and activity of insects (Weiss, 1990; Magarey et al., 2005). Like primitive microorganisms, fungal pathogens in air strongly rely on the presence of a layer of free water on the foliage surface to penetrate the host plant during the infection. Dew, which accumulates on leaves when their surface temperature is at or below dew-point temperature (condensation), is also known to influence plant growth and photosynthetic gas exchange in semi-arid regions for several reasons. Dew water drips off the leaves and wets the soil; it reduces the vapor pressure deficit within the canopy; foliar absorption of dew water occurs to maintain sufficient turgidity for growth; and dew water on surfaces delays the onset of evapotranspiration losses until after leaves dry. Duration, frequency, amount, and extent of leaf surface wetness are consequently important input variables for plant pathologists and agricultural meteorologists who have devised numerous measuring techniques such as changes in the electrical conductivity of the surfaces of natural and artificial leaves (electronic leaf), beta-ray gauge, weighing of the moisture deposits, and more recently optical sensors (e.g., Weiss, 1990; Griffioen et al., 1992).

The problem of light focusing by water drops on plants has fascinated scientists for a very long time. It was first described by Thomas Melvill (1726–1753), a Scottish natural philosopher who studied optics intending to verify Newton's theories. In the third section of his *Observations on Light and Colours* entitled *On the Silver-like Appearance of Drops of Water on the Leaves of Colewort* and delivered to the Medical Society of Edinburgh in 1752, he writes:

> It is common to admire the volubility and lustre of drops of rain that lie on the leaves of Colewort, and some other vegetables; but no philosopher, as far as I know, has put himself to the trouble of explaining this curious phenomenon. Upon inspecting them narrowly, I find, that the lustre of the drops arises from a copious reflection of light from the flattened part of its surface contiguous to the plant: I observe further, that, when the drop rolls along a part which has been wetted, it immediately loses all its lustre; the green plant being then seen clearly through it; whereas, in the other case, it is hardly to be discerned. *(Melvill, 1914)*

From these observations, he concluded that the drops may not be in contact with the leaf surface but hang in the air at some distance from it, by the force of a repulsive power. This was proved by the silverlike color, he said, which could only come by the reflection of light from the undersurface of the drop, which would not be possible if the drop actually touched the leaf. Another argument in favor of this theory was that drops left no trace of moisture after they rolled off. For this to be accomplished, Lommel (1874) asserted that the drop must be supported by hair about 1 focal length from the surface (Figure 12.11). If we treat the drop as a thin lens, however, it can be directly in contact with the leaf surface. Mattsson and Cavallin (1972) showed that light was focused 1 radius length past the drop and that dew shined brightly in the backward direction to form a "heiligenschein." This optical phenomenon, also called Cellini's halo, is similar to the hotspot; it creates a diffuse bright region surrounding the shadow when the Sun is low in the sky and when the surface is covered by many dewdrops (Figure 12.11). Heiligenschein should not be confused with the sylvanshine, another optical phenomenon occurring at the surface of some tree leaves and named by Fraser (1994), an expert in meteorological optics. Fraser was not the first person to document it but he appears to be the first to provide a detailed mathematical and biological explanation. The more or less spherical drops act like convergent lenses, which focus the light to a spot on the surface of the leaf, and the image of this spot is directed as rays in

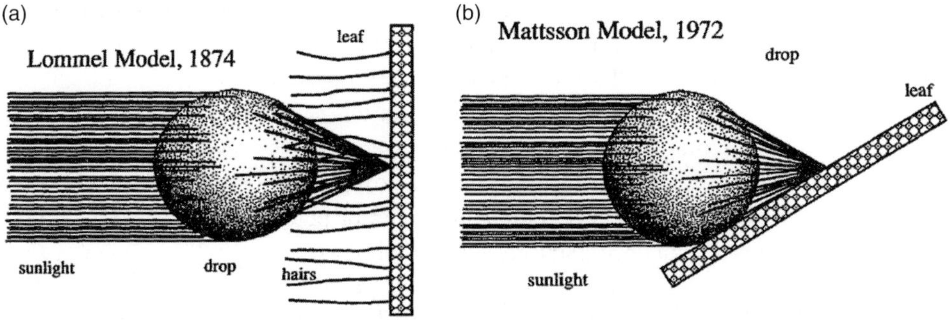

Figure 12.11 (a) Lommel's model relied on hairy leaves to hold the dewdrop one focal length from the surface. (b) Mattsson's model relied on oblique illumination to enable the light to be focused on the leaf. The light enters at an angle of $\pi/3$ from the normal. This is obtained by observing that at the drop center, the contact point with the leaf, and the focal point form a triangle with the focal distance for a water drop equal to twice its radius (from Fraser (1994), *Applied Optics*. Copyright © 1994 with permission from the Optical Society of America).

the opposite direction. The sylvanshine appears to be highly species-specific and it has been seen only on conifers. Contrary to the heiligenschein that usually extends over a very small solid angle and is seen on species with small leaves, such as grass, the sylvanshine extends over a very large solid angle. Note that the intensity of the coherent light transmitted by the dew pattern has been theoretically studied by Nikolayev et al. (1998).

Non-contact detection of dew or spray droplets on the leaf surface was assessed using a multispectral imaging system (Ramalingam et al., 2003). After compensating for background reflectance, the equivalent water thickness (EWT) was highly correlated with measured EWT but was consistently lower by 9%.

Heusinkveld et al. (2008) designed an active optical sensor where the emitter and detector are aligned so that the reflectance is measured in the specular direction. A modulated wideband light source allows reducing noise and discriminating reflected light from solar radiation. The reflectance is measured at 1.70 μm and 1.94 μm (water absorption bands) using two interference filters. Unfortunately, it is not clear how this sensor is practically used in the field to determine leaf surface wetness. A similar project was undertaken in 1997 by the Institut national de la recherche agronomique (INRA) in France, on the basis of preliminary work that consisted of analyzing the reflection of a coherent light beam off a leaf surface covered by dew droplets (Hubert et al., 2009, unpublished report). To simulate the bidirectional reflectance distribution function (BRDF) of such a surface, a Monte Carlo ray tracing model was developed; the image is made of spherical drops of variable size that intersect a plane, the leaf surface, at different contact angles (Figure 12.12).

These contact angles depend on the surface roughness, the nature of the liquid (in this case water), and the surface tension (Bhushan and Jung, 2006). For instance, Beysens (1995) showed that the presence of a waxy layer of cutin on *Iris* leaves led to partial wetting with a contact angle near 90°. Brewer et al. (1991) showed that leaf trichomes had a strong influence on leaf water repellence, as well as the retention of droplets on the leaf (see Chapter 2). It is a widespread belief that plants must not be watered at midday, because water drops adhering to leaves may cause sunburn as a result of the intense focused sunlight. Egri et al. (2010) conducted both computational and experimental studies of light focusing by water drops on three plant species. They came to the conclusion that the hydrophobic surface of floating watermoss (*Salvinia natans*) leaves, which produced spheroid water

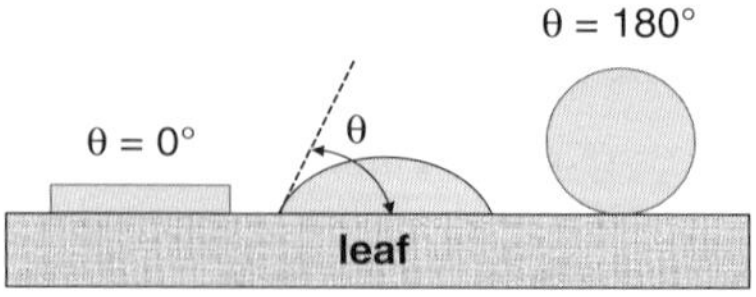

Figure 12.12 Wetting conditions for liquid water on a leaf, as defined by the contact angle θ. Complete wetting: $\theta = 0°$; partial wetting: $0° < \theta < 180°$; complete drying: $\theta = 180°$.

droplets, suffered more from sunburn than the hairless leaves of maidenhair ginkgo (*Ginkgo biloba*) and Norway maple (*Acer platanoides*) leaves, which produced flat lens-shaped water drops. But this effect was limited and only occurred in very special circumstances (Jones, 2010; Lee, 2011). Finally, leaf bidirectional optical properties can be calculated using a BRDF model (see Section 8.3). Preliminary results obtained with the model of Bousquet et al. (2005) showed that it was possible to simulate the effects of leaf wetness on visible reflectance and to measure it using optical methods.

12.4 Vegetation Remote Sensing

Remote sensing of vegetation covers a wide range of terrestrial environmental uses from land cover, agriculture, forestry, geology, hydrology, conservation and biodiversity, global change, and many others. These applications have been the subject of many reference books, therefore we will not extensively review them in this chapter (e.g., Chukhlantsev, 2006; Jones and Vaughan, 2010; Thenkabail et al., 2011). Environmental applications of remote sensing, at scales from local to the planet, are a logical extension of the principles of spectroscopy and leaf optics. However, to connect them one needs additional information about the spectral properties of the non-foliar elements in the canopy: understory canopy elements, plant debris and litter, and finally soils and other materials in the image. It includes the atmospheric conditions at the time of data collection. The three-dimensional arrangements of these elements affect the measured spectral properties due to their effects on photon scattering processes, which must be taken into account. Several good books on radiative transfer methods are available (e.g., Ross, 1981; Asrar, 1989; Myneni and Ross, 1991; Natsuyama et al., 1998), while books on leaf optics have not been previously covered in any detail, so we remain focused on leaf optics and limit our discussion about remote sensing applications to a few examples in a few areas.

12.4.1 Upscaling

Signatures at leaf scale derived from laboratory measurements may provide useful information on the influence of various scene components, especially in humid tropical forests where variation in the foliar chemistry is poorly understood (e.g., Asner and Martin, 2008b). The problem of upscaling has been questioned for a long time in remote sensing. Canopy architecture (LAI, leaf angle distribution), and leaf and soil optical properties are the primary factors that contribute to the reflectance of plant canopies during development, but the relative importance of the various components changes with time. Early in the season, only reflectance from leaves and soil need to be considered. As the vegetation develops, the importance of soil reflectance decreases, because the developing plant canopy obscures the background, and plant leaf response dominates. Therefore, at first sight, radiation reflected by vegetation canopies is similar to laboratory spectral measurements of individual leaves (e.g., Dean et al., 1986; Lorenzen and Jensen, 1991; Baynes, 2007). Sykioti et al. (2011) reported good correlations between CHRIS data and leaf absorption features in terms of magnitude, pattern, and seasonal variability using continuum removal or narrow band absorption features. Much work has attempted to directly link canopy chemistry and hyperspectral data through regression analysis or the use of narrow band spectral indices. Reality is obviously more complicated. Differences observed in the reflectance signature of leaves and canopies are mainly attributed to canopy structure and, to a lesser extent, to viewing and solar geometry. They are greater for pine trees than for broadleaf species, because the needles are clumped and this explains why spectral indices or statistical methods developed at leaf level cannot be directly transposed to canopy level. A controversy has recently arisen about the direct detection of canopy nitrogen from remote sensing observations. A strong canopy-scale positive correlation has been reported in the near-infrared between mass-based nitrogen and albedo in some temperate and boreal forests (Ollinger et al., 2008). However, such a relationship was not observed in leaf reflectance and transmittance (Bartlett et al., 2011). This theory was attacked by Knyazikhin et al. (2012) who explicitly tested whether assumptions that canopy structure can be ignored in quantifying biochemical composition, with a detailed analysis of the physical processes of photon scattering from leaves and plant canopies, using a three-dimensional radiative transfer model. They showed that the aforesaid results were an artifact and that structural variation was affecting the interpretation of canopy reflectance.

Disentangling the effects of canopy structure and leaf chemicals is difficult with empirical approaches. To get around this, radiative transfer models that began development in the 1960s, both at leaf and canopy levels, have turned out to be very powerful tools. Some canopy reflectance models have been coupled with leaf optical properties models to link the spectral variation of canopy reflectance, which is mainly related to leaf biochemical contents, with its directional variation, which is primarily related to canopy architecture and soil/vegetation contrast (Table 12.2). This link allows description of both the spectral and directional variation of canopy reflectance as a function of leaf biochemistry – mainly chlorophyll, water, and dry matter contents – and canopy architecture – primarily leaf area index, leaf angle distribution, and leaf size. It is also key to simultaneous estimation of canopy biophysical/structural variables at different scales. For example, PROSPECT was linked with the Scattering by Arbitrary Inclined Leaves (SAIL) family of models in PROSAIL (Jacquemoud et al., 2009b). It has been also integrated into other canopy reflectance models that address some specific types of canopy architectures: Fast Canopy Reflectance (FCR), New Advanced DIscrete Model (NADIM), and Markov–Chain Canopy Reflectance Model (MCRM) for homogeneous crop canopies; and Discrete Anisotropic Radiative Transfer (DART), Spreading of Photons for Radiation INTerception (SPRINT), Forest Light Interaction Model (FLIM), and Forest LIGHT (FLIGHT) for discontinuous forest canopies. Similar leaf-canopy coupled models were attempted with LCM2 (LEAFMOD-CANMOD), LIBERTY-FLIGHT, and LIBSAIL (LIBERTY-SAIL). Nonetheless, these models are simplifications of natural heterogeneity while, at the same time, require a large number of inputs that make them difficult to apply to image data except for experimental settings. The distribution of these codes in the remote sensing community has remained limited compared to PROSPECT and SAIL because the authors of these models have not made them freely available for everyone to use. The issues raised by the application of these models demonstrate the complexity of the physical interactions and need for more work in this area.

Table 12.2 *Coupled leaf-canopy radiative transfer models (*Author's name of the first citation).*

Leaf model	Canopy model	Leaf+canopy models	Reference*
LEAFMOD	CANMOD (1D)	LCM2	Ganapol et al. (1999)
LIBERTY	SAIL (1D)	LIBSAIL	Dash and Curran (2004)
	FLIGHT (3D)		Dawson et al. (1999)
PROSPECT	SAILH (1D)	PROSAIL	Jacquemoud (1993)
	GEOSAIL (1D)	PROGEOSAIL / PROGOSAIL	Bowyer and Danson (2004); Koetz et al. (2004)
	2 M-SAIL / GEOSAIL (1D)		Weiss et al. (2001); Verhoef and Bach (2003)
	4SAIL (1D)	SLC	Verhoef and Bach (2007)
	FCR (1D)	MSCRM	Kuusk (1994)
	NADIM (1D)		Jacquemoud et al. (2000)
	MCRM (1D)		Kuusk (1995b)
	DART (3D)		Demarez and Gastellu-Etchegorry (2000)
	SPRINT (3D)		Zarco-Tejada et al. (2004a)
	FLIM (3D)		Zarco-Tejada et al. (2004b)
	FLIGHT (3D)		Koetz et al. (2004)

With one or two exceptions, all canopy reflectance models assume that the individual scattering elements, mainly the leaves, act as perfect Lambertian diffusors. That hypothesis continued due to the lack of experimental data and to the complexity of the transport problem. The single-scattering albedo of an object is the ratio of its scattering coefficient (K_s) to its extinction coefficient ($K_e = K_s + K_a$):

$$\omega(\lambda) = \rho\ (\lambda) + \tau(\lambda) \tag{12.24}$$

In the case of planar objects, such as plant leaves, K_s is the sum of the hemispherical reflectance ρ and the hemispherical transmittance τ, and $K_a = 1 - K_s$ so their single scattering albedo is the sum of the two components: $\omega(\lambda) = \rho\ (\lambda) + \tau(\lambda)$. As seen in Section 5.1, most leaves are not isotropic diffusers except under perpendicular illumination. For oblique incidence angles, their reflectance displays a strong specular component and a weak, narrow hot spot component. To calculate plant canopy BRDF, Myneni et al. (1987) used reflection and transmission phase functions measured by Breece and Holmes (1971) on soybean (*Glycine max*) leaves, for both top and bottom incidences. They defined single scattering albedos of the leaves dependent on the incident beam direction, on its wavelength, and on which face the leaf was irradiated (upper and lower). Ross and Marshak (1989) designed a model where the leaf transmittance was isotropic and the leaf reflectance consisted of two components, a diffuse component and a specular component depending on the incidence angle and the refractive index of the wax (see Section 3.6). The calculation of the contribution of the specular component of leaf reflectance is not straightforward, as it mainly changes canopy BRDF in the red, and the assumption of Lambertian leaves apparently leads to small errors under most viewing/illumination geometries. The relative contribution of leaf reflectance and leaf transmittance to canopy reflectance has been studied by Otterman et al. (1995). Large values of leaf reflectance have a strong effect at nadir and produce strong backscattering, whereas high values of leaf transmittance produce strong forward scattering. With a view to the inversion of radiative transfer model to infer canopy biophysical variables, the sensitivity of canopy reflectance to leaf spectral optical properties has been investigated using the PROSAIL (e.g., Baret et al., 1994; Jacquemoud et al., 2000, 2009b; Bacour et al., 2002b), PROSPECT-NADIM (Ceccato et al., 2002), and LIBERTY-FLIGHT (Dawson et al., 1999) models.

12.4.2 Biodiversity

Biodiversity includes genetic variation within species, the variety of species in an area, and the variety of habitat types within a landscape. Recognition of different plant species based on foliar biochemistry is possible if there is sufficient spectral contrast between the species of interest and other species in the environment (see Section 6.5). However, we have very little knowledge of the distribution of leaf biochemical constituents in ecosystems. According to the Convention on Biological Diversity (www.cbd.int/Same here), the main causes of biodiversity loss on Earth are land degradation, climate change, and invasive species. There is growing concern about invasive plants that have aggressively spread outside their natural range in recent decades and transform entire ecosystems by altering hydrology, fire regime, nutrient cycle, and other ecological processes (Mooney and Cleland, 2001). Economic impacts stem from time and financial costs associated with both manual and mechanical controls. Their detection can be performed using airborne or spaceborne hyperspectral data and processing techniques based on their spectral signatures (e.g., Underwood et al., 2003; Lass et al., 2005; Lawrence et al., 2006; Pengra et al., 2007; Asner et al., 2008).

Assessing the diversity and functioning of terrestrial ecosystems is also a question of primary interest. Tropical forests are vast, mostly unexplored places. The Spectranomics project (http://spectranomics.stanford.edu/) initiated by the Carnegie Institution in the 2000s has showed the way to approach this problem (e.g., Asner, 2008; Asner and Martin, 2009). It consists of studying the variation in chemical, taxonomic, and spectral properties among tropical plant species over large regions. Asner et al. (2009) showed that the natural variation of leaf chemical and spectral properties is genetically driven, which is a discovery of the greatest importance in ecology. Last but not least, Ribeiro da Luz and Crowley (2010) also recently identified thermal infrared spectroscopy as a potential candidate to determine plant biodiversity.

12.4.3 Plant Functional Models

Vegetation provides the foundation for life on Earth through several ecological functions: regulation of climate and water, habitat for animals, supply of food, fiber, and other goods and services. Three striking examples illustrate its fundamental role: i) all life depends ultimately on primary production by plants; ii) there is a good correlation between the species number of vascular plants and the species

Table 12.3 *Linkage of ecophysiological variables with biophysical processes. In italics, the variables directly connected to plant leaves.*

Biophysical processes	Ecophysiological variables
Interception of solar radiation	fAPAR Green LAI Canopy structure
Photosynthesis/Respiration	LAI *Leaf pigment and water content* Leaf anatomy
Decomposition of organic matter	Soil water content and temperature *Leaf C:N ratio*
Evapotranspiration	fCover LAI Albedo Soil water content *Leaf water content*
Plant species/Biodiversity	Topography Soil water content and composition LAI Canopy structure *Leaf chemical content*

number of animals (insects, tetrapods, etc.); and iii), about 25% of the drugs prescribed worldwide come from plants. Physiological processes at different scales like photosynthesis, evapotranspiration, carbon storage, and decomposition of organic matter depend on plant species and age, canopy density, and architecture, but also leaf anatomy and chemical composition (Table 12.3).

Vegetation can be studied at multiple temporal, spatial, and spectral scales. It has been a predominant topic in remote sensing for decades (e.g., Wessman, 1992; Schaepman et al., 2009). As seen earlier, leaf reflectance and transmittance can be considered intermediate products carrying information about foliar biochemistry. When scaling from chloroplast to leaf level, from leaf to canopy level, from canopy to ecosystem level, from ecosystem to the whole Earth level, coupled radiative transfer models can be used (Figure 12.13).

Earth vegetation can be segmented into different biomes depending on climatic variables such as mean annual temperature and annual precipitation (Woodward et al., 2004). How do biomes respond to changes in natural and human-induced forcing such as climate or land use? What is the efficiency of vegetation feedback on climate? To model these complex processes it is necessary to understand vegetation biophysical processes that are partly controlled by ecophysiological variables including leaf area index (LAI), fractional cover (fCover), fraction of absorbed photosynthetically active radiation (fAPAR), and leaf chemical composition (Table 12.3). The latter that can be inferred from hyperspectral data and integrated into ecological (Rautiainen et al., 2010) or soil-vegetation-atmosphere-transfer (SVAT) models (Krinner et al., 2005). SVAT models such as Interactions

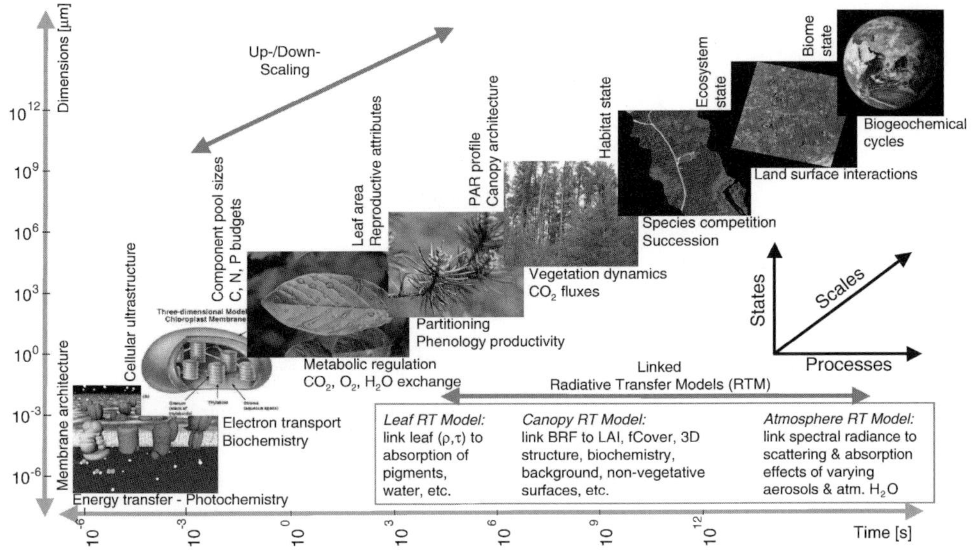

Figure 12.13 Coupled states, processes, and scales ranging from cellular architecture to global biogeochemical cycles. The contribution of linked radiative transfer models in down- and upscaling ranges from leaves to biomes (from Schaepman et al. (2009), *Remote Sensing of Environment*. Copyright © 2009 with permission from Elsevier).

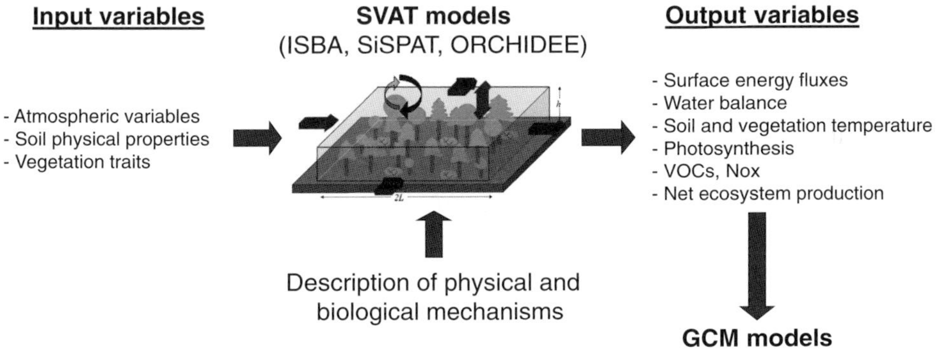

Figure 12.14 Principle of a SVAT model.

between Soil, Biosphere and Atmosphere (ISBA), Simple Soil-Plant-Atmosphere Transfer model (SISPT) model, or ORganizing Carbon and Hydrology In Dynamic EcosystEms (ORCHIDEE) are designed to calculate the flow of energy and mass between the surface and the atmosphere, trace gases fluxes, photosynthesis, and net ecosystem production, among others (Figure 12.14). There processes are closely linked to vegetation biophysical properties that can be determined by remote sensing at different scales. Although maps of LAI, fCover, and fAPAR are now routinely produced from space, the estimation of leaf chemical content is still in its infancy.

Le us take the example of organic matter decomposition, which was in the news in the 1990s. It is affected by the presence of carbon and nitrogen in the soil. When trees lose their leaves, these

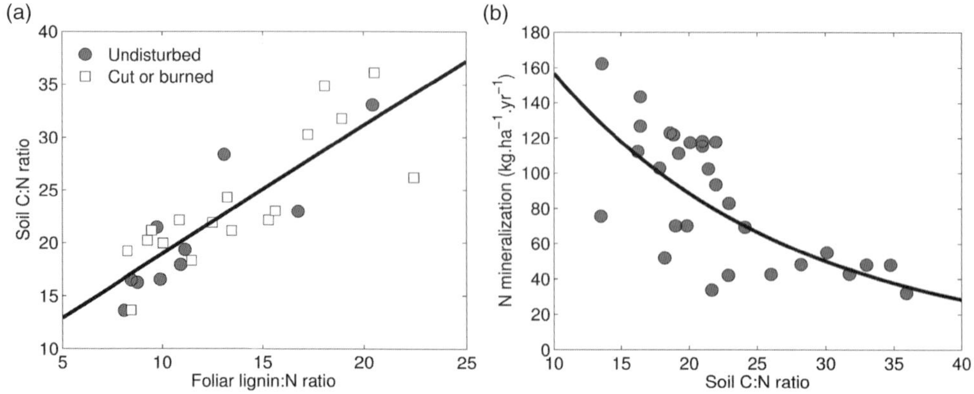

Figure 12.15 (a) Soil C:N ratio (organic plus mineral) in relation to mass-based foliar lignin:N ratio. (b) Soil C:N ratio in relation to net N mineralization (redrawn from Ollinger et al. (2002)).

constituents become part of the organic matter that is decomposed by soil micro-organisms. The C:N ratio that represents the relative proportion of the two elements is a good indicator of the rate of nitrogen mineralization, such as of soil respiration (Figure 12.15). The ratio is higher in Gymnosperms (which typically have thick needle-shape leaves) and decomposition is slower than in Angiosperms (monocots and dicots with laminar leaves). Since the release of carbon to the atmosphere through microbial respiration is an important component of the global carbon cycle, knowledge of C:N at different spatial scales would be particularly valuable to describe the functioning of terrestrial ecosystems (e.g., Wessman et al., 1988b; Martin and Aber, 1997; Ollinger et al., 2002).

The use of remote sensing to drive a SVAT model requires rethinking of traditional modeling approaches and the use of inverse modeling techniques. As seen in Section 12.4, the combined PROSPECT leaf optical properties model and SAIL canopy bidirectional reflectance model, also referred to as PROSAIL, has become one of the most popular radiative transfer tools to analyze hyperspectral data due to its ease of use, general robustness, and consistent validation by lab/field/space experiments over the years (Jacquemoud et al., 2009b). PROSAIL allows description of both the spectral and directional variation of canopy reflectance as a function of leaf biochemistry and canopy architecture. It can be run in direct mode to design and evaluate vegetation indexes, but also directly inverted on multispectral or hyperspectral data to retrieve and map canopy biophysical variables such as leaf pigments, water, and dry matter (e.g., Jacquemoud et al., 1995a; Bacour et al., 2002a; Combal et al., 2002; Zhang et al., 2005; Trombetti et al., 2008; le Maire et al., 2008).

12.4.4 Geobotany

Geobotany is defined as the study of plants that are indicators of their geological environment (e.g., Rose et al., 1980; Raines and Canney, 1980; Ustin et al., 1999). From the very first uses of imaging spectrometry, plants have been identified as potential indicators of the underlying geology or soil. Although plant community distributions are generally defined by climate relationships and ecological history, some species and community distributions are related to edaphic conditions, such as salt marshes and alkaline meadows (Schimper et al., 1903) and serpentine soils (Kruckeberg, 1951). For example, sites that have high concentrations of heavy metals (see Section 7.1), such as mine waste or mill tailings for lead, copper, zinc, nickel, and other metals, generally develop unique floras because conditions are toxic to many species (Antonovics et al., 1971). Species that are tolerant of one metal or nutrient condition are not necessarily tolerant of others. Thus, indicator species can be developed for certain conditions. For plants to act as indicators, the environment must be deficient in an essential nutrient (e.g., serpentine soils are deficient in calcium relative to magnesium) or contain minerals and nutrients in excess that cause injury or mortality. In remote sensing applications, the symptoms caused by the deficiency or excess must produce changes in plant growth or conditions that can be observed by optical sensors. When long-term, multi-generational impacts occur, ecological succession produces changes in the canopy density, and composition of the plant community. Therefore, under these circumstances it is anomalous species distributions or low leaf area that is the geobotanical signal. Short-term changes, due to pollution, drought, weather, or other temporary environmental stressors, often cause a change in leaf physiology like chlorosis or other pigment changes (increased anthocyanins, carotenoids, or change in chlorophyll *a:b* ratio). Reduced water content or increased presence of leaf dry biomass are observed from optical sensors. Horler et al. (1980) and Chang and Collins (1983) demonstrated experimentally that chlorophyll content and reflectance change when plants are grown in soils with heavy metals, which were reviewed by Horler and Barber (1981). These early methods generally could not identify specific stressors but in recent years some more specific applications using imaging spectrometry have been developed, for example, detection of oil spills (Li et al., 2005c), metals (Wang et al., 2010), cadmium and vanadium (Rosso et al., 2005), nuclear contamination (Davids and Tyler, 2003), TNT (Naumann et al., 2010), or micronutrient deficiency (Meggio et al., 2010).

12.5 Color Perception by Animals

While the coevolution of insects and flowers has been long studied, leaf–animal interactions are much less well known. Many biologists have interpreted some leaf characters, such as specific chemical, trichomes, shape, and color to have a detrimental effect on herbivores, indicating that their primary function might be to defend plants against them. Others believe that these characters serve ecological and physiological functions (Rausher, 1992). It is extremely difficult to identify and quantify evolutionary forces in nature, because direct proof is often lacking. However, progress has been made so far in identification of such relationships between animals and plant leaves.

12.5.1 Insects

Are some herbivore insects able to discriminate among plant species on the basis of one or more leaf color attributes such as hue, saturation, or intensity (see Section 5.3)? Our knowledge in this domain is still sketchy (Prokopy and Owens, 1983). The cabbage root fly (*Delia radicum*) is a serious pest affecting most cabbage (*Brassica oleracea*) plants and related species. To study the visual detection of plants by these insects, Prokopy et al. (1983) made artificial leaves having spectral reflectance properties equivalent to those of real leaves in the visible and showed that leaf color is used by flies to help landing. Red foliage seems to be less detectable (or less attractive to) than green foliage by herbivore insects, the majority of which apparently lack a red receptor. Similarly, Kelber (1999a) shows that females of *Papilio aegeus* preferred to oviposit on substrata that looks green to humans.

As seen in Section 5.1.2, reflected light by waxy biological materials such as plant leaves or petals of flowers is partially polarized, and the degree of polarization depends on the nature of the surface. Is this signal useful to insects? In complex environments such as the tropical forest understory, light polarization may provide additional information about surface orientation for oviposition. For instance, some insects have two kinds of receptors with different microvillar orientations: the first are sensitive to color, such as wavelength and light intensity, the second to the angle and degree of polarization. This may help them discriminate between shiny and matt leaves before landing. Kelber (1999b) and Horváth et al. (2002) studied the ability of *Papilio* butterflies to discern both signals separately and demonstrated that they could perceive false colors. Hegedüs and Horvath (2004a, 2004b) tested and invalidated that hypothesis, showing that polarization sensitivity was even disadvantageous for the identification of a preferred surface orientation.

The leaves of many temperate deciduous trees change color in autumn. This phenomenon is due to the degradation of chlorophylls which, in non-senescent leaves, mask the presence of carotenoids and flavonoids, and to the synthesis of new pigments. In the past decade, several hypotheses have emerged to explain the evolution of autumn colors (e.g., Archetti et al., 2009c; Jolivet, 2010). Evolutionary biologists recently hypothesized that bright colors of certain tree species could be an adaptive defense signal directed toward potential parasites (e.g., Archetti, 2000, 2009a, 2009b; Hamilton and Brown, 2001; Hagen et al., 2003; Archetti and Brown, 2004; Manetas, 2006; Lev-Yadun and Gould, 2007, 2009; Sinkkonen et al., 2012). This theory was labeled as "the coevolution theory" or "leaf signal theory." It is limited to leaf senescence in spite of the fact that coevolution between insects and plants has also been suggested as an explanation for the occurrence of anthocyanins in young unfolding leaves (Schaefer and Wilkinson, 2004; Karageorgou et al., 2008).

Aphids, a group of small sap-sucking insects that have color-sensitive vision, are a good illustration of this coevolution theory. At the approach of autumn, they migrate to their winter hosts, the trees, and lay eggs causing severe damage to the vegetation. Vigorous trees seem to produce brighter colors than weak ones, acting as a deterrent for these insects which prefer green or dull colored leaves. Archetti and Leather (2005) found a negative correlation between the percentage of red/yellow leaves and the number of aphid species, supporting the basic prediction of the coevolution theory (Figure 12.16). The more colorful trees in autumn suffer less insect damage the following spring.

However, more recent observations tend to invalidate some of those results. Holopainen et al. (2009) demonstrated that aphids prefer yellow leaves to green leaves, suggesting that yellow leaves are not a warning signal for them. This is confirmed by their extreme abundance on silver birch (*Betula pendula*) in northern latitudes. Moreover, aphids that lack a red receptor may not be able to distinguish between red and green leaves, contrary to humans or other animals (see Section 12.5.3). Döring et al. (2009) performed a catch experiment in the laboratory with painted water traps and autumnal leaves collected in the UK and Germany. They could build a color preference model for winged aphids in autumn (Figure 12.17). Archetti et al. (2009c), however, stress that a test with artificial targets cannot easily be extrapolated to leaf colors.

12.5.2 Other Animals

Burns et al. (2009) investigated whether fruit color diversity was higher in tropical regions, where birds are the primary consumers of these fruits, and therefore the primary seed dispersers, compared to temperate regions. They measured reflectance spectra of both fruits and leaves in the background,

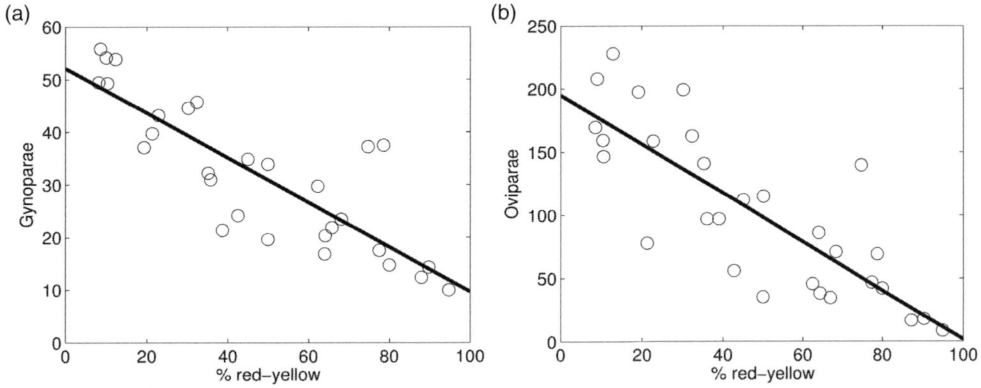

Figure 12.16 Correlations between the mean number of (a) gynoparae (winged morphs that fly to the overwintering host) and (b) oviparae (sexual morphs that mate with the males and produce the overwintering eggs) aphids per 100 leaves and the mean degree of red-yellow (redrawn from Archetti and Leaver (2005)).

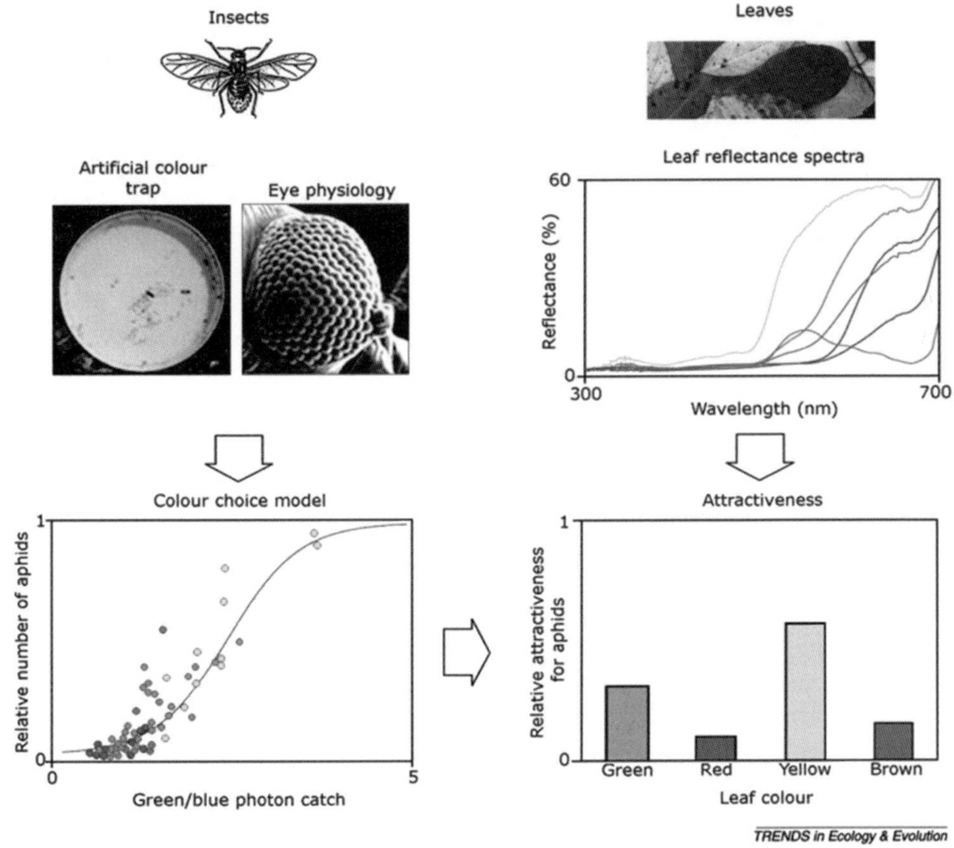

Figure 12.17 Measuring color preference in insects. An objective color measure (the leaf reflectance spectra) of leaves is integrated with a color choice model for aphids (derived from artificial color trap measurements and the physiology of aphid color vision) to obtain a measure of the attractiveness for aphids of leaf colors according to color names based on human vision (from Archetti et al. (2009c), *Trends in Ecology & Evolution*. Copyright © 2009 with permission from Elsevier). (A black and white version of this figure will appear in some formats. For the color version, please refer to the plate section.)

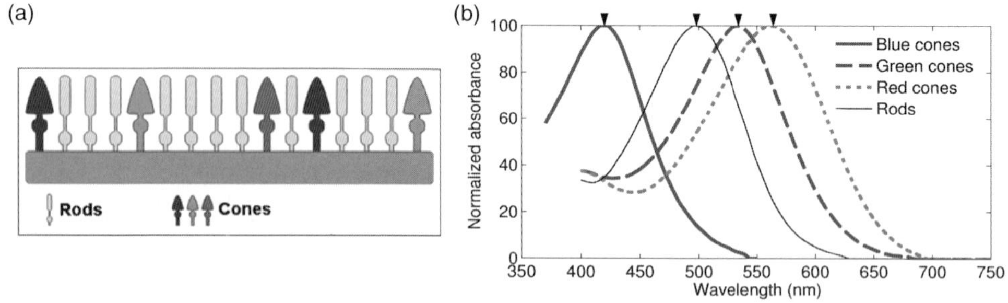

Figure 12.18 (a) Schematic view of the human retina including rod and cone light receptors. (b) Spectral absorption curves of the blue-sensitive cones (420 nm), green-sensitive cones (534 nm), red-sensitive cones (564 nm), and rods (498 nm). Both color and intensity are most sensitive in the green part of the visible spectrum (redrawn from Bowmaker and Dartnall (1980)).

and analyzed their color according to the visual system of birds. In contrast to their predictions, fruit colors in the tropics, an area generally characterized by higher bird diversity, are neither more diverse nor more conspicuous than in temperate regions. Moreover, they found that fruit color was positively correlated with leaf color across all regions. They hypothesized that both color patterns might be evolutionarily linked.

12.5.3 Humans

Generally, color vision is prevalent in diurnally active animals rather than nocturnal ones. The human eye is exceptionally sensitive to different shades of green leaves due to our trichromatic vision system (Figure 12.18a). We discriminate leaf color by combining information from the rod and cone cells in the retina. Rod cells only sense intensity of light therefore are most sensitive for low-light vision, while cone cells discriminate colors but function best in bright light. The three color cone cells are differentially sensitive to visible light between 400 nm and 700 nm (Figure 12.18b). Human cone cells have varying sensitivity across a broad range of wavelengths, with peak sensitivity in the blue, yellow-green, and red regions. Similarly, greatest sensitivity to luminosity (brightness) is the green part of the spectrum. Humans share an advanced trichromatic vision of colors with Old World primates and some New World primates, often females, contrary to most New World primates. The individualization of the red cone from the green cone occurred about 32 MYr ago, when the African and American continents started to diverge. Previously, this selective advantage was explained by the necessity of finding ripe colored fruits amongst foliage, and the fruits themselves have evolved to be salient to primates and so secure dissemination of their seeds (e.g., Osorio and Bossomaier, 1992; Osorio and Vorobyev, 1996; Regan et al., 2001). Seen from this point of view, it can be considered a coevolution process. By studying the behavior of primates in the Kibale National Park, Uganda, Dominy and Lucas (2001) hypothesize that the color of plant leaves may be more important. Although the color of young leaves in temperate regions is nearly always green, a substantial proportion of those in tropical forests are red: 50 ± 62% in the Old World and 18 ± 36% in the New World. They often correspond to young leaves that are tender, having higher protein content and lower fiber content than mature leaves. When food becomes scarce, arboreal primates preferentially forage for these developing leaves.

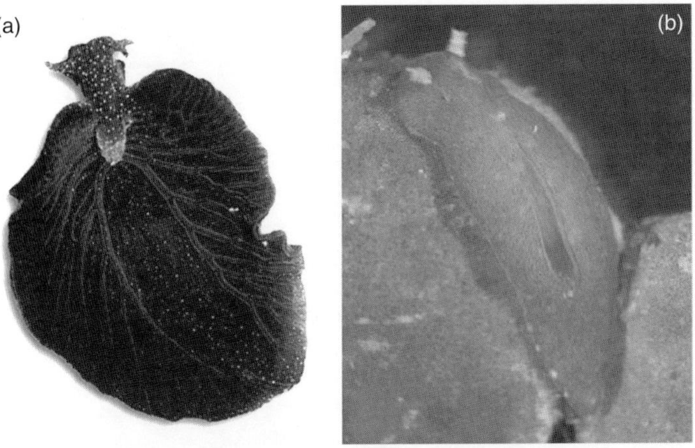

Figure 12.19 (a) Sea slug *Elysia chlorotica* about 6 cm in length (from Rumpho et al. (2000), *Plant Physiology*. Copyright © 2000 with permission from The American Society of Plant Biologists), and (b) *Elysia pusilla* about 3 cm in length, showing that chloroplasts are located throughout the digestive system (from Wägele and Klussmann-Kolb (2005), *Frontiers in Zoology*. Copyright © 2005 with permission from The Deutsche Zoologische Gesellschaft). The color depends on the algae being ingested, the age and condition of the sea slug.

12.6 Autotrophic Endosymbiosis: Animals That Photosynthesize

A small number of insects, sea slugs, and animals can capture chloroplasts from algae and can keep them intact and functioning within their bodies for extended periods in a process termed kleptoplasty (Waugh and Clark, 1986; Serôdio et al., 2014). Kleptoplasty differs from most symbiotic associations of intact free-living organisms, and defines instead a pseudo-symbiotic relationship where the plastids (mostly chloroplasts) from algae are captured and are retained in a functional state for periods of days to many months (Rumpho et al., 2000). It is now known in about 75 sacoglossan species (the sap-sucking sea slugs), some of which have long-term retention of the chloroplasts include *Elysia chlorotica*, *Elysia tuca*, *Elysia timida*, *Elysia crispata*, *Elysia clarki*, *Plakobranchus ocellatus*, and *Costasiella ocellifera*, which are multicellular eukaryotic marine mollusks in the opisthobranch of gastropods (Figure 12.19). The sea slug *Elysia chloritica* has been well studied for its ability to utilize photosynthesis from its captured chloroplasts. It harvests chloroplasts from the marine alga *Vaucheria litorea*, which it captures by siphoning off the cell constituents and retaining the chloroplasts in its digestive diverticula. The diverticula extend throughout the slug's body and are located one cell layer below the epidermis. The complete explanation for how the chloroplasts are kept alive is not yet known but at least some parts of the algal genome that support photosynthesis have been transferred to the slug (Rumpho et al., 2008; de Vries et al., 2014a,b). A similar mechanism has been reported for *Elysia timida*.

While not common, endosymbiont plastid relationships are known in diverse organisms stretching across the phylogenetic tree. Phyla include *Porifera* (marine and freshwater sponges), *Cnidaria* (corals, sea anemones, and freshwater hydra), *Acoelomorpha* (marine turbellaria), *Mollusca* (giant clams and nudibranchs), and *Chordata* (marine ascidians) (e.g., Trench, 1979; Brümmer et al., 2008; Venn et al., 2008; Serôdio et al., 2014). The unsegmented worms, plathelminth species, including *Convoluta roscoffensis* is in association with the algae *Tetraselmis convolutae* (Nissen et al., 2015).

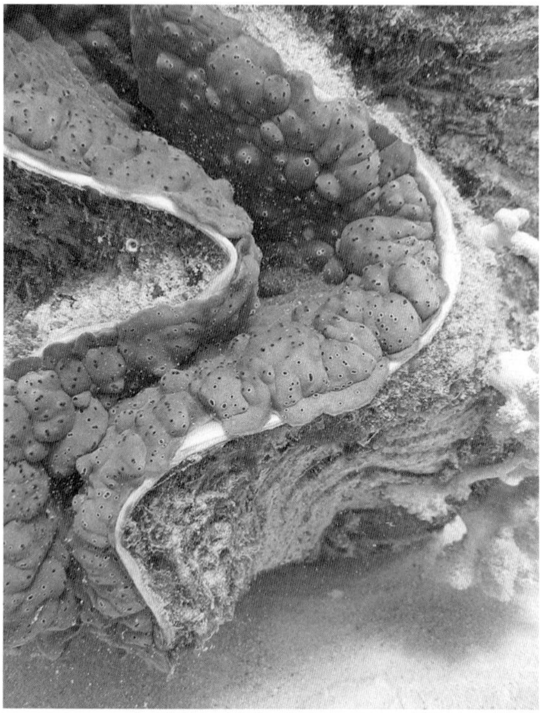

Figure 12.20 The mantle of the Giant clam, *Tridacna gigas*, has the color of the algae it has ingested. It is the largest of all mollusks and *T. gigas* is the largest of nine species of giant clam in the Great Barrier Reef of Australia (Penny and Willan, 2014). The iridescent blue eyespots aid light penetration farther into the mantle. (Photographed by Jan Derk in November 2002 on the Great Barrier Reef, Australia, and released into the public domain by its author, at English Wikipedia.)

Giant clams might not appear to be a good choice for an association with photosynthetic algae since their shells would prevent light penetration; however, the mantle is extruded and chloroplasts are concentrated there. In a study of three giant clam species in the genus *Tridacna* from Indonesia, de Boer et al. (2012) found that the clams had multiple symbionts from between and within dinoflagellate algal clades in the genus *Symbiodinium* (Lee et al., 2015), a common symbiont partner in coral reefs, anemones, jellyfish, nudibranch sea slugs, and others (Figure 12.20). de Boer et al. (2012) suggest that the different algal clades of *Symbiodinium* may provide differences in physiological tolerances (e.g., temperature) that would facilitate adaptation to variable environments. Another study of active endosymbiotic algal photosynthesis in oligotrophic seawater may be related to algal dependence on host-derived nitrogen and phosphorus (Tanaka et al., 2015). Symbiodinium algal types have specific association in the corals *Acropora downingi*, *Cyphastrea microphthalma*, and *Playgyra daedalea*. Smith et al. (2017) report that each coral species is associated with a specific type of algal symbiont. Some small bivalves, conchs, and snails have algae that survive on the low levels of light that penetrate the shell.

The yellow-spotted salamanders and some frogs are the only vertebrates with known endosymbiotic algae (Kim et al., 2014). Females of *Ambystoma maculatum* (spotted salamander) and *A. Gracile* (Northwestern salamander), and ranid frogs *Lithobates sylvatica* (wood frog) and *L. aurora*

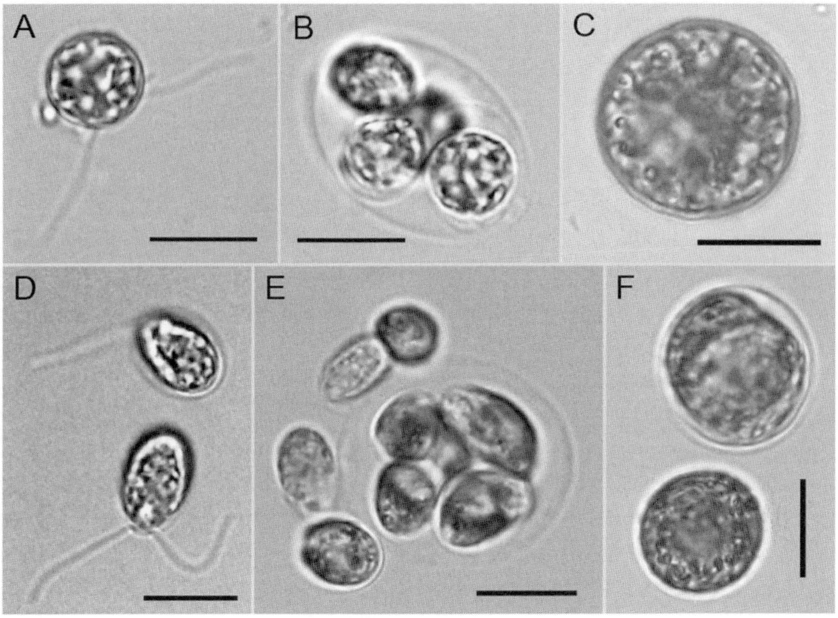

Figure 12.21 Eggs of *Ambystoma maculatum* with chloroplasts from *Oophilia*, illustrating morphological heterogeneity of cell types. Scale bars = 10 μm (from Kim et al. (2014), reproduced from *PLoS ONE*).

(red-legged frog) store algal cells in their egg capsules. The algae represent different strains of the free-living clamydomonad green algae (*Oophila amblystomatis*), with different strains forming partnerships with the amphibians (Figure 12.21).

12.7 Camouflage

Camouflage is used extensively in the animal realm to help in the struggle for survival. Predators use camouflage in stalking their prey, who in turn strive, in many ways, to escape detection (Overton, 1969). Fewer examples are known for plants. One should not confuse camouflage (or crypsis), the ability of an organism to blend in with its environment, with mimicry, the process where one species resembles another.

12.7.1 Relationships between Leaf Color and Animals

Many animal species have evolved camouflage defenses to dissuade their predators from attacking them. The most famous example is given by leaf insects, a group of tropical insects that belong to the order *Phasmatodea*, which are very similar in appearance and size to a leaf (Figure 12.22). They do this so ingeniously that they are practically invisible among tree foliage and predators cannot distinguish them from real leaves. In some species the edge of the leaf insect's body even has the appearance of bite marks. When leaf insects do move, they are even more cryptic in that they gently sway their whole body from side to side, looking just like a leaf blowing in the wind. It means that

Figure 12.22 Insect leaf mimicry. (a) *Tropidoderus childrenii* among eucalyptus leaves (© David Midgley, www.flickr.com/photos/petrichor/). (b) *Phyllium bioculatum* (http://en.wikipedia.org/wiki/Phyllium_bioculatum). (c) *Choeradodis*, commonly named leaf Mantis, in the Osa Peninsula (Costa Rica) (http://en.wikipedia.org/wiki/Mantis).

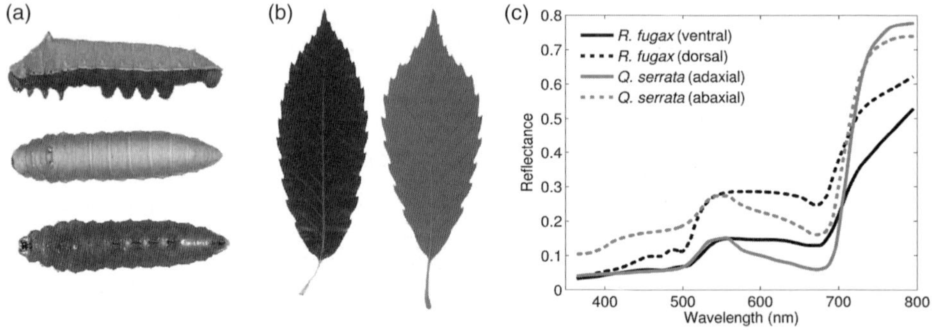

Figure 12.23 (a) Lateral (upper), dorsal (middle), and ventral (lower) views of *Rhodinia fugax* larvae. (b) Adaxial and abaxial views of bao li (*Quercus serrata*) leaves, its host plant. (c) Spectral reflectance of *R. fugax* larval integument (ventral/dorsal) and *Q. serrata* leaf (abaxial/adaxial) (redrawn from Saito (2001)).

they can keep up their disguise even when walking. And that thanks to their perfect camouflage, they remain active during the day.

Another example is given by the caterpillars of phytophagous insects. They are frequently green, a coloration that presumably serves as a protective camouflage against predators. Figure 12.23 shows the larva of a tropical moth, *Rhodinia fugax*. It is light yellowish-green on its dorsal surface and dark green on its ventral surface with a lateral line in-between. One of the host plants for the larvae is the bao li (*Quercus serrata*), a rare deciduous tree native to Eastern Asia. The adaxial surface of its leaves is dark green, whereas the abaxial surface is light green. This resemblance between *R. fugax* and *Q. serrata* was studied by Saito (2001), who measured the spectral reflectance of the dorsal/ventral larva integuments and of the adaxial/abaxial leaf surfaces. All reflectance spectra display a small peak from 540 nm to 669 nm and the reflectance of the upper surface of both the larvae and the leaves was less than that of the lower surface (Figure 12.23). The amount of blue biliproteins in the hemolymph and integument of the larva explains this difference in color between its ventral and dorsal surfaces, and is controlled by light intensity. Camouflage is not restricted to the visible region, and may extend to the near-infrared. An unusual example is provided by green leaf-sitting neotropical frogs (Schwalm et al., 1977). The reflectance of *Centrolenella fleischmanni*, the northern glass frog, increases from 700 nm to 900 nm so that it blends with foliage both in the visible and near-infrared; on the contrary, the reflectance of *Hyla cinerea*, the American green tree frog, remains low and is distinguished from the leaf surface in an infrared photograph.

If animals often use colors to hide from predators or advertise defenses, until now there has been little evidence of color-based defenses in plants. For instance, Givnish (1990) hypothesized that leaf mottling may serve to camouflage the foliage of certain groups of forest herbs, by disrupting their outline as perceived by color-blind vertebrate herbivores (e.g., deer) in the sun-dappled understory. Lev-Yadun et al. (2004) suggested that plant coloration undermines herbivorous insect camouflage. A leaf that has two different colors on the upper and lower sides is more likely to avoid herbivory than an all-green leaf. These authors claimed that the enormous variation in coloration of leaves, petioles, and stems as well as of flowers and fruits undermine the camouflage of invertebrate herbivores, especially insects. Lev-Yadun (2006) saw a defensive function in the white color of dune plants, mainly due their trichomes. Beside their protection role against radiation (see Section 5.1.3), sticky glandular trichomes may defend plants from insect herbivory; white coloration of leaves and branches may also mimic fungal infestation. Fadzly et al. (2009) discovered that trees in New Zealand evolved camouflage defenses to protect themselves from predators, providing the first evidence of such a strategy in plant life. They studied the leaves of lancewood (*Pseudopanax crassifolius*), a tree native to New Zealand and well adapted to its natural habitat. *P. crassifolius* is a heteroblastic species, that is, its lance-like foliage dramatically changes as the tree grows. This species goes through several strange color transitions during the process from germination to maturity. In fact, young trees are so different from adults that early botanists believed they were different species. The reason for these curious morphological changes is thought to be a defense strategy from the moa, a giant flightless bird which was hunted to extinction 750 years ago. Seedlings produce small narrow leaves that are mottled and similar in appearance to leaf litter on the forest floor, which would have made them difficult for a moa to see. Once above moa height, approximately 3 m for the largest known fossil skeleton, saplings produce larger, more elongated leaves that are ordinary in size, shape, and color, turning into a "regular" tree. To prove this, Fadzly et al. (2009) compared *P. crassifolius* leaves to samples from Chatham Island lancewood (*Pseudopanax chathamicus*), a similar species of tree that grows in the Chatham Islands, over 800 km east of New Zealand. In the absence of moa, *P. chathamicus* displays less morphological changes between adults and juveniles. Niu et al. (2014) investigated the coloration of *Corydalis benecincta*, an alpine plant growing in alpine screes and displaying remarkably dimorphic leaf colors; they can be gray or green. They showed that the leaves of the two morphs had similar chlorophyll content, but that the anthocyanin content of gray leaves was much higher than that of green leaves (~ 10-fold). The net photosynthetic rate of both leaves, however, was similar. Recently, the same authors Niu et al. (2017) quantified background matching between leaves of *Corydalis hemidicentra*, another alpine plant whose leaves match the color of rocks where it grows (Figure 12.24). They showed that plants exhibited remarkable color divergence between populations and that leaf color varied because of a different quantitative combination of chlorophyll and anthocyanin, plus different air spaces.

As noted by Lee (2011), variegation, a common character in plants growing in forest understory environments (see Section 5.3), may also be a strategy to avoid being eaten by insects. Such leaves

Figure 12.24 The leaves from four populations of *Corydalis hemidicentra* covering the main distribution area of this species: from the left to the right, Pujinlangba (PJ), Caishichang (CSC), Tianbaoshan (TBS), and Huangguanfeng (HGF) (from Niu et al. (2017), *Proceedings of the Royal Society of London. Series B*. Copyright © 2017 with permission from The Royal Society). (A black and white version of this figure will appear in some formats. For the color version, please refer to the plate section.)

are perceived as not leaf-like or as already containing herbivorous insects. For instance, Smith (1986) studied the juvenile leaves of a subcanopy liana, Espino hueco (*Byttneria aculeate*), and identified two distinct forms: plain green leaves that predominate in the forest and variegated leaves that predominate in open sites. Within a given habitat, the latter were less attacked by leaf miners. A similar mechanism was observed by Soltau et al. (2009) with the leaves of elephant ear (*Caladium steudneriifolium*) plants, an understory species of the Ecuador submontane rainforest. Burns (2010) assumes that, since cryptic coloration is widespread in animals, it might also be common in plants and he notes that we are far from determining if this is true (see also Lev-Yadun, 2003). Our understanding of color-based defenses in plants is in its infancy.

Although given much consideration in the animal literature, little work has been performed regarding aposematic coloration in plants. Some plants are thought to employ visual aposematism to warn herbivores. The warning signal may take the form of odors (unpalatable chemicals), physical defenses (prickled leaves or thorns), or conspicuous colors. In the latter case, herbivores associate color or a color pattern with unpleasantness. Simcha Lev-Yadun from the University of Haifa published a series of articles about this phenomenon (e.g., Lev-Yadun, 2001, 2009; Lev-Yadun and Ne'eman, 2004; Lev-Yadun et al., 2009; Lev-Yadun and Keasar, 2012).

Last but not least, some leaves can imitate (size, shape, color, orientation, and even the vein patterns) the leaves of one or, more exceptionally, several other plant species. A remarkable feature of many Australian types of mistletoe is their apparent close vegetative similarity to their usual hosts. This phenotypic plasticity seems to be more frequent in Australia than anywhere else. One of the most amazing examples of host resemblance is *Dendrophthoe homoplastica* that grows on Shirley's silver-leafed ironbark (*Eucalyptus shirleyi*) trees. Native to temperate forests of central and southern Chile and Argentina, pilpil (*Boquila trifoliolata*) is the first plant shown to imitate several hosts that it climbs upon (Gianoli and Carrasco-Urra, 2014; see also the dispatch by Pannell, 2014). It is a rare quality, known as a mimetic polymorphism, which was previously observed only in butterflies. According to these researchers, the deceit serves as a defense against plant-eating herbivores like weevils and leaf beetles.

12.7.2 Applications for Defense

There is little historical information on the use of camouflage by ancient man, but it was almost certainly used in hunting, by both the hunter and the hunted. Later, it became a well-developed art, mainly for military purposes. Camouflage techniques using natural foliage can be very effective but foliage must be replaced periodically since noticeable color changes may occur with leaf senescence within a relatively short period of time (Henderson et al., 1974). Consequently, more stable camouflage fabric simulating foliage has long been used. An object is visible in an image because its reflectance is different from that of the surrounding background. It may appear darker or brighter. Different hues can be employed to reduce this contrast. Those found in most camouflage patterns used in temperate and tropical forests are green, olive, khaki, brown, and black. With the development of near-infrared surveillance technology during the Second World War it became necessary to develop a new generation of camouflage materials that attempt to minimize the contrast between objects and their environment over the near-infrared region in addition to the visible region (Burkinshaw et al., 1996). A classical approach consists in dying cellulosic or polyethylene terephthalate, commonly abbreviated PET, fabrics to imitate the reflectance of a green leaf in the VIS-NIR (Hui and Jianchun, 2007; Zhang and Zhang, 2008). The choice of the dyes, the absorption peaks of which occur at different wavelengths, their proportion, and concentration are critical factors in the manufacture of the fabric. For example, 1–2% of Disperse Blue 56 matches well with the actual leaf red-edge. Other vat dyes were applied to cotton fabrics to provide reflectance spectra similar to a plant leaf (Goudarzi et al., 2013, 2014). The emergence of multi-spectral imaging cameras that operate across the whole optical domain, from the visible to the longwave infrared, questions conventional camouflage materials that work well in the visible domain but that are ineffective in the others. This is illustrated by Figure 12.25 that shows a VIS/NIR/SWIR image where a MARPAT (MARine PATtern, digital camouflage designed by the US Marines for their uniforms) hat and a leaf are side by side. The leaf is clearly distinguished from the hat because of its significantly lower reflectance in the SWIR spectrum due to a strong water absorption peak at 1.4 μm (Leong, 2008).

New multilayer composite materials imitating the structure of plant leaves and petals have recently been designed. Yang et al. (2010, 2011a,b) used PROSPECT to analyze the determinism of leaf spectral reflectance and build a bionic leaf. Then they hypothesized that a material having the same biophysical properties as a plant leaf in terms of anatomical structure, chlorophyll, and water content should display similar reflectance properties. They built a proper structure made of three layers, as shown in Figure 12.26.

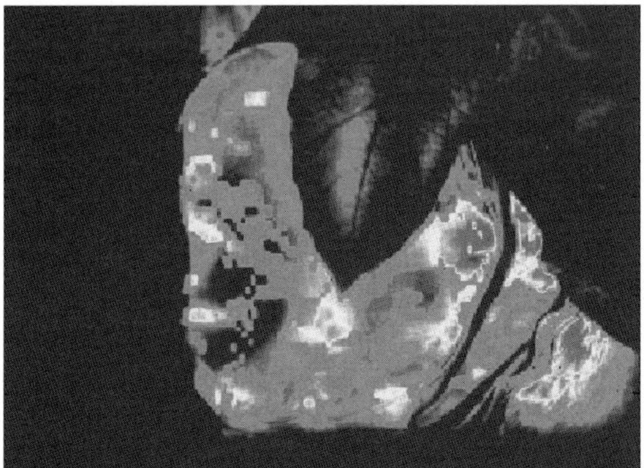

Figure 12.25 SWIR image showing reflectance contrast between foliage and non-foliage objects (from Leong (2008), *POINTER*. Copyright © 2008 with permission from the Singapore Government). (A black and white version of this figure will appear in some formats. For the color version, please refer to the plate section.)

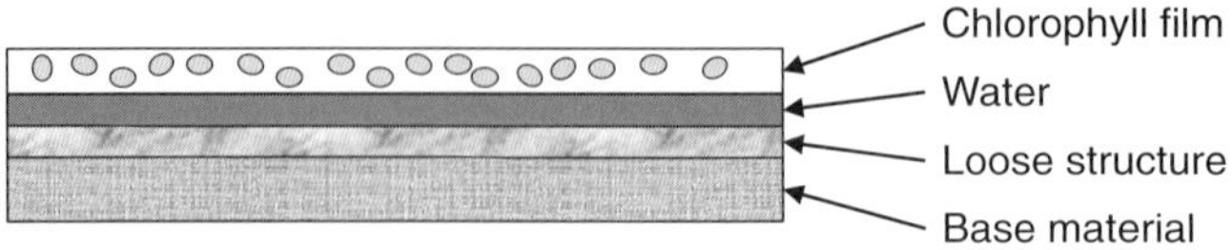

Figure 12.26 Structure of bionic composite material (redrawn from Yang et al. (2010)).

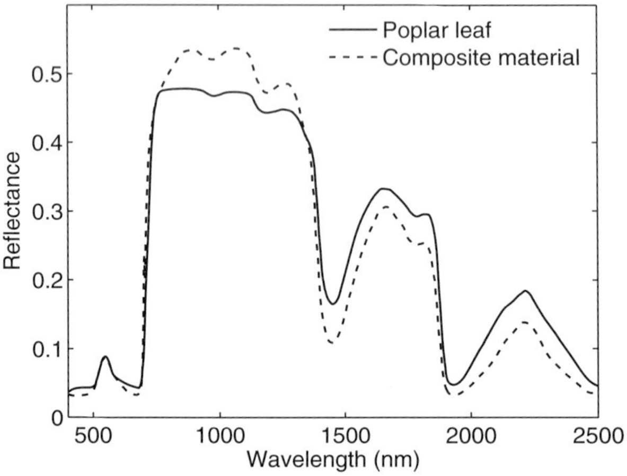

Figure 12.27 Reflectance spectra of poplar leaves and bionic composite material (redrawn from Yang et al. (2010)).

The upper layer is a chlorophyll polyvinyl alcohol film, which improves the photostability of chlorophyll by isolating it from oxygen; the middle layer is a sealed bag of polyvinylidene chloride containing pure liquid water to prevent evaporation; and the lower layer is a piece of paper that simulates the spongy mesophyll. The three layers are bound together using a polyurethane film to form the bionic composite material. The reflectance spectrum of the bionic leaf is very similar to that of a poplar (*Populus* sp.) leaf, even after three months of exposure to the rays of the Sun (Figure 12.27). Note that chlorophyll and water are not distributed into the porous structure like the chloroplasts and vacuoles in leaves. Liu et al. (2008b, 2009b) synthesized a more sophisticated biomimetic leaf model made of polyurethane foam containing plant pigments and water. They also compared its reflectance spectrum to that of an Oriental plane (*Platanus orientalis*) leaf and showed a good agreement, except around 550 nm where the reflectance of the bionic leaf was too high. The method may be refined by entrapping chlorophyll molecules in nanoporous silica gel (Manna et al., 2009). Qin et al. (2012) developed novel hydrophilic poly(ureaformaldehyde) (PUF) microcapsules that can reproduce reflectance spectra of plant leaves at different water contents from 300 nm to 2500 nm.

In order to maintain the advantage and effective intelligence on the battlefield, technologies in counter-camouflage, concealment, and deception (CC&D) need to permanently adapt by developing new systems that mimic leaf optical properties in extended wavelength domains, surface temperature, but also physiological processes and energy budget. Ye et al. (2013, 2015) and Yuan et al. (2014b) recently designed a bionic leaf, the upper side of which has surface roughness and emissivity properties similar to those of a real leaf and which can regulate its water content. Therefore, it can imitate the thermal infrared optical properties of vegetation. Based on leaf thermophysical properties, these authors

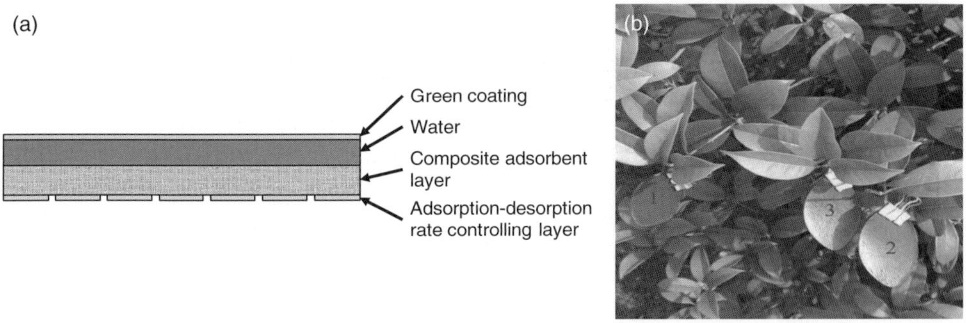

Figure 12.28 (a) Structure of the bionic leaf and heat exchange with its environment. (b) Field experiment (redrawn from Yuan et al. (2014b)).

imagined a green coating, a water holding layer, and an adsorbent layer made of carbon composite materials (Figure 12.28). The latter is the key material, containing calcium chloride ($CaCl_2$), a hygroscopic salt that absorbs water vapor during the night and releases it during the day. The bionic leaf has been successfully tested in the field using a thermography camera. The average radiative temperature difference between the bionic and natural leaves is lower than 1.0°C and we can hardly distinguish them.

12.8 Astrobiology

The search for extraterrestrial life has aroused much speculation throughout the centuries. Since we do not know anything about life-forms in the Universe, there is a great temptation to assume that it developed in similar forms as on Earth, for instance, like vegetation, which uses visible light as an external energy source. As seen in Chapter 2, the evolution of plants since the Ordovician, around 450 million years ago, has resulted in levels of complexity that defy imagination. In the mid-19th century, Darwin (1862) was already amazed by the extraordinary variability of vegetable kingdom:

> The more I study nature, the more I become impressed (.) that the contrivances and beautiful adaptations (acquired through natural selection) transcend in an incomparable degree (those) which the most fertile imagination of the most imaginative man could suggest with unlimited time at his disposal.

What would plants look like on another world? That is an even more difficult question. Specific atmospheric and environmental conditions on alien planets would lead to the evolution of life-forms with specific features, different from those found on Earth. The plants of Figure 12.29 are based on the work of scientists from NASA, SETI (Search for Extraterrestrial Intelligence), and leading British universities. Paleobotanists imagined that on a planet with low gravity and a thick atmosphere, vegetation could grow to the size of hot air balloons. The hydrogen gas produced by bacteria would allow some of these plants to float several hundreds of meters above the surface. Pagoda trees and stinger fans are other kinds of proposed extraterrestrial plants. Professor Jodie S. Holt, a plant physiologist from the University of California, Riverside, provided botanical concepts like these that were incorporated into the 2009 movie, *Avatar*. She gave scientific names and descriptions to 55 of the most bizarre plants in the film, which can be consulted in an on-line encyclopedia called Pandorapedia (www.pandorapedia.com/). Such plants that are a figment of scientists' imagination remain a mix of botany and science fiction. "The plants are fake, but the science is real," Professor Holt says.

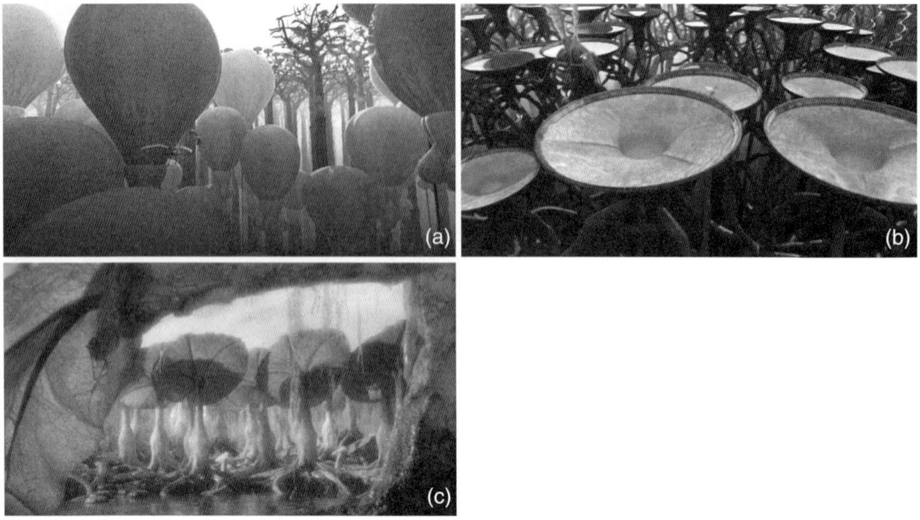

Figure 12.29 Computer artwork depicting a scientist's impression of alien plant forms called: (a) Balloon plants, 5 m in diameter and 80 m height; (b) Pagoda plants; and (c) Stinger fans (©2005 Big Wave Productions/Science Source. All rights reserved). (A black and white version of this figure will appear in some formats. For the color version, please refer to the plate section.)

Astrobotany is a branch of botany that investigates the possibility that plants grow on other planets. It is an integral component part of astrobiology, the science of extraterrestrial life. It began to develop roughly in 1945, in connection with the question of whether or not there was vegetation on Mars (Katterfeld and Suslov, 1969). Russian astrophysicist Gavriil A. Tikhov (1875–1960) was the founder of this new science (e.g., Omarov and Tashenov, 2005; Harvey, 2007; Briot, 2013a, 2013b). As a firm opponent of geocentrism, he believed that tiny organisms as well as plants existed on Mars, Venus, and possibly other planets. He insisted that life on other worlds might be very different from the forms with which we are familiar here on Earth and could appear under very different conditions, a view now commonly held by many scientists. Tikhov (1960, 1966) supposed that vegetation on Mars was made of blue-purple organisms like lichens or mosses growing in dense clumps or mats on a red soil (Figure 12.30).

Spectroscopy was the main tool of astrobotany, which consisted of comparing the reflectance spectra of plants that grow in Mars-like regions and of various segments of the Mars surface. During the 1909, 1918, and 1920 oppositions of Mars, when the distance from Earth to Mars was the closest, Tikhov carried out a large number of observations at the Pulkovo Observatory near Saint Petersburg to obtain spectra of the dark "sea" and light "desert" areas and perform a detailed comparison of their colors. Canadian astronomer Peter Millman (1906–1990) did the same in 1937 at the David Dunlap Observatory outside Toronto. Even though they both found that the dark areas appeared to be relatively greener than the light areas, the color did not match the reflection spectrum of chlorophyll observed on fresh green leaves. To solve this puzzle, Tikhov organized several scientific expeditions to regions where there are extreme climatic conditions, such as the high Pamir Mountains in the subarctic tundra. He studied how plants survive in low-temperature, high-altitude, and oxygen deficient regimes and measured their optical properties (Figure 12.31). He assumed that vegetation growing in the cold climatic conditions prevalent on Mars had a much greater need of solar heat than

Figure 12.30 Martian landscape as imagined by Tikhov: a low blue vegetation covering a red soil. A sandstorm, frequent on this planet, is rising on the horizon (from Tikhov (1960)). (A black and white version of this figure will appear in some formats. For the color version, please refer to the plate section.)

Figure 12.31 Tikhov measuring plant spectra (source: http://planlab.tripod.com/Tikhov1.htm).

vegetation on Earth. Therefore the main chlorophyll absorption bands become weaker and may even disappear completely in the sub-arctic regions. These plants could consequently exhibit unusual reflectance spectra consistent with Mars observations (Dadykin and Bedenko, 1960). This research became so important that in 1947 Tikhov established the world's first Department of Astrobotany (Kazakhstan Academy of Sciences), which he headed. This department attracted young graduate students in biology and physics from across the Soviet Union until 1960 when Tikhov died. Tens of papers about the optical properties of subarctic plants were published in Russian over these 15 years (e.g., Kutyreva et al., 1960; Perevertun, 1960; Semenenko, 1960b). Note that principal works in astrobotany have been translated by the American Department of Defense.

But the Martian story is not complete. During the 1956 and 1958 oppositions of Mars, American astronomer William Sinton (1925–2004) performed new spectroscopic observations covering the 2.7–3.8 μm wavelength range. He identified absorption features centered at 3.43 μm, 3.56 μm, and 3.67 μm that he attributed to organic compounds such as carbohydrates produced by photosynthesis in plants, analogous to tests done on plants on Earth (Sinton, 1957, 1959). This hypothesis was overcome a few years later when refined measurements showed that two of the so-called "Sinton bands" were due to heavy deuterated water in the Earth's atmosphere.

In 1995, at the Haute-Provence Observatory in the southeast of France, two Swiss astrophysicists, Michel Mayor and Didier Queloz, discovered the first extrasolar planets. It gave a boost to exploratory research and, to date, more than 4000 exoplanets have been identified, notably Earth-like or Super Earth-like planets located in the habitable zone (http://exoplanet.eu/). One can well imagine that, in a few decades, a number of techniques will emerge for direct detection and imaging of such planets around nearby stellar sources. There are two types of biosignatures: the first are indirect and consist of biological activity by-products, that is, atmospheric molecules such as oxygen, ozone, in association with water vapor, methane, and carbon dioxide; a second type is provided by indications that stellar light is transformed into biochemical energy, that is, photosynthetic pigments contained in plant leaves that produce the color of vegetation (e.g., Kiang et al., 2007a, 2007b; Kiang, 2008). Our Earth is an excellent model to test such techniques. American astronomer Carl Sagan (1934–1996) was the first to make this suggestion in a founding paper. Sagan et al. (1993) analyzed a Galileo image of the Amazon rainforest on Earth and identified a spectral signature of vegetation in the red-edge. They established a methodology to find evidence of life on Earth, called the "Sagan Criteria for Life." How to measure the spectrum of the Earth as observed from a very large distance, typically several parsecs? This could be done from a space probe traveling through the Solar System and looking back at the Earth, like Voyager 1 in 1990 or Mars Express in 2003. Unfortunately, no spectrum in the interesting wavelengths for life detection is available. An alternative method consists of measuring the Earthshine spectrum, that is the reflected earthlight visible on the Moon's night side (Briot et al., 2010). Figure 12.32 shows that, besides the strong signatures of atmospheric molecules (absorption and Rayleigh scattering), the vegetation red-edge that is typical of leaf reflectance and transmittance spectra is still detectable in the Earthshine spectrum (e.g., Arnold et al., 2002; Woolf et al., 2002).

A recent appealing idea concerns the measurement of circular polarization (see Section 4.1.2) as a signature of homochirality to detect biological activity on planets. Plant leaves are constructed from arrays of chiral molecules, for instance, photosynthetic pigments (chlorophyll) are either inherently chiral or embedded in a chiral matrix of proteins and carbohydrates. The identification of a universal biosignature that could be sensed remotely using circular polarization gave rise to a few studies since the end of the 1960s. Pospergelis (1969) first performed spectroscopic laboratory measurements of the complete Stokes vector on different rock samples and on green leaves. He notes that the transparency of leaf tissues and the presence of optically active substances are responsible for a distinctive behavior that may be useful for the detection of vegetation on planets.

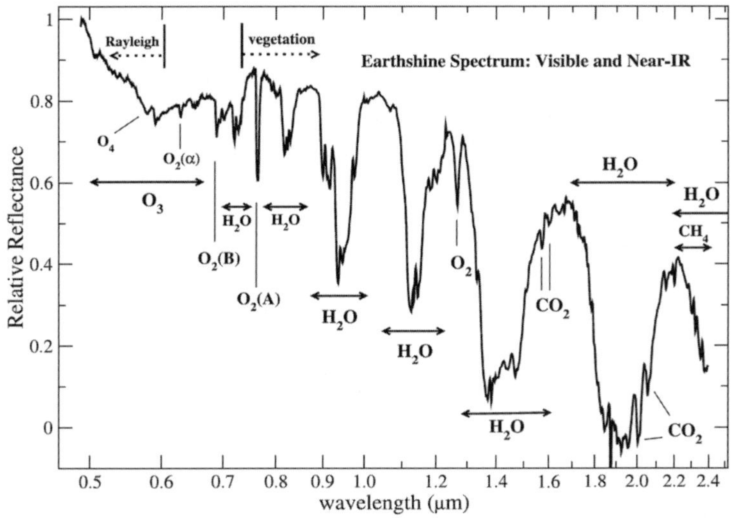

Figure 12.32 Earth's observed reflectance spectrum, at visible and near-infrared wavelengths (from Turnbull et al. (2006), *The Astrophysical Journal*. Copyright © 2006 with permission from IOP Publishing).

Wolstencroft et al. (2002) measured the DR-CD (diffuse reflectance circular dichroism) spectra of Hebe (*Hebe* sp.) and polka dot plant (*Hypoestes sanguinolenta*) leaves from 200 nm to 700 nm. They showed a strong polarization signal associated with photosynthetic pigments, in particular chlorophyll absorption features between 650 nm and 700 nm. This was confirmed by Sparks et al. (2009a, 2009b) on fresh leaves of red maple (*Acer rubrum*), oak (*Quercus* sp.), and red clover (*Trifolium pratense*) both on reflection and transmission spectra, and by Sterzik et al. (2010) on *Philodendron* leaves.

12.9 Image Synthesis

Many researchers have tried to produce realistic computer images of vegetation for a wide range of applications including ecology, agriculture, forestry, remote sensing, architecture, image synthesis, and the movie industry (e.g., Prusinkiewicz and Lindenmayer, 1990; Lehmann et al., 2005; Deussen and Lintermann, 2005). Obtaining natural visual impressions of a tree, a maize seedling, or a prairie is still very challenging because it depends on biological and environmental factors that are not always understood. Baranoski and Rokne (2004) pioneered the design of fast biophysically based models of light interaction with plant tissues combining computer graphics techniques and biological data (see Section 8.3.6 for a detailed description of these models and Wang et al., 2009b, for a review). If the physical processes driving leaf appearance such as color, gloss, and translucency are now well known, we desperately lack information about the natural range, distribution, and evolution of leaf biophysical variables. Although the natural progression of autumn coloring on a tree is a fascinating process, it has been studied by few researchers. We mention in Section 6.3.2 that autumn colors of many deciduous species are simply a consequence of the process of senescence. Accordingly, Chiba et al. (1996) modeled pigment evolution as an aging process controlled by the degeneration of the chloroplasts and the decomposition of the chlorophyll pigments to colorless low molecular weight

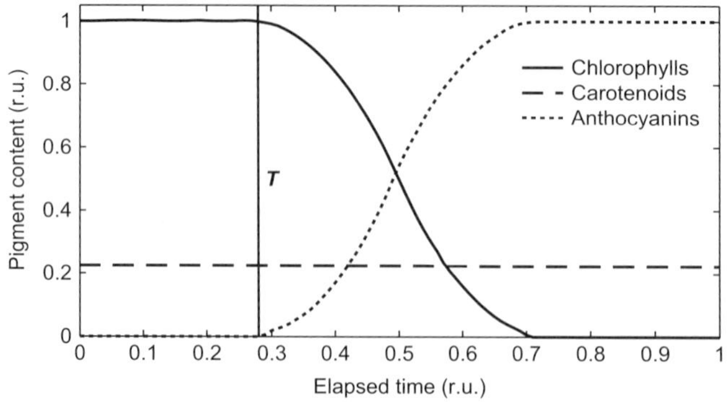

Figure 12.33 Model of autumn coloration. The time ***T*** stands for the beginning of aging (redrawn from Chiba et al. (1996)).

products. The preferential destruction of chlorophylls, that in green leaves disguise the presence of carotenoids and anthocyanins, allows the yellow and red hues of these pigments (Figure 12.33). Their model assumes a constant carotenoid content. It is also sophisticated enough to create variegated leaves (see Section 5.3) where colors vary as a function of the distance to major veins and the petiole, since aging definitely does not occur uniformly in leaves.

In addition, all the leaves do not turn red or yellow at the same time in a single tree or a forest stand. Mochizuki et al. (2001, 2007) proposed a model based on fractals, which preserves both temporal and spatial autumn coloring complexity. Although air temperature, humidity, and sunlight exposure are the main environmental factors of autumn coloring, they only consider the amount of sunlight which determines the amount of photosynthesized sugar in a leaf. They use experimental biological data to calibrate the model parameters and simulate the autumn coloring of a maple tree. The first image in Figure 12.34 shows the tree in summer with full green leaves. In the second image, the first changes in leaf color due to changes in the leaf pigment concentration appear and, in the next images, the red coloring of the leaves becomes dominant due to increasing anthocyanin content. In contrast to this approach, Braitmaier et al. (2004) assume the temperature as a triggering factor for seasonal changes. They also study them in two seasons, spring and autumn. Instead of focusing only on the effect of a single environmental factor at a time, Zhou et al. (2006) study the joint influence of several factors under a unified framework. They determine climate conditions such as sunlight, precipitation, and temperature on the basis of long-term observations to calculate foliar pigment content using mathematical models of chlorophyll and anthocyanin evolution. By moving the edge of these curves (see Figure 12.33) they can simulate the effect of abnormal climatic situations like precipitation or sunlight deficiency, or a warm autumn. Also note the existence of the Foliage Network (http://foliagenetwork.com/) that weekly analyzes and plots data collected in the northeastern United States by "foliage spotters" and produces reports which are used by travelers to find the best foliage conditions!

Real leaves display a wide range of shapes, margins, and venations (see Section 2.2) and, obviously, they are not flat or homogeneously colored. Several techniques have been proposed to capture their complexity under normal natural conditions, but also under stress or during senescence (Figure 12.35). Modeling such phenomena is a challenging and important problem in computer graphics. Different approaches including L-system grammar and other methods

Figure 12.34 Virtual autumn colors of a sugar maple (*Acer saccharum*) tree (from Mochizuki et al. (2007), *The Journal of the Society for Art and Science* © 2007 with permission from the Society for Art and Science). (A black and white version of this figure will appear in some formats. For the color version, please refer to the plate section.)

have been satisfactorily applied to create venation patterns and shapes of various kinds of leaves (e.g., Rodkaew et al., 2004; Peyrat et al., 2008). The natural deformation of leaf surface, for example, curling or wilting, can now be represented with great realism by constraining the movement of the venation skeleton that consists of segmented vertices (e.g., Basu, 2006). Special attention has also been paid to the reproduction of texture and color patterns that may be due to

Figure 12.35 (a) Leaf degradation of rose tree affected by color changes, insect attacks, and geometric shrinking (from Peyrat et al. (2008), *The Visual Computer* © 2009 with permission from Springer). (b) Autumn leaves stacking on the grass (from Desbenoit et al. (2006) © 2006 with permission from *The Eurographics Association*). (c) Simulated shape of medium wilting cucumber (*Cucumis sativus*) leaf (from Lu et al. (2009), *International Journal of Computer Games Technology* © 2009 with permission from Hindawi Publishing Corporation). (A black and white version of this figure will appear in some formats. For the color version, please refer to the plate section.)

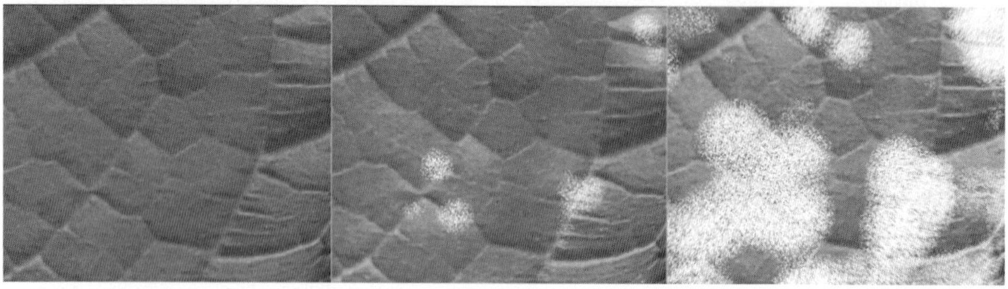

Figure 12.36 Simulation of leaf color changes with different powdery mildew infestation (from Miao et al. (2014), The *Transactions of the Chinese Society of Agricultural Engineering* © 2014 with permission from the Editorial Office of Transactions of the Chinese Society of Agricultural Engineering).

venation, variegation, or natural variability of pigment content (e.g., Desbenoit et al., 2006; Chi et al., 2009b; Binsfeld et al., 2011).

In addition to all these phenomena, the latest research tries to integrate meteorological and physiological factors, such as plant stress due to different causes listed in Chapter 7 (e.g., Lu et al., 2009; Tian et al., 2012a; Wang et al., 2013; Miao et al., 2013, 2014, Figure 12.36).

12.10 Science and Art

12.10.1 The Artificial Leaf

Bionics is the application of biological principles found in nature to the study and design of engineering systems. Why not look at what a leaf does and try to replicate it? Some examples have been already provided in this book: the hydrophobic properties of the lotus leaf (Section 12.3.3) have inspired, among other things, paint for mineral, cement, and concrete surfaces that is self cleaning, and a bionic leaf, the surface of which is coated with SiO_2 nanoparticles. In Section 12.7, we mentioned a bionic leaf to simulate the thermal effect of leaf transpiration. But designing a bionic leaf that converts sunlight into fuels is the Holy Grail of the artificial leaf because it would provide an abundant and cheap source of renewable energy (Birch, 2009; Regalado, 2010; Nocera, 2012). As seen in Section 2.3.3, photosynthesis is the process of converting solar energy to chemical energy and storing it in the bonds of sugar. It takes place in the chloroplasts and uses chlorophyll as a light-harvesting molecule. Photosynthesis feeds the world but it is an extremely inefficient process, as only about 1–2% of the solar energy is recovered as new photosynthetic products. The ultimate goal is to copy chlorophyll and design a bionic leaf that makes chemicals directly from solar radiation, more efficiently than a real leaf. It should use water as the sole source of electrons and protons for the reduction of CO_2 into high-energy density liquid fuels. The artificial leaf developed by Harvard researchers sandwiches a photovoltaic cell between two thin metal oxide catalysts. When submersed in a glass of water at room temperature and normal atmospheric pressure, it mimics photosynthesis and produces hydrogen, which is very expensive, and sometimes dangerous, to store and transport. A different approach involving bacteria that turn the hydrogen into something easier to handle has been taken recently (Torella et al., 2015). It is highly possible that in the future the world could be powered by photosynthesis from artificial leaves. The race for designing a commercially viable one is just beginning.

12.10.2 The Artistic Leaf

Plant leaves have always fascinated artists. Here are just a few examples in the domain of painting. As mentioned in Chapter 1, in the study of color perspective, Leonardo da Vinci was fascinated by the lighting conditions that produce the most beautiful colors in nature. He carefully observed trees and leaves: "different colors are found in the same tree, the leaves being more beautiful and larger on some branches than on other ones," seeking out new information on the way to represent them. The way light is transmitted by a leaf, for example, had important pictorial applications for Leonardo: "never paint leaves transparent to the sun, because they are confused; and this is because on the transparency of one leaf will be seen the shadow of another leaf which is above it." In the *Virgin of the Rocks*, a painting that has two versions, one in the Louvre, Paris, and another in the National Gallery, London, he has selectively exploited the effects he described to enhance the modeling and foreshortening of the leaves. The different orientation of the curved planes of a leaf on either side of its central vein is handled in a particularly emphatic manner; undulations in the surface of the leaves are modeled with great deliberation; and there is a thin scumble of lead-tin yellow to create the relatively lighter, denser edges of the individual leaves in the background (Bayer, 2004; Keith et al., 2011).

The French painter Claude Lorrain (1600–1682) was one of the first artists to practice painting outdoors. In an interview book, David Hockney, one of the most influential British painters of the 20th century, notes that much of the popularity of Lorrain was related to the delicacy of the foliage in

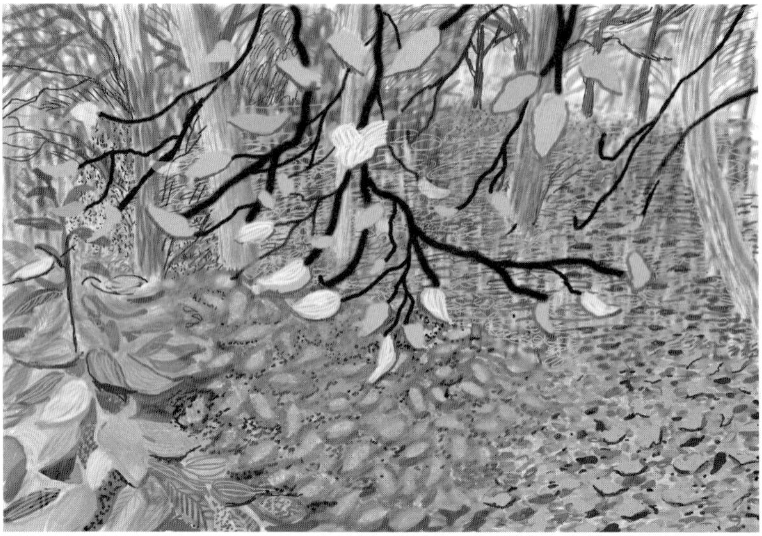

Figure 12.37 *"Autumn Leaves" 2008*, inkjet printed computer drawing on paper, 35 × 45 1/2" (© David Hockney, www.davidhockney.co/). (A black and white version of this figure will appear in some formats. For the color version, please refer to the plate section.)

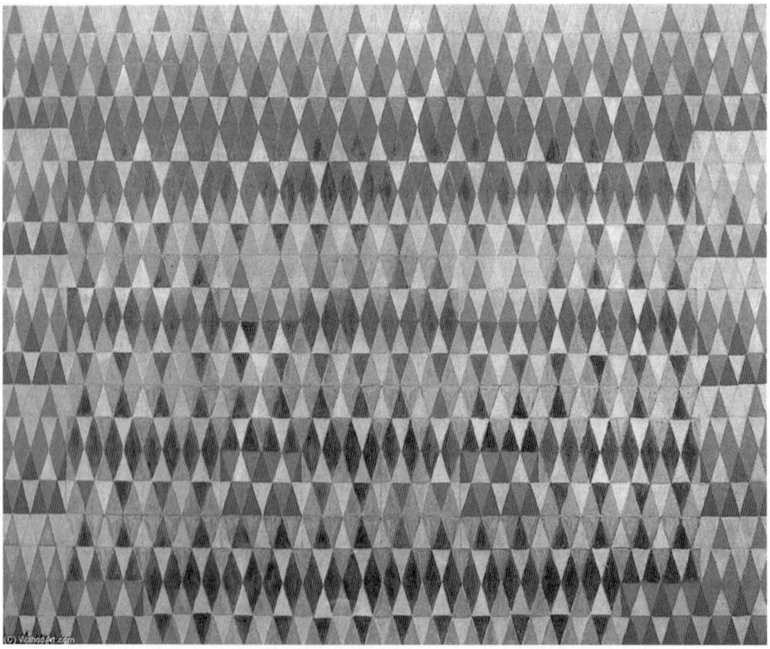

Figure 12.38 Eucalyptus, oil on canvas, 100.2 cm × 120.5 cm (G. Balla, 1914). (A black and white version of this figure will appear in some formats. For the color version, please refer to the plate section.).

his paintings (Gayford, 2011): "The details of the trees and the way he paints the leaves are beautiful. It probably isn't that natural, but it looks it. The trees are really like that." In *Autumn Leaves*, an image produced on a computer, Hockney shows the observation of the color change between the leaves in autumn (Figure 12.37).

Dewdrop, one of Escher's most famous mezzotint 1948 works, displays the leaf of a succulent plant. In the dewdrop, one can see the reflection of a window, which behaves like an optical lens, which shows the veins of the leaf and the bubbles of air between the leaf and the dewdrop (see Section 12.3.3).

Eucalyptus is a masterpiece from Giacomo Balla (1871–1958), an Italian painter who created a pictorial depiction of light, movement, and speed (Figure 12.38). It is entitled after the artist's favorite tree which he could admire from his studio window. Balla focuses particularly on the leaves of the tree, which are thin, pointed, and elongated, in several shades of green and yellow depending on which way you look at the surface. When the wind blows, the whole ensemble takes on a profusion of colors that seem to vibrate with vigorous energy. In this composition, constructed on the principle of formal repetition, the artist portrays the tree as if seen through a kaleidoscope. The leaves of the eucalyptus, aligned in head-to-tail fashion, appear to be viewed on both sides, offering up a color palette that ranges from green to pale yellow, while the contrast created by the repetition of small blue and light pink triangles in the background intensifies the visual effect with the foliage jumping out of the canvas at the viewer like a dazzling fireball.

Conclusion

A fundamental understanding of leaf optics has widespread applications ranging from botany, environmental sciences, physics, and astronomy, to applications in art, movies, and videogames. Improved understanding of how plant leaves interact with light permits a more complete understanding and quantifying of ecological processes and functions, not only on Earth, but also for exploration of other planets. Through primary production, plants provide the energy and nutrient resources that supports all living systems on Earth. In addition, the leaf energy budget, largely dictated by the amount of energy absorbed, impacts carbon sequestration, respiration, and transpiration rates, thus providing feedback mechanisms between biogeochemical cycles and the climate system. Plants have evolved diverse mechanisms to control their interactions with the environment, from changing turgor pressure in epidermal cells that allow light to be focused into the leaf's interior, and to motile stomata that move to the edge of palisade parenchyma cells or to their interior to enhance or limit light absorption. When present in excess, photosynthetic light is released through fluorescence, protecting the reaction centers. Even the clustering of chloroplasts or their dispersion in the cell extends or reduces the path length for transmitted light, affecting the efficiency of light absorption. Moreover, leaves change structural properties, such as increased cell wall thicknesses providing additional cellular support during periods of limited water availability, and increased leaf pubescence and glandular trichomes that reduce light absorption.

The color of the leaf can be quantified using spectroscopy to estimate the composition of pigments and nutrient concentrations in leaves. This information contributes to understanding of assimilation rates, while knowledge of the composition of pigments leads to knowledge of plant functioning. For example, xanthophyll pigments reduce the amount of energy reaching the photosynthetic reaction center through non-photochemical quenching, a mechanism that prevents photoinhibition under high light conditions. But many other environmental factors also affect leaf color from pubescence to anatomy, thus measuring color alone does not ensure correct estimate of the leaf's biochemical status or its physiology. Nonetheless, by combining reflectance information from across the solar spectrum, we can derive a more complete picture of the physiological status of the leaf and its functioning that let us place the pigment information in a more complete context. Besides pigments, other biochemicals, such as cellulose and lignin, allow retrieval of information about leaf dry weight, and derived properties like the dry mass per area of leaves, a property correlated with rates of growth. Improved understanding of plant functioning on Earth has been used as a guide to search for life on other planets. Finally, animals, including insects, often use leaf color (as measured across the UV to the VIS) or other structural clues to identify host plants while other organisms rely on leaf mimicry to avoid their predators. The vast wealth of interactions between plants and the rest of the biotic organisms on Earth began in the earliest stages of plant evolution, approximately 500 million years ago and continues today.

After presenting information about the physical and structural properties of leaves, and their optical properties and processes, we provide a review of the state of radiative transfer (RT) models and what types of information could lead to better or more reliable models in the future. Today's RT models could be improved with more detailed information about the range of variability in leaf anatomy and the chemistry associated with different cell types and cell wall materials, for example, better understanding of mesophyll heterogeneity (within and between palisade parenchyma and spongy mesophyll). How important is chlorophyll fluorescence when much of the energy lost is reabsorbed in the mesophyll tissues, which could improve model performance. The significance of leaf BRDF at the canopy scale is still unsettled as are mechanisms to measure it from space. How universal is the importance of epidermal cell structures for scattering and internal cell scattering for modifying the energy budget? Among foliar pigments there is promise to improve concentration estimates of additional pigments, including the so-called "brown pigments," which are the early degradative products of the breakdown of chlorophyll and carotenoid pigments. Absorption coefficients have recently been estimated for xanthophyll cycle pigments, making them accessible to RT modeling. Finally, measurement of new pigments, especially the flavonoids and polyphenols, may be possible.

Improved measurement of dynamic changes in leaves by better temporal modeling will greatly improve knowledge of fluxes of carbon, water, and energy. Little is known about the dynamic changes in leaf optical properties. New satellites with more frequent revisits and/or more spectral bands that measure the necessary wavelengths and are acquired at scales appropriate to trees or small groups of plants, perhaps about 10 m to 30 m pixels, will eventually provide information on rates of change in these properties.

It is possible that some new wavelength domains that have only recently become available could aid this growing understanding. For example, to elucidate the protective mechanisms of phenolic compounds requires measurements in the ultraviolet (UV) or in the longwave infrared (LWIR), the thermal emissive part of the spectrum. In the midwave infrared (MWIR) wavelengths, between 3 μm and 6 μm, the water content of the plants or other surface materials dominates energy absorption, thus their emissivity, while the signal is mostly independent of water and more closely related to the chemistry of the surface materials at longer wavelengths. These measurements combine both emitted infrared energy and the kinetic temperatures. Those wavelengths have reduced emissions due to absorption of energy in these regions. More accurate estimates of leaf surface temperatures, especially at higher spatial resolutions of 20 m to 50 m resolution, will significantly improve estimates of transpiration, a critical measurement for agricultural crops and for understanding surface temperatures. Several gases in the atmosphere are related to nutrient cycling and emissions from plants in the environment. These include gaseous oxygen and ozone, carbon dioxide and methane, nitrous oxide, and sulfur compounds. We provide examples of even longer wavelengths that relate to energy absorption and scattering by terahertz radiation. One characteristic that these data provide is more detailed structure of the within-leaf anatomy. Such information could lead to a better understanding of mesophyll heterogeneity. Polarization sensors may allow detection and quantification of some leaf surface properties, which are expected to lead to improved estimates of the fraction of photosynthetic active light (PAR) entering the leaf, thus better estimates of absorbed photosynthetic active radiation (APAR).

Appendix A

Glossary and Acronym List

A.1 Glossary

References by Gausman et al. (1973), Allaby (2006), Lambers et al. (2008), Stearn (2010), and Diah et al. (2014) were used for the definitions below. See also

http://en.wikipedia.org/wiki/Glossary_of_botanical_terms
www.ibiblio.org/pic/botanical_dictionary.htm
www.thefreedictionary.com/
http://dictionary.reference.com/
www.yourdictionary.com/
www.uri.edu/cels/bio/plant_anatomy/glossary.html
www.merriam-webster.com/dictionary/
https://goldbook.iupac.org/

Abaxial: lower leaf surface, directed outward from the axis.

Absorbance: the ability of a sample to absorb radiation, expressed mathematically as the negative natural or decimal logarithm of transmittance.

Absorptance: the ratio of the absorbed to the incident radiant power. It is also called absorption factor.

Absorption feature: A characteristic absorption pattern across a sequence of wavelengths, usually attributed to a particular absorbing compound and located at particular wavelengths.

Achlorophyllous: not having chlorophyll and, hence, unable to engage in photosynthesis.

Acrodomous: two or more primary veins that diverge from near the base of the blade that run the length of the blade converging at the apex.

Actinodromous: three or more primary veins diverging from a single point near the base of the blade and running radially toward the margin.

Acylation: introduction of an acyl group into a chemical compound.

Advection: the transport of a substance by the mass motion of a fluid, such as currents in wind or water.

Adaxial: upper leaf surface, directed toward the axis.

ADP (or **adenosine diphosphate**): high-energy phosphoric ester of the nucleoside adenosine.

Albedo: the fraction of incident radiation that is reflected by a surface over a specific set of wavelengths, typically the full range of the solar spectrum.

Allele: any of the alternative forms of a gene that may occur at a given locus.
Allelomorph (see **Allele**): either of a pair (or series) of alternative forms of a gene that can occupy the same position on a particular chromosome and that control the same character.
Allelopathic: relating to or characterized by allelopathy.
Allelopathy: a usually negative effect on the growth or development of an organism of one species, caused by a chemical released by an organism of another species.
Angiosperms: plants with seeds enclosed in an ovary; flowering plants.
Antheraxanthin: a yellow xanthophyll cycle pigment, a type of carotenoid pigment, is an intermediate in the reversible protective transformation from violaxanthin to zeaxanthin under high light conditions.
Anthocyanins: water-soluble vacuolar red to violet plant pigments, non-photosynthetic, related to the flavonoids.
Anticlinal wall: perpendicular to a surface.
Aposematic: colored or constructed in a way that indicates special capabilities for defense.
Arabinose: a pentose sugar, $C_5H_{10}O_5$, obtained from plant polysaccharides such as gums and hemicelluloses.
Arabinoxylan: a hemicellulose found in the primary and secondary cell walls of plants, consisting of copolymers of arabinose and xylose.
Areola: a small space or interstice in a tissue or part, such as the area bounded by small veins.
Ascomycete: a fungus in which meiotically produced spores are contained within asci.
ATP: a phosphorylated nucleotide composed of adenosine and three phosphate groups that supply energy for cellular processes.
Attenuation coefficient: (or **Extinction coefficient**): imaginary part of the complex refractive index.
Azimuth angle: an angle measured clockwise in the horizontal plane between a reference direction and any other line.
Band center: the central wavelength of a wavelength interval that defines a band.
Basidiomycete: a fungus in which meiotically produced spores are produced in basidia.
Betalains: a class of red and yellow indole-derived pigments found in some plants, where they replace anthocyanin pigments. They include betacyanins (betanin, isobetanin, probetanin, and neobetanin) and betaxanthins (amaranthine, vulgaxanthin, miraxanthin, portulaxanthin, and indicaxanthin).
Bidirectional: involving, moving, or taking place in two directions.
Bidirectional reflectance distribution function (BRDF): describes the reflectance properties of a surface by specifying the amount of radiation incident from one direction that is reflected into another direction, with respect to the surface normal. It has units of steradian.
Bidirectional transmittance distribution function (BTDF): describes the transmittance properties of a surface by specifying the amount of radiation incident from one direction that is transmitted into another direction, with respect to the surface normal. It has units of steradian.
Bifacial: upper and lower surfaces (abaxial and adaxial) of the blade are different (present two faces).
Biophysical: (see **Biophysics**).
Biophysics: a branch of science concerned with the application of physical principles and methods to biological processes and phenomena.
Blade (see **Lamina**): the flattened part of the lamina above the petiole.
Bloom: a fine powdery wax with a whitened or bluish appearance on the surface of some leaves.

Brewster's Angle: angle of incidence for which light, at a particular polarization, will be perfectly transmitted through a transparent dielectric surface with no reflection.

Broadleaf: (or **Broad-leaved**): having leaves with a broad blade, not needlelike or scalelike.

Bryophytes: plants with a dominant gametophyte, including mosses and liverworts.

Bulliform cells: enlarged epidermal cells occurring in longitudinal rows of similar cells in the *Gramineae*. They are thought to play a role in the rolling and unrolling of leaves.

Bundle sheath: the layer of tightly packed cells that surround the vascular tissue in leaves.

Campylodromous: a type of leaf venation with the veins being more or less parallel and where the veins extend in incurved arches from the base and converge at the apex.

Carotenoids: a generic term for water-insoluble pigments, which often function as accessory photosynthetic pigments in higher plants.

Cellulose: straight chain polysaccharides formed from many (100s) of β-glucose molecules that are the central material of cell walls.

Chalcones: the parent compound of a series of plant pigments related to the flavones.

Chemometrics: the science of extracting information from measurements made on chemicals in mixed environments (e.g., within a leaf) with the use of mathematical and statistical procedures.

Chlorenchyma: chloroplast-containing parenchyma tissue.

Chlorophyll: pigments that absorb in the blue and red parts of the spectrum, with a central magnesium-containing porphyrin ring that captures light energy, making them essential for photosynthesis. Green algae and higher plants contain chlorophylls *a* and *b* that differ in their side chains; other algae have other chlorophyll pigments.

Chloroplast: organelle in the mesophyll tissue of the leaf that captures light energy and converts it into carbohydrate and produces usable cellular energy in the form of ATP and NADPH molecules.

Chlorosis: loss of chlorophyll in the leaf (loss of green color) indicating environmental stress or leaf senescence.

Clade: a monophyletic group of organisms classified together taxonomically based on the presence of homologous features that can be traced to a common ancestor.

Collenchyma: a specialized type of living cell with unevenly thickened primary cell walls that provide support for leaf tissue.

Collinearity: in statistics, correlation between predictor variables (or independent variables), such that they express a linear relationship in a regression model.

Compact leaf: leaf with a mesophyll comprised of relatively compact chlorenchyma with few intercellular spaces (nonporous mesophyll).

Compound leaf: a leaf composed of several leaflets oppositely arranged along the rachis.

Conifer: the group of cone-bearing gymnospermous seed plants.

Continuum removal: a process to normalize reflectance spectra by fitting a convex hull to a spectrum. The "removal" is performed by subtracting the spectral feature values from the convex hull values at each wavelength.

Convex hull (or Convex envelope): given a set of points in the plane, the convex hull is the smallest convex polygon that contains all the points.

Cotyledon: the seed leaf or leaves that develop before germination of the seed of a seed plant, which may photosynthesize and provide nutrients in the early stage of growth.

Cuticle: a layer of fatty substance, cutin, on the outer epidermal cell walls that make the leaf almost impermeable to water.

Cutin: a waxy waterproof substance, consisting of hydroxy of fatty acids that are attached to each other by ester linkages, that is the main constituent of the plant cuticle.

Cytoplasm: within the cell walls, the matrix of the cell, containing cytosol in which the organelles are embedded, includes the nucleus, chloroplasts, mitochondria, vacuole, and other organelles.

Cytoplasmic: of or pertaining to the cytoplasm, within the cytoplasm.

Cytosol: the fluid portion of the cytoplasm, i.e., the cytoplasm exclusive of organelles and membranes.

Deciduous: a leaf that lives for one or part of one growing season.

Dendritic: a tree-like structure with branches arising from the main axis (e.g., branched trichomes from an epidermal cell).

Dicotyledons (or **Dicots**): an informal grouping of angiosperms other than monocots, with two seed leaves (cotyledons) that are part of the embryo. Dicots include about 80% of angiosperm species, representing trees, shrubs, and herbs.

Dielectric constant (or **Relative permittivity**): ratio of the electric field strength in vacuum to that in a given medium.

Diffuse light: light scattered onto or from a surface in all directions. A surface that scatters perfectly in all directions is termed a Lambertian (isotropic) surface.

Dimorphism: the occurrence of two distinct forms of the same parts in one plant, as in the juvenile and adult leaves of ivy.

Dioecious: possessing male and female reproductive organs on separate, unisexual, individual plants.

Dorsiventral leaf: a leaf with differentiated abaxial and adaxial sides, most commonly with palisade parenchyma cells on abaxial side of the blade and spongy parenchyma cells on the other.

Downy: covered with fine soft hairs.

Edaphic: influenced by the soil; edaphic factors include abiotic components like pH and chemical composition (e.g., nitrogen content).

Ellipsometry: an optical technique for investigating the dielectric properties (complex refractive index or dielectric function) of thin films.

Endodermis: the protective outermost cell layer, usually one cell thick, surrounds the ground tissue of the leaf.

Endosymbiont: an organism living inside another organism in an endosymbiotic relationship.

Endosymbiosis: a symbiotic relationship between two organisms in which one lives within the body of the other.

Enzyme: generally a protein molecule present in low concentrations that enhances the rate and efficiency of chemical reactions.

Epicuticle: the waxy layer on the outside of the cuticle.

Epidermis: the outer cellular layer of a leaf, primary in origin; if multiseriate (multiple layers of epidermis), only the outer layer differentiates epidermal characteristics.

Epoxidation: chemical reaction that yields an epoxide, a molecule with an oxygen atom and two carbon atoms.

Erectophyll: having the leaves largely hanging vertically.

Eucamptodromous: pinnate leaf venation in which the secondary veins gradually diminish inside the margin where they connect to adjacent secondary veins by a series of cross-veins.

Eudicotyledons (or **Eudicots**): the "true dicots" or the clade of dicotyledonous plants that have tricolpate and derived pollen, in contrast to the uniaperaturate pollen of other angiosperms, including monocots and "basal" groups such as magnolias.
Evapotranspiration: loss of water from the soil both by evaporation and by transpiration from the plants growing thereon.
Evergreen: leaves that are alive for more than one growing season, for some species possibly 5–10 years.
Excitation: process causing the transition of a system from one state to another of higher energy.
Extinction coefficient: a measure of the rate of diminution of transmitted light through a medium via scattering and absorption.
Fenestration: refers to holes or translucent area in the leaves of some species of plants. Can be natural or caused by insects.
Flavanone: any of various aromatic ketones that often occur in plants as glycosides and that constitute a subset of the flavonoids.
Flavone: any of the derivatives of the colorless crystalline aromatic ketone $C_{15}H_{10}O_2$, which occur as yellow plant pigments in the form of glycosides and are used as dyestuffs.
Flavonoid: molecules formed from a C-15 skeleton with an aromatic ring in positions 2, 3, or 4. Two benzene rings in the nucleus, one derived from condensation of acetate units and the other from the shikimic acid pathway with a C-3 structure.
Flavonol: any of various hydroxy derivatives of flavone.
Fructan: any polysaccharide composed of fructose residues.
Fructose: a crystalline sugar $C_6H_{12}O_6$ sweeter and more soluble than glucose.
Functional trait: morphological, biochemical, physiological, structural, phonological, or behavioral characteristics of a leaf that influence performance.
Galactan: a polysaccharide consisting of many galactose residues.
Galactose: a monosaccharide sugar $C_6H_{12}O_6$ less soluble and less sweet than glucose.
Genus (pl. **Genera**): a group of closely related species. Related genera constitute a family.
Glabrous: leaf surface without hairs, scales, bristles, or other protrusions; having a smooth surface.
Glaucous: having a whitish or bluish color of the leaf surface; caused by a waxy powder.
Glaucousness: shiny appearance of leaves, due to the presence of specific wax compounds.
Glossiness: the property of being smooth and shiny.
Glucomannan: a hemicellulose component in the cell walls of some plant species.
Glucose: a monosaccharide sugar $C_6H_{12}O_6$ that occurs widely in nature and is the usual form in which carbohydrate is assimilated by animals.
Glycosylation: the process by which sugars are chemically attached to proteins to form glycoproteins.
Granule: a small particle of a substance.
Granum (pl. **Grana**): the collective term for a stack of thylakoids, a membranous structure in the chloroplast where the light reactions of photosynthesis occur.
Guttation: the secretion of droplets of water from the pores of plants.
Gymnosperms: the informal group of seed-bearing plants with ovules not enclosed in a carpel. The term means naked (exposed) seeds; taxa include conifers (mostly northern hemisphere, with the family *Podocarpaceae* being dominant in the southern hemisphere), and *Ginkgo*, *Gnetum*, and cycads.

Heiligenschein: a diffuse bright region surrounding the shadow an observer's head casts on an irregular surface.

Hemicellulose: a class of substances that are part of the cell walls; they are polysaccharides that are simpler in structure then cellulose.

Heteroblastic: showing a marked difference in form between the juvenile and the adult structures.

Heterophylly: denoting plants that have more than one form of foliage leaf on the same individual.

Hirsute: covered with coarse hairs.

Hispid: covered with stiff or rough hairs.

Homochirality: describes a property of some materials made up of chiral units.

Hydrenchyma: large cells with large vacuoles, but lacking chloroplasts.

Hydrophilic: of, relating to, or having a strong affinity for water.

Hydrophytic: plants that grow partially or fully in water, such as water lilies.

Hydroxylation: chemical process that introduces a hydroxyl into an organic compound.

Hyperspectral data: of or relating to the measurements of objects or regions by combining information from a large number of detected spectral bands, as of visible light, infrared, and ultraviolet frequencies.

Hypostomatous: leaves with most of their stomata on the abaxial side.

Intercellular space: space among cells within the leaf.

Inverse model: a model that uses typical outputs to infer the inputs. For example, in radiative transfer (RT) modeling, ingesting of reflectance and transmittance data can be used to predict the biophysical and biochemical characteristics of the organisms (traits).

Iridescence: the property of certain surfaces that appear to change color as the viewing angle or the illumination angle changes.

Isoflavone: a colorless, crystalline, bioactive ketone $C_{15}H_{10}O_2$.

Isoflavonoid: any of a large group of water-soluble antioxidant polyphenolic plant compounds, including the isoflavones, which are found chiefly in leguminous plants.

Isolateral leaf (or **Isobifacial leaf**): a leaf that has palisade parenchyma cells on both sides of the blade.

Isotropy: uniformity of optical properties in all directions.

Kleptoplasty: a symbiotic phenomenon whereby plastids from algae are sequestered by host organisms.

Lacuna (pl. **Lacunae**): air space.

Lambert's cosine law: a law stating that the luminous intensity, in any direction from a plane surface, varies as the cosine of the angle between that direction and the perpendicular to the surface.

Lamina: the blade or flattened surface of a leaf.

Latent heat flux: energy released or absorbed by a body or a thermodynamic system, during a constant-temperature process. It is associated with evaporation or condensation of water vapor at the leaf surface; a component of the energy budget.

Leaf area index (LAI): The one-sided area of leaves above an area on the ground surface, typically based on leaf area (m^2) per m^2 ground area.

Leaf hairs (or **Trichomes**): single elongated cells or rows of cells on the surface of the leaf epidermis.

Liana: a woody plant that is rooted in the soil and grows toward the upper canopy of other plants by attaching itself to them with specialized stems and leaves (tendrils, twining stems, etc.) to reach the top of the canopy.

Libration: a type of reciprocating motion in which an object with a nearly fixed orientation repeatedly rotates slightly back and forth.

Lignin: a complex organic polymer deposited into the secondary cell walls, making them rigid and woody. Lignin fills the spaces between cellulose, hemicellulose, and protein components of the walls.

Litter decomposition: the process by which leaf litter and other plant substances are broken down into simple organic molecules, as a component of nutrient cycling.

Lutein: a type of xanthophyll pigment that is yellow and contributes to modulating the light energy in the chloroplast and contributes to non-photochemical quenching.

Macronutrient: essential nutrients required by plants in relatively high concentrations: nitrogen, potassium, calcium, magnesium, phosphorous, and sulfur.

Mannose: a monosaccharide found in some plants and produced by the oxidation of mannitol.

Maxwell's equations: a set of partial differential equations that, together with the Lorentz force law, form the foundation of classical electromagnetism, classical optics, and electric circuits.

Meristem (or **Meristematic cells**): undifferentiated cells that retain the capacity to divide and produce new cells.

Meristematic: consisting of or having the properties of meristem.

Meristemoid: the meristematic precursor cell of a stomate.

Mesic: characterized by, relating to, or requiring a moderate amount of moisture.

Mesophyll: parenchyma tissue of a leaf that is located between the epidermal layers.

Mesophytic: a plant that thrives under intermediate environmental conditions of moderate moisture levels, an example is a temperate meadow.

Methylation: an addition or a substitution of a methyl group ($R\text{-}CH_3$) to a substrate.

Micronutrient: essential elements needed in very small quantities by plants: chlorine, iron, boron, manganese, zinc, copper, molybdenum, and nickel.

Midrib: the central and most prominent vein in a leaf.

Model: a simplified description, usually conceptual or mathematical, of some aspect of the world and often how it works; a set of rules or mathematical functions.

Monocotyledons (or **Monocots**): a monophyletic group of flowing plants that have a single cotyledon or seed leaf that develops in the seed and becomes the first leaf at emergence. Monocots include grasses, sedges, and herbs but also include palms and other tree groups; the leaves are typically parallel veined.

Morphology: a branch of biology that deals with the form and structure of plants.

Multiseriate: consisting of many layers of cells.

Multispectral: of or relating to two or more ranges of frequencies or wavelengths in the electromagnetic spectrum. Multispectral bands typically have gaps between them, and therefore do not measure all contiguous bands in the spectral interval. For comparison: panchromatic (1 band), multispectral (2–20 bands), hyperspectral (>20-several hundred), ultraspectral (several thousand).

Nadir: the point of the celestial sphere that is directly opposite the zenith and vertically downward from the observer.

NADP (or **nicotinamide adenine dinucleotide phosphate**): a coenzyme, $C_{21}H_{28}N_7O_{17}P_3$, occurring in most living cells and utilized similarly to NAD but interacting with different metabolites.

NADPH: the reduced form of NADP.

Nectary: a multicellular glandular structure in leaves that secretes a sugary liquid.

Neoxanthin: a xanthophyll carotenoid pigment derived from violaxanthin when leaves are exposed to high light environments.

Neural network: a computing system (machine learning technique) inspired by the neural network in the brain. The neural network itself is not an algorithm, but rather a framework for different machine learning algorithms to process complex data inputs.

Nonphotochemical quenching: defensive mechanism to protect chloroplasts from high light intensity. The excited chlorophyll is quenched by transferring the energy through internal conversion back to the ground state.

Oligotrophic: lacking in plant nutrients and having a large amount of dissolved oxygen throughout.

Optical path length (or Optical distance): the product of the geometric length of the path followed by light through a given system, and the index of the refraction of the medium through which it propagates.

Organelle: a specialized part of a cell having some specific function.

Palinactinodromous: actinodromous, in which the primary veins have one or more subsidiary branch points above the primary one.

Palisade parenchyma: parenchyma tissue of a leaf mesophyll whose cells have an elongated form (palisade cells) perpendicular to the leaf surface.

Palmate: radially lobed or divided, with veins radiating from one point at the base of the leaf.

Panchromatic: sensitive to light of all colors in the visible spectrum.

Paradermal (or **Tangential**): refers to a section made parallel with the surface of a leaf.

Parallelodromous: having two or more primary veins originating at the base of the blade and running more or less in parallel to the apex, where they converge.

Parenchyma: unspecialized plant tissue consisting of simple thin-walled cells with intervening air spaces.

Parenchyma cell: thin-walled cell found in leaves that is capable of growth and division.

Pectin: a complex set of polysaccharides that are important parts of most primary cell walls in the non-woody parts of terrestrial plants. Pectin is a major component of the middle lamella, helping to bind cells together.

Periclinal wall: parallel to a surface.

Permittivity: permittivity of vacuum multiplied by the relative permittivity.

Petiole: the stalk that attaches a leaf to the stem.

Petiolule: the stalk of a leaflet on a compound leaf.

Phenolic compounds (or **Polyphenols**): the number and characteristics of the phenol structures determine the properties of particular members of the class. Compounds are derived from shikimate/phenylpropanoid and/or the polyketide pathways, featuring more than one phenolic unit and without nitrogen-based functions.

Phenotype: the characteristic state of an organism; its characteristics as determined by the environment and the genotype.

Phenylpropanoid: any of a class of plant-derived organic compounds that are biosynthesized from the amino acid phenylalanine.

Phloem: the vascular tissue that transports sugars and other metabolic products from the leaves to the rest of the plant.

Photochemical reflectance index (PRI): a narrow band normalized difference index, typically using a band at or near 531 nm that responds to changes in the relative amounts of xanthophyll cycle pigments and a reference band at or near 570 nm that does not respond to these changes. Under some conditions (e.g., diurnally varying illumination), PRI can provide an indicator of photosynthetic light-use efficiency. Under longer (seasonal) periods, it is often correlated with the relative concentrations of chlorophylls and carotenoids.

Photomorphogenesis: part of the plant's growth controlled by the light environment. For example, plants can growth toward or away from the light.

Photon: an elementary particle, the quantum of all forms of electromagnetic radiation including light.

Photosynthate: the sugars and starches that are produced by photosynthesis; these can be transported throughout the plant as an energy source.

Photosynthesis: process in which light energy is used to reduce CO_2 to organic compounds; occurs in chloroplasts in higher plants and algae.

Photosynthetic capacity: The maximum photosynthetic rate for a species under a set of environmental conditions, often defined as full sunlight and ambient carbon dioxide concentration.

Photosystem: functional and structural units of protein-pigment complexes located on the thylakoid membranes of the chloroplast that are involved in the absorption of light and the transfer of energy and electrons.

Phytophagous: herbivorous.

Pilose: covered with usually soft hair.

Pinnate: a compound leaf with leaflets arranged on either side of the common rachis.

Planck's law: describes the spectral density of electromagnetic radiation emitted by a black body in thermal equilibrium at a given temperature.

Planophyll: having the leaves more or less horizontal.

Plant litter: dead plant material (leaves, stems, bark) that has fallen to the ground.

Plasmodesmata: plasma membrane-lined pores that span the adjoining walls of plant cells.

Podocarpaceae: a large family of mainly southern hemisphere conifers.

Polymorphism: polymorphism is said to occur when two or more clearly genetically distinct forms exist in the same population of a species.

Polyphenols (or **Phenolic compounds**): a structural class of compounds characterized by the presence of large multiples of phenol structural units, which are multiples of benzene rings with attached hydroxyl groups.

Polysaccharide: a starch, cellulose, or glycogen polysaccharide composed of numerous glucose units bonded together by glycosidic linkages. Molecules may be linear to highly branched.

Porphyrin ring: a group of heterocyclic macrocycle organic compounds, composed of four modified pyrrole subunits interconnected at their α carbon atoms via methine bridges (=CH−). There are many types, including the chlorophyll ring with magnesium at the center.

Prenylation: the addition of hydrophobic molecules to a protein or chemical compound. It is usually assumed that prenyl groups (3-methyl-but-2-en-1-yl) facilitate attachment to cell membranes and are important for protein–protein binding.

Protochlorophyll: a green magnesium-containing pigment that is present in etiolated leaves and seedlings which develop in the dark and that is converted to chlorophyll by reduction under the influence of light.

Protochlorophyllide: any of several fluorescent pigments that is converted to chlorophyll by photoconversion.

Protoplast: a plant cell that has had its cell wall removed.

Pteridophytes: an informal term for vascular plants (such as ferns) that reproduce by spores rather than seeds.

Pterocarpan: any of various isoflavonoid derivatives found in the *Fabaceae* family.

Puberulent: covered with soft erect hairs.

Pubescent: covered with minute hairs or very fine down.

Pyrrole: a heterocyclic aromatic compound forming a five-membered four-carbon ring.

Quinone: class of organic compounds derived from aromatic compounds like benzene after conversion of an even number of –CH= groups into –C(=O)– groups with any necessary rearrangement of double bonds.

Radiative transfer: the physical phenomenon of energy transfer in the form of electromagnetic radiation. The equation of radiative transfer describes these interactions mathematically.

Rayleigh scattering: the scattering of light by particles and molecules which are much smaller than the wavelength of the light.

Reflectance: fraction of incident radiation reflected by a surface or discontinuity.

Reflectivity: the property of reflecting light or radiation, especially reflectance as measured independently of the thickness of a material.

Refraction: deflection from a straight path undergone by a light ray or energy wave in passing obliquely from one medium (such as air) into another (such as glass) in which its velocity is different.

Refractive index (or **Index of refraction**): the ratio of the speed of light in vacuum to that in a given medium.

Refractometry: the method of measuring substances' refractive index.

Relative permittivity (see **Dielectric constant**): ratio of the electric field strength in vacuum to that in a given medium.

Rodlet: any of various rod-shaped cells in the retina that respond to dim light.

RuBisCO (or **Ribulose-1,5-bisphosphate carboxylase/oxygenase**): an enzyme that catalyzes the first major step of carbon fixation.

Scabrous: having or covered with scales or small projections and rough to the touch

Scale: one of several types of cuticular materials on leaf surfaces, flattened waxy material.

Scattering: a process in which a change in direction or energy of an incident radiation is caused by interaction with a particle. Rayleigh scattering occurs when the particle size is much smaller than the wavelength of the light (molecules), while Mie scattering occurs when the particle size is about equal to the wavelength (aerosols, particles). Non-selective scattering occurs when the particle size is several times larger than the wavelength.

Sclereid: a short thick-walled plant cell that functions for support rather than conduction.

Sclerenchyma: a tissue of thick-walled cells (sclereids) whose principal function is strengthening plant parts.

Sclerophyll: a woody plant with small leathery evergreen leaves that is the dominant plant form in certain hot dry areas, especially the Mediterranean region.

Sclerophyllous: having the appearance of thick leathery leaves that are common to plants from arid and semiarid regions.

Senescence: the process of tissue death either because of unfavorable environmental conditions or because the leaf (or plant) is nearing the end of its life cycle.

Signal-to-noise ratio: a measure used in science and engineering that compares the level of a desired signal to the level of background noise.

Snell's Law (or **Snell–Descartes law** or **Law of refraction**): a formula used to describe the relationship between the angles of incidence and refraction, when referring to light or other waves passing through a boundary between two different isotropic media, such as water, glass, or air.

Softwood: the wood of conifers.

Spectral band: a defined wavelength interval of electromagnetic radiation in which, for example, a detector can absorb photons.

Spectral characteristics: number, position, and range of bands within the electromagnetic spectrum in which a sensor operates; the spectral properties (the absorption or scattering properties by wavelength) of a species.

Spectral index: a mathematical expression, often a ratio or normalized difference of measured reflectance or radiance from two or more spectral bands, with or without a constant or scaling factor that highlights a particular feature in the spectral signal.

Spectroscopy (or Optical spectroscopy): the study of systems by the electromagnetic radiation with which they interact or that they produce. By convention, spectroscopic properties are expressed in terms of wavelength.

Spectrum analysis: the interpretation of the information present in an energy spectrum in terms of radiation energy and intensity.

Spermatophytes: seed plants, including angiosperms and gymnospermous groups, the dominant land plants.

Spongy parenchyma: parenchyma tissue of a leaf mesophyll with conspicuous intercellular spaces (porous mesophyll).

Sporophyll: a modified leaf that bears the sporangia (reproductive tissue found in non-seed bearing plants like ferns).

Stellate: arranged or shaped like a star.

Stoichiometry: founded on the law of conservation of mass which has led to insights that the relations among quantities of reactants and products typically form a ratio of positive integers, which means if one reactant is known the rest can be estimated.

Stomata (singular **Stoma or Stomate**): minute apertures on the epidermis. They consist of two specialized guard cells that surround a pore. These valves let gases in and out of the leaf and are actively controlled by the osmotic pressure within the guard cells.

Storage cells: large thin-walled cells used for storage of water and mucilages.

Strigose: having stiff, straight, closely appressed hairs.

Stroma: the matrix within a chloroplast, containing various molecules and ions and in which the thylakoid membrane system is suspended.

Substomatal: located immediately below the stoma.

Succulent leaf: fleshy leaves (malacophyllous) with many cells that store water and mucilage.

Tannins: any of various water-soluble polyphenols found in plant tissues that bind proteins.

Thallophytes: any nonvascular plantlike organisms showing no clear distinction of roots, stem, or leaves and not producing flowers or seeds. These included the fungi, lichens, and most algae.

Thylakoid: elongated thick walled xylem cells for transport of water and support of the leaf.

Tomentose: covered with a dense layer of short, matted, woolly hairs.

Tracheids: elongated, tapering xylem cells having lignified, pitted, intact walls, adapted for water conduction and support.

Trait (see **Functional trait**): biochemical, physiological, morphological, structural, phonological, or functional characteristics of leaves and plants.

Transmission: movement of electromagnetic waves (visible light, radio waves, ultraviolet, etc.) through a material.

Transmissivity: ability of a material to transmit electromagnetic energy as light.

Transmittance: the ratio of the transmitted radiant power to that incident on the sample.

Transpiration: the process by which liquid water is transported to the stomata where it vaporizes and is released to the atmosphere.

Transverse: a cross-section; a section taken perpendicular to the longitudinal axis of the cell; also called transection.

Trichomes (or **Leaf hairs**): outgrowths of the epidermal cells, can be single cell or multicellular (branched) hairs, spines, scales, or with glands. Secretions produced by glands may be oils or other chemicals.

Tubule: a very small tube or tubular structure.

Turgid: swollen with water. Turgidity is required for leaf expansion and for stomatal opening of the guard cells.

Turgor: the state of turgidity in leaves and the resulting rigidity of cells (or tissues), due to the absorption of water.

Ultramafic: ultrabasic igneous rocks.

Understory forest: plant life growing beneath the forest canopy without penetrating it to any extent.

Vacuole: a cavity or vesicle in the cytoplasm of a cell usually containing fluid.

Vascular bundle: a strand of vascular tissue consisting of xylem and phloem.

Vein: a strand of vascular tissue in the leaf. Veins are ordered in descending size, down to the smallest. Most cells in the leaf are within a few cells of a vein or veinlet.

Venation: the arrangement of the veins in a leaf.

Vessel element: an elongated, water-conducting cell (a tracheary element) in the xylem of angiosperms. At maturity, the cells die leaving only their lignified cell walls. They are stacked one on top of another in long columns, called vessels. Water flows unimpeded from cell to cell along these columns through perforations in the cell walls.

Violaxanthin: a natural xanthophyll carotenoid pigment, orange in color, which is the state of the xanthophyll cycle pigments under low light conditions. Violaxanthin is transformed to antheraxanthin and zeaxanthin under higher light conditions.

Wien's displacement law: states that the black body radiation curve for different temperatures peaks at a wavelength inversely proportional to the temperature.

Xeric: characterized by, relating to, or requiring only a small amount of moisture.

Xerophytic: a plant adapted to low water status; often found in arid and semiarid regions.

Xylem: a tissue in vascular plants that transports water and nutrients from the roots to the leaves. The xylem also provides support and consists of tracheids, vessels, parenchyma cells, and woody fibers.

Xylose: a crystalline monosaccharide, $C_5H_{10}O_5$, that is a component of most hemicelluloses in plants.

Xyloglucan: the main hemicellulose in the primary cell wall of dicotyledonous plants.

Zeaxanthin: a yellow pigment and one of the most common carotenoids found in nature; it is important in the xanthophyll cycle.

Zenith angle: the angle between the direction to the zenith and the direction of a light ray.

A.2 Acronym List

ABM	algorithmic BDF model
ADP	adenosine diphosphate
ANMB	area under curve normalized to maximal band depth
ANN	artificial neural network
ANOVA	analysis of variance
APAR	absorbed photosynthetic active radiation
ARI	anthocyanin reflectance index
ATRMO	automated radiative transfer models operator
ATP	adenosine triphosphate
ATR	attenuated total reflection
BDT	bagged decision trees
BIEN	Botanical Information and Ecology Network
BMA	Bayesian model averaging
BP	back propagation
BRDF	bidirectional reflectance distribution function
BRF	bidirectional reflectance factor
BSDF	bidirectional scattering distribution function
BTDF	bidirectional transmittance distribution function
BTF	bidirectional transmittance function
CAM	Crassulacean acid metabolism
CARI	carotenoid index
CCD	charge coupled device
CC&D	counter-camouflage, concealment, and deception
CCI	chlorophyll concentration index
CCM	chlorophyll content meter
CCSM	cross correlogram spectral matching
ChlF	chlorophyll fluorescence
Chl NDI	chlorophyll normalized difference index
CHRIS	compact high resolution imaging spectrometer
CIE	Commission Internationale de l'Eclairage
CMF	color matching functions
COSINE	close-range spectral imaging of leaves
CRI	carotenoid reflectance index

CSG	constructive solid geometry
CWT	continuous wavelet transform
DCI	derivative chlorophyll index
DD	double difference
DGCI	dark green color index
DHR	directional-hemispherical reflectance
DHRF	directional-hemispherical reflectance factor
DHTF	directional-hemispherical transmittance factor
DI	derivative index
DLM	dorsiventral leaf model
DMC	dry matter content
DOES	design of experiments for simulation
DRCD	diffuse reflectance circular dichroism
DWI	difference water index
DWT	discrete wavelets transform
EcoSIS	ecological spectral information system
eFAST	extended Fourier amplitude sensitivity test
EHF	extremely high frequency
ELF	extremely low frequency
EMR	electromagnetic radiation
EMT	effective medium theory
EOSDIS	Earth Observing System Data and Information System
EPI	eucalyptus pigment index
ESDIS	Earth Science Data and Information System
EUV	extreme ultraviolet
eV	electron-volt
EWT	equivalent water thickness
fAPAR	fraction of absorbed photosynthetic active radiation
FAST	Fourier amplitude sensitivity test
FBA	foreground-background analysis
FMC	foliage moisture content
FRT	fluorescence-reflectance transmittance
FSM	foliar scattering model
FD	first derivative
FTIR	Fourier transform infrared
FUV	far ultraviolet
FWIR	farwave infrared
gNDVI	green NDVI
GAF	geometrical attenuation factor
GA-PLS	genetic algorithms with partial least squares
GRG	generalized Rayleigh–Gans
GUI	graphic user interface
GVMI	global vegetation moisture index
HATR	horizontal attenuated total reflection

HF	high frequency
HFBA	hierarchical foreground-background analysis
HLB	huanglongbing
H Lyman-α	hydrogen Lyman-α
INRA	Institut national de la recherche agronomique
IRRI	International Rice Research Institute
KM	Kubelka–Munk
LAI	leaf area index
LCC	leaf color chart
LDA	linear discriminant analysis
LDMC	leaf dry matter content
LEAFMOD	leaf experimental absorptivity feasibility mode
LED	light emitting diode
LHCPII	light-harvesting chlorophyll *a*/*b* binding protein of photosystem II
LIBERTY	leaf incorporating biochemistry exhibiting reflectance and transmittance yields
LLL	Landau, Lifshitz, Looyenga
LOPEX	leaf optical properties experiment
LMA	leaf mass per area
LNA	leaf nitrogen density
LRR	logarithm of reciprocal reflectance
LUE	light use efficiency
LUT	lookup table
LWC	leaf water content
LWCI	leaf water content index
LWIR	longwave infrared
mARI	modified anthocyanin reflectance index
MARPET	MARine PATtern
mCRI	modified carotenoid reflectance index
MESMA	multiple endmember mixture model
MLF	multi-layer feed-forward
mNDI	modified normalized difference index
MNDWI	mid-wave infrared normalized difference water index
MRI	magnetic resonance imaging
MSE	mean squared error
MSI	moisture stress index
mSR	modified simple ratio
MUV	middle ultraviolet
MWIR	midwave infrared
NASA	National Aeronautics and Space Administration
NC	number of combinations
ND	normalized difference
NDMI	normalized dry matter index
NDVI	normalized difference vegetation index
NDWI	normalized difference water index

NIR	near infrared
NIRS	near-infrared reflectance spectroscopy
NUV	near ultraviolet
ORNL DAAC	Oak Ridge National Laboratory Distributed Active Archive Center
PAL	phenylalanine ammonia-lyase
PAR	photosynthetic active radiation
PAS	photoacoustic spectroscopy
PCA	principal component analysis
PCR	principal components regression
PDA	penalized discriminant analysis
PDF	probability density function
PET	polyethylene terephthalate
PhilRice	Philippines Rice Research Institute
PLS	partial least square
PLSR	partial least square regression
PPO	polyphenol oxidase
PRESS	prediction residual error sum of squares
PRI	photosynthetic reflectance index
PROSPECT	leaf optical properties spectra
PSI/II	photosystem I/II
PSND	pigment specific normalized difference
PSSR	pigment specific simple ratio
PTFE	polytetrafluoroethylene
QDA	quadratic discriminant analysis
QSEU	quinine sulfate equivalent unit
QSPECT	quatre PROSPECT
RAMIS	radiomètre portatif de mesure *in situ*
RARS	ratio analysis ofreflectance spectra
RAYTRAN	ray racing model
REIP	red-edge inflexion point
REP	red-edge position
RES	red-edge symmetry
RGR	red: green ratio
RMSE	root mean square error
RMSEP	root mean square error of prediction
RPV	Rahman–Pinty–Verstraete
RT	radiative transfer
RWC	relative water content
SAIL	scattering by arbitrary inclined leaves
SCPO	surface current physical optics
SEM	scanning electron microscopy
SEP	standard error of prediction
SETI	search for extraterrestrial intelligence
SFY	spectral fluorescence yield

SHF	super-high frequency
SIPI	structure insensitive pigment index
SLA	specific leaf area
SLM	specific leaf mass
SLOP	stochastic model for leaf optical properties
SLRA	stepwise linear regression analysis
SLW	specific leaf weight
SMA	spectral mixture analysis
SNR	signal-to-noise ratio
SPAD	soil plant analysis development
SR	simple ratio
SRBM	simple yet realistic BRDF model
SRI	summed reflectance index
SVM	support vector machine
SVMR	support vector machine regression
SWIR	shortwave infrared
SXR	soft X-rays
TBRI	three-band ratio index
TIR	thermal infrared
TOMS	total ozone mapping spectrometer
UHF	ultra high frequency
UV	ultraviolet
UVA	ultraviolet A
UVB	ultraviolet B
UVC	ultraviolet C
VHF	very high frequency
VIPO	volume integral physical optics
VIS	visible
VLA	leaf vein length per unit leaf area
VLWIR	very longwave infrared
VUV	vacuum ultraviolet
WABI	water balance index
WI	water Index
YI	yellowness index

Appendix B
Leaf Molecules

Detailed information about the molecules hereafter can be found in the open internet accessible chemistry database PubChem (https://pubchem.ncbi.nlm.nih.gov/). All molecules have been created using the ACD/ChemSketch freeware, a drawing package that allows drawing chemical structures.

B.1 Phenolic Compounds

See Section 2.3.4.

B.1.1 Flavonoids/Flavonols

Flavonols are flavonoids with a ketone group. They occur abundantly in a variety of fruits, vegetables, and leaves, for example, green and black tea (*Camellia sinensis*) leaves, and the main representatives are quercetin and kaempferol. Intake of flavonols is found to be associated with a wide range of health benefits, which includes antioxidant potential and reduced risk of vascular disease (Table B.1 and Figure B.1).

B.1.2 Flavonoids/Isoflavonoids/Isoflavones

Isoflavones are derived from the phenylpropanoid pathway and widely distributed in the *Fabaceae* family. They are structurally similar to the mammalian estrogen, estradiol-17β. Because of their structural property, isoflavones have potential beneficial effects on various diseases such as cancer-preventive, antioxidative, antiosteoporosis, and anticarcinogenic, as well as a cardiovascular-enhancing agent (Table B.2 and Figure B.2).

B.1.3 Flavonoids/Isoflavonoids/Pterocarpans

Pterocarpans are derivatives of isoflavonoids found in the *Fabaceae* family. They possess antibiotic activity and are produced by plant tissues in response to biotic and abiotic stresses such as pathogen infection, ultraviolet irradiation, and heavy metal ions (Table B.3 and Figure B.3).

Table B.1 *Some flavonols present in plant leaves.*

Name	Formula	PubChem CID	Refractive index	Plant species
Kaempferol	$C_{15}H_{10}O_6$	5280863	1.784	*Brassica napus, Glycine max, Trifolium pratense*
Quercetin	$C_{15}H_{10}O_7$	5280343	1.823	*Brassica napus, Trifolium pratense*

Table B.2 *Some isoflavones present in plant leaves.*

Name	Formula	PubChem CID	Refractive index	Plant species
Daidzein	$C_{15}H_{10}O_4$	5281708	1.698	*Pueraria lobata, Pueraria mirifica, Pueraria thomsonii*
Genistein	$C_{15}H_{10}O_5$	5280961	1.732	*Trifolium pratense*
Pseudobaptigenin	$C_{16}H_{10}O_5$	5281805	1.690	*Pisum sativum, Trifolium pratense*
Formononetin	$C_{16}H_{12}O_4$	5280378	1.641	*Trifolium pratense*
Biochanin A	$C_{16}H_{12}O_5$	5280373	1.668	*Cicer arietinum, Dalbergia paniculata, Lupinus reflexus, Trifolium pratense, Trifolium subterraneum*
Calycosin	$C_{16}H_{12}O_5$	5280448	1.668	*Astragalus membranaceus, Trifolium pratense*
Prunetin	$C_{16}H_{12}O_5$	5281804	1.668	*Dalbergia hainanensis, Trifolium pratense*
Methylorobol	$C_{16}H_{12}O_6$	5319744	1.697	*Trifolium pratense*
Pratensein	$C_{16}H_{12}O_6$	5281803	1.697	*Trifolium pratense*
Afrormosin	$C_{17}H_{14}O_5$	5281704	1.621	*Trifolium pratense*

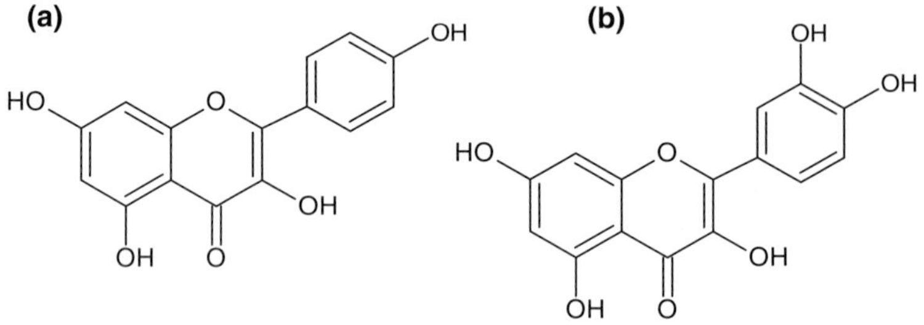

Figure B.1 Some flavonols present in plant leaves: (a) kaempferol; and (b) quercetin.

B.2 Waxes

See Section 2.3.5.

B.2.1 n-alkanes (Paraffin)

n-alkanes are epicuticular compounds of plant leaves that range in chain lengths from C_{17} to C_{35} and that can be used as a criterion for systematic classification. Dozens of studies have examined the

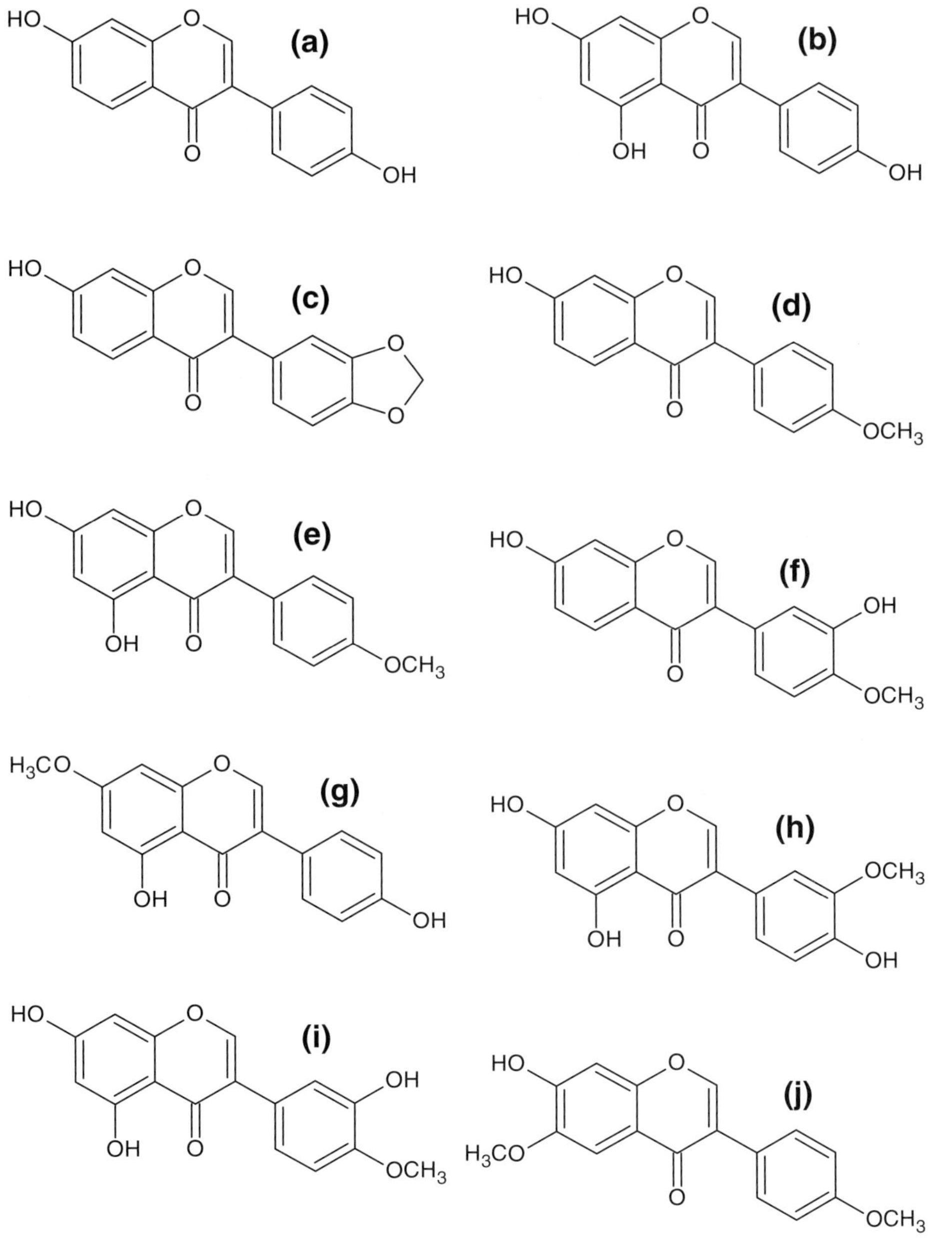

Figure B.2 Some isoflavones present in plant leaves: (a) daidzein; (b) genistein; (c) pseudobaptigenin; (d) formononetin; (e) biochanin A; (f) calycosin; (g) prunetin; (h) methylorobol; (i) pratensein; and (j) afrormosin.

Table B.3 *Some pterocarpans present in plant leaves.*

Name	Formula	PubChem CID	Refractive index	Plant species
Bitucarpin A	$C_{22}H_{24}O_4$	21576494	1.573	*Bituminaria morisiana*
Erybraedin A	$C_{25}H_{28}O_4$	362562	1.610	*Erythrina eriotricha*

Table B.4 *Some n-alkanes present in plant leaves.*

Name	Formula	PubChem CID	Refractive index	Plant species
Heptadecane	$C_{17}H_{36}$	12398	1.437	*Lepidagathis cristata, Pimenta racemosa*
Nonadecane	$C_{19}H_{40}$	12401	1.441	*Pimenta racemosa, Senna surattensis*
Nonacosane	$C_{29}H_{60}$	12409	1.453	*Acacia senegal, Brassica nigra, Brassica oleracea, Pisum sativum, Spinacia oleracea*
Hentriacontane	$C_{31}H_{64}$	12410	1.428	*Ipomoea aquatica, Nicotania tabacum, Pisum sativum, Prunus persica, Sambucus nigra*

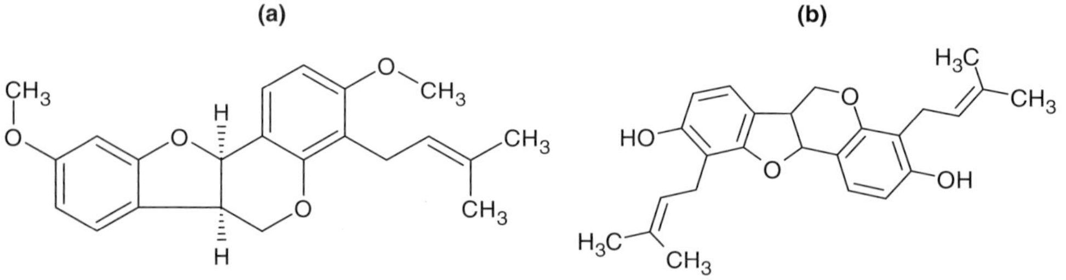

Figure B.3 Some pterocarpans present in plant leaves: (a) bitucarpin A; and (b) erybraedin A.

range and variation of n-alkane chain-length abundances in modern plants from around the world (Table B.4 and Figure B.4).

B.2.2 Alkyl Esters

Alkyl esters range in chain lengths from C_{36} to C_{72}. It has been shown that they are present in the leaf cuticles of European beech (*Fagus sylvati*ca), silver linden (*Tilia tomento*sa), and common oak (*Quercus robur*), at all developmental stages from buds to mature foliage (Table B.5 and Figure B.5).

B.2.3 Ketones

A ketone is an organic compound characterized by a carbonyl group RC(=O)R', where R and R' can be a variety of carbon-containing substituents (Table B.6 and Figure B.6).

Table B.5 *Some alkyl esters present in plant leaves.*

Name	Formula	PubChem CID	Refractive index	Plant species
Octadecyl hexacosanoate	$C_{44}H_{88}O_2$	13908832	1.459	*Quercus robur, Copernicia cerifera*
Octadecyl octacosanoate	$C_{46}H_{92}O_2$	13908836	1.460	*Musa paradisiaca*
Tetracosanyl docosanoate	$C_{46}H_{92}O_2$	13908825	1.460	*Aleurodicus dugesii*
Docosyl tetracosanoate	$C_{46}H_{92}O_2$	13908830	1.460	*Artemisia afra*

Table B.6 *Some ketones present in plant leaves.*

Name	Formula	PubChem CID	Refractive index	Plant species
Nonacosan-15-one	$C_{29}H_{58}O$	75997	1.453	*Allium porrum*

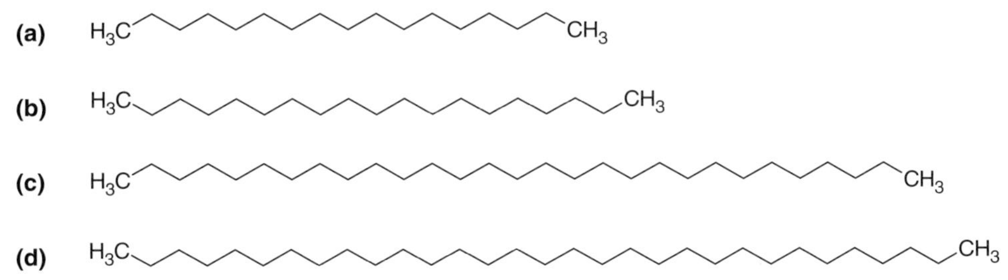

Figure B.4 Some n-alkanes present in plant leaves: (a) heptadecane; (b) nonadecane; (c) nonacosane; and (d) hentriacontane.

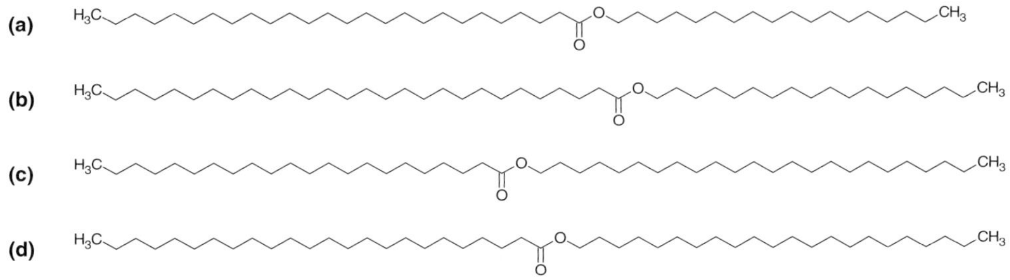

Figure B.5 Some alkyl esters present in plant leaves: (a) octadecyl hexacosanoate; (b) octadecyl octacosanoate; (c) tetracosanyl docosanoate; and (d) docosyl tetracosanoate.

B.2.4 β-diketones

A diketone is a molecule containing two ketone groups. β-diketones have a number of very interesting and specific properties due to the presence of two carbonyl groups separated with one carbon atom (Table B.7 and Figure B.7). They form hollow wax nanotubes in many grasses.

Table B.7 *Some β-diketones present in plant leaves.*

Name	Formula	PubChem CID	Refractive index	Plant species
Hentriacontane-14,16-dione	$C_{31}H_{60}O_2$	441489	1.457	*Eucalyptus globulus, Leymus arenarius, Secale cereale*
Hentriacontane-8,10-dione	$C_{31}H_{60}O_2$	13915231	1.457	*Buxus sempervirens*
Tritriacontane-16,18-dione	$C_{33}H_{64}O_2$	154874	1.458	*Eucalyptus*
Pentatriacontane-16,18-dione	$C_{35}H_{68}O_2$	116652	1.459	*Eucalyptus globulus*

Table B.8 *Some secondary alcohols present in plant leaves.*

Name	Formula	PubChem CID	Refractive index	Plant species
Nonacosan-10-ol	$C_{29}H_{60}O$	25240035	1.458	*Aquilegia vulgaris, Dawsonia superba*

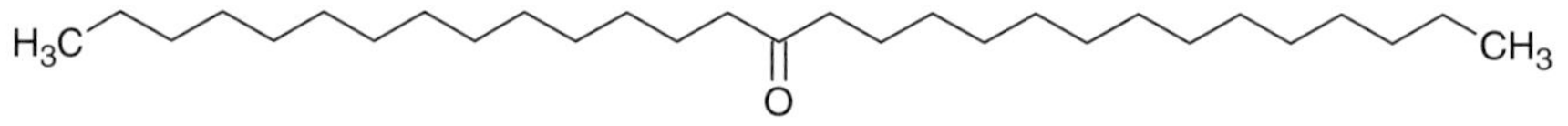

Figure B.6 Some ketones present in plant leaves: nonacosan-15-one.

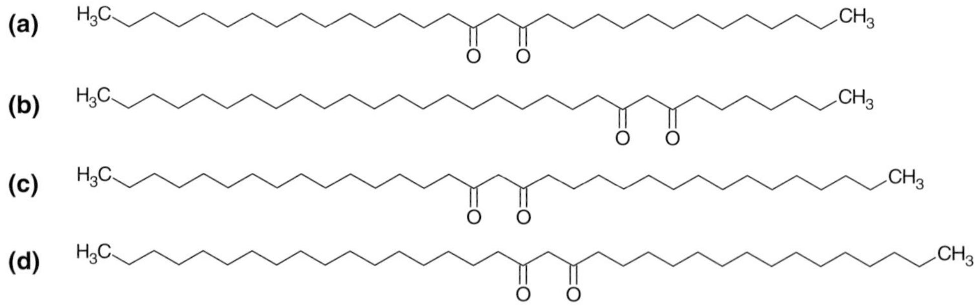

Figure B.7 Some β-diketones present in plant leaves: (a) hentriacontane-14,16-dione; (b) hentriacontane-8,10-dione; (c) tritriacontane-16,18-dione; and (d) pentatriacontane-16,18-dione.

B.2.5 Secondary Alcohols

A secondary alcohol is a compound in which a hydroxy group, –OH, is attached to a saturated carbon atom. which has two other carbon atoms attached to it (Table B.8 and Figure B.8).

B.2.6 Fatty Acids

Plants synthesize a huge variety of fatty acids, although only a few are major and common constituents. Hexadecanoic and octadecanoic acids are incorporated into leaf lipids. Researchers working on biofuels have recently taken an interest in lipids in leaves (Table B.9 and Figure B.9).

Table B.9 *Some fatty acids present in plant leaves.*

Name	Formula	PubChem CID	Refractive index	Plant species
Hexadecanoic acid	$C_{16}H_{32}O_2$	985	1.453	*Allium porrum*
Octacosanoic acid	$C_{28}H_{56}O_2$	10470	1.461	*Ceanothus velutinus*

Table B.10 *Some hydroxy acids present in plant leaves.*

Name	Formula	PubChem CID	Refractive index	Plant species
16-Hydroxy hexadecanoic acid	$C_{16}H_{32}O_3$	10466	1.468	*Vicia faba*
18-Hydroxy octadecanoic acid	$C_{18}H_{36}O_3$	5282915	1.469	*Rosmarinus officinalis*

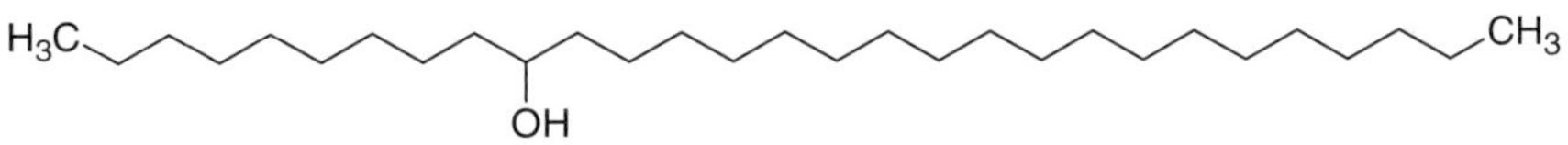

Figure B.8 Some secondary alcohols present in plant leaves: nonacosan-10-ol.

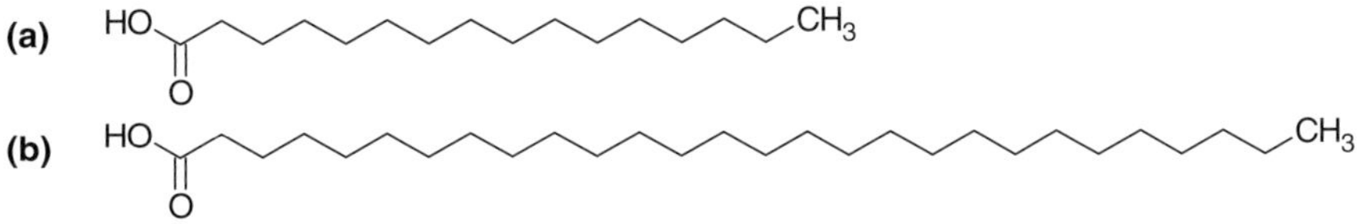

Figure B.9 Some fatty acids present in plant leaves: (a) hexadecanoic acid (palmitic acid); and (b) octacosanoic acid (montanic acid).

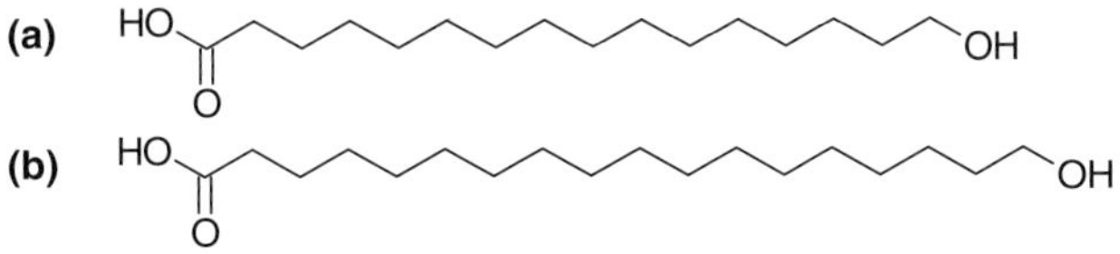

Figure B.10 Some hydroxy acids present in plant leaves: (a) 16-hydroxy hexadecanoic acid; and (b) 18-hydroxy octadecanoic acid.

B.2.7 Hydroxy Acids

Hydroxy acids are organic acids having one or more hydroxyl group attached directly to the carbon chain of an aliphatic or alicyclic carbon atom. Hydroxy acids occur in nature and can also be synthesized through chemical or enzymatic methods. Many of the hydroxy acids occur in plants (Table B.10 and Figure B.10).

Table B.11 *Some primary alcohols present in plant leaves.*

Name	Formula	PubChem CID	Refractive index	Plant species
Pentacosanol	$C_{25}H_{52}O$	92247	1.456	*Stellaria media*
Hexacosanol	$C_{26}H_{54}O$	68171	1.457	*Pisum sativum*
Heptacosanol	$C_{27}H_{56}O$	74822	1.457	*Encephalartos* sp.
Nonacosanol	$C_{29}H_{60}O$	243696	1.458	Encephalartos sp.
Octacosanol	$C_{28}H_{58}O$	68406	1.458	*Acacia* sp., *Pisum* sp., *Trifolium* sp., *Triticum aestivum*

Table B.12 *Some aldehydes present in plant leaves.*

Name	Formula	PubChem CID	Refractive index	Plant species
Hexacosanal	$C_{26}H_{52}O$	3084462	1.451	*Blumeria graminis, Fagus sylvatica, Hordeum vulgare*
Octacosanal	$C_{28}H_{56}O$	3084376	1.453	*Brassica oleracea, Citrus limon*

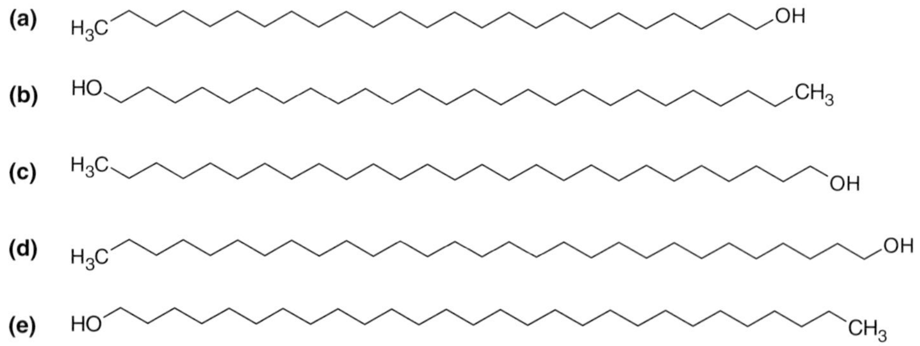

Figure B.11 Some primary alcohols present in plant leaves: (a) pentacosanol; (b) hexacosanol; (c) heptacosanol; (d) nonacosanol; and (e) octacosanol.

B.2.8 Primary Alcohols

Primary alcohols are present in plant and insect waxes. In the plant world, these waxes are cuticle excretions (Table B.11 and Figure B.11).

B.2.9 Aldehydes

Aldehydes are organic molecules containing a functional group with the structure –CHO, consisting of a carbonyl center (a carbon double-bonded to oxygen) with the carbon atom also bonded to hydrogen and to an R group, which is any generic alkyl or side chain. They are widespread in the plant community and responsible for many odors and fragrances and tastes and flavors.

Table B.13 *Some monoterpenes present in plant leaves.*

Name	Formula	PubChem CID	Refractive index	Plant species
α-Pinene	$C_{10}H_{16}$	6654	1.479	*Cannabis sativa, Rosmarinus officinalis, Satureja myrtifolia*
Sabinene	$C_{10}H_{16}$	18818	1.483	*Achillea millefolium, Sequoia sempervirens*

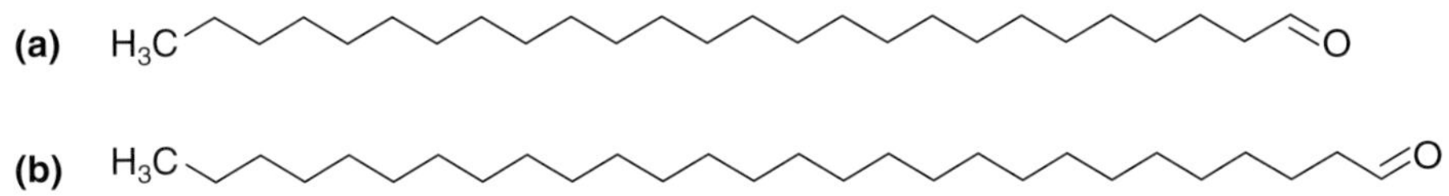

Figure B.12 Some aldehydes present in plant leaves: (a) hexacosanal; and (b) octacosanal.

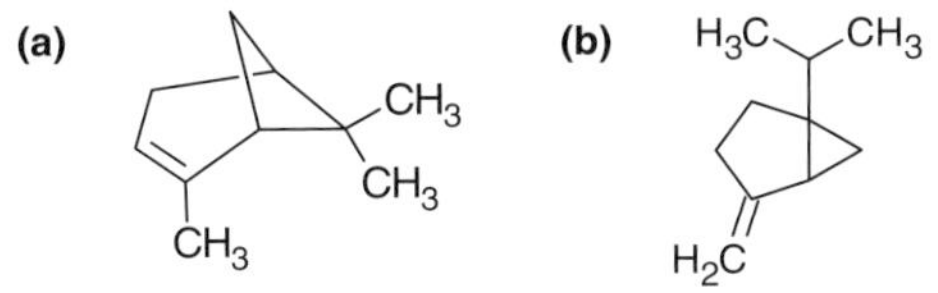

Figure B.13 Some monoterpenes present in plant leaves; (a) α-pinene; and (b) sabinene.

B.3 Oils

See Section 2.3.5.

B.3.1 Terpenoids, Monoterpenes

Monoterpenes are a class of terpenes that consist of two isoprene units and have the molecular formula $C_{10}H_{16}$. They are predominantly products of the secondary metabolism of plants, widely occurring, and are the primary contributors to the organoleptic properties associated with various herbs, spices, citrus, conifers, etc. (Table B.13 and Figure B.13).

B.3.2 Terpenoids, Triterpenes

Triterpenes are a class of chemical compounds composed of three terpene units with the molecular formula $C_{30}H_{48}$. These triterpenes are described to present different properties, such as antitumoral activity, cardioprotective activity, anti-inflammatory activity, and antioxidant protection (Table B.14 and Figure B.14).

B.4 Carbohydrates

Cellulose occurs as a crystalline polymer of D-glucose units; hemicelluloses are a family of polysaccharides that contain many different sugar monomers of D-glucose, D-xylose (e.g., xylans,

Table B.14 *Some triterpenes present in plant leaves.*

Name	Formula	PubChem CID	Refractive index	Plant species
Ursolic acid	$C_{30}H_{48}O_3$	64945	1.555	*Orthosiphon stamineus, Perilla frutescens, Rosmarinus officinalis, Terminalia catappa*
Oleanolic acid	$C_{30}H_{48}O_3$	10494	1.557	*Olea europaea, Orthosiphon stamineus, Perilla frutescens*

Table B.15 *Some carbohydrates present in plant leaves.*

Name	Formula	PubChem CID	Refractive index
D-arabinose	$C_5H_{10}O_5$	5460291	1.543
D-xylose	$C_5H_{10}O_5$	135191	1.646
Galacturonic acid	$C_6H_{10}O_7$	84740	1.591
Fructose	$C_6H_{12}O_6$	5984	1.574
D-galactose	$C_6H_{12}O_6$	6036	1.635
D-glucose	$C_6H_{12}O_6$	5793	1.635
D-mannose	$C_6H_{12}O_6$	18950	1.635
Sucrose	$C_{12}H_{22}O_{11}$	5988	1.656
Raffinose	$C_{18}H_{32}O_{16}$	439242	1.676

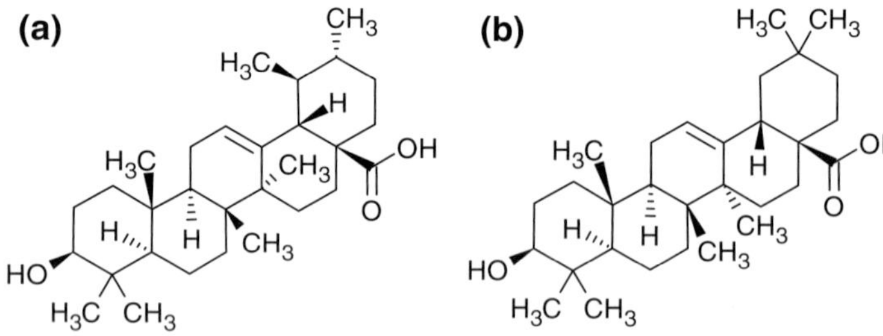

Figure B.14 Some triterpenes present in plant leaves: (a) ursolic acid; and (b) oleanolic acid.

xyloglucans), D-arabinose (e.g., arabinoxylans), D-mannose (e.g., mannans, glucomannans), D-galactose (e.g., galactomannans, galactans), etc. Pectin is composed of a chain of galacturonic acid and other molecules. Starch and cellulose have both the same glucose-based repeat units; in starch, they are oriented in the same direction while in cellulose, they are rotated 180 degrees around the axis of the polymer backbone chain, relative to the last repeat unit (Table B.15 and Figure B.15).

B.5 Carotenoids

See Section 2.3.3.

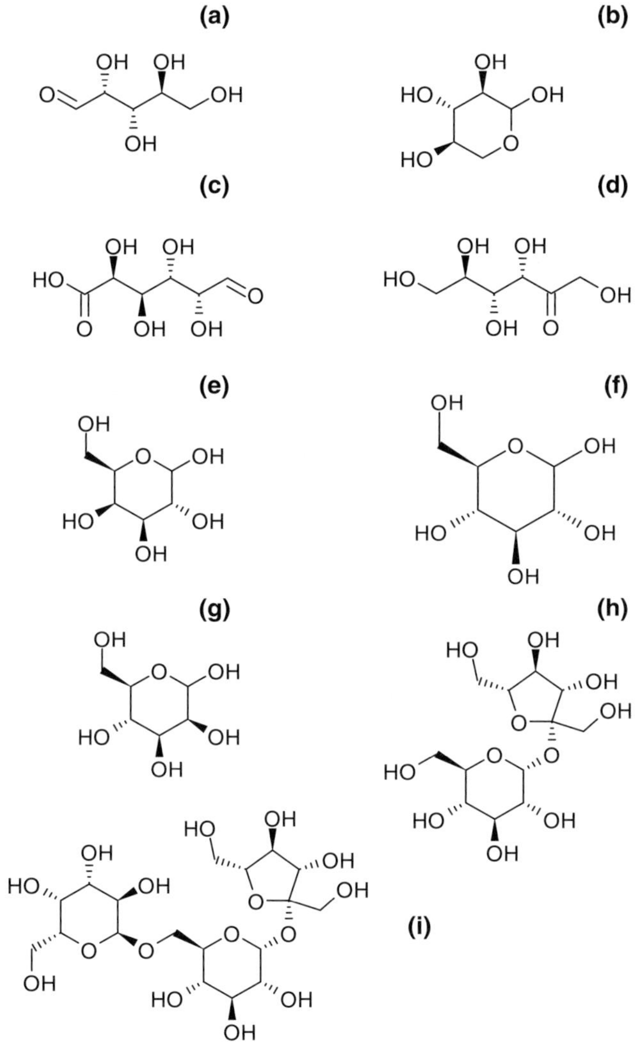

Figure B.15 Some carbohydrates present in plant leaves: (a) D-arabinose; (b) D-xylose; (c) galacturonic acid; (d) fructose; (e) D-galactose; (f) D-glucose; (g) D-mannose; (h) sucrose; and (i) raffinose.

Table B.16 *Some carotenoids present in plant leaves.*

Name	Formula	PubChem CID	Refractive index
α-carotene	$C_{40}H_{56}$	6419725	1.563
β-carotene	$C_{40}H_{56}$	5280489	1.565
Lutein	$C_{40}H_{56}O_2$	5281243	1.582
Zeaxanthin	$C_{40}H_{56}O_2$	5280899	1.585
Antheraxanthin	$C_{40}H_{56}O_3$	5281223	1.597
Neoxanthin	$C_{40}H_{56}O_4$	5281247	1.572
Violaxanthin	$C_{40}H_{56}O_4$	448438	1.610

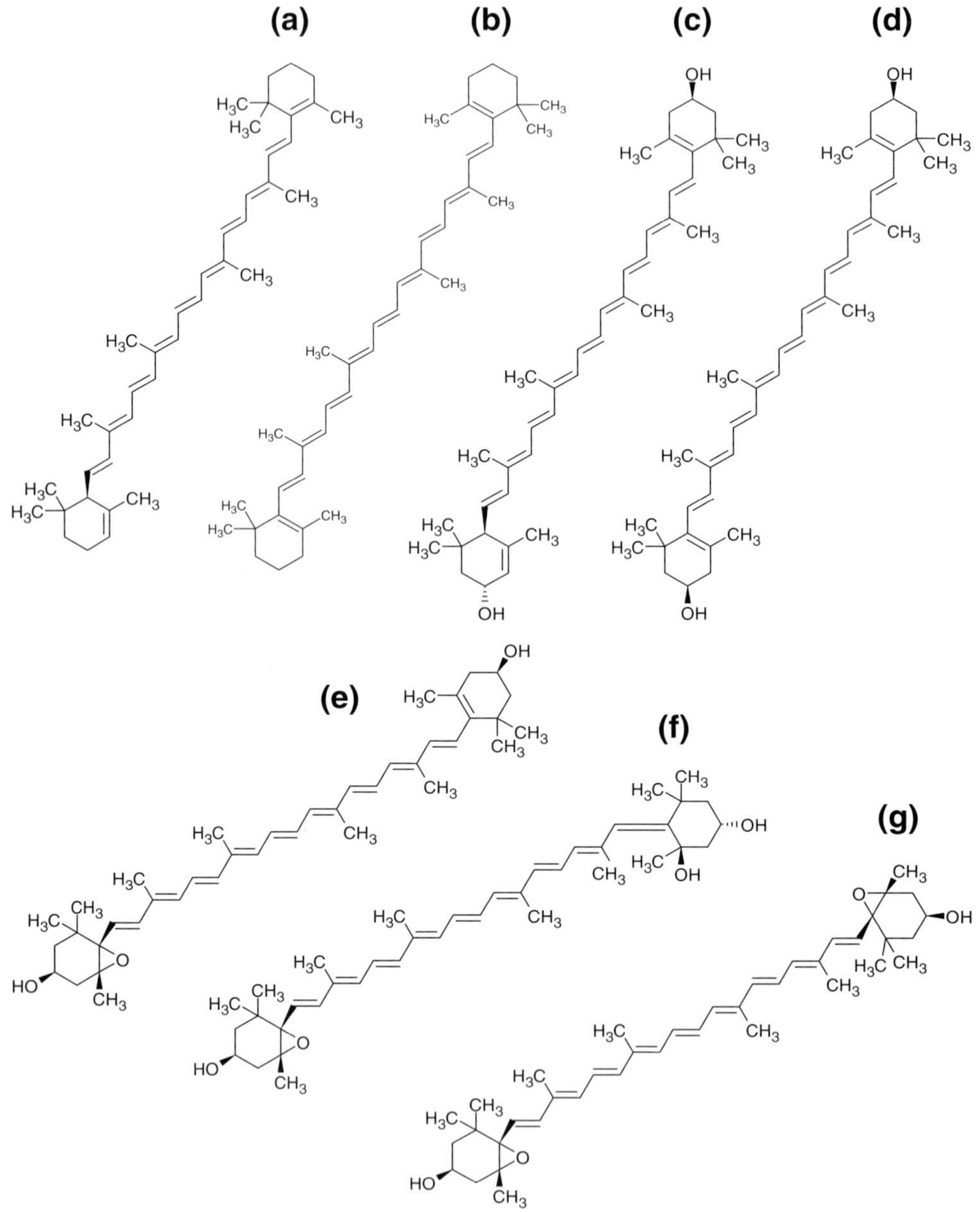

Figure B.16 Some carotenoids present in plant leaves: (a) α-carotene; (b) β-carotene; (c) lutein; (d) zeaxanthin; (e) antheraxanthin; (f) neoxanthin; and (g) violaxanthin.

Appendix C
Planck's Law

The blackbody spectral radiance in unit of power per unit of surface area per unit of wavelength per unit of solid angle (W m^{-2} mm^{-1} sr^{-1}) is given by the famous Planck's law:

$$L_e(\lambda, T) = \frac{2hc^2}{\lambda^5\left(e^{hc/k\lambda T} - 1\right)} \tag{C.1}$$

where λ is the wavelength (m), T is the absolute temperature (K), h is the Planck's constant (6.626×10^{-34} J s), c is the speed of light in a vacuum (2.998×10^8 m s^{-1}), and k is the Boltzmann's constant (1.38×10^{-23} J K^{-1}).

C.1 Wien's Displacement Law

For constant T, the maximum spectral radiance with respect to the wavelength is obtained by differentiating $L_e(\lambda, T)$ and by setting the first derivative equal to zero:

$$\frac{dL_e(\lambda, T)}{d\lambda} = 0 \Rightarrow 5\lambda^4\left(e^{hc/k\lambda T} - 1\right) - \lambda^5\frac{hc}{k\lambda^2 T}e^{hc/k\lambda T} = 0 \Rightarrow 5e^{hc/k\lambda T} - 5 - \frac{hc}{k\lambda T}e^{hc/k\lambda T} = 0 \tag{C.2}$$

If we define the single variable $x \equiv \frac{hc}{k\lambda T}$, then Eq. C.2 becomes

$$f(x) = e^x(5 - x) - 5 = 0 \tag{C.3}$$

Equation C.3 is a transcendental equation that cannot be solved in terms of elementary functions. Its solutions can be found iteratively using the Newton–Raphson method, where the zero of a real function is approximated considering the function, its derivative, and an initial x-value. Without going into details, we write the Taylor series of the function $f(x)$, that is n times continuously differentiable in the neighborhood of x_0:

$$f(x) = f(x_0) + \frac{\partial f}{\partial x}(x_0)\frac{(x - x_0)}{1!} + \frac{\partial^2 f}{\partial x^2}(x_0)\frac{(x - x_0)^2}{2!} + \ldots + \frac{\partial^n f}{\partial x^n}(x_0)\frac{(x - x_0)^n}{n!} \tag{C.4}$$

We search the value of x such that $f(x) = 0 \Rightarrow f(x_0) + \frac{\partial f}{\partial x}(x_0)(x - x_0) \approx 0 \Rightarrow x \approx x_0 - \frac{f(x_0)}{f'(x_0)}$. We start the process with an arbitrary initial value and the method will usually converge, provided this initial guess is close enough to the solution. For the function $f(x)$ of Eq. C.4, the recurrence relation is

$$x_{k+1} = x_k - \frac{e^{x_k}(5 - x_k) - 5}{e^x(4 - x_k)} = x_k - \frac{5 - x_k - 5e^{-x_k}}{4 - x_k} \tag{C.5}$$

If the initial guess is $x_0 = 6$, then $x_1 = 5.4938$, $x_2 = 5.1495$, $x_3 = 4.9942$, $x_4 = 4.9660$, $x_5 = 4.9651$, and $x_6 = 4.9651$, which is the solution. Then we obtain

$$\frac{hc}{k\lambda T} = 4.9651 \Rightarrow \lambda_{\max} \times T = 2.898 \times 10^{-3}\text{m K} \tag{C.6}$$

This law states that the product of the wavelength, where the spectral radiance is maximum, and the absolute temperature is a constant. Note that one can start with the form of Planck's law as a function of frequency.

C.2 Stefan–Boltzmann Law

Let now integrate $L_e(\lambda, T)$ over the whole spectrum and over the hemisphere into which the radiation is emitted. First, we integrate over the wavelength from zero to infinity to obtain the total radiance in unit of power per unit of surface area per unit of solid angle (W m^{-2} sr^{-1}):

$$L_e(T) = \int_0^\infty L_e(\lambda, T)\, d\lambda \tag{C.7}$$

For this calculation it is easier to present the Plank's law over a frequency scale instead of a wavelength scale. If we set $\lambda = \frac{c}{\nu} \Rightarrow d\lambda = -\frac{c}{\nu^2}d\nu$, Eq. C.7 becomes

$$L_e(T) = \frac{2h}{c^2}\int_0^\infty \frac{\nu^3}{e^{h\nu/kT} - 1} d\nu \tag{C.8}$$

We introduce a new variable $y = \frac{h\nu}{kT} \Rightarrow \nu = \frac{ykT}{h}$ and $d\nu = \frac{kT}{h}dy$. With this, the integration reads as

$$L_e(T) = \frac{2k^4T^4}{h^3c^2}\int_0^\infty \frac{y^3}{e^y - 1} dy \tag{C.9}$$

There are a number of ways to solve this integration; a simple one consists of using the infinite geometric series: for $|x| < 1$, $\frac{1}{1-x} = \sum_{n=0}^{\infty} x^n$. By replacing x with e^{-y}, one obtains $\frac{1}{1-e^{-y}} = \sum_{n=0}^{\infty}(e^{-y})^n = \sum_{n=0}^{\infty} e^{-ny}$. Then we can write that $\frac{1}{e^y-1} = \frac{e^{-y}}{1-e^{-y}} = \sum_{n=0}^{\infty} e^{-(n+1)y} = \sum_{m=1}^{\infty} e^{-my}$. Since the integral of a sum is equal to the sum of the integral, Eq. C.9 is transformed into

$$L_e(T) = \frac{2k^4T^4}{h^3c^2}\int_0^\infty y^3 \sum_{m=1}^{\infty} e^{-my} dy = \frac{2k^4T^4}{h^3c^2}\sum_{m=1}^{\infty}\int_0^\infty y^3 e^{-my} dy \tag{C.10}$$

and after three integrations by parts, we obtain

$$L_e(T) = \frac{2k^4T^4}{h^3c^2}\sum_{m=1}^{\infty}\frac{6}{m^4} \tag{C.11}$$

The Riemann zeta-function defined as $\zeta(4) = \sum_{m=1}^{\infty}\frac{1}{m^4}$ converges very quickly toward $\frac{\pi^4}{90}$, so that

$$L_e(T) = \frac{2\pi^4k^4}{15h^3c^2}T^4 \tag{C.12}$$

For the integration over the hemisphere, we assume that the emission from a plane radiating surface into the hemisphere is proportional to the projected surface area and that it is isotropic, that is, $L_e(T)$ does not depend on directions:

$$M(T) = \int_\Omega L_e(T)\cos\theta d\Omega = L_e(T)\int_0^{\pi/2}\int_0^{2\pi}\cos\theta\sin\theta\, d\theta\, d\phi = \pi L_e(T) \tag{C.13}$$

which leads to the famous Stefan–Boltzmann law:

$$M(T) = \frac{2\pi^5k^4}{15h^3c^2}T^4 = \sigma T^4 \tag{C.14}$$

where σ is the Stefan–Boltzmann constant (5.67×10^{-8} W m^{-2} K^{-4}). The total radiated power of a blackbody surface is proportional to the fourth power of the temperature.

C.3 Band Emission

It is often interesting to know the fraction of the total emission from a blackbody within a certain wavelength band. For instance, in the wavelength range between 0 and λ, this fraction may be expressed as

$$F(T) = \frac{\pi\int_0^\lambda L_e(\lambda, T)\, d\lambda}{\sigma T^4} \tag{C.15}$$

For a prescribed temperature of 5800 K (surface temperature of the Sun), $F = 0.2502$ at $\lambda = 0.5$ μm, $F = 0.7199$ at $\lambda = 1$ μm, $F = 0.9401$ at $\lambda = 2$ μm, and $F = 0.9790$ at $\lambda = 3$ μm.

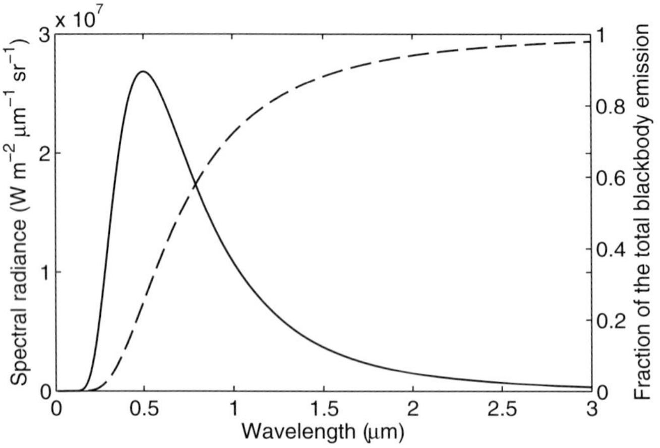

Figure C.1 Spectral radiance (plain line, left *y*-axis) and fraction of the total blackbody emission (dashed line, right *y*-axis) between 0 and $\lambda = 3$ μm for a surface temperature of 5800 K.

Appendix D
Radiometry

D.1 Lambertian Surface

A Lambertian surface is a surface that has a constant radiance L (radiant flux Φ that leaves a surface per unit solid angle and unit projected area of that emitting surface) (see Section 4.1.3). In accordance with the definition of radiance, a radiant (luminous) flux may be emitted, transmitted, and/or reflected by the surface (Figure D.1).

The radiant intensity I (radiant flux Φ proceeding from a point source per unit solid angle in the direction considered) varies as a Lambert's cosine law:

$$I_r = I_0 \cos\theta_v \tag{D.1}$$

where θ_v is the viewing zenith angle.

D.2 Perfect Reflector

Consider an elementary surface dA uniformly irradiated in the direction (θ_i, ϕ_i). Assume now that this elementary surface is observed in the direction (θ_r, ϕ_r) in the solid angle element $d\Omega_r = \sin\theta_r d\theta_r d\phi_r$. If $d\phi_R$ (W) is the radiant flux that leaves the surface, the radiance L_R (W m^{-2} sr^{-1}) is given by

$$L_R(\theta_i, \phi_i, \theta_r, \phi_r) = \frac{d^2\Phi_R}{dA\cos\theta_r d\Omega_r} \tag{D.2}$$

Then the radiant flux equals

$$d^2\Phi_R = L_R(\theta_i, \phi_i, \theta_r, \phi_r) dA \cos\theta_r d\Omega_r \tag{D.3}$$

The total flux is obtained by integrating Eq. D.3:

$$d\Phi_R = dA \int_0^{2\pi} d\phi_r \int_0^{\pi/2} L_R(\theta_i, \phi_i, \theta_r, \phi_r) \sin\theta_r \cos\theta_r d\theta_r \tag{D.4}$$

For a Lambertian surface:

$$L_R(\theta_i, \phi_i, \theta_r, \phi_r) = L_R(\theta_i, \phi_i) \tag{D.5}$$

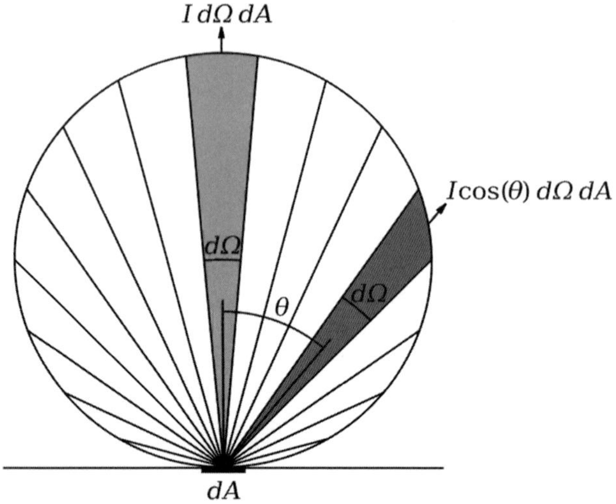

Figure D.1 Geometry of reflected elementary beams (source: Wikipedia).

then

$$\begin{aligned} d\Phi_R &= dA\ L_R(\theta_i,\phi_i)\int_0^{2\pi} d\phi_r \int_0^{\pi/2} \sin\theta_r \cos\theta_r d\theta_r \\ &= 2\pi dA\ L_R(\theta_i,\phi_i)\int_0^{\pi/2} \frac{\sin 2\theta_r}{2} d\theta_r \\ &= \pi dA\ L_R(\theta_i,\phi_i)\left[-\frac{\cos 2\theta_r}{2}\right]_0^{\pi/2} \\ &= \pi dA\ L_R(\theta_i,\phi_i) \end{aligned} \tag{D.6}$$

Therefore the relationship between irradiance and radiance is

$$E_I(\theta_i,\phi_i) = \frac{d\Phi_R}{dA} = \pi\ L_R(\theta_i,\phi_i) \tag{D.7}$$

D.3 Consequence for the BRDF

The bidirectional reflectance distribution function (BRDF) (sr^{-1}) of the surface is then defined as (see Section 4.2.1)

$$BRDF(\theta_i,\phi_i,\theta_r,\phi_r) = \frac{L_R(\theta_i,\phi_i,\theta_r,\phi_r)}{E_I(\theta_i,\phi_i)} \tag{D.8}$$

The bidirectional reflectance factor (BRF) is then defined as the ratio of the radiance reflected by the surface and the radiance reflected by a Lambertian surface, also called a perfect reflector (pr). In the same direction:

$$BRF(\theta_i, \phi_i, \theta_r, \phi_r) = \frac{L_R(\theta_i, \phi_i, \theta_r, \phi_r)}{L_{pr}(\theta_i, \phi_i, \theta_r, \phi_r)} \tag{D.9}$$

As seem in Eq. D.5:

$$L_{pr}(\theta_i, \phi_i, \theta_r, \phi_r) = L_{pr}(\theta_i, \phi_i) = \frac{E_I(\theta_i, \phi_i)}{\pi} \tag{D.10}$$

then

$$BRDF(\theta_i, \phi_i, \theta_r, \phi_r) = \frac{1}{\pi} \frac{L_R(\theta_i, \phi_i, \theta_r, \phi_r)}{L_{pr}(\theta_i, \phi_i, \theta_r, \phi_r)} = \frac{1}{\pi} BRF(\theta_i, \phi_i, \theta_r, \phi_r) \tag{D.11}$$

Appendix E
Fresnel's Equations

This appendix develops the reflection and refraction of light at a flat interface between two optical media (or dielectrics). For simplicity, we will assume non-absorbing media. Consider an electromagnetic wave propagating in a medium in the direction $\vec{k}$: the electric field vector $\vec{E}$ and the magnetic field vector $\vec{H}$ oscillate perpendicular to one another and in phase. $\vec{E}$, $\vec{H}$, and $\vec{k}$ vectors form a right-handed triad. A wave whose electric field oscillates in a particular direction is said to be polarized, and if the oscillation occurs in a plane, it is said to be plane polarized. We shall distinguish the case where $\vec{E}$ is perpendicular to the plane of incidence (the plane containing the normal to the interface and the propagation vector of the incident wave), identified by the subscript $\perp$, from the case where $\vec{E}$ is in the plane of incidence, identified by the subscript $\|$ (Figure E.1).

A wave that is vibrating in more than one plane is referred to as an unpolarized wave. This is typically the light emitted by the Sun before interaction with the atmosphere. It can be convenient to consider unpolarized light as an electromagnetic wave which has half of its vibrations in a horizontal plane and half of its vibrations in a vertical plane.

E.1 Snell's Law

When incident radiation moves from the medium having an index of refraction n_1 into the second medium with index n_2, both reflection and refraction may occur. The incident wave (subscript i) gives rise to a reflected wave (subscript r), which remains in the medium of index n_1 and to a transmitted wave (subscript t), which is propagated in the medium of index n_2. When the incident wave falls at an angle θ_i on the interface, it is reflected according to the angle $\theta_r = \theta_i$ and refracted according to the angle θ_t which is related to θ_i by the Snell–Descartes law of refraction:

$$n_1 \sin\theta_i = n_2 \sin\theta_t \tag{E.1}$$

This law is named after the Dutch scientist Willebrørd Snell (1591–1626) who first stated it in a manuscript in 1621, and after the French scientist René Descartes (1596–1650) who first put the law into widespread circulation in his *Discourse on Method*, published in 1637. But the origin of this law is probably much older (Kwan et al., 2002).

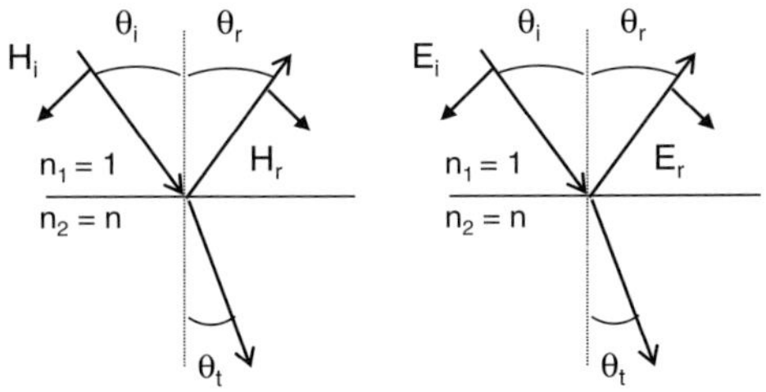

Figure E.1 Reflection and refraction of a light beam: (a) perpendicular polarization; and (b) parallel polarization.

E.2 Refractive Index

In an isotropic medium, the (electric) permittivity ε (F m^{-1}) and the (magnetic) permeability μ (H m^{-1}) determine the phase velocity v (m s^{-1}) of electromagnetic radiation through that medium:

$$v^2\varepsilon\mu = 1 \tag{E.2}$$

The refractive index (or index of refraction) is defined as the ratio of velocity of light c (m s^{-1}) in a vacuum to its phase velocity v (m s^{-1}) in the medium:

$$n = \frac{c}{v} \tag{E.3}$$

By combining Eq. E.2 and Eq. E.3 we find

$$n = c\sqrt{\varepsilon\mu} \tag{E.4}$$

Let us define the relative permittivity of a material:

$$\varepsilon_r = \frac{\varepsilon}{\varepsilon_0} \tag{E.5}$$

with ε_0 the vacuum permittivity (8.854×10^{-12} F m^{-1}). One can also define the relative permeability:

$$\mu_r = \frac{\mu}{\mu_0} \tag{E.6}$$

with μ_0 the vacuum permeability (1.256×10^{-6} H m^{-1}). The speed of light in a vacuum c is related to ε_0 and μ_0 by the relationship:

$$c = \frac{1}{\sqrt{\varepsilon_0\mu_0}} \tag{E.7}$$

The combination of Eq. E.4 and Eq. E.7 gives another expression for the refractive index:

$$n = \sqrt{\varepsilon_r\mu_r} \tag{E.8}$$

E.3 Fresnel's Equations

The electric field vector $\vec{E}$ and the magnetic field vector $\vec{H}$ components of the electromagnetic waves are related by the cross product:

$$\vec{H} = \frac{\vec{k} \wedge \vec{E}}{\omega\mu} \tag{E.9}$$

with $\vec{k} = k\,\vec{u}$ the wave vector, k the wavenumber, and ω the angular frequency. According to the triple product expansion, which states that $\vec{u}_1 \wedge (\vec{u}_2 \wedge \vec{u}_3) = (\vec{u}_1.\vec{u}_3)\vec{u}_2 - (\vec{u}_1.\vec{u}_2)\vec{u}_3$:

$$\vec{k} \wedge \vec{H} = \frac{\vec{k} \wedge \left(\vec{k} \wedge \vec{E}\right)}{\omega\mu} = \frac{\left(\vec{k}.\vec{E}\right)\vec{k} - \left(\vec{k}.\vec{k}\right)\vec{E}}{\omega\mu} = \frac{-\vec{E}}{\omega\mu} \tag{E.10}$$

Using Eq. 2, the relationship that gives $\vec{E}$ as a function of $\vec{H}$ is straightforward:

$$\vec{E} = \frac{-\vec{k} \wedge \vec{H}}{\omega\varepsilon} \tag{E.11}$$

We will concentrate on the electric field vector $\vec{E}$. The ratio of the reflected and transmitted electric field amplitude to initial electric field for electromagnetic radiation incident on a dielectric is given by the well-known Fresnel's equations, after the French physicist and engineer Augustin-Jean Fresnel (1788–1827), who was the first to establish them in 1918 (Lvovsky, 2013).

E.3.1 Reflection and Transmission Coefficients

E.3.1.1 Perpendicular Polarization (S Polarization)

Let us assume that the electric field vector $\vec{E}$ is perpendicular to the plane of incidence (Figure E.2).

The electric field vectors of the incident, reflected, and transmitted (or refracted) waves are

$$\vec{E}_i = E_i e^{-j\vec{k}_i.\vec{r}}\vec{e}_x \tag{E.12}$$

$$\vec{E}_r = E_r e^{-j\vec{k}_r.\vec{r}}\vec{e}_x \tag{E.13}$$

$$\vec{E}_t = E_t e^{-j\vec{k}_t.\vec{r}}\vec{e}_x \tag{E.14}$$

with

$$\begin{cases} \vec{k}_i = k_0\vec{u}_i = k_0\left(\sin\theta_i\vec{e}_y - \cos\theta_i\vec{e}_z\right) = k_0\sin\theta_i\vec{e}_y - k_0\cos\theta_i\vec{e}_z \\ \vec{k}_r = k_0\vec{u}_r = k_0\left(\sin\theta_i\vec{e}_y + \cos\theta_i\vec{e}_z\right) = k_0\sin\theta_i\vec{e}_y + k_0\cos\theta_i\vec{e}_z \\ \vec{k}_t = \vec{\beta}_t - j\vec{\alpha}_t = \beta_t\vec{u}_t + j\alpha_t\vec{e}_z = \beta_t\left(\sin\theta_t\vec{e}_y - \cos\theta_t\vec{e}_z\right) + j\alpha_t\vec{e}_z = \beta_t\sin\theta_t\vec{e}_y + (-\beta_t\cos\theta + j\alpha_t)\vec{e}_z \end{cases}$$

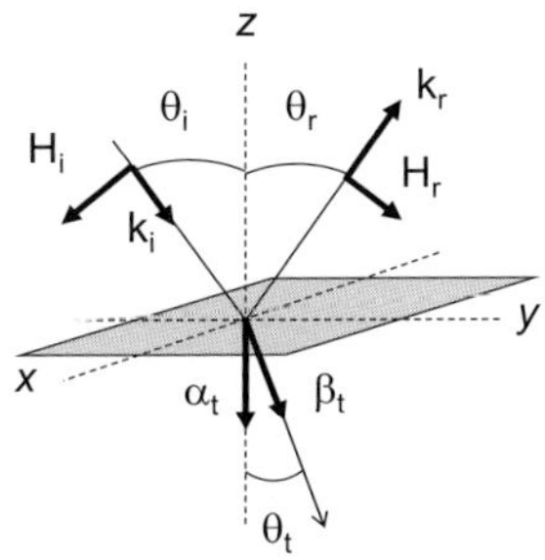

Figure E.2 Field vectors of the incident, transmitted, and reflected waves in the case where the electric field vectors are perpendicular to the plane of incidence (S polarization).

The phase velocity v is related to the wavenumber k and the angular frequency ω:

$$v = \frac{c}{n} = \frac{\omega}{k} \tag{E.15}$$

By combining Eq. E.7 and Eq. E.15, we find the so-called dispersion relation for electromagnetic waves in a vacuum:

$$k_0^2 = \omega^2 \varepsilon_0 \mu_0 \tag{E.16}$$

with k_0 the wavenumber in free space. The continuity of the electric field component $\vec{E}$ tells us that

$$\vec{E}_i + \vec{E}_r = \vec{E}_t \tag{E.17}$$

A projection along $\vec{e}_x$ leads to

$$E_i + E_r = E_t \Rightarrow 1 + \frac{E_r}{E_i} = \frac{E_t}{E_i} \Rightarrow 1 + R_\perp = T_\perp \tag{E.18}$$

The continuity of the magnetic field component $\vec{H}$ tells us that

$$\vec{H}_i + \vec{H}_r = \vec{H}_t \Rightarrow \frac{\vec{k}_i \wedge \vec{E}_i}{\omega\mu_0} + \frac{\vec{k}_r \wedge \vec{E}_r}{\omega\mu_0} = \frac{\vec{k}_t \wedge \vec{E}_t}{\omega\mu} \tag{E.19}$$

A projection along $\vec{e}_y$ leads to

$$\begin{aligned} &\frac{-k_0 \cos\theta_i E_i + k_0 \cos\theta_i E_r}{\mu_0} = \frac{-\beta_t \cos\theta_t + j\alpha_t}{\mu} E_t \\ &\Rightarrow \frac{k_0 \cos\theta_i}{\mu_0}\left(1 - \frac{E_r}{E_i}\right) = \frac{\beta_t \cos\theta_t - j\alpha_t}{\mu}\frac{E_t}{E_i} \\ &\Rightarrow \frac{k_0 \cos\theta_i}{\mu_0}(1 - R_\perp) = \frac{\beta_t \cos\theta_t - j\alpha_t}{\mu} T_\perp \end{aligned} \tag{E.20}$$

From Eq. E.18 and Eq. E.20, we deduce $R_\perp$ and $T_\perp$:

$$R_\perp = \frac{\mu k_0 \cos\theta_i - (\beta_t \cos\theta_t - j\alpha_t)\mu_0}{\mu k_0 \cos\theta_i + (\beta_t \cos\theta_t - j\alpha_t)\mu_0} \tag{E.21}$$

$$T_\perp = \frac{2k_0 \cos\theta_i \mu}{\mu k_0 \cos\theta_i + (\beta_t \cos\theta_t - j\alpha_t)\mu_0} \tag{E.22}$$

Now: $(\beta_t \cos\theta_t - j\alpha_t)^2 = \beta_t^2 \cos^2\theta_t - \alpha_t^2 - 2\,j\alpha_t\beta_t \cos\theta_t = \beta_t^2 - \alpha_t^2 - 2\,j\alpha_t\beta_t \cos\theta_t - \beta_t^2 \sin^2\theta_t = k_t^2 - k_o^2 \sin^2\theta_t$ because $\vec{k}_t = \vec{\beta}_t - j\vec{\alpha}_t$ and $\beta_t \sin\theta_t = k_0 \sin\theta_i$ (Snell's law). Then we obtain:

$$R_\perp = \frac{\mu \cos\theta_i - \sqrt{k_t^2/k_0^2 - \sin^2\theta_i}\mu_0}{\mu \cos\theta_i + \sqrt{k_t^2/k_0^2 - \sin^2\theta_i}\mu_0} \tag{E.23}$$

$$T_\perp = \frac{2\mu \cos\theta_i}{\mu \cos\theta_i + \sqrt{k_t^2/k_0^2 - \sin^2\theta_i}\mu_0} \tag{E.24}$$

By assuming that $\mu = \mu_0$ (non-magnetic medium) and by setting $n = k_t/k_0$:

$$R_\perp = \frac{\cos\theta_i - \sqrt{n^2 - \sin^2\theta_i}}{\cos\theta_i + \sqrt{n^2 - \sin^2\theta_i}} \tag{E.25}$$

$$T_\perp = \frac{2\cos\theta_i}{\cos\theta_i + \sqrt{n^2 - \sin^2\theta_i}} \tag{E.26}$$

E.3.1.2 Parallel Polarization (P Polarization)

Let us assume that the electric field vector $\vec{E}$ is parallel to the plane of incidence (Figure E.3).

The magnetic field vectors of the incident, reflected, and transmitted (or refracted) waves are

$$\vec{H}_i = H_i e^{-j\vec{k}_i.\vec{r}}\vec{e}_x \tag{E.27}$$

$$\vec{H}_r = H_r e^{-j\vec{k}_r.\vec{r}}\vec{e}_x \tag{E.28}$$

$$\vec{H}_t = H_t e^{-j\vec{k}_t.\vec{r}}\vec{e}_x \tag{E.29}$$

The continuity of the magnetic field component $\vec{H}$ tells us that

$$\vec{H}_i + \vec{H}_r = \vec{H}_t \tag{E.30}$$

A projection along $\vec{e}_x$ leads to

$$H_i + H_r = H_t \Rightarrow 1 + \frac{H_r}{H_i} = \frac{H_t}{H_i} \Rightarrow 1 + R_{||} = T_{||} \tag{E.31}$$

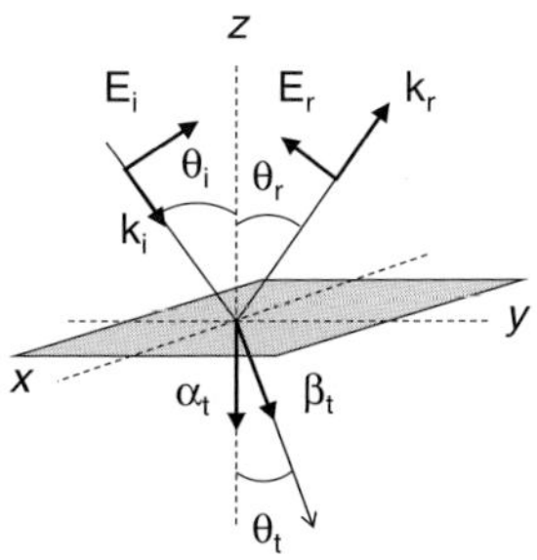

Figure E.3 Field vectors of the incident, transmitted, and reflected waves in the case where the electric field vectors lie within the plane of incidence (P polarization).

The continuity of the electric field component $\vec{E}$ tells us that

$$\vec{E}_i + \vec{E}_r = \vec{E}_t \Rightarrow \frac{\vec{k}_i \wedge \vec{H}_i}{\omega\varepsilon_0} + \frac{\vec{k}_r \wedge \vec{H}_r}{\omega\varepsilon_0} = \frac{\vec{k}_t \wedge \vec{H}_t}{\omega\varepsilon} \tag{E.32}$$

A projection along $\vec{e}_y$ leads to

$$\frac{k_0 \cos\theta_i}{\varepsilon_0}\left(1 - R_{||}\right) = \frac{\beta_t \cos\theta_t - j\alpha_t}{\mu} T_{||} \tag{E.33}$$

From Eq. E.31 and Eq. E.33, we deduce $R_{||}$ and $T_{||}$, which are similar to $R_\perp$ and $T_\perp$:

$$R_{||} = \frac{\varepsilon\cos\theta_i - \sqrt{n^2 - \sin^2\theta_i \varepsilon_0}}{\varepsilon\cos\theta_i + \sqrt{n^2 - \sin^2\theta_i \varepsilon_0}} \tag{E.34}$$

$$T_{||} = \frac{2\varepsilon\cos\theta_i}{\varepsilon\cos\theta_i + \sqrt{n^2 - \sin^2\theta_i \varepsilon_0}} \tag{E.35}$$

We multiply these two expressions by μ/ε_0. Now consider materials where $\mu = \mu_0$ (non-magnetic medium) and set $n^2 = \varepsilon\mu/\varepsilon_0\mu_0$ to obtain

$$R_{||} = \frac{n^2\cos\theta_i - \sqrt{n^2 - \sin^2\theta_i}}{n^2\cos\theta_i + \sqrt{n^2 - \sin^2\theta_i}} \tag{E.36}$$

$$T_{||} = \frac{2n^2\cos\theta_i}{n^2\cos\theta_i + \sqrt{n^2 - \sin^2\theta_i}} \tag{E.37}$$

E.3.2 Special Cases

One can remark that for normal angles of incidence ($\theta_i = 0$), then

$$R_\perp = \frac{1-n}{1+n} \tag{E.38}$$

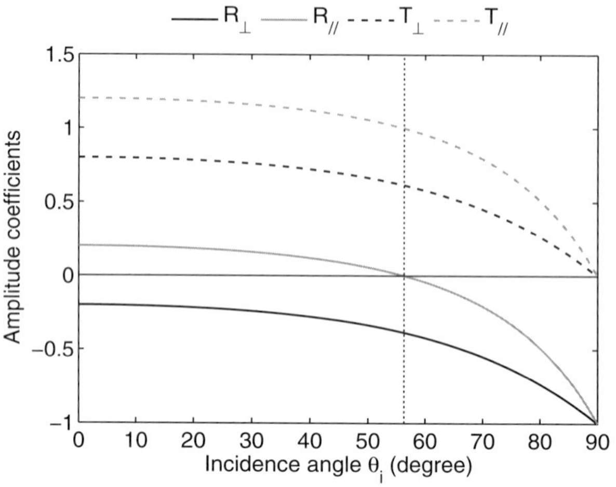

Figure E.4 Reflection and transmission coefficients calculated for perpendicular and parallel polarization and a refractive index $n = 1.5$. We can verify the relationships $1 + R_\perp = T_\perp$ and $1 + R_{||} = T_{||}$.

$$T_\perp = \frac{2}{1+n} \tag{E.39}$$

and

$$R_{||} = \frac{n-1}{n+1} \tag{E.40}$$

$$T_{||} = \frac{2n}{n+1} \tag{E.41}$$

For the glancing case of incidence $(\theta_i = \pi/2)$, $R_\perp = R_{||} = 1$ and $T_\perp = T_{||} = 1$. Moreover, Figure E.4 shows that there is a particular incident angle, called the Brewster's angle, where the parallel-polarized reflected light equals zero: $R_{||} = 0 \Rightarrow \theta_B = \arcsin\sqrt{n^2/n^2+1} = \arctan n$.

E.4 Reflectivity and Transmissivity

The transmissivity is the fraction of light transmitted by a non-absorbing medium. Let us define the Poynting vector which quantifies the energy flux (W m^{-2}) of an electromagnetic field in the direction of propagation. It is written as the cross product:

$$\vec{S} = \frac{1}{2}\vec{E} \wedge \vec{H}^* \tag{E.42}$$

The transmissivity for perpendicular polarization can be calculated using the real part of the complex Poynting vector:

$$t_\perp = |T_\perp|^2\, \Re\left(\frac{\sqrt{n^2 - \sin^2\theta_i}}{\cos\theta_i}\right) = \frac{4\cos^2\theta_i}{\left(\cos\theta_i + \sqrt{n^2 - \sin^2\theta_i}\right)^2} \times \frac{\sqrt{n^2 - \sin^2\theta_i}}{\cos\theta_i} \tag{E.43}$$

Eq. E.43 simplifies into

$$t_{\perp}(\theta_i, 1, n) = \frac{4\sqrt{1 - \sin^2\theta_i}\sqrt{n^2 - \sin^2\theta_i}}{\left(\sqrt{1 - \sin^2\theta_i} + \sqrt{n^2 - \sin^2\theta_i}\right)^2} \tag{E.44}$$

The same scheme applies to the transmissivity for parallel polarization:

$$t_{||} = |T_{||}|^2\, \Re\left(\frac{\sqrt{n^2 - \sin^2\theta_i}}{\cos\theta_i}\right) = \frac{4n^4\cos^2\theta_i}{\left(n^2\cos\theta_i + \sqrt{n^2 - \sin^2\theta_i}\right)^2} \times \frac{\sqrt{n^2 - \sin^2\theta_i}}{n^2\cos\theta_i} \tag{E.45}$$

Eq. E.45 simplifies into

$$t_{||}(\theta_i, 1, n) = \frac{4n^2\sqrt{1 - \sin^2\theta_i}\sqrt{n^2 - \sin^2\theta_i}}{\left(n^2\sqrt{1 - \sin^2\theta_i} + \sqrt{n^2 - \sin^2\theta_i}\right)^2} \tag{E.46}$$

If $\theta_i = 0$ (normal incidence), we obtain

$$t_{\perp}(0, 1, n) = t_{||}(0, 1, n) = \frac{4n}{(n+1)^2} \tag{E.47}$$

and since $r_{\perp} + t_{\perp} = 1$ and $r_{||} + t_{||} = 1$, we can deduce the reflectivity:

$$r_{\perp}(0, 1, n) = r_{||}(0, 1, n) = \left(\frac{n-1}{n+1}\right)^2 \tag{E.48}$$

If $\theta_i = \pi/2$ (grazing incidence), we obtain $t_{\perp}(0, 1, n) = t_{||}(0, 1, n) = 0$ and $r_{\perp}(0, 1, n) = r_{||}(0, 1, n) = 1$. Figure E.5 illustrates the variation of $t_{\perp}$, $t_{||}$, $r_{\perp}$ and $r_{||}$ as a function of θ_i, for a refractive index $n = 1.5$.

The average transmissivity in natural radiation is then given by

$$t(\theta_i, 1, n) = \frac{1}{2}\left(t_{\perp}(\theta_i, 1, n) + t_{||}(\theta_i, 1, n)\right) \tag{E.49}$$

Of particular interest is the transmissivity for incident radiation which impinges on the phase boundary surface at all angles θ_i between 0 and α, so that integration must be carried out over all these angles and the average obtained. Since the radiation flux impinging on a unit surface is proportional to the cosine of the angle of incidence, we have from the second mean value theorem:

$$t_{av}(\alpha, 1, n) = \frac{\displaystyle\int_0^{2\pi} d\varphi \int_0^{\alpha} t(\theta_i, 1, n)\cos\theta_i \sin\theta_i d\theta_i}{\displaystyle\int_0^{2\pi} d\varphi \int_0^{\alpha} \cos\theta_i \sin\theta_i d\theta_i} \tag{E.50}$$

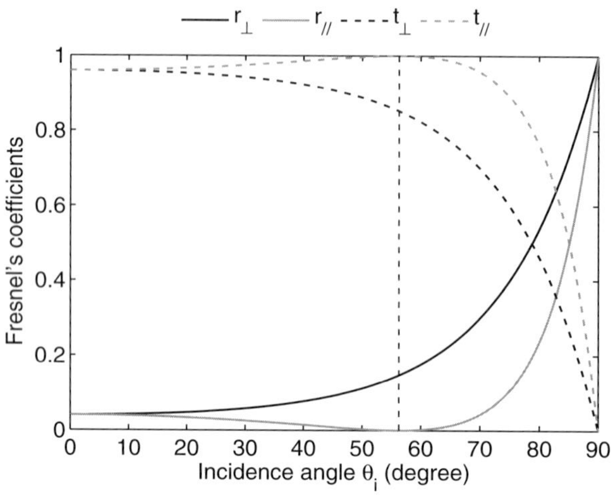

Figure E.5 Reflectivity and transmissivity coefficients calculated for perpendicular and parallel polarization and a refractive index $n = 1.5$. We can verify the relationships $r_\perp + t_\perp = 1$ and $r_{||} + t_{||} = 1$.

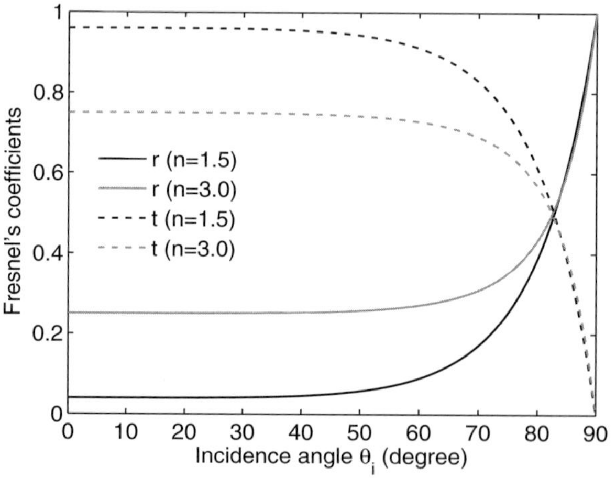

Figure E.6 Reflectivity and transmissivity coefficients calculated for unpolarized light and refractive indices $n = 1.5$ and $n = 3.0$.

If we make the substitution $x = \sin^2\theta_i$, we obtain $dx = 2\sin\theta_i \cos\theta_i$ and hence

$$t_{av}(\alpha, 1, n) = \frac{\displaystyle\int_0^{\sin^2\alpha} t(x, 1, n)dx}{\displaystyle\int_0^{\sin^2\alpha} dx} = \frac{1}{2}\left(t_{\perp,av}(\alpha, 1, n) + t_{||,av}(\alpha, 1, n)\right) \tag{E.51}$$

Simple formulations of $t_{av}(\alpha, 1, n)$ can be provided for $\alpha = 0$:

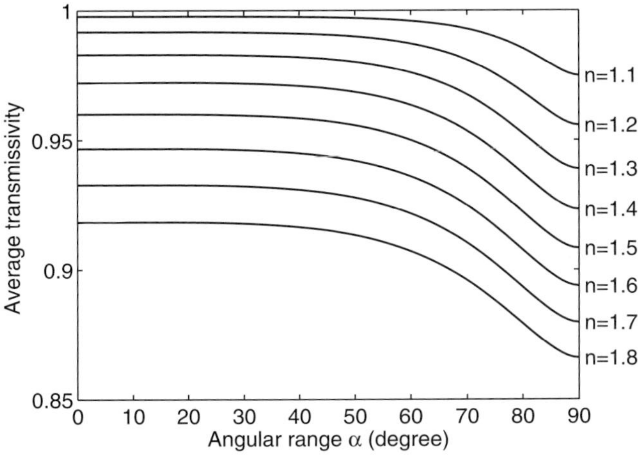

Figure E.7 Average transmissivity of a plane dielectric surface for different values of the refractive index over the angular range α measured from the surface.

$$t_{av}(0,1,n) = t_{\perp,av}(0,1,n) = t_{||,av}(0,1,n) = \frac{4n}{(n+1)^2} \tag{E.52}$$

The integration of Eq. E.51 has been carried out by Judd (1942) and later on by Stern (1964) for diffuse incident radiation ($\alpha = \pi/2$):

$$t_{\perp,av}\left(\frac{\pi}{2},1,n\right) = \frac{4(2n+1)}{3(n+1)^2} \tag{E.53}$$

$$t_{||,av}\left(\frac{\pi}{2},1,n\right) = \frac{4n^3(n^2+2n-1)}{(n^2+1)^2(n^2-1)} - \frac{2n^2(n^2+1)}{(n^2-1)^2}\log n + \frac{2n^2(n^2-1)^2}{(n^2+1)^3}\log\frac{n(n+1)}{n-1} \tag{E.54}$$

Allen (1973) extended it for any value of α. Figure E.6 shows that the transmissivity for diffuse incidence ($\alpha = \pi/2$) is smaller than that for perpendicular incidence ($\alpha = 0$).

Finally, when the radiation is incident from opposite directions, for example, from the material with the higher refractive index, Stern (1964) has shown that

$$t_{av}(\alpha,n,1) = n^{-2}t_{av}(\alpha,1,n) \tag{E.55}$$

Appendix F

Beer–Lambert Law

The Beer–Lambert law, sometimes called Beer's Law, the Bouguer–Lambert law, or Lambert's law of absorption, relates the amount of light absorbed and scattered in a medium and the distance it travels through this medium. Although Johann Heinrich Lambert (1728–1777) is generally given credit for this law that appears in 1760 in his treatise *Photometria*, Pierre Bouguer (1698–1758) stated empirically the same relationship much earlier in his *Essai d'optique sur la gradation de la lumière*, which appeared in Paris in 1729. In 1852, August Beer (1825–1863) indicated the mathematical relationship involved and extended this exponential absorption law to incorporate the concentration of solutions in the absorption coefficient (e.g., Pfeiffer and Liebhafsky, 1951; Malinin and Yoe, 1961).

Consider an energetic flux Φ (W) being propagated in a semi-transparent plate with an incident angle of $\theta_i = \theta$ (Figure F.1).

The medium is characterized by its extinction coefficient σ_e {m^{-1}), which is here assumed to be constant. The extinction coefficient can be written as the sum of the absorption coefficient σ_a {m^{-1}) and the scattering coefficient σ_s {m^{-1}):

$$\sigma_e = \sigma_a + \sigma_s \tag{F.1}$$

Moreover, if we assume that only absorption occurs in the medium, then we can set $\sigma_e = \sigma_a = k$. The variation of flux in the infinitesimal layer of thickness dx can be written as

$$\Phi(x + dx) - \Phi(x) = d\Phi(x) = -k\Phi(x)dx \tag{F.2}$$

Integrating Eq. F.2 over the whole thickness of the layer, we have the transmittance of the medium:

$$\tau(\theta, D) = \frac{\Phi_2}{\Phi_1} = \exp\left(-\frac{kD}{\cos\theta}\right) \tag{F.3}$$

Equation F.3 can be similarly written for any distance x:

$$\tau(\theta, x) = \exp\left(-\frac{kx}{\cos\theta}\right) \tag{F.4}$$

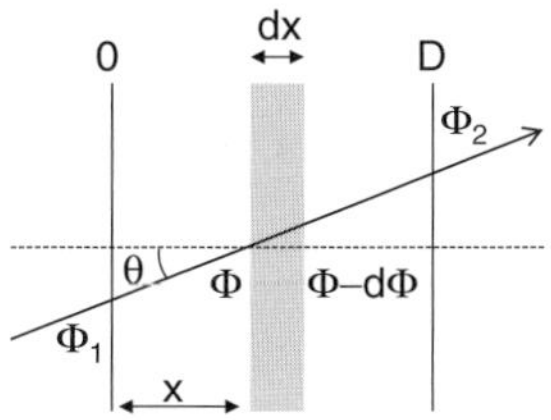

Figure F.1 Transfer of radiation through a plate of thickness *D*.

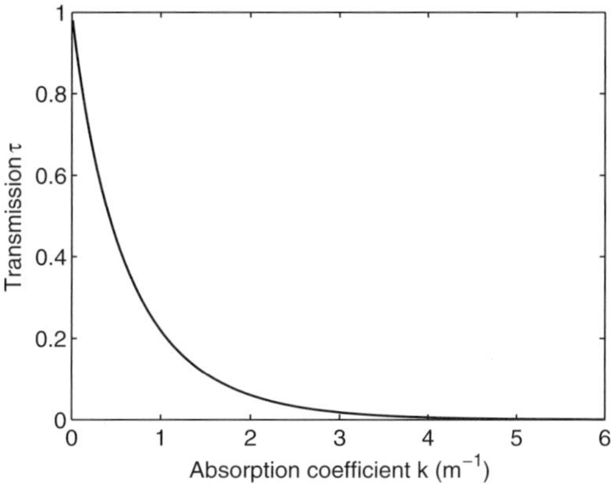

Figure F.2 Variation of the transmittance of a single layer for isotropic light as a function of its absorption coefficient, for a distance $x = 1\ m$.

The transmittance for isotropic light passing through the plate is obtained by straightforward integration of Eq. F.4 over the hemisphere:

$$\tau(x) = \frac{\int_0^{2\pi} d\varphi \int_0^{\pi/2} \tau(\theta, x) \cos\theta \sin\theta d\theta}{\int_0^{2\pi} d\varphi \int_0^{\pi/2} \cos\theta \sin\theta d\theta} = (1 - kx)\exp(-kx) + k^2x^2\Gamma(0, kx) \tag{F.5}$$

where $\Gamma(a, x)$ is the "upper" incomplete gamma function. By assuming a unit distance $x = 1\ m$, Eq. F.5 simplifies into $\tau = (1 - k)\exp(-k) + k^2\Gamma(0, k)$. Figure F.2 shows the variation of τ as a function of k: $\tau = 1$ when $k = 0$ and $\tau = 0$ when k tends to infinity.

Note: The "upper" incomplete gamma function should not be confused with the "lower" incomplete gamma function noted as $\gamma(a, x)$. They are given by

$$\Gamma(a, x) = \int_x^{\infty} t^{a-1} e^{-t} dt \ and, \tag{F.6}$$

$$\gamma(a, x) = \int_0^{x} t^{a-1} e^{-t} dt \tag{F.7}$$

By definition their sum satisfies:

$$\gamma(a,x)+\Gamma(a,x)=\int_0^x t^{a-1}e^{-t}dt+\int_x^\infty t^{a-1}e^{-t}dt=\int_0^\infty t^{a-1}e^{-t}dt=\Gamma(a)$$
$$\Rightarrow \frac{\gamma(a,x)}{\Gamma(a)}+\frac{\Gamma(a,x)}{\Gamma(a)}=1\Rightarrow P(a,x)+Q(a,x)=1 \tag{F.8}$$

where $P(a,x)$ and $Q(a,x)$ are the regularized "lower" and "upper" incomplete gamma function. Since the incomplete gamma function defined in Matlab is $P(a,x)$, one has to be very careful with the definitions! Finally, in the particular case when $a=0$, $\Gamma(a,x)$ is closely related to the exponential integral which will be used in the Matlab code:

$$\Gamma(0,x)=\int_x^\infty \frac{e^{-t}}{t}dt \tag{F.9}$$

The differentiation and integration of $\Gamma(0,x)$ are well known:

$$\frac{d\Gamma(0,x)}{dx}=-\frac{e^{-x}}{x} \text{ and } \frac{d\Gamma(0,kx)}{dx}=-\frac{e^{-kx}}{x} \tag{F.10}$$

$$\int\Gamma(0,x)=x\Gamma(0,x)-e^{-x} \text{ and } \int\Gamma(0,kx)=x\Gamma(0,kx)-\frac{e^{-kx}}{k} \tag{F.11}$$

Appendix G
Kubelka–Munk Theory

The Kubelka–Munk theory is one of the simplest ways to calculate leaf optical properties (Allen and Richardson, 1968). It was developed for the first time by astrophysicist Schuster (1905) who was interested in the propagation of sun radiation in a hazy atmosphere, but it takes its name from the Czech chemists Paul Kubelka and Franz Munk, the two authors who popularized it (Kubelka and Munk, 1931). It is still widely used in many fields (image synthesis, inks, biological tissues, etc.) even if there are now much more complex and accurate models (Kortüm, 1969). Consider a medium separated by two infinite parallel interfaces – there are no edge effects – and two diffuse light fluxes propagating perpendicularly to these interfaces: a downward flux I and an upward flux J (Figure G.1). The incident flux $I_0 = I(0)$ is assumed to be equal to 1.

The Kubelka–Munk theory is based on the following linear homogeneous system of differential equations with constant coefficients:

$$\begin{cases} dI = -(k+s)Idz + s\ Jdz \\ dJ = (k+s)Jdz - s\ Idz \end{cases} \tag{G.1}$$

with k and s the absorption and scattering coefficients of the medium (m^{-1}). Note that if there is only a downward flux, the system simplifies into $dI = -(k+s)Idz$, which is none other than the Beer–Lambert law (see Appendix F). Equation G.1 can be written in a matrix form:

$$\begin{pmatrix} I' \\ J' \end{pmatrix} = \begin{pmatrix} -(k+s) & s \\ -s & (k+s) \end{pmatrix} \begin{pmatrix} I \\ J \end{pmatrix} \tag{G.2}$$

By setting $X = \begin{pmatrix} I \\ J \end{pmatrix}$ and $A = \begin{pmatrix} -(k+s) & s \\ -s & (k+s) \end{pmatrix}$, we obtain

$$X' = AX \tag{G.3}$$

Homogeneous systems of equations with constant coefficients can be solved in different ways. The method of eigenvalues and eigenvectors provides a general solution of Eq. G.3:

$$X = c_1 e^{\lambda_1} U_1 + c_2 e^{\lambda_2} U_2 \text{ with } (c_1, c_2) \in \mathbb{R}^2 \tag{G.4}$$

(λ_1, λ_2) are the two eigenvalues of A and (U_1, U_2) are the two eigenvectors corresponding to the eigenvalues.

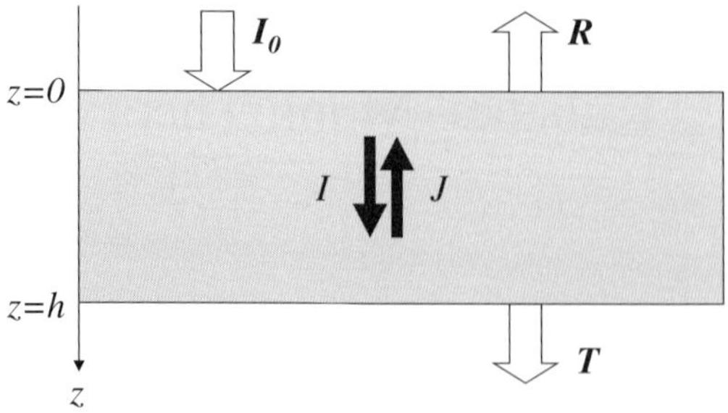

Figure G.1 Upward and downward fluxes.

G.1 Calculation of the Eigenvalues

The eigenvalues λ of the square matrix A of order 2 are the two solutions of the equation:

$$\det(A-\lambda I)=\begin{vmatrix}-(k+s)-\lambda & s\\ -s & (k+s)-\lambda\end{vmatrix}=0 \tag{G.5}$$

We easily find the following elementary quadratic equation:

$$\lambda^2=k(k+2s) \tag{G.6}$$

We set:

$$\alpha=\sqrt{k(k+2s)} \tag{G.7}$$

The values $\lambda_1=\alpha$ and $\lambda_2=-\alpha$ are the required eigenvalues of A.

G.2 Calculation of the Eigenvectors

For each of the eigenvalues calculated we have an individual eigenvector. To find them we simply plug in each eigenvalue into $(A-\lambda I)\vec{v}=\vec{0}$ and solve:

- Eigenvector $\vec{v}_1$ corresponding to the eigenvalue $\lambda_1=\alpha$

$$(A-\alpha I)\vec{v}_1=\vec{0}\Leftrightarrow\begin{pmatrix}-(k+s)-\alpha & s\\ -s & (k+s)-\alpha\end{pmatrix}\begin{pmatrix}x\\ y\end{pmatrix}=\begin{pmatrix}0\\ 0\end{pmatrix}\Leftrightarrow\begin{cases}(k+s+\alpha)x=sy\\ -sx=(\alpha-k-s)y\end{cases}$$

We sum the two equalities:

$$(\alpha+k)x=(\alpha-k)y\Leftrightarrow\begin{cases}x=\alpha-k\\ y=\alpha+k\end{cases}\Leftrightarrow\begin{cases}x=\alpha(1-k/\alpha)\\ y=\alpha(1+k/\alpha)\end{cases}$$

We set:

$$\beta=k/\alpha=\sqrt{k/(k+2s)} \tag{G.8}$$

Then $(A-\alpha I)\vec{v}_1=\vec{0} \Leftrightarrow \begin{cases} x=\alpha(1-\beta) \\ y=\alpha(1+\beta) \end{cases}$ and

$$\vec{v}_1=\begin{pmatrix} \alpha(1-\beta) \\ \alpha(1+\beta) \end{pmatrix}=\alpha\begin{pmatrix} 1-\beta \\ 1+\beta \end{pmatrix}=\alpha\vec{v}_{10} \tag{G.9}$$

- Eigenvector $\vec{v}_2$ corresponding to the eigenvalue $\lambda_2=-\alpha$

$$(A+\alpha I)\vec{v}_2=\vec{0} \Leftrightarrow \begin{pmatrix} -(k+s)+\alpha & s \\ -s & (k+s)+\alpha \end{pmatrix}\begin{pmatrix} x \\ y \end{pmatrix}=\begin{pmatrix} 0 \\ 0 \end{pmatrix} \Leftrightarrow \begin{cases} (k+s-\alpha)x=sy \\ -sx=-(\alpha+k+s)y \end{cases}$$

We sum the two equalities:

$$(\alpha-k)x=(\alpha+k)y \Leftrightarrow \begin{cases} x=\alpha+k \\ y=\alpha-k \end{cases} \Leftrightarrow \begin{cases} x=\alpha(1+k/\alpha) \\ y=\alpha(1-k/\alpha) \end{cases} \Leftrightarrow \begin{cases} x=\alpha(1+\beta) \\ y=\alpha(1-\beta) \end{cases}$$

Then

$$\vec{v}_2=\begin{pmatrix} \alpha(1+\beta) \\ \alpha(1-\beta) \end{pmatrix}=\alpha\begin{pmatrix} 1+\beta \\ 1-\beta \end{pmatrix}=\alpha\vec{v}_{20} \tag{G.10}$$

Note that the two eigenvectors are linearly independent.

G.3 General Solution of the System

A can be decomposed into

$$A=PBP^{-1} \tag{G.11}$$

with $B=\begin{pmatrix} \alpha & 0 \\ 0 & -\alpha \end{pmatrix}$, a diagonal matrix the diagonal entry of which are the eigenvalues of A, and $P=\begin{pmatrix} 1-\beta & 1+\beta \\ 1+\beta & 1-\beta \end{pmatrix}$, a square matrix composed of the eigenvectors of A. Therefore

$$X'=AX \Leftrightarrow X'=\left(PBP^{-1}\right)X \tag{G.12}$$

One multiplies the two members by P^{-1}:

$$P^{-1}X'=BP^{-1}X \tag{G.13}$$

Let us set $Y=P^{-1}X \Leftrightarrow Y'=P^{-1}X'=P^{-1}\left(PBP^{-1}\right)X=BP^{-1}X=BY$. The solution of $Y'=BY$ is straightforward and equal to $Y=\begin{pmatrix} c_1e^{\alpha z} \\ c_2e^{-\alpha z} \end{pmatrix}$ with $(c_1,c_2)\in\mathbb{R}^2$. $Y=P^{-1}X \Leftrightarrow X=PY=\begin{pmatrix} 1-\beta & 1+\beta \\ 1+\beta & 1-\beta \end{pmatrix}\begin{pmatrix} c_1e^{\alpha z} \\ c_2e^{-\alpha z} \end{pmatrix}$. Then the general solution of Eq. G.1 is

$$\begin{cases} I(z) = (1-\beta)c_1e^{\alpha z} + (1+\beta)c_2e^{-\alpha z} \\ J(z) = (1+\beta)c_1e^{\alpha z} + (1-\beta)c_2e^{-\alpha z} \end{cases} \text{with } (c_1, c_2) \in \mathbb{R}^2 \tag{G.14}$$

G.4 Calculation of the Constants

Let us assume that R_0 is the reflectance of the background. To calculate the constants we use the boundary conditions:

If $z = 0$, then $I(0) = (1-\beta)c_1 + (1+\beta)c_2 = 1$

If $z = h$, then $J(h) = (1+\beta)c_1e^{\alpha h} + (1-\beta)c_2e^{-\alpha h} = R_0I(h) = R_0(1-\beta)c_1e^{\alpha h} + R_0(1+\beta)\, c_2e^{-\alpha h}$

We obtain a system of two equations with two unknowns:

$$\begin{cases} (1-\beta)c_1 + (1+\beta)c_2 = 1 \\ \Big((1+\beta) - R_0(1-\beta)\Big)e^{\alpha h}c_1 + \Big((1-\beta) - R_0(1+\beta)\Big)e^{-\alpha h}c_2 = 0 \end{cases}$$

This system is easily solved by the Cramer's method:

$$D = \begin{vmatrix} (1-\beta) & (1+\beta) \\ \Big((1+\beta) - R_0(1-\beta)\Big)e^{\alpha h} & \Big((1-\beta) - R_0(1+\beta)\Big)e^{-\alpha h} \end{vmatrix}$$

$$= (1-\beta)^2e^{-\alpha h} - R_0(1-\beta^2)e^{-\alpha h} - (1+\beta)^2e^{\alpha h} + R_0(1-\beta^2)e^{\alpha h}$$

$$= (1-\beta)^2e^{-\alpha h} - (1+\beta)^2e^{\alpha h} - R_0(1-\beta^2)\left(e^{-\alpha h} - e^{\alpha h}\right)$$

$$c_1 = \frac{\begin{vmatrix} 1 & (1+\beta) \\ 0 & \Big((1-\beta) - R_0(1+\beta)\Big)e^{-\alpha h} \end{vmatrix}}{D} = \frac{\Big((1-\beta) - R_0(1+\beta)\Big)e^{-\alpha h}}{D} \text{ et}$$

$$c_2 = \frac{\begin{vmatrix} (1-\beta) & 1 \\ \Big((1+\beta) - R_0(1-\beta)\Big)e^{\alpha h} & 0 \end{vmatrix}}{D} = \frac{\Big(R_0(1-\beta) - (1+\beta)\Big)e^{\alpha h}}{D}$$

G.5 Exact Solution

We can calculate the analytical expressions of the upward and downward fluxes:

$$\begin{aligned} I(z) &= (1-\beta)c_1e^{\alpha z} + (1+\beta)c_2e^{-\alpha z} \\ &= \frac{(1-\beta)\Big((1-\beta) - R_0(1+\beta)\Big)e^{-\alpha(h-z)}}{D} + \frac{(1+\beta)\Big(R_0(1-\beta) - (1+\beta)\Big)e^{\alpha(h-z)}}{D} \\ &= \frac{(1+\beta)^2e^{\alpha(h-z)} - (1-\beta)^2e^{-\alpha(h-z)} - R_0(1-\beta^2)\left(e^{\alpha(h-z)} - e^{-\alpha(h-z)}\right)}{(1+\beta)^2e^{\alpha h} - (1-\beta)^2e^{-\alpha h} - R_0(1-\beta^2)\left(e^{\alpha h} - e^{-\alpha h}\right)} \end{aligned}$$

$$
\begin{aligned}
J(z) &= (1+\beta)c_1 e^{\alpha z} + (1-\beta)c_2 e^{-\alpha z} \\
&= \frac{(1+\beta)\Big((1-\beta) - R_0(1+\beta)\Big)e^{-\alpha(h-z)}}{D} + \frac{(1-\beta)\Big(R_0(1-\beta) - (1+\beta)\Big)e^{\alpha(h-z)}}{D} \\
&= \frac{(1-\beta^2)\left(e^{\alpha(h-z)} - e^{-\alpha(h-z)}\right) - R_0\left((1-\beta)^2 e^{\alpha(h-z)} - (1+\beta)^2 e^{-\alpha(h-z)}\right)}{(1+\beta)^2 e^{\alpha h} - (1-\beta)^2 e^{-\alpha h} - R_0\big(1-\beta^2\big)(e^{\alpha h} - e^{-\alpha h})}
\end{aligned}
$$

We recover the equations given by Allen and Richardson (1968):

$$
\begin{cases}
I(z) = \dfrac{(1+\beta)^2 e^{\alpha(h-z)} - (1-\beta)^2 e^{-\alpha(h-z)} - R_0\big(1-\beta^2\big)\left(e^{\alpha(h-z)} - e^{-\alpha(h-z)}\right)}{(1+\beta)^2 e^{\alpha h} - (1-\beta)^2 e^{-\alpha h} - R_0\big(1-\beta^2\big)(e^{\alpha h} - e^{-\alpha h})} \\
J(z) = \dfrac{(1-\beta^2)\left(e^{\alpha(h-z)} - e^{-\alpha(h-z)}\right) - R_0\left((1-\beta)^2 e^{\alpha(h-z)} - (1+\beta)^2 e^{-\alpha(h-z)}\right)}{(1+\beta)^2 e^{\alpha h} - (1-\beta)^2 e^{-\alpha h} - R_0\big(1-\beta^2\big)(e^{\alpha h} - e^{-\alpha h})}
\end{cases}
\tag{G.15}
$$

By setting $\alpha = \ln b \Leftrightarrow b = e^{\alpha}$ and $\beta = \frac{a-1}{a+1} \Leftrightarrow a = \frac{1+\beta}{1-\beta}$, one can simplify Eq. G.15:

$$
\begin{aligned}
I(z) &= \frac{\left(1+\dfrac{a-1}{a+1}\right)^2 b^{h-z} - \left(1-\dfrac{a-1}{a+1}\right)^2 b^{-(h-z)} - R_0\left(1-\left(\dfrac{a-1}{a+1}\right)^2\right)\left(b^{h-z} - e^{-(h-z)}\right)}{\left(1+\dfrac{a-1}{a+1}\right)^2 b^{h} - \left(1-\dfrac{a-1}{a+1}\right)^2 b^{-h} - R_0\left(1-\left(\dfrac{a-1}{a+1}\right)^2\right)(b^h - b^{-h})} \\
&= \frac{\dfrac{4a^2}{(a+1)^2} b^{h-z} - \dfrac{4}{(a+1)^2} b^{-(h-z)} - R_0 \dfrac{4a}{(a+1)^2}\left(b^{h-z} - b^{-(h-z)}\right)}{\dfrac{4a^2}{(a+1)^2} b^{h} - \dfrac{4}{(a+1)^2} b^{-h} - R_0 \dfrac{4a}{(a+1)^2}\left(b^{h} - b^{-h}\right)} \\
&= \frac{(a^2 - aR_0)b^{h-z} + (aR_0 - 1)b^{-(h-z)}}{(a^2 - aR_0)b^{h} + (aR_0 - 1)b^{-h}}
\end{aligned}
$$

$$
\begin{aligned}
J(z) &= \frac{\left(1-\left(\dfrac{a-1}{a+1}\right)^2\right)\left(b^{h-z} - b^{-(h-z)}\right) - R_0\left(\left(1-\dfrac{a-1}{a+1}\right)^2 b^{h-z} - \left(1+\dfrac{a-1}{a+1}\right)^2 b^{-(h-z)}\right)}{\left(1+\dfrac{a-1}{a+1}\right)^2 b^{h} - \left(1-\dfrac{a-1}{a+1}\right)^2 b^{-h} - R_0\left(1-\left(\dfrac{a-1}{a+1}\right)^2\right)(b^h - b^{-h})} \\
&= \frac{\dfrac{4a}{(a+1)^2}\left(b^{h-z} - b^{-(h-z)}\right)b^{h-z} - R_0\left(\dfrac{4}{(a+1)^2} b^{h-z} - \dfrac{4a^2}{(a+1)^2} b^{-(h-z)}\right)}{\dfrac{4a^2}{(a+1)^2} b^{h} - \dfrac{4}{(a+1)^2} b^{-h} - R_0 \dfrac{4a}{(a+1)^2}\left(b^{h} - b^{-h}\right)} \\
&= \frac{(a - R_0)b^{h-z} + (a^2 R_0 - a)b^{-(h-z)}}{(a^2 - aR_0)b^{h} + (aR_0 - 1)b^{-h}}
\end{aligned}
$$

Thus Eq. G.14 can be rewritten in simplified form:

$$\begin{cases} I(z) = \dfrac{a(a-R_0)b^{h-z} + (aR_0-1)b^{-(h-z)}}{a(a-R_0)b^h + (aR_0-1)b^{-h}} \\ J(z) = \dfrac{(a-R_0)b^{h-z} + a(aR_0-1)b^{-(h-z)}}{a(a-R_0)b^h + (aR_0-1)b^{-h}} \end{cases} \tag{G.16}$$

By definition, the reflectance is the upward flux at the top of the leaf:

$$R = J(0) = \frac{(a-R_0)b^h + a(aR_0-1)b^{-h}}{a(a-R_0)b^h + (aR_0-1)b^{-h}} \tag{G.17}$$

and the transmittance is the downward flux at the base of the leaf:

$$T = I(h) = \frac{a^2-1}{a(a-R_0)b^h + (aR_0-1)b^{-h}} \tag{G.18}$$

Note that the existence of $a = \dfrac{1+\beta}{1-\beta}$ implies that $\beta \neq 1 \Leftrightarrow \dfrac{k}{\alpha} \neq 1 \Leftrightarrow \dfrac{k}{k+2s} \neq 1 \Leftrightarrow s \neq 0$, that is, there is always scattering in the medium.

G.6 Special Case: The Background Reflectance Is Zero

$$R_0 = 0 \Rightarrow R = \frac{ab^h - ab^{-h}}{a^2b^h - b^{-h}} = \frac{b^h - b^{-h}}{ab^h - a^{-1}b^{-h}} \text{ and } T = \frac{a^2-1}{a^2b^h - b^{-h}} = \frac{a-a^{-1}}{ab^h - a^{-1}b^{-h}}$$

Note that the reflectance and the transmittance are linked by a simple relation:

$$\frac{R}{b^h - b^{-h}} = \frac{T}{a - a^{-1}} = \frac{1}{ab^h - a^{-1}b^{-h}} \tag{G.19}$$

Stokes (1862) obtains similar equations by determining the reflectance R and the transmittance T of a stack of N elementary layers, for which reflectance ρ and transmittance τ are known. The coefficients a and b, also known as Stokes parameters, are expressed as a function of ρ and τ and the height h is replaced by the number of elementary layers N. The PROSPECT model (Jacquemoud and Baret, 1990) is based on such formalism (see Chapter 9).

G.7 Special Case: The Medium Is Opaque

Consider an infinite medium, that is to say for which $h \to +\infty$. Rearranging the expression of R by dividing numerator and denominator by b^h:

$$R = \frac{(a-R_0) + a(aR_0-1)b^{-2h}}{a(a-R_0) + (aR_0-1)b^{-2h}} \tag{G.20}$$

then

$$\lim_{h \to +\infty} R = \frac{a - R_0}{a(a - R_0)} = a^{-1} \tag{G.21}$$

and

$$R_\infty = a^{-1} = \frac{1 - \beta}{1 + \beta} \tag{G.22}$$

Most articles and books on the theory of Kubelka–Munk refer to the reflectance function that connects R_∞ to the ratio k/s. Indeed it can be shown that $\frac{(1 - R_\infty)^2}{2R_\infty} = \frac{4\beta^2}{(1 + \beta)^2} \times \frac{1 + \beta}{2(1 - \beta)} = \frac{2\beta^2}{1 - \beta^2} = \frac{2}{1/\beta^2 - 1} = \frac{k}{s}$. So the reflectance of an opaque sample solely depends on this ratio. From Eq. G.20 one can extract b^{-2h} as a function of R, R_0 and R_∞:

$$\begin{aligned}
R &= \frac{(a - R_0) + a(aR_0 - 1)b^{-2h}}{a(a - R_0) + (aR_0 - 1)b^{-2h}} \\
\Leftrightarrow\ & R\Big(a(a - R_0) + (aR_0 - 1)b^{-2h}\Big) = (a - R_0) + a(aR_0 - 1)b^{-2h} \\
\Leftrightarrow\ & b^{-2h}\Big(R(aR_0 - 1) - a(aR_0 - 1)\Big) = (a - R_0) - aR(a - R_0) \\
\Leftrightarrow\ & b^{-2h} = \frac{(a - R_0)(1 - aR)}{(aR_0 - 1)(R - a)} \\
\Leftrightarrow\ & b^{-2h} = \frac{(1 - R_0R_\infty)(R_\infty - R)}{(R_0 - R_\infty)(RR_\infty - 1)}
\end{aligned}$$

By neglecting the reflectance products compared to 1, $R_0R_\infty \ll 1$ and $RR_\infty \ll 1$, one obtains

$$b^{-2h} = \frac{R_\infty - R}{R_\infty - R_0} \Leftrightarrow R = R_\infty + (R_0 - R_\infty)b^{-2h}$$

and then

$$R = R_\infty + (R_0 - R_\infty)e^{-2\alpha h} \tag{G.23}$$

Appendix H
Global Sensitivity Analysis

Let us denote $X = (X_1, X_2, \ldots, X_p) \in \mathbb{R}^p$ the model inputs, which are a vector of p parameters X_i, and $Y = f(X) \in \mathbb{R}$ the model output, which is a scalar. In the probabilistic setting, X is a random vector defined by a probability distribution and Y is a random variable. It is possible to decompose the function f into a sum of elementary functions:

$$Y = f(X) = f_0 + \sum_{i=1}^{p} f_i(X_i) + \sum_{i<j}^{p} f_{ij}(X_i, X_j) + \ldots + f_{12\ldots p}X \tag{H.1}$$

where f_0 is a constant. The total variance $V(Y)$ of the model output $Y = f(X)$ can be written as

$$V(Y) = \sum_{i=1}^{p} V_i + \sum_{i=1}^{p} \sum_{j>i} V_{ij} + \ldots + V_{12\ldots p} \tag{H.2}$$

where $V_i = V\big(E(Y|X_i)\big)$, $V_{ij} = V\big(E(Y|X_i, X_j)\big) - V_i - V_j$, and so on for higher-order interactions. V_i is the part of the variance that can be attributed to the input parameter X_i alone. Its expression ensues from the law of total variance, also known as variance decomposition formula: $V(Y) = V\big(E(Y|X_i)\big) + E\big(V(Y|X_i)\big)$. Therefore $V_i = V(Y) - E\big(V(Y|X_i)\big) = V\big(E(Y|X_i)\big)$. The Sobol' sensitivity indices are obtained as follows:

$$S_i = \frac{V_i}{V(Y)}, S_{ij} = \frac{V_{ij}}{V(Y)}, \ldots, S_{12\ldots p} = \frac{V_{12\ldots p}}{V(Y)} \tag{H.3}$$

The first-order sensitivity index S_i (between 0 and 1) is defined as the ratio of the output variance due to the ith parameter to the variance due to all variables. Therefore it measures the global sensitivity of Y with respect to X_i. Similarly, the second-order sensitivity index S_{ij} measures the pure interaction effect between X_i and X_j, that is, that part of the variation in Y due to X_i and X_j, which cannot be explained by the sum of the individual effects of X_i and X_j, and so on. Note that (Sobol', 1993):

$$\sum_{i=1}^{p} S_i + \sum_{i=1}^{p} \sum_{j>i} S_{ij} + \ldots + S_{12\ldots p} = 1 \tag{H.4}$$

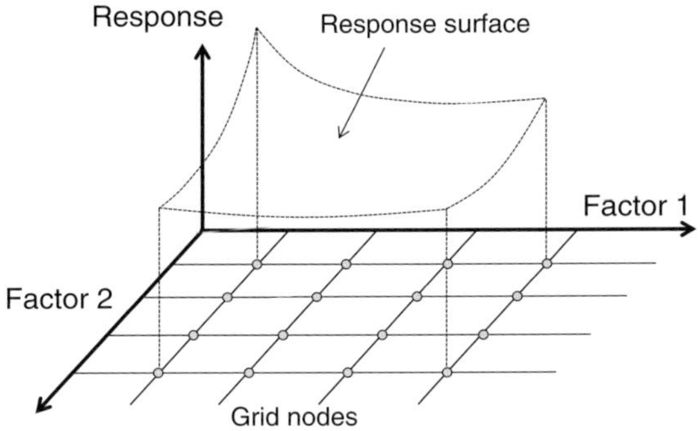

Figure H.1 Simulation at each node of the network (derived from Benoist et al., 1994).

For computational time and interpretation reasons, indices of an order higher than two are generally not considered. If the model is perfectly additive, that is, there are no interactions between the parameters, Eq. H.1 simplifies to

$$Y = f_0 + \sum_{i=1}^{p} f_i(X_i) + \varepsilon \tag{H.5}$$

In that case, $\sum S_i = 1$. It means that the total output variance is explained by the sum of the individual variances introduced by varying each parameter alone. In reality, models are not perfectly additive and $\sum S_i < 1$. The effect due to interactions among parameters are not captured by S_i. One can calculate the total-order effect through the total sensitivity index S_{Ti}, which estimates the sum of all effects involving the ith variable:

$$S_{Ti} = 1 - \frac{V_{\sim i}}{V} \tag{H.6}$$

Subscript $\sim i$ indicates all indices but i, unlike the S_i, $\sum S_{Ti} > 1$. Before quantifying the effects, contributions, and interactions of each of the input parameters, one must build an experiment table.

Consider a matrix of model runs generated to explore the parameter space. All simulations (*experiments*) are gathered with the input parameters varied in a structured pattern. A column of this matrix corresponds to the values (*levels*) taken by a particular parameter (*factor*) along the experiments. A line is a simulation to realize. Each line thus differs from others by the combination of the factor levels. As an illustration, consider a model with two input variables, each with four values distributed within their range of variation. The experimental design consists of $4^2 = 16$ runs (Figure H.1).

If the number of factors and levels remains reasonable, testing all possible combinations for sensitivity may be realistic. But when it increases, a full factorial design may lead to an unreasonable number of simulations. For instance, PROSPECT-D has 6 input parameters: considering 10 evenly spaced values for each parameter over it variation range, this experimental design would require $6^{10} =$ 60 466 176 combinations! It is evident from this example that a large number of parameters quickly prohibit a thorough examination of the model because of the tremendous number of model runs required, which would be very time-consuming and certainly demanding in computer resources. This problem can be bypassed by using an appropriate method (quasi-Monte Carlo method, Sobol sequence, or Latin hypercube sampling) for exploring multi-dimensional spaces by cutting down the number of trials with relatively little loss of information.

Appendix I
Leaf Three-Dimensional Reconstruction

I.1 Epidermis

Both the upper and lower epidermises are modeled as layers of compact ellipsoidal cells. Let a_e be the half-axis of an ellipsoid in the plane of the leaf (in two directions) and b_e the half-axis of the ellipsoid in the direction perpendicular to the leaf. The roughness of the leaf surface can be easily controlled by the oblateness $o_e = b_e/a_e$ of the ellipsoid. The distance between the centers of two cells along a row is $L_e = \sqrt{3}a_e$, and the distance between two rows of cells is given by $R_e = 3a_e/2$. Each cell is carved out at three symmetric points to leave space for the surrounding cells (Figure I.1). The volume of one cutout is expressed as

$$2\pi o_e \int_{L_e/2}^{a_e} (a_e^2 - x^2)dx = \frac{16 - 9\sqrt{3}}{12} \pi a_e^3 o_e \tag{I.1}$$

and the volume V_e of an epidermis cell is given by

$$V_e = \frac{4}{3}\pi a_e^3 o_e - 3\left(\frac{16 - 9\sqrt{3}}{12}\pi a_e^3 o_e\right) = \frac{27\sqrt{3} - 32}{12}\pi a_e^3 o_e \tag{I.2}$$

The dashed rectangle in Figure I.1 defines the size of an elementary lattice. Its total surface area is equal to $S_e = 3\sqrt{3}a_e^2$ and it contains $N_e = 2$ cells. Therefore, it is possible to define the volume per unit area $\overline{V_e} = N_e V_e / S_e$ and the fraction of intercellular air spaces in this tissue, that is, the fraction of space between the interior-side of the epidermis and the palisade parenchyma: $\xi_e = 1 - \overline{V_e}/(2a_e o_e) = 0.256$. Note that ξ_e is independent of the cell size. To define the volume of water V_{we}, the same reasoning can be applied. The vacuole membrane has a radius of $a_e - e_{ce}$ and is carved out with ellipsoids of radius $a_e + e_{ce}$, where e_{ce} is the cell wall thickness. The cell wall volume V_{ce} equals $V_{ce} = V_e - V_{we}$. It is also straightforward to express the cell wall volume per unit area $\overline{V_{ce}}$ and the water volume per unit area $\overline{V_{we}}$.

I.2 Palisade Parenchyma

The volume of a single palisade cell V_p is

$$V_p = \pi a_p^2 \left(h_p + \frac{4a_p o_p}{3}\right) \tag{I.3}$$

Table I.1 *Structural parameters defining the epidermis of* DICOT1. * *Values found in the literature.*

Characteristics	Parameter	Value (μm)
Cell radius	a_e	12.5*
Cell oblateness	o_e	0.7
Cell wall thickness	e_{ce}	1.0*

Table I.2 *Structural parameters defining the palisade parenchyma of* DICOT1. * *Values found in the literature.*

Characteristics	Parameter	Value (μm)
Cell radius	a_p	7.5*
Cell cap oblateness	o_p	1.0
Cell high	h_p	50*
Cell wall thickness	e_{cp}	1.0*
Chloroplast membrane thickness	e_{pp}	1.53*

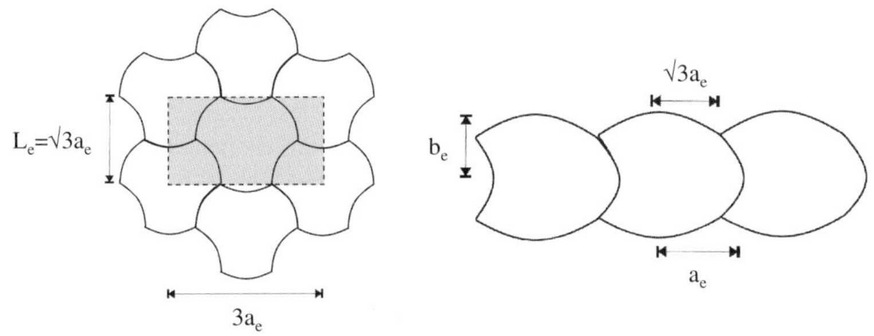

Figure I.1 (a) Top view of the cell arrangement for the epidermis. The shadowed area indicates the elementary lattice. (b) Transverse vertical section of the epidermis (from Govaerts et al. (1996), *Applied Optics*. Copyright © 1996 with permission from The Optical Society of America).

with a_p the radius of the palisade cell, o_p the oblateness of the spherical cap, and h_p the height of the cylinder. The oblateness o_p is defined as $o_p = b_p/a_p$ where b_p is the half-axis of the cap in the direction perpendicular to the cylinder radius. The triangle in Figure I.2 indicates the elementary lattice of area $S_p = \sqrt{3}a_p^2$ that contains the $N_p = 1/2$ cell. The volume per unit area $\overline{V_p}$ can be expressed as in the case of the epidermis and the fraction of intercellular air spaces ξ_p is

$$\xi_p = 1 - \frac{\sqrt{3}}{18}\frac{3h_p + 4a_p}{h_p + 2a_p} \tag{I.4}$$

One can determine the cell wall volume V_{cp}, as described above, by defining a smaller object whose dimensions are reduced by the cell wall thickness e_{cp}. To evaluate the thickness e_{pp} and volume V_{pp} of

Table I.3 *Structural parameters defining the spongy mesophyll of* DICOT1. ** Values found in the literature.*

Characteristics	Parameter	Value (μm)
Basic cell radius	a_s	7.5*
Fraction of air spaces	ξ_s	1.0
Tissue thickness	h_s	50*
Basic cell wall thickness	e_{cs}	1.0*
Basic chloroplast membrane thickness	e_{ps}	1.53*

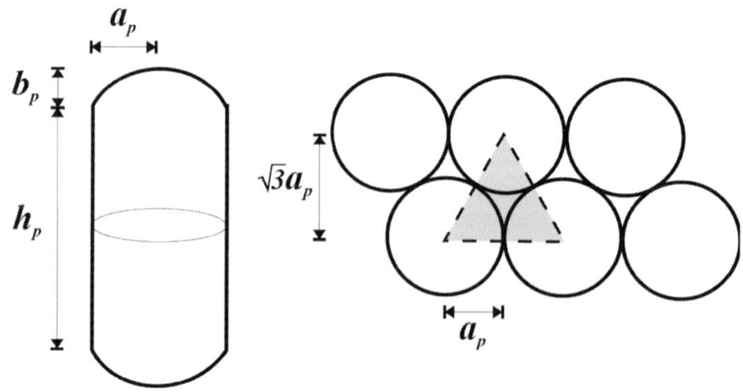

Figure I.2 (a) Palisade parenchyma cell is made of a cylinder with ellipsoid caps. (b) Paradermal section showing the arrangement in space. The shadowed triangle corresponds to the elementary lattice (from Govaerts et al. (1996), *Applied Optics*. Copyright © 1996 with permission from The Optical Society of America).

the layer containing chlorophyll, we assume that a typical palisade cell contains $n_{pp} = 40$ chloroplasts, each occupying a volume of 85 μm (Fukshansky et al., 1993). e_{pp} is numerically estimated using the analytical formula of V_{pp}. The remaining volume $V_{wp} = V_p - V_{cp} - V_{pp}$ is filled with water. The volume per unit area of $\overline{V_{cp}}$, $\overline{V_{pp}}$, $\overline{V_{wp}}$ are easily deduced.

I.3 Spongy Mesophyll

Consider a box of height h_s and base $S_s = 300$ μm × 300 μm, and assign the volumetric fraction of air spaces ξ_s as 45%. The spongy mesophyll is generated by filling the box at random with spheres of initial radius a_s without overlap. When no more space is available, the radius a_s is reduced by 10% and the filling process continues with smaller spheres. This iterative process ends when $1 - \xi_s$ of the space is occupied by cells. A total of 1139 spheres of 11 different sizes allowed us to fill the space available at the required density (Goverts et al., 1996). The thicknesses of all internal membranes of the smaller spheres are similarly decreased. To compute the chloroplast layer thickness of the initial spheres, we assume that the latter contain $n_{ps} = 25$ chloroplasts, each occupying a volume of 85 μm^3, and estimate e_{ps} numerically as for the palisade parenchyma. The remaining volume is then filled with

Table I.4 *Statistics for the biophysical properties of* DICOT1.

Cell properties	Parameter	Epidermis	Palisade parenchyma	Spongy mesophyll	Whole leaf
Whole tissue	h	17.5 μm	65 μm	50 μm	150 μm
	ξ	25.6%	16.3%	45.0%	
	N	246,336 cm^{-2}	513,200 cm^{-2}	1,265,556 cm^{-2}	2,271,428 cm^{-2}
Cell wall	α_c	27.0%	26.6%	23.0%	
	V	1427.51 μm^3	2815.91 μm^3	1662.95 μm^3	
	C_c	0.00052 g cm^{-2}	0.00213 g cm^{-2}	0.00093 g cm^{-2}	0.00410 g cm^{-2}
Pigments	n_p		40	25	
	α_p		32.1	21.4	
	V		3400.00 μm^3	2125.00 μm^3	
	C_p		41.1 μg cm^{-2}	11.0 μg cm^{-2}	60.1 μg cm^{-2}
Water	α_w	73.0%	41.4%	47.7%	
	V	3857.44 μm^3	4386.96 μm^3	3450.28 μm^3	
	C_w	0.00095 g cm^{-2}	0.00225 g cm^{-2}	0.00131 g cm^{-2}	0.00546 g cm^{-2}

water. This process is repeated for each class of spheres. The spongy mesophyll cell volume per unit area $\overline{V_s}$ can be expressed as a sum $\sum_k \overline{V_{s,k}}$ over the 11 different classes of spheres.

I.4 Whole Leaf

The leaf internal structure is fully defined by the 13 parameters summarized in Table I.4. The previous formula control the fraction of intercellular air spaces, the fractional volume of the different media, as well as the number of cells per square micrometer N_l which is given by

$$N_l = 2\frac{N_e}{S_e} + \frac{N_p}{S_p} + \frac{1}{S_s}\sum N_s \tag{I.5}$$

Assuming that α_c, α_p, and α_w are the fractional volumes of cell walls, photosynthetic pigments, and water respectively, one can calculate corresponding contents C_c, C_p, C_w expressed in g cm^{-2} or μg cm^{-2}. If the density of water is $v_w = 1.0$ g cm^{-3}, the water content of the whole leaf is

$$C_w = v_w\left(\alpha_{we}\overline{V_e} + \alpha_{wp}\overline{V_p} + \alpha_{ws}\overline{V_s} + \alpha_{we}\overline{V_e}\right) \tag{I.6}$$

The value $C_w = 0.00546$gcm^{-2} is low but within the range of laboratory measurements (0.0043–0.0713 g cm^{-2}, mean value 0.0129±0.0073 g cm^{-2}) (Féret et al., 2011). The case of cell walls is somewhat less straightforward, due to the lack of information about their biophysical properties. Cellulose (including hemicellulose) and lignin, the main constituents of the cell wall, are in proportions of 0.75 and 0.25, (Jacquemoud et al., 1996) and they have densities of 1.52 and 1.34 (Stamm and Sanders, 1966), respectively. The cell wall density has accordingly been set to $v_c = 1.47$ g cm^{-3}. The biochemical composition of cell walls is almost constant and so is its density. As for water, the cell wall content is expressed as

$$C_c = v_c\left(\alpha_{ce}\overline{V_e} + \alpha_{cp}\overline{V_p} + \alpha_{cs}\overline{V_s} + \alpha_{ce}\overline{V_e}\right) \tag{I.7}$$

Once again, the value $C_c = 0.00410\text{gcm}^{-2}$ agrees with experimental observations for dry matter (0.0008–0.0331 g cm^{-2}, mean value 0.0077±0.0035 g cm^{-2}) (Féret et al., 2011). Finally, the total leaf chlorophyll content C_p is given by

$$C_p = c_p\left(n_{pp}N_p + \sum n_{ps}N_s\right) \tag{I.8}$$

where n_{pp} is the average number of chloroplasts in a cell and the chlorophyll content of a chloroplast is $c_p = 2 \times 10^{-6}$ µg. The value of $C_p = 60.1\ \mu\text{g cm}^{-2}$ for this modeled leaf is given in Table I.4 and appears reasonable. This table also exhibits the other properties of our virtual leaf.

The current model omitted a number of leaf tissues and cell types that would make it more physically realistic but would require descriptive information not readily available. For example, the xylem and phloem conductive tissues are not included, nor are various specialized cells, for example, sclerids, glands, trichomes, and others.

Appendix J

Leaf Online Databases and Models

This appendix briefly presents some leaf online databases and model mirrors. Some databases contain leaf optical properties; others just leaf traits.

J.1 Databases with Leaf Optical Properties

The relationship between leaf biochemical and structural traits and leaf optical properties has long been investigated. Depending on physical environment and stress, plants invest in covarying traits, influencing spectral characteristics of leaves (e.g., Schweiger et al., 2017).

J.1.1 BOREAS

The Boreal Ecosystem-Atmosphere Study (BOREAS, https://daac.ornl.gov/cgi-bin/dataset_lister.pl?p=2) was a large-scale international interdisciplinary experiment in the boreal forests of central Canada. It aimed at improving our understanding of the exchanges of radiative energy, sensible heat, water, CO_2, and trace gases between the boreal forest and the lower atmosphere. A primary objective of BOREAS was to collect the data needed to improve computer simulation models of the important processes controlling these exchanges, so that scientists can anticipate the effects of global change on the biome. The data collected in the frame of that experiment are provided by the Oak Ridge National Laboratory Distributed Active Archive Center (ORNL DAAC), one of the NASA Earth Observing System Data and Information System (EOSDIS) data centers managed by the Earth Science Data and Information System (ESDIS) Project, which is responsible for providing scientists and other users access to data from NASA's Earth Science Missions. The spectral optical properties (reflectance and transmittance) and chemical composition of boreal forest conifers and broadleaf tree leaves were measured at several sites during the growing seasons of 1994 and 1996 (e.g., Chan and Middleton, 2000; Middleton and Sullivan, 2000).

J.1.2 LOPEX

The Leaf Optical Properties EXperiment (LOPEX, http://opticleaf.ipgp.fr/) was organized in the Joint Research Centre (Ispra, Italy) during the summer of 1993. A dataset associating visible/near infrared/shortwave infrared spectra of vegetation elements (leaves, conifer needles, stems, etc.) with physical measurements and biochemical analyses was constructed (Hosgood et al., 1995). In order to have a wide range of variation of leaf internal structure, pigmentation, water content, and biochemical components, plant species with different types of leaves were collected during two separate periods during the summer of 1993. About 70 leaf samples representative from more than 50 species were obtained from trees, crops, and plants in the area of the Joint Research Centre. This database is one of the most used by the scientific community to date.

J.1.3 Spectranomics

This is the case for the Spectranomics (http://spectranomics.stanford.edu/) project carried out by Dr Greg Asner at the Carnegie Institution of Washington's Department of Global Ecology (Stanford University). The information provided on plant leaves is extremely complete: pigments, nitrogen, potassium, etc. With the development of advanced spectrometers in laboratories and flown as airborne imaging spectrometer systems, it is likely that more comprehensive global data will become available over the next decade, making it likely that improved understanding of plant optical properties related to the biochemistry and structure of leaves will be possible (e.g., Asner and Martin, 2008a, 2016).

J.1.4 SPECCHIO

SPECCHIO (www.specchio.ch/), an Italian word meaning "mirror" or "looking glass," is an environmental database developed by Dr Andreas Hueni at the Remote Sensing Laboratories (University of Zurich) starting in the early 2000s. It provides a spectral library in various input formats of natural and artificial materials relevant for remote sensing applications. Most of the input data has come from special campaigns and experiments but it does not have comprehensive metadata to support interpretations of co-occurring leaf traits (as does the Spectranomics database). The strength of SPECCHIO is in the functionality of its JAVA application server, which performs various data transformations, manipulations, and corrections, and links to Matlab tools developed by the Zurich group (e.g., Bojinski et al., 2003; Hueni et al., 2009; Bojinski et al., 2003, 2011).

J.1.5 EcoSIS

The Ecological Spectral Information System (EcoSIS, http://labs.russell.wisc.edu/townsend/tag/ecosis/) is an ongoing program (started in late 2013 by NASA Terrestrial Ecology Program; expected release in 2016) that uses the open source CKAN web/database architecture that is widely used for public databases on the web. However, EcoSIS has adopted the SPECCHIO/Australia metadata schema to facilitate seamless data exchange between these two databases as it has received extensive vetting. EcoSIS is incorporating new field names for ancillary measurements that accompany the spectra, in particular to facilitate greater discovery and analysis of the source variation across spectra and different measurement campaigns. Such fields could include chemical concentrations of pigments or other leaf constituents and photosynthetic rates, as well as leaf and canopy structure.

J.1.6 OPTICLEAF

OPTICLEAF is a database on leaf optical properties (http://opticleaf.ipgp.fr/) designed and kept up to date by Professor Stéphane Jacquemoud (Université de Paris/Institut de Physique du Globe de Paris). It gathers more than 3000 references that are related to this topic.

J.2 Databases without Leaf Optical Properties

J.2.1 LEDA

The LEDA trait database (www.uni-oldenburg.de/en/landeco/research/leda/) provides information on plant traits that describe three key features of plant dynamics: persistence, regeneration, and dispersal (Kleyer et al., 2008). Among all the traits currently accessible via the query application are leaf distribution, leaf dry matter content, leaf mass, leaf size, and specific leaf area.

J.2.2 TRY Plant Trait Database

TRY is a network of vegetation scientists headed by Future Earth and the Max Planck Institute for Biogeochemistry, providing a global archive of plant traits (e.g., Kattge et al., 2011). To facilitate exchange data functionality, users are encouraged to use the field names and definitions in the TRY Plant Trait Database (www.try-db.org/). This database is expected to be housed at JPL when it is released.

J.2.3 BIEN

The Botanical Information and Ecology Network (BIEN, http://bien.nceas.ucsb.edu/bien/) is a network of ecologists, botanists, and computer scientists working together to document global patterns of plant diversity, function, and distribution (Engemann et al., 2016).

J.2.4 BIOPOP

BIOPOP was a database of plant traits and internet application for nature conservation sponsored by the German Federal Agency for Nature Conservation (BfN). It contains 104 584 datasets, including 60 traits and 4722 plant species (Poschlod et al., 2003).

J.3 Models

The Automated Radiative Transfer Models Operator (ARTMO, http://ipl.uv.es/artmo/) Graphic User Interface (GUI) is an in-house software package developed by the Image Processing Laboratory (IPL) of the University of Valencia (Spain). It provides essential tools for running and inverting a suite of plant radiative transfer models, both at the leaf and at the canopy level. ARTMO facilitates consistent and intuitive user interaction, thereby streamlining model set-up, running, storing, and spectra output plotting for any kind of optical sensor operating in the visible, near-infrared, and shortwave infrared range (400–2500 nm). The PROSPECT, LIBERTY, DLM, and FLUSPECT-B models detailed in Chapter 8 and Chapter 9 are available.

J.3.1 PROSPECT

The Leaf Optical Properties Spectra model (PROSPECT, Jacquemoud and Baret, 1990) has been used for about 30 years to study leaf optical properties in the solar and thermal infrared domain (see Chapter 9). The combined PROSPECT and SAIL canopy bidirectional reflectance model, also referred to as PROSAIL, allows to develop new methods for retrieval of vegetation biophysical properties (http://teledetection.ipgp.fr/prosail/). It links the spectral variation of canopy reflectance, which is mainly related to leaf biochemical contents, with its directional variation, which is primarily related to canopy architecture and soil/vegetation contrast. This link is key to simultaneous estimation of canopy biophysical/structural variables for applications in agriculture, plant physiology, and ecology at different scales. PROSAIL has become one of the most popular radiative transfer tools due to its ease of use, general robustness, and consistent validation by lab/field/space experiments over the years.

J.3.2 LIBERTY

The Leaf Incorporating Biochemistry Exhibiting Reflectance and Transmittance Yields (LIBERTY, http://rscl-grss.org/coderecord.php?id=491) model is a conifer leaf reflectance simulation model based on radiative transfer theory (see Section 8.1.2). LIBERTY provides a simulation of reflectance and transmittance of single leaf and quasi-infinite leaf (stacked leaves) at 400–2500 nm. The Matlab version was compiled based on the publication of Terence P. Dawson (Dawson et al., 1998a). The inputs include leaf structure and biochemical parameters. The outputs of the model include reflectance of single leaf and quasi-infinite leaf and transmittance. A user manual is provided, which contains the detailed description of input/output parameters. A GUI interface makes this program easy to operate.

J.3.3 ABM

The Algorithmic BDF models (ABM-U and ABM-B, Baranoski, 2006; Baranoski and Eng, 2007) employ an algorithmic Monte Carlo formulation to simulate light interactions with unifacial (e.g., maize and sugar cane) and bifacial (e.g., soybean and maple) plant leaves, respectively (www.npsg.uwaterloo.ca/models.php) (see Section 8.1.6 and Section 8.3.3). More specifically, radiation propagation is treated as a random walk process whose states correspond to the main tissue interfaces found in these leaves.

References

A.M.S. (1909), The temperature relations of foliage leaves, *New Phytologist*, 8(2):77–80.

Aalto T., Vesala T., Mattila T., Simbierowicz P. and Hari P. (1999), A three-dimensional stomatal CO_2 exchange model including gaseous phase and leaf mesophyll separated by irregular interfaces, *Journal of Theoretical Biology*, 196 (1):115–128.

Aalto T. and Juurola E. (2002), A three-dimensional model of CO_2 transport in airspaces and mesophyll cells of a silver birch leaf, *Plant, Cell & Environment*, 25(11):1399–1409.

Abbott L.A. and Lindenmayer A. (1981), Models for growth of clones in hexagonal cell arrangements: applications in *Drosophila* wing disc epithelia and plant epidermal tissues, *Journal of Theoretical Biology*, 90(4):495–544.

Abdel-Rahman E.M., Fethi F.B., van den Berg M. and Way M.J. (2010), Potential of spectroscopic data sets for sugarcane thrips (*Fulmekiola serrata* Kobus) damage detection, *International Journal of Remote Sensing*, 31(15):4199–4216.

Aber J.D., Bolster K.L., Newman S.D., Soulia M. and Martin M.E. (1994), Analyses of forest foliage. II: measurement of carbon fraction and nitrogen content by end-member analysis, *Journal of Near Infrared Spectroscopy*, 2(1):15–23.

Abera M.K., Retta M.A., Verboven P., Nicolai B.M., Berghuijs H. and Struik P. (2016), Virtual microstructural leaf tissue generation based on cell growth modeling, *Acta Horticulturae*, 1110:155–162.

Abou-Khaled A., Hagan R.M. and Davenport D.C. (1970), Effects of kaolinite as a reflective antitranspirant on leaf temperature, transpiration, photosynthesis, and water-use efficiency, *Water Resources Research*, 6(1):280–289.

Acem Z., Parent G., Monod B., Jeandel G. and Boulet P. (2010), Experimental study in the infrared of the radiative properties of pine needles, *Experimental Thermal and Fluid Science*, 34(7):893–899.

Acevedo M.F. and Ataroff M. (2012), Leaf spectra and weight of species in canopy, subcanopy, and understory layers in a Venezuelan Andean cloud forest, *Scientifica,* 2012: Article ID 839584, 14 pages.

Achard F.K. (1778), Mémoire sur les couleurs des végétaux. Première partie, *Nouveaux Mémoires de l'Académie Royale des Sciences et Belles-Lettres de Berlin*, 1:62–69 (Archiv der Berlin-Brandenburgischen Akademie der Wissenschaften).

Adams J. B. and Adams J.D. (1984), Geologic mapping using Landsat MSS and TM images: removing vegetation by modeling spectral mixtures, in *Proc. International Symposium on Remote Sensing of Environment Third Thematic Conference: Remote Sensing for Exploration Geology*, Ann Arbor, MI, 16–19 April 1984, ERIM, pp. 615–622.

Adams J.B., Smith M.O. and Johnson P.E. (1986), Spectral mixture modeling: a new analysis of rock and soil types at the Viking Lander I site, *Journal of Geophysical Research*, 91(B8):8098–8112.

Adams M.L., Norvell W.A., Peverly J.H. and Philpot W.D. (1993), Fluorescence and reflectance characteristics of manganese deficient soybean leaves: effects of leaf age and choice of leaflet, *Plant and Soil*, 155–156(1):235–238.

Adams M.L., Philpot W.D. and Norvell W.A. (1999), Yellowness index: an application of spectral second derivatives to estimate chlorosis of leaves in stressed vegetation, *International Journal of Remote Sensing*, 20(18):3663–3675.

Adams M.L., Norvell W.A., Philpot W.D. and Peverly J.H. (2000a), Spectral detection of micronutrient deficiency in "Bragg" soybean, *Agronomy Journal*, 92(2):261–268.

(2000b), Toward the discrimination of manganese, zinc, copper, and iron deficiency in "Bragg" soybean using spectral detection methods, *Agronomy Journal*, 92(2):268–274.

Adler P.M. (1992), *Porous Media: Geometry and Transports*, Butterworth-Heinemann, 560 pages.

Adler P.M., Jacquin C.G. and Thovert J.F. (1992), The formation factor of reconstructed porous media, *Water Resources Research*, 28(6):1571–1576.

Afandi S.D., Herdiyeni Y., Prasetyo L.B., Hasbi W., Arai K. and Okumurad H. (2016), Nitrogen content estimation of rice crop based on near infrared (NIR) reflectance using artificial neural network (ANN), *Procedia Environmental Sciences*, 33:63–69.

Afzal A. and Mousavi S.F. (2008), Estimation of moisture in maize leaf by measuring leaf dielectric constant, *International Journal of Agriculture & Biology*, 10(1):66–68.

Afzal A., Mousavi S.F. and Khademi M. (2010), Estimation of leaf moisture content by measuring the capacitance, *Journal of Agricultural Science and Technology*, 12(3):339–346.

Agarwal U.P. and Atalla R.H. (2010), Vibrational spectroscopy, in *Lignin and Lignans: Advances in Chemistry* (Heitner C., Dimmel D.R. and Schmidt J.A., Eds), CRC Press, pp. 103–136.

Agati G., Fusi F., Mazzinghi P. and Dipaolo M.L. (1993), A simple approach to the evaluation of the reabsorption of chlorophyll fluorescence spectra in intact leaves, *Journal of Photochemistry and Photobiology B*, 17(2):163–171.

Agati G., Mazzinghi P., Fusi F. and Ambrosini I. (1995), The F685/F730 chlorophyll fluorescence ratio as a tool in plant physiology: response to physiological and environmental factors, *Journal of Plant Physiology*, 145(3):228–238.

Agati G., Mazzinghi P., Lipucci di Paola, M., Fusi, F. and Cecchi, G. (1996), The F685/F730 chlorophyll fluorescence ratio as indicator of chilling stress in plants, *Journal of Plant Physiology*, 148(3–4):384–390.

Agati G. (1998), Response of the *in vivo* chlorophyll fluorescence spectrum to environmental factors and laser excitation wavelength, *Pure and Applied Optics*, 7(4):797–807.

Agati G., Cerovic Z.G. and Moya I. (2000), The effects of decreasing temperature up to chilling values on the *in vivo* F685/F735 chlorophyll fluorescence ratio in *Phaseolus vulgaris* and *Pisum sativum*: the role of photosystem I contribution to the 735 nm fluorescence band, *Photochemistry and Photobiology*, 72(1):75–84.

Agati G., Brunetti C., Di Ferdinando M., Ferrini F., Pollastri S. and Tattini M. (2013), Functional roles of flavonoids in photoprotection: new evidence, lessons from the past, *Plant Physiology and Biochemistry*, 72:35–45.

Agrios G. (2004), *Plant Pathology*, Elsevier, 952 pages.

Ahern F.J. (1988), The effects of bark beetle stress on the foliar spectral reflectance of lodgepole pine, *International Journal of Remote Sensing*, 9(9):1451–1468.

Ainsworth E.A., Serbin S.P., Skoneczka J.A. and Townsend P.A. (2014), Using leaf optical properties to detect ozone effects on foliar biochemistry, *Photosynthesis Research*, 119(1–2):65–76.

Al-Abbas A.H., Barr R., Hall J.D., Crane F.L. and Baumgardner M.F. (1974), Spectra of normal and nutrient-deficient maize leaves, *Agronomy Journal*, 66(1):16–20.

Alberts B. (2004), *Biologie moléculaire de la cellule*, Flammarion Médecine-Sciences, 1472 pages.

Alenius C.M., Vogelmann T.C. and Bornman J.F. (1995), A three-dimensional representation in the relationship between penetration of u.v.-B radiation and u.v.-screening pigments in leaves of *Brassica napus*, *New Phytologist*, 131 (3):297–302.

Ali A.M., Darvishzadeh R., Skidmore A.K., van Duren I., Heiden U. and Heurich M. (2015), PROSPECT inversion for indirect estimation of leaf dry matter content and specific leaf area, in *Proc. 36th International Symposium on Remote Sensing of Environment*, Berlin, Germany, 11–15 May 2015, ISPRS, pp. 277–284.

(2016), Estimating leaf functional traits by inversion of PROSPECT: assessing leaf dry matter content and specific leaf area in mixed mountainous forest, *International Journal of Applied Earth Observation and Geoinformation*, 45(part A):66–76.

Allaby M. (2006), *Dictionary of Plant Sciences*, Oxford University Press, 508 pages.

Allen E. (1964), Fluorescent white dyes: calculation of fluorescence from reflectivity values, *Journal of the Optical Society of America*, 54(4):506–515.

Allen L.H., Gausman H.W. and Allen W.A. (1975), Solar ultraviolet radiation in terrestrial plant communities, *Journal of Environmental Quality*, 4(3):285–294.

Allen W.A. and Richardson A.J. (1968), Interaction of light with a plant canopy, *Journal of the Optical Society of America*, 58 (8):1023–1028.

Allen W.A., Gausman H.W., Richardson A.J. and Thomas J.R. (1969), Interaction of isotropic light with a compact plant leaf, *Journal of the Optical Society of America*, 59(10):1376–1379.

Allen W.A., Gausman H.W. and Richardson A.J. (1970a), Mean effective optical constants of cotton leaves, *Journal of the Optical Society of America*, 60(4):542–547.

(1973), Willstätter-Stoll theory of leaf reflectance evaluated by ray tracing, *Applied Optics*, 12(10):2448–2453.

Allen W.A., Gausman H.W., Richardson A.J. and Wiegand C.L. (1970b), Mean effective optical constants of thirteen kinds of plant leaves, *Applied Optics*, 9(11):2573–2577.

Aluru M.R., Bae H., Wu D. and Rodermel S.R. (2001), The *Arabidopsis immutans* mutation affects plastid differentiation and the morphogenesis of white and green sectors in variegated plants, *Plant Physiology*, 127(1):67–77.

Alvarez-Añorve M.Y., Quesada M., Sánchez-Azofeifa G.A., Avila-Cabadilla L.D. and Gamon J.A. (2012), Functional regeneration and spectral reflectance of trees during succession in a highly diverse tropical dry forest ecosystem, *American Journal of Botany*, 99(5):816–826.

Álvarez-Arenas T.E.G., Sancho-Knapik D., Peguero-Pina J.J. and Gil-Pelegrín E. (2009a), Determination of plant leaves water status using air-coupled ultrasounds, in *Proc. 2009 IEEE International Ultrasonics Symposium (IUS)*, Rome, Italy, 20–23 September 2009, IEEE, pp. 771–774.

(2009b), Noncontact and noninvasive study of plant leaves using air-coupled ultrasounds, *Applied Physics Letters*, 95:193702.

Alves P.L.C.A., Magalhães A.C.N. and Barja P.R. (2002), The phenomenon of photoinhibition of photosynthesis and its importance in reforestation, *The Botanical Review*, 68(2):193–208.

Amigues S. (2003), *Théophraste. Recherches sur les plantes: Livres I-II*, Les Belles Lettres, 143 pages.

André Y.M. (1746), *Discours sur les merveilles de l'arc-en-ciel*, Ganeau Editeur., Paris, pp. 145–198.

Andrieu B., Baret F., Schellberg J. and Rinderle U. (1988), Estimation de spectres de feuilles à partir de mesures dans les bandes spectrales larges, in *Proc. 4th International Colloquium on Spectral Signatures of Objects in Remote Sensing*, Aussois, France, 18–22 January 1988, ESA, Vol. SP-287, pp. 351–356.

Andrieu B., Kiriakos S. and Jaggard K.W. (1992), Estimation de la concentration en chlorophylles de feuilles par mesure de leur réflectance ou par analyse numérique de photographies prises au laboratoire, *Agronomie*, 12(6):477–485.

Ångström A.J. (1853), Om växternas gröna färg, *Öfversigt af Kongl. Vetenskaps-akademiens forhandlingar*, 10:246–251. (1854), Ueber die grüne Farbe der Pflanzen, *Annalen der Physik*, 169(11):475–480.

Ansari A.Q. and Loomis W.E. (1959), Leaf temperatures, *American Journal of Botany*, 46(10):713–717.

Antonovics J., Bradshaw A.D. and Turner R.G. (1971), Heavy metal tolerance in plants, in *Advances in Ecological Research* (J.B. Cragg, Ed.), Academic Press, pp. 1–86.

Antunez A., Whiting M.D., Pierce F.J. and Stöckle C. (2008), Estimation of sweet cherry tree water status by spectral reflectance, *Acta Horticulturae*, 795:711–715.

Aoki M., Yabuki K. and Totsuka T. (1980), Remote sensing of the physiological functions of plants by infrared color aerial photography. I: Relation between leaf reflectivity ratio, bi-band ratio and photosynthetic function of leaves in several woody plants, *Research Report from the National Institute for Environmental Studies*, 11:225–237.

(1988), Effective spectral characteristics of leaf for the remote sensing of leaf water content *Journal of Agricultural Meteorology*, 44(2):111–117 (in Japanese).

Aoki M., Yabuki K., Totsuka T. and Nishida M. (1986), Remote sensing of chlorophyll content of leaf. I: Effective spectral reflection characteristics of leaf for the evaluation of chlorophyll content in leaves of Dicotyledons, *Environmental Control in Biology*, 24(1):21–26.

Archetti M. (2000), The origin of autumn colours by coevolution, *Journal of Theoretical Biology*, 205(4):625–630.

(2009a), Classification of hypotheses on the evolution of autumn colours, *Oikos*, 118(3):328–333.

(2009b), Phylogenetic analysis reveals a scattered distribution of autumn colours, *Annals of Botany*, 103(5):703–713.

Archetti M. and Brown S.P. (2004), The coevolution theory of autumn colours, *Proceedings of the Royal Society of London. Series B*, 271(1545):1219–1223.

Archetti M. and Leather S.R. (2005), A test of the coevolution theory of autumn colours: colour preference of *Rhopalosiphum padi on* Prunus padus, *Oikos*, 110(2):339–343.

Archetti M., Döring T.F., Hagen S.BHughes N.M., Leather S.R., Lee D.W., et al. (2009c), Unravelling the evolution of autumn colours: an interdisciplinary approach, *Trends in Ecology & Evolution*, 24(3):166–173.

Archetti M., Richardson A.D., O'Keefe J. and Delpierre N. (2013), Predicting climate change impacts on the amount and duration of autumn colors in a New England forest, *PLOS One* 8(3):e57373.

Arnold L., Gillet S., Lardière O., Riaud P. and Schneider J. (2002), A test for the search for life on extrasolar planets, *Astronomy & Astrophysics*, 392(1):231–237.

Arroyo-Mora J.P., Kalacska M., Caraballo B.L., Trujillo J.E. and Vargas O. (2008), Spectral expression of gender: a pilot study with two dioecious neotropical tree species, in *Hyperspectral Remote Sensing of Tropical and Sub-Tropical Forests* (Kalacska M. and Sánchez-Azofeifa G.A., Eds), CRC Press, pp. 125–140.

Ashourloo D., Mobasheri M.R. and Huete A. (2014), Developing two spectral disease indices for detection of wheat leaf rust (*Pucciniatriticina*), *Remote Sensing*, 6(6):4723–4740.

Askenasy E. (1875), Ueber die Temperatur welche Pflanzen im Sonnenlicht annehmen, *Botanische Zeitung*, 33:442–443.

Asmail C. (1991), Bidirectional Scattering Distribution Function (BSDF): a systematized bibliography, *Journal of Research of the National Institute of Standards and Technology*, 96(2):215–223.

Asner G.P., Wessman C.A., Schimel D.S. and Archer S. (1998), Variability in leaf and litter optical properties: implications for BRDF model inversions using AVHRR, MODIS, and MISR, *Remote Sensing of Environment*, 63(3):243–257.

Asner G.P. (2008), Hyperspectral remote sensing of canopy chemistry, physiology, and biodiversity in tropical rainforests, in *Hyperspectral Remote Sensing of Tropical and Sub-Tropical Forests* (Kalacska M. and Sánchez-Azofeifa G.A., Eds), CRC Press, pp. 261–296.

Asner G.P., Jones M.O., Martin R.E., Knapp D.E. and Hughes R.F. (2008a), Remote sensing of native and invasive species in Hawaiian forests, *Remote Sensing of Environment*, 112(5):1912–1926.

Asner G.P. and Martin R.E. 2008b), Spectral and chemical analysis of tropical forests: caling from leaf to canopy levels, *Remote Sensing of Environment*, 112(10):3958–3970.

(2009), Airborne spectranomics: Mapping canopy chemical and taxonomic diversity in tropical forests, *Frontiers in Ecology and the Environment*, 7(5):269–276.

(2011), Canopy phylogenetic, chemical and spectral assembly in a lowland Amazonian forest, *New Phytologist*, 189 (4):999–1012.

(2016), Spectranomics: Emerging science and conservation opportunities at the interface of biodiversity and remote sensing, *Global Ecology and Conservation*, 8:212–219.

Asner G.P., Martin R.E., Ford A.J., Metcalfe D.J. and Liddell M.J. (2009), Leaf chemical and spectral diversity in Australian tropical forests, *Ecological Applications*, 19(1):236–253.

Asner G.P., Martin R.E., Knapp D.E., Tupayachi R., Anderson C., Carranza L., et al. (2011a), Spectroscopy of canopy chemicals in humid tropical forests, *Remote Sensing of Environment*, 115(12):3587–3598.

Asner G.P., Martin R.E., Tupayachi R., Emerson R., Martinez P., Sinca F., et al. (2011b), Taxonomy and remote sensing of leaf mass per area (LMA) in humid tropical forests, *Ecological Applications*, 21(1):85–98.

Asner G.P., Martin R.E. and Bin Suhaili A. (2012), Sources of canopy chemical and spectral diversity in lowland Bornean forest, *Ecosystems*, 15(3):504–517.

Asner G.P. (2014), A chemical-evolutionary basis for remote sensing of tropical forest diversity, in *Forests and Global Change* (Coomes D.A., Burslem D.F.R.P. and Simonson D., Eds), Cambridge University Press, pp. 343–358.

Asner G.P., Martin R.E., Carranza-Jimenez L., Sinca F., Tupayachi R., Anderson C.B., et al. (2014), Functional and biological diversity of foliar spectra in tree canopies throughout the Andes to Amazon region, *New Phytologist*, 204(1):127–139.

Asrar G. (1989), *Theory and Applications of Optical Remote Sensing*, Wiley Interscience, 734 pages.

Atkinson M.D., Jervis A.P. and Sangha R.S. (1997), Discrimination between *Betula pendula, Betula pubescens*, and their hybrids using near-infrared reflectance spectroscopy, *Canadian Journal of Forest Research*, 27(11):1896–1900.

Atrashevskii Y.I., Sikorskii A.V., Sikorskii V.V., Stel'makh G.F. and Shuplyak V.I. (1998), Spectral polarization characteristics of optical radiation reflected from leaves subjected to unfavorable ecological factors, *Journal of Applied Spectroscopy*, 65 (1):110–116 (cover-to-cover translation from *Zhurnal Prikladnoi Spektroskopii*, 65(1):107–113, 1998).

Atrashevskii Y.I., Sikorskii A.V., Sikorskii V.V. and Stel'makh G.F. (1999), The reflection and scattering of light by a plant leaf, *Journal of Applied Spectroscopy*, 66(1):105–114 (cover-to-cover translation from *Zhurnal Prikladnoi Spektroskopii*, 66(1):101–108, 1999).

Ausmus B.S. and Hilty J.W. (1972), Reflectance studies of healthy, maize dwarf mosaic virus-infected, and *Helminthosporium maydis*-infected corn leaves, *Remote Sensing of Environment*, 2:77–81.

Avalos G., Mulkey S.S. and Kitajima K. (1999), Leaf optical properties of trees and lianas in the outer canopy of a tropical dry forest, *Biotropica*, 31(3):517–520.

Averill B. and Eldredge P. (2011), *General Chemistry: Principles, Patterns, and Applications*, Saylor Foundation, https://open.umn.edu/opentextbooks/BookDetail.aspx?bookId=69.

Azia F. and Stewart K.A. (2001), Relationships between extractable chlorophyll and SPAD values in muskmelon leaves, *Journal of Plant Nutrition*, 24(6):961–966.

Bacci L., Benincasa F. and Rapi B. (1996), Effect of growth temperature on the spectrocolorimetric characteristics of sorghum plants (*Sorghum bicolor* (L.) Moench.): indices of stress, *European Journal of Agronomy*, 5(1–2):45–57.

Bacci L., De Vincenzi M., Rapi B., Arca B. and Benincasa F. (1998), Two methods for the analysis of colorimetric components applied to plant stress monitoring, *Computers and Electronics in Agriculture*, 19(2):167–186.

Bacour C., Jacquemoud S., Leroy M., et al. (2002a), Reliability of the estimation of vegetation characteristics by inversion of three canopy reflectance models on airborne POLDER data, *Agronomie*, 22:555–565.

Bacour C., Jacquemoud S., Tourbier Y., Dechambre M. and Frangi J.P. (2002b), Design and analysis of numerical experiments to compare four canopy reflectance models, *Remote Sensing of Environment*, 79(1):72–83.

Baffioni C., Careri G. and Giansanti A. (1985), Leaf tissue desiccation process: a dielectric-gravimetric study, *Lettere Al Nuovo Cimento*, 42(6):295–298.

Bagard M., Le Thiec D., Delacote E., Hasenfratz-Sauder M.P., Banvoy J., Gerard J., et al. (2008), Ozone-induced changes in photosynthesis and photorespiration of hybrid poplar in relation to the developmental stage of the leaves, *Physiologia Plantarum*, 134(4):559–574.

Bailey L.H. (1963), *How Plants Get Their Names*, Dover Publications, 181. pages.

Bailey S., Walters R.G., Jansson S. and Horton P. (2001), Acclimation of *Arabidopsis thaliana* to the light environment: the existence of separate low light and high light responses, *Planta*, 213(5):794–801.

Baldini E., Facini O., Nerozzi F., Rossi F. and Rotondi A. (1997), Leaf characteristics and optical properties of different woody species, *Trees – Structure and Function*, 12(2):73–81.

Baldy R.W., de Benedictis J.A., Johnson L.F., Weber E., Baldy M.W., Osborn B.P et al. (1996), Leaf color and vine size are related to yield in a phylloxera-infested vineyard, *Vitis*, 35(4):201–205.

Bálint J., Nagy B.V. and Fail J. (2013), Correlations between colonization of onion thrips and leaf reflectance measures across six cabbage varieties, *PLOS One*, 8(9):e73848.

Ball A., Sanchez-Azofeifa A.G., Portillo-Quintero C., Rivard B., Castro-Contreras S. and Fernandes G. (2015), Patterns of leaf biochemical and structural properties of Cerrado life forms: implications for remote sensing, *PLOS One*, 10 (2):e0117659.

Baker E.A. (1982), Chemistry and morphology of plant epicuticular waxes, in *The Plant Cuticle* (Cutler D.F., Alvin K.L. and Price C.E., Eds), Academic Press, pp. 139–165.

Baltzer J.L. and Thomas S.C. (2005), Leaf optical responses to light and soil nutrient availability in temperate deciduous trees, *American Journal of Botany*, 92(2):214–223.

Banchoff T. (1996), *La quatrième dimension – Voyage dans les dimensions supérieures*, Pour la Science, Paris, 206 pages.

Bandaru V., Hansen D.J., Codling E.E., Daughtry C.S., White-Hansen S. and Green C.E. (2010), Quantifying arsenic-induced morphological changes in spinach leaves: implications for remote sensing, *International Journal of Remote Sensing*, 31 (15):4163–4177.

Baranoski G.V.B. and Rokne J.G. (1997), An algorithmic reflectance and transmittance model for plant tissue, *Computer Graphics Forum*, 16(3):C141–C150.

(1999), A non-deterministic reconstruction approach for isotropic reflectances and transmittances, *Journal of Visualization and Computer Animation*, 10(4):225–231.

(2001), Efficiently simulating scattering of light by leaves, *Visual Computer*, 17(8):491–505.

(2004), *Light Interaction with Plants – A Computer Graphics Perspective*, Horwood Publishing, 154 pages.

(2005), A practical approach for estimating the red edge position of plant leaf reflectance, *International Journal of Remote Sensing*, 26(3):503–521.

Baranoski G.V.G. (2006), Modeling the interaction of infrared radiation (750 to 2500 nm) with bifacial and unifacial plant leaves, *Remote Sensing of Environment*, 100(3):335–347.

Baranoski G.V.G. and Eng D. (2007), An investigation on sieve and detour effects affecting the interaction of collimated and diffuse infrared radiation (750 to 2500 nm) with plant leaves, *IEEE Transactions on Geoscience and Remote Sensing*, 45 (8):2593–2599.

Baret F., Andrieu B. and Guyot G. (1988), A simple model for leaf optical properties in visible and near infrared: application to the analysis of spectral shifts determinism, in *Applications of Chlorophyll Fluorescence in Photosynthesis Research, Stress Physiology, Hydrobiology and Remote Sensing* (Lichtenthaler H.K., Ed), Kluwer Academic Publishers, Dordrecht, pp. 345–351.

Baret F., Jacquemoud S., Guyot G. and Leprieur C. (1992), Modeled analysis of the biophysical nature of spectral shifts and comparison with information content of broad bands, *Remote Sensing of Environment*, 41(2–3):133–142.

Baret F., Vanderbilt V.C., Steven M.D. and Jacquemoud S. (1994), Use of spectral analogy to evaluate canopy reflectance sensitivity to leaf optical properties, *Remote Sensing of Environment*, 48(2):253–260.

Baret F. and Fourty T. (1997), Estimation of leaf water content and specific leaf weight from reflectance and transmittance measurements, *Agronomie*, 17(9–10):455–464.

Baret F. and Buis S. (2008), Estimating canopy characteristics from remote sensing observations: review of methods and associated problems, in *Advances in Land Remote Sensing – System, Modeling, Inversion and Application* (S. Liang, Ed.), Springer, pp. 173–201.

Barnes J.D. and Cardoso-Vilhena J. (1996), Interaction between electromagnetic radiation and the plant cuticle, in *Plant Cuticles: an Integrated Functional Approach* (Kerstiens G.K., Ed), BIOS Scientific Publishers Limited, Oxford, pp. 157–174.

Barnes P.W., Flint S.D., Slusser J.R., Gao W. and Ryel R.J. (2008), Diurnal changes in epidermal UV transmittance of plants in naturally high UV environments, *Physiologia Plantarum*, 133(2):363–372.

Barry K.M., Newnham G.J. and Stone C. (2009), Estimation of chlorophyll content in *Eucalyptus globulus* foliage with the leaf reflectance model PROSPECT, *Agricultural and Forest Meteorology*, 149(6–7):1209–1213.

Barry K.M., Corkrey R., Thi H.P., Ridge S. and Mohammed C.L. (2011), Spectral characterization of necrosis from reflectance of *Eucalyptus globulus* leaves with Mycosphaerella leaf disease or subjected to artificial lesions, *International Journal of Remote Sensing*, 32(24):9243–9259.

Barthlott W., Mutke J., Rafiqpoor D., Kier G. and Kreft H. (2005), Global centers of vascular plant diversity, *Nova Acta Leopoldina*, 92(342):61–83.

Bartlett M.K., Ollinger S.V., Hollinger D.Y., Wicklein H.F. and Richardson A.D. (2011), Canopy-scale relationships between foliar nitrogen and albedo are not observed in leaf reflectance and transmittance within temperate deciduous tree species, *Botany-Botanique*, 89(7):491–497.

Barton C.V.M. (2001), A theoretical analysis of the influence of heterogeneity in chlorophyll distribution on leaf reflectance, *Tree Physiology*, 21(12–13):789–795.

Basayigit L., Albayrak S. and Senol H. (2014), Analysis of VNIR reflectance for prediction of macro and micro nutrient and chlorophyll contents in apple trees (*Malus communis*), *Asian Journal of Chemistry*, 21(2):1302–1308.

Basu S.K. (2006), *Realistic Rendering of Plant Leaves*, University of North Carolina, Chapel Hill, NC, 9 May 2006, 7 pages.

Bauer S.D., Korč F. and Förstner W. (2011), The potential of automatic methods of classification to identify leaf diseases from multispectral images, *Precision Agriculture*, 12(3):361–377.

Bauerle W.L., Weston D.J., Bowdena J.D., Dudley J.B. and Toler J.E. (2004), Leaf absorptance of photosynthetically active radiation in relation to chlorophyll meter estimates among woody plant species, *Scientia Horticulturae*, 101(1–2):169–178.

Bawhey C.I. and Grant R.H. (2003), Effect of epicuticular wax on UV scattering of sorghum leaves and canopies, in *Proc. Ultraviolet Ground- and Space-based Measurements, Models, and Effects III* (Slusser J.R., Herman J.R. and Gao W., Eds), San Diego, CA, 4 August 2003, SPIE, Vol. 5156, pp. 236–244.

Bayer A. (2004), *Painters of Reality: the Legacy of Leonardo and Caravaggio in Lombardy*, Metropolitan Museum of Art, 257 pages.

Baynes J. (2007), The reflectance signature of canopy components: implications for the interpretation of remotely sensed images, *Annals of Tropical Research*, 29(1):21–31.

Becher J.J. (1669), *Physica subterranea*, Georg Ernst Stahl, Leipzig (1738), 503 pages.

Becquerel E. (1868), *La lumière, ses causes et ses effets*, Librairie de Firmin Didot Frères, 377 pages.

Behera R.K. and Choudhury N.K. (2002), High irradiance induced pigment degradation and loss of photochemical activity of wheat chloroplasts, *Biologia Plantarum*, 45(1):45–49.

Behmann J., Mahlein A.K., Rumpf T., Römer C. and Plümer L. (2015), A review of advanced machine learning methods for the detection of biotic stress in precision crop protection, *Precision Agriculture*, 16(3):239–260.

Belanger M.J. (1990), *A seasonal Perspective of several leaf developmental characteristics as related to the red edge of plant leaf reflectance*, Master of Science Thesis, Faculty of Graduate Studies, York University, North York, ON, 110 pages.

Belanger M.J., Miller J.R. and Boyer G. (1995), Comparative relationships between some red edge parameters and seasonal leaf chlorophyll concentrations, *Canadian Journal of Remote Sensing*, 21(1):16–21.

Bell I.E. and Baranoski G.V.G. (2004), Reducing the dimensionality of plant spectral databases, *IEEE Transactions on Geoscience and Remote Sensing*, 42(3):570–576.

Bellante G.J., Powell S.L., Lawrence R.L., Repasky K.S. and Dougher T. (2014), Hyperspectral detection of a subsurface CO_2 leak in the presence of water stressed vegetation, *PLOS One*, 9(10):e108299.

Belov N.P., Sherstobitova A.S., Yaskov A.D. (2011), Diffuse reflection of light by cellulose pulp and optical absorption of aqueous residual lignin solutions, *Journal of Applied Spectroscopy*, 78(1):138–140.

Belyaev B.I., Belyaev Y.V., Chumakov A.V., Nekrasov V.P. and Shuplyak V.I. (2000), Spectral and spectral-polarization characteristics of potato leaves, *Journal of Applied Spectroscopy*, 67(4):723–730 (cover-to-cover translation from *Zhurnal Prikladnoi Spektroskopii*, 67(4):524–529).

Ben-Dor E., Inbar Y. and Chen Y. (1997), The reflectance spectra of organic matter in the visible near-infrared and short wave infrared region (400–2500 nm) during a controlled decomposition process, *Remote Sensing of Environment*, 61(1):1–15.
Benedict H.M. and Swidler R. (1961), Nondestructive method for estimating chlorophyll content of leaves, *Science*, 133 (3469):2015–2016.
Benford F. (1923), Reflection and transmission by parallel plates, *Journal of the Optical Society of America*, 7(11):1017–1025.
Benford F. (1946), Radiation in diffuse medium, *Journal of the Optical Society of America*, 36(9):524–554.
Benoist D., Tourbier Y. and Germain-Tourbier S. (1994), *Plans d'expériences : construction et analyse*, Lavoisier, Paris, 693 pages.
Berdnik V.V. and Loiko V.A. (1999), Modeling of radiative transfer in disperse layers of a medium with a highly stretched phase function, *Journal of Quantitative Spectroscopy & Radiative Transfer*, 61(1):49–57.
Berdnik V.V. and Mukhamed'yarov R.D. (2001), Radiation transfer in plant leaves, *Optics and Spectroscopy*, 90(4):580–591 (cover-to-cover translation from *Optika i Spektroskopiya*, 90(4):652–663).
Berg V.S. (1985), Physical, chemical and spectral properties of leaf surfaces of *Berberis aquifolium* (Pursh.), *Plant, Cell & Environment*, 8(9):631–638.
Berghuijs H.N.C., Yin X.Y., Ho Q.T., Driever S.M., Retta M.A., Nicolai B.M. et al. (2016), Mesophyll conductance and reaction-diffusion for CO_2 transport in C3 leaves; needs, opportunities and challenges, *Plant Science*, 252:62–75.
Berghuijs H.N.C., Yin X.Y., Ho Q.T., Retta M.A., Verboven P., Nicolaï B.M., et al. (2017), Localization of (photo)respiration and CO2 re-assimilation in tomato leaves investigated with a reaction-diffusion model, *PLOS One*, 12(9):e0183746.
Bernays E.A. and Chapman R.F. (1994), *Host-Plant Selection by Phytophageous Insects*, Springer, 312 pages.
Berthier S. (2006), *Iridescences: the Physical Colors of Insects*, Springer, 160 pages.
Bertoluzza A., Bottura G., Lucchi P., Marchetti L. and Zechini D'Aulerio A. (1999), Molecular monitoring of horse chestnut leaves affected with biotic and abiotic disorders, *Journal of Plant Pathology*, 81(2):89–94.
Berzelius J. (1837a), Ueber die gelbe Farbe der Blätter im Herbste, *Annalen der Pharmacie*, 21(3):257–262.
(1837b), Einige Untersuchungen über die Farbe, welche das Laub verschiedener Baumgattungen im Herbste vor dem Abfallen annimmt, *Annalen der Physik*, 118(11):422–433.
(1837c), Ueber den rothen Farbstoff der Beeren und Blätter im Herbst, *Annalen der Pharmacie*, 21(3):262–267.
Beysens D. (1995), The formation of dew, *Atmospheric Research*, 39(1–3):215–237.
Bhushan B. and Jung Y.C. (2006), Micro- and nanoscale characterization of hydrophobic and hydrophilic leaf surfaces, *Nanotechnology*, 17(11):2758–2772.
Bianchi G. (1995), Plant waxes, in *Waxes: Chemistry, Molecular Biology and Functions* (Hamilton R.D., Ed), The Oily Press, pp. 177–222.
Bilger W., Björkman O. and Thayer S.S. (1989), Light-induced spectral absorbance changes in relation to photosynthesis and the epoxidation state of xanthophyll cycle components in cotton leaves, *Plant Physiology*, 91(2):542–551.
Biliouris D., Verstraeten W.W., Dutré P., van Aardt J.A.N., Muys B. and Coppin P. (2007), A compact laboratory spectro-goniometer (CLabSpeG) to assess the BRDF of materials. Presentation, calibration and implementation on *Fagus sylvatica* L. leaves, *Sensors*, 7(9):1846–1870.
Biliouris D., van der Zande D., Verstraeten W.W., et al. (2009), RPV model parameters based on hyperspectral bidirectional reflectance measurements of *Fagus sylvatica* L. leaves, *Remote Sensing*, 1(2):92–106.
Billings W.D. and Morris R.J. (1951), Reflection of visible and infrared radiation from leaves of different ecological groups, *American Journal of Botany*, 38(5):327–331.
Binsfeld R., Gamboa J. and Walter M. (2011), Visual patterns in the plant kingdom, in *Proc. 24th Graphics, Patterns and Images* (Lewiner T. and Torres R., Eds), Maceio, Brazil, 28–31 August 2011, IEEE, pp. 86–92.
Birch H. (2009), The artificial leaf, *Chemistry World*, May 2009, pp. 42–45.
Birkebak R. and Birkebak R. (1964), Solar radiation characteristics of tree leaves, *Ecology*, 45(3):646–649.
Birth G.S. and Hecht H.G. (1987), The physics of near-infrared reflectance, in *Near-Infrared Technology in the Agricultural and Food Industries* (Norris P.W.K., Ed.), American Association of Cereal Chemists, St Paul, MN, pp. 1–15.
Bisba A., Petropoulou Y. and Manetas Y. (1997), The transiently pubescent young leaves of plane (*Platanus orientalis*) are deficient in photodissipative capacity, *Physiologia Plantarum*, 101(2):373–378.
Biswal B. (1995), Carotenoid catabolism during leaf senescence and its control by light, *Journal of Photochemistry and Photobiology B: Biology*, 30(1):3–13.
Björn L.O. (1986), Reply to comment on "Measurement of light gradients and spectral regime in plant tissue with fibre optic probe", *Physiologia Plantarum*, 67(3):494–497.
Björn L.O. and Vogelmann T.C. (1996), Quantifying light and ultraviolet radiation in plant biology, *Photochemistry and Photobiology*, 64(3):403–406.
Björn L.O. and Li S. (2011), Near-surface silica does not increase radiative heat dissipation from plant leaves, *Applied Physics Letters*, 99(2):024104.
Blackburn G.A. (1998a), Quantifying chlorophylls and carotenoids at leaf and canopy scales: an evaluation of some hyperspectral approaches, *Remote Sensing of Environment*, 66(3):273–285.
(1998b), Spectral indices for estimating photosynthetic pigment concentrations: a test using senescent tree leaves, *International Journal of Remote Sensing*, 19(4):657–675.
(1999), Relationships between spectral reflectance and pigment concentrations in stacks of deciduous broadleaves, *Remote Sensing of Environment*, 70(2):224–237.

(2004), Wavelet decomposition of hyperspectral reflectance data for quantifying photosynthetic pigment concentrations in vegetation, in *Proc. XXth ISPRS Congress: Geo-Imagery Bridging Continents* (Altan O., Ed), Istanbul, Turkey, 12–23 July 2004, ISPRS, pp. 878–882.

(2007a), Hyperspectral remote sensing of plant pigments, *Journal of Experimental Botany*, 58(4):855–867.

(2007b), Wavelet decomposition of hyperspectral data: a novel approach to quantifying pigment concentrations in vegetation, *International Journal of Remote Sensing*, 28(12):2831–2855.

Blackburn, G.A. and Ferwerda, J.G. (2008), Retrieval of chlorophyll concentration from leaf reflectance spectra using wavelet analysis, *Remote Sensing of Environment*, 112(4):1614–1632.

Blackmer T.M., Schepers J.S. and Vigil M.F. (1993), Chlorophyll meter readings in corn as affected by plant spacing, *Communications in Soil Science and Plant Analysis*, 24(17–18):2507–2516.

Blackmer T.M., Shepers J.S. and Varvel G.A. (1994), Light reflectance compared with other nitrogen stress measurements in corn leaves, *Agronomy Journal*, 86(6):934–938.

Blackmer T.M. and Shepers J.S. (1995), Use of a chlorophyll meter to monitor nitrogen status and schedule fertigation for corn, *Journal of Production Agriculture*, 8(1):56–60.

Blanc P. (2002), *Etre plante à l'ombre des forêts tropicales*, Nathan, 432 pages.

Blanchfield A.L., Robinson S.A., Renzullo L.J. and Powell K.S. (2006), Phylloxera-infested grapevines have reduced chlorophyll and increased photoprotective pigment content-can leaf pigment composition aid pest detection? *Functional Plant Biology*, 33(5):507–514.

Blank F. (1947), The anthocyanin pigments of plants, *Botanical Review*, 13(5):241–317.

Blazquez C.H. and Edwards G.J. (1986), Spectral reflectance of healthy and diseased watermelon leaves, *Annals of Applied Biology*, 108(2):243–249.

Blinn J.F. (1977), Models of light reflection for computer synthesized pictures, *ACM SIGGRAPH Computer Graphics*, 11 (2):192–198.

Blonder B., De Carlo F., Moore J., Rivers M. and Enquist B.J. (2012), X-ray imaging of leaf venation networks, *New Phytologist*, 196(4):1274–1282.

Boeckx T., Winters A.L., Webb K.J. and Kingston-Smith A.H. (2015), Polyphenol oxidase in leaves: is there any significance to the chloroplastic localization? *Journal of Experimental Botany*, 66(12):3571–3579.

Bojinski S., Schaepman M., Schläpfer D., Itten K. (2003), SPECCHIO: a spectrum database for remote sensing applications, *Computers & Geosciences*, 29(1):27–38.

Bolster K.L., Martin M.E. and Aber J.D. (1996), Determination of carbon fraction and nitrogen concentration in tree foliage by near infrared reflectance: a comparison of statistical methods, *Canadian Journal of Forest Research*, 26 (4):590–600.

Bone R.A., Lee D.W. and Norman J.M. (1985), Epidermal cells functioning as lenses in leaves of tropical rain forest shade plants, *Applied Optics*, 24(10):1408–1414.

Bonfil D.J., Karnieli A., Raz M., et al. (2005), Rapid assessing of water and nitrogen status in wheat flag leaves, *Journal of Food, Agriculture & Environment*, 3(2):148–153.

Bonham J.S. (1986), Fluorescence and Kubelka-Munk theory, *Color Research and Applications*, 11(3):223–230.

Bonham-Carter G. F. (1988), Numerical procedures and computer program for fitting an inverted Gaussian model to vegetation reflectance data, *Computers & Geosciences*, 14(3):339–356.

Boochs F., Kupfer G., Dokter K. and Kühbauch W. (1990), Shape of the red edge as vitality indicator for plants, *International Journal of Remote Sensing*, 11(10):1741–1753.

Borel C.C. and McIntosh R.E. (1988), Leaf backscattering measurements and modelling at 94 GHz, in *Proc. 8th International Geoscience and Remote Sensing Symposium (IGARSS'88)*, Edinburgh, Scotland, 13–16 September 1988, ESA, Vol. SP-284, pp. 1277–1278.

Born N., Behringer D., Liepelt S., et al. (2014a), Monitoring plant drought stress response using terahertz time-domain spectroscopy, *Plant Physiology*, 164(4):1571–1577.

Born N., Gente R., Behringer D., et al. (2014b), Monitoring the water status of plants using THz radiation, in *Proc. 39th International Conference on Infrared, Millimeter, and THz Waves*, Tucson, AZ, 14–19 September 2014. 2 pages.

Bornman J.F. and Vogelmann T.C. (1988), Penetration of blue and UV radiation measured by fiber optics in spruce and fir needles, *Physiologia Plantarum*, 72(4):699–705.

(1991), Effect of UV-B radiation on leaf optical properties measured with fibre optics, *Journal of Experimental Botany*, 42 (237):547–554.

Bornman J.F., Vogelmann T.C. and Martin G. (1991), Measurement of chlorophyll fluorescence within leaves using a fibreoptic microprobe, *Plant, Cell & Environment*, 14(7):719–725.

Bortolot Z.J. and Wynne R.H. (2003), A method for predicting fresh green leaf nitrogen concentrations from shortwave infrared reflectance spectra acquired at the canopy level that requires no *in situ* nitrogen data, *International Journal of Remote Sensing*, 24(3):619–624.

Bostock J. and Riley H.T. (1855), *The Natural History of Pliny*, H.G. Bohn, London, 543 pages.

Botha E., Zebarth B. and Leblon B. (2006), Non-destructive estimation of potato leaf chlorophyll and protein contents from hyperspectral measurements using the PROSPECT radiative transfer model, *Canadian Journal of Plant Science*, 86 (1):279–291.

Boulet P., Parent G., Collin A., et al. (2009), Spectral emission of flames from laboratory-scale vegetation fires, *International Journal of Wildland Fire*, 18(7):875–884.

Boulet P., Parent G., Acem Z., Collin A. and Séro-Guillaume O. (2011), On the emission of radiation by flames and corresponding absorption by vegetation in forest fires, *Fire Safety Journal*, 46(1–2):21–26.

Bousquet L., Lachérade S., Jacquemoud S. and Moya I. (2005), Leaf BRDF measurement and model for specular and diffuse component differentiation, *Remote Sensing of Environment*, 98(2–3):201–211.

Bousquet L. (2007), *Mesure et modélisation des propriétés optiques spectrales et directionnelles des feuilles*, PhD Thesis, Departement of Physics, Paris Diderot University, Paris, 203 pages.

Bowmaker J.K. and Dartnall H.J. (1980), Visual pigments of rods and cones in a human retina, *The Journal of Physiology*, 298 (1):501–511.

Bowman W.D. (1989), The relationships between leaf water status, gas exchange, and spectral reflectance in cotton leaves, *Remote Sensing of Environment*, 30(3):249–255.

Bowyer P. and Danson F.M. (2004), Sensitivity of spectral reflectance to variation in live fuel moisture content at leaf and canopy level, *Remote Sensing of Environment*, 92(3):297–308.

Boye C.L. and Rife D.C. (1938), Genetic studies of *Coleus*. I: Leaf color I, *Journal of Heredity*, 29(2):55–60.

Boyer M., Miller J.R., Belanger M., Hare E. and Wu J. (1988), Senescence and spectral reflectance in leaves in northern pin oak (*Quercus palustris* Muenchh.), *Remote Sensing of Environment*, 25(1):71–87.

Boyle R. (1664), *Experiments and Considerations Touching Colours*, H. Herringman, London, 423 pages.

Braitmaier M., Diepstraten J. and Ertl T. (2004), Real-time rendering of seasonal influenced trees, in *Proc. Theory and Practice of Computer Graphics*, Bournemouth, UK, 10 June 2004, IEEE, pp. 152–159.

Brakke T.W. and Smith J.A. (1987), A ray tracing model for leaf bidirectional scattering studies, in *Proc. 7th International Geoscience and Remote Sensing Symposium (IGARSS'87)*, Ann Arbor, MI, 18–21 May 1987, IEEE, pp. 643–648.

Brakke T.W., Smith J.A. and Harnden J.M. (1989), Bidirectional scattering of light from tree leaves, *Remote Sensing of Environment*, 29(2):175–183.

Brakke T.W. (1992), Goniometric measurements of light scattered in the principal plane from leaves, in *Proc. 12th International Geoscience and Remote Sensing Symposium (IGARSS'92)*, Houston, TX, 26–29 May 1992, IEEE, pp. 508–510.

(1994), Specular and diffuse components of radiation scattered by leaves, *Agricultural and Forest Meteorology*, 71 (3–4):283–295.

Brakke T.W., Wergin W.P., Erbe E.F. and Harnden J.M. (1993), Seasonal variation in the structure and red reflectance of leaves from yellow poplar red oak, and red maple, *Remote Sensing of Environment*, 43(2):115–130.

Brandt A.B. and Tageyeva S.V. (1967), *Optical Parameters of Plant Organisms*, Nauka Publishers, Moskva (in Russian).

Breece H.T. and Holmes R.A. (1971), Bidirectional scattering characteristics of healthy green soybeans and corn leaves *in vivo*, *Applied Optics*, 10(1):119–127.

Breitenstein B., Scheller M., Shakfa M.K., Kinder T., Müller-Wirts T., Koch M., et al. (2011), Introducing terahertz technology into plant biology: a novel method to monitor changes in leaf water status, *Journal of Applied Botany and Food Quality*, 84(2):158–161.

Brewer C.A., Smith W.K. and Vogelmann T.C. (1991), Functional interaction between leaf trichomes, leaf wettability and the optical properties of water droplets, *Plant, Cell & Environment*, 14(9):955–962.

Brewster D.K.H. (1834), On the colours of natural bodies, *Transactions of the Royal Society of Edinburgh*, 12:538–545.

Brewster D.K.H. (1855), *Memoirs of the Life, Writings and Discoveries of Sir Isaac Newton*, Thomas Constable and Co., Edinburgh, 564 pages.

Briot D., Arnold L., Jacquemoud S., Schneider J., Agabi A., Aristidi E., et al. B2010), The LUCAS experiment: Spectroscopy of Earthshine in Antarctica for detection of life, *Proc. 3rd ARENA* Conference: *"An Astronomical Observatory at CONCORDIA (Dome C, Antarctica)"*, EAS Publications Series, Vol. 40, pp. 361–365.

Briot D. (2013a), The creator of astrobotany, Gavriil Adrianovich Tikhov, in *Astrobiology, History, and Society* (Vakoch D.A., Ed), Springer-Verlag, pp. 175–185.

(2013b), Elements for the history of a long quest: search for life in the Universe, International *Journal of Astrobiology*, 12 (3):254–258.

Briottet X., Hosgood B., Meister G., Sandmeier S. and Serrot G. (2004), Laboratory measurement of bi-directional reflectance, in *Reflection Properties of Vegetation and Soil – With a BRDF Database* (von Schönermark M., Geiger B. and Röser H.P., Eds), Wissenschaft und technik Verlag, Berlin, pp. 173–194.

Brito F.A. and Freire M.L.F. (2006), A domain wall model for spectral reflectance of plant leaves, *Quantitative Biology*, 4 pp.

Broadhurst M.G., Chiang C.K., Wahlstrand K.J., Hill R.M. and Dissado L.A. (1987), The dielectric properties of biological tissue (*Crassula portulacea*) from 102 to 109 Hz, *Journal of Molecular Liquids*, 36:65–73.

Brodersen C.R. and Vogelmann T.C. (2007), Do epidermal lens cells facilitate the absorptance of diffuse light? *American Journal of Botany*, 94(7):1061–1066.

Brodersen C.R., Vogelmann T.C., Williams W.E. and Gorton H.L. (2008), A new paradigm in leaf-level photosynthesis: direct and diffuse lights are not equal, *Plant, Cell & Environment*, 31(1):159–164.

Brodersen C.R. and Vogelmann T.C. (2010), Do changes in light direction affect absorption profiles in leaves? *Functional Plant Biology*, 37(5):403–412.

Brodersen C.R. and Roddy A.B. (2016), New frontiers in the three-dimensional visualization of plant structure and function, *American Journal of Botany*, 103(2):184–188.

Brodribb T.J. and Feild T.S. (2010), Leaf hydraulic evolution led a surge in leaf photosynthetic capacity during early angiosperm diversification, *Ecology Letters*, 13(2):175–183.

Broglia M. (1993), Blue-green laser-induced fluorescence from intact leaves: actinic light sensitivity and subcellular origins, *Applied Optics*, 32(3):334–338.

Brown F.B.H. (1920), The refraction of light in plant tissues, *Bulletin of the Torrey Botanic Club*, 47(6):243–260.

Brown H.T. (1905), The reception and utilization of energy by a green leaf, *Scientific American*, 60:24678–24679.

Brown H.T. and Escombe F. (1905), Researches on some of the physiological processes of green leaves, with special reference to the interchange of energy between the leaf and its surroundings, *Proceedings of the Royal Society of London. Series B*, 76(507):29–111.

Brown J.M., Thomas J.F., Cofer G.P. and Johnson G.A. (1988), Magnetic resonance microscopy of stem tissue of *Pelargonium hortorum*, *Botanical Gazette*, 149(3):253–259.

Brümmer F., Pfannkuchen M., Baltz A., Hauser T. and Thiel V. (2008), Light inside sponges, *Journal of Experimental Marine Biology and Ecology*, 367(2):61–64.

Brunner U. and Eller M. (1977), Spectral properties of juvenile and adult leaves of Piper beetle and their ecological significance, *Physiologia Plantarum*, 41(2):22–24.

Buddenbaum H., Pueschel P., Stellmes M., Werner W. and Hill J. (2011), Measuring water and chlorophyll content on the leaf and canopy scale, *EARSeL eProceedings*, 10(1):66–72.

Buddenbaum H. and Hill J. (2015), PROSPECT inversions of leaf laboratory imaging spectroscopy – A comparison of spectral range and inversion technique influences PROSPECT, *Photogrammetrie – Fernerkundung – Geoinformation*, 2015 (3):231–240.

Buriol G.A., Menoux Y. and de Parcevaux S. (1984a), Détermination de la masse d'eau et des propriétés optiques d'une feuille à partir de modifications de son bilan énergétique. II: Applications en conditions artificielles et naturelles, *Agronomie*, 4 (6):501–506.

Buriol G.A., Santibanez F., Menoux Y., de Parcevaux S. and Bertolini J.M. (1984b), Détermination de la masse d'eau et des propriétés optiques d'une feuille à partir de modifications de son bilan énergétique. I: Bases théoriques de la méthode et technique de mesure, *Agronomie*, 4(6):493–500.

Burkholder A., Warner T.A., Culp M. and Landenberger R. (2011), Seasonal trends in separability of leaf reflectance spectra for *Ailanthus altissima* and four other tree species, *Photogrammetric Engineering & Remote Sensing*, 77(8):793–804.

Burkinshaw S.M., Hallas G. and Towns A.D. (1996), Infrared camouflage, *Coloration Technology*, 26(1):47–53.

Burns D.A. and Ciurczak E.W. (2007), *Handbook of Near-infrared Analysis*, 3rd Edition, CRC Press, 834 pages.

Burns K.C., Cazetta E., Galetti M., Valido A. and Schaefer H.M. (2009), Geographic patterns in fruit colour diversity: do leaves constrain the colour of fleshy fruits? *Oecologia*, 159(2):337–343.

Burns K.C. (2010), Is crypsis a common defensive strategy in plants? Speculation on signal deception in the New Zealand flora, *Plant Signaling & Behavior*, 5(1):9–13.

Buscaglia H.J. and Varco J.J. (2002), Early detection of cotton leaf nitrogen status using leaf reflectance, *Journal of Plant Nutrition*, 25(9):2067–2080.

Buschmann C. and Lichtenthaler H.K. (1988), Reflectance and chlorophyll fluorescence signatures of leaves, in *Applications of Chlorophyll Fluorescence in Photosynthesis Research, Stress Physiology, Hydrobiology and Remote Sensing* (Lichtenthaler H.K., Ed), Kluwer Academic Publishers, Dordrecht, pp. 325–332.

Buschmann C., Nagel E., Rang S. and Stober F. (1990), Interpretation of reflectance spectra of terrestrial vegetation based on specifical plant test systems, in *Proc. 10th International Geoscience and Remote Sensing Symposium*, College Park, MD, 20–24 May 1990, IEEE, Vol. 3, pp. 1927–1930.

Buschmann C. and Nagel E. (1993a), *In vivo* spectroscopy and internal optics of leaves as basis for remote sensing of vegetation, *International Journal of Remote Sensing*, 14(4):711–722.

(1993b), Variation of reflectance signatures of a leaf as indication of physiological activity, in *Proc. 13th International Geoscience and Remote Sensing Symposium (IGARSS'93)*, Tokyo, Japan, 18–21 August 1993, IEEE, Vol. 2, pp. 522–524.

Buschmann C., Nagel E., Szabo K. and Kocsanyi L. (1994), Spectrometer for fast measurements of *in vivo* reflectance, absorptance, and fluorescence in the visible and near-infrared, *Remote Sensing of Environment*, 48(1):18–24.

Buschmann C., Lenk S. and Lichtenthaler H.K. (2012), Reflectance spectra and images of green leaves with different tissue structure and chlorophyll content, *Israel Journal of Plant Sciences*, 60(1–2):49–64.

Butler W.L. (1964), Absorption spectroscopy *in vivo*: theory and application, *Annual Review of Plant Physiology*, 15 (1):451–470.

Calcante A., Mena A. and Mazzetto F. (2012), Evaluation of "ground sensing" optical sensors for diagnosis of *Plasmopara viticola* on vines, *Spanish Journal of Agricultural Research*, 10(3):619–630.

Caldwell M.M. (1968), Solar ultraviolet radiation as an ecological factor for alpine plants, *Ecological Monographs*, 38 (3):243–268.

Caldwell M.M., Robberecht R. and Flint S.D. (1983), Internal filters: prospects for UV-acclimation in higher plants, *Physiologia Plantarum*, 58(3):445–450.

Calvo-Alvarado J.C., Kalacska M., Sánchez-Azofeifa G.A. and Bell L.S. (2008), Effect of soil type on plant growth leaf nutrient/chlorophyll concentration, and leaf reflectance of tropical tree and grass species, in *Hyperspectral Remote Sensing of Tropical and Sub-Tropical Forests* (Kalacska M. and Sanchez-Azofeifa G.A., Eds), CRC Press, pp. 87–123.

Cameron R.J. (1970), Light intensity and the growth of *Eucalyptus* seedlings. II: The effect of cuticular waxes on light absorption in leaves of *Eucalyptus* species, *Australian Journal of Botany*, 18(3):275–284.

Campbell G.S. (2000), Editorial. Monteith symposium: topics in environmental physics, *Agricultural and Forest Meteorology*, 104(1):1–4.

Campbell P.K.E., Middleton E.M., McMurtrey J.E., Corp L.A. and Chappelle E.W. (2007), Assessment of vegetation stress using reflectance or fluorescence measurements, *Journal of Environmental Quality*, 36(3):832–845.

Cannon H.L. (1971), The use of plant indicators in ground water surveys, geologic mapping, and mineral prospecting, *Taxon*, 20(2–3):227–256.

Cao K.F. (2000), Leaf anatomy and chlorophyll content of 12 woody species in contrasting light conditions in a Bornean heath forest, *Canadian Journal of Botany*, 78(10):1245–1253.

Card D.H., Peterson D.L., Matson P.A. and Aber J.D. (1988), Prediction of leaf chemistry by the use of visible and near infrared reflectance spectroscopy, *Remote Sensing of Environment*, 26(2):123–147.

Cardenas R., Gausman H.W., Allen W.A. and Schupp M. (1970), The influence of ammonia-induced cellular discoloration within cotton leaves (*Gossypium hirsutum* L.) on light reflectance, transmittance and absorptance, *Remote Sensing of Environment*, 1(3):199–202.

Cardenas R., Gausman H.W. and Thomas C.T. (1972), Photographic previsual detection of watermelon mosaic virus in cucumber, *Journal of the Rio Grande Valley Horticultural Society*, 26:73–75.

Carlson R.E. and Yarger D.N. (1971a), An evaluation of two methods for obtaining leaf transmissivity from leaf reflectivity measurements, *Agronomy Journal*, 63(1):78–81.

Carlson R.E., Yarger D.N. and Shaw R.H. (1971), Factors affecting the spectral properties of leaves with special emphasis on leaf water status, *Agronomy Journal*, 63(3):486–489.

Carpita N.C. and McCann M. (2000), The cell wall, in *Biochemistry and Molecular Biology of Plants* (Buchanan B., Gruissem W. and Jones R.L., Eds), American Society of Plant Physiologists, pp. 52–108.

Carter D.L. and Myers V.I. (1963), Light reflectance and chlorophyll and carotene contents of grapefruit leaves as affected by Na_2SO_4, NaCl and $CaCl_2$, *Proceeding of the American Society for Horticultural Science*, 82:217–221.

Carter G.A., Paliwal K., Pathre U., Green T.H., Mitchell R.J. and Gjerstad D.H. (1989), Effect of competition and leaf age on visible and infrared reflectance in pine foliage, *Plant, Cell and Environment*, 12(3):309–315.

Carter G.A. (1991), Primary and secondary effects of water-content on the spectral reflectance of leaves, *American Journal of Botany*, 78(7):916–924.

(1993), Response of leaf spectral reflectance to plant stress, *American Journal of Botany*, 80(3):239–243.

(1994), Ratios of leaf reflectances in narrow wavebands as indicators of plant stress, *International Journal of Remote Sensing*, 15(3):697–703.

Carter G.A., Mitchell R.J., Chappelka A.H. and Brewer C.H. (1992), Response of leaf spectral reflectance in loblolly pine to increased atmospheric ozone and precipitation acidity, *Journal of Experimental Botany*, 43(249):577–584.

Carter G.A. and McCain D.C. (1993), Relationship of leaf spectral reflectance to chloroplast water content determined using NMR microscopy, *Remote Sensing of Environment*, 46(3):305–310.

Carter G.A., Rebbeck J. and Percy K.E. (1995), Leaf optical properties in *Liriodendron tulipifera* and *Pinus strobus* as influenced by increased atmospheric ozone and carbon dioxide, *Canadian Journal of Forest Research*, 25(3):407–412.

Carter G.A., Bahadur R. and Norby R.J. (2000), Effects of elevated atmospheric CO_2 and temperature on leaf optical properties in *Acer saccharum*, *Environmental and Experimental Botany*, 43(3):267–273.

Carter G.A. and Spiering B.A. (2002), Optical properties of intact leaves for estimating chlorophyll concentration, *Journal of Environmental Quality*, 31(5):1424–1432.

Casa R., Castaldi F., Pascucci S. and Pignatti S. (2015), Chlorophyll estimation in field crops: an assessment of handheld leaf meters and spectral reflectance measurements, *Journal of Agricultural Science*, 153(5):876–890.

Castillo R., Contreras D., Freer J., Ruiz F. and Valenzuela S. (2008), Supervised pattern recognition techniques for classification of *Eucalyptus* species from leaves NIR spectra, *Journal of the Chilean Chemical Society*, 53(4):1709–1713.

Castle E.S. (1933), The refractive indices of whole cells, *Journal of General Physiology*, 17(1):41–47.

Castro-Díez P., Puyravaud J.P., and Cornelissen J.H.C. (2000), Leaf structure and anatomy as related to leaf mass per area variation in seedlings of a wide range of woody plant species and types, *Oecologia*, 124(4):476–486.

Castro-Esau K.L., Sánchez-Azofeifa G.A. and Caelli T. (2004), Discrimination of lianas and trees with leaf-level hyperspectral data, *Remote Sensing of Environment*, 90(3):353–372.

Castro-Esau K.L., Sánchez-Azofeifa G.A. and Rivard B. (2006a), Comparison of spectral indices obtained using multiple spectroradiometers, *Remote Sensing of Environment*, 103(3):276–288.

Castro-Esau K.L., Sánchez-Azofeifa G.A., Rivard B., Wright S.J. and Quesada M. (2006b), Variability in leaf optical properties of Mesoamerican trees and the potential for species classification, *American Journal of Botany*, 93 (4):517–530.

Castro K.L. and Sánchez-Azofeifa G.A. (2008), Changes in spectral properties, chlorophyll content and internal mesophyll structure of senescing *Populus balsamifera* and *Populus tremuloides* leaves, *Sensors*, 8(1):51–69.

Cavender-Bares J., Meireles J.E., Couture J.J., Kaproth M.A., Kingdon C.C., Singh A., et al. (2016), Associations of leaf spectra with genetic and phylogenetic variation in oaks: prospects for remote detection of biodiversity, *Remote Sensing*, 8 (3):221.

Ceccato P., Flasse S., Tarantola S., Jacquemoud S. and Grégoire J.M. (2001), Detecting vegetation water content using reflectance in the optical domain, *Remote Sensing of Environment*, 77(1):22–33.

Ceccato P., Gobron N., Flasse S., Pinty B. and Tarantola S. (2002), Designing a spectral index to estimate vegetation water content from remote sensing data. Part 1: theoretical approach, *Remote Sensing of Environment*, 82(2–3):188–197.

Cen Y.P. and Bornman J.F. (1993), The effect of exposure to enhanced UV-B radiation on the penetration of monochromatic and polychromatic UV-B radiation in leaves of *Brassica napus*, *Physiologia Plantarum*, 87(3):249–255.

Cerovic Z.G., Samson G., Morales F., Tremblay N. and Moya I. (1999), Ultraviolet-induced fluorescence for plant monitoring: present state and prospects, *Agronomie*, 19(7):543–578.

Cerovic Z.G., Ounis A., Cartelat A., et al (2002), The use of chlorophyll fluorescence excitation spectra for the non-destructive *in situ* assessment of UV-absorbing compounds in leaves, *Plant, Cell & Environment*, 25(12):1663–1676.

Cerovic Z.G., Masdoumier G., Ben Ghozlen N. and Latouche G. (2012), A new optical leaf-clip meter for simultaneous non-destructive assessment of leaf chlorophyll and epidermal flavonoids, *Physiologia Plantarum*, 146(3):251–260.

Chaerle L., van Caeneghem W., Messens E., Lambers H., van Montagu M. and van der Stracten D. (1999), Presymptomatic visualization of plant-virus interactions by thermography, *Nature Biotechnology*, 17(8):813–816.

Chaerle L. and van der Straeten D. (2000), Imaging techniques and the early detection of plant stress, *Trends in Plant Science*, 5 (11):495–501.

Chaerle L., De Boever F., Van Montagu M. and Van Der Straeten D. (2001), Thermographic visualization of cell death in tobacco and *Arabidopsis*, *Plant, Cell & Environment*, 24(1):15–25.

Chaerle L., vande Ven M., Valcke R. and Van Der Straeten D. (2002), Visualization of early stress responses in plant leaves, in *Proc. Thermosense XXIV* (Maldague X.P. and Rozlosnik A.E., Eds), Orlando, FL, 1–5 April 2002, SPIE, Vol. 4710, pp. 417–423.

Chaerle L., Hagenbeek D., De Bruyne E., Valcke R. and Van Der Straeten D. (2004), Thermal and chlorophyll-fluorescence imaging distinguish plant-pathogen interactions at an early stage, *Plant & Cell Physiology*, 45(7):887–896.

Chan S.S. and Middleton E.M. (2000), *BOREAS TE-10 Leaf Optical Properties*, NASA Goddard Space Flight Center, Greenbelt, MD, October 2000, NASA/TM-2000–209891/vol 162, 28 pages.

Chandrasekhar S. (1960), *Radiative Transfer*, Dover, New York, 416 pages.

Chang S.H. and Collins W. (1983), Confirmation of the airborne biogeophysical mineral exploration technique using laboratory methods, *Economic Geology*, 78(4):723–736.

Chappelle E.W., Kim M.S. and McMurtrey J.E. (1992), Ratio analysis of reflectance spectra (RARS): an algorithm for the remote estimation of the concentrations of chlorophyll A, chlorophyll B, and carotenoids in soybean leaves, *Remote Sensing of Environment*, 39(3):239–247.

Charney E. and Brackett F.S. (1961), The spectral dependence of scattering from a spherical alga and its implication for the state of organization of the light-accepting pigments, *Archives of Biochemistry and Biophysics*, 92(1):1–12.

Chautard J. (1872), Recherches sur les raies de la chlorophylle, *Comptes Rendus Hebdomadaires des Séances de l'Académie des Sciences*, 75:1836–1839.

Chavana-Bryant C., Malhi Y., Wu J., Asner G.P., Anastasiou A., Enquist B.J., et al. (2017), Leaf aging of Amazonian canopy trees as revealed by spectral and physiochemical measurements, *New Phytologist*, 214(3):1049–1063.

Chelle M. (2006), Could plant leaves be treated as Lambertian surfaces in dense crop canopies to estimate light absorption? *Ecological Modelling*, 198(1–2):219–228.

Chen B., Li S., Wang K., Zhou G. and Bai J. (2012a), Evaluating the severity level of cotton Verticillium using spectral signature analysis, *International Journal of Remote Sensing*, 33(9):2706–2724.

Chen C.T., Chen S., Hsieh K.W., Yang H.C., Hsiao S. and Yang I.C. (2007a), Estimation of leaf nitrogen content using artificial neural network with cross-learning scheme and significant wavelengths, *Transactions of the ASABE*, 50(1):295–301.

Chen L., Huang J.F., Wang F.M. and Tang Y.L. (2007b), Comparison between back propagation neural network and regression models for the estimation of pigment content in rice leaves and panicles using hyperspectral data, *International Journal of Remote Sensing*, 28(16):3457–3478.

Chen M., Glaz B., Gilbert R.A., Daroub S.H., Barton F.E. and Wan Y. (2002), Near-infrared reflectance spectroscopy analysis of phosphorus in sugarcane leaves, *Agronomy Journal*, 94(6):1324–1331.

Chen M. and Weng F. (2012), Kramers-Kronig analysis of leaf refractive index with the PROSPECT leaf optical property model, *Journal of Geophysical Research – Atmospheres*, 117:D18106.

Chen Q., Zhao J., Huang X., Zhang H. and Liu M. (2006), Simultaneous determination of total polyphenols and caffeine contents of green tea by near-infrared reflectance spectroscopy, *Microchemical Journal*, 83(1):42–47.

Chen Q., Zhao J., Liu M., Cai J. and Liu J. (2008), Determination of total polyphenols content in green tea using FT-NIR spectroscopy and different PLS algorithms, *Journal of Pharmaceutical and Biomedical Analysis*, 46(3):568–573.

Chen X., Han W. and Li M. (2012b), Spectroscopic determination of leaf water content using linear regression and an artificial neural network, *African Journal of Biotechnology*, 11(10):2518–2527.

Cheng Q., Tang D. and Wu X. (2008), Reflectance properties and physiological responses of Bulrush in Wetland Park to heavy metal contamination in *Proc. Remote Sensing for Environmental Monitoring, GIS Applications, and Geology VIII* (Michel U., Civco D.L., Ehlers M. and Kaufmann H.J., Eds), Cardiff, UK, 15 September 2008, SPIE, Vol. 71101 R, 9 pages.

Cheng T., Rivard B., Sánchez-Azofeifa G.A., Feng J. and Calvo-Polanco M. (2010), Continuous wavelet analysis for the detection of green attack due to mountain pine beetle infestation, *Remote Sensing of Environment*, 114(4):899–910.

Cheng T., Rivard B. and Sánchez-Azofeifa G.A. (2011), Spectroscopic determination of leaf water content using continuous wavelet analysis, *Remote Sensing of Environment*, 115(2):659–670.

Cheng T., Rivard B., Sánchez-Azofeifa G.A., Féret J.B., Jacquemoud S. and Ustin S.L. (2012), Predicting leaf gravimetric water content from foliar reflectance across a range of plant species using continuous wavelet analysis, *Journal of Plant Physiology*, 169(12):1134–1142.

(2014), Deriving leaf mass per area (LMA) from foliar reflectance across a variety of plant species using continuous wavelet analysis, ISPRS, *Journal of Photogrammetry and Remote Sensing*, 87:28–38.

Cheynier V. (2005), Polyphenols in foods are more complex than often thought, *American Journal of Clinical Nutrition*, 81 (1):223S–229S.
Chi G.Y., Chen X., Shi Y. and Liu X.H. (2009a), Spectral response of rice (*Oryza sativa* L.) leaves to Fe^{2+} stress, *Science in China Series C: Life Sciences*, 52(8):747–753.
Chi X.Y., Sheng B., Yang M., Chen Y.Y. and Wu E.H. (2009b), Simulation of autumn leaves, *Journal of Software*, 20 (3):702–712.
Chiba N., Ohshida K., Muraoka K. and Saito N. (1996), Visual simulation of leaf arrangement and autumn colours, *Journal of Visualization and Computer Animation*, 7(2):79–93.
Cho M.A. and Skidmore A.K. (2006), A new technique for extracting the red edge position from hyperspectral data: the linear extrapolation method, *Remote Sensing of Environment*, 101(2):181–193.
Christie C., Mann P. and Cloutis E.A. (2013), Spectral characteristics of natural and laboratory-induced leaf senescence in four common North American tree species, *International Journal of Remote Sensing Applications*, 3(2):75–85.
Christie J.M. (2007), Phototropin blue-light receptors, *Annual Review of Plant Biology*, 58:21–45.
Chuah H.T., Lee K.Y. and Lau T.W. (1995), Dielectric constants of rubber and oil palm leaf samples at X-band, *IEEE Transactions on Geoscience and Remote Sensing*, 33(1):221–223.
Chuah H.T., Kam S.W. and Chye Y.H. (1997), Microwave dielectric properties of rubber and oil palm leaf samples: measurement and modelling, *International Journal of Remote Sensing*, 18(12):2623–2639.
Chudek J.A. and Hunter G. (1997), Magnetic resonance imaging of plants, *Progress in Nuclear Magnetic Resonance Spectroscopy*, 31(1):43–62.
Chukhlantsev A.A. (2006), *Microwave Radiometry of Vegetation Canopies*, Springer, 287 pages.
Chung B.K. (2006), A convenient method for complex permittivity measurement of thin materials at microwave frequencies, *Journal of Physics D: Applied Physics*, 39(9):1926–1931.
(2007), Dielectric constant measurement for thin material at microwave frequencies, *Progress in Electromagnetics Research*, 75:239–252.
Chwirot S. and Slevin J. (1986), Comment on "Measurement of light gradients and spectral regime in plant tissue with a fibre optic probe", *Physiologia Plantarum*, 67(3):493–494.
Cibula W.G. and Carter G.A. (1992), Identification of a far-red reflectance response to ectomycorrhizae in slash pine, *International Journal of Remote Sensing*, 13(5):925–932.
Cipar J., Cooley T. and Lockwood R. (2008), Summer to autumn changes in vegetation spectral indices of deciduous trees, in *Proc. Remote Sensing and Modeling of Ecosystems for Sustainability V* (Gao W. and Wang H., Eds), San Diego, CA, 13 August 2008, SPIE, Vol. 7083, 708306.
Clark D.H., Mayland H.F. and Lamb R.C. (1987), Mineral analysis of forages with near infrared reflectance spectroscopy, *Agronomy Journal*, 79(3):485–490.
Clark J.B. and Lister G.R. (1975), Photosynthetic action spectra of trees. Part II: The relationship of cuticle structure to the visible and ultraviolet spectral properties of needles from four coniferous species, *Plant Physiology*, 55 (2):407–413.
Clark M.L., Roberts D.A. and Clark D.B. (2005), Hyperspectral discrimination of tropical rain forest tree species at leaf to crown scales, *Remote Sensing of Environment*, 96(3–4):375–398.
Clark R.N. and Roush T.L. (1984), Reflectance spectroscopy – Quantitative-analysis techniques for remote sensing applications, *Journal of Geophysical Research*, 89(B7):6329–6340.
Clarke F.J.J. and Larkin J.A. (1985), Measurement of total reflectance, transmittance and emissivity over the thermal IR spectrum, *Infrared Physics*, 25(1–2):359–367.
Clijsters H. and Van Assche F. (1985), Inhibition of photosynthesis by heavy metals, *Photosynthesis Research*, 7(1):31–40.
Close D.C. and Beadle C.L. (2003), The ecophysiology of foliar anthocyanin, *The Botanical Review*, 69(2):149–161.
Close D.C., Davidson N.J., Shields C.B. and Wiltshire R. (2007), Reflectance and phenolics of green and glaucous leaves of *Eucalyptus urnigera*, *Australian Journal of Botany*, 55(5):561–567.
Clum H.H. (1926), The effect of transpiration and environmental factors on leaf temperatures. I: Transpiration. *American Journal of Botany*, 13(3):194–216.
Coblentz W.W. (1913), The diffuse reflecting power of various substances, *Bulletin of the Bureau of Standards*, 9(2):283–325.
Cochran G.W., Welkie G.W. and Chidester J.L. (1960), Direct infra-red spectral analysis of the cucumber mosaic virus Infection process, *Nature*, 187:1049–1050.
Cochrane M.A. (2000), Using vegetation reflectance variability for species level classification of hyperspectral data, *International Journal of Remote Sensing*, 21(10):2075–2087.
Cockell S.S. and Knowland J. (1999), Ultraviolet radiation screening compounds, *Biological Reviews of the Cambridge Philosophical Society*, 74(3):311–345.
Cohen W.B. (1991), Temporal versus spatial variation in leaf reflectance under changing water stress conditions, *International Journal of Remote Sensing*, 12(9):1865–1876.
Cohen Y., Alchanatis V., Meron M., Saranga Y. and Tsipris J. (2005), Estimation of leaf water potential by thermal imagery and spatial analysis, *Journal of Experimental Botany*, 56(417):1843–1852.
Cohen Y., Alchanatis V., Zusman Y., et al. (2010), Leaf nitrogen estimation in potato based on spectral data and on simulated bands of the VENµS satellite, *Precision Agriculture*, 11(5):520–537.
Colombo R., Meroni M., Marchesi A., et al. (2008), Estimation of leaf and canopy water content in poplar plantations by means of hyperspectral indices and inverse modeling, *Remote Sensing of Environment*, 112(4):1820–1834.

Colwell R.N. (1956), Determining the prevalence of certain cereal crop diseases by means of aerial photography, *Hilgardia*, 26 (5):223–286.

Comar A., Baret F., Viénot F., Yan L. and de Solan B. (2012), Wheat leaf bidirectional reflectance measurements: description and quantification of the volume, specular and hot-spot scattering features, *Remote Sensing of Environment*, 121:26–35.

Comar A., Baret F., Obein G., et al. (2014), ACT: a leaf BRDF model taking into account the azimuthal anisotropy of monocotyledonous leaf surface, *Remote Sensing of Environment*, 143:112–121.

Combal B., Baret F., Weiss M., et al. (2002), Retrieval of canopy biophysical variables from bidirectional reflectance using prior information to solve the ill-posed inverse problem, *Remote Sensing of Environment*, 84(1):1–15.

Combes D., Bousquet L., Jacquemoud S., Sinoquet H., Varlet-Grancher C. and Moya I. (2007), A new spectrogoniophotometer to measure leaf spectral and directional optical properties, *Remote Sensing of Environment*, 109(1):107–117.

Conejo E., Frangi J.P. and de Rosny G. (2010), Biophotonic *in situ* sensor for plant leaves, *Applied Optics*, 49(10):1687–1697.

(2015), Neural network implementation for a reversal procedure for water and dry matter estimation on plant leaves using selected LED wavelengths, *Applied Optics*, 54(17):5453–5460.

Conel J.E., Bosch J.V.D. and Grove C.I. (1993a), Application of a two-stream radiative transfer model for leaf lignin and cellulose concentrations from spectral reflectance measurements. Part 1, in *Proc. 4th Annual JPL Geoscience Workshop: AVIRIS* (Green R.O., Ed), Washington, DC, 25–29 October 1993, JPL Publication, Vol. 93–26, pp. 39–43.

Conel J.E., Bosch J.V.D. and Grove C.I. (1993b), Application of a two-stream radiative transfer model for leaf lignin and cellulose concentrations from spectral reflectance measurements. Part 2, in *Proc. 4th Annual JPL Geoscience Workshop: AVIRIS* (Green R.O., Ed), Washington, DC, 25–29 October 1993, JPL Publication, Vol. 93–26, pp. 45–51.

Cook R.L. and Torrance K.E. (1981), A reflectance model for computer graphics, *Computer Graphics*, 15(3):307–316.

Coops N.C., Dury S., Smith M.L., Martin M. and Ollinger S. (2002), Comparison of green leaf eucalypt spectra using spectral decomposition, *Australian Journal of Botany*, 50(5):567–576.

Coops N.C. and Stone C. (2005), A comparison of field-based and modelled reflectance spectra from damaged *Pinus radiata* foliage, *Australian Journal of Botany*, 53(5):417–429.

Cordon G.B. and Lagorio M.G. (2007a), Absorption and scattering coefficients: a biophysical-chemistry experiment using reflectance spectroscopy, *Journal of Chemical Education*, 84(7):1167–1170.

(2007b), Optical properties of the adaxial and abaxial faces of leaves. Chlorophyll fluorescence, absorption and scattering coefficients, *Photochemical & Photobiological Sciences*, 6(8):873–882.

Cornard J.P. and Merlin J.C. (2002), Spectroscopic and structural study of complexes of quercetin with Al(III), *Journal of Inorganic Biochemistry*, 92(1):19–27.

Cortazar B., Koydemir H.C., Tseng D., Fenga S. and Ozcan A. (2015), Quantification of plant chlorophyll content using Google Glass, *Lab on a Chip*, 15:1708–1716.

Coward J.L. (2010), FTIR spectroscopy of synthesized racemic nonacosan-10-ol: a model compound for plant epicuticular waxes, *Journal of Biological Physics*, 36(4):405–425.

Croft H., Chen J.M., Zhang Y., et al. (2015), Evaluating leaf chlorophyll content prediction from multispectral remote sensing data within a physically-based modelling framework, *ISPRS Journal of Photogrammetry and Remote Sensing*, 102:85–95.

Cui M., Vogelmann T.C. and Smith W.K. (1991), Chlorophyll and light gradients in sun and shade leaves of *Spinacia oleracea, Plant, Cell & Environment*, 14(5):493–500.

Cunningham S.A. and Floyd R.B. (2004), Leaf compositional differences predict variation in *Hypsipyla robusta* damage to *Toona ciliata* in field trials, *Canadian Journal of Forest Research*, 34(3):642–648.

Curran P.J. (1989), Remote sensing of foliar chemistry, *Remote Sensing of Environment*, 30(3):271–278.

Curran P.J., Dungan J.L. and Gholz H.L. (1990), Exploring the relationship between reflectance red-edge and chlorophyll content in slash pine, *Tree Physiology*, 7(1–4):33–48.

Curran P.J., Dungan J.L., Macler B.A. and Plummer S.E. (1991), The effect of a red leaf pigment on the relationship between red edge and chlorophyll concentration, *Remote Sensing of Environment*, 35(1):69–76.

Curran P.J., Dungan J.L., Macler B.A., Plummer S.E. and Peterson D.L. (1992), Reflectance spectroscopy of fresh whole leaves for the estimation of chemical concentration, *Remote Sensing of Environment*, 39(2):153–166.

Curran P.J., Windham W.R. and Gholz H.L. (1995), Exploring the relationship between reflectance red edge and chlorophyll concentration in slash pine leaves, *Tree Physiology*, 15(3):203–206.

Curran, P.J., Dungan, J.L. and Peterson, D.L. (2001), Estimating the biochemical concentration of leaves with reflectance spectrometry. Testing the Kokaly and Clark methodologies, *Remote Sensing of Environment*, 76(3):349–359.

Cutler D.F., Alvin K.L. and Price C.E., Eds (1982), *The Plant Cuticle*, Academic Press, New York, 461 pages.

Dadykin V.P. and Bedenko V.P. (1960), Concerning the geographic variability of optical properties in plant leaves, *Doklady Botanical Sciences Sections*, 130(3):6–8 (cover-to-cover translation from *Doklady Akademii Nauk SSSR*, 130 (3):674–677).

Dahm D.J. and Dahm K.D. (1999), Representative layer theory for diffuse reflectance, *Applied Spectroscopy*, 53(6):647–654.

Danson F.M. and Bowyer P. (2004), Estimating live fuel moisture content from remotely sensed reflectance, *Remote Sensing of Environment*, 92(3):309–321.

Danson F.M., Steven M.D., Malthus T.J. and Clark J.A. (1992), High-spectral resolution data for determining leaf water content, *International Journal of Remote Sensing*, 13(3):461–470.

Darwin C.R. (1862), *On the Various Contrivances by which British and Foreign Orchids are Fertilised by Insects, and on the Good Effects of Intercrossing*, John Murray, London, 360 pages.

Dash J. and Curran P.J. (2004), The MERIS terrestrial chlorophyll index, *International Journal of Remote Sensing*, 25 (33):5403–5413.

Datt B. (1998), Remote sensing of chlorophyll a, chlorophyll b, chlorophyll a+b, and total carotenoid content in *Eucalyptus* leaves, *Remote Sensing of Environment*, 66(2):111–121.

(1999a), A new reflectance index for remote sensing of chlorophyll content in higher plants: tests using *Eucalyptus* leaves, *Journal of Plant Physiology*, 154(1):30–36.

(1999b), Remote sensing of water content in *Eucalyptus* leaves, *Australian Journal of Botany*, 47(6):909–923.

(1999c), Visible/near infrared reflectance and chlorophyll content in *Eucalyptus* leaves, *International Journal of Remote Sensing*, 20(14):2741–2759.

Daughtry C.S.T., Biehl L.L. and Ranson K.J. (1989), A new technique to measure the spectral properties of conifer needles, *Remote Sensing of Environment*, 27(1):81–91.

Daughtry C.S.T. and Walthall C.L. (1998), Spectral discrimination of *Cannabis sativa* L. leaves and canopies, *Remote Sensing of Environment*, 64(2):192–201.

Daughtry C.S.T., Walthall C.L., Kim M.S., Brown de Colstoun E. and McMurtrey J.E. (2000), Estimating corn leaf chlorophyll concentration from leaf and canopy reflectance, *Remote Sensing of Environment*, 74(2):229–239.

Dauriac F. (2004), *Suivi multi-échelle par télédétection et spectroscopie de l'état hydrique de la végétation méditerranéenne pour la prévention du risque de feu de forêt*, Thèse de doctorat Sciences de l'eau, École Nationale du Génie Rural, des Eaux et des Forêts, Montpellier, France, 288 pages.

Davids C. and Tyler A.N. (2003), Detecting contamination-induced tree stress within the Chernobyl exclusion zone, *Remote Sensing of Environment*, 85(1):30–38.

Davis P.A., Caylor S., Whippo C.W. and Hangarter R.P. (2011), Changes in leaf optical properties associated with light-dependent chloroplast movements, *Plant, Cell & Environment*, 34(12):2047–2059.

Dawson T.P., Curran P.J. and Plummer S.E. (1995), Modelling the spectral response of coniferous leaf structures for the estimation of biochemical concentrations, in *Proc. 21st Annual Conference of the Remote Sensing Society: Remote Sensing in Action*, 11–14 September 1995, University of Southampton, UK, Remote Sensing Society pp. 587–594.

(1998a), LIBERTY – Modeling the effects of leaf biochemical concentration on reflectance spectra, *Remote Sensing of Environment*, 65(1):50–60.

(1998b), The biochemical decomposition of slash pine needles from reflectance spectra using neural networks, *International Journal of Remote Sensing*, 19(7):1433–1438.

Dawson T.P. and Curran P.J. (1998), A new technique for interpolating the reflectance red edge position, *International Journal of Remote Sensing*, 19(11):2133–2139.

Dawson T.P., Curran P.J., North P.R.J. and Plummer S.E. (1999), The propagation of foliar biochemical absorption features in forest canopy reflectance: a theoretical analysis, *Remote Sensing of Environment*, 67(2):147–159.

Day T.A., Vogelmann T.C. and De Lucia E.H. (1992), Are some plant life forms more effective than others in screening out ultraviolet-B radiation? *Oecologia*, 92(4):513–519.

Day T.A., Howells B.W. and Rice W.J. (1994), Ultraviolet absorption and epidermal-transmittance spectra in foliage, *Physiologia Plantarum*, 92(2):207–218.

Deák S., Márk G.I., Szabó T. and Füleky G. (2007), Spectral properties of strawberry plants, *International Journal of Horticultural Science*, 13(2):17–22.

Dean K.G., Kodama Y. and Wendler G. (1986), Comparison of leaf and canopy reflectance of subarctic forests, *Photogrammetric Engineering & Remote Sensing*, 52(6):809–811.

De Boer T.S., Baker A.C., Erdmann M.V., Ambariyanto, Jones P.R. and Barber P.H. (2012), Patterns of *Symbiodinium* distribution in three giant clam species across the biodiverse Bird's Head region of Indonesia, *Marine Ecology Progress Series*, 444:117–132.

Degl'innocenti E., Guidi L., Pardossi A. and Tognoni F. (2005), Biochemical study of leaf browning in minimally processed leaves of lettuce (*Lactuca sativa* L. Var. Acephala), *Journal of Agricultural and Food Chemistry*, 53 (26):9980–9984.

Delafolie (1774), Examen d'une Terre verte que l'on trouve abondamment aux environs du Pont-Audemer en Normandie, & probablement en d'autres endroits : avec diverses expériences qui paraissent démontrer que les couleurs variées de toutes les plantes ne font que le résultat des précipités ferrugineux, Observations sur la physique, sur l'histoire naturelle et sur les arts, Tome IV, pp. 349–359.

Delalieux S., van Aardt J., Keulemans W., Schrevens E. and Coppin P. (2007), Detection of biotic stress (*Venturia inaequalis*) in apple trees using hyperspectral data: non-parametric statistical approaches and physiological implications, *European Journal of Agronomy*, 27(1):130–143.

Delalieux S., Somers B., Verstraeten W.W., van Aardt J.A.N., Keulemans W. and Coppin P. (2009), Hyperspectral indices to diagnose leaf biotic stress of apple plants, considering leaf phenology, *International Journal of Remote Sensing*, 30 (8):1887–1912.

Del'Arco Sanches I., Souza Filho C.R. and Kokaly R.F. (2014), Spectroscopic remote sensing of plant stress at leaf and canopy levels using the chlorophyll 680 nm absorption feature with continuum removal, *ISPRS Journal of Photogrammetry and Remote Sensing*, 97:111–122.

Delaval E.H. (1774), Recherches expérimentales sur la cause des changements de couleur dans les corps opaques naturellement colorés, *Nouveaux Mémoires de l'Académie Royale des Sciences et Belles-Lettres de Berlin*, 1:154–194 (Archiv der Berlin-Brandenburgischen Akademie der Wissenschaften).

Delozier G., Eckard K., Greene M. and Lord E.M. (1987), A computer graphics program for the three-dimensional reconstruction of plant organs from serial sections, *American Journal of Botany*, 74(1):136–140.

DeLucia E.H., Shenoi H.D., Naidu S.L. and Day T.A. (1991), Photosynthetic symmetry of sun and shade leaves of different orientations, *Oecologia*, 87(1):51–57.

DeLucia E.H., Day T.A. and Vogelman T.C. (1992), Ultraviolet-B and visible light penetration into needles of two species of subalpine conifers during foliar development, *Plant, Cell & Environment*, 15(8):921–929.

Demarez V. and Gastellu-Etchegorry J.-P. (2000), A modeling approach for studying forest chlorophyll content, *Remote Sensing of Environment*, 71(2):226–238.

Demetriades-Shah T.H., Steven M.D. and Clark J.A. (1990), High resolution derivative spectra in remote sensing, *Remote Sensing of Environment*, 33(1):55–64.

Demmig-Adams B., Winter K., Krüger A. and Czygan F.C. (1987), Photoinhibition and zeaxanthin formation in intact leaves: a possible role of the xanthophyll cycle in the dissipation of excess light energy, *Plant Physiology*, 84(2):118–224.

Demmig-Adams B. (1990), Carotenoids and photoprotection in plants: a role for the xanthophyll zeaxanthin, *Biochimica et Biophysica Acta (BBA)-Bioenergetics*, 1020(1):1–24.

Demmig-Adams B., Gilmore A.M. and Adams III W.W. (1996), Carotenoids 3: *in vivo* functions of carotenoids in higher plants, *The FASEB Journal*, 10(4):403–412.

Dengler N. and Kang J. (2001), Vascular patterning and leaf shape, *Current Opinion in Plant Biology*, 4(1):50–56.

Dennison P.E. and Roberts D.A. (2003a), Endmember selection for multiple endmember spectral mixture analysis using endmember average RMSE, *Remote Sensing of Environment*, 87(2–3):123–135.

(2003b), The effects of vegetation phenology on endmember selection and species mapping in southern California chaparral, *Remote Sensing of Environment*, 87(2–3):295–309.

Dennison P.E., Halligan K.Q. and Roberts D.A. (2004), A comparison of error metrics and constraints for multiple endmember spectral mixture analysis and spectral angle mapper, *Remote Sensing of Environment*, 93(3):359–367.

De Parcevaux S., Bertolini J.M. and Katerji N. (1995), Masse en eau des feuilles: détermination non destructive en mileu naturel, *Agronomie*, 15(9–10):547–556.

Derkacheva O. and Sukhov D. (2005), Investigation of lignins by FTIR spectroscopy, *Macromolecular Symposia*, 265 (1):61–68.

Deroin J.P. and Deroin T. (1996), The relationships among vegetation, geology and spectral reponse: a quantitative approach on the European beech (*Fagus sylvatica*, Fagaceae), *Fragmentu Floristica et Geobotanica*, 41(2):791–801.

(1990), Effet d'une déficience en manganèse des sols sur la réponse spectrale des feuilles de hêtre (*Fagus sylvatica* L.), Comptes Rendus de l'Académie des Sciences Paris, t. 311, Série II, pp. 605–611.

de Rosny G., Vanderhaghen R., Baret F., Equer B. and Frangi J.P. (1995), A device for in situ measurements of leaf chlorophyll and carotenoid concentrations, in *Proc. International Colloquium on Photosynthesis and Remote Sensing* (G. Guyot, Ed), Montpellier, France, 28–30 August 1995, pp. 135–141.

de Saussure N.T. (1804), *Recherches chimiques sur la végétation*, Nyon, Paris, 327 pages.

Desbenoit B., Galin E., Akkouche S. and Grosjean J. (2006), Modeling autumn sceneries, in *Proc. 26th International Conference on Eurographics* (Fellner D.W. and Hansen C., Eds), Vienna, Austria, 4–8 September 2006, The Eurographics Association, pp. 107–110.

Deussen O. and Lintermann B. (2005), *Digital Design of Nature*, Springer, 295 pages.

de Vries J., Rauch C., Christa G. and Gould S.B. (2014a), A sea slug's guide to plastid symbiosis, *Acta Societatis Botanicorum Poloniae*, 83(4):415–421.

de Vries J., Christa G. and Gould S.B. (2014b), Plastid survival in the cytosol of animal cells, *Trends in Plant Science*, 19 (6):347–350.

DeWitt D.P. and Nutter G.D. (1988), *Radiation Thermometry*, John Wiley & Sons, 1138 pages.

Dey P.M. and Harborne J.B. (1997), *Plant Biochemistry*, Academic Press, 554 pages.

Diago M.P., Fernandes A.M., Millan B., Tardaguila J. and Melo-Pinto P. (2013), Identification of grapevine varieties using leaf spectroscopy and partial least squares, *Computers and Electronics in Agriculture*, 99:7–13.

Diah S.Z.M., Karman S.B. and Gebeshuber I.C. (2014), Nanostructural colouration in Malaysian plants: lessons for biomimetics and biomaterials, *Journal of Nanomaterials*, 2014:878409.

Dijkstra P. (1989), Cause and effect of differences in specific leaf area, in *Causes and Consequences of Variation in Growth Rate and Productivity of Higher Plants* (Lambers H., Ed), Academic Publishing, The Hague, pp. 125–140.

Dillen S.Y., Op de Beeck M., Hufkens K., Buonanduci M. and Phillips N.G. (2012), Seasonal patterns of foliar reflectance in relation to photosynthetic capacity and color index in two co-occurring tree species, *Quercus rubra* and *Betula papyrifera*, *Agricultural and Forest Meteorology*, 160:60–68.

Di Vittorio A.V. (2009), Enhancing a leaf radiative transfer model to estimate concentrations and *in vivo* specific absorption coefficients of total carotenoids and chlorophylls a and b from single-needle reflectance and transmittance, *Remote Sensing of Environment*, 113(9):1948–1966.

Vittorio A.V. and Biging G.S. (2009), Spectral identification of ozone-damaged pine needles, *International Journal of Remote Sensing*, 30(12):3041–3073.

Djurišić A.B. and Stanić B.V. (1998), Modeling the wavelength dependence of the index of refraction of water in the range 200 nm to 200 μm, *Applied Optics*, 37(13):2696–2698.

Dlugunovich V.A., Zaitseva V.A. and Tsaryuk O.V. (2001a), Changes in the polarization characteristics of the He-Ne laser radiation reflected from plant leaves, *Journal of Applied Spectroscopy*, 68(1):94–100.

Dlugunovich V.A., Zaitseva V.A., Sergeichik S.A. and Tsaryuk O.V. (2001b), Polarization characteristics of the He–Ne laser radiation reflected from rhododendron leaves subjected to acid treatment, *Journal of Applied Spectroscopy*, 68 (6):1020–1025.

Dobek A., Paillotin G., Gapinski J., Breton J., Leibl W. and Trissl H.W. (1994), Amplitude and polarity of the light gradient photovoltage from chloroplasts, *Journal of Theoretical Biology*, 170(2):129–143.

Dobrowski S.Z., Pushnik J.C., Zarco-Tejeda P.J. and Ustin S.L. (2005), Simple reflectance indices track heat and water stressed induced changes in steady state chlorophyll fluorescence at the canopy scale, *Remote Sensing of Environment*, 97 (3):403–414.

Doi H. and Takahashi M. (2008), Latitudinal patterns in the phenological responses of leaf colouring and leaf fall to climate change in Japan, *Global Ecology and Biogeography*, 17(4):556–561.

Dominy N.J. and Lucas P.W. (2001), Ecological importance of trichromatic vision to primates, *Nature*, 410:363–366.

Donner C. and Jensen H.W. (2005), Light diffusion in multi-layered translucent materials, *ACM Transactions on Graphics*, 24 (3):1032–1039.

Döring T.F., Archetti M. and Hardie J. (2009), Autumn leaves seen through herbivore eyes, *Proceedings of the Royal Society of London. Series B, Biological Sciences*, 276(1654):121–127.

Doughty C.E., Asner G.P. and Martin R.E. (2011), Predicting tropical plant physiology from leaf and canopy spectroscopy, *Oecologia*, 165(2):289–299.

Du L.J. and Peake W.H. (1969), Rayleigh scattering from leaves, *Proceedings of the IEEE*, 57(6):1227–1229.

Dubis E.N., Dubis A.T., and Morzycki J.W. (1999), Comparative analysis of plant cuticular waxes using HATR FT-IR reflection technique, *Journal of Molecular Structure*, 511–512:173–179.

Dubis E.N., Dubis A.T, and Poplawski J. (2001), Determination of the aromatic compounds in plant cuticular waxes using FT-IR spectroscopy, *Journal of Molecular Structure*, 596(1–3):83–88.

Dunagan S.C., Gilmore M.S. and Varekamp J.C. (2007), Effects of mercury on visible/near-infrared reflectance spectra of mustard spinach plants (*Brassica rapa* P.), *Environmental Pollution*, 148(1):301–311.

Dunn A. and Richards-Kortum R. (1996), Three-dimensional computation of light scattering from cells, *IEEE Journal of Quantum Electronics*, 2(4):898–905.

Durgante F.M., Higuchi N., Almeida A. and Vicentini A. (2013), Species spectral signature: discriminating closely related plant species in the Amazon with near-infrared leaf-spectroscopy, *Forest Ecology and Management*, 291:240–248.

Dutton H.J., Manning W.M. and Duggar B.M. (1943), Chlorophyll fluorescence and energy transfer in the diatom, *Nitzschia closterium, Journal of Physical Chemistry*, 47(4):308–313.

Dwyer L.M., Tollenaar M. and Houwing L. (1991), A nondestructive method to monitor leaf greenness in corn, *Canadian Journal of Plant Science*, 71(2):505–509.

Earl H.J. and Tollenaar M. (1997), Maize leaf absorptance of photosynthetically active radiation and its estimation using a chlorophyll meter, *Crop Science*, 37(2):436–440.

Edwards G.J., Schehl T. and DuCharme E.P. (1975), Multispectral sensing of *Citrus* young tree decline, *Photogrammetric Engineering*, 41(5):653–657.

Egan W.G. (1970), Optical stokes parameters for farm crop identification, *Remote Sensing of Environment*, 1(3):165–180.

Egbert K.J. and Martin C.E. (2000), Light penetration via leaf windows does not increase photosynthesis in three species of desert succulents, *Journal of Plant Physiology*, 157(5):521–525.

Egbert K.J., Martin C.E. and Vogelmann T.C. (2008), The influence of epidermal windows on the light environment within the leaves of six succulents, *Journal of Experimental Botany*, 59(7):1863–1873.

Eggli U. and Newton L.E. (2004), *Etymological Dictionary of Succulent Plant Names*, Springer, 266 pages.

Egri A., Horváth A., Kriska G. and Horváth G. (2010), Optics of sunlit water drops on leaves: conditions under which sunburn is possible, *New Phytologist*, 185(4):979–987.

Ehleringer J.R. (1976), Leaf absorptance and photosynthesis as affected by pubescence in the genus *Encelia*, *Carnegie Institution of Washington Year Book*, 75:413–418.

Ehleringer J., Bjorkman O. and Mooney H.A. (1976a), Leaf pubescence: effects on absorptance and photosynthesis in a desert shrub, *Science*, 192(4237):376–377.

Ehleringer J.R. and Björkman O. (1978), Pubescence and leaf spectral characteristics in a desert shrub, *Encelia farinosa*, *Oecologia*, 36(2):151–162.

Ehleringer J.R. and Mooney H.A. (1978), Leaf hairs: effects on physiological activity and adaptive value to a desert shrub, *Oecologia*, 37(2):183–200.

Ehleringer J.R. (1981), Leaf absorptances of Mohave and Sonoran plants, *Oecologia*, 49(3):366–370.

(1984), Ecology and ecophysiology of leaf pubescence in North American desert plants, in *Biology and Chemistry of Plant Trichomes* (Rodriquez E., Healey P. and Mehta I., Eds), Plenum Press, New York, pp. 113–132.

(1986), Modifications of solar-radiation absorption patterns and implications for carbon gain at the leaf level, in *On the Economy of Plant Form and Function* (Givnish T.J., Ed), Cambridge University Press, London, pp. 57–81.

(1988), Changes in leaf characteristics of species along elevational gradients in the Wasatch Front, Utah, *American Journal of Botany*, 75(5):680–689.

Ehleringer J.R. and Cook C.S. (1990), Characteristics of *Encelia* species differing in leaf reflectance and transpiration rate under common garden conditions, *Oecologia*, 82(4):484–489.

Elias H. (1971), Three-dimensional structure identified from single sections, *Science*, 174(4013):993–1000.

Elias M., and Lafait J. (2006), *La couleur. Lumière, vision et matériaux*, Belin, 352 pages.

Eller B.M. and Brunner U. (1975), Der Einfluß von Straßenstaub auf die Strahlungsabsorption durch Blätter, *Archiv für Meteorologie, Geophysik und Bioklimatologie. Serie B: Klimatologie, Umweltmeteorologie, Strahlungsforschung*, 23 (1–2):137–146 (in German: Road-dust changed absorption of solar radiation by plant leaves).

Eller B.M. (1977a), Leaf pubescence: the significance of lower surface hairs for the spectral properties of the upper surface, *Journal of Experimental Botany*, 28(105):1054–1059.

(1977b), Road dust induced increase of leaf temperature, *Environmental Pollution*, 13(2):99–107.

(1977c), Beeinflussung der Energiebilanz von Blättern durch Straßenstaub, *Angewandte Botanik*, 51(1–2):9–15 (in German: Road dust changed energy balance of leaves).

Eller B.M. and Willi P. (1977a), Globalstrahlungsabsorption von *Hedera helix* L. unter Straßenstaubimmissionen, *Gartenbauwissenschaft*, 42(2):49–53 (in German: Road dust emissions on *Hedera helix* L. and changed absorption of global radiation).

(1977b), Die Bedeutung der Wachsausblühungen auf Blaättern von *Kalanchoe pumila* Baker für die Absorption der Globalstrahlung, *Flora*, 166(5):461–474 (in German: Significance of wax bloom on leaves of *Kalanchoe pumila* Baker for absorption of global radiation).

(1977c), The significance of leaf pubescence for the absorption of global radiation by *Tussilago farfara* L., *Oecologia*, 29 (2):179–187.

Eller B.M., Brinckmann E. and von Willert D.J. (1983), Optical properties and succulence of plants in the arid Richtersveld (Cp., Rep. South Africa), *Botanica Helvetica*, 93(1):47–55.

El-Rayes M.A. and Ulaby F.T. (1987), Microwave dielectric spectrum of vegetation-Part I: experimental observations, *IEEE Transactions on Geoscience and Remote Sensing*, 25(5):541–549.

Elser J.J., Fagan W.F., Kerkhoff A.J., Swenson N.G. and Enquist B. J. (2010), Biological stoichiometry of plant production: metabolism, scaling and ecological response to global change, *New Phytologist*, 186(3):593–608.

Elvidge C.D. (1988), Thermal infrared reflectance of dry plant materials: 2.5–20.0 μm, *Remote Sensing of Environment*, 26 (3):265–285.

(1990), Visible and near infrared reflectance characteristics of dry plant materials, *International Journal of Remote Sensing*, 11(10):1775–1795.

Emberson L. (2003), Air pollution impacts on crops and forests: an introduction, in *Air Pollution Impacts on Crops and Forests: a Global Assessment* (Emberson L., Ashmore M.R. and Murray F., Eds), World Scientific Publishing Company, pp. 3–34.

Emengini E.J., Blackburn G.A. and Theobald J.C. (2013), Early detection of oil-induced stress in crops using spectral and thermal responses, *Journal of Applied Remote Sensing*, 7(1):073596.

Emengini E.J. and Ugbelase V.N. (2013), Mapping the effects of hydrocarbon spillage on plant spectral properties, *International Journal of Environmental Science, Management and Engineering Research*, 2(1):30–38.

Emmel P., and Hersch (1998), Spectral colour prediction model for a transparent fluorescent ink on paper, in *Proc. 6th Color Imaging Conference: Color Science, Systems and Applications*, 17–20 November 1998, Scottsdale, AZ, pp. 116–122.

Emmel P. (2000), Nouvelle formulation du modèle de Kubelka et Munk avec applications aux encres fluorescentes, in *Actes de l'Ecole de Printemps 2000 – Le Pays d'Apt en Couleurs*, 14–18 mars 2000, Roussillon en Provence, France, pp. 87–96.

(2003), Physical models for color prediction, in *Digital Color Imaging Handbook* (Sharma G., Ed), CRC Press, 66 pages.

Engemann K., Sandel B., Morueta-Holme N., Enquist B.J., Peet R.K., Wiser S. et al. (2016), Patterns and drivers of plant functional group dominance across the Western Hemisphere – A macroecological re-assessment based on a massive botanical dataset, *Botanical Journal of the Linnaean Society*, 180(2):141–60.

Essery C.I. and Morse A.P. (1992), The impact of ozone and acid mist on the spectral reflectance of young Norway spruce trees, *International Journal of Remote Sensing*, 13(16):3045–3054.

Esteban R., Fernández-Marín B., Olano J.M., Becerril J.M. and García-Plazaola J.I. (2014), Does plant colour matter? Wax accumulation as an indicator of decline in *Juniperus thurifera*, *Tree Physiology*, 34(3):267–274.

Estrella N. and Menzel A. (2006), Responses of leaf colouring in four deciduous tree species to climate and weather in Germany, *Climate Research*, 32(3):253–267.

Evans J.R. (1989), Photosynthesis and nitrogen relationships in leaves of C_3 plants, *Oecologia*, 78(1):9–19.

(1999), Leaf anatomy enables more equal access to light and CO_2 between chloroplasts, *New Phytologist*, 143(1):93–104.

Evans J.R., Jakobsen I., Ogren E. (1993), Photosynthetic light-response curves. 2: gradients of light absorption and photosynthetic capacity, *Planta*, 189(2):191–200.

Evans J.R. and Vogelmann T.C. (2003), Profiles of 14C fixation through spinach leaves in relation to light absorption and photosynthetic capacity, *Plant, Cell & Environment*, 26(4):547–560.

(2006), Photosynthesis within isobilateral *Eucalyptus pauciflora* leaves, *New Phytologist*, 171(4):771–782.

Evans J.R., Vogelmann T.C., Williams W.E. and Gorton H.L. (2004), Chloroplast to leaf, in *Photosynthetic Adaptation: Chloroplast to Landscape* (Smith W.K., Vogelmann T.C. and Critchley C., Eds), *Ecological Studies 178*, Springer, pp. 15–41.

Fabre S., Lesaignoux A., Olioso A. and Briottet X. (2011), Influence of water content on spectral reflectance of leaves in the 3–15 μm domain, *IEEE Geoscience and Remote Sensing Letters*, 8(1):143–147.

Fadzly N., Jack C., Schaefer H.M. and Burns K.C. (2009), Ontogenetic colour changes in an insular tree species: signalling to extinct browsing birds? *New Phytologist*, 184(2):495–501.

Fan D.Y., Wang Q., Li M. and Gao R.F. (2002), The lens effect of the epidermic cell layer of the leaf of *Euonymus japonicus* T. on the light gradients within leaf, *Acta Phytoecologica Sinica*, 26(5):594–598.

Fan D.Y., Han T., Li J., Wang Q. and Gao R.F. (2003), Profile of absolute light utilization efficiency within leaves of *Euonymus japonicus*, *Acta Botanica Sinica*, 45(2):169–176.
Fan Q., Wang Y., Sun P., Liu S. and Li Y. (2010), Discrimination of *Ephedra* plants with diffuse reflectance FT-NIRS and multivariate analysis, *Talanta*, 80(3):1245–1250.
Fang M.H. and Ju W.M. (2015), A inversion model for remote sensing of leaf water content based on the leaf optical property, *Spectroscopy and Spectral Analysis*, 35 (1):167–171 (in Chinese).
Fariñas M.D., Sancho-Knapik D., Peguero-Pina J.J., Gil-Pelegrín E. and Gomez Álvarez-Arenas T.E. (2013), Shear waves in vegetal tissues at ultrasonic frequencies, *Applied Physics Letters*, 102(10):103702.
(2014a), Monitoring plant response to environmental stimuli by ultrasonic sensing of the leaves, *Ultrasound in Medicine & Biology*, 40(9):2183–2194.
Fariñas M.D. and Gomez Álvarez-Arenas T.E. (2014b), Ultrasonic assessment of the elastic functional design of component tissues of *Phormium tenax* leaves, *Journal of the Mechanical Behavior of Biomedical Materials*, 39:304–315.
Farmer A.M. (1993), The effects of dust on vegetation – A review, *Environmental Pollution*, 79(1): 63–75.
Farquhar G.D. and Roderick M.L. (2003), Pinatubo, diffuse light, and the carbon cycle, *Science*, 299(5615):1997–1998.
Farrar J.F. and Mapunda O.P. (1977), Optical properties of the leaves of some African crop plants, *Applied Optics*, 16 (1):248–251.
Faruque A., Bahadur R. and Carter G.A. (2001), Application of artificial neural networks for the classification of remote sensing spectral reflectance data of fungal infected soybean leaf, *Journal of the Mississippi Academy of Sciences*, 46 (2):119–119.
Favier J.F., Ross D.W., Tsheko R., Kennedy D.D., Muir A.Y. and Fleming J. (1998), Discrimination of weeds in brassica crops using optical spectral reflectance and leaf texture analysis, in *Proc. Precision Agriculture and Biological Quality* (Meyer G.E. and DeShazer J.A., Eds), Boston, MA, 3 November 1998, SPIE, Vol. 3543, pp. 311–318.
Federici J.F. (2012), Review of moisture and liquid detection and mapping using terahertz imaging, *Journal of Infrared, Millimeter, and Terahertz Waves*, 33(2):97–126.
Feilhauer H., Asner G.P. and Martin R.E. (2015), Multi-method ensemble selection of spectral bands related to leaf biochemistry, *Remote Sensing of Environment*, 164:57–65.
Feng H., Li Z., Jin X., Yang G., Wan P., Guo J., et al. (2016), Estimating equivalent water thickness of apple leaves using hyperspectral data based on EFAST and PLS, *Transactions of the Chinese Society of Agricultural Engineering*, 32 (12):165–171 (in Chinese).
Féret J.B., François C., Asner G.P., Gitelson A.A., Martin R.E., Bidel L.P.R., et al (2008), PROSPECT-4 and 5: advances in the leaf optical properties model separating photosynthetic pigments, *Remote Sensing of Environment*, 112(6):3030–3043.
Féret J.B. and Asner G.P. (2011), Spectroscopic classification of tropical forest species using radiative transfer modeling, *Remote Sensing of Environment*, 115(9):2415–2422.
Féret J.B., François C., Gitelson A., Asner G.P., Barry K.M., Panigada C., et al. (2011), Optimizing spectral indices and chemometric analysis of leaf chemical properties using radiative transfer modeling, *Remote Sensing of Environment*, 115 (10):2742–2750.
Féret J.B., Gitelson A.A., Noble S.D. and Jacquemoud S. (2017), PROSPECT-D: Towards modeling leaf optical properties through a complete lifecycle, *Remote Sensing of Environment*, 193:204–215.
Fernandes A.M., Melo-Pinto P., Millan B., Tardaguila J. and Diago M.P. (2015), Automatic discrimination of grapevine (*Vitis vinifera* L.) clones using leaf hyperspectral imaging and partial least squares, *The Journal of Agricultural Science*, 153 (3):455–465.
Field C.B. and Mooney H.A. (1986), The photosynthesis – nitrogen relationship, in *On the Economy of Plant Form and Function* (Givnish T.J., Ed), Cambridge University Press, New York, pp. 25–55.
Field C.B., Chapin F.S., Matson P.A. and Mooney H.A. (1992), Responses of terrestrial ecosystems to the changing atmosphere: A resource-based approach, *Annual Review of Ecology and Systematics*, 23:201–235.
Filella I. and Peñuelas J. (1999), Altitudinal differences in UV absorbance, UV reflectance and related morphological traits of *Quercus ilex* and *Rhododendron ferrugineum* in the Mediterranean region, *Plant Ecology*, 145(1):157–165.
Filella I., Porcar-Castell A., Munné-Bosch S., Bäck J., Garbulsky M.F. and Peñuelas J. (2009), PRI assessment of long-term changes in carotenoids/chlorophyll ratio and short-term changes in de-epoxidation state of the xanthophyll cycle, *International Journal of Remote Sensing*, 30(17):4443–4455.
Filhol M.E. (1865), Recherche sur les propriétés chimiques de la chlorophylle, *Comptes Rendus Hebdomadaires des Séances de l'Académie des Sciences*, 61:371–373.
Fisher D.G. (1986), Ultrastructure, plasmodesmatal frequency, and solute concentration in green areas of variegated *Coleus blumei* Benth. leaves, *Planta*, 169(2):141–152.
Flint S.D., Jordan P.W. and Caldwell M.M. (1985), Plant protective response to enhanced UV-B radiation under field conditions: leaf optical properties and photosynthesis, *Photochemistry and Photobiology*, 41(1):95–99.
Foley S., Rivard B., Sánchez-Azofeifa G.A. and Calvo J. (2006), Foliar spectral properties following leaf clipping and implications for handling techniques, *Remote Sensing of Environment*, 103(3):265–275.
Foley W.J., McIlwee A., Lawler I., Aragones L., Woolnough A.P. and Berding N. (1998), Ecological applications of near infrared reflectance spectroscopy – A tool for rapid, cost-effective prediction of the composition of plant and animal tissues and aspects of animal performance, *Oecologia*, 116(3):293–305.

Fondom N.Y., Castro-Nava S. and Huerta A.J. (2009), Photoprotective mechanisms during leaf ontogeny: cuticular development and anthocyanin deposition in two morphs of *Agave striata* that differ in leaf coloration, *Botany*, 87 (12):1186–1197.

Font R., del Río-Celestino M., and De Haro A. (2002), Use of near infrared spectroscopy to evaluate heavy metal content in *Brassica juncea* cultivated on the polluted soils of the Guadiamar River area, *Fresenius Environmental Bulletin*, 11 (10):777–781.

Font R., del Río-Celestino M., Vélez D., Montoro R., and de Haro-Bailón A. (2004), Use of near-infrared spectroscopy for determining the total arsenic content in prostrate amaranth, *Science of the Total Environment*, 327(1–3):93–104.

Font R., Vélez D., del Río-Celestino M., de Haro-Bailón A., and Montoro R. (2005), Screening inorganic arsenic in rice by visible and near-infrared spectroscopy, *Microchimica Acta*, 151(3):231–239.

Font R., del Río-Celestino M. and de Haro-Bailón A. (2007), Near-infrared reflectance spectroscopy: methodology and potential for predicting trace elements in plants. in *Phytoremediation, Methods and Reviews* (Willey N., Ed), Humana Press, Inc., pp. 205–217.

Fooshee W.C. and Henny R.J. (1990), Chlorophyll leaves and anatomy of variegated and nonvarigated areas of *Aglaonema nitidum* leaves, *Proceedings of the Florida State Horticultural Society*, 103:170–172.

Foster A.J., Kakani V.G., Ge J. and Mosali J. (2012), Discrimination of switchgrass cultivars and nitrogen treatments using pigment profiles and hyperspectral leaf reflectance data, *Remote Sensing*, 4(9):2576–2594.

Fourty T., Baret F., Jacquemoud S., Schmuck G. and Verdebout J. (1996), Leaf optical properties with explicit description of its biochemical composition: direct and inverse problems, *Remote Sensing of Environment*, 56(2):104–117.

Fourty T. and Baret F. (1998), On spectral estimates of fresh leaf biochemistry, *International Journal of Remote Sensing*, 19 (7):1283–1297.

Fox D.L. and Wells J.R. (1971), Schemochromic blue leaf-surfaces of *Selaginella*, *American Fern Journal*, 61(3):137–139.

Franklin B. (1751), *Experiments and Observations on Electricity*, E. Cave, London, 86 pages.

Franz E., Gebhardt M.R. and Unklesbay K.B. (1991), The use of local spectral properties of leaves as an aid for identifying weed seedlings in digital images, *Transactions of the ASAE*, 34(2):682–687.

Franzke O. and Deussen O. (2003), Accurate graphical representation of plant leaves, in *Proc. Plant Growth Modeling, Simulation, Visualization, and Their Applications* (Hu B.G. and Jaeger M., Eds), Beijing, China, 13–16 October 2003, 16 pages.

Fraser A.B. (1994), The sylvanshine: retroreflection from dew-covered trees, *Applied Optics*, 33(21):4539–4547.

Fremy M.E. (1860), Recherches sur la matière colorante verte des feuilles, *Comptes Rendus Hebdomadaires des Séances de l'Académie des Sciences*, 50:405–411.

(1865), Recherches chimiques sur la matière verte des feuilles, *Comptes Rendus Hebdomadaires des Séances de l'Académie des Sciences*, 61:188–192.

Fridgen J.L. and Varco J.J. (2004), Dependency of cotton leaf nitrogen, chlorophyll, and reflectance on nitrogen and potassium availability, *Agronomy Journal*, 96(1):63–69.

Fry E.S. (2000), Visible and near-ultraviolet absorption spectrum of liquid water: comment, *Applied Optics*, 39(16):2743–2744.

Fuchs L. (1542), *De historia stirpium commentarii insignes*, Isingrin, Basel, 896 pages.

Fuchs M. and Tanner C.B. (1966), Infrared thermometry of vegetation, *Agronomy Journal*, 58(6):597–601.

Fuhrer M., Jensen H.W. and Prusinkiewicz P. (2004), Modeling Hairy Plants, in *Proc. 12th Pacific Conference on Computer Graphics and Applications*, Seoul, Republic of Korea, 6–8 October 2004, pp. 217–226.

Fujino M., Endo R. and Omasa K. (2002), Nondestructive instrumentation of water-stressed cucumber leaves – Comparison between changes in spectral reflectance, stomatal conductance, PSII yield and shape, *Agricultural Information Research*, 11 (2):161–170 (in Japanese).

Fukshansky L. (1978), On the theory of light absorption in non-homogeneous objects: the sieve-effect in one-component suspensions, *Journal of Mathematical Biology*, 6(2):177–196.

(1981), Optical properties of plants, in *Plant and the Daylight Spectrum* (Smith H., Ed), Academic Press, London, pp. 21–40.

(1987), Absorption statistics in turbid media, *Journal of Quantitative Spectroscopy & Radiative Transfer*, 38(5):389–406.

Fukshansky L. and Kazarinova N. (1980), Extension of the Kubelka-Munk theory of light propagation in intensely scattering materials to fluorescent media, *Journal of the Optical Society of America*, 70(9):1101–1111.

Fukshansky L., Kazarinova N. and Martinez von Remisowsky A. (1991), Estimation of optical parameters in a living tissue by solving the inverse problem of the multiflux radiative transfer, *Applied Optics*, 30(22):3145–3153.

Fukshansky L. and Martinez von Remisowsky A. (1992), A theoretical study of the light microenvironment in a leaf in relation to photosynthesis, *Plant Science*, 86(2):167–182.

Fukshansky L., Martinez von Remisowsky A., McClendon J., Ritterbusch A., Richter T. and Mohr H. (1993), Absorption spectra of leaves corrected for scattering and distributional error: a radiative transfer and absorption statistics treatment, *Photochemistry and Photobiology*, 57(3):538–555.

Fukuhara M., Degawa T., Okushima L., Matsuo K. and Honma T. (2000), Physical characteristics of tea leaves by ultrasonic transmission method, *Tea Research Journal*, 90: 108–109 (in Japanese).

Fukuhara M. (2002), Acoustic characteristics of botanical leaves using ultrasonic transmission waves, *Plant Science*, 162 (4):521–528.

Fukuhara M., Okushima L., Matsuo K. and Homma T. (2005), Acoustic characteristics of fresh tea leaves, *Japan Agricultural Research Quarterly*, 39(1):45–49.

Fukuhara M., Dutta Gupta S. and Okushima L. (2006), Acoustic characteristics of plant leaves using ultrasonic transmission waves, in *Plant Tissue Culture Engineering* (Dutta Gupta S. and Ibaraki Y., Eds), Springer, pp. 427–439.

Fung T., Ma F.Y. and Siu W.L. (1998), Hyperspectral data analysis for subtropical tree species recognition, in *Proc. IEEE International Geoscience and Remote Sensing Symposium (IGARSS'98)*, Seattle, WA, 6–10 July 1998 IEEE, Vol. 3, pp. 1298–1300.

Fung T., Fung H., Ma Y. and Siu W.L. (2003), Band selection using hyperspectral data of subtropical tree species, *Geocarto International*, 18(4):3–11.

Furuya S. (1986), Growth diagnosis of rice plants by means of leaf color, *Japan Agricultural Research Quarterly*, 20 (3):147–153.

Gäb M., Hoffmann K., Lobe M., Metzger R., van Ooyen S., Elbers G. et al. (2006), NIR-spectroscopic investigation of foliage of ozone-stressed *Fagus sylvatica* trees, *Journal of Forest Research*, 11(2):69–75.

Gabrys-Mizera H. (1976), Model considerations of the light conditions in noncylindrical plant cells, *Photochemistry and Photobiology*, 24(5):453–461.

Gal A., Brumfeld V., Weiner S., Addadi L. and Oron D. (2012), Certain biominerals in leaves function as light scatterers, *Advanced Optical Materials*, 24(10):OP77–OP83.

Galtie J.F. and Lescure M. (2009), *Dispositif de mesure optoélectronique de l'hydratation d'un végétal dans son environnement naturel*, Patent WO/2009/007269, 15 January 2009.

Galvez-Sola L., García-Sánchez F., Pérez-Pérez J.G., et al. (2015), Rapid estimation of nutritional elements on citrus leaves by near infrared reflectance spectroscopy, *Frontiers in Plant Science*, 6:571.

Gamon J.A., Field C.B., Bilger W., Björkman O., Fredeen A.L. and Peñuelas J. (1990), Remote sensing of the xanthophyll cycle and chlorophyll fluorescence in sunflower leaves and canopies, *Oecologia*, 85(1):1–7.

Gamon J.A., Peñuelas J. and Field C.B. (1992), A narrow-waveband spectral index that tracks diurnal changes in photosynthetic efficiency, *Remote Sensing of Environment*, 41(1):35–44.

Gamon J.A., Filella I. and Peñuelas J. (1993), The dynamic 531-nanometer Delta reflectance signal: a survey of twenty Angiosperm species, in *Photosynthetic Responses to the Environment* (Yamamoto H.Y. and Smith C.M., Eds), American Society of Plant Physiologists, pp. 172–177.

Gamon J.A., Serrano L. and Surfus J.S. (1997), The photochemical reflectance index: an optical indicator of photosynthetic radiation use efficiency across species, functional types, and nutrient levels, *Oecologia*, 112(4):492–501.

Gamon J.A. and Qiu H.L. (1999), Ecological applications of remote sensing at different scales, in *Handbook of Functional Plant Ecology* (Pugnaire F. and Valladares F., Eds), CRC Press, pp. 805–846.

Gamon J.A. and Surfus J.S. (1999), Assessing leaf pigment content and activity with a reflectometer, *New Phytologist*, 143 (1):105–117.

Gamon J.A., Field C.B., Fredeen A.L. and Thayer S. (2001), Assessing photosynthetic downregulation in sunflower stands with an optically-based model, *Photosynthesis Research*, 67(1–2):113–125.

Gamon J.A. and Berry J.A. (2012), Facultative and constitutive pigment effects on the Photochemical Reflectance Index (PRI) in sun and shade conifer needles, *Israel Journal of Plant Sciences*, 60(1–2):85–95.

Ganapol B.D., Johnson L.F., Hammer P.D., Hlavka C.A. and Peterson D.L. (1998), LEAFMOD: a new within-leaf radiative transfer model, *Remote Sensing of Environment*, 63(2):182–193.

Ganapol B.D., Johnson L.F., Hlavka C.A., Peterson D.L. and Bond B. (1999), LCM2: A coupled leaf/canopy radiative transfer model, *Remote Sensing of Environment*, 70(2):153–166.

Gapinski J., Dobek A., Paillotin G., Breton J., Leibl W. and Trissl H.W. (1994), Light gradient in photosynthetic systems: theory and experiment, *Laser Physics*, 4:191–198.

Garbe C.S., Schurr U. and Jähne B. (2002), Thermographic measurements on plant leaves, in *Proc. Thermosense XXIV* (Maldague X.P. and Rozlosnik A.E., Eds), Orlando, FL, 1–5 April 2002, SPIE, Vol. 4710, pp. 407–416.

Garbulsky M.F., Peñuelas J., Gamon J., Inoue Y. and Filella I. (2011), The photochemical reflectance index (PRI) and the remote sensing of leaf, canopy and ecosystem radiation use efficiencies: a review and meta-analysis, *Remote Sensing of Environment*, 115(2):281–297.

Garcia Ciudad A., Fernandez Santos B., Vazquez de Aldana B.R., Zabalgogeazcoa I., Gutierrez M.Y. and Garcia Criado B. (2004), Use of near infrared reflectance spectroscopy to assess forage quality of a Mediterranean shrub, *Communications in Soil Science and Plant Analysis*, 35 (5–6):665–678.

Gardner C.M., Jacques S.L. and Welch A.J. (1996), Fluorescence spectroscopy of tissue: recovery of intrinsic fluorescence from measured fluorescence, *Applied Optics*, 35(10):1780–1792.

Garlaschi F.M., Zucchelli G. and Jennings R.C. (1989), Studies on light absorption and photochemical activity changes in chloroplast suspensions and leaves due to light scattering and light filtration across chloroplast and vegetation layers, *Photosynthesis Research*, 20(3):207–220.

Garnier E., Shipley B., Roumet C. and Laurent G. (2001), A standardized protocol for the determination of specific leaf area and leaf dry matter content, *Functional Ecology*, 15(5):688–695.

Garriga M., Retamales J.B., Romero-Bravo S., Caligari P.D.S. and Lobos G.A. (2014), Chlorophyll, anthocyanin, and gas exchange changes assessed by spectroradiometry in *Fragaria chiloensis* under salt stress, *Journal of Integrative Plant Biology*, 56(5):505–515.

Gates D.M. and Tantraporn W. (1952), The reflectivity of deciduous trees and herbaceous plants in the infrared to 25 microns, *Science*, 115(2997):613–616.

Gates D.M. (1962), Leaf temperature and energy exchange, *Archiv für Meteorologie, Geophysik und Bioklimatologie*, 12 (2):321–336.
(1964), Leaf temperature and transpiration, *Agronomy Journal*, 56(3):273–277.
(1965), Energy, plants, and ecology, *Ecology*, 46(1–2):1–13.
(1966), Transpiration and energy exchange, *The Quarterly Review of Biology*, 41(4):353–364.
(1967), Remote sensing for the biologist, *BioScience*, 17(5):303–307.
(1968), Energy exchange and ecology, *BioScience*, 18(2):90–95.
(1976), Energy exchange and transpiration, in *Water and Plant Life* (Lange O.L., Kappen L. and Schulze E.D., Eds), Springer-Verlag, pp. 137–147.
(1980a), *Biophysical Ecology*, Springer Verlag, 611 pages.
(1980b), Energy budgets of plants, in *Biophysical Ecology* (Reichle D.E., Ed), Springer-Verlag, New York, pp. 345–381.
(2003), *Biophysical Ecology*, Dover, 656 pages.
Gates D.M., Keegan H.J., Schleter V.R. and Weidner V.R. (1965), Spectral properties of plants, *Applied Optics*, 4(1):11–20.
Gates D.M., Alderfer R. and Taylor E. (1968), Leaf temperatures of desert plants, *Science*, 159(3818):994–995.
Gauslaa Y. (1984), Heat resistance and energy budget in different Scandinavian plants. 1. Infrared and visible reflectance in different alpine vascular plants, *Holarctic Ecology*, 7(1):7–12.
Gausman H.W., Allen W.A. and Cardenas R. (1969a), Reflectance of cotton leaves and their structure, *Remote Sensing of Environment*, 1(1):19–22.
Gausman H.W., Allen W.A., Cardenas R. and Richardson A.J. (1969b), Relation of light reflectance to cotton leaf maturity (*Gossypium hirsutum* L.), in *Proc. 6th International Symposium on Remote Sensing of Environment*, Ann Arbor, MI, 13–16 October 1969, ERIM, Vol. 2, pp. 1123–1141.
(1970c), Relation of light reflectance to histological and physical evaluations of cotton leaf maturity, *Applied Optics*, 9 (3):545–552.
(1971a), Effects of leaf nodal position on absorption and scattering coefficients and infinite reflectance of cotton leaves, *Gossypium hirsutum* L., *Agronomy Journal*, 63(1):87–91.
(1972), Effects of leaf age for four growth stages of cotton and corn plants on leaf reflectance, structure, thickness, water and chlorophyll concentrations and selection of wavelengths for crop discrimination, in *Proc. Remote Sensing of Earth Resources: Technical papers selected from the Conference on Earth Resources Observation and Information Analysis System* (Shahrokhi F., Ed), Tullahoma, TN, 13–14 March 1972, Vol. 1, pp. 25–51.
Gausman H.W., Allen W.A., Myers V.I. and Cardenas R. (1969c), Reflectance and internal structure of cotton leaves, *Gossypium hirsutum* L., *Agronomy Journal*, 61(3):374–376.
Gausman H.W. and Cardenas R. (1969), Effect of leaf pubescence of *Gynura aurantiaca* on light reflectance, *Botanical Gazette*, 130(3):158–162.
(1973), Light reflectance by leaflets of pubescent, normal, and glabrous soybean lines, *Agronomy Journal*, 65(5):837–838.
Gausman H.W., Allen W.A., Cardenas R. and Bowen R.L. (1970a), Color photos, cotton leaves, and soil salinity, *Photogrammetric Engineering*, 36(5):454–459.
(1970b), Detection of foot rot disease of grapefruit trees with infrared color film, *Journal of the Rio Grande Valley Horticultural Society*, 24:36–42.
Gausman H.W., Allen W.A., Myers V.I., Cardenas R. and Leamer R.W. (1970d), Reflectance of single leaves and field plots of Cycocel-treated cotton (*Gossypium hirsutum* L.) in relation to leaf structure, *Remote Sensing of Environment*, 1 (2):103–107.
Gausman H.W., Allen W.A., Escobar D.E., Rodriguez R.R. and Cardenas R. (1971b), Age effects of cotton leaves on light reflectance, transmittance, and absorptance and on water content and thickness, *Agronomy Journal*, 63(3):465–469.
Gausman H.W., Allen W.A., Wiegand C.L., Escobar D.E. and Rodriguez R.R. (1971c), Leaf light reflectance, transmittance, absorptance, and optical and geometrical parameters for eleven plant genera with different leaf mesophyll arrangements, in *Proc. 7th International Symposium on Remote Sensing of Environment*, Ann Arbor, MI, 17–21 May 1971, Vol. 3, pp. 1599–1625.
Gausman H.W., Allen W.A., Wiegand C.L., Escobar D.E., Rodriguez R.R. and Richardson A.J. (1973), *The Leaf Mesophylls of Twenty Crops, their Light Spectra, and Optical and Geometrical Parameters*, USDA, Technical Bulletin 1465, 59 pages.
Gausman H.W. (1974), Leaf reflectance of near-infrared, *Photogrammetric Engineering*, 40(2):183–191.
Gausman H.W., Allen W.A. and Escobar D.E. (1974a), Refractive index of plant cell walls, *Applied Optics*, 13(1):109–111.
Gausman H.W., Cardenas R. and Berumen A. (1974b), Effects of leaf age within growth stages of pepper and sorghum plants on leaf thickness, water, chlorophyll, and light reflectance, in *Proc. Remote Sensing of Earth Resources: Technical papers selected from the Conference on Earth Resources Observation and Information Analysis System* (Shahrokhi F., Ed), Tullahoma, TN, 25–27 March 1974, Vol. 3, pp. 39–56.
Gausman H.W. and Hart W.G. (1974a), Reflectance of four levels of sooty-mold deposits produced from the honeydew of three insect species, *Journal of the Rio Grande Valley Horticultural Society*, 28:131–136.
(1974b), Reflectance of sooty mold fungus on citrus leaves over the 2.5 to 40-micrometer wavelength interval, *Journal of Economic Entomology*, 67(4):479–480.
Gausman H.W., Heald C.M. and Escobar D.E. (1975a), Effect of *Rotylenchulus reniformis* on reflectance of cotton plant leaves, *Journal of Nematology*, 7(4):368–374.

(1975b), Ultraviolet radiation reflectance, transmittance, and absorptance by plant leaf epidermises, *Agronomy Journal*, 67 (5):720–724.

Gausman H.W., Rodriguez R.R. and Richardson A.J. (1976), Infinite reflectance of dead compared with live vegetation, *Agronomy Journal*, 68(2):295–296.

Gausman H.W., Escobar D.E. and Wiegand C.L. (1977), Reflectance and photographic characteristics of three citrus varieties for discrimination purpose, in *Proc. Remote Sensing of Earth Resources: Technical papers selected from the Sixth Annual Remote Sensing of Earth Resources Conference* (Shahrokhi F., Ed), Tullahoma, TN, 29–31 March 1977, Vol. 6, pp. 341–355.

Gausman H.W., Escobar D.E., Everitt J.H., Richardson A.J. and Rodriguez R.R. (1978a), Distinguishing succulent plants from crop and woody plants, *Photogrammetric Engineering & Remote Sensing*, 44(4):487–491.

Gausman H.W., Escobar D.E., Rodriguez R.R., Thomas C.E. and Bowen R.L. (1978b), Ozone damage detection in cantaloupe plants, *Photogrammetric Engineering & Remote Sensing*, 44(4):481–485.

Gausman H.W., Everitt J.H. and Escobar D.E. (1979), Seasonal nitrogen concentration and reflectance of seven woody plant species, *Journal of the Rio Grande Valley Horticultural Society*, 33:101–104.

Gausman H.W., Menges R.M., Richardson A.J., Walter H., Rodriguez R.R. and Tamez S. (1981), Optical parameters of leaves of seven weed species, *Weed Science*, 29(1):24–26.

Gausman H.W. (1984), Evaluation of factors causing reflectance differences between sun and shade leaves, *Remote Sensing of Environment*, 15(2):177–181.

Gauthier A. (1906), Sur la coloration rouge éventuelle de certaines feuilles et sur la couleur des feuilles d'automne, *Comptes Rendus Hebdomadaires des Séances de l'Académie des Sciences*, 143:490–491.

Gay A., Thomas H., Roca M., et al. (2008), Nondestructive analysis of senescence in mesophyll cells by spectral resolution of protein synthesis-dependent pigment metabolism, *New Phytologist*, 179(3):663–674.

Gayford M. (2011), *A Bigger Message: Conversations with David Hockney*, Thames & Hudson Ltd, 248 pages.

Gazala I.F.S., Sahoo R.N., Pandey R., Mandal B., Gupta V.K., Singh R., et al. (2013), Spectral reflectance pattern in soybean for assessing yellow mosaic disease, *Indian Journal of Virology*, 24(2):242–249.

Ge H., Lu S. and Zhao Y.S. (2012), Effects of leaf hair on leaf reflectance and hyperspectral vegetation indices, *Spectroscopy and Spectral Analysis*, 32 (2):439–444 (in Chinese).

Gebeshuber I.C. and Lee D.W. (2012), Nanostructures for coloration (Organisms other than animals), in *Encyclopedia of Nanotechnology*, Springer, pp. 1790–1803.

Geiger B. (1993), Three-dimensional modeling of human organs and its application to diagnosis and surgical planning, in *INRIA, Programme 4. Robotique*, Image et Vision, 119 pages.

Gemmell F.M. and Colls J.J. (1992), The effects of sulphur dioxide on the spectral characteristics of leaves of *Vicia faba* L., *International Journal of Remote Sensing*, 13(14):2547–2563.

Gente R., Born N., Voß N., Sannemann W., Léon J., Koch M. and Castro-Camus E. (2013), Determination of leaf water content from terahertz time-domain spectroscopic data, *Journal of Infrared, Millimeter, and Terahertz Waves*, 34(3–4):316–323.

Gente R. and Koch M. (2015), Monitoring leaf water content with THz and sub-THz waves, *Plant Methods*, 11:15.

Geoffroy C.J. (1707), Observations sur les huiles essentielles, avec quelques conjectures sur sur la cause des couleurs des feuilles et des fleurs des plantes, *Histoire de l'Académie royale des sciences*, Tome I, pp. 517–526.

Georgiev G.T., Gatebe C.K., Butler J.J. and King M.D. (2007), BRDF calibration of natural samples in support of remote sensing, in *Proc. International Geoscience and Remote Sensing Symposium (IGARSS'07)*, Barcelona, Spain, 23–28 July 2007, IEEE, pp. 2877–2880.

(2009), BRDF analysis of savanna vegetation and salt-pan samples, *IEEE Transactions on Geoscience and Remote Sensing*, 47(8):2546–2556.

Gerber F., Marion R., Olioso A., Jacquemoud S., Ribeiro da Luz B. and Fabre S. (2011), Modeling directional-hemispherical reflectance and transmittance of fresh and dry leaves from 0.4 μm to 5.7 μm with the PROSPECT-VISIR model, *Remote Sensing of Environment*, 115(1):404–414.

Gianoli E. and Carrasco-Urra F. (2014), Leaf mimicry in a climbing plant protects against herbivory, *Current Biology*, 24 (9):984–987.

Gibeaut D.M. and Thomson W.W. (1989), Stereology of the internal structures of leaves in *Peperomia obtusifolia, P. camptotricha*, and *P. scandens*, *Botanical Gazette*, 150(2):115–121.

Gibson L.J. and Ashby M.F. (1997), *Cellular Solids: Structure and Properties*, Cambridge University Press, 532 pages.

Gillon D., Joffre R. and Dardenne P. (1993), Predicting the stage of decay of decomposing leaves by near-infrared reflectance spectroscopy, *Canadian Journal of Forest Research*, 23(12):2552–2559.

Gillon D., Hernando C., Valette J.C. and Joffre R. (1997), Fast estimation of the calorific values of forest fuels by near-infrared reflectance spectroscopy, *Canadian Journal of Forest Research*, 27(5):760–765.

Gillon D., Houssard C. and Joffre R. (1999a), Using near-infrared reflectance spectroscopy to predict carbon, nitrogen and phosphorus content in heterogeneous plant material, *Oecologia*, 118(2):173–182.

Gillon D., Joffre R. and Ibrahima A. (1999b), Can litter decomposability be predicted by near infrared reflectance spectroscopy? *Ecology*, 80(1):175–186.

Gillon D., Dauriac F., Deshayes M., Valette J.C. and Moro C. (2002), Foliage moisture content and spectral characteristics using near infrared reflectance spectroscopy (NIRS), in *Proc. 4th International Conference on Forest Fire Research* (Viegas D.X., Ed), Coimbra, Portugal, 18–23 November 2002, 13 pages.

Gillon D., Dauriac F., Deshayes M., Valette J.C. and Moroc C. (2004), Estimation of foliage moisture content using near infrared reflectance spectroscopy, *Agricultural and Forest Meteorology*, 124(1–2):51–62.

Gilmore A.M., Hazlett T.L. and Govindjee (1995), Xanthophyll cycle-dependent quenching of photosystem II chlorophyll a fluorescence: formation of a quenching complex with a short fluorescence lifetime, *Proceedings of the National Academy of Sciences*, 92(6):2273–2277.

Gislum R., Micklander E. and Nielsen J.P. (2004), Quantification of nitrogen concentration in perennial ryegrass and red fescue using near-infrared reflectance spectroscopy (NIRS) and chemometrics, *Field Crops Research*, 88(2–3):269–277.

Gitelson A. and Merzlyak M.N. (1994a), Quantitative estimation of chlorophyll-a using reflectance spectra: experiments with autumn chestnut and maple leaves, *Journal of Photochemistry and Photobiology. B, Biology*, 22(3):247–252.

(1994b), Spectral reflectance changes associated with autumn senescence of *Aesculus hippocastanum* L. and *Acer platanoides* L. leaves. Spectral features and relation to chlorophyll estimation, *Journal of Plant Physiology*, 143 (3–4):286–292.

(1996), Signature analysis of leaf reflectance spectra: algorithm development for remote sensing of chlorophyll, *Journal of Plant Physiology*, 148(3–4):494–500.

(1997), Remote estimation of chlorophyll content in higher plant leaves, *International Journal of Remote Sensing*, 18 (12):2691–2697.

Gitelson A.A., Merzlyak M.N. and Lichtenthaler H.K. (1996), Detection of red edge position and chlorophyll content by reflectance measurements near 700 nm, *Journal of Plant Physiology*, 148(3–4):501–508.

(1998), Leaf chlorophyll fluorescence corrected for re-absorption by means of absorption and reflectance measurements, *Journal of Plant Physiology*, 152(2–3):283–296.

(1999), The chlorophyll fluorescence ratio R735/F700 as an accurate measure of the chlorophyll content in plants, *Remote Sensing of Environment*, 69(3):296–302.

Gitelson A.A., Merzlyak M.N. and Chivkunova O.B. (2001), Optical properties and nondestructive estimation of anthocyanin content in plant leaves, *Photochemistry and Photobiology*, 74(1):38–45.

Gitelson A.A., Zur Y., Chivkunova O.B. and Merzlyak M.N. (2002), Assessing carotenoid content in plant leaves with reflectance spectroscopy, *Photochemistry and Photobiology*, 75(3):272–291.

Gitelson A.A., Gritz Y. and Merzlyak M.N. (2003), Relationships between leaf chlorophyll content and spectral reflectance and algorithms for non-destructive chlorophyll assessment in higher plant leaves, *Journal of Plant Physiology*, 160 (3):271–282.

Gitelson A.A., Keydan G.P. and Merzlyak M.N. (2006), Three-band model for noninvasive estimation of chlorophyll, carotenoids, and anthocyanin contents in higher plant leaves, *Geophysical Research Letters*, 33:L11402.

Gitelson A.A., Gamon J.A. and Solovchenko A. (2017a), Multiple drivers of seasonal change in PRI: implications for photosynthesis. 1: Leaf level, *Remote Sensing of Environment*, 191:110–116.

(2017b), Multiple drivers of seasonal change in PRI: implications for photosynthesis 2. Stand, *Remote Sensing of Environment*, 190:198–206.

Giusti M.M. and Wrolstad R.R. (2001), Characterization and measurement of anthocyanins by UV-visible spectroscopy, in *Current Protocols in Food Analytical Chemistry*, John Wiley & Sons, pp. F1.2.1-F1.2.13.

Giusti M.M. and Jing P. (2008), Analysis of anthocyanins, in *Food Colorants: Chemical and Functional Properties*, CRC Press, pp. 479–506.

Givnish T.J. (1990), Leaf mottling: relation to growth form and leaf phenology and possible role as camouflage, *Functional Ecology*, 4(4):463–474.

Glassner A.S. (1989), Surface physics for ray tracing, in *An Introduction to Ray Tracing, The Morgan Kaufmann Series in Computer Graphics*, Morgan Kaufmann, pp. 121–160.

Gledhill D. (1985), *The Names of Plants*, 2nd Edition, Cambridge University Press, 202 pages.

Glover B.J. and Whitney H.M. (2010), Structural colour and iridescence in plants: the poorly studied relations of pigment colour, *Annals of Botany*, 105(4):505–511.

Goetz A.F.H., Gao B.C., Wessman C.A. and Bowman W.D. (1990), Estimation of biochemical constituents from fresh, green leaves by spectrum matching techniques, in *Proc. 10th International Geoscience and Remote Sensing Symposium (IGARSS'90)*, College Park, MD, 20–24 May 1990, IEEE, pp. 971–974.

Goldstein D.H. and Cox J.L. (2004), Spectropolarimetric properties of vegetation, in *Proc. Polarization: Measurement, Analysis, and Remote Sensing I* (Goldstein D.H. and Chenault D.B., Eds), Orlando, FL, 12 April 2004, SPIE, Vol. 5432, pp. 53–62.

Gonzales R., Paul N.D., Percy K.E., Ambrose M., McLaughlin C.K., Barnes J.D., et al. (1996), Responses to ultraviolet-B radiation (280–315 nm) of pea (*Pisum sativum*) lines differing in leaf surface wax, *Physiologia Plantarum*, 98 (4):852–860.

González-Martín I., Hernández-Hierro J.M. and González-Cabrera J.M. (2007), Use of NIRS technology with a remote reflectance fibre-optic probe for predicting mineral composition (Ca, K, P, Fe, Mn, Na, Zn), protein and moisture in alfalfa, *Analytical and Bioanalytical Chemistry*, 387(6):2199–2205.

Gorbunova G.S., Parshina Z.S. and Bedenko V.P. (1960),Optical properties and photosynthesis of certain species of cultivated and wild plants in relation to ecological conditions, *Trudy sektora astrobotaniki*, 8: 31–51 (in Russian).

Gorton H.L., Williams W.E. and Vogelmann T.C. (1999), Chloroplast movement in *Alocasia macrorrhiza*, *Physiologia Plantarum*, 106(4):421–428.

Gorton H.L., Brodersen C.R., Williams W.E. and Vogelmann T.C. (2010), Measurement of the optical properties of leaves under diffuse light, *Photochemistry and Photobiology*, 86(5):1076–1083.
Göttlicher D., Albert J., Nauss T. and Bendix J. (2011), Optical properties of selected plants from a tropical mountain ecosystem – Traits for plant functional types to parametrize a land surface model, *Ecological Modelling*, 222 (3):493–502.
Gottwald T.R., da Graça J.V. and Bassanezi R.B. (2007), *Citrus Huanglongbing*: the pathogen and its impact, in *Plant Health Progress*, 36 pages.
Goudarzi U., Mokhtari J. and Nouri M. (2013), Camouflage of cotton fabrics in visible and near infrared region using vat dyes, *Journal of Color Science and Technology*, 7 (1):35–25 (in Persian).
(2014), Camouflage of cotton fabrics in visible and NIR region using three selected vat dyes, *Color Research & Application*, 39(2):200–207.
Goulas Y., Cerovic Z.G., Cartelat A. and Moya I. (2004), Dualex: a new instrument for field measurements of epidermal ultraviolet absorbance by chlorophyll fluorescence, *Applied Optics*, 43(23):4488–4496.
Gould J.M. (1982), Characterization of lignin *in situ* by photoacoustic spectroscopy, *Plant Physiology*, 70(5):1521–1525.
Gould K.S. and Lee D.W. (1996), Physical and ultrastructural basis of blue leaf iridescence in four Malaysian understory plants, *American Journal of Botany*, 83(1):45–50.
Gould K.S., Vogelmann T.C., Han T. and Clearwater M.J. (2002), Profiles of photosynthesis within red and green leaves of *Quintinia serrata*, *Physiologia Plantarum*, 116(1):127–133.
Gould K.S., Davies K.M. and Winefield C., Eds (2009), *Anthocyanins: Biosynthesis, Functions, and Applications*, Springer, 329 pages.
Govaerts Y., Jacquemoud S., Verstraete M.M. and Ustin S.L. (1996), Three-dimensional radiation transfer modeling in a dicotyledon leaf, *Applied Optics*, 35(33):6585–6598.
Govaerts Y. and Verstraete M.M. (1998), Raytran: a Monte Carlo ray-tracing model to compute light scattering in three-dimensional heterogeneous media, *IEEE Transactions on Geoscience and Remote Sensing*, 36(2):493–505.
Govindjee and Krogmann D. (2004), Discoveries in oxygenic photosynthesis (1727–2003): a perspective, *Photosynthesis Research*, 80(1–3):15–57.
Graeff S., Steffens D. and Schubert S. (2001), Use of reflectance measurements for the early detection of N, P, Mg, and Fe deficiencies in *Zea mays* L., *Journal of Plant Nutrition and Soil Science*, 164(4):445–450.
Graeff S. and Claupein W. (2003), Quantifying nitrogen status of corn (*Zea mays* L.) in the field by reflectance measurements, *European Journal of Agronomy*, 19(4):611–618.
Graeff S., Link J. and Claupein W. (2006), Identification of powdery mildew (*Erysiphe graminis* sp. tritici) and take-all disease (*Gaeumannomyces graminis* sp. tritici) in wheat (*Triticum aestivum* L.) by means of leaf reflectance measurements, *Central European Journal of Biology*, 1(2):275–288.
Graham R.M., Lee D.W. and Norstog K. (1993), Physical and ultrastructural basis of blue leaf iridescence in two neotropical ferns, *American Journal of Botany*, 80(2):198–203.
Grant L. (1985), *Polarized and non-polarized components of leaf reflectance*, PhD Thesis, Purdue University, West Lafayette, IN, 142 pages.
(1987), Diffuse and specular characteristics of leaf reflectance, *Remote Sensing of Environment*, 22(2):309–322.
Grant L., Daughtry C.S.T. and Vanderbilt V.C. (1987a), Polarized and non-polarized leaf reflectance of *Coleus blumei*, *Environmental and Experimental Botany*, 27(2):139–145.
(1987b), Variations in the polarized leaf reflectance of *Sorghum bicolor*, *Remote Sensing of Environment*, 21(3):333–339.
(1993), Polarized and specular reflectance variation with leaf surface features, *Physiologia Plantarum*, 88(1):1–9.
Grant R.H., Heisler G.M., Gaoa W. and Jenks M. (2003), Ultraviolet leaf reflectance of common urban trees and the prediction of reflectance from leaf surface characteristics, *Agricultural and Forest Meteorology*, 120(1–4):127–139.
Graβmann J. (2005), Terpenoids as plant antioxidants. In *Vitamins and Hormones*, Vol. 72, Elsevier, pp. 505–535.
Gratani L., Covone F. and Larcher W. (2006), Leaf plasticity in response to light of three evergreen species of the Mediterranean maquis, *Trees – Structure and Function*, 20(5):549–558.
Greiner M.A., Duncan B.D. and Dierking M.P. (2007), Bidirectional scattering distribution functions of maple and cottonwood leaves, *Applied Optics*, 46(25):6485–6494.
(2009), Monte Carlo simulation of multiple photon scattering in sugar maple tree canopies, *Applied Optics*, 48 (32):6159–6171.
Grew N. (1682), *The Anatomy of Plants with an Idea of a Philosophical History of Plants*, W. Rawlins, London, 561 pages.
Griffioen H., Kornet J.G. and Schurer K. (1992), An optical leaf wetness sensor, *Acta Horticulturae*, 304:127–135.
Griffiths P.R. and Dahm D.J. (2008), Continuum and discontinuum theories of diffuse reflection, in *Handbook of Near-Infrared Analysis* (Burns D.A. and Ciurczak E.W., Eds), CRC Press, pp. 21–64.
Grime J.P. (1961), Measurement of leaf colour, *Nature*, 191(4788):614–615.
Grisham M.P., Johnson R.M. and Zimba P.V. (2010), Detecting sugarcane yellow leaf virus infection in asymptomatic leaves with hyperspectral remote sensing and associated leaf pigment changes, *Journal of Virological Methods*, 167(2):140–145.
Gronwall T.H. (1926), Reflection of radiation from a finite number of equally spaced parallel planes, *Physical Review*, 27 (3):277–285.
Grossman Y.L., Ustin S.L., Jacquemoud S., Sanderson E., Schmuck G. and Verdebout J. (1996), Critique of stepwise multiple linear regression for the extraction of leaf biochemistry information from leaf reflectance data, *Remote Sensing of Environment*, 56(3):182–193.

Gruninger J.H., Robertson D.C. and Pervaiz M.M. (1992), *Data Analysis for Bark and Leaf Reflectance Measurements*, Smart Weapons Operability Enhancement, June 1992, SWOE Report 92–6, 52 pages.

Gudi G., Krähmer A., Krüger H. and Schulz H. (2015), Attenuated total reflectance–Fourier transform infrared spectroscopy on intact dried leaves of sage (*Salvia officinalis* L.): accelerated chemotaxonomic discrimination and analysis of essential oil composition, *Journal of Agricultural and Food Chemistry*, 63(39):8743–8750.

Guo H., Jin L.S. and Lin G.L. (2008), Rapid determination of nitrogen and potassium contents in tobacco leaves by near infrared reflectance spectroscopy, *Heilongjiang Agricultural Sciences*, 4: 103–104 (in Chinese).

Guo H., Chen J., Pan T., Wang J. and Cao G. (2014), Vis-NIR wavelength selection for non-destructive discriminant analysis of breed screening of transgenic sugarcane, *Analytical Methods*, 6:8810–8816.

Guyot G., Baret F. and Jacquemoud S. (1992), Imaging spectroscopy for vegetation studies, in *Imaging Spectroscopy: Fundamentals and Prospective Application* (Toselli F. and Bodechtel J., Eds), pp. 145–165.

Gupta R.K. and Woolley J.T. (1971), Spectral properties of soybean leaves, *Agronomy Journal*, 63(1):123–126.

Guyon E. and Troadec J.P. (1994), *Du sac de billes au tas de sable*, Editions Odile Jacob, Paris, 306 pages.

Habel R., Kusternig A. and Wimmer M. (2007), Physically based real-time translucency for leaves, in *Proc. Eurographics Symposium on Rendering* (Kautz J. and Pattanaik S., Eds), pp. 253–263.

Habel R. (2009), *Real-time Rendering and Animation of Vegetation*, PhD Thesis, Fakultät für Technische Naturwissenschaften und Informatik, Technischen Universität Wien, Wien (Osterreich), 137 pages.

(2010), Real-time rendering and animation of vegetation, *Buletinul Institutului Politehnic din Iaşi*, 4:115–130.

Haberlandt G. (1914), Optical sense-organs, in *Physiological Plant Anatomy* (Haberlandt G., Ed), MacMillan and Co., London, pp. 613–631.

Hadjiloucas S., Karatzas L.S. and Bowen J.W. (1999), Measurements of leaf water content using terahertz radiation, *IEEE Transactions on Microwave Theory and Techniques*, 47(2):142–149.

Hadjiloucas S., Galvão R.K.H. and Bowen J.W. (2002), Analysis of spectroscopic measurements of leaf water content at terahertz frequencies using linear transforms, *Journal of the Optical Society of America A*, 19(12):2495–2509.

Hadjiloucas S., Walker G.C., Bowen J.W. and Zafiropoulos A. (2009), Propagation of errors from a null balance terahertz reflectometer to a sample's relative water content, *Journal of Physics: Conference Series*, 178(1):012012.

Hagen S.B., Folstad I. and Jakobsen S.W. (2003), Autumn colouration and herbivore resistance in mountain birch (*Betula pubescens*), *Ecology Letters*, 6(9):807–811.

Hagenbach E. (1870), Untersuchung über die optischen Eigenschaften des Blattgrüns, *Annalen der Physik*, 217(10):245–275.

Hale G.M. and Querry M.R. (1973), Optical constants of water in the 200-nm to 200-μm wavelength region, *Applied Optics*, 12(3):555–562.

Hamilton R.J. (1995), *Waxes: Chemistry, Molecular Biology and Function*, Oily Press Ltd, 360 pages.

Hamilton W.D. and Brown S.P. (2001), Autumn tree colours as a handicap signal, *Proceedings of the Royal Society of London. Series B*, 268(1475):1489–1493.

Han S., Yu B., Sun M., Huang W., Liu L. and Sun G. (2009), Development of a portable meter for detecting chlorophyll, nitrogen and water contents of plants, *Transactions of the Chinese Society of Agricultural Machinery*, 40(z1):256–259 (in Chinese).

Hana M., McClure W.F., Whitaker T.B., White M.W. and Bahler D.R. (1997), Applying neural networks. Part 2. Using near infrared to classify tobacco types and identify native grown tobacco, *Journal of Near Infrared Spectroscopy*, 5 (1):19–25.

Hanrahan P. and Krueger W. (1993), Reflection from layered surfaces due to subsurface scattering, in *Proc. Computer Graphics Proceedings (SIGGRAPH 93)* (Kajiya J.T., Ed), Anaheim, CA, Vol. 27, pp. 165–174.

Hara N. (1957), Study of the variegated leaves with special reference to those caused by air spaces, *Japanese Journal of Botany*, 16(1):86–101.

Harbinson J. and Woodward I. (1987), The use of microwaves to monitor the freezing and thawing of water in plants, *Journal of Experimental Botany*, 38(183):1325–1335.

Hardacre A.K., Nicholson H.F. and Boyce M.L.P. (1984), A portable photometer for the measurement of chlorophyll in intact leaves, *New Zealand Journal of Experimental Agriculture*, 12(4):357–362.

Hardin J.A., Jones C.L., Maness N.O., Weckler P.R. and Dillwith J.W. (2011), Rapid *in situ* quantification of leaf cuticular wax using FTIR-ATR and DSC, in *Proc. 2011 ASAE Annual Meeting*, Louisville, KY, 7–10 August 2011, ASABE, 18 pages.

Hardin J.A., Jones C.L., Weckler P.R., Maness N.O., Dillwith J.W. and Madden R.D. (2013), Rapid quantification of spinach leaf cuticular wax using Fourier transform infrared attenuated total reflectance spectroscopy, *Transactions of the ASABE*, 56(1):331–339.

Hardwick K. and Baker N.R. (1973), *In vivo* measurement of chlorophyll content of leaves, *New phytologist*, 72(1):51–54.

Harrington D.F. and Clark C. (1989), Reduction in light reflectance of leaves of *Encelia densifolia* (Asteraceae) by trichome wetting, *Madroño*, 36(3):180–186.

Harron J. (2000), *Optical properties of phytoelements in conifers*, MSc Thesis, Department of Earth and Space Science, York University, North York, ON, 208 pages.

Hart W.G. and Myers V.I. (1968), Infrared aerial color photography for detection of populations of brown soft scale in citrus groves, *Journal of Economic Entomology*, 61(3):617–624.

Hart W.G., Gausman H.W. and Rodriguez R.R. (1976), Citrus blackfly (Hemiptera: Aleyrodidae), feeding injury and its influence on the spectral properties of citrus foliage, *Journal of the Rio Grande Valley Horticultural Society*, 30:37–43.

Harting P. (1855), Ueber das Absorptionsvermögen des reinen und des unreinen Chlorophylls für die Strahlen der Sonne, *Annalen der Physik*, 172(12):543–550.

Harvey B. (2007), *Russian Planetary Exploration: History, Development, Legacy, Prospects*, Springer, New York, 354 pages.

Hashimoto Y., Ino T., Kramer P.J., Naylor A.W. and Strain B.R. (1984), Dynamic analysis of water stress of sunflower leaves by means of a thermal image processing system, *Plant Physiology*, 76(1):266–269.

Hatier J.H.B. and Gould K.S. (2007), Black coloration in leaves of *Ophiopogon planiscapus* 'Nigrescens'. Leaf optics, chromaticity, and internal light gradients, *Functional Plant Biology*, 34(2):130–138.

Hattey J.A., Sabbe W.E., Baten G.D. and Blakeney A.B. (1994), Nitrogen and starch analysis of cotton leaves using near infrared reflectance spectroscopy (NIRS), *Communications in Soil Science and Plant Analysis*, 25(9–10):1855–1863.

Hawkins S.A., Park B., Poole G.H., Gottwald T., Windham W.R. and Lawrence K.C. (2010), Detection of Citrus Huanglongbing by Fourier transform infrared-attenuated total reflection spectroscopy, *Applied Spectroscopy*, 64 (1):100–103.

Heath R. (1994), Possible mechanisms for the inhibition of photosynthesis by ozone, *Photosynthesis Research*, 39 (3):439–451.

Heath S.M., Southworth D. and D'Allura J.A. (1997), Localization of nickel in epidermal subsidiary cells of leaves of *Thlaspi montanum* var. Siskiyouense (Brassicaceae) using energy-dispersive X-ray microanalysis, *International Journal of Plant Sciences*, 158(2):184–188.

Hébant C. and Lee D.W. (1984), Ultrastructural basis and developmental control of blue iridescence in *Selaginella* leaves, *American Journal of Botany*, 71(2):216–219.

Hecht H.G. (1976), The interpretation of diffuse reflectance spectra, *Journal of Research of the National Bureau of Standards-A. Physics and Chemistry*, 80A(4):567–583.

Hegedüs R. and Horvath G. (2004a), How and why are uniformly polarization-sensitive retinae subject to polarization-related artefacts? Correction of some errors in the theory of polarization-induced false colours, *Journal of Theoretical Biology*, 230(1):7 SPIE, 7–87.

(2004b), Polarizational colours could help polarization-dependent colour vision systems to discriminate between shiny and matt surfaces, but cannot unambiguously code surface orientation, *Vision Research*, 44(20):2337–2348.

Heil M., Baumann B., Andary C., Linsenmair K.E. and McKey D. (2002), Extraction and quantification of "condensed tannins" as a measure of plant anti-herbivore defence? Revisiting an old problem, *Naturwissenschaften*, 89(11):519–524.

Helhel S., Colak B. and Özen S. (2009), Measurement of dielectric constant of thin leaves by moisture content at 4 mm band, *Progress in Electromagnetics Research Letters*, 7:183–191.

Hellicar A.D., Li L., Greene K., Hislop G., Hanham S., Nikolic N. and Du J. (2007), A 500–700 GHz system for exploring the THz frequency regime, in *Proc. 2nd International Conference on Wireless Broadband and Ultra Wideband Communications*, Sydney, Australia, 27–30 August 2007, IEEE, 6 pages.

Hellmann C., Große-Stoltenberg A., Lauströer V., Oldeland J. and Werner C. (2015), Retrieving nitrogen isotopic signatures from fresh leaf reflectance spectra: disentangling δ15N from biochemical and structural leaf properties, *Frontiers in Plant Science*, 6:307.

Hemenger R.P. (1977), Optical properties of turbid media with specularly reflecting boundaries: applications to biological problems, *Applied Optics*, 16(7):2007–2012.

Henckel J.F. (1760), *Flora saturnisans, ou preuve de l'alliance qui existe entre le règne végétal et le règne minéral*, Chez Jean-Thomas Hérissant, Paris, 284 pages (Traduit de l'Allemand par M. Charas, Apothicaire à Paris).

Henderson N.C., Dick R.J., Grandinetti P.M., Riley G.R. and Roop D.E. (1974), *Camouflage by Artificial Foliage*, Battelle Colombus Laboratories, Colombus, OH, June 1974, 64 pages.

Hendry G.A.F., Houghton J.D. and Brown S.B. (1987), The degradation of chlorophyll – A biological enigma, *New Phytologist*, 107(2):255–302.

Henry W.B., Shaw D.R., Reddy K.R., Bruce L.M. and Tamhankar H.D. (2004), Spectral reflectance curves to distinguish soybean from common Cocklebur (*Xanthium strumarium*) and Sicklepod (*Cassia obtusifolia*) grown with varying soil moisture, *Weed Science*, 52(5):788–796.

Heredia A. (2003), Biophysical and biochemical characteristics of cutin, a plant barrier biopolymer, *Biochimica et Biophysica Acta*, 1620(1–3):1–7.

Hernández E.I., Melendez-Pastor I., Navarro-Pedreño J. and Gómez I. (2014), Spectral indices for the detection of salinity effects in melon plants, *Scientia Agricola (Piracicaba, Braz.)*, 71(4):324–330.

Hernández-Clemente R., Navarro-Cerrillo R.M., Suárez L., Morales F. and Zarco-Tejada P.J. (2011), Assessing structural effects on PRI for stress detection in conifer forests, *Remote Sensing of Environment*, 115(9):2360–2375.

Herrmann I., Berenstein M., Sade A., Karnieli A., Bonfil D.J. and Weintraub P.G. (2012), Spectral monitoring of two-spotted spider mite damage to pepper leaves, *Remote Sensing Letters*, 3(4):277–283.

Hesketh M. and Sánchez-Azofeifa G.A. (2012), The effect of seasonal spectral variation on species classification in the Panamanian tropical forest, *Remote Sensing of Environment*, 118:73–82.

Hesketh M. and Sánchez-Azofeifa G.A. (2013), A review of remote sensing of tropical dry forests, in *Tropical Dry Forests in the Americas* (Quesada M., Ed), CRC Press, pp. 83–100.

Hettinger J.W., de la Peña Mattozzi M., Myers W.R., et al. (2000), Optical coherence microscopy. A technology for rapid, *in vivo*, non-destructive visualization of plants and plant cells, *Plant Physiology*, 123(1):3–15.

Heusinkveld B.G., Berkowicz S.M., Jacobs A.F.G., Hillen W. and Holtslag A.A.M. (2008), A new remote optical wetness sensor and its applications, *Agricultural and Forest Meteorology*, 148(4):580–591.

Hickey L.J. (1973), Classification of the architecture of dicotyledonous leaves, *American Journal of Botany*, 60(1):17–33.

Hildebrandt S. and Tromba A. (1995), *The Parsimonious Universe – Shape and Form in the Natural World*, Springer-Verlag, New York, 330 pages.

Hill J.F. (2012), Early pioneers of photosynthesis research, in *Photosynthesis: Plastid Biology, Energy Conversion and Carbon Assimilation, Advances in Photosynthesis and Respiration, Vol 34* (Eaton-Rye J.J., Tripathy B.C. and Sharkey T.D., Eds), Springer, Dordrecht, pp. 771–800.

Hill R.M., Dissado L.A., Pugh J., Broadhurst M.G., Chiang C.K. and Wahlstrand K.J. (1986), The dielectric response of *Portulacaceae* (Jade) leaves over an extended frequency range, *Journal of Biological Physics*, 14(4):133–135.

Hill R.M., Dissado L.A. and Pathmanathan K. (1987), The low-frequency dielectric properties of leaves, *Journal of Biological Physics*, 15(1):2–16.

Hillnhütter C., Mahlein A.K., Sikora R.A. and Oerke E.C. (2012), Use of imaging spectroscopy to discriminate symptoms caused by *Heterodera schachtii* and *Rhizoctonia solani* on sugar beet, *Precision Agriculture*, 13(1):17–32.

Himmelsbach D.S., Boer H., Akin D.E. and Barton F.E. (1988), Solid-state 13C NMR, FTIR, and NIRS spectroscopic studies of ruminant silage digestion, in *Analytical Applications in Spectroscopy* (Creaser C.S. and Davies A.M.C., Eds), The Royal Society of Chemistry, London, pp. 410–413.

Hirano T., Kiyota M., Seki K. and Aiga I. (1992), Effects of volcanic ashes from Mt Unzen-Fugendake and Mt Sakurajima on leaf temperature and stomatal conductance of cucumber, *Journal of Agricultural Meteorology*, 48(2):139–145.

Hirata M., Ishii R., Kumura A. and Murata Y. (1983), Photoinhibition of photosynthesis in soybean leaves. III: Leaf transmittance change in response to incident light intensity, *Japanese Journal of Crop Science*, 52(4):430–434.

Hiukka R. (1998), A multivariate approach to the analysis of pine needle samples using NIR, *Chemometrics and Intelligent Laboratory Systems*, 44(1–2):395–401.

Hlavka C.A., Peterson D.L., Johnson L.F. and Ganapol B. (1997), Analysis of forest foliage spectra using a multivariate mixture model, *Journal of Near Infrared Spectroscopy*, 5(3):167–173.

Hmimina G., Dufrêne E. and Soudani K. (2014), Relationship between photochemical reflectance index and leaf ecophysiological and biochemical parameters under two different water statuses: towards a rapid and efficient correction method using real-time measurements, *Plant, Cell & Environment*, 37(2):473–487.

Ho Q.T., Berghuijs H.N.C., Watté R., Verboven P., Herremans E., Yin X., et al. (2016), Three-dimensional microscale modelling of CO_2 transport and light propagation in tomato leaves enlightens photosynthesis, *Plant, Cell & Environment*, 39(1):50–61.

Hoch W.A., Zeldin E.L. and McCown B.H. (2001), Physiological significance of anthocyanins during autumnal leaf senescence, *Tree Physiology*, 21(1):1–8.

Hoel B.O. and Solhaug K.A. (1998), Effect of irradiance on chlorophyll estimation with the Minolta SPAD-502 leaf chlorophyll meter, *Annals of Botany*, 82(3):389–392.

Holdridge L.R. (1967), *Life Zone Ecology*, Tropical Science Center, San Jose, Costa Rica, 149 pages.

Holloway P.J. (1982), The chemical constitution of plant cutins, in *The Plant Cuticle* (Cutler D.F., Alvin K.L., and Price C.E., Eds), Academic Press, London, pp. 45–85.

Holmes M.G. and Keiller D.R. (2002), Effects of pubescence and waxes on the reflectance of leaves in the ultraviolet and photosynthetic wavebands: a comparison of a range of species, *Plant, Cell & Environment*, 25(1):85–93.

Holopainen J.K., Semiz G. and Blande J.D. (2009), Life-history strategies affect aphid preference for yellowing leaves, *Biology Letters*, 5(5):603–605.

Hongo C., Kobayashi T. and Arita Y. (1998), The response of water status in tree seedlings through spectral reflectance, *Journal of the Japan Society of Photogrammetry and Remote Sensing*, 37(4):43–50 (in Japanese).

Hoque E. and Hutzler P.J.S. (1992), Spectral blue-shift of red edge monitors damage class of beech trees, *Remote Sensing of Environment*, 39(1):81–84.

Horler D.N.H., Barber J. and Barringer A.R. (1980), Effects of heavy metals on the absorbance and reflectance spectra of plants, *International Journal of Remote Sensing*, 1(2):121–136.

Horler D.H.N. and Barber J. (1981), Principles of remote sensing of plants, in *Plants and the Daylight Spectrum* (Smith H., Ed), Academic Press, London, pp. 43–63.

Horler D.N.H., Barber J., Darch J.P., Ferns D.C. and Barringer A.R. (1983a), Approaches to detection of geochemical stress in vegetation, *Advances in Space Research*, 3(2):175–179.

Horler D.N.H., Dockray M. and Barber J. (1983b), The red edge of plant leaf reflectance, *International Journal of Remote Sensing*, 4(2):273–288.

Horler D.N.H., Dockray M., Barber J. and Barringer A.R. (1983c), Red edge measurements for remotely sensing plant chlorophyll content, *Advances in Space Research*, 3(2):273–277.

Hormaetxe K., Becerril J.M., Fleck I., Pinto M. and Garcia-Plazaola J.I. (2005), Functional role of red (retro)-carotenoids as passive light filters in the leaves of *Buxus sempervirens* L.: increased protection of photosynthetic tissues? *Journal of Experimental Botany*, 56(420):2629–2636.

Hort A.F. (1916), *Theophrastus. Enquiry into plants: Books 1–5*, Harvard University Press, Cambridge, 475 pages.

Hörtensteiner S. and Feller U. (2002), Nitrogen metabolism and remobilization during senescence, *Journal of Experimental Botany*, 53(370):927–937.

Horváth G., Gál J., Labhart T. and Wehner R. (2002), Does reflection polarization by plants influence colour perception in insects? Polarimetric measurements applied to a polarization-sensitive model retina of Papilio butterflies, *Journal of Experimental Biology*, 205:3281–3298.

Hosgood B., Jacquemoud S., Andreoli G., Verdebout J., Pedrini G. and Schmuck G. (1995), *Leaf Optical Properties EXperiment 93 (LOPEX93)*, European Commission, Joint Research Centre, Ispra, Italy, EUR 16095 EN, 20 pages.
Howard J.A. (1966), Spectral energy relations of isobilateral leaves, *Australian Journal of Biological Sciences*, 19 (5):757–766.
(1969), Increased luminance in the direction of reflex reflexion – A recently observed natural phenomenon, *Nature*, 224 (5224):1102–1103.
(1971a), Luminance and luminous intensity indicatrices of isolateral leaves, *Applied Optics*, 10(10):2354–2360.
Howard J.A., Watson R.D. and Hessin T.D. (1971), Spectral reflectance properties of *Pinus ponderosa* in relation to copper content of the soil-Malachite mine, Jefferson County, Colorado, in *Proc. 7th International Symposium on Remote Sensing of Environment*, Ann Arbor, MI, 17–21 May 1971, ERIM, Vol. 1, pp. 285–297.
Hsieh C.Y. (2003), Dependence of backscattering from leaves on dry-matter fraction and permittivity of saline water of leaves, *Journal of Microwaves and Optoelectronics*, 3(3):1–4.
Hu B.B. and Nuss M.C. (1995), Imaging with terahertz waves, *Optics Letters*, 20(16):1716–1718.
Huang J.F. and Apan A. (2006), Detection of Sclerotinia rot disease on celery using hyperspectral data and partial least squares regression, *Journal of Spatial Science*, 51(2):129–142.
Huang J.F. and Blackburn G.A. (2011), Optimizing predictive models for leaf chlorophyll concentration based on continuous wavelet analysis of hyperspectral data, *International Journal of Remote Sensing*, 32(24):9375–9396.
Hubert L., Flura D., Saint-Jean S., Chelle M., Beysens D. and Jacquemoud S. (2009), *L'eau liquide à la surface des feuilles: développement d'une méthode de mesure non intrusive de la condensation en vue d'applications à des processus biophysiques ou chimiques superficiels*, Projet Innovant 2007–2008, INRA, Département Environnement et Agronomie, 18 pages.
Hueni A., Nieke J., Schopfer J., Kneubühler M. and Itten K.I. (2009), The spectral database SPECCHIO for improved long-term usability and data sharing, *Computers & Geosciences*, 35(3):557–565.
Hueni A., Malthus T., Kneubuehler M. and Schaepman M. (2011), Data exchange between distributed spectral databases, *Computers & Geosciences*, 37:861–873.
Hui Z. and Jianchun Z. (2007), Near-infrared green camouflage of PET fabrics using disperse dyes, *Sen'i Gakkaishi*, 63 (10):223–229.
Huibers P.D.T. (1997), Models for the wavelength dependence of the index of refraction of water, *Applied Optics*, 36 (16):3785–3787.
Hulbary R.L. (1944), The influence of air spaces on the three-dimensional shapes of cells in *Elodea* stems, and a comparison with pith cells of *Ailanthus, American Journal of Botany*, 31(9):561–580.
Hunt E.R., Rock B.N. and Nobel P.S. (1987), Measurement of leaf relative water content by infrared reflectance, *Remote Sensing of Environment*, 22(3):429–435.
Hunt E.R. and Rock B.N. (1989), Detection of changes in leaf water content using near and middle-infrared reflectances, *Remote Sensing of Environment*, 30(1):43–54.
Hunt E.R. and Daughtry C.S.T. (2014), Chlorophyll meter calibrations for chlorophyll content using measured and simulated leaf transmittances, *Agronomy Journal*, 106(3):931–939.
Idso S.B., Baker D.G. and Gates D.M. (1966), The energy environment of plants, *Advances in Agronomy*, 18:171–218.
Idso S.B., Jackson R.D., Ehrler W.L. and Mitchell S.T. (1969), A method for determination of infrared emittance of leaves, *Ecology*, 50(5):899–902.
Inada K. (1963), Studies on a method for determining the deepness of green color and chlorophyll content of intact crop leaves and its practical applications. 1: Principal for estimating the deepness of green color and chlorophyll content of whole leaves, *Japanese Journal of Crop Science*, 32(2):157–162.
(1965), Studies on a method for determining the deepness of green color and chlorophyll content of intact crop leaves and its practical applications. 2: Photoelectric characters of chlorophyllo-meter and correlation between the reading and chlorophyll content in leaves, *Japanese Journal of Crop Science*, 33 (4):301–308 (in Japanese).
(1976), Action spectra for photosynthesis in higher plants, *Plant & Cell Physiology*, 17(2):355–365.
(1980), Spectral absorption property of pigments in living leaves and its contribution to photosynthesis, *Japanese Journal of Crop Science*, 49(2):286–294.
(1985), Spectral ratio of reflectance for estimating chlorophyll content of leaf, *Japanese Journal of Crop Science*, 54 (3):261–265.
Ingen-Housz J. (1779), *Experiments upon Vegetables, Discovering Their Great Power of Purifying the Common Air in the Sunshine and of Injuring It in the Shade and at Night; to Which Is Joined a New Method of Examining the Accurate Degree of Salubrity of the Atmosphere*, Elmsley and Payne, London, 402 pages.
Inoue Y. and Shibata K. (1973), Light-induced chloroplasts rearrangements and their action spectra as measured by absorption spectrophotometry, *Planta*, 114(4):341–358.
Inoue Y., Kimball B.A., Jackson R.D., Pinter P.J. and Reginato R.J. (1990), Remote estimation of leaf transpiration rate and stomatal resistance based on infrared thermometry, *Agricultural and Forest Meteorology*, 51(1):21–33.
Inoue Y., Morinaga S. and Shibayama M. (1993), Non-destructive estimation of water status of intact crop leaves based on spectral reflectance measurements, *Japanese Journal of Crop Science*, 62(3):462–469.
Inoue Y. and Peñuelas J. (2006), Relationship between light use efficiency and photochemical reflectance index in soybean leaves as affected by soil water content, *International Journal of Remote Sensing*, 27(22):5109–5114.

Intaravanne Y. and Sumriddetchkajorn S. (2015), Android-based rice leaf color analyzer for estimating the needed amount of nitrogen fertilizer, *Computers and Electronics in Agriculture*, 116:228–233.

Irvine W.M. and Pollack J.B. (1968), Infrared optical properties of water and ice spheres, *Icarus*, 8(1–3):324–360.

Ismail R., Mutanga O. and Ahmed F. (2008), Discriminating *Sirex noctilio* attack in pine forest plantations in South Africa using high spectral resolution data, in *Hyperspectral Remote Sensing of Tropical and Sub-Tropical Forests* (Kalacska M. and Sánchez-Azofeifa G.A., Eds), CRC Press, pp. 161–175.

Ivanova L.A., Petrov M.S. and Kadushnikov R.M. (2006), Determination of mesophyll diffusion resistance in *Chamaerion angustifolium* by the method of three-dimensional reconstruction of the leaf cell packing, *Russian Journal of Plant Physiology*, 53(3):316–324.

Jackson R.D. (1982), Canopy temperature and crop water stress, *Advances in Irrigation*, 1:43–85.

(1986), Remote sensing of biotic and abiotic plant stress, *Annual Review of Phytopathology*, 24:265–287.

Jacquemoud S. and Baret F. (1990), PROSPECT: a model of leaf optical properties spectra, *Remote Sensing of Environment*, 34(2):75–91.

Jacquemoud S. (1992), *Utilisation de la haute résolution spectrale pour l'étude des couverts végétaux : développement d'un modèle de réflectance spectrale*, Thèse de Doctorat de Physique, Spécialité Méthodes physiques en télédétection, Université Paris Diderot, 92 pages.

(1993), Inversion of the PROSPECT+SAIL canopy reflectance model from AVIRIS equivalent spectra: theoretical study, *Remote Sensing of Environment*, 44(2–3):281–292.

(2004), Leaf optical properties, in *Reflection Properties of Vegetation and Soil – with a BRDF Data Base* (von Schönermark M., Geiger B. and Röser H.P., Eds), Wissenschaft & Technik Verlag, Berlin, Germany, pp. 56–70.

Jacquemoud S., Baret F., Andrieu B., Danson F.M. and Jaggard K. (1995a), Extraction of vegetation biophysical parameters by inversion of the PROSPECT + SAIL model on sugar beet canopy reflectance data – Application to TM and AVIRIS sensors, *Remote Sensing of Environment*, 52(3):163–172.

Jacquemoud S., Verdebout J., Schmuck G., Andreoli G. and Hosgood B. (1995b), Investigation of leaf biochemistry by statistics, *Remote Sensing of Environment*, 54(3):180–188.

Jacquemoud S., Ustin S.L., Verdebout J., Schmuck G., Andreoli G. and Hosgood B. (1996), Estimating leaf biochemistry using the PROSPECT leaf optical properties model, *Remote Sensing of Environment*, 56(3):194–202.

Jacquemoud S., Frangi J.P., Govaerts Y. and Ustin S.L. (1997), Three-dimensional representation of leaf anatomy – Application to the study of photon transport, in *Proc. 7th International Symposium on Physical Measurements and Signatures in Remote Sensing* (Guyot G. and Phulpin T., Eds), Courchevel, France, 7–11 April 1997, Balkema, pp. 295–302.

Jacquemoud S., Bacour C., Poilvé H. and Frangi J.P. (2000), Comparison of four radiative transfer models to simulate plant canopies reflectance – Direct and inverse mode, *Remote Sensing of Environment*, 74(3):471–481.

Jacquemoud S. (2004), Leaf optical properties, in *Reflection Properties of Vegetation and Soil – with a BRDF Data Base* (von Schönermark M., Geiger B. and Röser H.P., Eds), Wissenschaft & Technik Verlag, Berlin, Germany, pp. 56–70.

Jacquemoud S., Féret J.B. and Ustin S.L. (2009a), Compréhension et modélisation de la couleur des feuilles, in *Proc. Ecole thématique interdisciplinaire du CNRS – Couleur, question d'échelle : l'espace*, Roussillon en Provence, France, 23–27 mars 2009, pp. 181–190.

Jacquemoud S., Verhoef W., Baret F., Bacour C., Zarco-Tejada P.J., Asner G.P., et al. (2009b), PROSPECT + SAIL Models: a review of use for vegetation characterization, *Remote Sensing of Environment*, 113(1):S56–S66.

Jay S., Bendoula R., Hadoux X. and Gorretta N. (2015), Mapping of foliar content using radiative transfer modeling and VIS-NIR hyperspectral close-range remote sensing, in *Proc. ISPRS Geospatial Week 2015*, La Grande Motte, France, 28 September–2 October 2015, ISPRS, 6 pages.

Jay S., Bendoula R., Hadoux X., Féret J.B. and Gorretta N. (2016), A physically-based model for retrieving foliar biochemistry and leaf orientation using close-range imaging spectroscopy, *Remote Sensing of Environment*, 177:220–236.

Jeje A. and Zimmermann M. (1983), The anisotropy of the mesophyll and CO_2 capture sites in *Vicia faba* L. leaves at low light intensities, *Journal of Experimental Botany*, 34(12):1676–1694.

Jenkins M.W., Krofcheck D.J., Teasdale R., Houpis J. and Pushnik J. (2012), Exploring the edge of a natural disaster, *Open Journal of Ecology*, 2(4):222–232.

Jensen H.W., Marschner S.R., Levoy M. and Hanrahan P. (2001), A practical model for subsurface light transport, in *Proc. 28th Annual Conference on Computer Graphics and Interactive Techniques*, Los Angeles, CA, 12–17 August 2001, ACM, pp. 511–518.

Jensen H.W. and Buhler J. (2002), A rapid hierarchical rendering technique for translucent materials, *ACM Transactions on Graphics*, 21(3):576–581.

Jernshøj K.D. and Hassing S. (2009), Analysis of reflectance and transmittance measurements on absorbing and scattering small samples using a modified integrating sphere setup, *Applied Spectroscopy*, 63(8):879–888.

Ji L. and Peters A.J. (2007), Performance evaluation of spectral vegetation indices using a statistical sensitivity function, *Remote Sensing of Environment*, 106(1):59–65.

Jiang H., Ying Y. and Bao Y. (2005a), Study on the water content measurement of tomatoes by near infrared technique, in *Proc. Optical Sensors and Sensing Systems for Natural Resources and Food Safety and Quality* (Chen Y.R., Meyer G.E. and Tu S.I., Eds), Boston, MA, 23–24 October 2005, SPIE, Vol. 5996, 599612.

Jiang H., Bao Y. and Ying Y. (2005b), Measure the chlorophyll content in leaves by near infrared analysis, in *Proc. Optical Sensors and Sensing Systems for Natural Resources and Food Safety and Quality* (Chen Y.R., Meyer G.E. and Tu S.I., Eds), Boston, MA, 23–24 October 2005, SPIE, Vol. 5996, 599617.

Jiang H. and Lu Y.Y.H. (2006), Near-infrared diffuse reflection systems for chlorophyll content of tomato leaves measurement, in Proc. Optics for Natural Resources, Agriculture, and Foods (Chen Y.R., Meyer G.E. and Tu S.I., Eds), Boston, MA, SPIE, Vol. 6381, 638112.

Jin J., Jiang H., Zhang X., Wang Y., Cheng M. and Song X. (2013), Using multivariate analysis to detect the hyperspectral response of Chinese fir to acid stress, *International Journal of Remote Sensing*, 34(11):3775–3786.

Joffre R., Gillon D., Dardenne P. and Agneessens R. (1992), The use of near-infrared reflectance spectroscopy in litter decomposition studies, *Annales des Sciences Forestières*, 49(5):481–488.

Johnson H.B. (1975), Plant pubescence: an ecological perspective, *The Botanical Review*, 41(3):233–258.

Johnson L.F. (2001), Nitrogen influence on fresh-leaf NIR spectra, *Remote Sensing of Environment*, 78(3):314–320.

Jolivet P. (2010), Vert, jaune, blanc ou rouge . . . Pourquoi les feuilles changent-elles de couleur? Incidence sur les insectes, *L' Entomologiste*, 66(3):177–188.

Jones C.D., Jones J.B., and Lee W.S. (2010), Diagnosis of bacterial spot of tomato using spectral signatures, *Computers and Electronics in Agriculture*, 74(2):329–335.

Jones H.G. (1999), Use of thermography for quantitative studies of spatial and temporal variation of stomatal conductance over leaf surfaces, *Plant, Cell & Environment*, 22(9):1043–1055.

(2010), Can water droplets on leaves cause leaf scorch? *New Phytologist*, 185(4):865–867.

Jones H.G. and Vaughan R.A. (2010), *Remote Sensing of Vegetation: Principles, Techniques, and Applications*, Oxford University Press, Oxford, 353 pages.

Jordan G.J., Dillon R.A. and Weston P.H. (2005), Solar radiation as a factor in the evolution of scleromorphic leaf anatomy in *Proteaceae*, *American Journal of Botany*, 92(5):789–796.

Jördens C., Scheller M., Breitenstein B., Selmar D. and Koch M. (2009), Evaluation of leaf water status by means of permittivity at terahertz frequencies, *Journal of Biological Physics*, 35(3):255–264.

Jørgensen R.N., Christensen L.K. and Bro R. (2007), Spectral reflectance at sub-leaf scale including the spatial distribution discriminating NPK stress characteristics in barley using multiway partial least squares regression, *International Journal of Remote Sensing*, 28(5):943–962.

Jovanic B.R. and Jovanic S.B. (2001), The effect of high concentrations of negative ions in the air on the chlorophyll contents in plant leaves, *Water, Air, and Soil Pollution*, 129(1–4):259–265.

Ju C.H., Tian Y.C., Yao X., Cao W.X., Zhu Y. and Hannaway D. (2010), Estimating leaf chlorophyll content using red edge parameters, *Pedosphere*, 20(5):633–644.

Judkins W.P. and Wander I.W. (1950), Correlation between leaf color, leaf nitrogen content, and growth of apple, peach, and grape plants, *Plant Physiology*, 25(1):78–85.

Juniper B.E. and Jeffree C.E. (1983), *Plant Surfaces*, Hodder Arnold, London, 93 pages.

Jusoff K., Yusoff M.M. and Mohd Ali N.H. (2010), Spectral signatures of leaf fall diseases in *Hevea brasiliensis* using a handheld spectroradiometer, *Modern Applied Science*, 4(2):78–84.

Justice C., Belward A., Morisette J., Lewis P., Privette J. and Baret F. (2000), Developments in the "validation" of satellite sensor products for the study of the land surface, *International Journal of Remote Sensing*, 21(17):3383–3390.

Kaasalainen S. and Rautiainen M. (2007), Backscattering measurements from individual Scots pine needles, *Applied Optics*, 46(22):4916–4922.

Kaiser J., Reale L., Ritucci A., Tomassetti G., Poma A., Spanò L., et al. (2005), Mapping of the metal intake in plants by large-field X-ray microradiography and preliminary feasibility studies in microtomography, *The European Physical Journal D*, 32(1):113–118.

Kaiser J., Samek O., Reale L., Liška M., Malina R., Ritucci A., et al. (2007), Monitoring of the heavy-metal hyperaccumulation in vegetal tissues by X-ray radiography and by femto-second laser induced breakdown spectroscopy, *Microscopy Research and Technique*, 70(2):147–153.

Kakani V.G., Reddy K.R., Zhao D. and Gao W. (2004), Senescence and hyperspectral reflectance of cotton leaves exposed to ultraviolet-B radiation and carbon dioxide, *Physiologia Plantarum*, 121(2):250–257.

Kakani V.G., Reddy K.R. and Zhao D. (2007), Deriving a simple spectral reflectance ratio to determine cotton leaf water potential, *Journal of New Seeds*, 8(3):11–27.

Kallel A. (2018), Leaf polarized BRDF simulation based on Monte Carlo 3-D vector RT modeling, *Journal of Quantitative Spectroscopy and Radiative Transfer*, 221: 202–224.

Kalacska M., Bohlman S., Sánchez-Azofeifa G.A., Castro-Esau K. and Caellic T. (2007), Hyperspectral discrimination of tropical dry forest lianas and trees: comparative data reduction approaches at the leaf and canopy levels, *Remote Sensing of Environment*, 109(4):406–415.

Karabourniotis G., Papadopoulos K., Papamarkou M. and Manetas Y. (1992), Ultraviolet-B radiation absorbing capacity of leaf hairs, *Physiologia Plantarum*, 86(3):414–418.

Karabourniotis G., Kyparissis A. and Manetas Y. (1993), Leaf hairs of *Olea europaea* L. protect underlying tissues against ultraviolet-B radiation damage, *Environmental and Experimental Botany*, 33:341–345.

Karabourniotis G., Papastergiou N., Kabanopoulou E. and Fasseas C. (1994), Foliar sclereids of *Olea europaea* may function as optical fibres, *Canadian Journal of Botany*, 72(3):330–336.

Karabourniotis G., Kotsabassidis D. and Manetas Y. (1995), Trichome density and its protective potential against ultraviolet-B radiation damage during leaf development, *Canadian Journal of Botany*, 73(3):376–383.

Karabourniotis G. (1998), Light-guiding function of foliar sclereids in the evergreen sclerophyll *Phillyrea latifolia*: a quantitative approach, *Journal of Experimental Botany*, 49(321):739–746.

Karabourniotis G. and Bornman J.F. (1999), Penetration of UV-A, UV-B and blue light through the leaf trichome layers of two xeromorphic plants, olive and oak, measured by optical fibre microprobes, *Physiologia Plantarum*, 105 (4):655–661.

Karabourniotis G., Bornman J.F. and Liakoura V. (1999a), Different leaf surface characteristics of three grape cultivars affect leaf optical properties as measured with fibre optics: possible implication in stress tolerance, *Australian Journal of Plant Physiology*, 26(1):47–53.

Karabourniotis G., Bornman J.F. and Nikolopoulos D. (2000), A possible optical role of the bundle sheath extensions of the heterobaric leaves of *Vitis vinifera* and *Quercus coccifera*, *Plant, Cell & Environment*, 23(4):423–430.

Karageorgou P., Buschmann C. and Manetas Y. (2008), Red leaf color as a warning signal against insect herbivory: honest or mimetic? *Flora*, 203(8):648–652.

Karam M.A., Fung A.K., Blanchard A.J. and Shen G.X. (1988), The leaf-shape effect on electromagnetic scattering from vegetated media, in *Proc. IEEE Geoscience and Remote Sensing Symposium (IGARSS'88)*, Edinburgh, Scotland, 12–16 September 1988 IEEE, Vol. 2, pp. 677–680.

Karam M.A. and Fung A.K. (1989), Leaf-shape effects in electromagnetic wave scattering from vegetation, *IEEE Transactions on Geoscience and Remote Sensing*, 27(6):687–697.

Karvaly B. (1970), Investigations on the connections between the bulk absorption and diffuse reflectance spectra of powdered solids, *Acta Physica Academiae Scientiarum Hungaricae*, 28(4):381–399.

Katterfeld G.N. and Suslov A.K. (1969), *Bibliography on problems of Astrobiology*, NASA, 1 April 1969, 56 pages (Translation of: "Literatura po Problemam Astrobiologii" Subcommission on Astrobiology and Planetary Physics, Commission on Planetology, Geographic Society of the USSR, Leningrad, 1967, 76 pp.).

Kattge J. et al. (2011), TRY – a global database of plant traits, *Global Change Biology*, 17(9):2905–2935.

Kattge J., Díaz S., Lavorel S., Prentice I.C., Leadley P., Boenisch G., et al. and The TRY Consortium (2015), TRY 3.0 – a substantial upgrade of the global database of plant traits: more data, more species, largely open access, *Geophysical Research Abstracts*, 17, EGU2015-15745–2.

Kaufmann W.F. and Hartmann K.M. (1988), Internal brightness of disk-shaped samples, *Journal of Photochemistry and Photobiology B: Biology*, 1(3):337–360.

Kautsky H. and Hirsch A. (1931), Neue Versuche zur Kohlensäureassimilation, *Die Naturwissenschaften*, 19(48):964–964.

Kawashima S. and Nakatani M. (1998), An algorithm for estimating chlorophyll content in leaves using a video camera, *Annals of Botany*, 81(1):49–54.

Kazarinova-Fukshansky N., Seyfried M. and Schafer E. (1985), Distortion of action spectra in photomorphogenesis by light gradients within the plant tissue, *Photochemistry and Photobiology*, 41(6):689–702.

Keegan H.J., Schleter J.C., Hall W.A. and Haas G.M. (1956), Spectrophotometric and colorimetric study of diseased and rust resisting cereal crops, National Bureau of Standards, July 1956, NBS Report 4591, 128 pages. https://archive.org/details/spectrophotometr4591keeg/page/n145

Keith L., Roy A., Morrison R. and Schade P. (2011), Leonardo da Vinci's "Virgin of the Rocks": treatment, technique and display, *National Gallery Technical Bulletin*, 32:32–56.

Kelber A. (1999a), Ovipositing butterflies use a red receptor to see green, *Journal of Experimental Biology*, 202 (19):2619–2630.

(1999b), Why "false" colours are seen by butterflies, *Nature*, 402(6759):251–251.

Kempeneers P., De Backer S., Debruyn W., Coppin P. and Scheunders P. (2005), Generic wavelet-based hyperspectral classification applied to vegetation stress detection, *IEEE Transactions on Geoscience and Remote Sensing*, 43 (3):610–614.

Kendal D., Hauser C.E., Garrard G.E., Jellinek S., Giljohann K.M. and Moore J.L. (2013), Quantifying plant colour and colour difference as perceived by humans using digital images, *PLOS One*, 8(8):e72296.

Kerstiens G., Ed (1996), *Plant Cuticles: An Integrated Functional Approach*, BIOS Scientific Publishers Limited, Oxford, UK, 337 pages.

Kestner J.M., Leidecker H.W., Irons J.R., Smith J.A., Brakke T.W. and Horning N.A. (1988), Goniometric observations of light scattered from soils and leaves, *Journal of Wave-Material Interaction*, 3(2):189–198.

Khamis S., Lamaze T., Lemoine Y. and Foyer C. (1990), Adaptation of the photosynthetic apparatus in maize leaves as a result of nitrogen limitation: relationships between electron transport and carbon assimilation, *Plant Physiology*, 94 (3):1436–1443.

Khanchi A.R., Mahani M.K., Hajihosseini M., Maragheh M.G., Chaloosi M. and Bani F. (2007), Simultaneous spectrophotometric determination of caffeine and theobromine in Iranian tea by artificial neural networks and its comparison with PLS, *Food Chemistry*, 103(3):1062–1068.

Khanna S., Santos M.J., Ustin S.L. and Haverkamp P.J. (2011), An integrated approach to a biophysiologically based classification of floating aquatic macrophytes, *International Journal of Remote Sensing*, 32(4):1067–1094.

Kharuk V.I. and Yegorov V.V. (1990), Polarimetric indication of plant stress, *Remote Sensing of Environment*, 33(1):35–40.

Khavanin Zadeh A.R., Veroustraete F., Wuyts K., Kardel F. and Samson R. (2012), Dorsi-ventral leaf reflectance properties of *Carpinus betulus* L.: an indicator of urban habitat quality, *Environmental Pollution*, 162:332–337.

Khavanin Zadeh A.R., Veroustraete F., Buytaert J.A.N., Dirckx J. and Samson R. (2013), Assessing urban habitat quality using spectral characteristics of *Tilia* leaves, *Environmental Pollution*, 178:7–14.

Khavanin Zadeh A.R., Veroustraete F., Buytaert J.A.N. and Samson R. (2014), Leaf injury symptoms of *Tilia* sp. as an indicator of urban habitat quality, *Ecological Indicators*, 41:58–64.

Kiang N.Y., Siefert J., Govindjee and Blankenship R.E. (2007a), Spectral signatures of photosynthesis. I: review of Earth organisms, *Astrobiology*, 7(1):222–251.

Kiang N.Y., Segura A., Tinetti G., Govindjee G., Blankenship R.E., Cohen M., et al. (2007b), Spectral signatures of photosynthesis. II: coevolution with other stars and the atmosphere on extrasolar worlds, *Astrobiology*, 7 (1):252–274.

Kiang N.Y. (2008), The color of plants on other worlds, *Scientific American*, April 2008, pp. 48–55.

Kim D.M., Zhang H., Zhou H., Du T., Wu Q., Mockler T.C. et al. (2015), Highly sensitive image-derived indices of water-stressed plants using hyperspectral imaging in SWIR and histogram analysis, *Scientific Reports*, 5:15919.

Kim E., Lin Y., Kerney R., Blumenberg L. and Bishop C. (2014), Phylogenetic analysis of algal symbionts associated with four North American amphibian egg masses, *PLOS One*, 9(11):e108915.

Kim M.S., McMurtrey J.E., Mulchi C.L., Daughtry C.S.T., Chappelle E.W. and Chen Y.R. (2001), Steady-state multispectral fluorescence imaging system for plant leaves, *Applied Optics*, 40(1):157–166.

Kira O., Linker R. and Gitelson A. (2015), Non-destructive estimation of foliar chlorophyll and carotenoid contents: focus on informative spectral bands, *International Journal of Applied Earth Observation and Geoinformation*, 38:251–260.

Klančnik K., Mlinar M. and Gaberščik A. (2012), Heterophylly results in a variety of "spectral signatures" in aquatic plant species, *Aquatic Botany*, 98(1):20–26.

Klančnik K., Vogel-Mikuš K. and Gaberščik A. (2014a), Silicified structures affect leaf optical properties in grasses and sedge, *Journal of Photochemistry and Photobiology B: Biology*, 130:1–10.

Klančnik K., Pančić M. and Gaberščik A. (2014b), Leaf optical properties in amphibious plant species are affected by multiple leaf traits, Hydrobiologia, 737(1):121–130.

Klančnik K., Zelnik I., Gnezda P. and Gaberščik A. (2015), Do reflectance spectra of different plant stands in wetland indicate species properties? in *The Role of Natural and Constructed Wetlands in Nutrient Cycling and Retention on the Landscape* (Vymazal J., Ed), Springer, pp. 73–86.

Kleyer M., Bekker R.M., Knevel I.C., Bakker J.P., Thompson K. et al. (2008), The LEDA traitbase: a database of life-history traits of the Northwest European flora, *Journal of Ecology*, 96(6):1266–1274.

Knapp A.K., Vogelmann T.C., McClean T.M. and Smith W.K. (1988), Light and chlorophyll gradients within *Cucurbita* cotyledons, *Plant, Cell & Environment*, 11(4):257–263.

Knapp A.K. and Carter G.A. (1998), Variability in leaf optical properties among 26 species from a broad range of habitats, *American Journal of Botany*, 85(7):940–946.

Knipling E.B. (1970), Physical and physiological basis for the reflectance of visible and near-infrared radiation from vegetation, *Remote Sensing of Environment*, 1(3):155–159.

Knoerr K.R. and Gay L.W. (1965), Tree leaf energy balance, *Ecology*, 46(1–2):17–24.

Knudson L.L., Tibbitts T.W. and Edwards G.E. (1977), Measurement of ozone injury by determination of leaf chlorophyll concentration, *Plant Physiology*, 60(4):606–608.

Knyazikhin Y., Schull M.A., Yang Y., Stenberg P., Mõttus M., Rautiainen M., et al. (2012), Hyperspectral remote sensing of foliar nitrogen content, *Proceedings of the National Academy of Sciences*, 110(3):E185–E192.

Kobayashi S., Rokugawa S., Yamagata Y. and Oguma H. (2002), A study on predicting biochemical contents of larch needles – A study on LIBERTY model, *Journal of the Remote Sensing Society of Japan*, 22 (5):571–587 (in Japanese).

Kobayashi T., Kanda E., Naito S., Nakajima T., Arakawa I., Nemoto K., et al. (2003), Ratio of rice reflectance for estimating leaf blast severity with a multispectral radiometer, *Journal of General Plant Pathology*, 69(1):17–22.

Koch, K. and W. Barthlott (2009), Superhydrophobic and superhydrophilic plant surfaces: an inspiration for biomimetic materials, *Philosophical Transactions of the Royal Society A: Mathematical, Physical and Engineering Sciences*, 367 (1893):1487–1509.

Kochubey S.M., Kobets N.I. and Shadchina T.M. (1987), The shape of reflectance spectra of leaves as informative basis of the remote sensing of crop states, *Physiology and Biochemistry of Cultivated Plants*, 19(6):539–545 (in Russian).

(1988), Quantitative analysis of the spectral distribution curve shape for reflection of plant leaves as a method for testing their state, *Physiology and Biochemistry of Cultivated Plants*, 20(6):535–540 (in Russian).

Kochubey S.M. and Kazantsev T.A. (2007), Changes in the first derivatives of leaf reflectance spectra of various plants induced by variations of chlorophyll content, *Journal of Plant Physiology*, 164(12):1648–1655.

(2012), Derivative vegetation indices as a new approach in remote sensing of vegetation, *Frontiers of Earth Science*, 6 (2):188–195.

Koetz B., Schaepman M., Morsdorf F., Bowyer P., Itten K. and Allgöwer B. (2004), Radiative transfer modeling within heterogeneous canopy for estimation of forest fire fuel properties, *Remote Sensing of Environment*, 92(3):332–344.

Koizumi M., Takahashi K., Mineuchi K., Nakamura T. and Kano H. (1998), Light gradients and the transverse distribution of chlorophyll fluorescence in mangrove and *Camellia* leaves, *Annals of Botany*, 81(4):527–533.

Kokalj M., Kolar J., Trafela T. and Kreft S. (2011), Differences among *Epilobium* and *Hypericum* species revealed by four IR spectroscopy modes: transmission, KBr tablet, diffuse reflectance and ATR, *Phytochemical Analysis*, 22(6):541–546.

Kokaly, R.F. and Clark, R.N. (1999), Spectroscopic determination of leaf biochemistry using band-depth analysis of absorption features and stepwise multiple linear regression, *Remote Sensing of Environment*, 67(3):267–287.

Kokaly R.F. (2001), Investigating a physical basis for spectroscopic estimates of leaf nitrogen concentration, *Remote Sensing of Environment*, 75(2):153–161.

Kokaly R.F. and Skidmore A.K. (2015), Plant phenolics and absorption features in vegetation reflectance spectra near 1.66 μm, *International Journal of Applied Earth Observation and Geoinformation*, 43:55–83.

Kolattukudy P.E. (1976), *Chemistry and Biochemistry of Natural Waxes*, Elsevier, Amsterdam, 459 pages.

Kolb C.A. and Pfündel E.E. (2005), Origins of non-linear and dissimilar relationships between epidermal UV absorbance and UV absorbance of extracted phenolics in leaves of grapevine and barley, *Plant, Cell & Environment*, 28(5):580–590.

Kondo A., Kaikawa J., Funaguma T. and Ueno O. (2004), Clumping and dispersal of chloroplasts in succulent plants, *Planta*, 219(3):500–506.

Korn R.W. (1974), The three-dimensional shape of plant cells and its relationship to pattern of tissue growth, *New Phytologist*, 73(5):927–935.

Körner C. (2003), *Alpine Plant Life: Functional Plant Ecology of High Mountain Ecosystems*, Springer Verlag, Berlin, 344 pages.

Kortüm G. (1969), *Reflectance Spectroscopy*, Springer Verlag, Berlin, 365 pages.

Kováč D., Navrátil M., Malenovský Z., Štroch M., Špunda V. and Urban O. (2012), Reflectance continuum removal spectral index tracking the xanthophyll cycle photoprotective reactions in Norway spruce needles, *Functional Plant Biology*, 39 (12):987–998.

Kováč D., Malenovský Z., Urban O., Špunda V., Kalina J., Ač A., et al. (2013), Response of green reflectance continuum removal index to the xanthophyll de-epoxidation cycle in Norway spruce needles, *Journal of Experimental Botany*, 64 (7):1817–1827.

Kozlova K.I. (1957), Experiment of spectrophotometric investigation of the reflection by plants of close ultraviolet rays, *Trudy sektora astrobotaniki*, 5: 110–117 (in Russian).

Kraft M., Weigel H.J., Mejer G.J. and Brandes F. (1996), Reflectance measurements of leaves for detecting visible and non-visible ozone damage to crops, *Journal of Plant Physiology*, 148(1–2):148–154.

Krauss P., Markstädter C. and Riederer M. (1997), Attenuation of UV radiation by plant cuticles from woody species, *Plant, Cell & Environment*, 20(8):1079–1085.

Krekov G.M., Krekova M.M., Lisenko A.A. and Sukhanov A.Y. (2009), Radiative characteristics of plant leaf, *Atmospheric and Oceanic Optics*, 22(2):241–256 (original text published in Russian in *Optika Atmosfery i Okeana*, 22(4):397–410).

Krezhova D.D., Yanev T.K., Alexieva V.S. and Ivanov C.V. (2005), Early detection of changes in leaf reflectance of pea plants (*Pisum sativum* L.) under herbicide action in *Proc. 2nd International Conference on Recent Advances in Space Technologies (RAST'05)*, Istanbul, Turkey, 9–11 June 2005, IEEE, pp. 636–641.

Krezhova D.D. and Yanev T.K. (2007), Use of a remote sensing method to estimate the influence of anthropogenic factors on the spectral reflectance of plant species, in *Proc. 6th International Conference of the Balkan Physical Union* (Cetin S. A. and Hikmet I., Eds), Istanbul, Turkey, 22–26 August 2006, American Institute of Physics, Vol. 899, pp. 738–738.

Krezhova D.D., Alexieva V.S., Yanev T.K. and Ivanov C.V. (2007a), Remote sensing study of the influence of herbicides Fluridone and Acifluorfen on the spectral reflectance of pea plant leaves (*Pisum sativum* L.), in *Proc. 3rd International Conference on Recent Advances in Space Technologies,* Istanbul, Turkey, 14–16 June 2007, IEEE, pp. 326–330.

Krezhova D.D. and Yanev T.K., Ivanov C.V. and Alexieva V.S. (2007b), Remote sensing of the effect of the herbicide glyphosate on the leaf spectral reflectance of pea plants (*Pisum sativum* L.), in *New Developments and Challenges in Remote Sensing* (Bochenek Z., Ed.), Millpress, pp. 45–52.

Krezhova D.D., Iliev I.T., Yanev T.K. and Kirova E.B. (2009a), Assessment of the effect of salinity on the early growth stage of soybean plants (*Glycine max* L.), in *Proc. 4th International Conference on Recent Advances in Space Technologies*, Istanbul, Turkey, 11–13 June 2009, IEEE, pp. 397–402.

Krezhova D.D., Kirova E.B., Yanev T.K. and Iliev I.T. (2009b), Effects of salinity on leaf spectral reflectance and biochemical parameters of nitrogen fixing soybean plants (*Glycine max* L.), in *Proc. 7th International Conference of the Balkan Physical Union* (Angelopoulos A. and Fildisis T., Eds), Alexandroupolis, Greece, 9–13 September 2009, American Institute of Physics, Vol. 1203, pp. 694–696.

Krezhova D.D., Hristova D.P. and Yanev T.K. (2010), Spectral remote sensing of tomato plants (*Lycopersicon esculentum* L.) infected with tomato mosaic virus (ToMV), in *Proc. Remote Sensing for Science, Education, and Natural and Cultural Heritage* (Reuter R., Ed), Paris, France, 31 May 2010–3 June 2010, pp. 715–722.

Krezhova D.D. (2011), Spectral remote sensing of the responses of soybean plants to environmental stresses, in *Soybean – Genetics and Novel Techniques for Yield Enhancement* (Krezhova D.D., Ed), InTech, pp. 216–256.

Krinner G., Viovy N., de Noblet-Ducoudré N., Ogée J., Polcher J., Friedlingstein P., et al. (2005), A dynamic global vegetation model for studies of the coupled atmosphere-biosphere system, *Global Biogeochemical Cycles*, 19: GB1015.

Krishnan S. and Nordine P.C. (1994), Mueller-matrix ellipsometry using the division-of-amplitude photopolarimeter: a study of depolarization effects, *Applied Optics*, 33(19):4184–4192.

Kruckeberg A.R. (1951), Intraspecific variability in the response of certain native plant species to serpentine soil, *American Journal of Botany*, 38(6):408–419.

Krulik G.A. (1980), Light transmission in window-leaved plants, *Canadian Journal of Botany*, 58(14):1591–1600.

Kubelka P. and Munk F. (1931), Ein beitrag zur optik der farbanstriche, *Zeitschrift fur Technische Physik*, 12:593–601.

Kubinova L., Janacek J., Albrechtova J. and Karen P. (2004), Stereological and digital methods for estimating geometrical characteristics of biological structures using confocal microscopy, in *Proc. NATO Advanced Study Institute on From Cells to Proteins: Imaging Nature across Dimensions* (Evangelista V., Barsanti L., Passarelli V. and Gualtieri P., Eds), Pisa, Italy, 12–23 September 2004, Springer, Vol. 3/2006, pp. 271–321.

Kudo M., Watt R.A. and Moffat A.C. (2000), Rapid identification of *Digitalis purpurea* using near-infrared reflectance spectroscopy, *Journal of Pharmacy and Pharmacology*, 52(10):1271–1277.

Kumar L. (2007a), A comparison of reflectance characteristics of some Australian *Eucalyptus* species based on high spectral resolution data – Discriminating using the visible and NIR regions, *Journal of Spatial Science*, 52(2):51–64.

(2007b), High-spectral resolution data for determining leaf water content in *Eucalyptus* species: leaf level experiments, *Geocarto International*, 22(1):3–16.

Kumar L., Skidmore A.K. and Mutanga O. (2010), Leaf level experiments to discriminate between eucalyptus species using high spectral resolution reflectance data: use of derivatives, ratios and vegetation indices, *Geocarto International*, 25 (4):327–344.

Kumar R. and Silva L. (1973), Light ray tracing through a leaf cross-section, *Applied Optics*, 12(12):2950–2954.

Kümmerlen B., Dauwe S., Schmundt D. and Schurr U. (1999), Thermography to measure relations of plant leaves, in *Handbook of Computer Vision and Applications* (Jahne B., Haussecker H. and Geissler P., Eds), Academic Press, pp. 763–781.

Kupková L., Potůčková M., Zachová K., Lhotáková Z., Kopačková V. and Albrechtová J. (2012), Chlorophyll determination in silver birch and scots pine foliage from heavy metal polluted regions using spectral reflectance data, *EARSeL eProceedings*, 11(1):64–73.

Kursar T.A. and Coley P.D. (1992), Delayed greening in tropical leaves: an anti-herbivore defense? *Biotropica*, 24 (2):256–262.

Kutschera U. (2008), The growing outer epidermal wall: design and physiological role of a composite structure, *Annals of Botany*, 101(5):615–621.

Kutyreva A.P., Intykbayeva B.B. and Kuatova Z. (1960), Characteristics of the optical properties of alpine plants of eastern Pamir, *Trudy sektora astrobotaniki*, 8:65–121 (English translation).

Kuusk A. (1994), A multispectral canopy reflectance model, *Remote Sensing of Environment*, 50(2):75–82.

(1995a), A fast, invertible canopy reflectance model, *Remote Sensing of Environment*, 51(3):342–350.

(1995b), A Markov chain model of canopy reflectance, *Agricultural and Forest Meteorology*, 76(3–4):221–236.

Kwan A., Dudley J. and Lantz E. (2002), Who really discovered Snell's law? *Physics World*, April 2002, pp. 64–64.

Kwolek W.F. (1982), Variations in leaf coloration using a reflectance colorimeter, *Journal of Arboriculture*, 8(6):157–159.

Labovitz M.L., Masuoka E.J., Bell R., Siegrist A.W. and Nelson R.F. (1983), The application of remote sensing to geobotanical exploration for metal sulfides – Results from the 1980 field season at Mineral, Virginia, *Economic Geology*, 78(4):750–760.

Lacaze B. and Joffre R. (1994), Extracting biochemical information from visible and near infrared reflectance spectroscopy of fresh and dried leaves, *Journal of Plant Physiology*, 144(3):277–281.

Lambers H., Chapin F.S. and Pons T.L. (2008), *Plant Physiological Ecology*, Springer, 605 pages.

Landrin A. (1864), *Notice historique et analytique sur les travaux relatifs a la coloration des végétaux. Rédigée sur la demande du Comité des Sciences appliquées de la Société d'Horticulture de Seine-et-Oise*, Imprimerie de E. Aubert, Versailles, France, 15 pages.

Lao C., Li B., Guo Y. and Yan T. (2005), Design and evaluation of a new device for measuring scattering light distribution of leaf, *Transactions of the Chinese Society of Agricultural Engineering*, 21 (9):85–89 (in Chinese).

Lao C., Guo Y. and Li B. (2006), Parameterization of bidirectional reflection from maize leaves with measurements in the principal plane, in *Proc. 2nd International Symposium on Plant Growth Modeling, Simulation,* Visualization *and Applications*, Beijing, China, 13–17 November 2006, pp. 109–115.

(2009), Simulating the distribution of R/FR in maize canopies with Monte Carlo ray tracing approach, in *Proc. 3rd International Symposium on Plant Growth Modeling, Simulation, Visualization and Applications,* Beijing, China, 9–13 November 2009, IEEE, pp. 65–71.

Lass L.W., Prather T.S., Glenn N.F., Weber K.T., Mundt J.T. and Pettingill J. (2005), A review of remote sensing of invasive weeds and example of the early detection of spotted knapweed (*Centaurea maculosa*) and babysbreath (*Gypsophila paniculata*) with a hyperspectral sensor, *Weed Science*, 53(2):242–251.

Lautenschlager-Fleury D. (1955), Über die Ultraviolettdurchlässigkeit von Blattepidermen, *Berichte der Schweizerischen Botanischen Gesellschaft*, 65:343–386.

Lawrence R.L., Wood S.D. and Sheley R.L. (2006), Mapping invasive plants using hyperspectral imagery and Breiman Cutler classifications (randomForest), *Remote Sensing of Environment*, 100(3):356–362.

Leadbetter J. (2005), *Environmental Microbiology*, Academic Press, Amsterdam, 568 pages.

Leaf Architecture Working Group (1979), *Manual of leaf architecture: morphological description and categorization of dicotyledonous and net-veined monocotyledonous angiosperms*, Smithsonian Institution, Washington, DC, 65 pages.

Lecointre G., Le Guyader H., Visset D. and McCoy K. (2007), *The Tree of Life: A Phylogenetic Classification*, Harvard University Press, Cambridge, MA, 560 pages.

Lee D.W. and Lowry J.B. (1975), Physical basis and ecological significance of iridescence in blue plants, *Nature*, 254 (5495):50–51.

(1980), Young-leaf anthocyanin and solar ultraviolet, *Biotropica*, 12(1):75–76.

Lee D.W. (1977), On iridescent plants, *The Gardens' Bulletin Singapore*, 30:21–29.

(1986), Unusual strategies of light absorption in rain-forest herbs, in *On the Economy of Plant Form and Function* (Givnish T.J., Ed), Cambridge University Press, New York, pp. 105–131.

(1997), Iridescent blue plants, *American Scientist*, 85(1):56–58.

(2007), *Nature's Palette. The science of Plant Color*, The University of Chicago Press, 409 pages.

(2011), Biomimicry of the ultimate optical device – The plant, in *Biomimetics: Nature-Based Innovation* (Bar-Cohen Y., Ed), CRC Press, Boca Raton, pp. 307–330.
Lee D.W. and Graham R. (1986), Leaf optical properties of rainforest sun and extreme shade plants, *American Journal of Botany*, 73(8):1100–1108.
Lee D.W., Bone R.A., Tarsis S.L. and Storch D. (1990), Correlates of leaf optical properties in tropical forest sun and extreme-shade plants, *American Journal of Botany*, 77(3):370–380.
Lee S.Y., Jeong H.J., Kang N.S., Jang T.Y., Jang S.H. and Lajeunesse T.C. (2015), *Symbiodinium tridacnidorum* sp. nov., a dinoflagellate common to Indo-Pacific giant clams, and a revised morphological description of *Symbiodinium microadriaticum* Freudenthal, emended Trench & Blank, *European Journal of Phycology*, 50(2):155–172.
Lehmann A.S., Pont S.C. and Geusebroek J.M. (2005), Tree textures: modern techniques in art-historical context, in *Proc. 4th International Workshop on Texture Analysis and Synthesis*, Beijing, China, 31 October 2005, pp. 43–48.
Lehmann J.R.K., Große-Stoltenberg A., Römer M. and Oldeland J. (2015), Field spectroscopy in the VNIR-SWIR region to discriminate between Mediterranean native plants and exotic-invasive shrubs based on leaf tannin content, *Remote Sensing*, 7(2):1225–1241.
Lei T.T., Tabuchi R., Kitao M. and Koike T. (1996), Functional relationship between chlorophyll content and leaf reflectance, and light-capturing efficiency of Japanese forest species, *Physiologia Plantarum*, 96(3):411–418.
Le Maire G., François C. and Dufrêne E. (2004), Towards universal broadleaf chlorophyll indices using PROSPECT simulated database and hyperspectral reflectance measurements, *Remote Sensing of Environment*, 89(1):1–28.
Le Maire G., François C., Soudani K., Berveiller D., Pontailler J.Y., Bréda N., et al. (2008), Calibration and validation of hyperspectral indices for the estimation of broadleaved forest leaf chlorophyll content, leaf mass per area, leaf area index and leaf canopy biomass, *Remote Sensing of Environment*, 112(10):3846–3864.
Lemery (1706), Que les plantes contiennent réellement du fer, & que ce métal entre nécessairement dans leur composition naturelle, *Histoire de l'Académie royale des sciences*, Tome I, pp. 411–418.
Lenk S., Chaerle L., Pfündel E., Langsdorf G., Hagenbeek D., Lichtenthaler H.K., et al. (2007), Multispectral fluorescence and reflectance imaging at the leaf level and its possible applications, *Journal of Experimental Botany*, 58(4):807–814.
Leon C.T., Shaw D.R., Bruce L.M. and Watson C. (2003), Effect of purple (*Cyperus rotundus*) and yellow nutsedge (*C. esculentus*) on growth and reflectance characteristics of cotton and soybean, *Weed Science*, 51(4):557–564.
Léon L. and Downey G. (2006), Preliminary studies by visible and near-infrared reflectance spectroscopy of juvenile and adult olive (*Olea europaea* L.) leaves, *Journal of the Science of Food and Agriculture*, 86(6):999–1004.
Leong H.C. (2008), Imaging and reflectance spectroscopy for the evaluation of effective camouflage in the SWIR, *POINTER*, 34(3):77–84.
Lestari W.A., Herdiyeni Y., Prasetyo L.B., Hasbi W., Arai K. and Okumura H. (2015), Nitrogen estimation of paddy based on leaf reflectance using artificial neural network, in *Proc. 7th International Conference of Soft Computing and Pattern Recognition*, Fukouka, Japan, 13–15 November 2015, IEEE, pp. 224–229.
Leuning R. (1989), Leaf energy balances: developments and applications, *Philosophical Transactions of the Royal Society of London. Series B, Biological Sciences*, 324(1223):191–206.
Lev-Yadun S. (2001), Aposematic (warning) coloration associated with thorns in higher plants, *Journal of Theoretical Biology*, 210(3):385–388.
(2003), Why do some thorny plants resemble green zebras? *Journal of Theoretical Biology*, 224(4):483–489.
(2006), Defensive functions of white coloration in coastal and dune plants, *Israel Journal of Plant Sciences*, 54(4):317–325.
(2009), Aposematic (warning) coloration in plants, in *Plant-Environment Interactions: From Sensory Plant Biology to Active Plant Behavior* (Baluška F., Ed), Springer, Heidelberg, Germany, pp. 167–202.
Lev-Yadun S. and Ne'eman G. (2004), When may green plants be aposematic? *Biological Journal of the Linnean Society*, 81 (3):413–416.
Lev-Yadun S., Dafni A., Flaishman M.A., et al. (2004), Plant coloration undermines herbivorous insect camouflage, *BioEssays*, 26(10):1126–1130.
Lev-Yadun S. and Gould K.S. (2007), What do red and yellow autumn leaves signal? *The Botanical Review*, 73(4):279–289.
(2009), Role of anthocyanins in plant defence, in Anthocyanins: Biosynthesis, Functions, and Applications (Gould K.S., Davies K. and Winefield C., Eds), Springer-Verlag, New York, pp. 21–28.
Lev-Yadun S. and Keasar T. (2012), Prerequisites for evolution: variation and selection in yellow autumn birch leaves, *New Phytologist*, 195(2):282–284.
Lewis F.T. (1923), The typical shape of polyhedra cells in vegetable parenchyma and the restoration of that shape following cell division, *Proceedings of the American Academy of Arts and Sciences*, 58(15):537–552.
Lewis M. (2002), Spectral characterization of Australian arid zone plants, *Canadian Journal of Remote Sensing*, 28 (2):219–230.
Lewis P. and Disney M. (2007), Spectral invariants and scattering across multiple scales from within-leaf to canopy, *Remote Sensing of Environment*, 109(2):196–206.
Lhotáková Z., Albrechtová J., Janáček J. and Kubinova L. (2008), Advantages and pitfalls of using free-hand sections of frozen needles for three-dimensional analysis of mesophyll by stereology and confocal microscopy, *Journal of Microscopy*, 232(1):56–63.
Li B., Liew O.W. and Asundi A.K. (2005a), Early detection of calcium deficiency in plants using red edge position, in *Proc. Optical Sensors and Sensing Systems for Natural Resources and Food Safety and Quality* (Chen Y.R., Meyer G.E. and Tu S.I., Eds), Boston, MA, 8 November 2005, SPIE, Vol. 5996, 599609.

(2006), Pre-visual detection of iron and phosphorus deficiency by transformed reflectance spectra, *Journal of Photochemistry and Photobiology B: Biology*, 85(2):131–139.

Li B., Wah L.O. and Asundi A.K. (2005b), Use of reflectance spectroscopy for early detection of calcium deficiency in plants, in *Proc. 3rd International Conference on Experimental Mechanics* (Quan C., Chau F.S., Asundi A., Wong B.S. and Lim C.T., Eds), Seville, Spain, SPIE, Vol. 5852, pp. 693–697.

Li B., Liu Z., Huang J., Zhang L., Zhou W. and Shi J. (2009), Hyperspectral identification of rice diseases and pests based on principal component analysis and probabilistic neural network, Transactions of the CSAE, 25(9):143–147 (in Chinese).

Li D., Xu B., Shi X., Zhang C. and Wu R. (1996), Influence of simulated acid rain on characteristics of reflectance spectra of rice leaves, *Remote Sensing of Environment China*, 11(4):241–247 (in Chinese).

Li D., Cheng T., Yao X., Zhang Z., Tian Y., Zhu Y. et al. (2016), Wavelet-based PROSPECT inversion for retrieving leaf mass per area (LMA) and equivalent water thickness (EWT) from leaf reflectance, in *Proc. International Geoscience and Remote Sensing Symposium (IGARSS'16)*, Beijing, 10–15 July 2016, IEEE, pp. 6910–6913.

Li G., Alchanatis V. and Shmilovitch Z. (1999), Nitrogen status detection of corn leaves by reflectance technique, in *Proc. International Conference on Agricultural Engineering*, Beijing, China,14–17 December 1999, Vol. 5, pp. 19–26.

Li L., Ustin S.L. and Lay M. (2005c), Application of AVIRIS data in detection of oil-induced vegetation stress and cover change at Jornada, New Mexico, *Remote Sensing of Environment*, 94(1):1–16.

Li L., Ustin S.L. and Riano D. (2007), Retrieval of fresh leaf fuel moisture content using genetic algorithm partial least squares (GA-PLS) modeling, *IEEE Geoscience and Remote Sensing Letters*, 4(2):216–220.

Li L., Cheng Y.B., Ustin S.L., Hua X.T. and Riaño D. (2008), Retrieval of vegetation equivalent water thickness from reflectance using genetic algorithm (GA)-partial least squares (PLS) regression, *Advances in Space Research*, 41 (11):1755–1763.

Li P. and Wang Q. (2011), Retrieval of leaf biochemical parameters using PROSPECT inversion: a new approach for alleviating ill-posed problems, *IEEE Transactions on Geoscience and Remote Sensing*, 49(7):2499–2506.

Li P. and Wang Q. (2013), Retrieval of chlorophyll for assimilating branches of a typical desert plant through inversed radiative transfer models, *International Journal of Remote Sensing*, 34(7):2402–2416.

Li X. and He Y. (2008), Discriminating varieties of tea plant based on Vis/NIR spectral characteristics and using artificial neural networks, *Biosystems Engineering*, 99(3):313–321.

Li Y., Ni S. and Huang J. (2003), Simulation rice leaf reflectance and its inversion, in *Proc. Multispectral and Hyperspectral Remote Sensing Instruments and Applications* (Larar A.M., Tong Q. and Suzuki M., Eds), Hangzhou, China, Vol. 4897, pp. 171–178.

Liakoura V., Bornman J.F. and Karabourniotis G. (2003), The ability of abaxial and adaxial epidermis of sun and shade leaves to attenuate UV-A and UV-B radiation in relation to the UV absorbing capacity of the whole leaf methanolic extracts, *Physiologia Plantarum*, 117(1):33–43.

Liao Q., Wang J., Yang G., Zhang D., Li H., Fu Y. et al. (2013), Comparison of spectral indices and wavelet transform for estimating chlorophyll content of maize from hyperspectral reflectance, *Journal of Applied Remote Sensing*, 7(1):073575.

Lichtenthaler H.K. (1987), Chlorophylls and carotenoids: pigments of photosynthetic biomembranes, *Methods in Enzymology*, 148:350–382.

Lichtenthaler H. K. and Rinderle U. (1988), The role of chlorophyll fluorescence in the detection of stress conditions in plants, *CRC Critical Reviews in Analytical Chemistry*, 19(1):S29–S85.

Lichtenthaler H.K., Gitelson A. and Lang M. (1996), Non-destructive determination of chlorophyll content of leaves of a green and an aurea mutant of tobacco by reflectance measurements, *Journal of Plant Physiology*, 148(3–4):483–493.

Lichtenthaler H.K., Wenzel O., Buschmann C. and Gitelson A. (1998), Plant stress detection by reflectance and fluorescence, *Annals of the New York Academy of Sciences*, 851(1):271–285.

Liew O.W., Boey W.S.L., Asundi A.K., Chen J.W. and He D.M. (1999), Fibre optic spectrophotometry monitoring of plant nutrient deficiency under hydroponic culture conditions, in *Proc. Optical Engineering for Sensing and Nanotechnology* (Yamaguchi I., Ed), Yokohama, Japan, 7 May 1999, SPIE, Vol. 3740, pp. 186–190.

Lillesaeter O. (1982), Spectral reflectance of partly transmitting leaves: laboratory measurements and mathematical modeling, *Remote Sensing of Environment*, 12(3):247–254.

Lin B., Sun W., Min Q. and Hu Y. (2008), Numerical studies of scattering properties of leaves and leaf moisture influences on the scattering at microwave wavelengths, *IEEE Transactions on Geoscience and Remote Sensing*, 46(2):353–360.

Lin Z.F. and Ehleringer J.R. (1983), Epidermis effects on spectral properties of leaves of four herbaceous species, *Physiologia Plantarum*, 59(1):91–94.

Linacre E.T. (1964), Determinations of the heat transfer coefficient of a leaf, *Plant Physiology*, 39(4):687–690.

Lindenmayer A. (1984), Models for plant tissue development with cell division orientation regulated by preprophase bands of microtubules, *Differentiation*, 26(1–3):1–10.

Lindenthal M., Steiner U., Dehne H.W. and Oerke E.C. (2005), Effect of downy mildew development on transpiration of cucumber leaves visualized by digital infrared thermography, *Phytopathology*, 95(3):233–240.

Linke R., Richter K., Haumann J., Schneider W. and Weihs P. (2008), Occurrence of repeated drought events: can repetitive stress situations and recovery from drought be traced with leaf reflectance? *Periodicum Biologorum*, 110(3):219–229.

Lintern M., Anand R., Ryan C. and Paterson D. (2013), Natural gold particles in *Eucalyptus* leaves and their relevance to exploration for buried gold deposits, *Nature Communications*, 4:2614.

Lister S.J., Dhanoa M.S., Stewart J.L. and Gill M. (2000), Classification and comparison of *Gliricidia* provenances using near infrared reflectance spectroscopy, *Animal Feed Science and Technology*, 86(3–4):221–238.

Litjens R.A.J., Quickenden T.I. and Freeman C.G. (1999), Visible and near-ultraviolet absorption spectrum of liquid water, *Applied Optics*, 38(7):1216–1223.

Liu C., Guo J., Cui Y., Lü T., Zhang X. and Shi G. (2011a), Effects of cadmium and salicylic acid on growth, spectral reflectance and photosynthesis of castor bean seedlings, *Plant Soil*, 344(1–2):131–141.

Liu M., Liu X., Li T. and Xiu L. (2010a), Analysis of hyperspectral singularity of rice under Zn pollution stress, *Transactions of the Chinese Society of Agricultural Engineering*, 26(3):191–197 (in Chinese).

Liu N., Peng C.L., Lin Z.F., Lin G.Z., Zhang L.L. and Pan X.P. (2006), Changes in photosystem II activity and leaf reflectance features of several subtropical woody plants under simulated SO_2 treatment, *Journal of Integrative Plant Biology*, 48 (11):1274–1286.

Liu N., Lin Z.F., Van Devender A., Lin G.Z., Peng C.L., Pan X.P., et al. (2009a), Spectral reflectance indices and pigment functions during leaf ontogenesis in six subtropical landscape plants, *Plant Growth Regulation*, 58(1):73–84.

Liu P., Shi R., Wang H., Bai K. and Gao W. (2014), Estimating leaf photosynthetic pigments information by stepwise multiple linear regression analysis and a leaf optical model, in *Proc. Remote Sensing and Modeling of Ecosystems for Sustainability XI* (Gao W., Chang N.B. and Wang J., Eds), San Diego, CA, 17 August 2014, SPIE, Vol. 9221, 92211A.

Liu P., Zhou J., Shi R., Zhang C., Liu C., Sun Z. et al. (2016), Identification of *Phragmites australis* and *Spartina alterniflora* in the Yangtze Estuary between Bayes and BP neural network using hyper-spectral data, in *Proc. Remote Sensing and Modeling of Ecosystems for Sustainability XIII* (Gao B.C. and Chang N.B., Eds), San Diego, CA, 28 August–1 September 2016, SPIE, Vol. 9975, 99750H.

Liu S.H., Liu X.H., Hou J., Chi G.Y. and Cui B.S. (2008a), Study on the spectral response of *Brassica campestris* L. leaf to the copper pollution, *Science in China Series E: Technological Sciences*, 51(2):202–208.

Liu W. and Lao C. (2007), The application of three B-spline interpolation in measuring scattering flux from leaves, *Control & Management*, 6: 181–182 (in Chinese).

Liu W., Chang Q.R., Guo M., Xing D.X. and Yuan Y.S. (2011b), Diagnosis of phosphorus nutrition in winter wheat based on first derivative spectra and radial basis function neural network, *Spectroscopy and Spectral Analysis*, 31(4):1092–1096 (in Chinese).

Liu Y., Chen H., Wu G. and Wu X. (2010b), Feasibility of estimating heavy metal concentrations in *Phragmites australis* using laboratory-based hyperspectral data-A case study along Le'an River, China, *International Journal of Applied Earth Observation and Geoinformation*, 12(2):S166–S170.

Liu Z.M., Wu W.J. and Hu B.R. (2008b), Design of biomimetic camouflage materials based on angiosperm leaf organs, *Science in China Series E: Technological Sciences*, 51(11):1902–1910.

Liu Z.M., Hu B.R., Wu W.J. and Zhang Y. (2009b), Spectral imaging of green coating camouflage under hyperspectral detection, *Acta Photonica Sinica*, 38(4):885–890 (in Chinese).

Liu Z.Y., Huang J.F., Shi J.J., Tao R.X., Zhou W. and Zhang L.L. (2007), Characterizing and estimating rice brown spot disease severity using stepwise regression, principal component regression and partial least-square regression, *Journal of Zhejiang University SCIENCE B*, 8(10):738–744.

Liu Z.Y., Huang J.F. and Tao R.X. (2008c), Characterizing and estimating fungal disease severity of rice brown spot with hyperspectral reflectance data, *Rice Science*, 15(3):232–242.

Liu Z., Cheng J., Huang W., Li C., Xu X., Ding X., et al. (2012), Hyperspectral discrimination and response characteristics of stressed rice leaves caused by rice leaf folder, in *Computer and Computing Technologies in Agriculture V* (Li D. and Chen Y., Eds), Springer, pp. 528–537.

Lolli L., Pisani M., Rajteri M., Widlowski J.L., Bialek A., Greenwell C. et al. (2014), Phytos: a portable goniometer for *in situ* spectro-directional measurements of leaves, *Metrologia*, 51(6):S309–S313.

Lommel E. (1874), Ueber den Lichtschein um den Schatten des Kopfe, in *Annalen der Physik und Chemie*., Jubelband, Leipzig, pp. 10–21.

Lorenzen B. and Jensen A. (1989), Changes in leaf spectral properties induced in barley by cereal powdery mildew, *Remote Sensing of Environment*, 27(2):201–209.

(1991), Spectral properties of barley canopy in relation to the spectral properties of single leaves and the soil, *Remote Sensing of Environment*, 37(1):23–34.

Lorenzen B., Skovhus K. and Jensen A. (1990), Spectral properties and net photosynthesis of *Aster tripolium* L. and *Halimione portulacoides* (L.) Aellen leaves under saline and hypoxic conditions, *New Phytologist*, 116(2):255–262.

Lovchikova L.P., Nikonenko S.V., Plyuta V.E. and Sergeichik S.A. (1997), Spectral reflection characteristics of spruce needles and leaves of trees in the zones with industrially polluted air, *Journal of Applied Spectroscopy*, 64(6):804–808 (cover-to-cover translation from *Zhurnal Prikladnoi Spektroskopii*, 64(6):789–792).

Lovelock C.E., Clough B.F. and Woodrow I.E. (1992), Distribution and accumulation of ultraviolet-radiation absorbing compounds in leaves of tropical mangroves, *Planta*, 188(2):143–154.

Lu C., Hen S.B. and Liu W.S. (2013), Research of PROSPECT leaf optical property model, *Global Geology*, 32(1):177–188 (in Chinese).

Lu S., Zhao C. and Guo X. (2009), Venation skeleton-based modeling plant leaf wilting, *International Journal of Computer Games Technology*, ID 890917, 8 pages.http://dx.doi.org/10.1155/2009/890917

Lucarini V., Saarinen J.J., Peiponen K.E. and Vartiainen E.M. (2005), *Kramers–Kronig Relations in Optical Materials Research*, Springer-Verlag, Berlin 162 pages.

Lück J., Lindenmayer A. and Lück H.B. (1988), Models of cell tetrads and clones in meristematic cell layers, *Botanical Gazette*, 149(2):127–141.

Lüdeker W. and Günther K.P. (1990), Leaf reflectance: a stochastic model for analysing the blue shift, in *Proc. Symposium on Global and Environmental Monitoring Techniques and Impacts*, Victoria, BC, 17–21 September 1990, ISPRS, Vol. 28, pp. 475–480.
Lukeš P., Stenberg P., Rautiainen M., Mõttus M. and Vanhatalo K.M. (2013), Optical properties of leaves and needles for boreal tree species in Europe, *Remote Sensing Letters*, 4(7):667–676.
Lunadei L., Diezma B., Lleó L., Ruiz-Garcia L., Cantalapiedra S. and Ruiz-Altisent M. (2012), Monitoring of fresh-cut spinach leaves through a multispectral vision system, *Postharvest Biology and Technology*, 63(1):74–84
Lunagaria M.M., Patel H.R. and Pandey V.Y.A.S. (2015), Evaluation and calibration of noninvasive leaf chlorophyll meters for wheat, *Journal of Agrometeorology*, 17(1):51–54.
Luo N.N., Zhao W.J. and Yan X. (2013), Impact of dust-fall on spectral features of plant leaves, *Spectroscopy and Spectral Analysis*, 33(10):2715–20 (in Chinese).
Lvovsky A.I. (2013), Fresnel equations, in *Encyclopedia of Optical Engineering*, CRC Press, 7 pp.
Ma K., Baret F., Barroy P. and Bousquet L. (2007), A leaf optical properties model accounting for differences between two faces, in *Proc. 10th International Symposium on Physical Measurements and Signatures in Remote Sensing* (Schaepman M., Liang S., Groot N. and Kneubühler M., Eds), Davos, Switzerland, 12–14 March 2007, ISPRS, pp. 89–95.
Ma Q., Ishimaru A., Phu P. and Kuga Y. (1990), Transmission, reflection, and depolarization of an optical wave for a single leaf, *IEEE Transactions on Geoscience and Remote Sensing*, 28(5):865–872.
Ma Z.G., Chen X., Wang Q., Li P.H. and Jiapaer G. (2012), Retrieval of leaf biochemical properties by inversed PROSPECT model and hyperspectral indices: an application to *Populus euphratica* polymorphic leaves, *Journal of Arid Land*, 4 (1):52–62.
Maas S.J. and Dunlap J.R. (1989), Reflectance, transmittance, and absorptance of light by normal, etiolated, and albino corn leaves, *Agronomy Journal*, 81(1):105–110.
Macaire-Princep I. (1828), Mémoire sur la coloration automnale des feuilles, *Mémoires de la Société de physique et d'histoire naturelle de Genève*, 4:43–53.
MacArthur A. and Malthus T. (2012), *Calluna vulgaris* foliar pigments and spectral reflectance modelling, *International Journal of Remote Sensing*, 33(16):5214–5239.
Maccioni A., Agati G. and Mazzinghi P. (2001), New vegetation indices for remote measurement of chlorophylls based on leaf directional reflectance spectra, *Journal of Photochemistry and Photobiology. B, Biology*, 61(1–2):52–61.
Machado M.L., Pinto F.A.C., Paula T.J., Queiroz D.M. and Cerqueira O.A.T. (2015), White mold detection in common beans through leaf reflectance spectroscopy, *Engenharia Agrícola*, 35(6):1117–1126.
Macior W.A. and Matzke E.B. (1951), An experimental analysis of cell-wall curvatures, and approximations to minimal tetrakaidecahedra in the leaf parenchyma of *Rhoeo discolor*, *American Journal of Botany*, 38(10):783–793.
MacNeil J.D., Downing R.S. and Hikichi M. (1974), Diffuse reflectance spectroscopy – Applications in measuring mite-feeding injury to orchard leaves, *Canadian Journal of Plant Science*, 54(3):505–509.
MacNeil J.D., Hikichi M. and Downing R.S. (1987), An investigation of the effects of seasonal changes, leaf maturity, nitrogen deficiency and leafhopper injury on the chlorophyll content and diffuse reflectance spectroscopic properties of orchard leaves, *International Journal of Environmental and Analytical Chemistry*, 31(1):55–62.
Macnicol P.K., Dudzinski M.L. and Condon B.N. (1976), Estimation of chlorophyll in tobacco leaves by direct photometry, *Annals of Botany*, 40(1):143–152.
Madeira A.C., Mendonca A., Ferreira M.E. and Taborda M.D. (2000), Relationship between spectroradiometric and chlorophyll measurements in green beans, *Communications in Soil Science and Plant Analysis*, 31(5–6):631–643.
Madeira A.C., Ferreira A., de Varennes A. and Vieira M.I. (2003), SPAD meter versus tristimulus colorimeter to estimate chlorophyll content and leaf color in sweet pepper, *Communications in Soil Science and Plant Analysis*, 34 (17–18):2641–2470.
Magarey R.D., Russo J.M., Seem R.C. and Gadoury D.M. (2005), Surface wetness duration under controlled environmental conditions, *Agricultural and Forest Meteorology*, 128(1–2):111–122.
Magnussen S., Coops N., Luther J.E. and Carroll A.L. (2004), An approach for the analysis of vegetation spectra using non-linear mixed modeling of truncated power spectra, *Annals of Forest Science*, 61(6):515–523.
Mahlein A.K., Steiner U., Dehne H.W. and Oerke E.C. (2010), Spectral signatures of sugar beet leaves for the detection and differentiation of diseases, *Precision Agriculture*, 11(4):413–431.
Mahlein A.K., Steiner U., Hillnhütter C., Dehne H.W. and Oerke E.C. (2012), Hyperspectral imaging for small-scale analysis of symptoms caused by different sugar beet diseases, *Plant Methods*, 8:3.
Mahlein A.K., Rumpf T., Welke P., Dehne H.W., Plümer L., Steiner U. et al. (2013), Development of spectral indices for detecting and identifying plant diseases, *Remote Sensing of Environment*, 128:21–30.
Maier S.W., Lüdeker W. and Günther K.P. (1999), SLOP: a revised version of the stochastic model for leaf optical properties, *Remote Sensing of Environment*, 68(3):273–280.
Maier S.W. (2000), *Modeling the radiative transfer in leaves in the 300 nm to 2.5 μm wavelength region taking into consideration chlorophyll fluorescence – The leaf model SLOPE*, PhD Thesis, Deutsches Fernerkundungstagsdatenzentrum, Technische Universität München, Oberpfaffenhofen, Germany, 110 pages.
Majeke B., van Aardt J.A.N. and Cho M.A. (2008), Imaging spectroscopy of foliar biochemistry in forestry environments, *Southern Forests*, 70(3):275–285.
Major D.J., McGinn S.M., Gillespie T.J. and Baret F. (1993), A technique for determination of single leaf reflectance and transmittance in field studies, *Remote Sensing of Environment*, 43(2):209–215.

Maki M., Ishiahra M. and Tamura M. (2004), Estimation of leaf water status to monitor the risk of forest fires by using remotely sensed data, *Remote Sensing of Environment*, 90(4):441–450.

Makrides S.C. and Goldthwaite J. (1981), Biochemical changes during bean leaf growth, maturity, and senescence, *Journal of Experimental Botany*, 32(129):725–735.

Main R., Cho M.A., Mathieu R., O'Kennedy M.M., Ramoelo A. and Koch S. (2011), An investigation into robust spectral indices for leaf chlorophyll estimation, *ISPRS Journal of Photogrammetry and Remote Sensing*, 66(6):751–761.

Male E.J., Pickles W.L., Silver E.A., Hoffmann G.D., Lewicki J., Apple M., et al. (2010), Using hyperspectral plant signatures for CO_2 leak detection during the 2008 ZERT CO_2 sequestration field experiment in Bozeman, Montana, *Environmental Earth Sciences*, 60(2):251–261.

Malenovsky Z., Albrechtova J., Lhotakova Z., Zurita-Milla R., Clevers J.G.P.W., Schaepman M.E. et al. (2006), Applicability of the PROSPECT model for Norway spruce needles, *International Journal of Remote Sensing*, 27(24):5315–5340.

Malhotra S.S. (1977), Effects of aqueous sulphur dioxide on chlorophyll destruction in *Pinus contorta*, *New physiologist*, 78 (1):101–109.

Malinin D.R. and Yoe J.H. (1961), Development of the laws of colorimetry: a historical sketch, *Journal of Chemical Education*, 38(3):129–131.

Malthus T.J. and Madeira A.C. (1993), High resolution spectroradiometry: spectral reflectance of field bean leaves infected by *Botrytis fabae*, *Remote Sensing of Environment*, 45(1):107–116.

Manabe S.I., Fujioka R., Iigima H. and Yaginuma Y. (2001), Change in refractive index of cellulose particle with particle size, *Bulletin of the Faculty of Human Environmental Science*, 32:65–69.

Mandelis A., Boroumand F. and vanden Bergh H. (1990), Quantitative diffuse reflectance spectroscopy of large powders: the Melamed model revisited, *Applied Optics*, 29(19):2853–2860.

Mandelis A. and Grossmann J.P. (1992), Perturbation theoretical approach to the generalized Kubelka-Munk problem in nonhomogeneous optical media, *Applied Spectroscopy*, 46(5):737–745.

Mandoli D.F. and Briggs W.R (1982), Optical properties of etiolated plant tissues, *Proceedings of the National Academy of Sciences of the United States of America*, 79(9):2902–2906.

(1984a), Fiber optics plant tissues: spectral dependence in dark-grown and green tissues, *Photochemistry and Photobiology*, 39(3):419–424.

(1984b), Fiber optics in plants, *Scientific American*, 251(2):90–98.

Mandre M. and Tuulmets L. (1997), Pigment changes in Norway spruce induced by dust pollution, *Water, Air, & Soil Pollution*, 94(3–4):247–258.

Manetas Y., Grammatikopoulos G. and Kyparissis A. (1998), The use of the portable, non-destructive, SPAD-502 (Minolta) chlorophyll meter with leaves of varying trichome density and anthocyanin content, *Journal of Plant Physiology*, 153 (3–4):513–516.

Manetas Y. (2003), The importance of being hairy: the adverse effects of hair removal on stem photosynthesis of *Verbascum speciosum* are due to solar UV-B radiation, *New Phytologist*, 158(3):503–508.

(2006), Why some leaves are anthocyanic and why most anthocyanic leaves are red? *Flora*, 201(3):163–177.

Manna J.S., Basu S., Mitra M.K., Mukherjee S. and Chandra Das G. (2009), Study on the biostability of chlorophyll a entrapped in silica gel nanomatrix, *Journal of Materials Science: Materials in Electronics*, 20(11):1068–1072.

Maquenne E. (1880), Recherches sur la détermination des pouvoirs absorbants et diffusifs des feuilles, *Annales Agronomiques*, 6:321–390.

Maréchal Y., and Chanzy H. (2000), The hydrogen bond network in Iβ cellulose as observed by infrared spectrometry, *Journal of Molecular Structure*, 523(1–3):183–196.

Marenco R.A., Antezana-Vera S.A. and Nascimento H.C.S. (2009), Relationship between specific leaf area, leaf thickness, leaf water content and SPAD-502 readings in six Amazonian tree species, *Photosynthetica*, 47(2):184–190.

Mariotti M., Ercoli L. and Masoni A. (1996), Spectral properties of iron-deficient corn and sunflower leaves, *Remote Sensing of Environment*, 58(3):282–288.

Markel V.A. (2016), Introduction to the Maxwell Garnett approximation: tutorial, *Journal of the Optical Society of America A*, 33(7):1244–1256.

Markwell J., Osterman J.C. and Mitchell J.L. (1995), Calibration of the Minolta SPAD-502 leaf chlorophyll meter, *Photosynthesis Research*, 46(3):467–472.

Markwell J. and Blevins D. (1999), The Minolta SPAD-502 leaf chlorophyll meter: an exciting new tool for education in the plant sciences, *American Biology Teacher*, 61(9):672–676.

Markwell J. (2002), Nondestructive chlorophyll assessment, *New Phytologist*, 153(1):7–8.

Marquart L.C. (1835), *Die Farben der Blüthen: Eine Chemisch-Physiologische Abhandlung*, T. Habicht, Bonn, 75 pages.

Martin C.E., Brandmeyer E.A. and Ross R.D. (2013), Ecophysiological function of leaf 'windows' in Lithops species – "Living Stones" that grow underground, *Plant Biology*, 15(1):243–247.

Martin G., Josserand S.A., Bornman J.F. and Volgemann T.C. (1989), Epidermal focussing and the light microenvironment within leaves of *Medicago sativa*, *Physiologia Plantarum*, 76(4):485–492.

Martin G., Myers D.A. and Vogelmann T.C. (1991), Characterization of plant epidermal lens effects by a surface replica technique, *Journal of Experimental Botany*, 42(238):581–587.

Martin J.T. and Juniper B.E. (1970), *The Cuticles of Plants*, E. Arnold, London, 347 pages.

Martin M.E. and Aber J.D. (1994), Analyses of forest foliage III: determining nitrogen, lignin and cellulose in fresh leaves using near infrared reflectance data, *Journal of Near Infrared Spectroscopy*, 2(1):25–32.

(1997), High spectral resolution remote sensing of forest canopy lignin, nitrogen, and ecosystem processes, *Ecological Applications*, 7(2):431–443.
Martin R.E., Asner G.P. and Sack L. (2007), Genetic variation in leaf pigment, optical and photosynthetic function among diverse phenotypes of *Metrosideros polymorpha* grown in a common garden, *Oecologia*, 151(3):387–400.
Martin R.E. and Asner G.P. (2009), Leaf chemical and optical properties of *Metrosideros polymorpha* across environmental gradients in Hawaii, *Biotropica*, 41(3):292–301.
Martin W.E., Hesse E., Hough J.H., Sparks W.B., Cockell C.S., Ulanowski Z., et al. (2010), Polarized optical scattering signatures from biological materials, *Journal of Quantitative Spectroscopy & Radiative Transfer*, 111 (16):2444–2459.
Martin W.K. and Thomas S.B. (1887), The autumnal changes in maple leaves, *Scientific American*, 24:9813–9814.
Martinez D.E. and Guiamet J.J. (2004), Distortion of the SPAD 502 chlorophyll meter readings by changes in irradiance and leaf water status, *Agronomie*, 24(1):41–46.
Martinez von Remisowsky A., McClendon J.H. and Fukshansky L. (1992), Estimation of the optical parameters and light gradients in leaves: multi-flux versus two-flux treatment, *Photochemistry and Photobiology*, 55(6):857–865.
Masaitis G., Mozgeris G. and Augustaitis A. (2013), Spectral reflectance properties of healthy and stressed coniferous trees, *iForest*, 6:30–36.
Masoni A., Ercoli L., Mariotti M. and Barberi P. (1994), Changes in spectral properties of ageing and senescing maize and sunflower leaves, *Physiologia Plantarum*, 91(2):334–338.
Masoni A., Ercoli L. and Mariotti M. (1996), Spectral properties of leaves deficient in iron, sulfur, magnesium, and manganese, *Agronomy Journal*, 88(6):937–943.
Massantini F., Masoni A., Mariotti M. and Volterrani M. (1992), Influenza della concimazione azotata sulla riflettanza, assorbanza e trasmittanza della foglia a bandiera del frumento tenero (*Triticum aestivum* L.), *Rivista di Agronomia*, 26 (4):489–497 (in Italian).
Matile P., Flach B.M.P. and Eller B.M. (1992), Autumn leaves of *Ginkgo biloba* L. optical properties, pigments and optical brighteners, *Botanica Acta*, 105(1):13–17.
Matile P. (2000), Biochemistry of Indian summer: physiology of autumnal leaf coloration, *Experimental Gerontology*, 35 (2):145–158.
Mattsson J.O. and Cavallin C. (1972), Retroreflection of light from drop-covered surfaces and an image-producing device for registration of this light, *Oikos*, 23(3):285–294.
Mätzler C. (1994), Microwave (1–100 GHz) dielectric model of leaves, *IEEE Transactions on Geoscience and Remote Sensi*ng, 32(5):947–949.
Mätzler C. and Sume A. (1989), Microwave radiometry of leaves, in *Microwave Radiometry and Remote Sensing Applications* (Pampaloni P., Ed), VSP Books, Utrech, The Netherlands, pp. 133–148.
Maxwell K. and Johnson G.N. (2000), Chlorophyll fluorescence-a practical guide, *Journal of Experimental Botany*, 51(345): 659–668.
Maxwell Garnett J.C. (1904), Colours in metal glasses and in metallic films, *Philosophical Transactions of the Royal Society of London, Series A*, 203:385–420.
Mayer J.R. (1845), *Die organische Bewegung in ihrem Zusammenhange mit dem Stoffwechsel* : Ein *Beitrag zur Naturkunde*, Heilbronn, 112 pages (in German).
McClendon J.H. (1984), The micro-optic of leaves. I: Patterns of reflection from the epidermis, *American Journal of Botany*, 71(10):1391–1397.
McClendon J.H. and Fukshansky L. (1990a), On the interpretation of absorption spectra of leaves. I. Introduction and the correction of leaf spectra for surface reflection, *Photochemistry and Photobiology*, 51(2):203–210.
(1990b), On the interpretation of absorption spectra of leaves. II: The non-absorbed ray of the sieve effect and the mean optical pathlength in the remainder of the leaf, *Photochemistry and Photobiology*, 51(2):211–216.
McCree K.J. (1972), The action spectrum, absorptance and quantum yield of photosynthesis in crop plants, *Agricultural Meteorology*, 9:191–216.
McLellan T.M., Aber J.D., Martin M.E., Melillo J.M. and Nadelhoffer K.J. (1991a), Determination of nitrogen, lignin, and cellulose content of decomposing leaf material by near infrared reflectance spectroscopy, *Canadian Journal of Forest Research*, 21(11):1684–1688.
McLellan T.M., Martin M.E., Aber J.D., Melillo J.M., Nadelhoffer K.J. and Dewey B. (1991b), Comparison of wet chemistry and near infrared reflectance measurements of carbon-fraction chemistry and nitrogen concentration of forest foliage, *Canadian Journal of Forest Research*, 21(11):1689–1693.
McMurtrey III J.E., Chappelle E.W., Kim M.S., Meisinger J.J. and Corp L.A. (1994), Distinguishing nitrogen fertilization levels in field corn (*Zea Mays* L.) with actively induced fluorescence and passive reflectance measurements, *Remote Sensing of Environment*, 47(1):36–44.
McPherson S. (2010), Iridescent plants of the world, *The Plantsman*, 9(2):120–125.
Meggio F., Zarco-Tejada P.J., Núñez L.C., Sepulcre-Cantó G., González M.R. and Martín P. (2010), Grape quality assessment in vineyards affected by iron deficiency chlorosis using narrow-band physiological remote sensing indices, *Remote Sensing of Environment*, 114(9):1968–1986.
Meinander O., Somersalo S., Holopainen T. and Strasser R.J. (1996), Scots pines after exposure to elevated ozone and carbon dioxide probed by reflectance spectra and chlorophyll a fluorescence transients, *Journal of Plant Physiology*, 148 (1–2):229–236.

Meinzer F. and Goldstein G. (1985), Some consequences of leaf pubescence in the Andean giant rosette plant *Espeletia timotensis*, *Ecology*, 66(2):512–520.

Meir S., Philosoph-Hadasa S., Glotera P. and Aharonia N. (1992), Nondestructive assessment of chlorophyll content in watercress leaves by a tristimulus reflectance colorimeter, *Postharvest Biology and Technology*, 2(2):117–124.

Meissner T. and Wentz F.J. (2004), The complex dielectric constant of pure and sea water from microwave satellite observations, *IEEE Transactions on Geoscience and Remote Sensing*, 42(9):1836–1849.

Melamed N.T. (1963), Optical properties of powders. Part I. Optical absorption coefficients and the absolute value of the diffuse reflectance. Part II: Properties of luminescent powders, *Journal of Applied Physics*, 34(3):560–570.

Melvill T. (1914), Observations on light and colours, *Journal of the Royal Astronomical Society of Canada*, 8:231–272.

Mendelssohn I.A., McKee K.L. and Kong T. (2001), A comparison of physiological indicators of sublethal cadmium stress in wetland plants, *Environmental and Experimental Botany*, 46(3):263–275.

Menesatti P., Antonucci F., Pallottino F., Roccuzzo G., Allegra M., Stagno F. et al. (2010), Estimation of plant nutritional status by Vis–NIR spectrophotometric analysis on orange leaves (*Citrus sinensis* (L) Osbeck cv Tarocco], *Biosystems Engineering*, 105(4):448–454.

Merk S., Blume A. and Riederer M. (1998), Phase behaviour and crystallinity of plant cuticular waxes studied by Fourier transform infrared spectroscopy, *Planta*, 204(1):44–53.

Meroni M., Picchi V., Rossini M., et al. (2008), Leaf level early assessment of ozone injuries by passive fluorescence and photochemical reflectance index, *International Journal of Remote Sensing*, 29(17): 5409–5422.

Meroni M., Panigada C., Rossini M., Picchi V., Cogliati S. and Colombo R. (2009), Using optical remote sensing techniques to track the development of ozone-induced stress, *Environmental Pollution*, 157(5):1413–1420.

Merzlyak M.N. and Gitelson A. (1995), Why and what for the leaves are yellow in autumn? On the interpretation of optical spectra of senescing leaves (*Acer platanoides* L.), *Journal of Plant Physiology*, 145(3):315–320.

Merzlyak M.N., Gitelson A.A., Pogosyan S.I., Chivkunova O.B., Lehimena L., Garson M., et al. (1997), Reflectance spectra of leaves and fruits during their development and senescence and under stress, *Russian Journal of Plant Physiology*, 44 (5):614–622.

Merzlyak M.N., Gitelson A.A., Chivkunova O.B. and Rakitin V.Y. (1999), Non-destructive optical detection of pigment changes during leaf senescence and fruit ripening, *Physiologia Plantarum*, 106(1):135–141.

Merzlyak M.N. and Chivkunova O.B. (2000), Light-stress-induced pigment changes and evidence for anthocyanin photoprotection in apples, *Journal of Photochemistry and Photobiology B: Biology*, 55(2–3):155–163.

Merzlyak M.N., Chivkunova O.B. and Razi Naqvi K. (2002), Does a leaf absorb radiation in the near infrared (780–900 nm) region? A new approach to quantifying optical reflection, absorption and transmission of leaves, *Photosynthesis Research*, 72(3):263–270.

Pogosyan S.I. (2003), Application of reflectance spectroscopy for analysis of higher plant pigments, *Russian Journal of Plant Physiology*, 50(5):704–710 (cover-to-cover translation from *Fiziologiya Rastenii*, 50(5):785–792).

Merzlyak M.N., Melo T.B. and Razi Naqvi K. (2004), Estimation of leaf transmittance in the near infrared region through reflectance measurements, *Journal of Photochemistry and Photobiology. B, Biology*, 74(2–3):145–150.

Merzlyak M.N., Chivkunova O.B., Zhigalova T.V. and Naqvi K.R. (2009), Light absorption by isolated chloroplasts and leaves: effects of scattering and “packing”, *Photosynthesis Research*, 102(1):31–41.

Mesarch M.A., Walter-Shea E.A., Asner G.P., Middleton E.M. and Chan S.S. (1999), A revised measurement methodology for conifer needles spectral optical properties: evaluating the influence of gaps between elements, *Remote Sensing of Environment*, 68(2):177–192.

Messier C. and Bellefleur P. (1987), Variations des propriétés spectrales des feuilles au cours de la croissance du bouleau jaune, *Canadian Journal of Botany*, 65(8):1682–1686.

Metzner P. (1957), Zur Optik der Blattoberflächen, *Die Kulturpflanze*, 5(1):221–239.

Meyer S., Cerovic Z.G., Goulas Y., Montpied P., Demotes-Mainard S., Bidel L.P.R., et al. (2006), Relationships between optically assessed polyphenols and chlorophyll contents, and leaf mass per area ratio in woody plants: a signature of the carbon-nitrogen balance within leaves? *Plant, Cell & Environment*, 29(7):1338–1348.

Meyer-Arendt J.R. (1968), Radiometry and photometry: units and conversion factors, *Applied Optics*, 7(10):2081–2084.

Meziane D. and Shipley B. (1999), Interacting determinants of specific leaf area in 22 herbaceous species: effects of irradiance and nutrient availability, *Plant, Cell & Environment*, 22(5):447–459.

Meziane D. and Shipley B. (2001), Direct and indirect relationships between specific leaf area, leaf nitrogen and leaf gas exchange. Effects of irradiance and nutrient supply, *Annals of Botany*, 88(5):915–927.

Miao Q., Zhao W., Guo X., Liu K., Han J. and Wang Z. (2011), Inversion of reed leaf chlorophyll content based on PROSPECT model, in *Proc. 19th International Conference on Geoinformatics*, Shanghai, China, 24–26 June 2011, IEEE, pp. 1–4.

Miao T., Zhao C.J., Guo X.Y. and Lu S.L. (2013), A framework for plant leaf modeling and shading, *Mathematical and Computer Modelling*, 58(3–4):710–718.

Miao T., Zhao C., Guo X., Lu S. and Wen W. (2014), Visual simulating appearance of plant leaves infected by disease and insect pests, *Transactions of the Chinese Society of Agricultural Engineering*, 30 (2):169–175 (in Chinese).

Middleton E.M. and Walter-Shea E.A. (1995), Optical properties of canopy elements in the boreal forest, in *Proc. 15th International Geoscience and Remote Sensing Symposium (IGARSS'95)*, Florence, Italy, 10–14 July 1995, IEEE, Vol. 1, pp. 789–793.

Middleton E.M., Chan S.S., Mesarch M.A. and Walter-Shea E.A. (1996), A revised measurement methodology for spectral optical properties of conifer needles, in *Proc. 16th International Geoscience and Remote Sensing Symposium (IGARSS'96)*, Lincoln, NB, 27–31 May 1996, IEEE, Vol. 2, pp. 1005–1009.

Middleton E.M., Sullivan J.H., Bovard B.D., DeLucas A.J., Chan S.S. and Cannon T.A. (1997), Seasonal variability in foliar characteristics and physiology for boreal forest species at the five Saskatchewan tower sites during the 1994 Boreal Ecosystem-Atmosphere Study (BOREAS), *Journal of Geophysical Research – Atmospheres*, 102(D24):28831–28844.

Middleton E.M. and Sullivan J. (2000), *BOREAS TE-10 Leaf Chemistry Data*, NASA Goddard Space Flight Center, Greenbelt, MD, 1 October 2000, NASA/TM-2000–209891/VOL161, 30 pages.

Middleton E.M., Campbell P.K.E., McMurtrey J.E., Corp L.A., Butcher L.M. and Chappelle E.W. (2002), "Red edge" optical properties of corn leaves from different nitrogen regimes, in *Proc. International Geoscience and Remote Sensing Symposium (IGARSS'02)*, Toronto, ON, 24–28 June 2002, IEEE, Vol. 4, pp. 2208–2210.

Middleton E.M., McMurtrey J.E., Campbell P.K.E., Corp L.A., Butcher L.M. and Chappelle E.W. (2003), Optical and fluorescence properties of corn leaves from different nitrogen regimes, in *Proc. Remote Sensing for Agriculture, Ecosystems, and Hydrology IV* (Owe M., D'Urso G. and Toulios L., Eds), Agia Pelagia, Greece, 23 September 2002 SPIE, Vol. 4879, pp. 72–83.

Milanowska J. and Gruszecki W.I. (2005), Heat-induced and light-induced isomerization of the xanthophyll pigment zeaxanthin, *Journal of Photochemistry and Photobiology B: Biology*, 80(3):178–186.

Miller C.E. (2001), Chemical principles of near infrared technology, in *Near-Infrared Technology in the Agricultural and Food Industries* (Williams P. and Norris K., Eds), American Association of Cereal Chemist, pp. 19–37.

Miller J.R., Hare E.W. and Wu J. (1990), Quantitative characterization of the vegetation red edge reflectance. 1: An inverted-Gaussian model, *International Journal of Remote Sensing*, 11(10):1755–1773.

Miller J.R., Wu J., Boyer M.G., Belanger M. and Hare E.W. (1991), Seasonal patterns in leaf reflectance red-edge characteristics, *International Journal of Remote Sensing*, 12(7):1509–1523.

Miller J.R., Steven M.D. and Demetriades-shah T.H. (1992), Reflection of layered bean leaves over different soil backgrounds: measured and simulated spectra, *International Journal of Remote Sensing*, 13(17):3273–3286.

Miller J., Berger M., Goulas Y., et al. (2005), *Development of A Vegetation Fluorescence Canopy Model,* ESTEC Contract No. 16365/02/NL/FF, 138 pages.

Milton N.M., Ager C.M. and Power M.S. (1988), *Spectral Reflectance Changes in Greenhouse-Grown Metal-Doped Plants*, USGS, Reston, VA, 13 pages.

Milton N.M., Ager C.M., Eiswerth B.A. and Power M.S. (1989a), Arsenic- and selenium-induced changes in spectral reflectance and morphology of soybean plants, *Remote Sensing of Environment*, 30(3):263–269.

Milton N.M. and Mouat D.A. (1989b), Remote sensing of vegetation responses to natural and cultural environmental conditions, *Photogrammetric Engineering & Remote Sensing*, 55(8):1167–1173.

Milton N.M., Eiswerth B.A. and Ager C.M. (1991), Effect of phosphorus deficiency on spectral reflectance and morphology of soybean plants, *Remote Sensing of Environment*, 36(2):121–127.

Min M. and Lee W.S. (2003), Spectral-based nitrogen sensing for Citrus, in *Proc. 2003 ASAE Annual Meeting*, Las Vegas, NV, 27–30 July 2003, ASAE, 031137.

(2005), Determination of significant wavelengths and prediction of nitrogen content for Citrus, *Transactions of the ASAE*, 48(2):455–461.

Min M., Lee W.S. and Bogrekci I. (2004), The effect of water and variety on nitrogen sensing of *Citrus* leaf, in *Proc. 2004 ASAE Annual Meeting*, Ottawa, ON, 1–4 August 2004, ASAE, 041080.

Mirzaie M., Darvishzadeh R., Shakiba A., Matkan A.A., Atzberger C. and Skidmore A.K. (2014), Comparative analysis of different uni- and multi-variate methods for estimation of vegetation water content using hyper-spectral measurements, *International Journal of Applied Earth Observation and Geoinformation*, 26:1–11.

Młodzińska E. (2009), Survey of plant pigments: molecular and environmental determinants of plant colors, *Acta Biologica Cracovienca Series Botanica*, 51(1):7–16.

Mobasheri M.R. and Fatemi S.B. (2013), Leaf equivalent water thickness assessment using reflectance at optimum wavelengths, *Theoretical and Experimental Plant Physiology*, 25(3):196–202.

Mochizuki S., Cai D., Komori T., Kimura H. and Hori R. (2001), Virtual autumn coloring system based on biological and fractal model, in *Proc. 9th Pacific Conference on Computer Graphics and Applications*, Tokyo, Japan, 16–18 October 2001, IEEE, pp. 348–354.

Mochizuki S., Horie D. and Cai D. (2007), Fractal shading for autumn coloring, *The Journal of the Society for Art and Science*, 6(2):76–87.

Mohammed G.H., Noland T.L., Irving D., Sampson P.H., Zarco-Tejada P.J. and Miller J.R. (2000), *Natural and Stress-Induced Effects on Leaf Spectral Reflectance in Ontario Species*, Forest Research Report No.156, 42 pages.

Mohl H. (1837), Untersuchungen über die winterliche Färbung der Blätter, *Flora oder allgemeine botanische Zeitung*, 20 (2):673–720.

(1838), Recherches sur la coloration hibernale des feuilles, *Annales des sciences Naturelles*, 9:212–235.

Mohr H. and Schopfer P. (1995), *Plant Physiology*, Springer-Verlag, Berlin, 629 pages.

Moldau H. (1965), *On the Use of Polarized Radiation to Analyse the Reflection Indicatrixes of Leaves, 7: Questions on Radiation Regime of Plant Stand.* Acad Sci ESSR Inst Phys Astron, Tartu, pp. 96–101 (in Russian).

(1967), Optical model of plant leaf, in *Photoactinometric Investigations of Plant Canopy*, Valgus Publishers, Tallinn, pp. 89–109 (in Russian).

Monje O.A. and Bugbee B. (1992), Inherent limitations of nondestructive chlorophyll meters: a comparison of two types of meters, *HortScience*, 27(1):69–71.

Monod B., Collin A., Parent G. and Boulet P. (2009), Infrared radiative properties of vegetation involved in forest fires, *Fire Safety Journal*, 44(1):88–95.

Monson R. and Baldocchi D. (2014), *Terrestrial Biosphere-Atmosphere Fluxes*, Cambridge University Press, Cambridge, 507 pages.

Mooney H.A. and Cleland E.E. (2001), The evolutionary impact of invasive species, *Proceedings of the National Academy of Sciences*, 98(10):5446–5451.

Moorthy I., Miller J.R., Noland T.L., Nielsen U. and Zarco-Tejada P.J. (2003a), Chlorophyll content estimation of Boreal conifers using hyperspectral remote sensing, in *Proc. International Geoscience and Remote Sensing Symposium (IGARSS'03)*, Toulouse, France, 21–25 July 2003, IEEE, Vol. 4, pp. 2568–2570.

Moorthy I., Miller J.R., Zarco-Tejada P.J. and Noland T.L. (2003b), Needle chlorophyll content estimation: a comparative study of PROSPECT and LIBERTY, in *Proc. International Geoscience and Remote Sensing Symposium (IGARSS'03)*, Toulouse, France, 21–25 July 2003, IEEE, Vol. 3, pp. 1676–1678.

Moorthy I., Miller J.R. and Noland T.L. (2008), Estimating chlorophyll concentration in conifer needles with hyperspectral data: an assessment at the needle and canopy level, *Remote Sensing of Environment*, 112(6):2824–2838.

Mora C., Tittensor D.P., Adl S., Simpson A.G.B. and Worm B. (2011), How many species are there on Earth and in the ocean? *PLoS Biology*, 9(8):e1001127.

Moran J.A., Mitchell A.K., Goodmanson G. and Stockburger K.A. (2000), Differentiation among effects of nitrogen fertilization treatments on conifer seedlings by foliar reflectance: a comparison of methods, *Tree Physiology*, 20 (16):1113–1120.

Moran, M.S., Maas S.J., Vanderbilt V.C., Barnes E.M., Miller S.N. and Clark T.R. (2004), Application of image-based remote sensing to irrigated agriculture, in *Remote Sensing for Natural Resource Management and Environmental Monitoring* (Ustin S.L., Ed.), *Manual of Remote Sensing Third Edition*, Vol. 4, Wiley and Sons, pp. 617–676.

Moreno N., Bougourd S., Haseloff J. and Feijó J.A. (2006), Imaging plant cells, in *Handbook of Biological Confocal Microscopy* (Pawley J.B., Ed), Springer, US, pp. 769–787.

Morozzo (1782), Lettre de M. le Comte Morozzo à M. L'Abbé Mongez, Auteur du Journal de Physique, Sur les expériences de M. Achard, sur la couleur des végétaux, *Journal de physique, de chimie, d'histoire naturelle et des arts*, 20: 385–389 .

Morren E. (1858a), *Dissertation sur les feuilles vertes et colorées*, Imprimerie et Lithographie de C. Annoot-Braeckman, Gand, Belgium, 249 pages.

(1858b), Notice sur les changements de couleur des feuilles pendant l'automne, l'hiver et le printemps, *La Belgique Horticole*, 8:53–58.

(1858c), Notice sur les changements de couleur des feuilles pendant l'automne, l'hiver et le printemps (suite et fin), *La Belgique Horticole*, 8:81–85.

Mortenson M.E. (1985), *Geometric Modeling*, Industrial Press, South Norwalk, 452 pages.

Moss D.M. and Rock B.N. (1991), Analysis of red edge spectral characteristics and total chlorophyll values for red spruce (*Picea rubens*) branch segments from Mt Moosilauke, NH, in *Proc. 11th International Geoscience and Remote Sensing Symposium (IGARSS'91)*, Helsinki, Finland, 3–6 June 1991, IEEE, Vol. 3, pp. 1529–1532.

Moss D.N. (1964), Optimum lighting of leaves, *Crop Science*, 4:131–136.

Moss R.A. and Loomis W.E. (1952), Absorption spectra of leaves. I: The visible spectrum, *Plant Physiology*, 27(2):370–391.

Mõttus M., Sulev M. and Hallik L. (2014), Seasonal course of the spectral properties of alder and birch leaves, *IEEE Journal of Selected Topics in Applied Earth Observations and Remote Sensing*, 7(6):2496–2505.

Mouroux (1770), Examen physico-chimique sur les couleurs des fleurs & de quelques autres substances végétales, *Miscellanea philosophico-mathematica Societatis privatae Taurinensis*, 6:11–51.

Muir C.D. (2013), How did the Swiss cheese plant get its holes? *The American Naturalist*, 181(2):273–281.

Muller C. and Riederer M. (2005), Plant surface properties in chemical ecology, *Journal of Chemical Ecology*, 31 (11):2621–2651.

Mulroy T.W. (1979), Spectral properties of heavily glaucous and non-glaucous leaves of a succulent rosette-plant, *Oecologia*, 38(3):349–357.

Murchie E.H and Horton P. (1997), Acclimation of photosynthesis to irradiance and spectral quality in British plant species: chlorophyll content, photosynthetic capacity and habitat preference, *Plant, Cell & Environment*, 20:438–448.

Murray I. and Williams P.C. (1987), *Chemical Principles of Near-Infrared Technology*, American Association of Cereal Chemists, St Paul, MN, 312 pages.

Mutanga O., Ismail R., Ahmed F. and Kumar L. (2007), Using *in situ* hyperspectral remote sensing to discriminate pest attacked pine forests in South Africa, in *Proc. 28th Asian Conference on Remote Sensing*, Kuala Lumpur, Malaysia, 12–16 November 2007, AARS, pp. PS3.G5.4.

Myers D.A., Vogelmann T.C. and Bornman J.F. (1994), Epidermal focussing and effects on light utilization in *Oxalis acetosella*, *Physiologia Plantarum*, 91(4):651–656.

Myers V.I. and Allen W.A. (1968), Electrooptical remote sensing methods as nondestructive testing and measuring techniques in agriculture, *Applied Optics*, 7(9):1819–1838

Myneni R.B., Asrar G., Burnett R.B. and Kanemasu E.T. (1987), Radiative transfer in an anisotropically scattering vegetative medium, *Agricultural and Forest Meteorology*, 41(1–2):97–121.

Myneni R.B. and Ross J., Eds (1991), *Photon-Vegetation Interactions: Applications in Optical Remote Sensing and Plant Ecology*, Springer-Verlag, Berlin, 560 pages.

Naidu R.A., Perry E.M., Pierce F.J. and Mekuria T. (2009), The potential of spectral reflectance technique for the detection of grapevine leafroll-associated virus-3 in two red-berried wine grape cultivars, *Computers and Electronics in Agriculture*, 66(1):38–45.

Nakatani M. and Kawashima S. (1994), An attempt to assess leaf color of winter cereals using a simple video system, Japanese *Journal of Crop Science*, 63(1):42–47 (in Japanese).

Naqvi K.R., Merzlyak M.N. and Melo T.B. (2004), Absorption and scattering of light by suspensions of cells and subcellular particles: an analysis in terms of Kramers-Kronig relations, *Photochemical & Photobiological Sciences*, 3(1):132–137.

Narayanan R.M., Green S.E. and Alexander D.R. (1991), Mid-infrared backscatter spectra of selected agricultural crops, in *Proc. Optics in Agriculture* (DeShazer J.A. and Meyer G.E., Eds), Boston, MA, SPIE, Vol. 1379, pp. 116–122.

Narayanan R.M., Mielke L.N. and Schirmer T.J. (1992), Mid-infrared Laser reflectance of crop leaves subjected to water stress, in *Proc. 12th International Geoscience and Remote Sensing Symposium (IGARSS'92)*, Houston, TX, 26–29 May 1992, IEEE, pp. 339–341.

National Academy of Sciences (1982), *Causes and Effects of Stratospheric Ozone Reduction: An Update*, The National Academies Press, Washington, DC, 352 pages.

Natsuyama H., Ueno S., and Wang A.P. (1998), *Terrestrial Radiative Transfer: Modeling, Computation, and Data Analysis*, Springer Verlag, Tokyo, 279 pages.

Naumann J.C., Anderson J.E. and Young D.R. (2010), Remote detection of plant physiological responses to TNT soil contamination, *Plant and Soil*, 329(1–2):239–248.

Nauš J., Klinkovský T. and Korčáková B. (1993), A model of one chloroplast in a cell for evaluation of reabsorption in the fluorescence spectrum, *Acta Universitatis Palackianae Olomucensis*, 111:47–60.

Nauš J., Klinkovský T., Ilik P. and Cikanek D. (1994), Model studies of chlorophyll fluorescence reabsorption at chloroplast level under different exciting conditions, *Photosynthesis Research*, 40(1):67–74.

Neill S. and Gould K.S. (1999), Optical properties of leaves in relation to anthocyanin concentration and distribution, *Canadian Journal of Botany*, 77(12):1777–1782.

Neinhuis C. and Barthlott W. (1997), Characterization and distribution of water-repellent, self-cleaning plant surfaces, *Annals of Botany*, 79(6):667–677.

Neitzke M. and Therburg A. (2003), Seasonal changes in UV-B absorption in beech leaves (*Fagus sylvatica* L.) along an elevation gradient, *Forstwissenschaftliches Centralblatt*, 122(1):1–21.

Nelson V.N., Gjerstad D.H. and Glover G.R. (1986), Determining nitrogen status of young loblolly pine by leaf reflectance, *Tree Physiology*, 1(3):333–339.

Neto A.J.S., Lopes D.C. and Borges J.C.F. (2017), Assessment of photosynthetic pigment and water contents in intact sunflower plants from spectral indices, *Agriculture*, 7:8.

Neuwirthová E., Lhotáková Z. and Albrechtová J. (2017), The effect of leaf stacking on leaf reflectance and vegetation indices measured by contact probe during the season, *Sensors*, 17(6):1202.

Newnham G.J. and Burt T. (2001), Validation of a leaf reflectance and transmittance model for three agricultural crop species, in *Proc. International Geoscience and Remote Sensing Symposium (IGARSS'01),* Sydney, Australia, 9–13 July 2001, IEEE, Vol. 7, pp. 2976–2978.

Newton I. (1704), *Optics : or a Treatise of the Reflexions, Refractions Inflexions and Colours of Light*, W. Innys, London, 382 pages.

Ngo V.D., Ryu D.K., Chung S.O., Park S.U., Kim S.J. and Park J.T. (2013), Comparison of optical reflectance spectrum at blade and vein parts of cabbage and kale leaves, *CNU Journal of Agricultural Science*, 40(2):163–167.

Nicodemus F.E. (1968), Emissivity of isothermal spherical cavity with gray Lambertian walls, *Applied Optics*, 7 (7):1359–1362.

Nicodemus F.E., Richmond J.C., Hsia J.J., Ginsberg I.W. and Limperis T. (1977), *Geometrical Consideration and Nomenclature for Reflectance*, NBS Monograph, October 1977, NBS MN-160, 52 pages.

Nicoletti S. and Adams A.J. (2000), Optical transmittance of a rice leaf via ray tracing, *Journal of the Arkansas Academy of Science*, 54:81–87.

Niinemets Ü. (1999), Components of leaf dry mass per area – thickness and density – alter leaf photosynthetic capacity in reverse directions in woody plants, *New Phytologist*, 144(1):35–47.

Niklas K.J. (1992), *Plant Biomechanics: An Engineering Approach to Plant Form and Function*, University of Chicago Press, Chicago, IL, 622 pages.

Nikolayev V.S., Sibille P., and Beysens D.A. (1998), Coherent light transmission by a dew pattern, *Optics Communication*, 150 (1–6):263–269.

Nilson T. and Kuusk A. (1989), A reflectance model for the homogeneous plant canopy and its inversion, *Remote Sensing of Environment*, 27(2):157–167.

Nishida K., Kosugi Y. and Ohte N. (2000), Spectral reflectance, photosynthesis, and water deficit stress of tree leaves, *Journal of the Remote Sensing Society of Japan*, 20 (3):230–240 (in Japanese).

Nishio J.N., Sun J. and Vogelmann T.C. (1993), Carbon fixation gradients across spinach leaves do not follow internal light gradients, *The Plant Cell*, 5(8):953–961.

Nissen M., Shcherbakov D., Heyer A., Brümmer F. and Schill R.O. (2015), Behaviour of the platehelminth *Symsagittifera roscoffensis* under different light conditions and the consequences for the symbiotic algae *Tetraselmis convolutae*, *Journal of Experimental Biology*, 218(11):1693–1698.

Niu Y., Chen G., Peng D.L., Song B., Yang Y., Li Z.M. et al. (2014), Grey leaves in an alpine plant: a cryptic colouration to avoid attack? *New Phytologist*, 203(3):953–963.
Niu Y. and Su H. (2014), Alpine scree plants benefit from cryptic coloration with limited cost, *Plant Signaling & Behavior*, 9 (9):e29698.
Niu Y., Chen Z., Stevens M. and Sun H. (2017), Divergence in cryptic leaf colour provides local camouflage in an alpine plant, *Proceedings of the Royal Society of London. Series B*, 284(1864):20171654.
Niyogi K.K., Björkman O. and Grossman A.R. (1997), The roles of specific xanthophylls in photoprotection, *Proceedings of the National Academy of Sciences*, 94(25):14162–14167.
Nobel P.S. (2009), *Physicochemical and Environmental Plant Physiology*, Academic Press, 600 pages.
Noble S.D. and Crowe T.G. (2001a), Background effects on apparent leaf reflectance, in *Proc. ASAE Meeting*, ASAE, Guelph, ON, 8–11 July 2001, 01–1177.
(2001b), Plant discrimination based on leaf reflectance, in *Proc. 2001 ASAE Annual Meeting*, ASAE, Sacramento, CA, 30 July–1 August 2001, 01–1150.
(2005), Analysis of crop and weed leaf diffuse reflectance spectra, *Transactions of the ASAE*, 48(6):2379–2387.
(2007), Sample holder and methodology for measuring the reflectance and transmittance of narrow-leaf samples, *Applied Optics*, 46 (22):4968–4976.
Nocera D.G. (2012), The artificial leaf, *Accounts of Chemical Research*, 45(5):767–776.
Noda H.M., Motohka T., Murakami K., Muraoka H. and Nasahara K.N. (2013), Accurate measurement of optical properties of narrow leaves and conifer needles with a typical integrating sphere and spectroradiometer, *Plant, Cell & Environment*, 36 (10):1903–1909.
Noomen M.F., Skidmore A.K., van der Meer F.D. and Prins H.H.T. (2006), Continuum removed band depth analysis for detecting the effects of natural gas, methane and ethane on maize reflectance, *Remote Sensing of Environment*, 105 (3):262–270.
Noomen M.F. and Skidmore A.K. (2009), The effects of high soil CO_2 concentrations on leaf reflectance of maize plants, *International Journal of Remote Sensing*, 30(2):481–497.
Nørgaard L., Saudland A., Wagner J., Nielsen J.P., Munck L., and Engelsen S.B. (2000), Interval partial least-squares regression (iPLS): a comparative chemometric study with an example from near-infrared spectroscopy, *Applied Spectroscopy*, 54(3):413–419.
Norris K.H., Barnes R.F., Moore J.E. and Shenk J.S. (1976), Predicting forage quality by infrared reflectance spectroscopy, *Journal of Animal Science*, 43(4):889–897.
Nostell P., Roos A. and Rönnow D. (1999), Single-beam integrating sphere spectrophotometer for reflectance and transmittance measurements versus angle of incidence in the solar wavelength range on diffuse and specular samples, *Review of Scientific Instruments*, 70:2481–2494.
Ntefidou M. and Manetas Y. (1996), Optical properties of hairs during the early stages of leaf development in *Platanus orientalis*, *Australian Journal of Plant Physiology*, 23(4):535–538.
Nybakken L., Aubert S. and Bilger W. (2004a), Epidermal UV-screening of arctic and alpine plants along a latitudinal gradient in Europe, *Polar Biology*, 27(7):391–398.
Nybakken L., Bilger W., Johanson U., Björn L.O., Zielke M. and Solheim B. (2004b), Epidermal UV-screening in vascular plants from Svalbard (Norwegian Artic), *Polar Biology*, 27(7):383–390.
Nybakken L. and Bilger W. (2007), Effects of enhanced UV-B radiation and epidermal UV screening in arctic and alpine plants, in *Arctic Alpine Ecosystems and People in a Changing Environment* (Ørbæk J.B., Kallenborn R., Tombre I., Hegseth E.N., Falk-Petersen S. and Hoel A.H., Eds), Springer-Verlag, Berlin, pp. 195–209.
Oerke E.C., Steiner U., Dehne H.W. and Lindenthal M. (2006), Thermal imaging of cucumber leaves affected by downy mildew and environmental conditions, *Journal of Experimental Botany*, 57(9):2121–2132.
Oerke E.C. and Steiner U. (2010), Potential of digital thermography for disease control, in *Precision Crop Protection – The Challenge and Use of Heterogeneity* (Oerke E.C., Gerhards R., Menz G. and Sikora R.A., Eds), Springer-Verlag, Berlin, pp. 167–182.
Oerke E.C., Fröhling P. and Steiner U. (2011), Thermographic assessment of scab disease on apple leaves, *Precision Agriculture*, 12(5):699–715.
Ögren E. and Evans J.R. (1993), Photosynthetic light-response curves. I: The influence of CO_2 partial pressure and leaf inversion, *Planta*, 189(2):182–190.
Oh Y. and Hong J.Y. (2007), Re-examination of analytical models for microwave scattering from deciduous leaves, *IET Microwaves, Antennas & Propagation*, 1(3):617–623.
Okayama H. (1996), How different are the indicatrixes of the leaves of various woody plant species? *Applied Optics*, 35 (18):3250–3254.
Okayama H. and Li C. (2007), Evaluation of the surface of objects by use of Minnaert constants, in *Proc. Remote Sensing for Agriculture, Ecosystems, and Hydrology IX* (Neale C.M.U., Owe M. and D'Urso G., Eds), Florence, Italy, 19 September 2007, SPIE, Vol. 6742, 674204.
Olf H.G. (1988), Stokes's pile of plates revisited, *Journal of the Optical Society of America*, 5(10):1620–1625.
Olioso A., Sòria G., Sobrino J. and Duchemin B. (2007), Evidence of low land surface thermal infrared emissivity in the presence of dry vegetation, *IEEE Geoscience and Remote Sensing Letters*, 4(1):112–116.
Ollinger S.V., Smith M.L., Martin M.E., Hallett R.A., Goodale C.L. and Aber J.D. (2002), Regional variation in foliar chemistry and N cycling among forests of diverse history and composition, *Ecology*, 83(2):339–355.

Ollinger S.V., Richardson A.D., Martin M.E., Hollinger D.Y., Frolking S.E., Reich P.B., et al. (2008), Canopy nitrogen, carbon assimilation, and albedo in temperate and boreal forests: functional relations and potential climate feedbacks, *Proceedings of the National Academy of Sciences of the United States of America*, 105(49):19336–19341.

Olson C.E. and Good R.E. (1962), Seasonal changes in light reflectance from forest vegetation, *Photogrammetric Engineering*, 28(3):107–114.

Omarov T.B. and Tashenov B.T. (2005), Tikhov's astrobotany as a prelude to modern astrobiology. in *Perspectives in Astrobiology* (Hoover R.B., Rozanov A.Y. and Paepeet R.R., Eds), IOS Press, pp. 86–87.

Omasa K., Abo F., Hashimoto Y. and Aiga I. (1980a), Evaluation of air pollution injury to plants by image processing, *Research Report from the National Institute for Environmental Studies*, 11: 249–254.

(1980b), Measurement of the thermal pattern of plant leaves under fumigation with air pollutant, *Research Report from the National Institute for Environmental Studies*,11: 239–247.

Ono K., Hiraide M. and Amari M. (2003), Determination of lignin, holocellulose, and organic solvent extractives in fresh leaf, litterfall, and organic material on forest floor using near-infrared reflectance spectroscopy, *Journal of Forest Research*, 8 (3):191–198.

Ordoñez C., Martínez J., Matías J.M., Reyes A.N. and Rodríguez-Pérez J.R. (2010), Functional statistical techniques applied to vine leaf water content determination, *Mathematical and Computer Modelling*, 52(7–8):1116–1122.

Ordóñez C., Rodríguez-Pérez J.R., Moreira J.J. and Sanz E. (2013), Using hyperspectral spectrometry and functional models to characterize vine-leaf composition, *IEEE Transactions on Geoscience and Remote Sensing*, 51 (5):2610–2618.

Oren M. and Nayar S.K. (1995), Generalization of the Lambertian model and implications for machine vision, *International Journal of Computer Vision*, 14(3):227–251.

Osborne B.A. and Raven J.A. (1986), Light absorption by plants and its implications for photosynthesis, *Biological Reviews*, 61(1):1–61.

Osnas J.L., Lichstein J.W., Reich P.B. and Pacala S.W. (2013), Global leaf trait relationships: mass, area, and the leaf economics spectrum, *Science*, 340(6133):741–744.

Osorio S. and Bossomaier T.R.J. (1992), Human cone-pigment spectral sensitivities and the reflectances of natural surfaces, *Biological Cybernetics*, 67:217–222.

Osorio D. and Vorobyev M. (1996), Colour vision as an adaptation to frugivory in primates, *Proceedings of the Royal Society of London. Series B, Biological Sciences*, 263(1370):593–599.

Otterman J., Brakke T. and Smith J. (1995), Effects of leaf-transmittance versus leaf-reflectance on bidirectional scattering from canopy/soil surface: an analytical study, *Remote Sensing of Environment*, 54(1):49–60.

Ounis A., Cerovic Z.G., Briantais J.M. and Moya I. (2001), Dual excitation FLIDAR for the estimation of epidermal UV absorption in leaves and canopies, *Remote Sensing of Environment*, 76(1):33–48.

Ourcival J.M., Joffre R. and Rambal S. (1999), Exploring the relationships between reflectance and anatomical and biochemical properties of *Quercus ilex* leaves, *New Phytologist*, 143(2):351–364.

Overton T.K. (1969), Camouflage colours, *Journal of the Society of Dyers and Colourists*, 85(4):152–154.

Pachepsky L.B. and Acock B. (1996), A model 2DLEAF of leaf gas exchange: development, validation, and ecological application, *Ecological Modelling*, 93(1–3):1–18.

Pachepsky L.B. and Acock B. (1998), Effects of leaf anatomy on hypostomatous leaf gas exchange: a theoretical study with the 2DLEAF model, *Biotronics*, 27:1–14.

Pacumbaba R.O. and Beyl C.A. (2011), Changes in hyperspectral reflectance signatures of lettuce leaves in response to macronutrient deficiencies, *Advances in Space Research*, 48(1):32–42.

Paillotin G., Dobek A., Breton J., Leibl W. and Trissl H.W. (1993), Why does the light-gradient photovoltage from photosynthetic organelles show a wavelength-dependent polarity? *Biophysical Journal*, 65(1):379–385.

Paillotin G., Leibl W., Gapinski J., Breton J. and Dobek A. (1998), Light gradients in spherical photosynthetic vesicles, *Biophysical Journal*, 75(1):124–133.

Palamaryuk V.E. and Guminetskii S.G. (1968), Dynamics of the change in spectral coefficients of plant leaves under illumination, *Journal of Applied Spectroscopy*, 9 (1):743–745 (cover-to-cover translation from Zhurnal Prikladnoi Spektroskopii, 9(1):152–155, 1968).

Palmer J.M. (1997), The measurement of transmission, absorption, emission, and reflection, in *Handbook of Optics*, American Institute of Physics, New York, Vol. II, pp. 25.1–25.25.

Palmer T. (1877a), The various changes caused on the spectrum by different vegetable colouring matters, *The Monthly Microscopical Journal*, 17(5):225–235.

(1877b), Vegetable coloring matters, *Scientific American*, 3:1214–1214.

Pan G. and Narayanan R.M. (2002), Comparison of microwave scattering models for leaf, in *Proc. International Geoscience and Remote Sensing Symposium (IGARSS'02)*, Toronto, ON, 24–28 June 2002, IEEE, Vol. 1, pp. 653–655.

Panigrahy S., Kumar T. and Manjunath K.R. (2012), Hyperspectral leaf signature as an added dimension for species discrimination: case study of four tropical mangroves, *Wetlands Ecology and Management*, 20(2):101–110.

Pannell J.R. (2014), Leaf mimicry: chameleon-like leaves in a Patagonian vine, *Current Biology*, 24(9):R357–R359.

Paradiso R., Meinen E., Snel J.F.H., De Visser P., Van Ieperen W., Hogewoning S.W. et al. (2011), Spectral dependence of photosynthesis and light absorptance in single leaves and canopy in rose, *Scientia Horticulturae*, 127(4):548–554.

Park Y.I., Chow W.S. and Anderson J.M. (1996), Chloroplast movement in the shade plant *Tradescantia albiflora* helps protect photosystem II against light stress, *Plant Physiology*, 111(3):867–875.

Parkhurst D.F. (1976), Effects of *Verbascum thapsus* leaf hairs on heat and mass transfer: a reassessment, *New Phytologist*, 76 (3):453–457.

(1982), Stereological methods for measuring internal leaf structure variables, *American Journal of Botany*, 69(1):31–39.

(1986), Internal leaf structure: a three-dimensional perspective, in *On the Economy of Plant Form and Function* (Givnish T. J., Ed.), Cambridge University Press, Cambridge, pp. 215–249.

Parmesan C. (2006), Ecological and evolutionary responses to recent climate change, *Annual Review of Ecology, Evolution, and Systematics*, 37:637–669.

Patil S.B. and Bodhe S.K. (2011), Leaf disease severity measurement using image processing, *International Journal of Engineering and Technology*, 3(5):297–301.

Pavan G., Jacquemoud S., Bidel L., François C., de Rosny G., Rambaut J.P. et al. (2004), RAMIS: a new portable field radiometer to estimate leaf biochemical content, in *Proc. 7th International Conference on Precision Agriculture and Other Precision Resources Management*, Minneapolis, MN, 25–28 July 2004, 15 pages.

Pearman G.I. (1966), The reflection of visible radiation from leaves of some Western Australian species, *Australian Journal of Biological Sciences*, 19(1):97–103.

Pedrini A. (1991), *Methodology for Fluorescence Yield Determination of Vegetation*, Unit for Advanced Techniques, Joint Research Centre, Ispra, Italy, 45 pages.

Pedrós R., Moya I., Goulas Y. and Jacquemoud S. (2008), Chlorophyll fluorescence emission spectrum inside a leaf, *Photochemical & Photobiological Sciences*, 7(4):498–502.

Pedrós R., Goulas Y., Jacquemoud S., Louis J. and Moya I. (2010), FluorMODleaf: a new leaf fluorescence emission model based on the PROSPECT model, *Remote Sensing of Environment*, 114(1):155–167.

Pelletier J. and Caventou J. (1817), Notice sur la matière verte des feuilles, *Journal de Pharmacie et des Sciences Accessoires*, 3:486–491.

Peltoniemi J.I., Gritsevich M. and Puttonen E. (2015), *Reflectance and polarization characteristics of various vegetation types*, in *Light Scattering Reviews 9* (Kokhanovsky A., Ed), Springer-Verlag, Berlin, pp. 257–293.

Pengra B.W., Johnston C.A. and Loveland T.R. (2007), Mapping an invasive plant, *Phragmites australis*, in coastal wetlands using the EO-1 Hyperion hyperspectral sensor, *Remote Sensing of Environment*, 108(1):74–81.

Penny S.S. and Willan R.C. (2014), Description of a new species of giant clam (Bivalvia: Tridacnidae) from Ningaloo Reef, Western Australia, *Molluscan Research*, 34(4):201–211.

Peñuelas J., Filella I., Biel C., Serrano L. and Savé R. (1993), The reflectance at the 950–970 nm region as an indicator of plant water status, *International Journal of Remote Sensing*, 14(10):1887–1905.

Peñuelas J., Gamon J.A., Fredeen A.L., Merino J. and Field C.B. (1994), Reflectance indices associated with physiological changes in nitrogen- and water-limited sunflower leaves, *Remote Sensing of Environment*, 48(2):135–146.

Peñuelas J., Filella I. and Gamon J.A. (1995), Assessment of photosynthetic radiation-use efficiency with spectral reflectance, *New Phytologist*, 131(3):291–296.

Peñuelas J., Llusia J., Piñol J. and Filella I. (1997), Photochemical reflectance index and leaf photosynthetic radiation-use-efficiency assessment in Mediterranean trees, *International Journal of Remote Sensing*, 18(13):2863–2868.

Peñuelas J. and Filella I. (1998), Visible and near-infrared reflectance techniques for diagnosing plant physiological status, *Trends in Plant Science*, 3(4):151–156.

Peñuelas J. and Inoue Y. (1999), Reflectance indices indicative of changes in water and pigment contents of peanut and weat leaves, *Photosynthetica*, 36(3):355–360.

Peñuelas J., Munné-Bosch S., Llusià J. and Filella I. (2004), Leaf reflectance and photo- and antioxidant protection in field-grown summer-stressed *Phillyrea angustifolia*. Optical signals of oxidative stress? *New Phytologist*, 162 (1):115–124.

Perevertun M.P. (1960), Optical properties of certain plant species in the infrared region of the spectrum in transmitted light, *Trudy sektora astrobotaniki*, 8: 59–73 (English translation).

Pérez-Harguindeguy N., Díaz S., Garnier E., Lavorel S., Poorter H., Jaureguiberry P., et al. (2013), New handbook for standardised measurement of plant functional traits worldwide, *Australian Journal of Botany*, 61(3):167–234.

Peterson D.L., Aber J.D., Matson P.A., et al. (1988), Remote sensing of forest canopy and leaf biochemical contents, *Remote Sensing of Environment*, 24(1):85–108.

Petisco C., Garcia-Criado B., Vazquez de Aldana B.R., Zabalgogeazcoa I., Mediavilla S. and Garcia-Ciudad A. (2005), Use of near-infrared reflectance spectroscopy in predicting nitrogen, phosphorus and calcium contents in heterogeneous woody plant species, *Analytical and Bioanalytical Chemistry*, 382(2):458–465.

Peynado A., Gausman H.W., Escobar D.E., Rodriguez R.R. and Garza M.V. (1979), Evidence of cell membrane injury detected by reflectance measurements, *Cryobiology*, 16(1):63–68.

Peyrat A., Terraz O., Merillou S. and Galin E. (2008), Generating vast varieties of realistic leaves with parametric 2Gmap L-systems, *The Visual Computer*, 24(7–9):807–816.

Pfeiffer H.G. and Liebhafsky H.A. (1951), The origin of Beer's law, *Journal of Chemical Education*, 28(3):123–125.

Pfündel E.E., Agati G. and Cerovic Z.G. (2006), Optical properties of plant surfaces, in *Biology of the Plant Cuticle* (Riederer M. and Müller C., Eds), Blackwell Publishing, pp. 216–249.

Phan C.T., Brach E.J. and Jasmin J.J. (1979), Studies of the detection of lettuce maturity: anatomical observations and reflectance measurements in the visible range (350–650 nm), *Canadian Journal of Plant Science*, 59(4):1067–1075.

Philips-Invernizzi B., Dupont D. and Cazé C. (2001), Biographical review for reflectance of diffusing media, *Optical Engineering*, 40(6):1082–1092.

Phipson M.T.L. (1858), Sur la couleur des feuilles, *Comptes Rendus Hebdomadaires des Séances de l'Académie des Sciences*, 47:912–913.
Pickering J.W., Moes C.J.M., Sterenborg H.J.C.M., Prahl S.A. and van Gemert M.J.C. (1992), Two integrating spheres with an intervening scattering sample, *Journal of the Optical Society of America*, 9(4):621–631.
Pickering J.W., Prahl S.A., van Wieringen N., Beek J.F., Sterenborg H.J.C.M. and van Gemert M.J.C. (1993), Double-integrating sphere system for measuring the optical properties of tissue, *Applied Optics*, 32(4):399–410.
Pieruschka R., Huber G. and Berry J.A. (2010), Control of transpiration by radiation, *Proceedings of the National Academy of Sciences of the United States of America*, 107(30):13372–13377.
Pietrzykowski E., Stone C., Pinkard E. and Mohammed C. (2006), Effects of *Mycosphaerella* leaf disease on the spectral reflectance properties of juvenile *Eucalyptus globulus* foliage, *Forest Pathology*, 36(5):334–348.
Pilarski J. (1999), Gradient of photosynthetic pigments in the bark and leaves of lilac (*Syringa vulgaris* L.), *Acta Physiologiae Plantarum*, 21(4):365–373.
Pilarski J. and Rajba S. (2004), Measurement of light gradient in plant organs with a fiber optic microprobe, *Acta Physiologiae Plantarum*, 26(4):405–410.
Pincebourde S. and Casas J. (2006), Leaf miner-induced changes in leaf transmittance cause variations in insect respiration rates, *Journal of Insect Physiology*, 52(2):194–201.
Pinzón J.E., Ustin S.L., Castañeda C.M. and Smith M.O. (1998), Investigation of leaf biochemistry by hierarchical foreground/background analysis, *IEEE Transactions on Geoscience and Remote Sensing*, 36(6):1913–1927.
Platt J.R. (1956), *Radiation Biology* (Hollaender A., Ed), McGraw-Hill, Vol. 3, pp. 71–123.
Pliny the Elder (1855), *Natural History* (Bostock J. and Riley H.T., translators), H.G.Bohn, London, 536 pages.
Polischuk V.P., Shadchina T.M., Kompanetz T.I., Budzanivskaya I.G., Boyko A.L. and Sozinov A.A. (1997), Changes in reflectance spectrum characteristic of *Nicotiana debneyi* plant under the influence of viral infection, *Archives of Phytopathology and Plant Protection*, 31(1):115–119.
Polívka T., Herek J.L., Zigmantas D., Åkerlund H.E. and Sundström V. (1999), Direct observation of the (forbidden) S_1 state in carotenoids, *Proceedings of the National Academy of Sciences of the United States of America*, 96(9):4914–4917.
Ponzoni F.J. and Gonçalves J.L. (1999), Spectral features associated with nitrogen, phosphorous, and potassium deficiencies in *Eucalyptus saligna* seedling leaves, *International Journal of Remote Sensing*, 20(11):2249–2264.
Poorter H., Niinemets Ü., Poorter L., Wright I.J. and Villar R. (2009), Causes and consequences of variation in leaf mass per area (LMA): a meta-analysis, *New Phytologist*, 182(3):565–588.
Poorter L., Oberbauer S.E. and Clark D.B. (1995), Leaf optical properties along a vertical gradient in a tropical rain forest canopy in Costa Rica, *American Journal of Botany*, 82(10):1257–1263.
Poorter L., Kwant R., Hernandez R., Medina E. and Werger M.J.A. (2000), Leaf optical properties in Venezuelan cloud forest trees, *Tree Physiology*, 20(8):519–526.
Poschlod P., Kleyer M., Jackel A.K., Dannemann A. and Tackenberg O. (2003), BIOPOP – A database of plant traits and internet application for nature conservation, *Folia Geobotanica*, 38(3):263–271.
Pospergelis M.M. (1969), Spectroscopic measurements of the four Stokes parameters for light scattered by natural objects, *Soviet Astronomy*, 12(6):973–977.
Poulson M.E. and Vogelmann T.C. (1990), Epidermal focussing and effects upon photosynthetic light-harvesting in leaves of *Oxalis*, *Plant, Cell & Environment*, 13(8):803–811.
Poulson M.E. and DeLucia E.H. (1993), Photosynthetic and structural acclimation to light direction in vertical leaves of *Silphium terebinthinaceum*, *Oecologia*, 95(3):393–400.
Pourcel L., Routaboul J.M., Cheynier V., Lepiniec L. and Debeaujon I. (2007), Flavonoid oxidation in plants: from biochemical properties to physiological functions, *Trends in Plant Science*, 12(1):29–36.
Powell K., Lamb D., Barret D., Blanchfield A., Held A., Renzullo L., et al. (2006), *Early detection of phylloxera in grapevines* (Vitis vinifera) *– A test-bed for spectro-optical and chemical fingerprinting of pest-induced grapevine stress*, Cooperative Research Centre for Viticulture, Australia, April 2006, 153 pages.
Powles S.B. (1984), Photoinhibition of photosynthesis induced by visible light, *Annual Review of Plant Physiology*, 35:15–44
Prahl S. (2001), *Optical Absorption of Water*. Oregon Medical Laser Center. http://omlc.ogi.edu/spectra/water/
Prance G.T. and Sandved K.B. (1985), *Leaves. The formation, Characteristics, and Uses of Hundreds of Leaves Found in all Parts of the World*, Thames & Hudson, London. 244 pages.
Prioul J.L. and Chartier P. (1977), Partitioning of transfer and carboxylation components of intracellular resistance to photosynthetic CO_2 fixation: a critical analysis of the method used, *Annals of Botany*, 41(174):789–800.
Priestley J. (1774), *Experiments and Observations on Different Kinds of Air*, Vol, I. Printed for J Johnson, London, 324 pages.
Prokopy R.P. and Owens E.D. (1983), Visual detection of plants by herbivorous insects, *Annual Review Entomology*, 28:337–364.
Prokopy R.P., Collier R.H. and Finch S. (1983), Visual detection of host plants by cabbage root flies, *Entomologia Experimentalis et Applicata*, 34(1):85–89.
Prusinkiewicz P. and Lindenmayer A. (1990), *The Algorithmic Beauty of Plants*, Springer-Verlag, New York, 228 pages.
Pu R., Ge S., Kelly N.M. and Gong P. (2002), Correlation analysis of hyperspectral absorption features with the water status of coast live oak leaves, in *Proc. Imaging Spectrometry VII* (Descour M.R. and Shen S.S., Eds), San Diego, CA, 1–3 August 2001, SPIE, Vol. 4480, pp. 147–153.
Pu R., Ge S., Kelly N.M. and Gong P. (2003), Spectral absorption features as indicators of water status in coast live oak (*Quercus agrifolia*) leaves, *International Journal of Remote Sensing*, 24(9):1799–1810.

Pu R., Foschi L. and Gong P. (2004), Spectral feature analysis for assessment of water status and health level in coast live oak (*Quercus agrifolia*) leaves, *International Journal of Remote Sensing*, 25(20):4267–4286.
Pu R., Kelly M., Chen Q. and Gong P. (2008), Spectroscopic determination of health levels of coast live oak (*Quercus agrifolia*) leaves, *Geocarto International*, 23(1):3–20.
Purcell D.E., O'Shea M.G., Johnson R.A. and Kokot S. (2009), Near-infrared spectroscopy for the prediction of disease ratings for Fiji leaf gall in sugarcane clones, *Applied Spectroscopy*, 63(4):450–457.
Pydipati R., Burks T.F. and Lee W.S. (2005), Statistical and neural network classifiers for Citrus disease detection using machine vision, *Transactions of the ASAE*, 48(5):2007–2014.
(2006), Identification of Citrus disease using color texture features and discriminant analysis, *Computers and Electronics in Agriculture*, 52(1–2):49–59.
Qi Y., Bai S., Vogelmann T.C., Heisler G.M. and Qin J. (2002), Methodology for comprehensive evaluation of UV-B tolerance in trees, in *Proc. Ultraviolet Ground- and Space-based Measurements, Models, and Effects* (Slusser J.R., Herman J.R. & Gao W., Eds), San Diego, CA, 29 July 2001, SPIE, Vol. 4482, pp. 367–380.
Qi Y., Bai S., Gao W. and Heisler G.M. (2003a), Intra- and inter-specific comparisons of leaf UV-B absorbing-compound concentration of southern broadleaf trees in the United States, in *Proc. Ultraviolet Ground- and Space-based Measurements, Models and Effects II* (Gao W., Herman J.R., Shi G., Shibasaki K. and Slusser J.R., Eds), Hangzhou, China, SPIE, Vol. 4896, pp. 120–129.
Qi Y., Bai S. and Heisler G.M. (2003b), Changes in ultraviolet-B and visible optical properties and absorbing pigment concentrations in pecan leaves during a growing season, *Agricultural and Forest Meteorology*, 120(1–4):229–240.
Qi Y., Bai S., Vogelmann T.C. and Heisler G.M. (2003c), Penetration of UV-A, UV-B, blue and red light into leaf tissues of pecan measured by a fiber optic microprobe system, in *Proc. Ultraviolet Ground- and Space-based Measurements, Models, and Effects III* (Slusser J.R., Herman J.R. and Gao W., Eds), San Diego, CA, SPIE, Vol. 5156, pp. 281–290.
Qi Y., Heisler G.M., Gao W., Vogelmann T.C. and Bai S. (2010), Characteristics of UV-B radiation tolerance in broadleaf trees in southern USA, in *UV Radiation in Global Climate Change* (Gao W., Slusser J.R. and Schmoldt D.L., Eds), Springer–Verlag, Berlin, pp. 509–530.
Qin J.L., Rundquist D., Gitelson A., Steele M., Harkins C. and Briles R. (2010), A non-linear model for measuring grapevine leaf thickness by means of red-edge/near-infrared spectral reflectance, *Acta Ecologica Sinica*, 30(6):297–303.
Qin R., Xu G., Guo L., Jiang Y. and Ding R. (2012), Preparation and characterization of a novel poly(urea-formaldehyde) microcapsules with similar reflectance spectrum to leaves in the UV-Vis-NIR region of 300–2500 nm, *Materials Chemistry and Physics*, 136(2–3):737–743.
Qu Y., Zhu Y. and Ge X. (2014), A reflectance spectra model of heavy metal stressed leaves: advances in the PROSPECT model adding specific absorption coefficients of heavy metal ion, in *Proc. 35th International Symposium on Remote Sensing of Environment*, Beijing, China, 22–26 April 2013, 012041.
Quickenden T.I., Freeman C.G. and Litjens R.A. (2000), Some comments on the paper by Edward S. Fry on the visible and near-ultraviolet absorption spectrum of liquid water, *Applied Optics*, 39(16):2740–2742.
Rabideau G.S., French C.S. and Holt A.S. (1946), The absorption and reflection spectra of leaves, chloroplasts suspensions, and chloroplast fragments as measured in the Ulbricht sphere, *American Journal of Botany*, 33(10):769–777.
Rabinowitch E.I. (1951), Light absorption by pigments in the living cell, in *Photosynthesis and Related Processes* (Rabinowitch E.I., Ed.), Interscience Publishers, New York, pp. 672–739.
Rahman H., Pinty B. and Verstraete M.M. (1993), Coupled surface-atmosphere reflectance (CSAR) model. 2. Semiempirical surface model usable with NOAA Advanced Very High Resolution Radiometer Data, *Journal of Geophysical Research*, 98 (D11):20,791–20,801.
Raines G.L. and Canney F.C. (1980), Vegetation and geology, in *Remote Sensing in Geology* (Siegal B.S. and Gillespie A.R., Eds), John Wiley & Sons, pp. 365–380.
Ramalingam N., Ling P.P. and Derksen R.C. (2001), Leaf surface moisture detection by multi-spectral imaging, in *Proc. 2001 ASAE Annual Meeting*, Sacramento, CA, 30 July–1 August 2001, ASAE, 01–3004.
Ramalingam N., Ling P.P. and Derksen R.C. (2003), Leaf surface wetness detection using ground based multispectral imaging, in *Proc. 2003 ASAE Annual Meeting*, Las Vegas, NV, 27–30 July 2003, ASAE, 033130.
Ramos M.E. and Lagorio M.G. (2004), True fluorescence spectra of leaves, *Photochemical & Photobiological Sciences*, 3 (11–12):1063–1066.
Ramsey E.W. and Rangoonwala A. (2004), Remote sensing and the optical properties of the narrow cylindrical leaves of *Juncus roemerianus*, *IEEE Transactions on Geoscience and Remote Sensing*, 42(5):1064–1075.
Rapaport T., Hochberg U., Shoshany M., Karnieli A. and Rachmilevitch S. (2015), Combining leaf physiology, hyperspectral imaging and partial least squares-regression (PLS-R) for grapevine water status assessment, *ISPRS Journal of Photogrammetry and Remote Sensing*, 109:88–97.
Rathod P.H., Rossiter D.G., Noomen M.F. and van der Meer F.D. (2013), Proximal spectral sensing to monitor phytoremediation of metal-contaminated soils, *International Journal of Phytoremediation*, 15(5):405–426.
Raunkiaer C. (1934), *The Life Forms of Plants and Statistical Plant Geography*, Oxford University Press, New York, 632 pages.
Rausher M.D. (1992), Natural selection and the evolution of plant-insect interactions, in *Insect Chemical Ecology – An Evolutionary Approach* (Roitberg B.D. and Isman M.B., Eds), Springer, US, 376 pages.
Rautiainen M., Heiskanen J., Eklundh L., Mõttus M., Lukeš P. and Stenberg P. (2010), Ecological applications of physically based remote sensing methods, *Scandinavian Journal of Forest Research*, 25(4):325–339.

Rautiainen M., Lukeš P., Homolová L., Hovi A., Pisek J. and Mõttus M. (2018), Spectral properties of coniferous forests: a review of *in situ* and laboratory measurements, Remote Sensing, 10(2):207.

Raven P.N., Jordan D.L. and Smith C.E. (2002), Polarized directional reflectance from laurel and mullein leaves, *Optical Engineering*, 41(5):1002–1012.

Ravi J., Lee S.T., Paulraj M. and Hernandez R. (2008), Feasibility of neural network approach in spectral mixture analysis of reflectance spectra, *International Journal of Remote Sensing*, 29(10):2981–2992.

Ray J. (1686), *Historia Plantarum*, Vol. 1, Clark, London, 862 pages.

Raymond C.A. and Schimleck L.R. (2002), Development of near infrared reflectance analysis calibrations for estimating genetic parameters for cellulose content in *Eucalyptus globulus*, *Canadian Journal of Forest Research*, 32(1):170–176.

Read J.J., Tarpley L., McKiniona J.M. and Reddyc K.R. (2002), Narrow-waveband reflectance ratios for remote estimation of nitrogen status in cotton, *Journal of Environmental Quality*, 31(5):1442–1452.

Reale L., Lai A., Tucci A., Poma A., Faenov A., Pikuz T., et al. (2004), Differences in X-ray absorption due to cadmium treatment in *Saponaria officinalis* leaves, *Microscopy Research and Technique*, 64(1):21–29.

Reale L., Lai A., Bellucci I., Faenov A., Pikuz T., Flora F., et al. (2006), Microradiography as a tool to detect heavy metal uptake in plants for phytoremediation applications, *Microscopy Research and Technique*, 69(8):666–674.

Reale L., Kaiser J., Reale A., Lai A., Flora F., Balerna A., et al. (2008a), Mapping the intake of different elements in vegetal tissues by dual-energy X-ray imaging at DaΦne synchrotron light source, *Microscopy Research and Technique*, 71 (3):179–185.

Reale L., Lai A., Sighicelli M., Faenov A., Pikuz T., Flora F., et al. (2008b), Qualitative detection of Mg content in a leaf of *Hedera helix* by using X-ray radiation from a laser plasma source, *Microscopy Research and Technique*, 71(6):459–468.

Regalado A. (2010), Reinventing the leaf, *Scientific American*, 303(4):86–89.

Regan B.C., Julliot C., Simmen B., Viénot F., Charles-Dominique P. and Mollon J.D. (2001), Fruits, foliage and the evolution of primate colour vision, *Philosophical Transactions of the Royal Society of London. Series B*, 356(1407):229–283.

Reich P.B., Walters M.B. and Ellsworth D.S. (1997), From tropics to tundra: global convergence in plant functioning, *Proceedings of the National Academy of Sciences*, 94(25):13730–13734.

Reicosky D.A. and Hanover J.W. (1978), Physiological effects of surface waxes. I: Light reflectance for glaucous and nonglaucous *Picea pungens*, *Plant Physiology*, 62:101–104.

Reisig D. and Godfrey L. (2007), Spectral response of cotton aphid – (Homoptera: Aphididae) and spider mite – (Acari: Tetranychidae) infested cotton: controlled studies, *Environmental Entomology*, 36(6):1466–1474.

Renzullo L.J., Blanchfield A.L. and Powell K.S. (2005), Insights into the early detection of grapevine phylloxera from *in situ* hyperspectral data, in *Proc. ISHS Acta Horticulturae: III International Grapevine Phylloxera Symposium*, Fremantle, Australia, Vol. 733, pp. 59–74.

Renzullo L.J., Blanchfield A.L., Guillermin R., Powell K.S. and Held A.A. (2006a), Comparison of PROSPECT and HPLC estimates of leaf chlorophyll contents in a grapevine stress study, *International Journal of Remote Sensing*, 27 (4):817–823.

Renzullo L.J., Blanchfield A.L. and Powell K.S. (2006b), A method of wavelength selection and spectral discrimination of hyperspectral reflectance spectrometry, *IEEE Transactions on Geoscience and Remote Sensing*, 44(7):1986–1994.

Reum D. and Zhang Q. (2007), Wavelet based multi-spectral image analysis of maize leaf chlorophyll content, *Computers and Electronics in Agriculture*, 56(1):60–71.

Reyna E. and Badhwar G.D. (1985), Inclusion of specular reflectance in vegetative canopy models, *IEEE Transactions on Geoscience and Remote Sensing*, 23(5):731–736.

Riaño D., Ustin S.L., Usero L. and Patricio M.A. (2005a), Estimation of fuel moisture content using neural networks, in *Proc. International Work-conference on the Interplay between Natural and Artificial Computation* (Mira J. and Alvarez J.R., Eds), Las Palmas de Gran Canaria, Canary Islands, 15–18 June 2005, pp. 489–498.

Riaño D., Vaughan P., Chuvieco E., Zarco-Tejada P.J. and Ustin S.L. (2005b), Estimation of fuel moisture content by inversion of radiative transfer models to simulate equivalent water thickness and dry matter content: analysis at leaf and canopy level, *IEEE Transactions on Geoscience and Remote Sensing*, 43(4):819–826.

Ribeiro da Luz B. (2006), Attenuated total reflectance spectroscopy of plant leaves: a tool for ecological and botanical studies, *New Phytologist*, 172(2):305–318.

Ribeiro da Luz B. and Crowley J.K. (2007), Spectral reflectance and emissivity features of broadleaf plants: prospects for remote sensing in the thermal infrared (8.0–14.0 μm), *Remote Sensing of Environment*, 109(4):393–405.

(2010), Identification of plant species by using high spatial and spectral resolution thermal infrared (8.0–13.5 μm) imagery, *Remote Sensing of Environment*, 114(2):404–413.

Richards-Kortum R., Rava R.P., Fitzmaurice M., Tong L.L., Ratliff N.B., Kramer J.R. et al. (1989), A one-layer model of laser-induced fluorescence for diagnosis of disease in human tissue: applications to atherosclerosis, *IEEE Transactions on Biomedical Engineering*, 36(12):1222–1232.

Richardson A.D. and Berlyn G.P. (2002), Spectral reflectance and photosynthetic properties of *Betula papyrifera* (Betulaceae) leaves along an elevational gradient on Mt Mansfield, Vermont, USA, *American Journal of Botany*, 89(1):88–94.

Richardson A.D., Duigan S.P. and Berlyn G.P. (2002), An evaluation of noninvasive methods to estimate foliar chlorophyll content, *New Phytologist*, 153(1):185–194.

Richardson A.D., Berlyn G.P. and Duigan S.P. (2003a), Reflectance of Alaskan black spruce and white spruce foliage in relation to elevation and latitude, *Tree Physiology*, 23:537–544.

Richardson A.D., Reeves J.B. and Gregoire T.G. (2003b), Multivariate analyses of visible/near infrared (VIS/NIR) absorbance spectra reveal underlying spectral differences among dried, ground conifer needle samples from different growth environments, *New Phytologist*, 161(1):291–301.

Richardson A.J. and Gausman H.W. (1982), Reflectance differences between untreated and Mepiquat Chloride-treated, field-grown cotton through a growing season, *Remote Sensing of Environment*, 12(6):501–507.

Richter J.P. (1970), *The Notebooks of Leonardo da Vinci Compiled and Edited from the Original Manuscripts*, Vol. I, Dover, 367 pages.

Richter T. and Fukshansky L. (1994), Authentic *in vivo* absorption spectra for chlorophyll in leaves as derived from *in situ* and *in vitro* measurements, *Photochemistry and photobiology*, 59(2):237–247.

(1996a), Optics of a bifacial leaf. 1: A novel combined procedure for deriving the optical parameters, *Photochemistry and photobiology*, 63(4):507–516.

(1996b), Optics of a bifacial leaf. 2. Light regime as affected by the leaf structure and the light source, *Photochemistry and Photobiology*, 63(4):517–527.

Ridley M. (2003), *Evolution*, Wiley-Blackwell, 784 pages.

Riedell W.E. and Blackmer T.M. (1999), Leaf reflectance spectra of cereal aphid-damaged wheat, *Crop Science*, 39 (6):1835–1840.

Riederer M. and Muller C. (2006), *Biology of the Plant Cuticle*, Blackwell Publishing, 456 pp.

Ripple W.J. (1986), Spectral reflectance relationships to leaf water stress, *Photogrammetric Engineering & Remote Sensing*, 52(10):1669–1675.

Ripullone F., Rivelli A.R., Baraldi R., et al. (2011), Effectiveness of the photochemical reflectance index to track photosynthetic activity over a range of forest tree species and plant water statuses, *Functional Plant Biology*, 38 (3):177–186.

Rivard B., Feng J., Gallie A. and Sanchez-Azofeifa A. (2008a), Continuous wavelets for the improved use of spectral libraries and hyperspectral data, *Remote Sensing of Environment*, 112(6):2850–2862.

Rivard B., Sánchez-Azofeifa G.A., Foley S. and Calvo-Alvarado J.C. (2008b), Species classification of tropical tree leaf reflectance and dependence on selection of spectral bands, in *Hyperspectral Remote Sensing of Tropical and Sub-Tropical Forests* (Kalacska M. and Sánchez-Azofeifa G.A., Eds), CRC Press, pp. 141–159.

Robberecht R. and Caldwell M.M. (1978), Leaf epidermal transmittance of ultraviolet radiation and its implications for plant sensitivity to ultraviolet-radiation induced injury, *Oecologia*, 32(3):277–287.

Robberecht R., Caldwell M.M. and Billings W.D. (1980), Leaf ultraviolet optical properties along a latitudinal gradient in the arctic-alpine life zone, *Ecology*, 61(3):612–619.

Robberecht R. and Caldwell M.M. (1983), Protective mechanisms and acclimation to solar ultraviolet-B radiation in *Oenothera stricta*, *Plant, Cell & Environment*, 6(6):477–485.

(1986), Leaf UV optical properties of *Rumex patientia* L. and *Rumex obtusifolius* L. in regard to a protective mechanism against solar UV-B radiation injury, in *Stratosphere Ozone Reduction, Solar Ultraviolet Radiation and Plant Life* (Worrest R.C. and Caldwell M.M., Eds), Springer-Verlag, Berlin, pp. 251–259.

Roberts D.A., Adams J.B. and Smith M.O. (1990a), Predicted distribution of visible and near-infrared radiant flux above and below a transmittant leaf, *Remote Sensing of Environment*, 34(1):1–17.

(1990b), Transmission and scattering of light by leaves: effects on spectral mixtures in *Proc. 10th International Geoscience and Remote Sensing Symposium (IGARSS'90)*, College Park, MD, 20–24 May 1990, IEEE, Vol. 3, pp. 1381–1384.

Roberts D.A., Gardner M., Church R., Ustin S., Scheer G. and Green, R.O. (1998a), Mapping chaparral in the Santa Monica Mountains using multiple endmember spectral mixture models, *Remote Sensing of Environment*, 65(3):267–279.

Roberts D.A., Nelson B.W., Adams J.B. and Palmer F. (1998b), Spectral changes with leaf aging in Amazon caatinga, *Trees – Structure and Function*, 12(6):315–325.

Robinson G.W., Cho C.H. and Gellene G.I. (2000), Refractive index mysteries of water, *Journal of Physical Chemistry B*, 104 (30):7179–7182.

Rock B.N., Hoshizaki T. and Miller J.R. (1988), Comparison of *in situ* and airborne spectral measurements of the blue shift associated with forest decline, *Remote Sensing of Environment*, 24(1):109–127.

Rock B.N., Williams D.L., Moss D.M., Lauten G.N. and Kim M. (1994), High-spectral resolution field and laboratory optical reflectance measurements of red spruce and eastern hemlock needles and branches, *Remote Sensing of Environment*, 47 (2):176–189.

Rodkaew Y., Chongstitvatana P., Siripant S. and Lursinsap C. (2004), Modeling plant leaves in marble-patterned colours with particle transportation system, in *Proc. 4th International Workshop on Functional-Structural Plant Models* (Godin C., Ed), Montpellier, France, 7–11 June 2004, pp. 391–397.

Rodriguez R.R. and Gausman H.W. (1977), Leaf ultraviolet radiation reflectance, transmittance, and absorptance of ten crop species, *Journal of the Rio Grande Valley Horticultural Society*, 31:175–180.

Rodríguez-Pérez J.R., Riaño D., Carlisle E., Ustin S. and Smart D.R. (2007), Evaluation of hyperspectral reflectance indexes to detect grapevine water status in vineyards, *American Journal of Enology and Viticulture*, 58(3):302–317.

Roelofsen H.D., van Bodegom P.M., Kooistra L. and Witte J.P.M. (2014), Predicting leaf traits of herbaceous species from their spectral characteristics, *Ecology and Evolution*, 4(6):706–719.

Rogge D., Rivard B., Deyholos M.K., Lévesque J., Ardouin J.P. and Faust A.A. (2012), Potential discrimination of toxic industrial chemical effects on poplar, canola and wheat, detectable in optical wavelengths 400–2450 nm, *IEEE Journal of Selected Topics in Applied Earth Observations and Remote Sensing*, 5(2):563–573.

Romberger J.A., Hejnowicz Z. and Hill J.F. (2004), *Plant Structure: Function and Development*, Springer-Verlag, Berlin, 524 pages.

Romero A., Aguado I., Chuvieco E. and Yebra M. (2007), Evaluation of dry foliage matter through normalised indexes and inversion of reflectivity models, in *Proc. 6th International Workshop of the European Association of Remote Sensing Laboratories* (Gitas I.Z. and Carmona-Moreno C., Eds), Thessaloniki, Greece, 27–29 September 2007, EARSeL, pp. 87–90.

Romero A., Aguado I. and Yebra M. (2012), Estimation of dry matter content in leaves using normalized indexes and PROSPECT model inversion, *International Journal of Remote Sensing*, 33(2):396–414.

Rose A.H. and Lindquist O.H. (1994), *Insects of Eastern Spruces, Fir and, Hemlock*, Canadian Forest Service, Ottawa, 162 pages.

Rose A.W., Hawkes H.E. and Webb J.S. (1980), *Geochemistry in Mineral Exploration*, Academic Press, 657 pages.

Rosema A., Verhoef W., Schroote J. and Snel J.F.H. (1991), Simulating fluorescence light-canopy interaction in support of laser-induced fluorescence measurements, *Remote Sensing of Environment*, 37(2):117–130.

Rosevear M.J., Young A.J. and Johnson G.N. (2001), Growth conditions are more important than species origin in determining leaf pigment content of British plant species, *Functional Ecology*, 15(4):474–480.

Ross J. (1981), *The Radiation Regime and Architecture of Plant Stands*, Springer, The Netherlands, 420 pages.

Ross J. and Marshak A. (1989), The influence of leaf orientation and the specular component of leaf reflectance on the canopy bidirectional reflectance, *Remote Sensing of Environment*, 27(3):251–260.

Ross J., Meinander O. and Sulev M. (1994), Spectral scattering properties of Scots pine shoots, in *Proc. 14th International Geoscience and Remote Sensing Symposium (IGARSS'94)*, Pasadena, CA, 8–12 August 1994, IEEE, Vol. 3, pp. 1451–1454.

Rosso P.H., Pushnik J.C., Lay M. and Ustin S.L. (2005), Reflectance properties and physiological responses of *Salicornia virginica* to heavy metal and petroleum contamination, *Environmental Pollution*, 137(2):241–252.

Roth-Nebelsick A., Uhl D., Mosbrugger V. and Kerp H. (2001), Evolution and function of leaf venation architecture: a review, *Annals of Botany*, 87(5):553–566.

Rozema J., Chardonnens A., Tosserams M., Hafkenscheid R. and Bruijnzeel S. (1997), Leaf thickness and UV-B absorbing pigments of plants in relation to an elevational gradient along the Blue Mountains, Jamaica, *Plant Ecology*, 128 (1–2):151–159.

Rudorff B.F., Mulchi C.L., Lee E.H., Rowland R. and Daughtry C. (1995), Effects of O_3 and SO_2 on leaf characteristics in soybeans grown under ambient- and enriched-carbon dioxide atmosphere, in *Proc. Air Toxics and Water Monitoring*, (Russwurm G.M., Ed), SPIE, Munich, Germany, Vol. 2503, pp. 89–100.

Ruhland C.T. and Day T.A. (1996), Changes in UV-B radiation screening effectiveness with leaf age in *Rhododendron maximum*, *Plant, Cell & Environment*, 19(6):740–746.

Rühle W. and Wild A. (1979), The intensification of absorbances in leaves by light-dispersion, *Planta*, 146(5):551–557.

Rumpf T., Mahlein A.K., Steiner U., Oerke E.C., Dehne H.W. and Plümer L. (2010), Early detection and classification of plant diseases with Support Vector Machines based on hyperspectral reflectance, *Computers and Electronics in Agriculture*, 74 (1):91–99.

Rumpho M.E., Summer E.J. and Manhart J.R. (2000), Solar-powered sea slugs. Mollusc/algal chloroplast symbiosis, *Plant Physiology*, 123(1):29–38.

Rumpho M.E., Worful J.M., Lee J., Kannan K., Tyler M.S., Bhattacharya D., et al. (2008), Horizontal gene transfer of the algal nuclear gene psbO to the photosynthetic sea slug *Elysia chlorotica*, *Proceedings of the National Academy of Sciences of Sciences of the United States of America*, 105(46):17867–17871.

Runeckles V.C. and Resh H.M. (1975), The assessment of chronic ozone injury to leaves by reflectance spectrophotometry, *Atmospheric Environment*, 9(4):447–452.

Rupert C.S. and Latarjet R. (1978), Toward a nomenclature and dosimetric scheme applicable to all radiations, *Photochemistry and Photobiology*, 28(1):3–5.

Ruth B., Hoque E., Weisel B. and Hutzler P.J.S. (1991), Reflectance and fluorescence parameters of needles of Norway spruce affected by forest decline, *Remote Sensing of Environment*, 38(1):35–44.

Rvachev V.P. and Guminetskii S.G. (1966), The structure of light beams reflected by plant leaves, *Journal of Applied Spectroscopy*, 4(5):303–307 (cover-to-cover translation from *Zhurnal Prikladnoi Spektroskopii*, 4(5):415–421).

Sack L. and Scoffoni C. (2013), Leaf venation: structure, function, development, evolution, ecology and applications in the past, present and future, *New Phytologist*, 198(4):983–1000.

Sack L., Caringella M., Scoffoni C., Mason C., Rawls M., Markesteijn L. et al. (2014), Leaf vein length per unit area is not intrinsically dependent on image magnification: avoiding measurement artifacts for accuracy and precision, *Plant Physiology*, 166(2):829–838.

Safir G.R., Svits G.H. and Ellingboe A.H. (1972), Spectral reflectance and transmittance of corn leaves infected with *Helminthosporium maydis*, *Phytopathology*, 62(10):1210–1213.

Sagan C., Thompson W.R., Carlson R. and Gurnett D. (1993), A search for life on Earth from the Galileo spacecraft, *Nature*, 365:715–721.

Sage R. (2005), Atmospheric CO_2, environmental stress, and evolution of C_4 photosynthesis, in *A history of Atmospheric CO_2 and Its Effects on Plants, Animals, and Ecosystems* (Ehleringer J.R., Cerling T.E. and Dearing M.D., Eds), Springer, New York, pp. 185–214.

Saito H. (2001), Blue biliprotein as an effective factor for cryptic colouration in *Rhodinia fugax* larvae, *Journal of Insect Physiology*, 47(2):205–212.

Salisbury J.W. (1986), Preliminary measurements of leaf spectral reflectance in the 8–14 μm region, *International Journal of Remote Sensing*, 7(12):1879–1886.

Salisbury J.W., Milton N.M. and Walsh P.A. (1987), Significance of non-isotropic scattering from vegetation for geobotanical remote sensing, *International Journal of Remote Sensing*, 8(7):997–1009.

Salisbury J.W. and Milton N.M. (1988), Thermal infrared (2.5- to 13.5-μm) directional-hemispherical reflectance of leaves, *Photogrammetric Engineering & Remote Sensing*, 54(9):1301–1304.

Salisbury J.W. and D'Aria D.M. (1992), Emissivity of terrestrial materials in the 8–14 μm atmospheric window, *Remote Sensing of Environment*, 42(2):83–106.

Salisbury J.W., Wald A. and D'Aria D.M. (1994), Thermal infrared remote sensing and Kirchhoff's law. 1: Laboratory measurements, *Journal of Geophysical Research-Solid Earth*, 99 (B6):11,897–11,911.

Salm-Horstmar F. (1854), Substanz der grünen Infusorien, *Annalen der Physik*, 169(9):159–159.

(1855a), Ueber das dispergirte rothe Licht in der Auflösung des Chlorophylls, *Annalen der Physik*, 170(3):467–468.

(1855b), Untersuchung des grünen Stoffes, den die kleinsten grünen Infusorien enthalten, *Annalen der Physik*, 170 (3):466–467.

(1856), Untersuchung des grünen Stoffes wahrer Infusorien, *Annalen der Physik*, 173(2):331–333.

Saltas V., Triantis D., Manios T. and Vallianatos F. (2007), Biomonitoring of environmental pollution using dielectric properties of tree leaves, *Environmental Monitoring and Assessment*, 133(1–3):69–78.

Saltelli A. and Bolado R. (1998), An alternative way to compute Fourier amplitude sensitivity test (FAST), *Computational Statistics & Data Analysis*, 26(4):445–460.

Saltelli A., Tarantola S. and Chan K.P.S. (1999), A quantitative, model independent method for global sensitivity analysis of model output, *Technometrics*, 41(1):39–56.

Saltelli A., Chan K. and Scott E.M. (2000), *Sensitivity Analysis*, John Wiley & Sons, 475 pages.

Sanches I.D., Souza Filho C.R. and Kokaly R.F. (2014), Spectroscopic remote sensing of plant stress at leaf and canopy levels using the chlorophyll 680 nm absorption feature with continuum removal, *ISPRS Journal of Photogrammetry and Remote Sensing*, 97:111–122.

Sanches M.C. and Válio I.F.M. (2006), Leaf optical properties of two liana species *Canavalia parviflora* Benth. and *Gouania virgata* Reissk in different light conditions, *Revista Brasileira de Botânica*, 29(2):319–330.

Sánchez-Azofeifa G.A., Castro K., Wright S.J., Gamon J., Kalacska M., Rivard B., et al. (2009), Differences in leaf traits, leaf internal structure, and spectral reflectance between two communities of lianas and trees: implications for remote sensing in tropical environments, *Remote Sensing of Environment*, 113(10):2076–2088.

Sánchez-Azofeifa G.A., Oki Y., Fernandes G.W., Ball R.A. and Gamon J. (2012), Relationships between endophyte diversity and leaf optical properties, *Trees – Structure and Function*, 26(2):291–299.

Sancho-Knapik D., Alvarez-Arenas T.E.G., Peguero-Pina J.J. and Gil-Pelegrin E. (2010), Air-coupled broadband ultrasonic spectroscopy as a new non-invasive and non-contact method for the determination of leaf water status, *Journal of Experimental Botany*, 61(5):1385–1391.

Sancho-Knapik D., Gismero, J. Asensio A., Peguero-Pina J.J., Fernández V., Álvarez-Arenas T.E.G., et al. (2011a), Microwave L-band (1730 MHz) accurately estimates the relative water content in poplar leaves. A comparison with a near infrared water index (R(1300)/R(1450)), *Agricultural and Forest Meteorology*, 151(7):827–832.

Sancho-Knapik D., Alvarez-Arenas T.E.G., Peguero-Pina J.J., Fernandez V. and Gil-Pelegrin E. (2011b), Relationship between ultrasonic properties and structural changes in the mesophyll during leaf dehydration, *Journal of Experimental Botany*, 62(10):3637–3645.

Sancho-Knapik D., Peguero-Pina J.J., Medrano H., Fariñas M.D., Álvarez-Arenas T.E.G. and Gil-Pelegrín E. (2013), The reflectivity in the S-band and the broadband ultrasonic spectroscopy as new tools for the study of water relations in *Vitis vinifera* L., *Physiologia Plantarum*, 148(4):512–521.

Sandquist D.R. and Ehleringer J.R. (1997), Intraspecific variation of leaf pubescence and drought response in *Encelia farinosa* associated with contrasting desert environments, *New Phytologist*, 135:635–644.

Sandwald E.F. (1981), Laboratory-determined spectral signatures of leaves of healthy and rizomania-diseased sugar beets and disease interpretability from aerial IRC photographs, in *Proc. Signatures spectrales d'objets en télédétection*, Avignon, France, 8–11 September 1981, INRA, pp. 201–208.

Sankaran S., Ehsani R. and Etxeberria E. (2010a), Mid-infrared spectroscopy for detection of Huanglongbing (greening) in citrus leaves, *Talanta*, 83(2):574–581.

Sankaran S., Mishra A., Ehsani R., and Davis C. (2010b), A review of advanced techniques for detecting plant diseases, *Computers and Electronics in Agriculture*, 72(1):1–13.

Sankaran S., Mishra A., Maja J.M. and Ehsani R. (2011), Visible-near infrared spectroscopy for detection of Huanglongbing in citrus orchards, *Computers and Electronics in Agriculture*, 77(2):127–134.

Sankaran S., Ehsani R., Inch S.A. and Ploetz R.C. (2012), Evaluation of visible-near infrared reflectance spectra of avocado leaves as a non-destructive sensing tool for detection of Laurel Wilt, *Plant Disease*, 96(11):1683–1689.

Santos A.O. and Kaye O. (2009), Grapevine leaf water potential based upon near infrared spectroscopy, *Scientia Agricola (Piracicaba, Braz.)*, 66(3):287–292.

Santos M.J., Hestir E.L., Khanna S. and Ustin S.L. (2012), Image spectroscopy and stable isotopes elucidate functional dissimilarity between native and nonnative plant species in the aquatic environment, *New Phytologist*, 193:683–695.

Sanz C. (1994), *Mesure et modélisation de la variation directionnelle des propriétés optiques des feuilles*, DEA Dynamique de la Biosphère Continentale, Université Paul Sabatier, Toulouse, France, 32 pp.

Sarabandi K. and Ulaby F.T. (1988), Technique for measuring the dielectric constant of thin materials, *IEEE Transactions on Instrumentation and Measurement*, 37(4):631–636.

Sarabandi K., Ulaby F.T. and Senior T.B.A. (1990), Millimeter wave scattering model for a leaf, *Radio Science*, 25 (1):9–18.

Sardans J., Peñuelas J. and Rodà F. (2006), Plasticity of leaf morphological traits, leaf nutrient content, and water capture in the Mediterranean evergreen oak *Quercus ilex* subsp. ballota in response to fertilization and changes in competitive conditions, *Ecoscience*, 13(2):258–270.

Sarto A.W., Woldemar C.M. and Vanderbilt V.C. (1989), Polarized Light Angle Reflectance instrument I polarized incidence (POLAR:I), in *Proc. Polarization Considerations for Optical Systems II* (Chipman R.A., Ed), San Diego, CA, 9–11 August 1989, SPIE, Vol. 1166, pp. 220–230.

Savenkov S.N., Muttiah R.S. and Oberemok Y.A. (2003), Transmitted and reflected scattering matrices from an English oak leaf, *Applied Optics*, 42(24):4955–4962.

Savenkov S.N., Mishchenko L.T., Muttiah R.S., Oberemok Y.A. and Mishchenko I.A. (2004), Mueller polarimetry of virus-infected and healthy wheat under field and microgravity conditions, *Journal of Quantitative Spectroscopy & Radiative Transfer*, 88(1–3):327–343.

Savenkov S.N. and Muttiah R.S. (2004a), Inverse polarimetry, and light scattering from leaves, in *Photopolarimetry in Remote Sensing* (Videen G., Ed), Kluwer Academic Publishers, pp. 243–264.

Savenkov S.N., Muttiah R.S., Yakubchak V.V. and Klimov A.S. (2007), Anisotropy parameters for *Chlorophytum* leaf epidermis, in *Proc. 10th International Conference on Electromagnetic & Light Scattering* (Videen G., Mishchenko M., Mengüç M.P. and Zakharova N., Eds), Bodrum, Turkey, 17–22 June 2007, pp. 185–188.

Savenkov S.N. (2015), Principles of the Mueller matrix measurements, in *Light Scattering Reviews 9* (Kokhanovsky A., Ed), Springer Praxis Books, pp. 213–255.

Saviranta N.M.M., Julkunen-Tiitto R., Oksanen E. and Karjalainen R.O. (2010), Leaf phenolic compounds in red clover (*Trifolium pratense* L.) induced by exposure to moderately elevated ozone, *Environmental Pollution*, 158(2):440–446.

Schade U., Holldack K., Kuske P., Wüstefeld G. and Hübers H.W. (2004), THz near-field imaging employing synchrotron radiation, *Applied Physics Letters*, 84(8):1422–1424.

Schade U., Holldack K., Martin M.C. and Fried D. (2005), THz near-field imaging of biological tissues employing synchrotron radiation, in *Proc. Ultrafast Phenomena in Semiconductors and Nanostructure Materials IX* (Tsen K.T., Song J.J. and Jiang H., Eds), San Jose, CA, 24 January 2005, SPIE, Vol. 5725, pp. 46–52.

Schaefer H.M. and Wilkinson D.M. (2004), Red leaves, insects and coevolution: a red herring? *Trends in Ecology & Evolution*, 19(12):616–618.

Schaepman M.E., Ustin S.L., Plaza A.J., Painter T.H., Verrelst J. and Liang S. (2009), Earth system science related imaging spectroscopy – An assessment, *Remote Sensing of Environment*, 113(1):S123–S137.

Schaepman-Strub G., Schaepman M.E., Painter T.H., Dangel S. and Martonchik J.V. (2006), Reflectance quantities in optical remote sensing – Definitions and case studies, *Remote Sensing of Environment*, 103(1):27–42.

Schanda R. (1986), *Physical Fundamentals of Remote Sensing*, Springer Verlag, Berlin, 187 pages.

Schaper H. and Chacko E.K. (1991), Relation between extractable chlorophyll and portable chlorophyll meter readings in leaves of eight tropical and subtropical fruit-tree species, *Journal of Plant Physiology*, 138(6):674–677.

Schellekens J.H., Gilbes F., Rivera G.D., Ysa Y.C., Chardon S. and Fong Y. (2005), Reflectance spectra of tropical vegetation as a response to metal enrichment in the substrate of west-central Puerto Rico, *Caribbean Journal of Earth Science*, 39:9–12.

Scheller M., Jansen C., and Koch M. (2010), Applications of effective medium theories in the terahertz regime, in *Recent Optical and Photonic Technologies* (Kim K.Y., Ed.), INTECH, Croatia, pp. 231–250.

Schepers J.S., Blackmer T.M., Wilhelm W.W. and Resende M. (1996), Transmittance and reflectance measurements of corn leaves from plants with different nitrogen and water supply, *Journal of Plant Physiology*, 148(5):523–529.

Schepers J.S., Blackmer T.M. and Francis D.D. (1997), Chlorophyll meter method for estimating nitrogen content in plant tissue, in *Handbook of Reference Methods for Plant Analysis* (Kalra Y.P., Ed), CRC Press, pp. 129–135.

Schimleck L.R., Doran J.C. and Rimbawanto A. (2003), Near infrared spectroscopy for cost effective screening of foliar oil characteristics in a *Melaleuca cajuputi* breeding population, *Journal of Agricultural and Food Chemistry*, 51 (9):2433–2437.

Schimper A.F.W., Fisher W.R., Groom P. and Balfour I.B. (1903), *Plant-Geography Upon a Physiological Basis*, Clarendon Press, 839 pages.

Schlemmer M.R., Francis D.D., Shanahan J.F. and Schepers J.S. (2005), Remotely measuring chlorophyll content in corn leaves with differing nitrogen levels and relative water content, *Agronomy Journal*, 97(1):106–112.

Schlerf M., Atzberger C., Udelhoven T., Jarmer T., Mader S., Werner W. et al. (2003), Spectrometric estimation of leaf pigments in Norway spruce needles using band-depth analysis, partial least-square regression and inversion of a conifer leaf model, in *Proc. 3rd EARSeL Workshop on Imaging Spectroscopy* (Habermeyer M., Müller A. and Holzwarth S., Eds), Herrsching, Germany, 13–16 May 2003, EARSeL, pp. 559–568.

Schmidt J.A. (2010), Electronic spectroscopy of lignins, in *Lignin and Lignans: Advances in Chemistry* (Heitner C., Dimmel D.R., and Schmidt J.A., Eds), CRC Press, pp. 49–102.

Schmidt K.S. and Skidmore A.K. (2001), Exploring spectral discrimination of grass species in African rangelands, *International Journal of Remote Sensing*, 22(17):3421–3434.

Schönn L. (1872), Ueber die Absorptionsstreifen des Blattgrüns, *Annalen der Physik*, 221(1):166–167.

Schreiber U., Vidaver W., Runeckles V.C. and Rosen P. (1978), Chlorophyll fluorescence assay for ozone injury in intact plants, *Plant Physiology*, 61(1):80–84.

Schubert E.F. (2006), *Light Emitting Diodes*, Cambridge University Press, Cambridge, 434 pages.

Schuepp P.H. (1993), Tansley Review N°59 – Leaf boundary layers, *New Phytologist*, 125(3):477–507.

Schultz H.R. (1996), Leaf absorptance of visible radiation in *Vitis vinifera* L.: estimates of age and shade effects with a simple field method, *Scientia Horticulturae*, 66(1–2):93–102.

Schulz H., Engelhardt U.H., Wegent A., Drews H.H. and Lapczynski S. (1999), Application of near-infrared reflectance spectroscopy to the simultaneous prediction of alkaloids and phenolic substances in green tea leaves, *Journal of Agricultural and Food Chemistry*, 47(12):5064–5067.

Schulze E.D., Eller B.M., Thomas D.A., Willert D.J. and Brinckmann E. (1980), Leaf temperatures and energy balance of *Welwitschia mirabilis* in its natural habitat, *Oecologia*, 44(2):258–262.

Schuster A. (1905), Radiation through a foggy atmosphere, *The Astrophysical Journal*, 21(1):1–22.

Schutt J.B., Rowland R.R. and Heartly W.H. (1984a), A laboratory investigation of a physical mechanism for the extended infrared absorption ("red shift") in wheat, *International Journal of Remote Sensing*, 5(1):95–102.

Schutt J.B., Rowland R.A. and Heggestad H.E. (1984b), Identification of injury resulting from atmospheric pollutants using reflectance measurements, *Journal of Environmental Quality*, 13(4):605–608.

Schwaller M.R., Schnetzler C.C. and Marshall P.E. (1981), *The Changes in Leaf Reflectance of Sugar Maple (*Acer saccharum *Marsh) Seedlings in Response to Heavy Metal Stress*, NASA, Greenbelt, MD, June 1981, Technical Memorandum 82150, 17 pages.

Schwaller M.R., Schnetzler C.C. and Marshall P.E. (1983), The changes in leaf reflectance of sugar maple (Acer saccharum Marsh) seedlings in response to heavy metal stress, *International Journal of Remote Sensing*, 4(1):93–100.

Schwalm P.A., Starrett P.H. and McDiarmid R.W. (1977), Infrared reflectance in leaf-sitting neotropical frogs, *Science*, 196 (4295):1225–1227.

Schweiger A.K., Schütz M., Risch A.C., Kneubühler M., Haller R. and Schaepman M.E. (2017), How to predict plant functional types using imaging spectroscopy: linking vegetation community traits, plant functional types and spectral response, *Methods in Ecology and Evolution*, 8(1):86–95.

Seelig H.D., Adams W.W., Hoehn A., Stodieck L.S., Klaus D.M. and Emery W.J. (2008a), Extraneous variables and their influence on reflectance-based measurements of leaf water content, *Irrigation Science*, 26(5):407–414.

Seelig H.D., Hoehn A., Stodieck L.S., Klaus D.M., Adams III W.W. and Emery W.J. (2008b), Relations of remote sensing leaf water indices to leaf water thickness in cowpea, bean, and sugarbeet plants, *Remote Sensing of Environment*, 112 (2):445–455.

(2008c), The assessment of leaf water content using leaf reflectance ratios in the visible, near-, and short-wave-infrared, *International Journal of Remote Sensing*, 29(13):3701–3713.

Seely G.R. and Jensen R.G. (1965), Effect of solvent on the spectrum of chlorophyll, *Spectrochimica Acta*, 21(10):1835–1845.

Segelstein, D.J. (1981), *The Complex Refractive Index of Water*, MSc Thesis, Department of Physics. University of Missouri, Kansas City, MO, 167 pages.

Semenenko A.D. (1960a), Optical properties of etiolated oats, *Trudy sektora astrobotaniki*, 8: 54–66 (in Russian).

(1960b), Optical properties of various species of wheat, *Trudy sektora astrobotaniki*, 8: 46–55 (English translation).

Senebier J. (1783), *Recherches sur l'influence de la lumière solaire pour métamorphoser l'air fixe en air pur par la végétation*, B. Chirol, Genève, 385 pages.

Senior T.B.A., Sarabandi K. and Ulaby F.T. (1987), Measuring and modeling the backscattering cross-section of a leaf, *Radio Science*, 22(6):1109–1116.

Serbin S.P., Dillaway D.N., Kruger E.L. and Townsend P.A. (2012), Leaf optical properties reflect variation in photosynthetic metabolism and its sensitivity to temperature, *Journal of Experimental Botany*, 63(1):489–502.

Serbin S.P., Singh A., McNeil B.E., Kingdon C.C. and Townsend P.A. (2014), Spectroscopic determination of leaf morphological and biochemical traits for northern temperate and boreal tree species, *Ecological Applications*, 24 (7):1651–1669.

Senn G. (1908), *Die Gestalts- und Lageveränderung der Pflanzen-Chromatophoren*, Wilhelm Engelmann, Leipzig, 397 pages.

Serôdio J., Cruz S., Cartaxana P., and Calado R. (2014), Photophysiology of kleptoplasts: photosynthetic use of light by chloroplasts living in animal cells, *Philosophical Transactions of The Royal Society B*, 369(1640):20130242.

Serrano L. (2008), Effects of leaf structure on reflectance estimates of chlorophyll content, *International Journal of Remote Sensing*, 29(17):5265–5274.

Seybold A. (1933a), Uber die Optischen Eigenschaften der Laubblätter. III: *Planta*, 20(3):577–601.

(1933b), Uber die Optischen Eigenschaften der Laubblätter. IV: *Planta*, 21(2):251–265.

(1956), Hat die Chloroplastenverlagerung in Laubblättern eine Bedeutung? *Naturwissenschaften*, 43(4):90–91.

Seyfried M. and Fukshansky L. (1983), Light gradients in plant tissue, *Applied Optics*, 22(9):1402–1408.

Seyfried M. and Schafer E. (1983), Changes in the optical properties of cotyledons of *Cucurbita pepo* during the first seven days of their development, *Plant, Cell & Environment*, 6(8):633–640.

Shacklette H.T., Lakin H.W., Hubert A.E. and Curtin G.C. (1970), Absorption of gold by plants, *USGS Bulletin*, 1314-B, 23 pages.

Shakespeare T. and Shakespeare J. (2003), A fluorescent extension to the Kubelka-Munk model, *Color Research and Applications*, 28(1):4–14.

Sharifi M.R., Gibson A.C. and Rundel P.W. (1997), Surface dust impacts on gas exchange in Mojave Desert shrubs, *Journal of Applied Ecology*, 34(4):837–846.

Shashar N., Cronin T.W., Wolff L.B. and Condon M.A. (1998), The polarization of light in a tropical rain forest, *Biotropica*, 30 (2):275–285.

Shenk J.S., Westerhaus M.O. and Hoover M.R. (1979), Analysis of forage by infrared reflectance, *Journal of Dairy Science*, 63 (5):807–812.

Shenk J.S., Landa I., Hoover M.R. and Westerhaus M.O. (1981), Description and evaluation of near infrared reflectance spectro-computer for forage and grain analysis, *Crop Science*, 21(3):355–358.

Shepherd K.D., Palm C.A., Gachengo C.N. and Vanlauwe B. (2003), Rapid characterization of organic resource quality for soil and livestock management in tropical agroecosystems using near-infrared spectroscopy, *Agronomy Journal*, 95(5):1314–1322.

Sheue C.R., Pao S.H., Chien L.F., Chesson P. and Peng C.I. (2012), Natural foliar variegation without costs? The case of Begonia, *Annals of Botany*, 109(6):1065–1074.

Shi R.H., Zhuang D.F. and Li S. (2005), Study on the extraction of plant biochemical information from canopy reflectance spectra, in *Proc. IEEE International Geoscience and Remote Sensing Symposium (IGARSS '05)*, Seoul, Republic of Korea, 25–29 July 2005, IEEE, Vol. 5, pp. 3132–3134.

Shi R.H., Zhuang D.F. and Niu Z. (2006), Simulated hyperspectral data analysis using continuum removal: case study on leaf chlorophyll prediction, in *Proc. 8th International Conference on Signal Processing*, Beijing, China, 16–20 November 2006Vol. 4, 3 pages.

(2007), Estimation of optimal mesophyll structure parameter of rice leaves, *Journal of Remote Sensing – Beijing*, 11 (5):626–631.

Shi R.H. and Sun J. (2007), Estimating leaf biochemical information from leaf reflectance spectrum using artificial neural network, in *Proc. 6th International Conference on Machine Learning and Cybernetics*, Hong Kong, 19–22 August 2007, IEEE, Vol. 4, pp. 2224–2228.

Shi R.H., Zhang H., Sun J., Gao W., Zhuang D.F. and Niu Z. (2008), Responses of plant biochemical substances to reflectance spectra at leaf and canopy scales, in *Proc. Remote Sensing and Modeling of Ecosystems for Sustainability V* (Gao W. and Wang H., Eds), San Diego, CA, 13 August 2008, SPIE, Vol. 7083, 70830X.

Shibghatallah M.A.H., Khotimah S.N., Suhandono S., Viridi S. and Kesuma T. (2013), Measuring leaf chlorophyll concentration from its color: a way in monitoring environment change to plantations, *AIP Conference Proceedings*, 1554 (1):210–213.

Shimazaki K., Sakaki T., Kondo N. and Sugahara K. (1980), Active oxygen participation in chlorophyll destruction and lipid peroxidation in SO_2-fumigated leaves of spinach, *Plant and Cell Physiology*, 21(8):1193–1204.

Shimshi D. and Livne A. (1967), The estimation of the osmotic potential of plant sap by refractometry and conductimetry: a field method, *Annals of Botany*, 31(3):505–511.

Shipley B. and Vu T.T. (2002), Dry matter content as a measure of dry matter concentration in plants and their parts, *New Phytologist*, 153(2):359–364.

Shul'gin I.A. and Kleshnin A.F. (1959), Correlation between optical properties of plant leaves and their chlorophyll content, *Doklady Botanical Sciences Sections*, 125:119–121 (cover-to-cover translation from *Doklady Akademii Nauk SSSR)*.

Shul'gin I.A., Khazanov V.S. and Kleshnin A.F. (1960a), On the reflection of light as related to leaf structure, *Doklady Botanical Sciences Sections*, 134:197–199 (cover-to-cover translation from *Doklady Akademii Nauk SSSR*, 134 (2):471–474).

Shul'gin I.A., Kleshnin A.F. and Podol'nyi V.Z. (1960b), Optical properties of plant leaves in the ultraviolet region, *Soviet Plant Physiology*, 7: 116–118 (cover-to-cover translation from *Fiziologiya Rastenii*, 7(2):141–144).

Shul'gin I.A. and Khzanov V.S. (1961), On the problem of light conditions in plant associations, *Doklady Botanical Sciences Sections*, 141: 210–212 (cover-to-cover translation from *Doklady Akademii Nauk SSSR*, 141(6):1493–1496).

Shul'gin I.A. and Moldau K.H. (1965), On coefficients of brightness of leaves in natural and polarized light, *Doklady Botanical Sciences Sections*, 162(6):99–101 (cover-to-cover translation from *Doklady Akademii Nauk SSSR*, 162(6):1430–1433).

Shull C.A. (1928), Reflection of light from the surface of leaves, *Science*, 67(1726):107–108.

(1929), A spectrophotometric study of reflection of light from leaf surfaces, *Botanical Gazette*, 87(5):583–607.

Shuplyak V.I., Atrashevskii Y.I., Sikorskii V.V., Stel'makh G.F. and Fomichev A.Y. (1994), Spectral and polarization characteristics of light reflected from the leaves of plants infected with bacterial cancer, *Journal of Applied Spectroscopy*, 60 (3–4):175–179 (cover-to-cover translation from *Zhurnal Prikladnoi Spektroskopii*, 60 (3–4):226–232).

Shuplyak V.I., Belyaev B.I., Belyaev Y.V., Chumakov A.V., Kurikina T.M. and Nekrasov V.P. (1997), Investigation of the spectropolarization characteristics angular dependence of radiation reflected by potato leaves, in *Proc. Earth Surface Remote Sensing*, London, UK, 21–25 September 1997, SPIE, Vol. 3222, pp. 88–95.

Sieghardt H. (1990), Effects of dust pollution on optical properties of leaves of *Erysimum sylvestre*, *Phyton*, 30(2):305–311.

Simon R., Holderied M.W., Koch C.U., von Helversen O. (2011), Floral acoustics: conspicuous echoes of a dish-shaped leaf attract bat pollinators, *Science*, 333:631–633.

Simpson L. and Moore R. (1984), Cellular structure and light absorption in leaves of *Frithia pulchra* (Mesembryanthemaceae), *Annals of Botany*, 53(3):413–420.

Simpson N. (2011), Colour and contemporary digital botanical illustration, *Optics & Laser Technology*, 43(2):330–336.

Sims D.A. and Gamon J.A. (2002), Relationships between leaf pigment content and spectral reflectance across a wide range of species, leaf structures and developmental stages, *Remote Sensing of Environment*, 81(2–3):337–354.

Sinclair R. and Thomas D.A. (1970), Optical properties of leaves of some species in arid south Australia, *Australian Journal of Botany*, 18(3):261–273.

Sinclair T.R., Hoffer R.M. and Schreiber M.M. (1971),Reflectance and internal structure of leaves from several crops during a growing season, *Agronomy Journal*, 63(6):864–867.

Sinclair T.R., Schreiber M.M. and Hoffer R.M. (1973), Diffuse reflectance hypothesis for the pathway of solar radiation through leaves, *Agronomy Journal*, 65(2):276–283.

Singh A., Yadav K., Sen A.K. (2012), Sal (*Shorea robusta*) leaves lignin epoxydation and its use in epoxy based coatings, *American Journal of Polymer Science*, 2(1):14–18.

Singhroy V., Kenny F. and Springer J. (1989), Reflectance spectra of vegetation growing on mine site in the Canadian Shield, in *Proc. International Geoscience and Remote Sensing Symposium (IGARSS'89)*, Vancouver, BC, 10–14 July 1989, IEEE, Vol. 2, pp. 665–669.

Singhroy V., Saint-Jean R., Levesque J. and Barnett P. (2000), Reflectance spectra of the boreal forest over mineralized sites, in *Proc. International Geoscience and Remote Sensing Symposium (IGARSS 2000)*, Honolulu, Hawaii, 24–28 July 2000, IEEE, Vol. 4, pp. 1379–1381.

Sinkkonen A., Somerkoski E., Paaso U., Holopainen J.K., Rousi M. and Mikola J. (2012), Genotypic variation in yellow autumn leaf colours explains aphid load in silver birch, *New Phytologist*, 195(2):461–469.

Sinton W.M. (1957), Spectroscopic evidence for vegetation on Mars, *The Astrophysical Journal*, 126(2):231–239.

(1959), Further evidence of vegetation on Mars, *Science*, 130(3384):1234–1237.

Slaton M.R., Hunt E.R. and Smith W.K. (2001), Estimating near-infrared leaf reflectance from leaf structural characteristics, *American Journal of Botany*, 88(2):278–284.

Slavik B. (1959), The relation of the refractive index of plant cell sap to its osmotic pressure, *Biologia Plantarum*, 1(1):48–53.

Sliney D.H. (2007), Radiometric quantities and units used in photobiology and photochemistry: recommendations of the Commission Internationale de l'Eclairage (International Commission on Illumination), *Photochemistry and Photobiology*, 83(2):425–432.

Slonecker T., Haack B. and Price S. (2009), Spectroscopic analysis of arsenic uptake in Pteris ferns, *Remote Sensing*, 1(4):644–675.

Smith A.P. (1986), Ecology of a leaf color polymorphism in a tropical forest species: habitat segregation and herbivory, *Oecologia*, 69(2):283–287.

Smith A.M. and Blackshaw R.E. (2002), Crop/weed discrimination using remote sensing, in *Proc. IEEE International Geoscience and Remote Sensing Symposium (IGARSS'02)*, Toronto, ON, 24–28 June 2002, IEEE, Vol. 4, pp. 1962–1964.

(2003), Weed-crop discrimination using remote sensing: a detached leaf experiment, *Weed Technology*, 17(4):811–820.

Smith E.G., Ketchum R.N. and Burt J.A. (2017), Host specificity of Symbiodinium variants revealed by an ITS2 metahaplotype approach, *The ISME Journal*, 11(6):1500–1503.

Smith J.H.C., Shibata K. and Hart R.W. (1957), A Spectrophotometer accessory for measuring absorption spectra of light-scattering samples: spectra of dark-grown albino leaves and of adsorbed chlorophylls, *Archives of Biochemistry and Biophysics*, 72(2):457–464.

Smith J.L. II and Hare J.D. (2004), Spectral properties, gas exchange, and water potential of leaves of glandular and non-glandular trichome types in *Datura wrightii* (Solanaceae), *Functional Plant Biology*, 31(3):267–273.

Smith K.F. and Flinn P.C. (1991), Monitoring the performance of a broad-based calibration for measuring the nutritive value of two independent populations of pasture using near infrared reflectance (NIR) spectroscopy, *Australian Journal of Experimental Agriculture*, 31(2):205–210.

Smith K.F., Willis S.E. and Flinn P.C. (1991), Measurement of the magnesium concentration in perennial ryegrass (*Lolium perenne*) using near infrared reflectance spectroscopy, *Australian Journal of Agricultural Research*, 42(8):1399–1404.

Smith K.L., Steven M.D. and Colls J.J. (2004a), Use of hyperspectral derivative ratios in the red-edge region to identify plant stress responses to gas leaks, *Remote Sensing of Environment*, 92(2):207–217.

(2004b), Spectral responses of pot-grown plants to displacement of soil oxygen, *International Journal of Remote Sensing*, 25(20):4395–4410.

(2005), Plant spectral responses to gas leaks and other stresses, *International Journal of Remote Sensing*, 18 (26):4067–4081.

Smith M.O., Ustin S.L., Adams J.B. and Gillespie A.R. (1990), Vegetation in deserts. I: A regional measure of abundances from multispectral images, *Remote Sensing of Environment*, 31(1):1–26.

Smith M.O., Roberts D.A., Hill J., Mehl W., Hosgood B., Venderbout J., et al. (1994), A new approach to quantifying abundancies of materials in multispectral images, in *Proc. International Geoscience and Remote Sensing Symposium (IGARSS'94)*, Pasadena, CA, 8–12 August 1994, IEEE, Vol. 4, pp. 2372–2374.

Smith S.E. and Read D.J. (2008), *Mycorrhizal Symbiosis*, Academic Press, 800 pages.

Smith W.K., Vogelmann T.C., DeLucia E.H., Bell D.T. and Shepherd K.A. (1997), Leaf form and photosynthesis, *Bioscience*, 47(11):785–793.

Snyder L.R. (1974), Classification of the solvent properties of common liquids, *Journal of Chromatographic A*, 92 (2):223–234.

Sobol' I.M. (1993), Sensitivity analysis for nonlinear mathematical models, *Mathematical Modeling and Computational Experiment*, 1:407–414.

Solovchenko A. (2010), *Photoprotection in Plants – Optical Screening-based Mechanisms*, Springer-Verlag, Berlin, 168 pages.

Soltau U., Dötterl S. and Liede-Schumann S. (2009), Leaf variegation in *Caladium steudneriifolium* (Araceae): a case of mimicry? *Evolutionary Ecology*, 23(4):503–512.

Song X.D., Jiang H., Yu S.Q., Zhou G.M. and Jiang Z.S. (2010), Relationship between simulated acid rain stress and leaf reflectance, *Spectroscopy and Spectral Analysis*, 30(1):165–169 (in Chinese).

Sonobe R. and Wang Q. (2016), Assessing the xanthophyll cycle in natural beech leaves with hyperspectral reflectance, *Functional Plant Biology*, 43(5):438–447.

(2017a), Hyperspectral indices for quantifying leaf chlorophyll concentrations performed differently with different leaf types in deciduous forests, *Ecological Informatics*, 37:1–9.

(2017b), Towards a universal hyperspectral index to assess chlorophyll content in deciduous forests, *Remote Sensing*, 9 (3):191.

Sorby H.C. (1871a), Memoirs: on the colour of leaves at different seasons of the year, *Journal of Cell Sciences*, 2–11 (43):215–234.

(1871b), On the various tints of autumnal foliage, *The Quarterly Journal of Science*, 8(29–32):64–77.

(1872), On comparative vegetable chromatology, *Proceedings of the Royal Society of London*, 21(139–147):442–483.

(1884), On the autumnal tints of foliage, *Nature*, 31:105–106.

Soukupova J., Rock B.N. and Albrechtova J. (2002), Spectral characteristics of lignin and soluble phenolics in the near infrared – A comparative study, *International Journal of Remote Sensing*, 23(15):3039–3055.

Souza R.P. and Valio I.F.M. (2003), Leaf optical properties as affected by shade in saplings of six tropical tree species differing in successional status, *Revista Brasileira de Fisiologia Vegetal*, 15(1):49–54.

Sparks W.B., Hough J.H., Germer T.A., Chen F., DasSarma S., DasSarma P., et al. (2009a), Detection of circular polarization in light scattered from photosynthetic microbes, *Proceedings of the National Academy of Sciences of the United States of America*, 106(19):7816–7821.

Sparks W.B., Hough J.H., Kolokolova L., Germer T.A., Chen F., DasSarma S., et al. (2009b), Circular polarization in scattered light as a possible biomarker, *Journal of Quantitative Spectroscopy & Radiative Transfer*, 110(14–16):1771–1779.

Spomer L.A., M.A.L. Smith and Sawwan J.S. (1988), Rapid nondestructive measurement of chlorophyll content in leaves with nonuniform chlorophyll distribution, *Photosynthesis Research*, 16(3):277–284.

Sridhar B.B.M., Han F.X., Diehl S.V., Monts D.L. and Su Y. (2007a), Spectral reflectance and leaf internal structure changes of barley plants due to phytoextraction of zinc and cadmium, *International Journal of Remote Sensing*, 28(5):1041–1054.

(2007b), Monitoring the effects of arsenic and chromium accumulation in Chinese brake fern, *International Journal of Remote Sensing*, 28(5):1055–1067.

Stahl E. (1896), Uber bunte Laubblätter. Ein Beitrag zur Pflanzenbiologie. II, *Annales du Jardin Botanique de Buitenzorg*, 13:137–216.

Stahl G.E. (1715), *Opusculum chymico-physico-medicum, Typis & impsensis Orphanotrophei, Halae Magdeburgi*cae, Orphanotrophei, Halle Madeburg, 896 pages.

Stahl U., Tusov V.B., Paschenko V. and Voigt J. (1989), Spectroscopic investigation of fluorescence behaviour, role and function of the long-wavelength pigments of Photosystem I: *Biochimica et Biophysica Acta*, 973(2):198–204.

Stamm A.J. and Sanders H.T. (1966), Specific gravity of the wood substance of loblolly pine as affected by chemical composition, *Tappi*, 49:397–400.

Stearn W.T. (1967), The anatomy of plants with an idea of a philosophical history of plants and several other lectures read before the Royal Society by Nehemiah Grew, *The British Journal for the History of Science*, 3(3):291–293.

Stearn W.T. (2010), *Botanical Latin*, 4th Edition, Timber Press, 546 pages.

Steddom K., Heidel G., Jones D. and Rush C.M. (2003), Remote detection of rhizomania in sugar beets, *Phytopathology*, 93 (6):720–726.

Sterner R.W. and Elser J.J. (2002), *Ecological Stoichiometry: The Biology of Elements from Molecules to the Biosphere*, Princeton University Press, Princeton, NJ, 584 pages.

Sterzik M., Bagnulo S., Azua A., Salinas F., Alfaro J. and Vicuna R. (2010), Astronomy meets biology: EFOSC2 and the chirality of life, *The Messenger*, 142:25–27.

Stokes G.G. (1852a), On the change of refrangibility of light, *Philosophical Transactions of the Royal Society of Londo*n, 142:463–562.

(1852b), On the composition and resolution of streams of polarized light from different sources, *Transactions of the Cambridge Philosophical Society*, 9:399–416.

(1862), On the intensity of the light reflected from or transmitted through a pile of plates, *Proceedings of the Royal Society of London. Series B*, 11:545–556.

Stone C., Chisholm L. and Coops N. (2001), Spectral reflectance characteristics of eucalypt foliage damaged by insects, *Australian Journal of Botany*, 49(6):687–698.

Stone C., Chisholm L.A. and McDonald S. (2003), Spectral reflectance characteristics of *Pinus radiata* needles affected by dothistroma needle blight, *Canadian Journal of Botany*, 81(6):560–569.

(2005), Effects of leaf age and psyllid damage on the spectral reflectance properties of *Eucalyptus saligna* foliage, *Australian Journal of Botany*, 53(1):45–54.

Strout G., Russell S.D., Pulsifer D.P., Erten S., Lakhtakia A. and Lee D.W. (2013), Silica nanoparticles aid in structural leaf coloration in the Malaysian tropical rainforest understorey herb *Mapania caudata*, *Annals of Botany*, 112(6):1141–1148.

Stuckens J., Verstraeten W.W., Delalieux S., Swennen R. and Coppin P. (2009), A dorsiventral leaf radiative transfer model: development, validation and improved model inversion techniques, *Remote Sensing of Environment*, 113(12):2560–2573.

Stuppy W.H., Maisano J.A., Colbert M.W., Rudall P.J. and Rowe T.B. (2003), Three-dimensional analysis of plant structure using high-resolution X-ray computed tomography, *Trends in Plant Science*, 8(1):2–6.

Stylinski C.D., Oechel W.C., Gamon J.A., Tissue D.T., Miglietta F. and Raschi A. (2000), Effects of lifelong [CO_2] enrichment on carboxylation and light utilization of *Quercus pubescens* Willd. examined with gas exchange, biochemistry and optical techniques, *Plant, Cell & Environment*, 23(12):1353–1362.

Stylinski C.D., Gamon J.A. and Oechel W.C. (2002), Seasonal patterns of reflectance indices, carotenoid pigments and photosynthesis of evergreen chaparral species, *Oecologia*, 131(3):366–374.

Su Y., Maruthi Sridhar B.B., Han F.X., Diehl S.V. and Monts D.L. (2007), Effect of bioaccumulation of Cs and Sr natural isotopes on foliar structure and plant spectral reflectance of Indian mustard (*Brassica juncea*), *Water, Air, and Soil Pollution*, 180(1–4):65–74.

Sui X. Y., Li S., Zhang X., et al. (2010), Measurement of cotton leaf thickness with hyper spectrum, *Transactions of the Chinese Society of Agricultural Engineering*, 26(1):262–266 (in Chinese).

Sui X., Zhang X., Wang Y. and Li S. (2012), Estimation of leaf thickness with remote sensing, *Applied Mechanics and Materials*, 263–266:339–345.

Sullivan J.H., Howells B.W., Ruhland C.T. and Day T.A. (1996), Changes in leaf expansion and epidermal screening effectiveness in *Liquidambar styraciflua* and *Pinus taeda* in response to UV-B radiation, *Physiologia Plantarum*, 98 (2):349–357.

Sultanova N., Kasarova S. and Nikolov I. (2009), Dispersion properties of optical polymers, *Acta Physica Polonica A*, 116 (4):585–587.

Sumriddetchkajorn S. and Intaravanne Y. (2014), Single-wavelength based rice leaf color analyzer for nitrogen status estimation, *Optics and Lasers in Engineering*, 53:179–184.

Sun C.X., Yuan F., Zhang Y.L., Chen Z.H., Chen L.J. and Wu Z.J. (2012), Study of photosynthetic characteristics of transgenic barley based on reflectance of single leaf, *Spectroscopy and Spectral Analysis*, 32(1):204–208 (in Chinese).

Sun C.X., Yuan F., Zhang Y.L., Cui Z.B., Chen Z.H., Chen L.J. et al. (2013), Unintended effects of genetic transformation on photosynthetic gas exchange, leaf reflectance and plant growth properties in barley (*Hordeum vulgare* L.), *Photosynthetica*, 51(1):22–32.

Sun J., Nishio J.N. and Vogelmann T.C. (1996), High-light effects on CO_2 fixation gradients across leaves, *Plant, Cell & Environment*, 19(11):1261–1270.

(1998), Green light drives CO_2 fixation deep within leaves, *Plant & Cell Physiology*, 39(10):1020–1026.

Sun P., Grignetti A., Liu S., Casacchia R., Salvatori R., Pietrini F., et al. (2008), Associated changes in physiological parameters and spectral reflectance indices in olive (*Olea europaea* L.) leaves in response to different levels of water stress, *International Journal of Remote Sensing*, 29(6):1725–1743.

Sušila, P. and Nauš, J.A. (2007), Monte Carlo study of the chlorophyll fluorescence emission and its effect on the leaf spectral reflectance and transmittance under various conditions, *Photochemical & Photobiological Sciences*, 6(8):894–902.

Sykioti O., Paronis D., Stagakis S. and Kyparissis A. (2011), Band depth analysis of CHRIS/PROBA data for the study of a Mediterranean natural ecosystem. Correlations with leaf optical properties and ecophysiological parameters, *Remote Sensing of Environment*, 115(2):752–766.

Syvertsen J.P. and Cunningham G.L. (1979), The effects of irradiating adaxial or abaxial leaf surface on the rate of net photosynthesis of *Perezia nana* and *Helianthus annuus*, *Photosynthetica*, 13:287–293.

Szalay L., Tombácz E. and Singhal G.S. (1974), Effect of solvent on the absorption spectra and Stokes' shift of absorption and fluorescence of chlorophylls, *Acta Physica Academiae Scientiarum Hungaricae*, 35(1–4):29–36.

Tageyeva S.V. and Brandt A.B. (1960a), Study of the optical properties of leaves in relation to the angle of incidence of light, *Biophysics*, 5: 354–365 (cover-to-cover translation from *Biofizika*, 5(3):308–317).

(1961a), Study of optical properties of leaves depending on the angle of light incidence, in *Progress in Photobiology* (Christensen B.C. and Buchmann B., Eds), Elsevier Publishing Company, Amsterdam, pp. 163–169.

Tageyeva S.V., Brandt A.B. and Derevyanko V.G. (1960), Changes in optical properties of leaves in the course of the growing season, *Doklady Botanical Sciences Sections*, 135:266–268.

(1961), Peculiarities of the optical properties of leaves during vegetation, in *Progress in Photobiology* (Christensen B.C. and Buchmann B., Eds), Elsevier Publishing Company, Amsterdam, pp. 158–162.

Taiz L. and Zeiger E. (2010), *Plant Physiology*, 5th Edition, Sinauer Associates, Inc., 782 pages.

Takahashi K., Mineuchi K., Nakamura T., Koizumi M. and Kano H. (1994), A system for imaging transverse distribution of scattered light and chlorophyll fluorescence in intact leaves, *Plant, Cell & Environment*, 17(1):105–110.

Takahashi T., Fujii T. and Yasuoka Y. (2004), Estimation and comparison of acid detergent lignin and acetyl bromide lignin in fallen leaves using near-infrared spectroscopy, *International Journal of Remote Sensing*, 25(24):5585–5600.

Takayama K., Nishina H. and Sakai Y. (2008), Detection of water stress induced photosynthetic dysfunction in tomato plant leaf with PRI measurement, *Acta Horticulturae*, 801:1213–1219.

Takiuchi M. and Hashimoto Y. (1977), Measurement of leaf temperature by means of infrared thermometer in connection with plant physiological information, *Transactions of the Society of Instrument and Control Engineers*, 13(5):482–488 (in Japanese).

Talmage D.A. and Curran P.J. (1986), Remote sensing using partially polarized light, *International Journal of Remote Sensing*, 7(1):47–64.

Tan H.S. (1981), Microwave measurements and modelling of the permittivity of tropical vegetation samples, *Applied Physics*, 25(3):351–355.

Tanaka M. (1968), Some leaf characters of flue-cured Tobacco leaves sprayed with transpiration suppressant, *Japanese Journal of Crop Science*, 37(3):436–441 (in Japanese).

Tanaka Y., Grottoli A.G., Matsui Y., Suzuki A. and Sakai K. (2015), Partitioning of nitrogen sources to algal endosymbionts of corals with long-term 15N-labelling and a mixing model, *Ecological Modelling*, 309–310:163–169.

Tanner V. and Eller B.M. (1986), Veränderungen der spektralen Eigenschaften der Blätter der Buche (*Fagus silvatica* L.) von Laubaustrieb bis Laubfall, *Allgemeine Forst und Jagdzeitung*, 157(6):108–117 (in German): Variations of the optical properties of the leaves of the European beech (*Fagus silvatica* L.) during the growing season).

Taoutaou A., Socaciu C., Pamfil D., Fetea F., Balazs E., Botez C., et al. (2010), Fourier-transformed infrared spectroscopy applied for studying compatible interaction in the pathosystem *Phytophtora infestans-Solanum tuberosum*, *Notulae Botanicae Horti AgrobotaniciCluj-Napoca*, 38(3):69–75.

Taoutaou A., Socaciu C., Pamfil D., Fetea F., Balazs E. and Botez C. (2012), New markers for potato late blight resistance and susceptibility using FTIR spectroscopy, *Notulae Botanicae Horti AgrobotaniciCluj-Napoca*, 40(1):150–154.

Tarantola A. (2005), *Inverse Problem Theory and Model Parameter Estimation*, Society for Industrial and Applied Mathematic, 352 pages.

Tarpley L., Reddy K.R. and Sassenrath Cole G.F. (2000), Reflectance indices with precision and accuracy in predicting cotton leaf nitrogen concentration, *Crop Science*, 40(6):1814–1819.

Taylor S.E. (1975), Optimal leaf form. In *Perspectives of Biophysical Ecology* (Gates D.M. and Schmerl R.B., Eds), Springer-Verlag, New York, pp. 73–86.

Taylor-Hell J.F., Baranoski G.V.G. and Rokne J.G. (2005), State of the art in the realistic modeling of plant venation systems, *International Journal of Image and Graphics*, 5(3):663–678.

Temizel K.E., Odabas M.S., Senyer N., Kayhan G., Bajwa S.G., Caliskan O. et al. (2014), Comparison of some models for estimation of reflectance of hypericum leaves under stress conditions, *Central European Journal of Biology*, 9 (12):1226–1234.

Teng P.S. and Close R.C. (1977), Spectral reflectance of healthy and leaf rust-infected barley leaves, *Australian Plant Pathology Society Newsletter*, 6(1):7–9.

Terashima I. and Saeki T. (1983), Light environment within a leaf. I: Optical properties of paradermal sections of *Camellia* leaves with a special reference to differences in the optical properties of palisade and spongy tissues, *Plant & Cell Physiology*, 24(8):1493–1501.

(1985), A new model for leaf photosynthesis incorporating the gradients of light environment and of photosynthetic properties of chloroplasts within a leaf, *Annals of Botany*, 56(4):489–499.

Terashima I. and Inoue Y. (1984), Comparative photosynthetic properties of palisade tissue chloroplasts and spongy tissue chloroplasts of *Camellia japonica* L.: functional adjustment of the photosynthetic apparatus to light environment within a leaf, *Plant & Cell Physiology*, 25(4):555–563.

Terashima I. (1986), Dorsiventrality in photosynthetic light response curves of a leaf, *Journal of Experimental Botany*, 37 (176):399–405.

(1989), Productive structure of a leaf, in *Photosynthesis* (Briggs W.R., Ed), Alan R. Liss, New York, pp. 207–226.

Terashima I. and Hikosaka K. (1995), Comparative ecophysiology of leaf and canopy photosynthesis, *Plant, Cell & Environment*, 18(10):1111–1128.

Terashima I., Fujita T., Inoue T., Chow W.S. and Oguchi R. (2009), Green light drives leaf photosynthesis more efficiently than red light in strong white light: revisiting the enigmatic question of why leaves are green, *Plant Cell Physiology*, 50 (4):684–697.

Terashima I., Hanba Y.T., Tholen D. and Niinemets U. (2011), Leaf functional anatomy in relation to photosynthesis, *Plant Physiology*, 155(1):108–116.

Thenkabail P.S., Lyon J.G. and Huete A. (2011), *Hyperspectral Remote Sensing of Vegetation*, CRC Press, 781 pages.

Thenot F., Méthy M. and Winkel T. (2002), The Photochemical Reflectance Index (PRI) as a water-stress index, *International Journal of Remote Sensing*, 23(23):5135–5139.

Thérézien M., Palmroth S., Brady R. and Oren R. (2007), Estimation of light interception properties of conifer shoots by an improved photographic method and a 3D model of shoot structure, *Tree Physiology*, 27(10):1375–1387.

Thom R. (1980), *Paraboles et catastrophes*, Flammarion, 193 pages.

Thomas J.R., Wiegan C.L. and Myer V.I. (1967), Reflectance of cotton leaves and its relation to yield, *Agronomy Journal*, 59 (6):551–554.

Thomas J.R. and Oerther G.F. (1972), Estimating nitrogen content of sweet pepper leaves by reflectance measurements, *Agronomy Journal*, 64(1):11–13.

Thomas J.R. and Gausman H.W. (1977), Leaf reflectance vs. leaf chlorophyll and carotenoid concentrations for eight crops, *Agronomy Journal*, 69(5):799–802.

Thomas K.R., Kolle M., Whitney H.M., Glover B.J. and Steiner U. (2010), Function of blue iridescence in tropical understorey plants, *Journal of the Royal Society Interface*, 7(53):1699–1707.

Thomas S.C. (2005), Increased leaf reflectance in tropical trees under elevated CO_2, *Global Change Biology*, 11(2):197–202.

Thomas S., Wahabzada M., Kuska M.T., Rascher U. and Mahlein A.K. (2017), Observation of plant–pathogen interaction by simultaneous hyperspectral imaging reflection and transmission measurements, *Functional Plant Biology*, 44(1):23–34.

Thompson D.W. (1992), *On Growth and Form: The Complete Revised Edition*, Cambridge University Press, Cambridge. 1116 pages.

Thompson J.A., Schweitzer L.E. and Nelson R.L. (1996), Association of specific leaf weight, an estimate of chlorophyll, and chlorophyll concentration with apparent photosynthesis in soybean, *Photosynthesis Research*, 49(1):1–10.

Thygesen L.G. (1994), Determination of dry matter content and the basic density of Norway spruce by near infrared reflectance and transmittance spectroscopy, *Journal of Near Infrared Spectroscopy*, 2(3):127–135.

Tian Q., Tong Q., Pu R., Guo X. and Zhao C. (2001), Spectroscopic determination of wheat water status using 1650–1850 nm spectral absorption features, *International Journal of Remote Sensing*, 22(12):2329–2338.

Tian Y., Zhao C., Lu S. and Guo X. (2010), Separating pigment components of leaf color image using FastICA, in *Advanced Intelligent Computing Theories and Applications* (Huang D.S., Zhao Z., Bevilacqua V. and Figueroa J.C., Eds), Springer-Verlag, Berlin, pp. 430–437.

(2011), Multiple classifier combination for recognition of wheat leaf diseases, *Intelligent Automation and Soft Computing*, 17(5):519–529.

(2012a), Advances in appearance modeling and photorealistic rendering of plant leaf, *Journal of Image and Graphics*, 17 (5):613–620 (in Chinese).

(2012b), SVM-based multiple classifier system for recognition of wheat leaf diseases, in *Proc. World Automation Congress*, Puerto Vallarta, Mexico, 24–28 June 2012, pp. 189–193.

Tidestrom I. (1909), Notes on *Populus*, Plinius, *American Midland Naturalist*, 1(5):113–118.

Tikhov G. (1960), *L'énigme des planètes*, Editions en langues étrangères, 183 pages.

Tikhov G.A. (1966), *Observations of the Moon, Mars, Uranus, and the stars. Optical properties of plants*, NASA, 320 pages.

Tilley D.R., Ahmed M., Son J.H. and Badrinarayanan H. (2007), Hyperspectral reflectance response of freshwater macrophytes to salinity in a Brackish subtropical marsh, *Journal of Environmental Quality*, 36(3):780–789.

Timiriazeff C. (1903), Croonian lecture: the cosmical function of the green plant, *Proceedings of the Royal Society of London*, 72:424–461.

Toomey M., Roberts D. and Nelson B. (2009), The influence of epiphylls on remote sensing of humid forests, *Remote Sensing of Environment*, 113(8):1787–1798.

Torella J.P., Gagliardi C.J., Chen J.S., et al. (2015), Efficient solar-to-fuels production from a hybrid microbial–water-splitting catalyst system, *Proceedings of the National Academy of Science of the United States of America*, 112(8):2337–2342.

Torii T., Okamoto T. and Kitani O. (1988), Non-destructive measurement of water content of a plant using ultrasonic technique, *Acta Horticulturae*, 230:389–396.

Torrance K.E. and Sparrow E.M. (1967), Theory for off-specular reflection from roughened surfaces, *Journal of the Optical Society of America*, 57(9):1105–1114.

Torres Netto A., Campostrini E., Gonçalves de Oliveira J. and Yamanishi O.K. (2002), Portable chlorophyll meter for the quantification of photosynthetic pigments, nitrogen and the possible use for assessment of the photochemical process in *Carica papaya* L., *Brazilian Journal of Plant Physiology*, 14(3):203–210.

Toth R. (1982), An introduction to morphometric cytology and its application to botanical research, *American Journal of Botany*, 69(10):1694–1706.

Trench R.K. (1979), The cell biology of plant-animal symbiosis, *Annual Review of Plant Physiology*, 30:485–531.

Trigui A. (1983), Evolution des propriétés biophysiques et optiques des jeunes feuilles d'olivier (*Olea europaea* L.) au cours de leur phase de croissance, *Comptes Rendus de l'Académie des Sciences Paris*, 297(3):543–545.

Trojan A. and Gabrys H. (1996), Chloroplast distribution in *Arabidopsis thaliana* (L.) depends on light conditions during growth, *Plant Physiology*, 111(2):419–425.

Trombetti M., Riaño D., Rubio M.A., Cheng Y.B. and Ustin S.L. (2008), Multi-temporal vegetation canopy water content retrieval using artificial neural networks for the USA, *Remote Sensing of Environment*, 112(1):203–215.

Tsel'niker Y.L. (1975), Effect of light intensity on optical properties of chloroplasts and leaf tissues in trees, Soviet Plant Physiology, 22:592–597 (cover-to-cover translation from *Fiziologiya rastenii*, 22(4):695–701).

Tsukaya H., Okada H. and Mohamed M. (2004), A novel feature of structural variegation in leaves of the tropical plant *Schismatoglottis calyptrata, Journal of Plant Research*, 117(6):477–480.

Tucker C.J. and Garratt M.W. (1977), Leaf optical system modeled as a stochastic process, *Applied Optics*, 16(3):635–642.

Tucker C.J. (1980), Remote sensing of leaf water content in the near infrared, *Remote Sensing of Environment*, 10(1):23–32.

Tuckerman L.B. (1947), On the intensity of the light reflected from or transmitted through a pile of plates, *Journal of the Optical Society of America*, 37(10):818–825.

Turnbull M.C., Traub W.A., Jucks K.W., et al. (2006), Spectrum of a habitable world: Earthshine in the near-infrared, *The Astrophysical Journal*, 644(1):551–559.

Turunen M.T., Vogelmann T.C. and Smith W.K. (1999), UV screening in lodgepole pine (*Pinus contorta* ssp. latifolia) cotyledons and needles, *International Journal of Plant Sciences*, 160(2):315–320.

Tyree M.T. and Hammel H.T. (1972), The measurement of the turgor pressure and the water relations of plants by the pressure-bomb technique, *Journal of Experimental Botany*, 23(1):267–282.

Tyree M.T. and Jarvis P.G. (1982), Water in tissues and cells, in *Physiological Plant Ecology II. Water Relations and Carbon Assimilation* (Lange O.L., Nobel P.S., Osmond C.B. and Ziegler H., Eds), Springer-Verlag, Berlin, pp. 36–77.

Uddling J., Gelang-Alfredsson J., Piikki K. and Pleijel H. (2007), Evaluating the relationship between leaf chlorophyll concentration and SPAD-502 chlorophyll meter readings, *Photosynthesis Research*, 91:37–46.

Ulaby F.T. and Jedlicka R.P. (1984), Microwave dielectric properties of plant materials, *IEEE Transactions on Geoscience and Remote Sensing*, 22(4):406–415.

Ulissi V., Antonucci F., Benincasa P., et al. (2011), Nitrogen concentration estimation in tomato leaves by VIS-NIR non-destructive spectroscopy, *Sensors*, 11(6):6411–6424.

Ullah S., Groen T.A., Schlerf M., Skidmore A.K., Nieuwenhuis W. and Vaiphasa C. (2012a), Using a genetic algorithm as an optimal band selector in the mid and thermal infrared (2.5–14 μm) to discriminate vegetation species, *Sensors*, 12 (7):8755–8769.

Ullah S., Schlerf M., Skidmore A.K. and Hecker C. (2012b), Identifying plant species using mid-wave infrared (2.5–6 μm) and thermal infrared (8–14 μm) emissivity spectra, *Remote Sensing of Environment*, 118:95–102.

Ullah S., Skidmore A.K., Naeem M. and Schlerf M. (2012c), An accurate retrieval of leaf water content from mid to thermal infrared spectra using continuous wavelet analysis, *Science of the Total Environment*, 437:145–152.

(2012d), Estimation of leaf water content from far infrared (2.5–14 μm) spectra using continuous wavelet analysis, in *Proc. IEEE International Geoscience and Remote Sensing Symposium (IGARSS12)*, Munich, Germany, IEEE, pp. 4817–4820.

Ullah S., Skidmore A.K., Groen T.A. and Schlerf M. (2013), Evaluation of three proposed indices for the retrieval of leaf water content from the mid-wave infrared (2–6 μm) spectra, *Agricultural and Forest Meteorology*, 171–172:65–71.

Ullah S., Skidmore A.K., Ramoelo A., Groen T.A., Naeem M. and Ali A. (2014), Retrieval of leaf water content spanning the visible to thermal infrared spectra, *ISPRS Journal of Photogrammetry and Remote Sensing*, 93:56–64.

Umezaki E. and Shimadaa T. (1999), Measurement of temperature on *Scindapsus* leaves subjected to ultraviolet radiation using infrared thermography techniques, in *Proc. Optical Engineering for Sensing and Nanotechnology* (Yamaguchi I., Ed), Yokohama, Japan, 16–18 June 1999, SPIE, Vol. 3740, pp. 358–361.

Underwood E., Ustin S.L. and DiPietro D. (2003), Mapping nonnative plants using hyperspectral imagery, *Remote Sensing of Environment*, 86(2):150–161.

Upchurch G.R. (1995), Dispersed angiosperm cuticles: their history, preparation, and application to the rise of angiosperms in Cretaceous and Paleocene coals, southern western interior of North America, *International Journal of Coal Geology*, 28 (2–4):161–227.

Ursprung A. (1918), Über die Absorptionskurve des grünen Farbstoffes lebender Blätter, *Berichte der Deutschen Botanischen Gesellschaft*, 36(2):73–85.

Ustin S.L., Adams J.B., Elvidge C.D., Rejmanek M., Rock B.N., Smith M.O., et al. (1986), Thematic Mapper studies of semiarid shrub communities, *BioScience*, 36:446–452.

Ustin S.L. and Curtiss B. (1990), Spectral characteristics of ozone-treated conifers, *Environmental and Experimental Botany*, 30(3):293–308.

Ustin, S.L., Smith, S. Jacquemoud, M.M. Verstraete and Y. Govaerts (1999), Geobotany: vegetation mapping for earth sciences, in *Remote Sensing for the Earth Sciences, Manual of Remote Sensing*, 3rd Edition, Vol. 3 (Rencz A., Ed), John Wiley & Sons, New York. pp. 189–248.

Ustin S.L., Jacquemoud S. and Govaerts Y. (2001), Simulation of photon transport in a three-dimensional leaf: implications for photosynthesis, *Plant, Cell & Environment*, 24(10):1095–1103.

Ustin S.L., Jacquemoud S., Zarco-Tejada P.J. and Asner G.P. (2004), Remote sensing of the environment: state of the science and new directions, in *Manual of Remote Sensing, Vol. 4: Remote Sensing for Natural Resource Management and Environmental Monitoring* (S.L. Ustin, Ed), John Wiley & Sons, pp. 679–729.

Ustin S.L., Gitelson A.A., Jacquemoud S., Schaepman M.E., Asner G.P., Gamon J.A., et al. (2009), Retrieval of foliar information about plant pigment systems from high resolution spectroscopy, *Remote Sensing of Environment*, 113(S1): S67–S77.

Ustin S.L. and Gamon J.A. (2010), Remote sensing of plant functional types, *New Phytologist*, 186(4):795–816.

Uto K. and Kosugi Y. (2012), Extraction of Lambert parameter from leaf scale hyperspectral images, in *Proc. 4th Workshop on Hyperspectral Image and Signal Processing*, Shanghai, China, 4-7 June 2012, IEEE, 4 pages.

(2013a), Estimation of Lambert parameter based on leaf-scale hyperspectral images using dichromatic model-based PCA, *International Journal of Remote Sensing*, 34(4):1386–1412.

(2013b), Leaf parameter estimation based on leaf scale hyperspectral imagery, *IEEE Journal of Selected Topics in Applied Earth Observations and Remote Sensing*, 6(2):699–707.

van den Berg A.K. and Perkins T.D. (2004), Evaluation of a portable chlorophyll meter to estimate chlorophyll and nitrogen contents in sugar maple (*Acer saccharum* Marsh.) leaves, *Forest Ecology and Management*, 200(1–3):113–117.

(2005), Nondestructive estimation of anthocyanin content in autumn sugar maple leaves, *HortScience*, 40(3):685–686.

Vanderbilt V.C. and Grant L. (1985), Plant canopy specular reflectance model, *IEEE Transactions on Geoscience and Remote Sensing*, 23(5):722–730.

(1986), Polarization photometer to measure bidirectional reflectance factor R(55°,0°,55°,180°) of leaves, *Optical Engineering*, 25(4):566–571.

Vanderbilt V.C., DeWitt D.P. and Robinson B.F. (1987), Integrating sphere transmissiometer for field measurement of leaf transmittance, *Optical Engineering*, 26(12):1191–1196.

Vanderbilt V.C., Grant L. and Ustin S.L. (1991), Polarization of light by vegetation. 2. Scattering by single leaves, in *Photon-Vegetation Interactions* (Myneni R.B. and Ross J., Eds), Springer-Verlag, New York, pp. 191–228.

Vanderbilt V.C. and Daughtry C.S.T. (2012), Mueller matrix of a dicot leaf, in *Proc. Polarization: Measurement, Analysis, and Remote Sensing X* (Chenault D.B. and Goldstein D.H., Eds), Baltimore, MD, 23 April 2012, SPIE, Vol. 8364, 83640R-1.

Vanderbilt V.C., Daughtry C.S.T. and Biehl L.L. (2014), Is there spectral variation in the polarized reflectance of leaves? in *Proc. Polarization: Measurement, Analysis, and Remote Sensing XI* (Chenault D.B. and Goldstein D.H., Eds), Baltimore, MD, 5 May 2014, SPIE, Vol. 9099, 909916.

van Deventer H. and Cho M.A. (2014), Assessing leaf spectral properties of *Phragmites australis* impacted by acid mine drainage, *South African Journal of Science*, 110(7–8):2013–0184.

van Deventer H., Cho M.A., Mutanga O., Naidoo L. and Dudeni-Tlhone N. (2015), Reducing leaf-level hyperspectral data to 22 components of biochemical and biophysical bands optimizes tree species discrimination, *IEEE Journal of Selected Topics in Applied Earth Observations and Remote Sensing*, 8(6):3161–3171.

van Maarschalkerweerd M., Bro R., Egebo M. and Husted S. (2013), Diagnosing latent copper deficiency in intact barley leaves (*Hordeum vulgare*, L.) using near infrared spectroscopy, *Journal of Agricultural and Food Chemistry*, 61 (46):10901–10910.

Van Wittenberghe S., Verrelst J., Rivera J.P., Alonso L., Moreno J. and Samson R. (2014), Gaussian processes retrieval of leaf parameters from a multi-species reflectance, absorbance and fluorescence dataset, *Journal of Photochemistry and Photobiology B: Biology*, 134:37–48.

Vargas W.E. and Niklasson G.A. (1997), Applicability conditions of the kubelka-Munk theory, *Applied Optics*, 36 (22):5580–5586.

Velázquez-Rosas N., Barradas V.L., Vázquez-Santana S., Cruz-Ortega R., García-Jiménez F., Toledo-Alvarado E. et al. (2010), Optical and morpho-functional traits of the leaves of tree species growing in a mountain cloud forest, *Acta Oecologica*, 36(6):587–598.

Verboven P., Herremans E., Helfen L., Ho Q.T., Abera M., Baumbach T., et al. (2015), Synchrotron X-ray computed laminography of the three-dimensional anatomy of tomato leaves, *The Plant Journal*, 81(1):169–182.

Veres J.S., Cofer G.P. and Johnson G.A. (1991), Distinguishing plant tissues with magnetic resonance microscopy, *American Journal of Botany*, 78(12):1704–1711.

(1993), Magnetic resonance imaging of leaves, *New Phytologist*, 123(4):769–774.

Verhoef W. and Bach H. (2003), Simulation of hyperspectral and directional radiance images using coupled biophysical and atmospheric radiative transfer models, *Remote Sensing of Environment*, 87(1):23–41.

(2007), Coupled soil-leaf-canopy and atmosphere radiative transfer modeling to simulate hyperspectral multi-angular surface reflectance and TOA radiance data, *Remote Sensing of Environment*, 109(2):166–182.

Verhoef W. (2011), Modelling vegetation fluorescence observations, in *Proc. 7th EARSEL workshop*, Edinburgh, Scotland, 11–13 April 2011, EARSEL, pp. 41–42.

Verdeil M.F. (1851), Recherches sur la matière colorante verte des plantes et sur la matière rouge su sang, *Comptes Rendus Hebdomadaires des Séances de l'Académie des Sciences*, 33:689–690.

Verrelst J., Rivera J.P., van der Tol C., Magnani F., Mohammed G. and Moreno J. (2015), Global sensitivity analysis of the SCOPE model: what drives simulated canopy-leaving sun-induced fluorescence? *Remote Sensing of Environment*, 166:8–21.

Vidaver W.E., Lister G.R., Brooke R.C. and Binder W.D. (1991), *A Manual for the Use of Variable Chlorophyll Fluorescence in the Assessment of the Ecophysiology of Conifer Seedlings*, FRDA Report 163, 60 pages.

Vieira Santos C. (2004), Regulation of chlorophyll biosynthesis and degradation by salt stress in sunflower leaves, *Scientia Horticulturae*, 103(1):93–99.

Vigneron J.P., Rassart M., Vértesy Z., et al (2005), Optical structure and function of the white filamentary hair covering the edelweiss bracts, *Physical Review E*, 71:011906.

Vignolini S., Moyroud E., Glover B.J. and Steiner U. (2013), Analysing photonic structures in plants, *Journal of the Royal Society Interface*, 10(87):20130394.

Vignolini S., Moyroud E., Hingant T., et al. (2015), The flower of *Hibiscus trionum* is both visibly and measurably iridescent, *New Phytologist*, 205(1):97–101.

Vile D., Garnier E., Shipley B., et al. (2005), Specific leaf area and dry matter content estimate thickness in laminar leaves, *Annals of Botany*, 96(6):1129–1136.

Vilfan N., van der Tol C., Muller O., Rascher U. and Verhoef W. (2016), Fluspect-B: A model for leaf fluorescence, reflectance and transmittance spectra, *Remote Sensing of Environment*, 186:596-615.

Vilfan N., van der Tol C., Yang P., Wyber R., Malenovský Z., et al. (2018), Extending Fluspect to simulate xanthophyll driven leaf reflectance dynamics, *Remote Sensing of Environment*, 211:345-356.

Ville G. (1889), Recherches sur les relations qui existent entre la couleur des plantes et la richesse des terres en agents de fertilité, *Comptes Rendus Hebdomadaires des Séances de l'Académie des Sciences*, 109:397–400.

Villena J.F., Domínguez E., and Heredia A. (2000), Monitoring biopolymers present in plant cuticles by FT-IR spectroscopy, *Journal of Plant Physiology*, 156(3):419–422.

Vincent J.F.V. (1999), From cellulose to cell, *The Journal of Experimental Biology*, 202:3263–3268.

Visser A.J., Tosserams M., Groen M.W., Kalis G., Kwant R., Magendas G.W.H. et al. (1997), The combined effects of CO_2 concentration and enhanced UV-B radiation on faba bean. 3: Leaf optical properties, pigments, stomatal index and epidermal cell density, *Plant Ecology*, 128(1–2):209–222.

Vitale A.T. (1997), *Leaves in Myth, Magic and Medicine*, Stewart, Tabori & Chang, 352 pages.

Vogelmann J.E., Rock B.N. and Moss D.M. (1993), Red edge spectral measurements from sugar maple leaves, *International Journal of Remote Sensing*, 14(8):1563–1575.

Vogelmann T.C. and Björn L.O. (1984), Measurement of light gradients and the spectral regime in plant tissue with a fiber optic probe, *Physiologia Plantarum*, 60(3):361–368.

(1986), Plants as light traps, *Physiologia Plantarum*, 68(4):704–708.

Vogelmann T.C. (1986), Light within the plant, in *Photomorphogenesis in Plants* (Kendrick R.E. and Kronenberg G.H.M., Eds), Martinus Nijhoff Publishers, Dordrecht, pp. 307–337.

1989), Penetration of light into plants, *Photochemistry and Photobiology*, 50(6):895–902.

Vogelmann T.C. (1993), Plant tissue optics, *Annual Review of Plant Physiology and Plant Molecular Biology*, 44:231–251.

Vogelmann T.C., Knapp A.K., McClean T.M. and Smith W.K. (1988), Measurements of light within thin plant tissues with fiber optic microprobes, *Physiologia Plantarum*, 72(3):623–630.

Vogelmann T.C., Bornman J.F. and Josserand S. (1989), Photosynthetic light gradients and spectral regime within leaves of *Medicago sativa*, *Philosophical Transactions of the Royal Society of London. Series B*, 323(1216):411–421.

Vogelmann T.C., Martin G., Chen G. and Buttry B. (1991), Fibre optic microprobes and measurement of the light micro-environment within plant tissues, *Advances in Botanical Research*, 18:255–295.

Vogelmann T.C. and Martin G. (1993), The functional significance of palisade tissue: penetration of directional versus diffuse light, *Plant, Cell & Environment*, 16(1):65–72.

Vogelmann T.C., Bornman J.F. and Yates D.J. (1996a), Focusing of light by leaf epidermal cells, *Physiologia Plantarum*, 98 (1):43–56.

Vogelmann T.C., Nishio N. and Smith W.K. (1996b), Leaves and light capture: light propagation and gradients of carbon fixation within leaves, *Trends in Plant Science*, 1(2):65–70.

Vogelmann T.C. and Han T. (2000), Measurement of gradients of absorbed light in spinach leaves from chlorophyll fluorescence profiles, *Plant, Cell & Environment*, 23(12):1303–1312.

Vogelmann T.C. and Evans J.R. (2002), Profiles of light absorption and chlorophyll within spinach leaves from chlorophyll fluorescence, *Plant, Cell & Environment*, 25(10):1313–1323.

Vogelmann T.C. and Gorton H.L. (2014), Leaf: light capture in the photosynthetic organ, in *The Structural Basis of Biological Energy Generation* (Hohmann-Marriott M.F., Ed), Springer, The Netherlands, pp. 363–377.

Von Schönermark M., Geiger B., Röser H.P. (2004), *Reflection Properties of Vegetation and Soils*, Wissenschaft und Technik Verlag, 352 pages.

Voshchula I.V., Zhumar A.Y. and Tsaryuk O.V. (2007), Elliptical polarization of laser light reflected from plant leaves and characteristics of the leaf cuticle, *Biophysics*, 52(4):418–422.

Vrindts E. and De Baerdemaeker J. (1998), Optical weed detection and evaluation using reflection measurements, in *Proc. Conference on Precision Agriculture and Biological Quality* (Meyer G.E. and DeShazer J.A., Eds), Boston, MA, 01 November 1998, SPIE, Vol. 3543, pp. 279–289.

Vukusic P. and Sambles J.R. (2003), Photonic structures in biology, *Nature*, 424:852–855.

Wada M., Kagawa T. and Sato Y. (2003), Chloroplast movement, *Annual Review of Plant Biology*, 54:455–468.

Wägele H. and Klussmann-Kolb A. (2005), Opisthobranchia (Mollusca, Gastropoda) – more than just slimy slugs. Shell reduction and its implications on defense and foraging, *Frontiers in Zoology*, 2: 3.

Walczak T. and Gabrys H. (1980), New type of photometer for measurements of transmission changes corresponding to chloroplast movements in leaves, *Photosynthetica*, 14(1):65–72.

Waller C. and Seibert A. (1955), Studies of the refractive indices of binary wax mixtures, *Journal of the American Oil Chemists' Society*, 32(12):709–712.

Wallihan E.F. (1973), Portable reflectance meter for estimating chlorophyll concentrations in leaves, *Agronomy Journal*, 65 (4):659–662.

Walter H. and Koch W. (1981), Optical parameters of leaves of crops and weeds, in *Proc. Signatures spectrales d'objets en télédétection*, (Guyot G. and Verbrugghe M., Eds), Avignon, France, 8–11 September 1981, INRA, pp. 225–232.

Walter L., Balling A., Zimmermann U., Haase A. and Kuhn W. (1989), Nuclear-magnetic-resonance imaging of leaves of *Mesembryanthemum crystallinum* L. plants grown at high salinity, *Planta*, 178(4):524–530.

Walter-Shea E.A., Norman J.M. and Blad B.L. (1989), Leaf bidirectional reflectance and transmittance in corn and soybean, *Remote Sensing of Environment*, 29(2):161–174.

Walter-Shea E.A. and Biehl L.L. (1990), Measuring vegetation spectral properties, *Remote Sensing Reviews*, 5(1):179–205.

Walter-Shea E.A., Norman J.M., Blad B.L. and Robinson B.F. (1991), Leaf reflectance and transmittance in soybean and corn, *Agronomy Journal*, 83(3):631–636.

Walthall C., Daughtry C.S.T., Pachepsky L., Erbe E., Lydon J., Higgins M., et al. (2006), Detection of illegal Cannabis cultivation using remote sensing in *Proc. IEEE International Conference on Geoscience and Remote Sensing Symposium (IGARSS'06)*, Denver, CO, 31 July–4 August 2006, IEEE, pp. 2281–2284.

Wang B.J. and Ju W. (2017), Limitations and improvements of the leaf optical properties model Leaf Incorporating Biochemistry Exhibiting Reflectance and Transmittance Yields (LIBERTY), *Remote Sensing*, 9(5):431.

Wang H., Shi R.H., Liu P.D. and Gao W. (2016), Dual NDVI ratio vegetation index: a kind of vegetation index assessing leaf carotenoid content based on leaf optical properties, *Spectroscopy and Spectral Analysis*, 36(7):2189–2194 (in Chinese).

Wang J., Xu R., Ma Y., Miao L., Cai R. and Chen Y. (2008), The research of air pollution based on spectral features in leaf surface of *Ficus microcarpa* in Guangzhou, China, *Environmental Monitoring and Assessment*, 142(1–3):73–83.

Wang K., Shen Z.Q., Abou-Ismail O., Yaghi A. and Wang R.C. (1997), Preliminary study on canopy and leaf reflectance characteristics of rice with various potassium levels, *Bulletin of Science and Technology*, 13(4):211–214 (in Chinese).

Wang L. and Bai Y. (2005), Correlation between corn leaves spectra reflectance and nutrient content under different potassium levels, in *Proc. 9th International Symposium on Physical Measurements & Signatures in Remote Sensing* (Liang S., Liu J., Li X., Liu R. and Schaepman M.E., Eds), Beijing, China, 17–19 October 2005, ISPRS, 3 pages.

Wang L., Wang W., Dorsey J., Yang X., Guo B. and Shum H.Y. (2005), Real-time rendering of plant leaves, in *Proc. SIGGRAPH 2005*, Los Angeles, CA, 31 July–4 August 2005, ACM, pp. 167–174.

Wang L., Bai Y.L. and Yang L.P. (2007), Spectral response and diagnosis of phosphorus nutrition in corn, *Plant Nutrition and Fertilizer Science*, 13 (5):802–808 (in Chinese).

Wang L., Luo Y.Q., Huang H.G., Shi J., Keliövaara K., Teng W.X. et al. (2009a), Reflectance features of water stressed *Larix gmelinii* needles, *Forestry Studies in China*, 11(1):28–33.

Wang L. and Sousa W.P. (2009), Distinguishing mangrove species with laboratory measurements of hyperspectral leaf reflectance, *International Journal of Remote Sensing*, 30(5):1267–1281.

Wang L., Hunt E.R., Qu J.J., Hao X. and Daughtry C.S.T. (2011a), Estimating dry matter content of fresh leaves from the residuals between leaf and water reflectance, *Remote Sensing Letters*, 2(2):137–145.

Wang L., Qu J.J., Hao X. and Hunt E.R. (2011b), Estimating dry matter content from spectral reflectance for green leaves of different species, *International Journal of Remote Sensing*, 32(22):7097–7109.

Wang P., Liu X.N. and Huang F. (2010), Retrieval model for subtle variation of contamination stressed maize chlorophyll using hyperspectral data, *Spectroscopy and Spectral Analysis*, 30(1):197–201.

Wang Q., Chen J. and Li Y. (2004), Nondestructive and rapid estimation of leaf chlorophyll and nitrogen status of peace lily using a chlorophyll Meter, *Journal of Plant Nutrition*, 27(3):557–569.

Wang T., Liu Y., Wu H.Y. and Zuo Y.M. (2012), Influence of foliar dust on crop reflectance spectrum and nitrogen monitoring, *Spectroscopy and Spectral Analysis*, 32(7):1895–1898 (in Chinese).

Wang Q. and Li P. (2012), Hyperspectral indices for estimating leaf biochemical properties in temperate deciduous forests: comparison of simulated and measured reflectance data sets, *Ecological Indicators*, 14(1):56–65.

Wang Q. and Jin J. (2015), Leaf transpiration of drought tolerant plant can be captured by hyperspectral reflectance using PLSR analysis, *iForest*, 9:30–37.

Wang X., Zhao C., Lu S. and Guo X. (2009b), Survey on modeling and visualization of plant leaf color, in *Proc. Third International Symposium on Plant Growth Modeling, Simulation, Visualization and Applications*, Beijing, China, 9–13 November 2009, IEEE, pp. 417–424.

Wang Y. and Cuitiño A.M. (2000), Three-dimensional nonlinear open-cell foams with large deformations, *Journal of the Mechanics and Physics of Solids*, 48(5):961–988.

Wang Y., Hao W., Wang G., Ning X., Tang J., Shi Z., et al. (2013), A method of realistic leaves modeling based on point cloud, in *Proc. 12th ACM SIGGRAPH International Conference on Virtual-Reality Continuum and Its Applications in Industry*, Hong Kong, 17–19 November 2013, pp. 123–130.

Wang Z., Skidmore A., Darvishzadeh R., Heiden U., Heurich M. and Wang T. (2015a), Leaf nitrogen content indirectly estimated by leaf traits derived from the PROSPECT model, *IEEE Journal of Selected Topics in Applied Earth Observations and Remote Sensing*, 8(6):3172–3182.

Wang Z., Skidmore A.K., Wang T., Darvishzadeh R. and Hearne J. (2015b), Applicability of the PROSPECT model for estimating protein and cellulose + lignin in fresh leaves, *Remote Sensing of Environment*, 168:205–218.

Waters K.R., Mobley J. and Miller J.G. (2005), Causality-imposed (Kramers-Kronig) relationships between attenuation and dispersion, *IEEE Transactions on Ultrasonics, Ferroelectrics, and Frequency Control*, 52(5):822–833.

Watté R., Aernouts B., Van Beers R., Herremans E., Ho Q.T., Verboven P., et al. (2015), Modeling the propagation of light in realistic tissue structures with MMC-fpf: a Meshed Monte Carlo method with free phase function, *Optics Express*, 23(13):17467–17486.

Waugh G.R. and Clark K.B. (1986), Seasonal and geographic variation in chlorophyll level of *Elysia tuca* (Ascoglossa: Opisthobranchia), *Marine Biology*, 92(4):483–387.

Weiss A. (1990), Leaf wetness: measurements and models, *Remote Sensing Reviews*, 5(1):215–224.

Weiss M., Troufleau D., Baret F., Chauki H., Prévot L., Olioso A., et al. (2001), Coupling canopy functioning and radiative transfer models for remote sensing data assimilation, *Agricultural and Forest Meteorology*, 108(2):113–128.

Welch A.J., Gardner C., Richards-Kortum R., Chan E., Criswell G., Pfefer J. et al. (1997), Propagation of fluorescent light, *Lasers in Surgery and Medicine*, 21(2):166–178.

Venn A.A., Loram J.E., Trapido-Rosenthal H.G., Joyce D.A. and Douglas A.E. (2008), Importance of time and place: patterns in abundance of symbiodinium Clades A and B in the tropical sea anemone *Condylactis gigantean*, *The Biological Bulletin*, 215(3):243–252.

Wessman C.A., Aber J.D., Peterson D.L. and Melillo J.M. (1988a), Foliar analysis using near infrared reflectance spectroscopy, *Canadian Journal of Forest Research*, 18(1):6–11.

(1988b), Remote sensing of canopy chemistry and nitrogen cycling in temperate forest ecosystems, *Nature*, 335 (6186):154–156.

Wessman C.A., Aber J.D. and Peterson D.L. (1989), An evaluation of imaging spectrometry for estimating forest canopy chemistry, *International Journal of Remote Sensing*, 10(8):1293–1316.

Wessman C.A. (1990), Evaluation of canopy biochemistry. In *Remote Sensing of Biosphere Functioning* (Hobbs R.J. and Mooney H.A., Eds), Springer-Verlag, New York, pp. 135–156.

(1992), Spatial scales and global change: bridging the gap from plots to CGM grid cells, *Annual Review of Ecology and Systematics*, 23:175–200.
Westman W.E. and Price C.V. (1988), Spectral changes in conifers subjected to air pollution and water stress: experimental studies, *IEEE Transactions on Geoscience and Remote Sensing*, 26(1):11–21.
Weybrew J.A. and Green P.E. (1952), Variation in coloring rates of tobacco, *Science*, 115(2991):466–468.
White J.D., Trotter C.M., Brown C.J. and Scott N. (2000), Nitrogen concentration in New Zealand vegetation foliage derived from laboratory and field spectrometry, *International Journal of Remote Sensing*, 21(12):2525–2531.
White N.S., Errington R.J., Fricker M.D. and Wood J.L. (1996), Aberration control in quantitative imaging of botanical specimens by multidimensional fluorescence microscopy, *Journal of Microscopy*, 181(2):99–116.
Wickham J., Chesley M., Lancaster J. and Mouat D. (1993), *Remote Sensing for the Geobotanical and Biogeochemical Assessment of Environmental Contamination*, Desert Research Institute, Reno, NV, DOE/NV/10845–27, 79 pages.
Wilkinson D.M., Sherratt T.N., Phillip D.M., Wratten S.D., Dixon A.F.G. and Young A.J. (2002), The adaptive significance of autumn leaf colours, *Oikos*, 99(2):402–407.
Williams D.L. (1991), A comparison of spectral reflectance properties at the needle, branch, and canopy level for selected conifer species, *Remote Sensing of Environment*, 35(2–3):79–93.
Williams J.A. and Ashenden T.W. (1992), Differences in the spectral characteristics of white clover exposed to gaseous pollutants and acid mist, *New Phytologist*, 120(1):69–75.
Williams P. and Norris K. (1987), *Near-Infrared Technology in the Agricultural and Food Industries*, American Association of Cereal Chemists, St Paul, MN, 330 pages.
Williams W.E., Gorton H.L. and Witiak S.M. (2003), Chloroplast movements in the field, *Plant, Cell & Environment*, 26(12):2005–2014.
Willstätter R. and Stoll A. (1918), *Untersuchungen über die Assimilation der Kohlensäure*, Verlag von Julius Springer, Berlin, 448 pages.
Wilson P.S. and Dunton K.H. (2009), Laboratory investigation of the acoustic response of seagrass tissue in the frequency band 0.5–2.5 kHz, *The Journal of the Acoustical Society of America*, 125(4):1951–1959.
Wilts B.D., Whitney H.M., Glover B.J., Steiner U. and Vignolini S. (2014), Natural helicoidal structures: morphology, self-assembly and optical properties, *Materials Today: Proceedings*, 1:177–185.
Windham W.R., Poole G.H., Park B., Heitschmidt G., Hawkins S.A., Albano J.P., et al. (2011), Rapid screening of Huanglongbing-infected citrus leaves by near infrared reflectance spectroscopy, *Transactions of the ASABE*, 54 (6):2253–2258.
Wing S.L. (1992), High-resolution leaf X-radiography in systematics and paleobotany, *American Journal of Botany*, 79 (11):1320–1324.
Winkel-Shirley B. (2001), Flavonoid biosynthesis. A colorful model for genetics, biochemistry, cell biology, and biotechnology, *Plant Physiology*, 126(2):485–493.
Witkowski E.T.F. and Lamont B.B. (1991), Leaf specific mass confounds leaf density and thickness, *Oecologia*, 88:486–493.
Witt C., Pasuquin J.M.C.A., Mutters R. and Buresh R.J. (2005), New leaf color chart for effective nitrogen management in rice, *Better Crops*, 89(1):36–39.
Woessner P. and Hapke B. (1987), Polarization of light scattered by clovers, *Remote Sensing of Environment*, 21 (3):243–261.
Wold S., Ruhe A., Wold H. and Dunn W.J. (1984), The collinearity problem in linear regression. The partial leastsquares (PLS) regression approach to generalized inverses, *Siam Journal on Scientific and Statistical Computing*, 5(3):735–743.
Wold S., Sjöström M. and Eriksson L. (2001), PLS-regression: a basic tool of chemometrics, *Chemometrics and Intelligent Laboratory Systems*, 58(2):109–130.
Wolfe J.A. and Upchurch G.R. (1986), Vegetation, climatic and floral changes at the Cretaceous-Tertiary boundary, *Nature*, 324:148–152.
(1987), Leaf assemblages across the Cretaceous-Tertiary boundary in the Raton Basin, New Mexico and Colorado, *Proceedings of the National Academy of Sciences of the United States of America*, 84(15):5096–5100.
Wolstencroft R.D., Tranter G.E. and Le Pevelen D.D. (2002), Diffuse reflectance circular dichroism for the detection of molecular chirality: an application in remote sensing of flora, in *Proc. Bioastronomy 2002: Life Among the Stars* (Norris R. & Stootman F., eds), Astronomical Society of the Pacific, pp. 149–153.
Wong C.L. and Blevin W.R. (1967), Infrared reflectances of plant leaves, *Australian Journal of Biological Sciences*, 20 (3):501–508.
Woodall G.S., Dodd I.C. and Stewart G.R. (1998), Contrasting leaf development within the genus *Syzygium*, *Journal of Experimental Botany*, 49(318):79–87.
Woodward F.I., Lomas M.R. and Kelly C.K. (2004), Global climate and the distribution of plant biomes, *Proceedings of the Royal Society B*, 359(1450):1465–1476.
Woolf N.J., Smith P.S., Traub W.A. and Jucks K.W. (2002), The spectrum of earthshine: a pale blue dot observed from the ground, *The Astrophysical Journal*, 574(1):430–433.
Woolley J.T. (1971), Reflectance and transmittance of light by leaves, *Plant Physiology*, 47(5):656–662.
(1975), Refractive index of soybean leaf cell walls, *Plant Physiology*, 55(2):172–174.
Wozniak B. and Dera J. (2007), *Light Absorption in Sea Water*, Springer-Verlag, New York, 453 pages.
Wright I.J, Reich P.B., Westoby M., Ackerly D.D., Baruch Z., Bongers F., et al. (2004). The worldwide leaf economics spectrum, *Nature*, 428:821–827.

Wu C. and Wang X. (2014), Effects of foliar dust on plant reflectance spectra and physiological ecology: a review, *Chinese Journal of Applied & Environmental Biology*, 20 (6):1132–1138 (in Chinese).

Wu D., Feng L., Zhang C. and He Y. (2008), Early detection of *Botrytis cinerea* on eggplant leaves based on visible and near-infrared spectroscopy, *Transactions of the ASABE*, 51(3):1133–1139.

Wu J., Feld M.S. and Rava R.P. (1993), An analytical model for extracting intrinsic fluorescence in a turbid media, *Applied Optics*, 32(19):3585–3595.

Wu J., Zhang J., Lü A. and Zhou L. (2012), An exploratory analysis of spectral indices to estimate vegetation water content using sensitivity function, *Remote Sensing Letters*, 3(2):161–169.

Wu J., Chavana-Bryant C., Prohaska N., Serbin S.P., Guan K., Albert L.P., et al. (2017), Convergence in relationships between leaf traits, spectra and age across diverse canopy environments and two contrasting tropical forests, *New Phytologist*, 214 (3):1033–1048.

Wu T., Zhang L., Cen Y., Huang C., Sun X., Zhao H. et al. (2013), Polarized spectral measurement and analysis of Sedum spectabile Boreau using a field imaging spectrometer system, *IEEE Journal of Selected Topics in Applied Earth Observations and Remote Sensing*, 6(2):724–730.

Wuenscher J.E. (1970), The effect of leaf hairs of *Verbascum thapsus* on leaf energy exchange, *New Phytologist*, 69(1):65–73.

Wuhrmann-Meyer K. and Wuhrmann-Meyer M. (1941), Untersuchungen über die Absorption ultravioletter Strahlen durch Kutikular- und Wachsschichten von Blättern. I, *Planta*, 32(1):43–50.

Wuyts N., Palauqui J.C., Conejero G., Verdeil J.L., Granier C. and Massonnet C. (2010), High-contrast three-dimensional imaging of the *Arabidopsis* leaf enables the analysis of cell dimensions in the epidermis and mesophyll, *Plant Methods*, 6:17.

Xiao Y., Zhao W., Zhou D. and Gong H. (2014), Sensitivity analysis of vegetation reflectance to biochemical and biophysical variables at leaf, canopy, and regional scales, *IEEE Transactions on Geoscience and Remote Sensing*, 5(7):4014–4024.

Xiao Y., Tholen D. and Zhu X.G. (2016), The influence of leaf anatomy on the internal light environment and photosynthetic electron transport rate: exploration with a new leaf ray tracing model, *Journal of Experimental Botany*, 67 (21):6021–6035.

Xie D.H., Wang P.J. and Zhu Q.J. (2010), Modeling polarimetric BRDF of leaves surfaces, *Spectroscopy and Spectral Analysis*, 30(12):3324–3328 (in Chinese).

Xie D.H., Qin W.H., Wang P.J., Shuai Y.M., Zhou Y.Y. and Zhu Q.J. (2017), Influences of leaf-specular reflection on canopy BRF characteristics: a case study of real maize canopies with a 3-D scene BRDF model, *IEEE Transactions on Geoscience and Remote Sensing*, 55(2):619–631.

Xie L.J., Ying Y.B. and Ying T.J. (2007), Quantification of chlorophyll content and classification of nontransgenic and transgenic tomato leaves using visible/near-infrared diffuse reflectance spectroscopy, *Journal of Agricultural and Food Chemistry*, 55(12):4645–4650.

Xie X., Jiang H. and Yua S. (2009), Study on leaf reflectance and red edge characteristics of Chinese fir (*Cunninghamia lanceolata*) caused by acid rain with hyperspectral, in *Proc. Remote Sensing and GIS Data Processing and Other Applications* (Maître H., Sun H., Lei B. and Feng J., Eds), Yichang, China, 30 October 2009, SPIE, Vol. 7498, 74980Q.

Xu C. and Gertner G. (2011), Understanding and comparisons of different sampling approaches for the Fourier Amplitudes Sensitivity Test (FAST), *Computational Statistics & Data Analysis*, 55(1):184–198.

Xu H.R., Ying Y.B. and Ye Z. (2005), Application of near infrared spectroscopy to predict plant diseases, in *Proc. Optical Sensors and Sensing Systems for Natural Resources and Food Safety and Quality* (Chen Y.R., Meyer G.E. and Tu S.I., Eds), Boston, MA, 23–26 October 2005, SPIE, Vol. 5996, 59960A.

Xu H.R., Ying Y.B., Fu X.P. and Zhu S.P. (2007), Near-infrared spectroscopy in detecting leaf miner damage on tomato leaf, *Biosystems Engineering*, 96(4):447–454.

Xu H.R., Yu P., Fu X.P. and Ying Y.B. (2009), On-site variety discrimination of tomato plant using visible-near infrared reflectance spectroscopy, *Journal of Zhejiang University SCIENCE B*, 10(2):126–132.

Xu J.H. and Yu J.T. (2013), Air dustfall impact on spectrum of *Ficus microcarpa*'s leaf, *Advanced Materials Research*, 655–657:813–815.

Xu R. and Ma Y. (2004), Remote sensing research in biogeochemistry of the Hetai gold deposit, Guangdong Province, China, *International Journal of Remote Sensing*, 25(2):437–453.

Xue L. and Yang L. (2009), Deriving leaf chlorophyll content of green-leafy vegetables from hyperspectral reflectance, *ISPRS Journal of Photogrammetry and Remote Sensing*, 64(1):97–106.

Xue Z., Gao H. and Zhao S. (2014), Effects of cadmium on the photosynthetic activity in mature and young leaves of soybean plants, *Environmental Science and Pollution Research*, 21(6):4656–4664.

Yadava U.L. (1986), A rapid nondestructive method to determine chlorophyll in intact leaves, *HortScience*, 21(6):1449–1450.

Yamada N. and Fujimura S. (1991), Nondestructive measurement of chlorophyll pigment content in plant leaves from three-color reflectance and transmittance, *Applied Optics*, 30(27):3964–3973.

Yamamoto A., Nakamura T., Adu-Gyamfi J.J. and Saigusa M. (2002), Relationship between chlorophyll content in leaves of sorghum and pigeonpea determined by extraction method and by chlorophyll meter (SPAD-502), *Journal of Plant Nutrition*, 25(10):2295–2301.

Yamamoto H., Suzuki Y. and Hayakawa S. (1993), Estimation of amount of volcanic ashes piled on plant leaves by spectral reflectance, *Journal of the Remote Sensing Society of Japan*, 13(3):240–248 (in Japanese).

Yamamoto H., Suzuki Y., Kojima T., Hayakawa S., Inoue Y. and Tanaka M. (1994), Estimation of leaf water content of plants by spectral reflectance of near infrared range, *Journal of the Remote Sensing Society of Japan*, 14(4):293–301 (in Japanese.

Yamamoto H., Suzuki Y. and Hayakawa S. (1995), The effects of overlapping, thickness and water content of plant leaves in spectral reflectance, *Journal of the Remote Sensing Society of Japan*, 15(5):463–470 (in Japanese).

Yamazaki K. (2010), Leaf mines as visual defensive signals to herbivores, *Oikos*, 119(5):796–801.

Yan X., Shi W., Zhao W. and Luo N. (2014), Estimation of atmospheric dust deposition on plant leaves based on spectral features, *Spectroscopy Letters*, 47(7):536–542.

Yáñez-Rausell L., Schaepman M.E. and Clevers J.G.P.W. (2014a), Minimizing measurement uncertainties of coniferous needle-leaf optical properties, Part I: Methodological review, *IEEE Journal of Selected Topics in Applied Earth Observations and Remote Sensing*, 7(2):399–405.

Yáñez-Rausell L., Malenovský Z., Clevers J.G.P.W. and Schaepman M.E. (2014b), Minimizing measurement uncertainties of coniferous needle-leaf optical properties, Part II: Experimental setup and error analysis, *IEEE Journal of Selected Topics in Applied Earth Observations and Remote Sensing*, 7(2):406–420.

Yang B., Knyazikhin Y., Lin Y., Yan K., Chen C., Park T., et al. (2016a), Analyses of impact of needle surface properties on estimation of needle absorption spectrum: case study with coniferous needle and shoot samples, *Remote Sensing*, 8(7):563.

Yang F., Lu Y., Zhou G., Pan Y. and Hu H. (2009), Heavy metal content estimation in leaf by spectrum features of plant in De-Xing copper mining area, in *Proc. 2008 International Conference on Optical Instruments and Technology: Optoelectronic Measurement Technology and Applications* (Ye S. and Zhang G., Eds), Beijing, China, 16–19 November 2008, SPIE, Vol. 7160, 71601B.

Yang G., Zhao C., Pu R., Feng H., Li Z., Li H. et al. (2015), Leaf nitrogen spectral reflectance model of winter wheat (*Triticum aestivum*) based on PROSPECT: simulation and inversion, *Journal of Applied Remote Sensing*, 9(1):095976–1.

Yang W.H., Peng S., Huang J., Sanico A.L., Buresh R.J. and Witt C. (2003), Using leaf color charts to estimate leaf nitrogen status of rice, *Agronomy Journal*, 95(1):212–217.

Yang X., Tang J. and Mustard J.F. (2014), Beyond leaf color: comparing camera-based phenological metrics with leaf biochemical, biophysical, and spectral properties throughout the growing season of a temperate deciduous forest, *Journal of Geophysical Research: Biogeosciences*, 119(3):181–191.

Yang X., Tang J., Mustard J.F., Wu J., Zhao K., Serbin S. et al. (2016), Seasonal variability of multiple leaf traits captured by leaf spectroscopy at two temperate deciduous forests, *Remote Sensing of Environment*, 179:1–12.

Yang X.S., Heisler G.M., Montgomery M.E., Sullivan J.H., Whereat E.B. and Miller D.R. (1995), Radiative properties of hardwood leaves to ultraviolet-irradiation, *International Journal of Biometeorology*, 38(2):60–66.

Yang Y., Liu Z., Hu B., Man Y. and Wu W. (2010), Bionic composite material simulating the optical spectra of plant leaves, *Journal of Bionic Engineering*, 7(S1):S43–S49.

Yang Y., Liu Z., Hu B. and Wu W. (2011a), Design of plant leaf bionic camouflage materials based on spectral analysis, *Spectroscopy and Spectral Analysis*, 31(6):1668–1672 (in Chinese).

Yang Y., Hu B. and Wu W. (2011b), Design and preparation of bionic camouflage materials by simulating plant leaves, *Journal of National University of Defense Technology*, 33(5):50–53 (in Chinese).

Yanovskaya E.A., Yanovskii A.F. and Dlugunovich V.A. (1991), Polarizational characteristics of radiation reflected from potato leaves, *Journal of Applied Spectroscopy*, 55(4):1029–1032 (Translated from *Zhurnal Prikladnoi Spektroskopii*, 55 (A):630–634).

Yao X., Tang S.P., Cao W.X., Tian Y.C. and Zhu Y. (2011), Estimating the nitrogen content in wheat leaves by near-infrared reflectance spectroscopy, *Chinese Journal of Plant Ecology*, 35(8):844–852 (in Chinese).

Yao Y., Li W., Wen S. and Zhao Y. (2012), Vegetational spectral characteristics in Hongtoushan mining area, Liaoning Province, in *Proc. 2nd International Conference on Remote Sensing, Environment and Transportation Engineering*, Nanjing, China, 1–3 June 2012, IEEE, 4 pages.

Yates D.J. (1981), Effect of the angle of incidence of light on the net photosynthesis rates of *Sorghum almum* leaves, *Australian Journal of Plant Physiology*, 8(3):335–346.

Ye H., Yuan Z. and Zhang S. (2013), The heat and mass transfer analysis of a leaf, *Journal of Bionic Engineering*, 10 (2):170–176.

Ye H., Gao Y., Li S. and Guo L. (2015), Bionic leaves imitating the transpiration and solar spectrum reflection characteristics of natural leaves, *Journal of Bionic Engineering*, 12(1):109–116.

Yeh P. (1988), *Optical Waves in Layered Media*, John Wiley & Sons, 406 pages.

Yi Q.X., Huang J.F., Wang F.M., Wang X.Z. and Liu Z.Y. (2007), Monitoring rice nitrogen status using hyperspectral reflectance and artificial neural network, *Environmental Science & Technology*, 41(19):6770–6775.

Yi S.L., Deng L., He S.L., Zheng Y.Q. and Mao S.S. (2010a), A spectrum based models for monitoring leaf potassium content of *Citrus sinensis* (L) cv. Jincheng orange, *Scientia Agricultura Sinica*, 43(4):780–786.

Yi S.L., Deng L., He S.L., Zheng Y.Q., Wang L. and Zhao X.A. (2010b), Research on zinc content in leaf of Olinda Valencia orange using visible near infrared spectroscopy model, *Spectroscopy and Spectral Analysis*, 30(11):2927–2931 (in Chinese).

Yoder B.J. and Daley L.S. (1990), Development of a visible spectroscopic method for determining chlorophyll a and b *in vivo* in leaf samples, *Spectroscopy*, 5(8):44–50.

Yoder B.J. and Pettigrew-Crosby R.E. (1995), Predicting nitrogen and chlorophyll content and concentrations from reflectance spectra (400–2500 nm) at leaf and canopy scales, *Remote Sensing of Environment*, 53(3):199–211.

Yoshimura H., Kobasi S., Ohte K. and Senoo T. (1991a), Characteristics and colorimetric analysis of the spectral reflectance of tree leaves, *Journal of the Remote Sensing Society of Japan*, 11(2):199–211 (in Japanese).

Yoshimura H., Tanaka S., Kobasi S., Ohte K., Senoo T. and Kunitomo M. (1991b), Seasonal change of near-infrared reflectance of single-leaf and multiple-leaf, *Journal of the Remote Sensing Society of Japan*, 11(4):607–621 (in Japanese).

Yoshimura H. (1998), Spectral properties of tree leaves with ageing and the effect of leaf stacking in the near-infrared region, *Journal of the Remote Sensing Society of Japan*, 18(1):42–56 (in Japanese).

Yoshimura H. (2001), The mechanism of remote detection concerning the xanthophyll cycle pigments, *Journal of the Remote Sensing Society of Japan*, 21(4):373–376 (in Japanese).

Yoshimura H., Zhu H., Wu Y. and Ma R. (2010), Spectral properties of plant leaves pertaining to urban landscape design of broad-spectrum solar ultraviolet radiation reduction, *International Journal of Biometeorology*, 54(2):179–191.

Younes H.A., Abdel-Aal R.M., Khodair M.M. and Abdel-Samie A.G. (1974), Spectral reflectance studies on mineral deficiency in corn plants, in *Proc. 9th International Symposium on Remote Sensing of Environment*, Ann Arbor, MI, 15–19 April 1974, ERIM, Vol. 2, pp. 1105–1125.

Young A.J. (1991), The photoprotective role of carotenoids in higher plants, Physiologia Plantarum, 83:702–708.

Yu G.R., Miwa T., Nakayama K., Matsuoka N. and Kon H. (2000), A proposal for universal formulas for estimating leaf water status of herbaceous and woody plants based on spectral reflectance properties, *Plant and Soil*, 227(1–2):47–58.

Yuan L., Zhang J., Zhao J., Cai S. and Wang J. (2013), Selection of leaf orientation insensitive bands for yellow rust detection, in *Computer and Computing Technologies in Agriculture VI* (Li D. and Chen Y., Eds), Springer-Verlag, Berlin, pp. 78–84.

Yuan L., Huang Y., Loraamm R.W., Nie C., Wang J. and Zhang J. (2014a), Spectral analysis of winter wheat leaves for detection and differentiation of diseases and insects, *Field Crops Research*, 156:199–207.

Yuan Z., Ye H. and Li S. (2014b), Bionic leaf simulating the thermal effect of natural leaf transpiration, *Journal of Bionic Engineering*, 11(1):90–97.

Yusuf B.L. and He Y. (2011), Application of hyperspectral imaging sensor to differentiate between the moisture and reflectance of healthy and infected tobacco leaves, *African Journal of Agricultural Research*, 6(29):6267–6280.

Yuzhu H., Xiaomei W. and Shuyao S. (2011), Nitrogen determination in pepper (*Capsicum frutescens* L.) plants by color image analysis (RGB), *African Journal of Biotechnology*, 10(77):17737–17741.

Zakaluk R. and Ranjan R.S. (2007), Artificial neural network modelling of leaf water potential for potatoes using RGB digital images: a greenhouse study, *Potato Research*, 49(4):255–272.

Zakharov V.P., Bratchenko I.A., Sindyaeva A.R. and Timchenko E.V. (2009), Modeling of optical radiation energy distribution in plant tissue, *Optics and Spectroscopy*, 107(9):903–908.

Zakharov V.P., Bratchenko I.A. and Timchenko E.V. (2010), Optical model of plant tissue, *Optics and Spectroscopy*, 109 (2):232–236.

Zamblé Fidèle T.B., Sadaiou Sabas B.Y., Maxime A.D. and Dongui B.K. (2014), Biomonitoring de la pollution urbaine en zone tropicale à partir des caractéristiques spectrales et anatomiques des feuilles de *Ficus polita* Vahl, *International Journal of Innovation and Applied Studies*, 8(2):861–870.

Zarco-Tejada P.J., Miller J.R., Mohammed G.H. and Noland T.L. (2000), Chlorophyll fluorescence effects on vegetation apparent reflectance. I: Leaf-level measurements and model simulation, *Remote Sensing of Environment*, 74(3):582–595.

Zarco-Tejada P.J., Miller J.R., Mohammed G.H., Noland T.L. and Sampson P.H. (2002), Vegetation stress detection through chlorophyll a + b estimation and fluorescence effects on hyperspectral imagery, *Journal of Environmental Quality*, 31 (5):1433–1441.

Zarco-Tejada P.J., Pushnik P.J., Dobrowski J.C. and Ustin S.L. (2003a), Steady-state chlorophyll a fluorescence detection from canopy derivative reflectance and double-peak red-edge effects, *Remote Sensing of Environment*, 84(2):283–294.

Zarco-Tejada P.J., Rueda C.A. and Ustin S.L. (2003b), Water content estimation in vegetation with MODIS reflectance data and model inversion methods, *Remote Sensing of Environment*, 85(1):109–124.

Zarco-Tejada P.J., Miller J.R., Harron J., Hu B., Noland T.L., Goel N., et al. (2004a), Needle chlorophyll content estimation through model inversion using hyperspectral data from Boreal conifer forest canopies, *Remote Sensing of Environment*, 89(2):1989–1999.

Zarco-Tejada P.J., Miller J.R., Morales A., Berjona A. and Agüerad J. (2004b), Hyperspectral indices and model simulation for chlorophyll estimation in open-canopy tree crops, *Remote Sensing of Environment*, 90(4):463–476.

Zeeman S.C., Smith S.M., snd Smith A.M. (2007), The diurnal metabolism of leaf starch, *Biochemical Journal*, 401(1):13–28.

Zhai Y., Cui L., Zhou X., Gao Y., Fei T. and Gao W. (2013), Estimation of nitrogen, phosphorus, and potassium contents in the leaves of different plants using laboratory-based visible and near-infrared reflectance spectroscopy: comparison of partial leastsquare regression and support vector machine regression methods, *International Journal of Remote Sensing*, 34 (7):2502–2518.

Zhang C., Kovacs J.M., Wachowiak M.P. and Flores-Verdugo F. (2013), Relationship between hyperspectral measurements and mangrove leaf nitrogen concentrations, *Remote Sensing*, 5(2):891–908.

Zhang H. and Zhang J.C. (2008), Near-infrared green camouflage of cotton fabrics using vat dyes, *Journal of the Textile Institute*, 99(1):83–88.

Zhang H., Hu H., Zhang X.B., et al. (2011), Estimation of rice neck blasts severity using spectral reflectance based on BP-neural network, *Acta Physiologiae Plantarum*, 33(6):2461–2466.

Zhang J., Sokhansanj S., Wu S., Fang R., Yang W. and Winter P. (1998), A transformation technique from RGB signals to the Munsell system for color analysis of tobacco leaves, *Computers and Electronics in Agriculture*, 19(2):155–166.

Zhang J.C., Pu R.L., Wang J.H., Huang W.J., Yuan L. and Luo J.H. (2012a), Detecting powdery mildew of winter wheat using leaf level hyperspectral measurements, *Computers and Electronics in Agriculture*, 85:13–23.

Zhang J.C., Yuan L., Wang J.H., Huang W.J., Chen L.P. and Zhang D.Y. (2012b), Spectroscopic leaf level detection of powdery mildew for winter wheat using continuous wavelet analysis, *Journal of Integrative Agriculture*, 11(9):1474–1484.

Zhang J.C., Yuan L., Pu R., Loraamm R.W., Yang G. and Wang J. (2014), Comparison between wavelet spectral features and conventional spectral features in detecting yellow rust for winter wheat, *Computers and Electronics in Agriculture*, 100:79–87.

Zhang Q., Xiao X., Braswell B., Linder E., Baret F. and Moore B. (2005), Estimating light absorption by chlorophyll, leaf and canopy in a deciduous broadleaf forest using MODIS data and a radiative transfer model, *Remote Sensing of Environment*, 99(3):357–371.

Zhang Q., Li Q. and Zhang G. (2012c), Rapid determination of leaf water content using VIS/NIR spectroscopy analysis with wavelength selection, *Spectroscopy*, 27(2):93–105.

Zhang S. and Wang Q. (2015), Inverse retrieval of chlorophyll from reflected spectra for assimilating branches of drought-tolerant *Tamarix ramosissima*, *IEEE Journal of Selected Topics in Applied Earth Observations and Remote Sensing*, 8(4):1498–1505.

Zhang X.J. and Li M.Z. (2008), Analysis and estimation of the phosphorus content in cucumber leaf in greenhouse by spectroscopy, *Spectroscopy and Spectral Analysis*, 28(10):2404–2408.

Zhang X. and He Y. (2013), Rapid estimation of seed yield using hyperspectral images of oilseed rape leaves, *Industrial Crops and Products*, 42:416–420.

Zhang Y., Chen J.M. and Thomas S.C. (2007), Retrieving seasonal variation in chlorophyll content of overstory and understory sugar maple leaves from leaf-level hyperspectral data, *Canadian Journal of Remote Sensing*, 33(5):406–415.

Zhang Y., Cong Q., Xie Y.F., Yang J.X. and Zhao. B. (2008), Quantitative analysis of routine chemical constituents in tobacco by near-infrared spectroscopy and support vector machine, *Spectrochimica Acta Part A: Molecular and Biomolecular Spectroscopy*, 71(4):1408–1413.

Zhang Y., Hayashi T., Hosokawa M., Yazawa S. and Li Y. (2009), Metallic lustre and the optical mechanism generated from the leaf surface of *Begonia rex* Putz., *Scientia Horticulturae*, 121(2):213–217.

Zhang Y., Huang J., Wang F., Blackburn G.A., Zhang H.K., Wang X., et al. (2017), An extended PROSPECT: advance in the leaf optical properties model separating total chlorophylls into chlorophyll *a* and *b*, *Scientific Reports*, 7:6429.

Zhao F., Guo Y., Huang Y., Reddy K.N., Lee M.A., Fletcher R.S. et al. (2014a), Early detection of crop injury from herbicide glyphosate by leaf biochemical parameter inversion, *International Journal of Applied Earth Observation and Geoinformation*, 31:78–85.

Zhao F., Huang Y., Guo Y., Reddy K.N., Lee M.A., Fletcher R.S., et al. (2014b), Early detection of crop injury from glyphosate on soybean and cotton using plant leaf hyperspectral data, *Remote Sensing*, 6(2):1538–1563.

Zhao F., Guo Y., Huang Y., Verhoef W., van der Tol C., Dai B., et al. (2015), Quantitative estimation of fluorescence parameters for crop leaves with Bayesian inversion, *Remote Sensing*, 7(10):14179–14199.

Zhao J., Yuan L., Luo J., Du S., Huang L. and Huang W. (2012), Spectral differences of opposite sides of stripe rust infested winter wheat leaves using ASD's Leaf Clip, in *Proc. IEEE International Geoscience and Remote Sensing Symposium (IGARSS'12)*, Munich, Germany, 22–27 July 2012, IEEE, pp. 6585–6588.

Zhao J., Huang L., Huang W., Zhang D., Yuan L., Zhang J. et al. (2014c), Hyperspectral measurements of severity of stripe rust on individual wheat leaves, *European Journal of Plant Pathology*, 139(2):407–417.

Zhao K., Valle D., Popescu S., Zhang X. and Mallick B. (2013), Hyperspectral remote sensing of plant biochemistry using Bayesian model averaging with variable and band selection, *Remote Sensing of Environment*, 132:102–119.

Zhou N., Dong W. and Mei X. (2006), Realistic simulation of seasonal variant maples, in *Proc. 2nd International Symposium on Plant Growth Modeling and Applications*, Beijing, China, 13–17 November 2006, pp. 295–301.

Zhou X., Huang W., Kong W., Ye H., Dong Y. and Casa R. (2017), Assessment of leaf carotenoids content with a new carotenoid index: development and validation on experimental and model data, *International Journal of Applied Earth Observation and Geoinformation*, 57:24–35.

Zhu J., Tremblay N. and Liang Y. (2012), Comparing SPAD and atLEAF values for chlorophyll assessment in crop species, *Canadian Journal of Soil Science*, 92(4):645–648.

Zhu Y., Qu Y., Liu S. and Chen S. (2014), A reflectance spectra model for copper-stressed leaves: advances in the PROSPECT model through addition of the specific absorption coefficients of the copper ion, *International Journal of Remote Sensing*, 35(4):1356–1373.

Zhumar A.Y. (1998), Effect of the structure of mesophyll in a leaf on the form of the spectral curve for the first derivatives of the reflection coefficients of leaves over the range 0.68–0.75 μm, *Journal of Applied Spectroscopy*, 65(6):967–972 (cover-to-cover translation from *Zhurnal Prikladnoi Spektroskopii*, 65(6):921–925).

Zhumar A.Y. and Zaitseva V.A. (2003a), Influence of the concentration of aqueous solutions of sulfuric acid on the reflection spectra of pine needles, *Journal of Applied Spectroscopy*, 70(2):292–297 (cover-to-cover translation from *Zhurnal Prikladnoi Spektroskopii*, 70(2):260–264).

(2003b), Influence of sulfuric acid solutions on the pine needles optical characteristics, in *Proc. International Geoscience and Remote Sensing Symposium (IGARSS'03)*, Toulouse, France, 21–25 July 2003, IEEE, Vol. 4, pp. 2891–2893.

Zhumar A.Y. and Tsaryuk A.V. (2005), Ellipticity of the linearly polarized He-Ne laser radiation reflected by the leaves of plants, *Journal of Applied Spectroscopy*, 72 (2):249–254 (translated from *Zhurnal Prikladnoi Spektroskopii*, 72 (2):236–240).

Ziechmann W. (1964), Spectroscopic investigations of lignin, humic substances and peat, *Geochimica et Cosmochimica Acta*, 28(10–11):1555–1566.

Zolotarev V.M. and Demin A.V. (1977), Optical constants of water over a broad range of wavelengths, 0.1 Å-1 m, *Optics and Spectroscopy*, 43(2):157–161.

Zuo J., Zhang Z.W., He J., Zhang L.L., Mu K.J. and Zhang C. (2011), The experimental research of leaf water content using terahertz time-domain spectroscopy, in *Proc. International Symposium on Photoelectronic Detection and Imaging 2011: Terahertz Wave Technologies and Applications* (Zhang X.C., Yao J., Zhang C. and Wang Z., Eds), Beijing, China, 24 May 2011, SPIE, Vol. 8195, 81951F.

Zuppiroli L. and Bussac M.N. (2001), *Traité des couleurs*, Presses polytechniques et universitaires romandes, Lausanne, Switzerland, 442 pages.

Zurzycki J. (1961), The influence of chloroplast displacements on the optical properties of leaves, *Acta Societatis Botanicorum Poloniae*, (3–4):503–527.

Zwiggelaar R. (1998), A review of spectral properties of plants and their potential use for crop/weed discrimination in row-crops, *Crop Protection*, 17(3):189–206.

Zygielbaum A.I., Gitelson A.A., Arkebauer T.J. and Rundquist D.C. (2009), Non-destructive detection of water stress and estimation of relative water content in maize, *Geophysical Research Letters*, 36:L12403.

Zygielbaum A.I., Arkebauer T.J., Walter-Shea E.A. and Scoby D.L. (2012), Detection and measurement of vegetation photoprotection stress response using PAR reflectance, *Israel Journal of Plant Sciences*, 60(1–2):37–47

Index